D1658074

52,-

35,-

Technik

ex libris '13

Identification of Dynamic Systems

Rolf Isermann · Marco Münchhof

Identification of Dynamic Systems

An Introduction with Applications

Springer

Prof. Dr. -Ing. Dr. h.c. Rolf Isermann
Technische Universität Darmstadt
Institut für Automatisierungstechnik
Landgraf-Georg-Straße 4
64283 Darmstadt
Germany
RIsermann@iat.tu-darmstadt.de

Dr. -Ing. Marco Münchhof, M.S./SUNY
Technische Universität Darmstadt
Institut für Automatisierungstechnik
Landgraf-Georg-Straße 4
64283 Darmstadt
Germany
MMuenchhof@iat.tu-darmstadt.de

Additional Material to this book can be downloaded from http://extras.springer.com/2011/978-3-540-78879-7

ISBN 978-3-540-78878-2 e-ISBN 978-3-540-78879-9
DOI 10.1007/978-3-540-78879-9
Springer Heidelberg Dordrecht London New York

© Springer-Verlag Berlin Heidelberg 2011
This work is subject to copyright. All rights are reserved, whether the whole or part of the material is concerned, specifically the rights of translation, reprinting, reuse of illustrations, recitation, broadcasting, reproduction on microfilm or in any other way, and storage in data banks. Duplication of this publication or parts thereof is permitted only under the provisions of the German Copyright Law of September 9, 1965, in its current version, and permission for use must always be obtained from Springer. Violations are liable to prosecution under the German Copyright Law.
The use of general descriptive names, registered names, trademarks, etc. in this publication does not imply, even in the absence of a specific statement, that such names are exempt from the relevant protective laws and regulations and therefore free for general use.

Cover design: WMXDesign GmbH, Heidelberg

Printed on acid-free paper

Springer is part of Springer Science+Business Media (www.springer.com)

Preface

For many problems of the design, implementation, and operation of automatic control systems, relatively precise mathematical models for the static and dynamic behavior of processes are required. This holds also generally in the areas of natural sciences, especially physics, chemistry, and biology, and also in the areas of medical engineering and economics. The basic static and dynamic behavior can be obtained by theoretical or physical modeling, if the underlying physical laws (first principles) are known in analytical form. If, however, these laws are not known or are only partially known, or if significant parameters are not known precisely enough, one has to perform an experimental modeling, which is called process or system identification. Then, measured signals are used and process or system models are determined within selected classes of mathematical models.

The scientific field of system identification was systematically developed since about 1960 especially in the areas of control and communication engineering. It is based on the methods of system theory, signal theory, control theory, and statistical estimation theory and was influenced by modern measurement techniques, digital computations and the need for precise signal processing, control, and automation functions. The development of identification methods can be followed in wide spread articles and books. However, a significant influence had the IFAC-symposia on system identification, which were since 1967 organized every three years around the world, in 2009 a 15th time in Saint-Malo.

The book is intended to give an introduction to system identification in an easy to understand, transparent, and coherent way. Of special interest is an application-oriented approach, which helps the user to solve experimental modeling problems. It is based on earlier books in German, published in 1971, 1974, 1991 and 1992, and on courses taught over many years. It includes own research results within the last 30 years and publications of many other research groups.

The book is divided into eight parts. After an introductory chapter and a chapter on basic mathematical models of linear dynamic systems and stochastic signals, part I treats *identification methods with non-parametric models and continuous time signals*. The classical methods of determining frequency responses with non-periodic and periodic test signals serve to understand some basics of identification and lay ground for other identifications methods.

Part II is devoted to the determination of *impulse responses with auto- and cross-correlation functions*, both in continuous and discrete time. These correlation methods can also be seen as basic identification methods for measurements with stochastic disturbances. They will later appear as elements of other estimation methods and allow directly the design of binary test signals.

The *identification of parametric models* in discrete time like difference equations in Part III is based mainly on least squares parameter estimation. These estimation methods are first introduced for static processes, also known as regression analysis, and then expanded to dynamic processes. Both, non-recursive and recursive parameter estimation methods are derived and various modifications are described, like methods of extended least squares, total least squares, and instrumental variables. The Bayes and maximum likelihood methods yield a deeper theoretical background, also with regard to performance bounds. Special chapters treat the parameter estimation of time-variant processes and under closed-loop conditions.

Part IV now looks at *parameter estimation methods for continuous-time models*. First parameter estimation is extended to measured frequency responses. Then, the parameter estimation for differential equations and subspace methods operating with state variable filters are considered.

The *identification of multi-variable systems (MIMO)* is the focus of Part V. First basic structures of linear transfer functions and state space models are considered. This is followed by correlation and parameter estimation methods, including the design of special uncorrelated test signals for the simultaneous excitation of several inputs. However, sometimes it is easier to identify single-input multiple outputs (SIMO) processes sequentially.

Of considerable importance for many complex processes is the *identification of non-linear systems*, treated in Part VI. Special model structures, like Volterra series, Hammerstein- and Wiener-models allow applying parameter estimation methods directly. Then, iterative optimization methods are treated, taking into account multi-dimensional, non-linear problems. Powerful methods were developed based on non-linear net models with parametric models like neural networks and their derivations and look-up tables (maps) as non-parametric representations. Also, extended Kalman filters can be used.

Some *miscellaneous issues*, which are common to several identification methods, are summarized in Part VII, as e.g. numerical aspects, practical aspects of parameter estimation and a comparison of different parameter estimation methods.

Part VIII then shows the *application* of several treated identification methods *to real processes* like electrical and hydraulic actuators, machine tools and robots, heat exchangers, internal combustion engines and the drive dynamic behavior of automobiles.

The *Appendix* as Part IX then presents some mathematical aspects and a description of the three mass oscillator process, which is used as a practical example throughout the book. Measured data to be used for applications by the reader can be downloaded from the Springer web page in the Internet.

The wide topics of dynamic system identification are based on the research performed by many experts. Because some early contributions lay the ground for many other developments we would just like to mention a few authors from early seminal contributions. The determination of characteristic parameters of step responses was published by V. Strejc (1959). First publications on frequency response measurement with orthogonal correlation go back to Schaefer and Feissel (1955) and Balchen (1962). The field of correlation methods and ways to design pseudo-random-binary signals was essentially brought forward by e.g. Chow, Davies (1964), Schweitzer (1966), Briggs (1967), Godfrey (1970) and Davies (1970). The theory and application of parameter estimation for dynamic processes was around 1960 until about 1974 essentially promoted by works of J. Durbin, R.C.K. Lee, V. Strejc, P. Eykhoff, K.J. Åström, V. Peterka, H. Akaike, P. Young, D.W. Clarke, R.K. Mehra, J.M. Mendel, G. Goodwin, L. Ljung, and T. Söderström.

This was followed by many other contributions to the field which are cited in the respective chapters, see also Table 1.3 for an overview over the literature in the field of identification.

The authors are also indebted to many contributions for developing and applying identifications methods from researchers at our own group since 1973 until now, like M. Ayoubi, W. Bamberger, U. Baur, P. Blessing, H. Hensel, R. Kofahl, H. Kurz, K.H. Lachmann, O. Nelles, K.H. Peter, R. Schumann, S. Toepfer, M. Vogt, and R. Zimmerschied. Many other developments with regard to special dynamic processes are referenced in the chapters on applications.

The book is dedicated as an introduction to system identification for undergraduate and graduate students of electrical and electronic engineering, mechanical and chemical engineering and computer science. It is also oriented towards practicing engineers in research and development, design and production. Preconditions are basic undergraduate courses of system theory, automatic control, mechanical and/or electrical engineering. Problems at the end of each chapter allow to deepen the understanding of the presented contents.

Finally we would like to thank Springer-Verlag for the very good cooperation.

Darmstadt,
June 2010

Rolf Isermann
Marco Münchhof

Contents

Part I IDENTIFICATION OF NON-PARAMETRIC MODELS IN THE FREQUENCY DOMAIN — CONTINUOUS TIME SIGNALS

List of Symbols

Only frequently used symbols and abbreviations are given.

Letter symbols

a	parameters of differential of difference equations, amplitude
b	parameters of differential or difference equations
c	spring constant, constant, stiffness, parameters of stochastic difference equations, parameters of physical model, center of Gaussian function
d	damping coefficient, direct feedthrough, parameters of stochastic difference equations, dead time, drift
e	equation error, control deviation $e = w - y$
e	number $\mathrm{e} = 2.71828\ldots$ (Euler's number)
f	frequency ($f = 1/T_\mathrm{p}$, T_p period time), function $f(\ldots)$
f_S	sample frequency
g	function $g(\ldots)$, impulse response
h	step response, undisturbed output signal for method IV, $h \approx y_\mathrm{u}$
i	index
$\mathrm{i} = \sqrt{-1}$	imaginary unit
j	integer, index
k	discrete number, discrete-time $k = t/T_0 = 0, 1, 2, \ldots$ (T_0: sample time)
l	index
m	mass, order number, model order, number of states
n	order number, disturbance signal
p	probability density function, process parameter, order number of a stochastic difference equation, parameter of controller difference equation, number of inputs, $p(x)$ probability density function
q	index, parameter of controller difference equation, time shift operator with $x(k)q^{-1} = x(k-1)$

r	number of outputs
r_{P}	penalty multiplier
s	Laplace variable $s = \delta + \mathrm{i}\omega$
t	continuous time
u	input signal change ΔU, manipulated variable
w	reference value, setpoint, weight, $w(t)$ window function
x	state variable, arbitrary signal
y	output signal change ΔY, signal
y_{u}	useful signal, response due to u
y_{z}	response due to disturbance z
z	disturbance variable change ΔZ, $\mathfrak{Z}$-transform variable $z = \mathrm{e}^{-T_0 s}$
A	denominator polynomial of process transfer function
B	numerator polynomial of process transfer function
$\mathcal{A}$	denominator polynomial of closed-loop transfer function
$\mathcal{B}$	numerator polynomial of closed-loop transfer function
C	denominator polynomial of stochastic filter equation, covariance function
D	numerator polynomial of stochastic filter equation, damping ratio
F	filter transfer function
G	transfer function
I	second moment of area
J	moment of inertia
K	constant, gain
M	torque
N	discrete number, number of data points
P	probability
Q	denominator polynomial of controller transfer function
R	numerator polynomial of controller transfer function, correlation function
S	spectral density, sum
T	time constant, length of a time interval
T_0	sample time
T_{M}	measurement time
T_{P}	period time
U	input variable, manipulated variable (control input)
V	cost function
W	complex rotary operator for DFT and FFT
Y	output variable, control variable
Z	disturbance variable
$\boldsymbol{a}$	vector
$\boldsymbol{b}$	bias

$\boldsymbol{b}, \boldsymbol{B}$	input vector/matrix
$\boldsymbol{c}, \boldsymbol{C}$	output vector/matrix
$\boldsymbol{e}$	error vector
$\boldsymbol{g}$	vector of inequality constraints with $\boldsymbol{g}(\boldsymbol{x}) \leq \boldsymbol{0}$
$\boldsymbol{h}$	vector of equality constraints with $\boldsymbol{h}(\boldsymbol{x}) = \boldsymbol{0}$
$\boldsymbol{n}$	noise vector
$\boldsymbol{s}$	search vector
$\boldsymbol{u}$	manipulated variables for neural net
$\boldsymbol{v}$	output noise
$\boldsymbol{w}$	state noise
$\boldsymbol{x}$	vector of design variables
$\boldsymbol{y}$	output vector
$\boldsymbol{z}$	operating point variables for neural net
$\boldsymbol{A}$	arbitrary matrix, state matrix
$\boldsymbol{C}$	covariance matrix, matrix of measurements for TLS
$\boldsymbol{D}$	direct feedthrough matrix
$\boldsymbol{G}$	transfer function matrix
$\boldsymbol{G}_\nu$	noise transfer function matrix
$\boldsymbol{H}$	Hessian matrix, Hadamard matrix
$\boldsymbol{I}$	identity matrix
$\boldsymbol{K}$	gain matrix
$\boldsymbol{P}$	correlation matrix, $\boldsymbol{P} = \boldsymbol{\Psi}^{\mathrm{T}}\boldsymbol{\Psi}$
$\boldsymbol{S}$	Cholesky factor
$\boldsymbol{T}$	similarity transform
$\boldsymbol{U}$	input matrix for subspace algorithms
$\boldsymbol{W}$	weighting matrix
$\boldsymbol{X}$	state matrix
$\boldsymbol{Y}$	output matrix for subspace algorithms
$\boldsymbol{A}^{\mathrm{T}}$	transposed matrix
α	factor, coefficients of closed-loop transfer function
β	factor, coefficients of closed-loop transfer function
γ	activation function
δ	decay factor, impulse function, time shift
ε	correlation error signal, termination tolerance, small positive number
ζ	damping ratio
η	noise-to-signal ratio
θ	parameter
λ	forgetting factor, cycle time of PRBS generator
μ	membership function, index, time scaling factor for PRBS, order of controller transfer function
ν	index, white noise (statistically independent signal), order of controller transfer function

ξ	measurement disturbance
π	number $\pi = 3.14159\ldots$
ϱ	Step width factor for stochastic approximation algorithms
τ	time, time difference
φ	angle, phase
ω	angular frequency, $\omega = 2\pi/T_{\mathrm{P}}$; T_{P} period, rotational velocity $\omega(t) = \dot{\varphi}(t)$
ω_0	undamped natural frequency
Δ	change, deviation
Π	product
Σ	sum
Φ	validity function, activation function, weighting function
Ψ	wavelet
$\boldsymbol{\gamma}$	correction vector
$\boldsymbol{\psi}$	data vector
$\boldsymbol{\theta}$	parameter vector
$\boldsymbol{\Delta}$	augmented error matrix
$\boldsymbol{\Sigma}$	covariance matrix of a Gaussian distribution, matrix of singular values
$\boldsymbol{\Phi}$	transition matrix
$\boldsymbol{\Psi}$	data matrix

Mathematical abbreviations

$\exp(x) = \mathrm{e}^x$	exponential function
dim	dimension
adj	adjoint
$\angle$	phase (argument)
arg	argument
cond	condition number
cov	covariance
det	determinant
lim	limit
max	maximum (also as index)
min	minimum (also as index)
plim	probability limit
tr	trace of a matrix
var	variance
Re{...}	real part
Im{...}	imaginary part
$\boldsymbol{Q}_{\mathrm{S}}$	controllability matrix
$\boldsymbol{Q}_{\mathrm{Sk}}$	extended reversed controllability matrix

$\mathrm{E}\{\ldots\}$	expected value of a statistical variable
$\mathfrak{F}$	Fourier transform
H	Hermitian matrix
$\boldsymbol{\mathcal{H}}$	Hankel matrix
$\mathfrak{H}(f(x))$	Hilbert transform
$\mathcal{H}$	Heaviside function
$\mathfrak{L}$	Laplace transform
$\boldsymbol{Q}_{\mathrm{B}}$	observability matrix
$\boldsymbol{Q}_{\mathrm{Bk}}$	extended observability matrix
$\mathcal{Z}$	z transform directly from s transform
$\boldsymbol{\mathcal{T}}$	Markov parameter matrix, Töplitz matrix
$\mathfrak{Z}$	z transform
$G(-\mathrm{i}\omega)$	conjugate complex, sometimes denoted as $G^*(\mathrm{i}\omega)$
$\\|\cdot\\|_2$	2-norm
$\\|\cdot\\|_{\mathrm{F}}$	Frobenius norm
$V_{\boldsymbol{\theta}}$	first derivative of V with respect to $\boldsymbol{\theta}$
$V_{\boldsymbol{\theta\theta}}$	second derivative of V with respect to $\boldsymbol{\theta}$
$\nabla f(\boldsymbol{x})$	gradient of $f(\boldsymbol{x})$
$\nabla^2 f(\boldsymbol{x})$	Hessian matrix of $f(\boldsymbol{x})$
$\hat{x}$	estimated or observed variable
$\tilde{x}$	estimation error
$\bar{x}$	average, steady-state value
$\dot{x}$	first derivative with respect to time t
$x^{(n)}$	n-th derivative with respect to time t
x_0	amplitude or true value
x_{00}	value in steady state
$\overline{x}$	mean value
x_{S}	sampled signal
x_δ	Dirac series approximation
x^*	normalized, optimal
x_{d}	discrete-time
x_{00}	steady state or DC value
$\boldsymbol{A}^\dagger$	pseudo-inverse
$\boldsymbol{f}/\boldsymbol{A}$	orthogonal projection
$\boldsymbol{f}/_{\boldsymbol{B}}\boldsymbol{A}$	oblique projection

Abbreviations

ACF	auto-correlation function, e.g. $R_{\mathrm{uu}}(\tau)$
ADC	analog digital converter
ANN	artificial neural network
AGRBS	amplitude modulated GRBS
APRBS	amplitude modulated PRBS
AR	auto regressive
ARIMA	auto regressive integrating moving average process

ARMA	auto regressive moving average process
ARMAX	auto regressive moving average with external input
ARX	auto regressive with external input
BLUE	best linear unbiased estimator
CCF	cross-correlation function, e.g. $R_{uy}(\tau)$
CDF	cumulative distribution function
CLS	bias corrected least squares
COR-LS	correlation analysis and method of least squares
CWT	continuous-time wavelet transform
DARE	differential algebraic Riccatti equation
DFT	discrete Fourier transform
DSFC	discrete square root filter in covariance form
DSFI	discrete square root filter in information form
DTFT	discrete time Fourier transform
DUDC	discrete UD-factorization in covariance form
EIV	errors in variables
EKF	extended Kalman filter
ELS	extended least squares
FFT	Fast Fourier Transform
FIR	finite impulse response
FLOPS	floating point operations
FRF	frequency response function
GLS	generalized least squares
GRBS	generalized random binary signal
GTLS	generalized total least squares
IIR	infinite impulse response
IV	instrumental variables
KW	Kiefer-Wolfowitz algorithm
LLM	local linear model
LPM	local polynomial model
LOLIMOT	local linear model tree
LPVM	linear parameter variable model
LQR	linear quadratic regulator
LRGF	locally recurrent global feedforward net
LS	least squares
M	model
MA	moving average
MIMO	multiple input, multiple output
ML	maximum likelihood
MLP	multi layer perceptron
MOESP	Multi-variable Output Error State sPace
N4SID	Numerical algorithms for Subspace State Space IDentification
NARX	non-linear ARX model
NDE	non-linear difference equation
NFIR	non-linear FIR model

NN	neural net
NOE	non-linear OE model
ODE	ordinary differential equation
OE	output error
P	process
PCA	principal component analysis
PDE	partial differential equation
PDF	probability density function $p(x)$
PE	prediction error
PEM	prediction error method
PRBS	pseudo-random binary signal
RBF	radial basis function
RCOR-LS	recursive correlation analysis and method of least squares
RGLS	recursive generalized least squares
RIV	recursive instrumental variables
RLS	recursive least squares
RLS-IF	recursive least squares with improved feedback
RML	recursive maximum likelihood
SISO	single input, single output
SNR	signal to noise ratio
SSS	strict sense stationary
STA	stochastic approximation
STFT	short time Fourier transform
STLS	structured total least squares
SUB	subspace
SUMT	sequential unconstrained minimization technique
SVD	singular value decomposition
TLS	total least squares
WLS	weighted least squares
WSS	wide sense stationary
ZOH	zero order hold

1

Introduction

The temporal behavior of systems, such as e.g. technical systems from the areas of electrical engineering, mechanical engineering, and process engineering, as well as non-technical systems from areas as diverse as biology, medicine, chemistry, physics, economics, to name a few, can uniformly be described by mathematical models. This is covered by *systems theory*. However, the application of systems theory requires that the mathematical models for the static and dynamic behavior of the systems and their elements are known. The process of setting up a suitable model is called *modeling*. As is shown in the following section, two general approaches to modeling exist, namely *theoretical* and *experimental modeling*, both of which have their distinct advantages and disadvantages.

1.1 Theoretical and Experimental Modeling

A *system* is understood as a confined arrangement of mutually affected entities, see e.g. DIN 66201. In the following, these entities are processes. A *process* is defined as the conversion and/or the transport of material, energy, and/or information. Here, one typically differentiates between individual (sub-)processes and the entire process. Individual processes, i.e. (sub-)processes, can be the generation of mechanical energy from electric energy, the metal-cutting machining of workpieces, heat transfer through a wall, or a chemical reaction. Together with other sub-processes, the entire process is formed. Such aggregate processes can be an electrical generator, a machine tool, a heat exchanger, or a chemical reactor. If such a process is understood as an entity (as mentioned above), then multiple processes form a system such as e.g. a power plant, a factory, a heating system, or a plastic material production plant. The behavior of a system is hence defined by the behavior of its processes.

The derivation of mathematical system and process models and the representation of their temporal behavior based on measured signals is termed *system analysis* respectively *process analysis*. Accordingly, one can speak of *system identification* or *process identification* when applying the experimental system or process analysis techniques described in this book. If the system is excited by a stochastic signal,

R. Isermann, M. Münchhof, *Identification of Dynamic Systems*,
DOI 10.1007/978-3-540-78879-9_1, © Springer-Verlag Berlin Heidelberg 2011

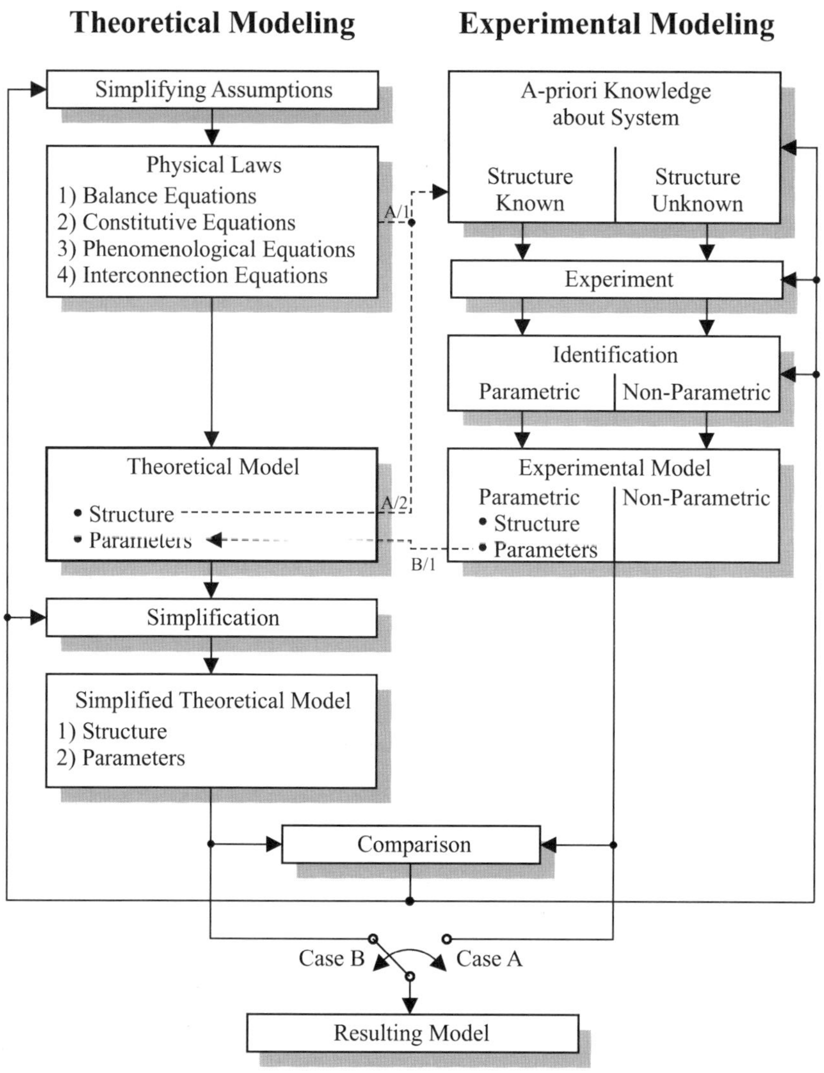

Fig. 1.1. Basic procedure for system analysis

one also has to analyze the signal itself. Thus the topic of *signal analysis* will also be treated. The title *Identification of Dynamic Systems* or simply *Identification* shall thus embrace all areas of identification as listed above.

For the derivation of mathematical models of dynamic systems, one typically discriminates between *theoretical* and *experimental* modeling. In the following, the basic approach of the two different ways of modeling shall be described shortly. Here, one has to distinguish *lumped parameter systems* and *distributed parameter systems*.

The states of *distributed parameter systems* depend on both the time and the location and thus their behavior has to be described by *partial differential equations (PDEs)*. *Lumped parameter systems* are easier to examine since one can treat all storages and states as being concentrated in single points and not spatially distributed. In this case, one will obtain *ordinary differential equations (ODEs)*.

For the *theoretical analysis*, also termed *theoretical modeling*, the model is obtained by applying methods from calculus to equations as e.g. derived from physics. One typically has to apply simplifying assumptions concerning the system and/or process, as only this will make the mathematical treatment feasible in most cases. In general, the following types of equations are combined to build the model, see also Fig. 1.1 (Isermann, 2005):

1. *Balance equations*: Balance of mass, energy, momentum. For distributed parameter systems, one typically considers infinitesimally small elements, for lumped parameter systems, a larger (confined) element is considered
2. *Physical or chemical equations of state*: These are the so-called constitutive equations and describe reversible events, such as e.g. inductance or the second Newtonian postulate
3. *Phenomenological equations*: Describing irreversible events, such as friction and heat transfer. An entropy balance can be set up if multiple irreversible processes are present
4. *Interconnection equations* according to e.g. Kirchhoff's node and mesh equations, torque balance, etc.

By applying these equations, one obtains a set of ordinary or partial differential equations, which finally leads to a theoretical model with a certain structure and defined parameters if all equations can be solved explicitly. In many cases, the model is too complex or too complicated, so that it needs to be simplified to be suitable for subsequent application. Figure 1.2 shows the order of the execution of individual simplifying actions. The first steps of this simplification procedure can already be carried out as the fundamental equations are set up by making appropriate simplifying assumptions. It is very tempting to include as many physical effects into the model as possible, especially nowadays, where simulation programs offer a wide variety of pre-build libraries of arbitrary degrees of complexity. However, this often occludes the predominant physical effects and makes both the understanding and the work with such a model a very tiresome, if not infeasible, endeavor.

But even if the resulting set of equations cannot be solved explicitly, still the individual equations give important hints concerning the model structure. Balance equations are always linear, some phenomenological equations are linear in a wide range. The physical and chemical equations of state often introduce non-linearities into the system model.

In case of an *experimental analysis*, which is also termed *identification*, a mathematical model is derived from measurements. Here, one typically has to rely on certain a priori assumptions, which can either stem from theoretical analysis or from previous (initial) experiments, see Fig. 1.1. Measurements are carried out and the input as well as the output signals are subjected to some identification method in order

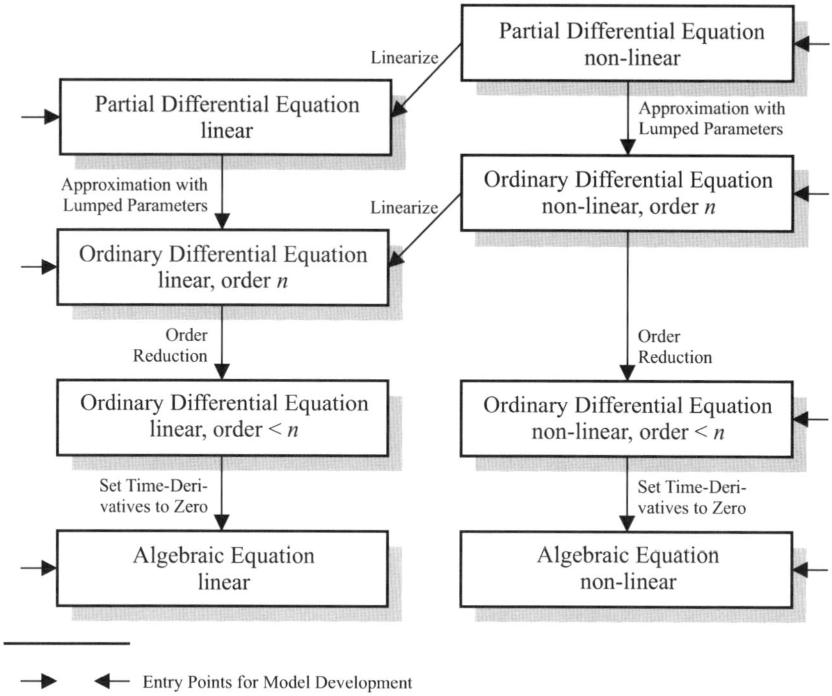

Fig. 1.2. Basic approach for theoretical modeling

to find a mathematical model that describes the relation between the input and the output. The input signals can either be a section of the the natural signals that act on the process during normal operation or can be an artificially introduced test signal with certain prespecified properties. Depending on the application, one can use *parametric* or *non-parametric models*, see Sect. 1.2. The resulting model is termed *experimental model.*

The theoretically and the experimentally derived models can be compared if both approaches can be applied and have been pursued. If the two models do not match, then one can get hints from the character and the size of the deviation, which steps of the theoretical or the experimental modeling have to be corrected, see Fig. 1.1.

Theoretical and experimental models thus complement one another. The analysis of the two models introduces a first feedback loop into the course of action for system analysis. Therefore, system analysis is typically an iterative procedure. If one is not interested in obtaining both models simultaneously, one has the choice between the experimental model (case A in Fig. 1.1) and the theoretical model (case B in Fig. 1.1). The choice mainly depends on the purpose of the derived model:

The theoretical model contains the functional dependencies between the physical properties of a system and its parameters. Thus, this model will typically be preferred if the system shall already be optimized in its static and dynamic behavior during the design phase or if its temporal behavior shall be simulated prior to the construction, respectively completion of the system.

On the contrary, the experimental model does only contain numbers as parameters, whose functional relations to the process properties remain unknown. However, this model can describe the actual dynamics of the system better and can be derived with less effort. One favors these experimental models for the adaptation of controllers (Isermann, 1991; Isermann et al, 1992; Åström et al, 1995; Åström and Wittenmark, 1997) and for the forecast of the respective signals or fault detection (Isermann, 2006).

In case B (Fig. 1.1), the main focus is on the theoretical analysis. In this setting, one employs the experimental modeling only once to validate the fidelity of the theoretical model or to determine process parameters, which can otherwise not be determined with the required accuracy. This is noted with the sequence B/1 in Fig. 1.1.

In contrast to case B, the emphasis is on the experimental analysis in case A. Here, one tries to apply as much a priori knowledge as possible from the theoretical analysis, as the model fidelity of the experimental model normally increases with the amount of a priori knowledge exploited. In the ideal case, the model structure is already known from the theoretical analysis (path A/2 in Fig. 1.1). If the fundamental equations of the model cannot be solved explicitly, if they are too complicated, or if they are not even completely known, one can still try to obtain information about the model structure from this incomplete knowledge about the process (sequence A/1 in Fig. 1.1).

The preceding paragraphs already pointed out that the system analysis can typically neither be completely theoretical nor completely experimental. To benefit from the advantages of both approaches, one does rarely use only theoretical modeling (leading to so-called *white-box models*) or only experimental modeling (leading to so-called *black-box models*), but rather a mixture of both leading to what is called *gray-box models*, see Fig. 1.3. This is a rather suitable combination of the two approaches, which is determined by the scope of application of the model and the system itself. The scope of application defines the required model accuracy and hence the effort that has to be put into the analysis. This introduces a second feedback loop into the schematic diagram presented in Fig. 1.1, which starts at the resulting models (either theoretical or experimental) and goes back to the individual modeling steps, hence one is confronted with a second iteration loop.

Despite the fact that the theoretical analysis can in principle deliver more information about the system, provided that the internal behavior is known and can be described mathematically, experimental analysis has found ever increasing attention over the past 50 years. The main reasons are the following:

- Theoretical analysis can become quite complex even for simple systems
- Mostly, model coefficients derived from the theoretical considerations are not precise enough
- Not all actions taking place inside the system are known

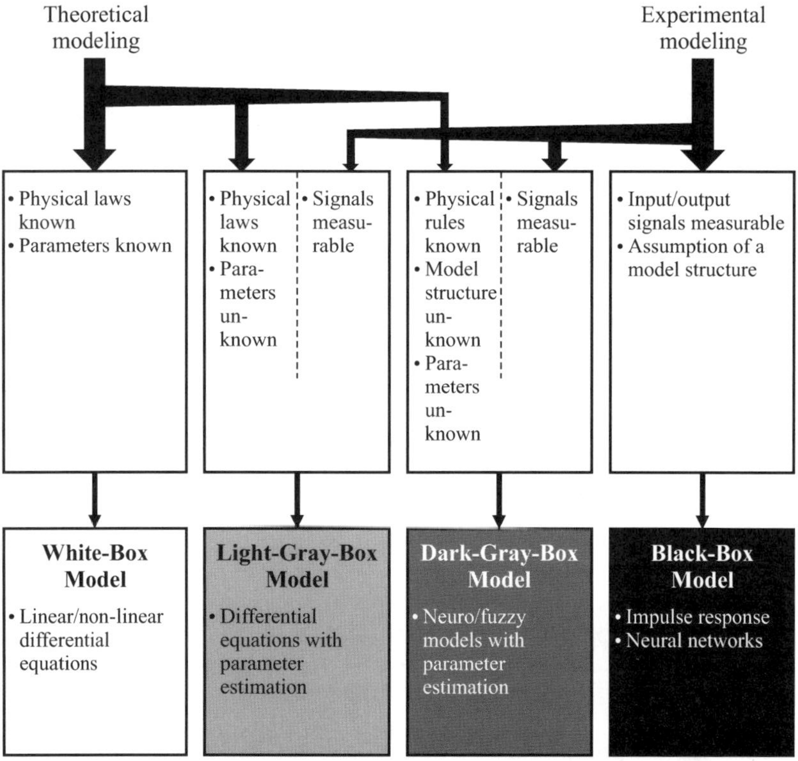

Fig. 1.3. Different kinds of mathematical models ranging from white box models to black box models

- The actions taking place cannot be described mathematically with the required accuracy
- Some systems are very complex, making the theoretical analysis too time-consuming
- Identified models can be obtained in shorter time with less effort compared to theoretical modeling

The experimental analysis allows the development of mathematical models by measurement of the input and output of systems of arbitrary composition. One major advantage is the fact that the same experimental analysis methods can be applied to diverse and arbitrarily complex systems. By measuring the input and output only, one does however only obtain models governing the input-output behavior of the system, i.e. the models will in general not describe the precise internal structure of the system. These input-output models are approximations and are still sufficient for many areas of application. If the system also allows the measurement of internal states, one can obviously also gather information about the internal structure of the

Table 1.1. Properties of theoretical modeling and identification

Theoretical Modeling	Identification
Model structure follows from laws of nature	Model structure must be assumed
Modeling of the input/output behavior as well as the internal behavior	Only the input/output behavior is identified
Model parameters are given as function of system properties	Model parameters are "numbers" only, in general no functional dependency to system properties known
Model is valid for the entire class of processes of a certain type and for different operating conditions	Model is only valid for investigated system and within operating limits
Model coefficients are not known exactly	Model coefficients are more precise for the given system within operating limits
Models can be build for non-existing systems	Model can only be identified for an existing system
The internal behavior of the system must be known and must be describable mathematically	Identification methods are independent of the investigated system and can thus be applied to many different systems
Typically lengthy process which takes up much time	Fast process if identification methods exist already
Models may be rather complex and detailed	Model size can be adjusted according to the area of application of the model

system. With the advent of digital computers starting in the 1960s, the development of capable identification methods has started. The different properties of theoretical modeling and identification have been summarized and set in contrast in Table 1.1.

1.2 Tasks and Problems for the Identification of Dynamic Systems

A process with a single input and a single output (SISO) is considered in the following. The process shall be stable to ensure a unique relation between input and output. Both the input and the output shall be measured without error. The task of identifying the process P is to find a mathematical model for the temporal behavior of the process from the measured input $u(t) = u_{\mathrm{M}}(t)$, the measured output $y(t) = y_{\mathrm{M}}(t)$ and optionally additional measured signals, see Fig. 1.4. This task is made more complicated, if *disturbances* $z_1 \dots z_i$ are acting on the process and are influencing the output signal. These disturbances can have various causes. The disturbances seen in the measured signals often stem from noise and hence will also be included in the term *noise* in the remainder of this book. The output is thus corrupted by a noise $n(t)$. In this case, one has to apply suitable techniques to separate the wanted signal

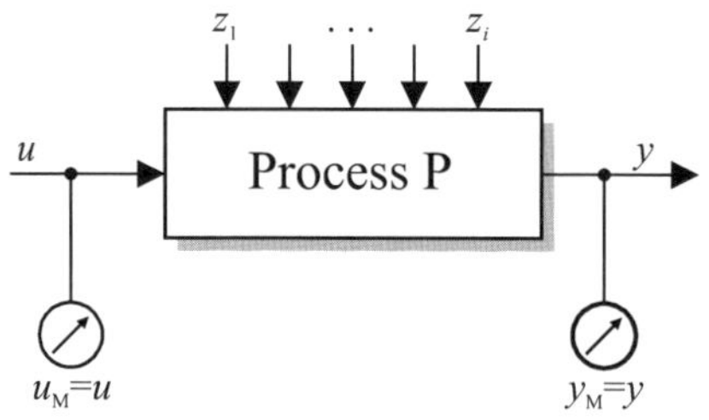

Fig. 1.4. Dynamic process with input u, output y and disturbances z_i

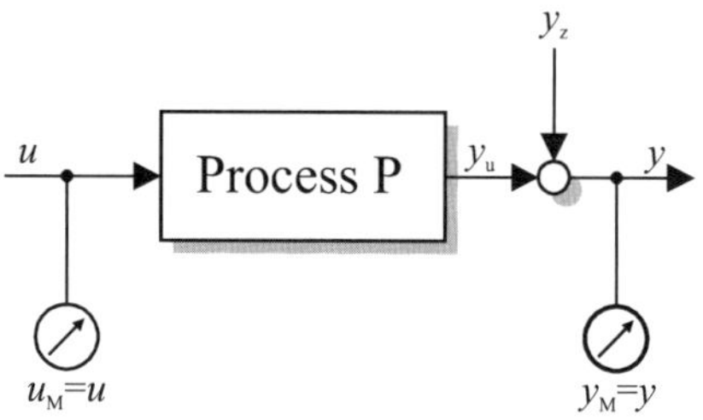

Fig. 1.5. Disturbed dynamic process with input u, output y, and noise n

$y_u(t)$, i.e. the response of the system due to the input $u(t)$, from the disturbances $n(t)$.

The term *identification* and the required subsequent tasks can thus be defined as follows:

Identification is the experimental determination of the temporal behavior of a process or system. One uses measured signals and determines the temporal behavior within a class of mathematical models. The error (respectively deviation) between the real process or system and its mathematical model shall be as small as possible.

This definition stems from Zadeh (1962), see also (Eykhoff, 1994). The measured signals are typically only the input to the system and the output from the system. However, if it is also possible to measure states of the process, then one can also gather information about the internal structure of the process.

In the following, a linear process is considered. In this case, the individual disturbance components $z_1, \dots, z_i$ can be combined into one representative disturbance $n(t)$, which is added to the wanted signal $y_u(t)$, see Fig. 1.5. If this disturbance $n(t)$ is not negligibly small, then its counterfeiting influence must be eliminated by the identification method as much as possible. For decreasing signal-to-noise ratios, the measurement time T_M must typically be increased.

For the identification itself, the following limitations have to be taken into consideration:

1. The available *measurement time* T_M is always limited, either due to technical reasons, due to time variance of the process parameters or due to economical reasons (i.e. budget), thus
$$T_M \le T_{M,max} \quad (1.2.1)$$
2. The maximum allowable change of the input signal, i.e. the *test signal height* u_0 is always limited, either due to technical reasons or due to the assumption of linear process behavior which is only valid within a certain operating regime
$$u_{min} \le u(t) \le u_{max} \quad (1.2.2)$$
3. The maximum allowable change of the output signal, *output signal*, y_0, may also be limited due to technical reasons or due to the assumption of linear process behavior

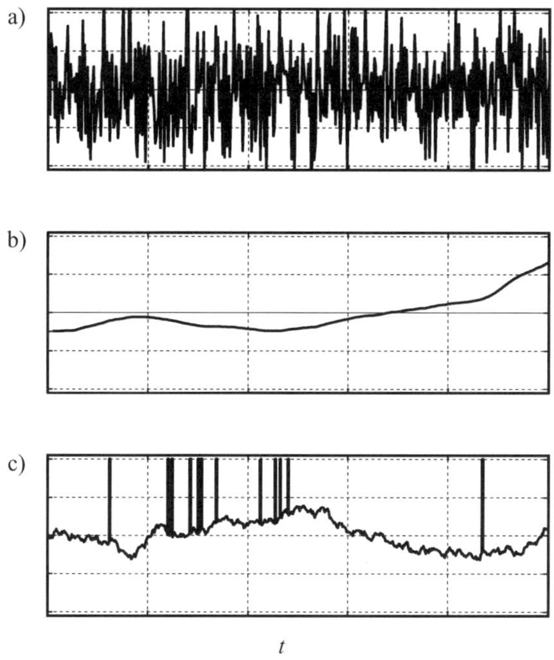

Fig. 1.6. Examples of disturbance components. (**a**) high frequent quasi-stationary stochastic disturbance. (**b**) low-frequent non-stationary stochastic disturbance. (**c**) disturbance with unknown character

which again is only valid within a certain operating regime

$$y_{\min} \leq y(t) \leq y_{\max} \tag{1.2.3}$$

4. The *disturbance* $n(t)$ typically consists of different components, which can be classified according to the following groups, see also Fig. 1.6:
   a) High-frequent quasi-stationary stochastic noise $n(t)$ with $\mathrm{E}\{n(t)\} = 0$. Higher frequent deterministic signal with $\overline{n(t)} = 0$.
   b) Low-frequent non-stationary stochastic or deterministic signal (e.g. drift, periodic signals with period times of one day or one year) $d(t)$
   c) Disturbance signal of unknown character (e.g. outliers) $h(t)$

It is assumed that within the limited measurement time, the disturbance component $n(t)$ can be treated as a stationary signal. The low-frequent component $d(t)$ must be treated as non-stationary, if it has stochastic character. Low-frequent deterministic disturbances can be drift and periodic signals with long period times such as one day or one year. Disturbance components with unknown character $h(t)$ are random signals, which cannot be described as stationary stochastic signals even for long measurement periods. This can be e.g. suddenly appearing, persistent, or disappearing disturbances and so-called *outliers*. These disturbances can e.g. stem from electromagnetic induction or malfunctions of the measurement equipment.

Typical identification methods can only eliminate the noise $n(t)$ as the measurement time is prolonged. Simple averaging or regression methods are often sufficient in this application. The components $d(t)$ require more specifically tailored measures such as special filters or regression methods which have been adapted to the very particular type of disturbance. Almost no general hints can be given concerning the elimination of the influence of $h(t)$. Such disturbances can only be eliminated manually or by special filters.

Effective identification methods must thus be able to determine the temporal behavior as precisely as possible under the constraints imposed by

- the given disturbance $y_z(t) = n(t) + d(t) + h(t)$
- the limited measurement time $T_M \leq T_{M,max}$
- the confined test signal amplitude $u_{min} \leq u(t) \leq u_{max}$
- the constrained output signal amplitude $y_{min} \leq y(t) \leq y_{max}$
- the purpose of the identification.

Figure 1.7 shows a general *sequence of an identification*. The following steps have to be taken:

First, the *purpose* has to be defined as the purpose determines the type of model, the required accuracy, the suitable identification methods and such. This decision is typically also influenced by the available budget, either the allocated financial resources or the expendable time.

Then, *a priori knowledge* must be collected, which encompasses all readily available information about the process to be identified, such as e.g.

- recently observed behavior of the process
- physical laws governing the process behavior
- rough models from previous experiments
- hints concerning linear/non-linear, time-variant/time-invariant as well as proportional/integral behavior of the process
- settling time
- dead time
- amplitude and frequency spectrum of noise
- operating conditions for conduction of measurements.

Now, the *measurement can be planned* depending on the purpose and the available a priori knowledge. One has to select and define the

- input signals (normal operating signals or artificial test signals and their shape, amplitude and frequency spectrum)
- sampling time
- measurement time
- measurements in closed-loop or open-loop operation of the process
- online or offline identification
- real-time or not

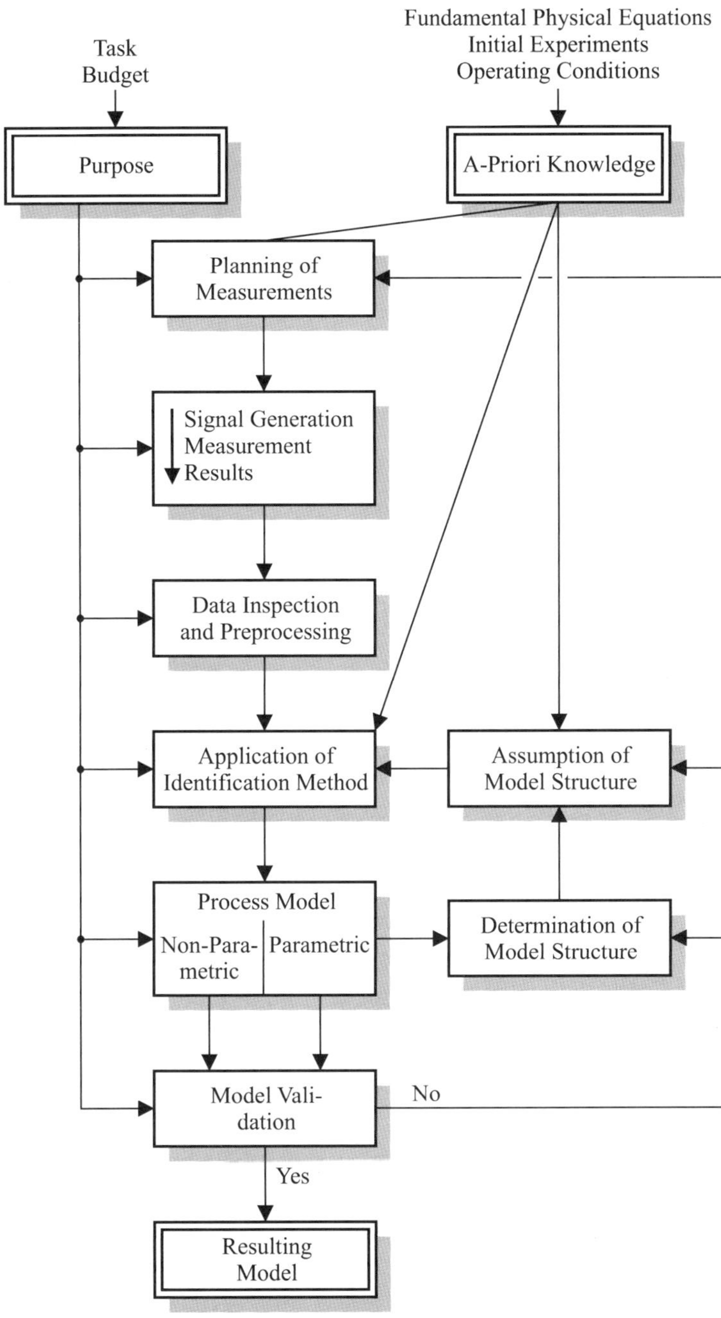

Fig. 1.7. Basic sequence of the identification

- necessary equipment (e.g. oscilloscope, PC, ...)
- filtering for elimination of noise
- limitations imposed by the actuators (saturation, ...).

Once these points have been clarified, the *measurements can be conducted.* This includes the signal generation, measurement, and data storage.

The collected data should undergo a first *visual inspection* and outliers as well as other easily detectable measurement errors should be removed. Then, as part of the further *pre-processing*, derivatives should be calculated, signals be calibrated, high-frequent noise be eliminated by e.g. low-pass filtering, and drift be removed. Some aspects of disturbance rejection and the removal of outliers by graphical and analytical methods are presented in Chap. 23. Methods to calculate the derivatives from noisy measurements are shown in Chap. 15.

After that, the *measurements will be evaluated* by the application of identification techniques and determination of model structure.

A very important step is the *performance evaluation of the identified model*, the so-called *validation* by comparison of model output and plant output or comparison of the experimentally established with the theoretically derived model. Validation methods are covered in Chap. 23. Typically, an identified model with the necessary model fidelity will not be derived in the first iteration. Thus, additional iteration steps might have to be carried out to obtain a suitable model.

Therefore, the last step is the possible *iteration*, i.e. the repeated conduction of measurements and evaluation of the measurements until a model meeting the imposed requirements has been found. One often has to conduct initial experiments, which allow to prepare and conduct the main experiments with better suited parameters or methods.

1.3 Taxonomy of Identification Methods and Their Treatment in This Book

According to the definition of identification as presented in the last section, the different identification methods can be classified according to the following criteria:

- Class of mathematical model
- Class of employed test signals
- Calculation of error between process and model

It has proven practical to also include the following two criteria:

- Execution of experiment and evaluation (online, offline)
- Employed algorithm for data processing

Mathematical models which describe the dynamic behavior of processes can be given either as functions relating the input and the output or as functions relating internal states. They can furthermore be set up as analytical models in the form of mathematical equations or as tables or characteristic curves. In the former case, the

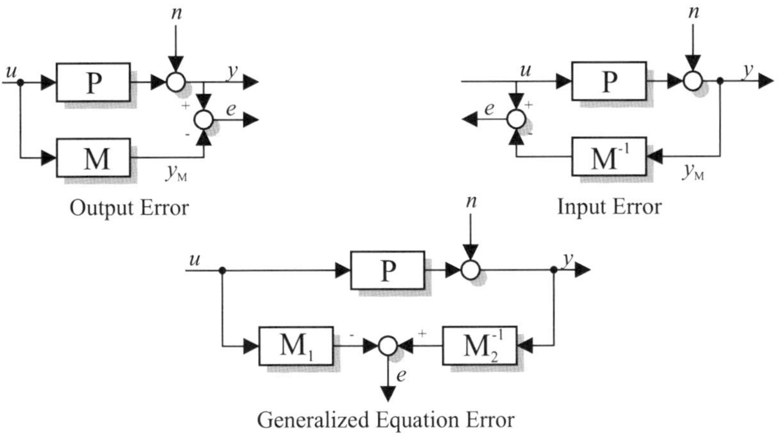

Fig. 1.8. Different setups for calculating the error between model M and process P

parameters of the model are explicitly included in the equation, in the latter case, they are not. Since the parameters of the system play a dominant role in identification, mathematical models shall first and foremost be classified by the model type as:

- Parametric models (i.e. models with structure and finite number of parameters)
- Non-parametric models (i.e. models without specific structure and infinite number of parameters)

Parametric models are equations, which explicitly contain the process parameters. Examples are differential equations or transfer functions given as an algebraic expression. *Non-parametric models* provide a relation between a certain input and the corresponding response by means of a table or sampled characteristic curve. Examples are impulse responses, step responses, or frequency responses presented in tabular or graphical form. They implicitly contain the system parameters. Although one could understand the functional values of a step response as "parameters", one would however need an infinite number of parameters to fully describe the dynamic behavior in this case. Consequently, the resulting model would be of infinite dimension. In this book, parametric models are thus understood as models with a finite number of parameters. Both classes of models can be sub-divided by the type of input and output signals as continuous-time models or discrete-time models.

The *input signals* respectively *test signals* can be deterministic (analytically describable) stochastic (random), or pseudo-stochastic (deterministic, but with properties close to stochastic signals).

As a measure for the *error* between model and process, one can choose between (see Fig. 1.8) the following errors:

- Input error
- Output error
- Generalized equation error

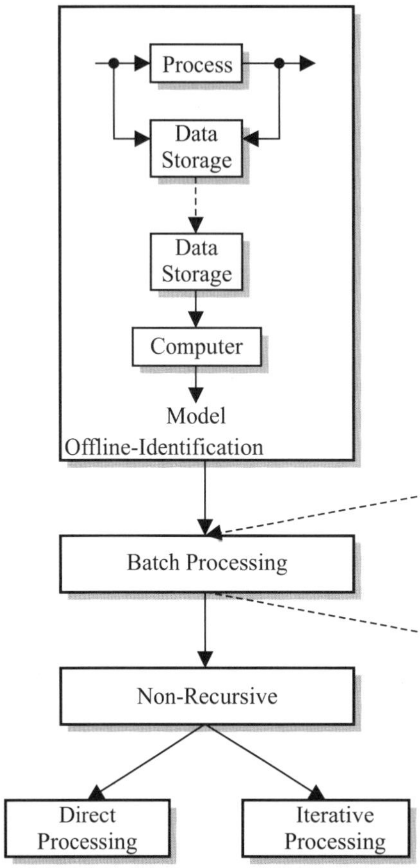

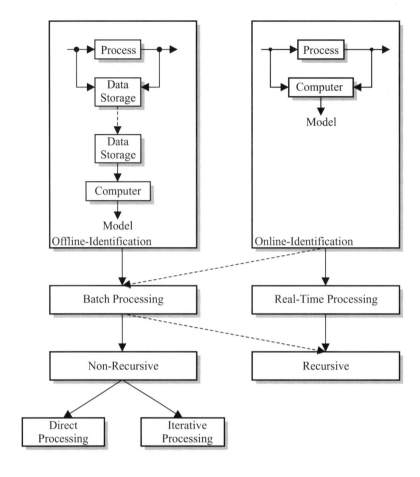

Fig. 1.9. Different setups for the data processing as part of the identification

Because of mathematical reasons, typically those errors are preferred, which depend linearly on the process parameters. Thus, one uses the output error if e.g. impulse responses are used as models and the generalized equation error if e.g. differential equations, difference equations, or transfer functions are employed. However, also output errors are used in the last case.

If digital computers are utilized for the identification, then one differentiates between two types of *coupling between process and computer*, see Fig. 1.9:

- Offline (indirect coupling)
- Online (direct coupling)

For the *offline identification*, the measured data are first stored (e.g. data storage) and are later transferred to the computer utilized for data evaluation and are processed there. The *online identification* is performed parallelly to the experiment. The computer is coupled with the process and the data points are operated on as they become available.

The identification with digital computers also allows to discern the identification according to the *type of algorithm* employed:

- Batch processing
- Real-time processing

In case of *batch processing*, the previously stored measurements will be processed in one shot, which is typically the case for offline applications. If the data are processed immediately after they become available, then one speaks of *real-time processing*, which necessitates a direct coupling between the computer and the process, see Fig. 1.9. Another feature is the processing of the data. Here, one can discern:

- Non-recursive processing
- Recursive processing

The *non-recursive methods* determine the model from the previously stored measurements and are thus a method of choice for offline processing only. On the contrary, the *recursive method* updates the model as each measurement becomes available. Hence, the new measurement is always used to improve the model derived in the previous step. The old measurements do not need to be stored. This is the typical approach for real-time processing and is called *real-time identification*. As not only the parameters, but also a measure of their accuracy (e.g. variance) can be calculated online, one can also think about running the measurement until a certain accuracy of the parameter estimates has been achieved (Åström and Eykhoff, 1971).

Finally, the non-recursive method can further be subdivided into:

- Direct processing
- Iterative processing

The *direct processing* determines the model in one pass. The *iterative processing* determines the model step-wise. Thus, iteration cycles are emerging and the data must be processed multiple times.

1.4 Overview of Identification Methods

The most important identification methods shall be described shortly. Table 1.2 compares their most prominent properties. A summary of the important advantages and disadvantages of the individual methods can be found in Sect. 23.4.

1.4.1 Non-Parametric Models

Frequency response measurements with periodic test signals allow the direct determination of discrete points of the frequency response characteristics for linear processes. The orthogonal correlation method has proven very effective for this task and is included in all frequency response measurement units. The necessary measurement

Table 1.2. Overview of the most prominent identification methods

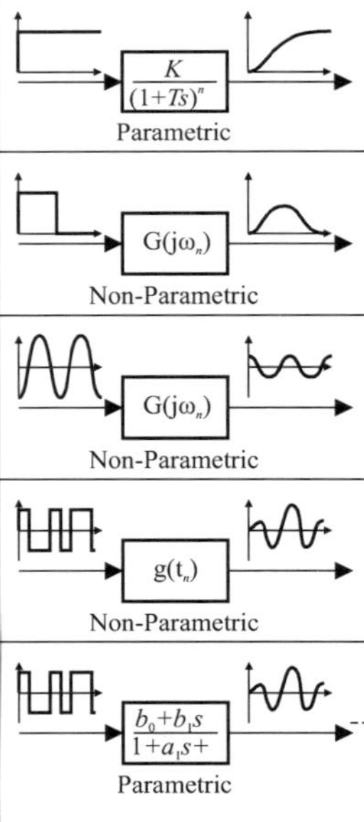

Method	Input / Model / Output	Linear Process	Nonlinear Process	Allowable Signal-to-Noise ratio	Online Processing	Offline Processing	Batch Processing	Real-Time Processing	Time-Variant Systems	MIMO Systems	Resulting Model Fidelity	Scope of Application
Determination of Characteristic Values	$\frac{K}{(1+Ts)^n}$ Parametric	✓	-	must be very large	-	✓	✓	-	-	-	Average	• Rough model • Controller tuning
Fourier Analysis	$G(j\omega_n)$ Non-Parametric	✓	-	must be large	✓	✓	✓	✓	-	✓	Average	• Validation of theoretically derived models
Frequency Response Measurement	$G(j\omega_\nu)$ Non-Parametric	✓	-	average	-	✓	✓	-	-	✓	Very good	• Validation of theoretically derived models • Design of classical (linear) controllers
Correlation Analysis	$g(t_\nu)$ Non-Parametric	✓	-	can be small	✓	✓	✓	✓	✓	✓	Good	• Determination of signal relations • Determination of time delays
Model Adjustment	$\frac{b_0+b_1 s}{1+a_1 s+}$ Parametric	✓	✓	must be large	✓	✓	-	✓	✓	-	Average	• Parameterization of models
Parameter Estimation	Parametric	✓	✓	can be small	✓	✓	✓	✓	✓	✓	Good	• Design of adaptive controllers • Adaptive controllers • Fault detection

✓ = Applicable; - = Not applicable

Table 1.2. Overview of the most prominent identification methods *(continued)*

Method	Input Model Output	Linear Process	Nonlinear Process	Allowable Signal-to-Noise ratio	Online Processing	Offline Processing	Batch Processing	Real-Time Processing	Time-Variant Systems	MIMO Systems	Resulting Model Fidelity	Scope of Application
Iterative Optimization	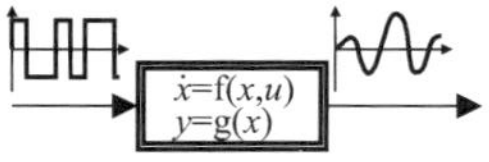 Parametric	-	✓	can be small	-	✓	✓	-	-	✓	Bad to Very Good	• Design of nonlinear controllers • Fault detection • Parameterization of models
Extended Kalman Filter	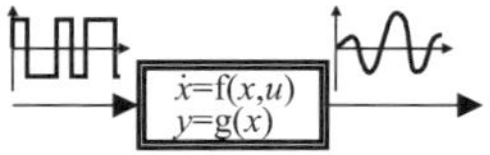 Parametric	✓	✓	average	✓	✓	✓	(✓)	✓	✓	Average	• Combined state and parameter estimation (e.g. no measurement of intermediate quantities • State estimation of nonlinear systems
Subspace Methods	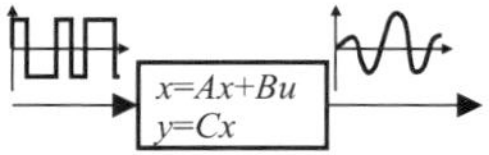 Parametric	✓	-	can be small	-	✓	✓	-	-	✓	Good	• Used for Modal Analysis
Neural Network	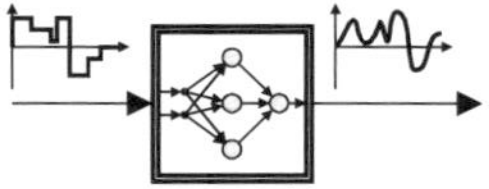 Parametric	✓	✓	can be small	-	✓	✓	-	-	✓	Average	• Design of nonlinear controllers • Fault detection • Modeling with little to no knowledge about the underlying process physics

✓= Applicable; (✓)= Possible, but not well suited; - = Not applicable

time is long if multiple frequencies shall be evaluated, but the resulting accuracy is very high. These methods are covered in this book in Chap. 5.

Fourier analysis is used to identify the frequency response from step or impulse responses for linear processes. It is a simple method with relatively small computational expense and short measurement time, but at the same time is only suitable for processes with good signal-to-noise ratios. A full chapter is devoted to Fourier analysis, see Chap. 3.

Correlation analysis is carried out in the time domain and works with continuous-time as well as discrete-time signals for linear processes. Admissible input signals are both stochastic and periodic signals. The method is also suitable for processes with bad signal-to-noise ratios. The resulting models are correlation functions or in special cases impulse responses for linear processes. In general, the method has a small computational expense. Correlation analysis is discussed in detail in Chap. 6 for the continuous-time case and Chap. 7 for the discrete-time case.

For all non-parametric identification techniques, it must only be ensured a priori that the process can be linearized. A certain model structure does not have to be assumed, what makes these methods very well suited for both lumped as well as distributed parameter systems with any degree of complexity. They are favored for the validation of theoretical models derived from theoretical considerations. Non-parametric models are favored since in this particular area of application, one is not interested in making any a priori assumptions about the model structure.

1.4.2 Parametric Models

For these methods, a dedicated model structure must be assumed. If assumed properly, more precise results are expected due to the larger amount of a priori knowledge.

The most simple method is the *determination of characteristic values*. Based on measured step or impulse responses, characteristic values, such as the delay time, are determined. With the aid of tables and diagrams, the parameters of simple models can then be calculated. These methods are only suitable for simple processes and small disturbances. They can however be a good starting point for a fast and simple initial system examination to determine e.g. approximate time constants, which allow the correct choice of the sample time for the subsequent application of more elaborate methods of system identification. The determination of characteristic values is discussed in Chap. 2.

Model adjustment methods were originally developed in connection with analog computers. However, they have lost most of their appeal in favor of parameter estimation methods.

Parameter estimation methods are based on difference or differential equations of arbitrary order and dead time. The methods are based on the minimization of certain error signals by means of statistical regression methods and have been complemented with special methods for dynamic systems. They can deal with an arbitrary excitation and small signal-to-noise ratios, can be utilized for manifold applications, work also in closed-loop, and can be extended to non-linear systems. A main focus of the book is placed on these parameter estimation methods. They are discussed e.g. in

Chap. 8, where static non-linearities are treated, Chap. 9, which discusses discrete-time dynamic systems, and Chap. 15, which discusses the application of parameter estimation methods to continuous-time dynamic systems.

Iterative optimization methods have been separated from the previously mentioned parameter estimation methods as these iterative optimization methods can deal with non-linear systems easily at the price of employing non-linear optimization techniques along with all the respective disadvantages.

Subspace-based methods have been used successfully in the area of modal analysis, but have also been applied to other areas of application, where parameters must be estimated. They are discussed in Chap. 16.

Also, *neural networks* as universal approximators have been applied to experimental system modeling. They often allow to model processes with little to no knowledge of the physics governing the process. Their main disadvantage is the fact that for most neural networks, the net parameters can hardly be interpreted in a physical sense, making it difficult to understand the results of the modeling process. However, local linear neural nets mitigate these disadvantages. Neural nets are discussed in detail in Chap. 20.

The *Kalman* filter is not used for parameter estimation, but is rather used for *state estimation* of dynamic systems. Some authors suggest to use the Kalman filter to smoothen the measurements as part of applying parameter estimation methods. A more general framework, the *extended Kalman Filter* allows the parallel estimation of states and parameters of both linear and non-linear systems. Its use for parameter estimation is reported in many citations. Chapter 21 will present the derivation of the Kalman filter and the extended Kalman filter and outline the advantages and disadvantages of the use of the extended Kalman filter for parameter estimation.

1.4.3 Signal Analysis

The signal analysis methods shown in Table 1.2 are employed to obtain parametric or non-parametric models of signals. Often, they are used to determine the frequency content of signals. The methods differ in many aspects.

A first distinction can be made depending on whether the method is used for *periodic, deterministic signals* or for *stochastic signals*. Also, not all methods are suited for *time-variant signals*, which in this context shall refer to signals, whose parameters (e.g. frequency content) change over time. There are methods available that work entirely in the *time domain* and others that analyze the signal in the *frequency domain*.

Not all methods are capable of making explicit statements on the *presence or absence of single spectral components*, i.e. oscillations at a *certain single frequency*, thus this capability represents another distinguishing feature. While many methods are capable of detecting periodic components in a signal, many methods can still not make a statement whether the recorded section of the signal is in itself *periodic* or not. Also, not all methods can determine the *amplitude* and the *phase* of the periodic signal components. Some methods can only determine the amplitude and some

Table 1.2. Overview of the most prominent signal analysis methods

Method	Periodic Signal	Stochastic Signal	Time-Variant	Time Domain	Frequency Domain	Single Frequency	Detection Periodicity	Amplitude	Phase	Comments
Bandpass Filtering	✓	✓	✓	-	(✓)	(✓)	-	✓	-	• Accuracy depends on filter passband • Old values do not need to be stored
Fourier Analysis	✓	-	-	-	✓	✓	✓	✓	✓	• Classical and easy to understand tool • Fast Fourier Transform existst in many implementations
Parametric Spectral Estimation	✓	-	-	✓	✓	✓	-	✓	✓	• Implementations available that do not suffer from windowing
Correlation Analysis	✓	✓	-	✓	-	-	✓	-	-	• Time domain method that allows to detect periodicity of signals and determine the period length
Spectrum Analysis	-	✓	-	-	✓	-	✓	✓	-	• Can use FFT as for Fourier Analysis
ARMA Parameter Estimation	✓	✓	-	-	✓	-	-	-	-	• Provides coefficients of a form filter that generates the signal • Typically slow convergence of the parameter estimation method
Short Time Fourier Transform	✓	-	✓	-	✓	✓	✓	✓	✓	• Blockwise application of the Fourier analysis
Wavelet Analysis	✓	-	✓	✓	-	✓	✓	✓	(✓)	• Well suited for signals with steep transients, e.g. rectangular waves and pulse type oscillations

✓= Applicable; - = Not Applicable

methods can neither determine the amplitude nor the phase without a subsequent analysis of the results delivered by the signal analysis method.

Bandpass filtering uses a bank of bandpass filters to analyze different frequency bands. The biggest advantage of this setup is that past values do not need to be stored. The frequency resolution depends strongly on the width of the filter passband.

Fourier analysis is a classical tool to analyze the frequency content of signals and is treated in detail in Chap. 3. The biggest advantage of this method is the fact that many commercial as well as non-commercial implementations of the algorithms exist.

Parametric spectral estimation methods can provide signal models as a form filter shaping white noise. They can also decompose a signal into a sum of sinusoidal oscillations. These methods are much less sensitive to the choice of the signal length than e.g. the Fourier analysis, where the sampling interval length typically has to be an integer multiple of the period length. These methods are discussed in Sect. 9.2.

Correlation analysis is discussed in detail in Chaps. 6 and 7. It is based on the correlation of a time signal with a time-shifted version of the same signal and is extremely well suited to determine whether a time signal is truly periodic and determine its period length.

Spectrum analysis examines the Fourier transform of the auto-correlation function, while the *ARMA parameter estimation* determines the coefficients of an ARMA form filter that generates the stochastic content of the signal. This will be presented in Sect. 9.4.2.

Finally, methods have been developed that allow a joint time-frequency analysis and can be used to check for changes in the signal properties. The *short time Fourier transform* applies the Fourier transform to small blocks of the recorded signals. The *wavelet analysis* calculates the correlation of the signal with a mother wavelet that is shifted and/or scaled in time. Both methods are presented in Chap. 3.

1.5 Excitation Signals

For identification purposes, one can supply the system under investigation either with the operational input signals or with artificially created signals, so-called *test signals*. Such test signals must in particular be applied, if the operational signals do not excite the process sufficiently (e. g. due to small amplitudes, non-stationarity, adverse frequency spectrum), which is often the case in practical applications. The favorable signals typically satisfy the following criteria:

- Simple and reproducible generation of the test signal with or without signal generator
- Simple mathematical description of the signal and its properties for the corresponding identification method
- Realizable with the given actuators
- Applicable to the process
- Good excitation of the interesting system dynamics

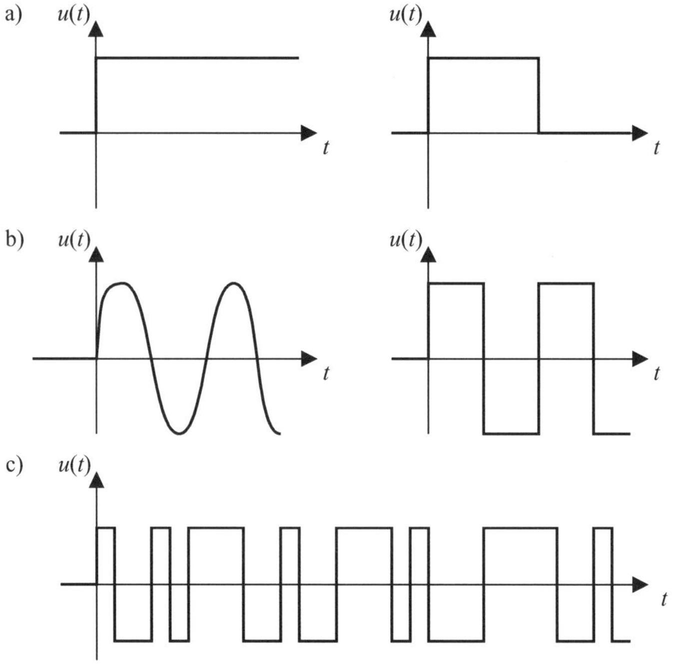

Fig. 1.10. Overview of some excitation signals. (**a**) non-periodic: step and square pulse. (**b**) periodic: sine wave and square wave. (**c**) stochastic: discrete binary noise

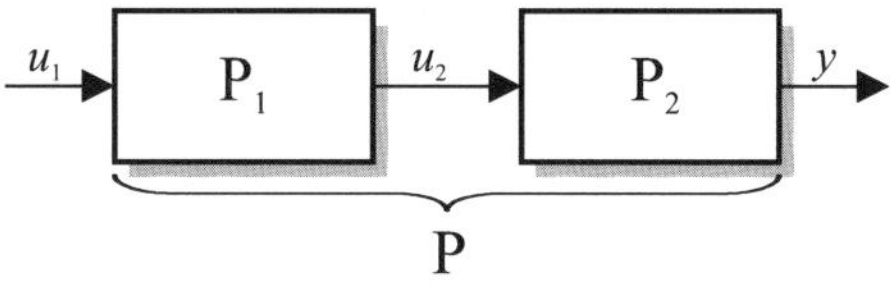

Fig. 1.11. Process P consisting of the subprocesses P_1 and P_2

Often, one cannot influence the input $u_2(t)$ that is directly acting on the subprocess P_2, which is to be identified. The input can only by influenced by means of the preceding subprocess P_1 (e.g. actuator) and its input $u_1(t)$, see Fig. 1.11. If $u_2(t)$ can be measured, the subprocess P_2 can be identified directly, if the identification method is applicable for the properties of $u_2(t)$. Is the method applicable for a special test signal $u_1(t)$ only, then one has to identify the entire process P and the sub-process P_1 and calculate P_2, which for linear systems is given as

$$G_{\text{P2}}(s) = \frac{G_{\text{P}}(s)}{G_{\text{P1}}(s)}, \tag{1.5.1}$$

where the $G(s)$ are the individual transfer functions.

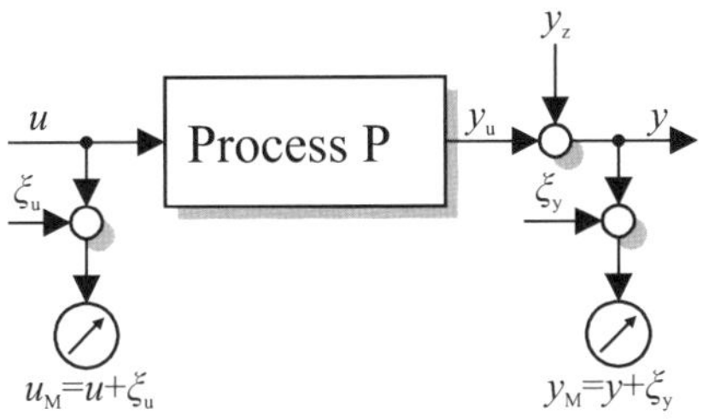

Fig. 1.12. Disturbed linear process with disturbed measurements of the input and output

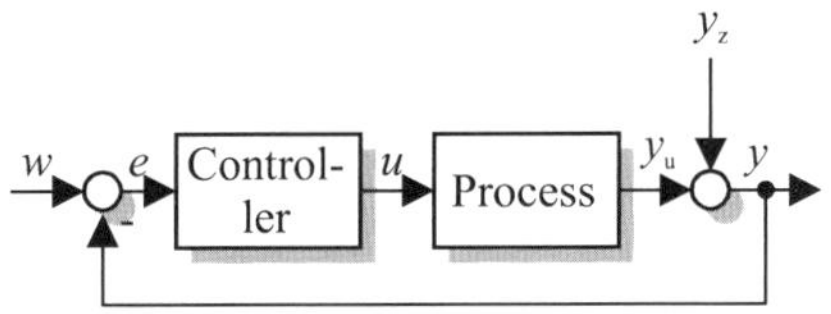

Fig. 1.13. Identification of a process in closed-loop

1.6 Special Application Problems

There are a couple of application problems, which shall be listed here shortly to sensitize the reader to these issues. They will be treated in more detail in later chapters.

1.6.1 Noise at the Input

So far, it was assumed that the disturbances acting on the process can be combined into a single additive disturbance at the output, y_z. If the measurement is disturbed by a disturbance $\xi_y(t)$, see Fig. 1.12, then this can be treated together with the disturbance $y_z(t)$ and thus does not pose a significant problem. More difficult is the treatment of a disturbed input signal $u(t)$, being counterfeit by $\xi_u(t)$. This is denoted as errors in variables, see Sect. 23.6. One approach to solve this problem is the method of total least squares (TLS) or the principal component analysis (PCA), see Chap. 10.

Proportional acting processes can in general be identified in open-loop. Yet, this is often not possible for processes with integral action as e.g. interfering disturbance signals may be acting on the process such that the output drifts away. Also, the process may not allow a longer open loop operation as the operating point may start to drift. In these cases as well as for unstable processes, one has to identify the process in closed-loop, see Fig. 1.13. If an external signal such as the setpoint is measurable, the process can be identified with correlation or parameter estimation methods. If there is no measurable external signal acting on the process (e.g. regulator settings with constant setpoint) and the only excitation of the process is by $y_z(t)$, then one is restricted in the applicable methods as well as the controller structure. Chapter 13 discusses some aspects that are proprietary to identification in closed-loop.

1.6.2 Identification of Systems with Multiple Inputs or Outputs

For linear systems with multiple input and/or output signals, see Fig. 1.14, one can also employ the identification methods for SISO processes presented in this book. For a system with one input and r outputs and one test signal, one can obtain r

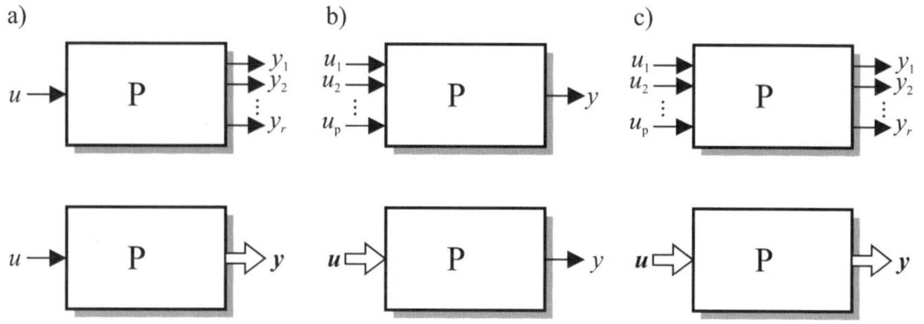

Fig. 1.14. Identification of (**a**) SIMO system with 1 input and r outputs, (**b**) MISO system with p inputs and 1 output, (**c**) MIMO system with p inputs and r outputs

input/output models by applying the identification method r times to the individual input/output combinations, see Fig. 1.14. A similar approach can be pursued for systems with r inputs and one output (MISO). One can excite one input after the other or one can excite all inputs at the same time with non-correlated input signals. The resulting model does not have to be the minimum realizable model, though.

For a system with multiple inputs and outputs (MIMO), one has three options: One can excite one input after the other and evaluate all outputs simultaneously, or one can excite all inputs at the same time and evaluate one output after the other, or one can excite all inputs simultaneously and also evaluate all outputs at the same time. If a model for the input/output behavior is sufficient, then one can successfully apply the SISO system identification methods. If however, one has p inputs which are excited simultaneously and r outputs, one should resort to methods specifically tailored to the identification of MIMO systems, as the assumed model structure plays an important role here. Parameter estimation of MIMO systems is discussed in Chap. 17.

1.7 Areas of Application

As already mentioned, the application of the resulting model has significant influence on the choice of the model classes, the required model fidelity, the identification method and the required identification hardware and software. Therefore, some sample areas of application shall be sketched in the following.

1.7.1 Gain Increased Knowledge about the Process Behavior

If it proves impossible to determine the static and dynamic behavior by means of theoretical modeling due to a lack of physical insight into the process, one has to resort to experimental modeling. Such complicated cases comprise technical processes as e.g. furnaces, combustion engines, bio-reactors, biological and economical processes. The choice of the identification method is mainly influenced by the questions whether or not special test signals can be inserted, whether the measurements can be taken continuously or only at discrete points in time, by the number of inputs and outputs, by the signal-to-noise ratio, by the available time for measurements and by the existence of feedback loops. The derived model must typically only be of good/medium fidelity. Often, it is sufficient to apply simple identification methods, however, one also applies parameter estimation methods quite frequently.

1.7.2 Validation of Theoretical Models

Due to the simplifying assumptions and the imprecise knowledge of process parameters, one quite frequently needs to validate a theoretically derived model with experiments conducted at a real process. For a (linear) model given in the form of a transfer function, the measurement of the frequency response provides a good tool to validate the theoretical model. The Bode diagram provides a very transparent representation of the dynamics of the process, such as resonances, the negligence of the higher frequent dynamics, dead time and model order. The major advantage of the frequency response measurement is the fact that no assumptions must be made about the model structure (e.g. model order, dead time,...). The most severe disadvantage is the long measurement time especially for processes with long settling times and the necessary assumption of linearity.

In the presence of mild disturbances only, it may also be sufficient to compare step responses of process and model. This is, of course, very transparent and natural. In the presence of more severe disturbances, however, one has to resort to correlation methods or parameter estimation methods for continuous-time models. The required model fidelity is medium to high.

1.7.3 Tuning of Controller Parameters

The rough tuning of parameters, e.g. for a PID controller, does not necessarily require a detailed model (like *Ziegler-Nichols experiment*). It is sufficient to determine some characteristic values from the step response measurement. For the fine-tuning however, the model must be much more precise. For this application, parameter estimation methods are favorable, especially for self-tuning digital controllers, see e.g. (Åström and Wittenmark, 1997, 2008; Bobál et al, 2005; O'Dwyer, 2009; Crowe et al, 2005; Isermann et al, 1992). These techniques should gain more momentum in the next decades as the technicians are faced with more and more controllers installed in plants and nowadays more than 50% of all controllers are not commissioned correctly, resulting in slowly oscillating control loops or inferior control performance (Pfeiffer et al, 2009).

1.7.4 Computer-Based Design of Digital Control Algorithms

For the design of model-based control algorithms, for e.g. internal model or predictive controllers or multi-variable controllers, one needs models of relatively high fidelity. If the control algorithms as well as the design methods are based on parametric, discrete-time models, parameter estimation methods, either offline or online, are the primary choice. For non-linear systems, either parameter estimation methods or neural nets are suitable (Isermann, 1991).

1.7.5 Adaptive Control Algorithms

If digital adaptive controllers are employed for processes with slowly time-varying coefficients, parametric discrete-time models are of great benefit since suitable models can be determined in closed loop and online by means of recursive parameter estimation methods. By the application of standardized controller design methods, the controller parameters can be determined easily. However, it is also possible to employ non-parametric models. This is treated e.g. in the books (Sastry and Bodson, 1989; Isermann et al, 1992; Ikonen and Najim, 2002; Åström and Wittenmark, 2008). Adaptive controllers are another important subject due to the same reasons already stated for the automatic tuning of controller parameters. However, the adaptation depends very much on the kind of excitation and has to be supervised continuously.

1.7.6 Process Supervision and Fault Detection

If the structure of a process model is known quite accurately from theoretical considerations, one can use continuous-time parameter estimation methods to determine the model parameters. Changes in the process parameters allow to infer on the presence of faults in the process. The analysis of the changes also allows to pinpoint the type of fault, its location and size. This task however imposes high requirements on the model fidelity. The primary choice are online identification methods with real time data processing or block processing. For a detailed treatment of this topic, see e.g. the book by Isermann (2006). Fault detection and diagnosis play an important role for *safety critical systems* and in the context of *asset management*, where all production equipment will be incorporated into a company wide network and all equipment will permanently assess its own state of health and request maintenance service autonomously upon the detection of tiny, incipient faults, which can cause harmful system behavior or stand-still of the production in the future.

1.7.7 Signal Forecast

For slow processes, such as e.g. furnaces or power plants, one is interested in forecasting the effect of the operator intervention by means of a simulation model to support the operator and enable him/her to judge the effects of his/her intervention. Typically, recursive online parameter estimation methods are exploited for the task of deriving a plant model. These methods have also been used to the prediction of economical markets as described e.g. by (Heij et al, 2007) as well as Box et al (2008).

Table 1.3. Bibliographical list for books on system identification since 1992 with no claim on completeness. ✓=Yes, (✓)=Yes, but not covered in depth, C=CD-ROM, D=Diskette, M=MatLab code or toolbox, W=Website. For a reference on books before 1992, see (Isermann, 1992). Realization theory based methods in (Juang, 1994; Juang and Phan, 2006) are sorted as subspace methods.

Citation (see references, pp. 29ff)	Characteristic parameters	Fourier transform	Freq. resp. measurement	Correlation analysis	LS parameter estimation	Numerical optimization	Standard Kalman filter	Extended Kalman filter	Neural nets / Neurofuzzy	Subspace methods	Frequency domain	Input signals	Volterra, Hammerst., Wiener	Continuous-time systems	MIMO systems	Closed loop identification	Real Data	Other
Chui and Chen (2009)							✓	✓						(✓)	✓			
Grewal and Andrews (2008)							✓	✓						(✓)	✓			C,M
Box et al (2008)		(✓)		✓	✓	(✓)						(✓)			✓		✓	
Garnier and Wang (2008)		✓			✓	✓				✓	✓			✓	✓	✓	✓	M
van den Bos (2007)					✓	✓						(✓)						
Heij et al (2007)		✓			(✓)		✓											C
Lewis et al (2008)							✓	✓			(✓)			✓	✓			
Mikleš and Fikar (2007)	✓				✓									✓			✓	W
Verhaegen and Verdult (2007)	✓	✓			✓	✓	✓			✓		✓			✓	(✓)		M
Bohlin (2006)						✓		✓						✓	✓		✓	W
Eubank (2006)							✓											
Juang and Phan (2006)										✓								
Goodwin et al (2005)						✓	(✓)											
Raol et al (2004)		✓			✓	✓	✓	✓	✓		(✓)			✓		✓	✓	W,M
Doyle et al (2002)											✓	✓	✓	✓			✓	
Harris et al (2002)					✓				✓									
Ikonen and Najim (2002)					✓	✓	✓	✓					✓					
Söderström (2002)							✓	✓						✓	✓	✓		
Pintelon and Schoukens (2001)		✓			✓	✓				✓	✓	✓	(✓)	✓	✓	✓	✓	M
Nelles (2001)					✓	✓			✓			✓	✓		✓	✓	✓	M
Unbehauen (2000)	(✓)		✓	✓	✓							✓			✓	✓		
Kamen and Su (1999)							✓	✓										M
Koch (1999)					✓													
Ljung (1999)		✓		✓	✓	✓	✓		✓	✓	✓	✓	✓	✓	✓	✓	✓	W,M
van Overshee and de Moor (1996)							(✓)			✓					✓		✓	D,M
Brammer and Siffling (1994)							✓							✓	✓	✓		
Juang (1994)		✓		✓	(✓)					✓	(✓)				✓	✓	✓	
Isermann (1992)	✓	✓	✓	✓	✓	(✓)					✓	✓	✓	✓	✓	✓	✓	

1.7.8 On-Line Optimization

If the task is to operate a process in its optimal operating point (e.g. for large Diesel ship engines or steam power plants), parameter estimation methods are used to derive an online non-linear dynamic model which then allows to find the optimal operating point by means of mathematical optimization techniques. As the price for energy and production goods, such as crude oil and chemicals, is increasing at a rapid level, it will become more and more important to operate the process as efficiently as possible.

From this variety of examples, one can clearly see the strong influence of the intended application on the choice of the system identification methods. Furthermore, the user is only interested in methods that can be applied to a variety of different problems. Here, parameter estimation methods play an important role since they can easily be modified to not only include linear, time-invariant SISO processes, but also cover non-linear, time-varying, and multi-variable processes.

It has also been illustrated that many of these areas of application of identification techniques will present attractive research and development fields in the future, thus creating a demand for professionals with a good knowledge of system identification. A selection of applications of the methods presented in this book is presented in Chap. 24.

1.8 Bibliographical Overview

The development of system identification has been pushed forward by new developments in diverse areas:

- System theory
- Control engineering
- Signal theory
- Time series analysis
- Measurement engineering
- Numerical mathematics
- Computers and micro-controllers

The published literature is thus spread across the different above-mentioned areas of research and their subject-specific journals and conferences. A systematic treatment of the subject can be found in the area of automatic control, where the *IFAC Symposia on System Identification (SYSID)* have been established in 1967 as a triennial platform for the community of scientists working in the area of identification of systems. The symposia so far have taken place in Prague (1967, *Symposium on Identification in Automatic Control Systems*), Prague (1970, *Symposium on Identification and Process Parameter Estimation*), The Hague (1973, *Symposium on Identification and System Parameter Estimation*), Tbilisi (1976), Darmstadt (1979), Washington, DC, (1982), York (1985), Beijing (1988), Budapest (1991), Copenhagen (1994), Kitakyushu (1997), Santa Barbara, CA (2000), Rotterdam (2003), Newcastle (2006) and Saint-Malo (2009). The *International Federation of Automatic Control* (IFAC) has also devoted the work of the Technical Committee 1.1 to the area of modeling, identification, and signal processing.

Due to the above-mentioned fact that system identification is under research in many different areas, it is difficult to give an overview over all publications that have appeared in this area. However, Table 1.3 tries to provide a list of books that are devoted to system identification, making no claim on completeness. As can be seen from the table many textbooks concentrate on certain areas of system identification.

Problems

1.1. Theoretical Modeling
Describe the theoretical modeling approach. Which equations can be set up and combined into a model? Which types of differential equations can result? Why is the application of purely theoretical modeling approaches limited?

1.2. Experimental Modeling
Describe the experimental modeling approach. What are its advantages and disadvantages?

1.3. Model Types
What are white-box, gray-box, and black-box models?

1.4. Identification
What are the tasks of the identification?

1.5. Limitations in Identification
Which limitations are imposed on a practical identification experiment?

1.6. Disturbances
Which typical disturbances are acting on the process? How can their effect be eliminated?

1.7. Identification
Which steps have to be taken in the sequence of system identification?

1.8. Taxonomy of Identification Methods
According to which features can identification methods be classified?

1.9. Non-Parametric/Parametric Models
What is the difference between a non-parametric and a parametric model? Give examples.

1.10. Areas of Application
Which identification methods are suitable for validation of theoretical linear models and the design of digital control algorithms.

References

Åström KJ, Eykhoff P (1971) System identification – a survey. Automatica 7(2):123–162

Åström KJ, Wittenmark B (1997) Computer controlled systems: theory and design, 3rd edn. Prentice-Hall information and system sciences series, Prentice Hall, Upper Saddle River, NJ

Åström KJ, Wittenmark B (2008) Adaptive control, 2nd edn. Dover Publications, Mineola, NY

Åström KJ, Goodwin GC, Kumar PR (1995) Adaptive control, filtering, and signal processing. Springer, New York

Bobál V, Böhm J, Fessl J, Machácek J (2005) Digital self-tuning controllers — Algorithms, implementation and applications. Advanced Textbooks in Control and Signal Processing, Springer, London

Bohlin T (2006) Practical grey-box process identification: Theory and applications. Advances in Industrial Control, Springer, London

van den Bos A (2007) Parameter estimation for scientists and engineers. Wiley-Interscience, Hoboken, NJ

Box GEP, Jenkins GM, Reinsel GC (2008) Time series analysis: Forecasting and control, 4th edn. Wiley Series in Probability and Statistics, John Wiley, Hoboken, NJ

Brammer K, Siffling G (1994) Kalman-Bucy-Filter: Deterministische Beobachtung und stochastische Filterung, 4th edn. Oldenbourg, München

Chui CK, Chen G (2009) Kalman filtering with real-time applications, 4th edn. Springer, Berlin

Crowe J, Chen GR, Ferdous R, Greenwood DR, Grimble MJ, Huang HP, Jeng JC, Johnson MA, Katebi MR, Kwong S, Lee TH (2005) PID control: New identification and design methods. Springer, London

DIN Deutsches Institut für Normung e V (Juli 1998) Begriffe bei Prozessrechensystemen

Doyle FJ, Pearson RK, Ogunnaike BA (2002) Identification and control using Volterra models. Communications and Control Engineering, Springer, London

Eubank RL (2006) A Kalman filter primer, Statistics, textbooks and monographs, vol 186. Chapman and Hall/CRC, Boca Raton, FL

Eykhoff P (1994) Identification in Measurement and Instrumentation. In: Finkelstein K, Grattam TV (eds) Concise Encyclopaedia of Measurement and Instrumentation, Pergamon Press, Oxford, pp 137–142

Garnier H, Wang L (2008) Identification of continuous-time models from sampled data. Advances in Industrial Control, Springer, London

Goodwin GC, Doná JA, Seron MM (2005) Constrained control and estimation: An optimisation approach. Communications and Control Engineering, Springer, London

Grewal MS, Andrews AP (2008) Kalman filtering: Theory and practice using MATLAB, 3rd edn. John Wiley & Sons, Hoboken, NJ

Harris C, Hong X, Gan Q (2002) Adaptive modelling, estimation and fusion from data: A neurofuzzy approach. Advanced information processing, Springer, Berlin
Heij C, Ran A, Schagen F (2007) Introduction to mathematical systems theory : linear systems, identification and control. Birkhäuser Verlag, Basel
Ikonen E, Najim K (2002) Advanced process identification and control, Control engineering, vol 9. Dekker, New York
Isermann R (1991) Digital control systems, 2nd edn. Springer, Berlin
Isermann R (1992) Identifikation dynamischer Systeme: Grundlegende Methoden (Vol. 1). Springer, Berlin
Isermann R (2005) Mechatronic Systems: Fundamentals. Springer, London
Isermann R (2006) Fault-diagnosis systems: An introduction from fault detection to fault tolerance. Springer, Berlin
Isermann R, Lachmann KH, Matko D (1992) Adaptive control systems. Prentice Hall international series in systems and control engineering, Prentice Hall, New York, NY
Juang JN (1994) Applied system identification. Prentice Hall, Englewood Cliffs, NJ
Juang JN, Phan MQ (2006) Identification and control of mechanical systems. Cambridge University Press, Cambridge
Kamen EW, Su JK (1999) Introduction to optimal estimation. Advanced Textbooks in Control and Signal Processing, Springer, London
Koch KR (1999) Parameter estimation and hypothesis testing in linear models, 2nd edn. Springer, Berlin
Lewis FL, Xie L, Popa D (2008) Optimal and robust estimation: With an introduction to stochastic control theory, Automation and control engineering, vol 26, 2nd edn. CRC Press, Boca Raton, FL
Ljung L (1999) System identification: Theory for the user, 2nd edn. Prentice Hall Information and System Sciences Series, Prentice Hall PTR, Upper Saddle River, NJ
Mikleš J, Fikar M (2007) Process modelling, identification, and control. Springer, Berlin
Nelles O (2001) Nonlinear system identification: From classical approaches to neural networks and fuzzy models. Springer, Berlin
O'Dwyer A (2009) Handbook of PI and PID controller tuning rules, 3rd edn. Imperial College Press, London
van Overshee P, de Moor B (1996) Subspace identification for linear systems: Theory – implementation – applications. Kluwer Academic Publishers, Boston
Pfeiffer BM, Wieser R, Lorenz O (2009) Wie verbessern Sie die Performance Ihrer Anlage mit Hilfe der passenden APC-Funktionen? Teil 1: APC-Werkzeuge in Prozessleitsystemen. atp 51(4):36–44
Pintelon R, Schoukens J (2001) System identification: A frequency domain approach. IEEE Press, Piscataway, NJ
Raol JR, Girija G, Singh J (2004) Modelling and parameter estimation of dynamic systems, IEE control engineering series, vol 65. Institution of Electrical Engineers, London

Sastry S, Bodson M (1989) Adaptive control : stability, convergence, and robustness. Prentice Hall Information and System Sciences Series, Prentice Hall, Englewood Cliffs, NJ

Söderström T (2002) Discrete-time stochastic systems: Estimation and control, 2nd edn. Advanced Textbooks in Control and Signal Processing, Springer, London

Unbehauen H (2000) Regelungstechnik Bd. 3: Identifikation, Adaption, Optimierung, 6th edn. Vieweg + Teubner, Wiesbaden

Verhaegen M, Verdult V (2007) Filtering and system identification: A least squares approach. Cambridge University Press, Cambridge

Zadeh LA (1962) From circuit theory to system theory. Proc IRE 50:850–865

2

Mathematical Models of Linear Dynamic Systems and Stochastic Signals

The main task of identification methods is to derive mathematical models of processes and their signals. Therefore, the most important mathematical models of linear, time-invariant SISO processes as well as stochastic signals shall shortly be presented in the following. It is assumed that the reader is already familiar with time- and frequency domain based models and methods. If this is not the case, the reader is referred to the multitude of textbooks dealing with control engineering and covering this topic in much more breadth (Åström and Murray, 2008; Chen, 1999; Dorf and Bishop, 2008; Franklin et al, 2009; Goodwin et al, 2001; Nise, 2008; Ogata, 2009). The following short discussion is only meant to agree upon the notation and allow the reader to recall the most important relations.

Systems are termed *linear*, if the superposition principle can be applied. The system output due to multiple input signals is then given as the superposition of the corresponding output signals. In the most simple case, the behavior of the linear system is described by means of a linear ordinary differential equation (ODE). If the coefficients do not change, the system is termed *time-invariant*, otherwise, i.e. if the system parameters change over time, one has to deal with a *time-variant* system. Suitable models for non-linear processes are introduced with the respective identification methods in later chapters. Systems are termed *affine*, if they have a constant term added to the output.

The *taxonomy* of the models described in the following leans on criteria which have been found useful with respect to the taxonomy of identification methods as well as the latter scope of application of the resulting model. In general, a distinction is made between parametric and non-parametric models, models in input/output or state space representation, time domain and frequency domain based models.

2.1 Mathematical Models of Dynamic Systems for Continuous Time Signals

First, the theory of mathematical models for dynamic systems in continuous-time shall be reviewed shortly as the understanding of these fundamentals is indispensable

R. Isermann, M. Münchhof, *Identification of Dynamic Systems*,
DOI 10.1007/978-3-540-78879-9_2, © Springer-Verlag Berlin Heidelberg 2011

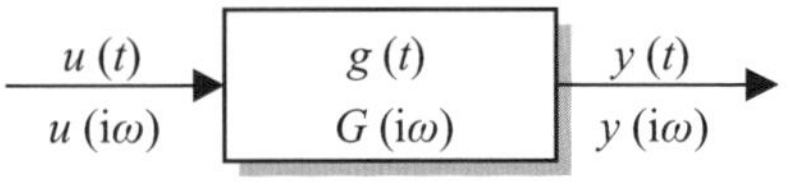

Fig. 2.1. Dynamic process with input u and output y

for the understanding and application of the identification methods presented in the remainder of this book.

2.1.1 Non-Parametric Models, Deterministic Signals

Mathematical models of processes, but also of signals, can be either non-parametric or parametric. *Non-parametric models* represent the relation between the input and the output by means of a table or curve. They do not exhibit a certain structure, are typically of infinite dimension and lay the foundation for so-called *black-box methods*. Thus, they shall be called *black models* in the following. The most prominent non-parametric models of time-invariant, linear processes are the impulse response, step response and the frequency response, see Fig. 2.1.

Impulse Response

The *impulse response* $g(t)$ is defined as the output of a process being excited by an *impulse* (Dirac's delta function) $\delta(t)$. This impulse function is defined as

$$\delta(t) = \begin{cases} \infty & \text{for } t = 0 \\ 0 & \text{for } t \neq 0 \end{cases} \tag{2.1.1}$$

$$\int_{-\infty}^{\infty} \delta(t)\,\mathrm{d}t = 1\,\mathrm{sec}\,. \tag{2.1.2}$$

By means of the impulse response, one can determine the output of a linear process for an arbitrary, deterministic input by employing the *convolution integral* as

$$y(t) = \int_0^t g(t-\tau)\,u(\tau)\mathrm{d}\tau = \int_0^t g(\tau)\,u(t-\tau)\mathrm{d}\tau\,. \tag{2.1.3}$$

The *step function* $\sigma(t)$ is also called the *Heaviside function* $\mathcal{H}(t)$. It is defined as

$$\sigma(t) = \begin{cases} 1 \text{ for } t \geq 0 \\ 0 \text{ for } t < 0\,. \end{cases} \tag{2.1.4}$$

A step can be obtained by integrating the impulse with respect to time t. The system output is defined as the *step response* $h(t)$ and can be calculated by convoluting the input signal with the impulse response $g(t)$ as

$$h(t) = \int_0^{\infty} g(\tau)\,\sigma(t-\tau)\mathrm{d}\tau = \int_0^t g(\tau)\mathrm{d}\tau\,. \tag{2.1.5}$$

The impulse response is thus the time-derivative of the step response, i.e.

$$g(t) = \frac{\mathrm{d}h(t)}{\mathrm{d}t} . \tag{2.1.6}$$

Note that the Heaviside function can also be defined as

$$\mathcal{H}_c(t) = \begin{cases} 1 \text{ for } t > 0 \\ c \text{ for } t = 0 \\ 0 \text{ for } t < 0 , \end{cases} \tag{2.1.7}$$

where $c = 0$ (Föllinger, 2010), $c = 1/2$, which increases the symmetry (Bracewell, 2000; Bronstein et al, 2008) or $c = 1$ which makes the definitions of the continuous-time and the discrete-time step function (which is also 1 for $k = 0$) quite similar.

Frequency Response, Transfer Function

The *frequency response* is the equivalent of the impulse response in the frequency domain. It is defined as the ratio of the vectors of the input and output quantity, if the process is excited by a harmonic oscillation and one waits until the steady-state response is fully developed,

$$G(\mathrm{i}\omega) = \frac{\boldsymbol{y}(\omega t)}{\boldsymbol{u}(\omega t)} = \frac{y_0(\omega)\mathrm{e}^{\mathrm{i}(\omega t+\varphi(\omega))}}{u_0(\omega)\mathrm{e}^{\mathrm{i}\omega t}} = \frac{y_0(\omega)}{u_0(\omega)}\mathrm{e}^{\mathrm{i}\varphi(\omega)} . \tag{2.1.8}$$

By means of the Fourier transform, which is treated in detail e.g. in (Papoulis, 1962; Föllinger and Kluwe, 2003), the frequency response can also be determined for non-periodic signals. The Fourier transform maps the function $x(t)$ in the time domain to the function $x(\mathrm{i}\omega)$ in the frequency domain as

$$\mathfrak{F}\{x(t)\} = x(\mathrm{i}\omega) = \int_{-\infty}^{\infty} x(t)\mathrm{e}^{-\mathrm{i}\omega t}\mathrm{d}t . \tag{2.1.9}$$

The corresponding *inverse Fourier transform* is given as

$$\mathfrak{F}^{-1}\{x(\mathrm{i}\omega)\} = x(t) = \frac{1}{2\pi}\int_{-\infty}^{\infty} x(\mathrm{i}\omega)\mathrm{e}^{\mathrm{i}\omega t}\mathrm{d}\omega . \tag{2.1.10}$$

If $f(t)$ is piecewise continuous and absolutely integrable, i.e.

$$\int_{-\infty}^{\infty} |x(t)|\,\mathrm{d}t < \infty , \tag{2.1.11}$$

then the Fourier transform exists and is a bounded continuous function (Poularikas, 1999) The frequency response is defined for non-periodic signals as the ratio of the Fourier transform of the output and the input,

$$G(\mathrm{i}\omega) = \frac{\mathfrak{F}\{y(t)\}}{\mathfrak{F}\{u(t)\}} = \frac{y(\mathrm{i}\omega)}{u(\mathrm{i}\omega)} . \tag{2.1.12}$$

The relation of the input and output in the time domain by means of the convolution is given as a simple multiplication

$$y(\mathrm{i}\omega) = G(\mathrm{i}\omega)\, u(\mathrm{i}\omega) \tag{2.1.13}$$

in the frequency domain. Since the Fourier transform of the Dirac delta impulse is

$$\mathfrak{F}\{\delta(t)\} = 1\,\mathrm{sec}\ , \tag{2.1.14}$$

one gets from (2.1.12)

$$G(\mathrm{i}\omega) = \frac{\mathfrak{F}\{g(t)\}}{\mathfrak{F}\{\delta(t)\}} = \int_0^\infty g(t)\,\mathrm{e}^{-\mathrm{i}\omega t}\,\mathrm{d}t\,\frac{1}{1\,\mathrm{sec}}\ , \tag{2.1.15}$$

which shows that the frequency response is the Fourier transform of the impulse response. The Fourier transform is treated again in Chap. 3, where the implementation of the Fourier transform on digital computers and the effect of applying the Fourier transform to data sequences of finite length are discussed in detail.

Since the Fourier transform does not exist for certain, often encountered input signals, such as e.g. the step function or the ramp function, one is interested in a way to determine the transfer function for these non-periodic signals as well. For this task, the *Laplace transform* is given as

$$\mathfrak{L}\{x(t)\} = x(s) = \int_0^\infty x(t)\mathrm{e}^{-st}\,\mathrm{d}t\ , \tag{2.1.16}$$

assuming that $x(t) = 0$ for $t < 0$, with the Laplace variable $s = \delta + \mathrm{i}\omega,\ \delta > 0$ and the *inverse Laplace transform*

$$\mathfrak{L}^{-1}\{x(s)\} = x(t) = \frac{1}{2\pi\mathrm{i}}\int_{\delta-\mathrm{i}\infty}^{\delta+\mathrm{i}\infty} x(s)\mathrm{e}^{st}\,\mathrm{d}s\ . \tag{2.1.17}$$

Now, the transfer function is given as the ratio of the Laplace transform of the output and the input as

$$G(s) = \frac{\mathfrak{L}\{y(t)\}}{\mathfrak{L}\{u(t)\}} = \frac{y(s)}{u(s)} \tag{2.1.18}$$

and in analogy to (2.1.15),

$$G(s) = \frac{\mathfrak{L}\{g(t)\}}{\mathfrak{L}\{\delta(t)\}} = \int_0^\infty g(t)\,\mathrm{e}^{-st}\,\mathrm{d}t\,\frac{1}{1\,\mathrm{sec}}\ . \tag{2.1.19}$$

For $\delta \to 0$ and thus $s \to \mathrm{i}\omega$, the transfer function evolves into the frequency response

$$\lim_{s\to\mathrm{i}\omega} G(s) = G(\mathrm{i}\omega)\ . \tag{2.1.20}$$

This concludes the composition of the most important fundamental equations for non-parametric, linear models and deterministic signals.

2.1.2 Parametric Models, Deterministic Signals

Parametric models represent the relation between the input and output by means of equations. In general, they contain a finite number of explicit parameters. These equations can be set up by the application of theoretical modeling techniques as was described in Sect. 1.1. By means of formulating balance equations for stored quantities, physical or chemical equations of state and phenomenological equations, a system of equations is constructed, which contains the physically defined parameters c_i, which shall be called *process coefficients* (Isermann, 2005). This system of equations reveals the *elementary model structure* and can be represented by means of a detailed block diagram. Models that exhibit such an elementary model structure can be called *white models* (white box) in contrast to the non-parametric, so-called black models that have been introduced in the previous section, recall also Fig. 1.3 for a comparison of the different modeling approaches ranging from white-box to black-box modeling.

Differential Equations

If only the input/output behavior of the process is of interest, then the system states will be eliminated (if possible). The resulting mathematical model assumes the form of an ordinary differential equation (ODE) for a lumped parameter system. In the linear case, this ODE is given as

$$\begin{aligned} & y^{(n)}(t) + a_{n-1}y^{(n-1)}(t) + \ldots + a_1\dot{y}(t) + a_0 y(t) \\ =& b_m u^{(m)}(t) + b_{m-1}u^{(m-1)}(t) + \ldots + b_1\dot{u}(t) + b_0 u(t) \,. \end{aligned} \tag{2.1.21}$$

The model parameters a_i and b_i are determined by the process coefficients c_i. For the transition from the physical process to the input/output model, the underlying model structure may be lost. For processes with distributed parameters, one can obtain similar partial differential equations (PDEs).

Transfer Function and Frequency Response

By application of the Laplace transform to the ODE in (2.1.21) and setting all initial conditions to zero, one obtains the (parametric) *transfer function*

$$G(s) = \frac{y(s)}{u(s)} = \frac{b_0 + b_1 s + \ldots + b_m s^m}{a_0 + a_1 s + \ldots + a_n s^n} = \frac{B(s)}{A(s)} \,. \tag{2.1.22}$$

By determining the limit $s \to \mathrm{i}\omega$, the (parametric) *frequency response* is obtained as

$$G(\mathrm{i}\omega) = \lim_{s \to \mathrm{i}\omega} G(s) = \big|G(\mathrm{i}\omega)\big| \mathrm{e}^{\mathrm{i}\varphi(\omega)} \tag{2.1.23}$$

with the magnitude $|G(\mathrm{i}\omega)|$ and the phase (argument) $\varphi(\mathrm{i}\omega) = \angle G(\mathrm{i}\omega)$, which can be expressed in dependence of the model parameters.

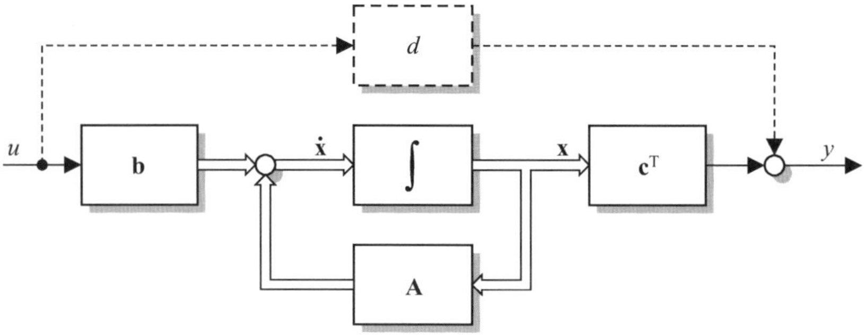

Fig. 2.2. State space representation of a SISO system

State Space Representation

If one is not only interested in the behavior of the output, but also the internal states of the system, one has to represent the system in its state space form. For a linear time-invariant process with one input and one output, the equations are given as

$$\dot{\boldsymbol{x}}(t) = \boldsymbol{A}\boldsymbol{x}(t) + \boldsymbol{b}u(t) \tag{2.1.24}$$

$$y(t) = \boldsymbol{c}^{\mathrm{T}}\boldsymbol{x}(t) + du(t)\ . \tag{2.1.25}$$

The elements of these equations are called *state vector* ($\boldsymbol{x}(t)$), *state matrix* ($\boldsymbol{A}$), *input vector* ($\boldsymbol{b}$), *output vector* ($\boldsymbol{c}^{\mathrm{T}}$) and *direct feedthrough* (d). The first equation is termed *state equation* and the second *output equation.* A block diagram representation is shown in Fig. 2.2.

The time solution of (2.1.24) and (2.1.25), which is important for the evaluation of e.g. the Kalman filter (see Chap. 21) is given as

$$\boldsymbol{x}(t) = \boldsymbol{\Phi}(t-t_0)\boldsymbol{x}(t_0) + \int_{t_0}^{t} \boldsymbol{\Phi}(t-\tau)\boldsymbol{b}u(\tau)\mathrm{d}\tau \tag{2.1.26}$$

with the transition matrix $\boldsymbol{\Phi}$ being determined by the matrix exponential

$$\boldsymbol{\Phi}(t) = \mathrm{e}^{\boldsymbol{A}t} = \lim_{n\to\infty}\left(\boldsymbol{I} + \boldsymbol{A}t + \boldsymbol{A}^2\frac{t^2}{2!} + \ldots + \boldsymbol{A}^n\frac{t^n}{n!}\right)\ . \tag{2.1.27}$$

Apart from the direct evaluation of the series as in (2.1.27), there are several other ways to calculate the matrix exponential (e.g. Moler and van Loan, 2003). Using the transition matrix, the output can be calculated as

$$y(t) = \boldsymbol{c}^{\mathrm{T}}\boldsymbol{\Phi}(t-t_0)\boldsymbol{x}(t_0) + \boldsymbol{c}^{\mathrm{T}}\int_{t_0}^{t} \boldsymbol{\Phi}(t-\tau)\boldsymbol{b}u(\tau)\mathrm{d}\tau + du(t)\ . \tag{2.1.28}$$

From the state space representation, one can also determine the transfer function in continuous-time by

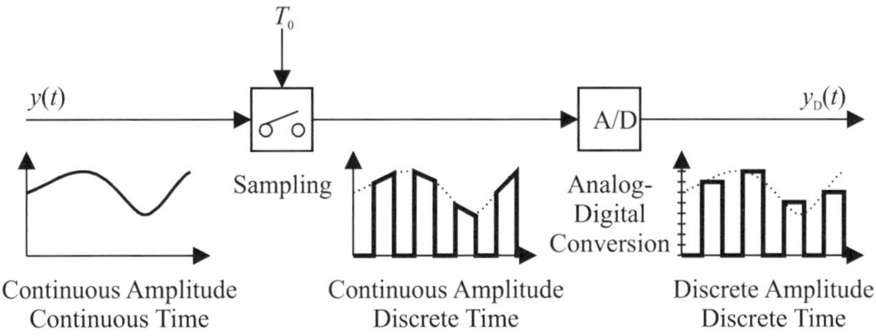

Fig. 2.3. Analog-digital conversion process and the subsequent generation of an amplitude-modulated discrete-time and discrete-amplitude, discrete-time signal

$$G(s) = \frac{y(s)}{u(s)} = \boldsymbol{c}^{\mathrm{T}}(s\boldsymbol{I} - \boldsymbol{A})^{-1}\boldsymbol{b} , \quad (2.1.29)$$

leading to a rational transfer function as in the form of (2.1.22).

2.2 Mathematical Models of Dynamic Systems for Discrete Time Signals

For the digital processing of measurements and thus also for the identification with digital computers, the measurements are sampled and digitized in the analog-digital converter (ADC). By this sampling and discretizing process, discrete signals are generated, which are both quantized in time and in amplitude. It is assumed that the quantization error for the amplitude is so small that the amplitude values can be assumed to be quasi-continuous. If the sampling process is periodic with the sample time T_0, then an amplitude modulated train of pulses, apart by the sample time T_0, results, see Fig. 2.3. This sampled signal can then be processed inside the digital computer, e.g. for control purposes (Franklin et al, 1998; Isermann, 1991) or for other purposes such as process identification. It is important to recognize that the process model inevitable also contains the sampling process at the input of the digital computer and the subsequent holding element as well as the sampling process at the output of the computer. A detailed description of discrete-time signals can be found in textbooks on digital control (Franklin et al, 1998; Isermann, 1991; Phillips and Nagle, 1995; Söderström, 2002). Therefore, only a short synopsis shall be provided in the following.

2.2.1 Parametric Models, Deterministic Signals

δ Impulse Series, z-Transform

If the continuous-time input and output of a process are sampled with a sufficiently high sample rate (compared to the process dynamics), one can obtain difference equations describing the process behavior by discretizing the differential equation using finite differences as a replacement for the continuous-time derivatives. However, a more suitable treatment, which is valid also for large sampling times is to approximate the pulses of the sampled signal $x_S(t)$ which have the width h by an area-equivalent δ impulse,

$$x_S(t) \approx x_\delta(t) = \frac{h}{1\,\text{sec}} \sum_{k=0}^{\infty} x(kT_0)\delta(t - kT_0)\,. \tag{2.2.1}$$

With the normalization $h = 1\,\text{sec}$, one gets

$$x^*(t) = \sum_{k=0}^{\infty} x(kT_0)\delta(t - kT_0)\,. \tag{2.2.2}$$

This expression can be subjected to the Laplace transform, yielding

$$x^*(s) = \mathfrak{L}\{x^*(t)\} = \sum_{k=0}^{\infty} x(kT_0)\mathrm{e}^{-kT_0 s}\,. \tag{2.2.3}$$

The Laplace transform $x^*(s)$ is periodic with

$$x^*(s) = x^*(s + \mathrm{i}\nu\omega_0),\ \nu = 0, 1, 2, \ldots \tag{2.2.4}$$

with the *sampling frequency* $\omega_0 = 2\pi/T_0$. Introducing the short hand notation

$$z = \mathrm{e}^{T_0 s} = \mathrm{e}^{T_0(\delta + \mathrm{i}\omega)}\,, \tag{2.2.5}$$

one obtains the *z-transform*

$$x(z) = \mathfrak{Z}\{x(kT_0\} = \sum_{k=0}^{\infty} x(kT_0)z^{-k}\,. \tag{2.2.6}$$

If $x(kT_0)$ is bounded, $x(z)$ converges for $|z| > 1$, which can be achieved for most interesting signals by an appropriate choice of δ. Similarly to the Laplace transform, it is assumed that $x(kT_0) = 0$ for $k < 0$ and $\delta > 0$ (Poularikas, 1999; Föllinger and Kluwe, 2003). $x(z)$ is in general a series of infinite length. For many test signals however, one can provide closed-form expressions.

Discrete Impulse Response

Since the response of a system due to the excitation with a δ-impulse is the impulse response $g(t)$, one obtains the convolution sum

$$y(kT_0) = \sum_{\nu=0}^{\infty} u(\nu T_0)g\big((k - \nu)T_0\big) \tag{2.2.7}$$

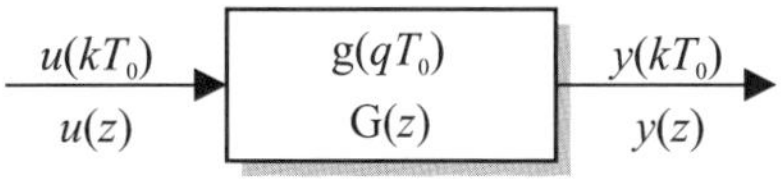

Fig. 2.4. Dynamic process with sampled input u and output y being characterized by its (discrete) impulse response and its z transfer function

which can be used to calculate the system output due to an input $u(kT_0)$, where

$$u^*(t) = \sum_{k=0}^{\infty} u(kT_0)\delta(t - kT_0) \tag{2.2.8}$$

is the δ-impulse approximation of the input. If the output is sampled synchronously to the input, then the convolution sum is given as

$$y(kT_0) = \sum_{\nu=0}^{\infty} u(\nu T_0) g\big((k-\nu)T_0\big) = \sum_{\nu=0}^{\infty} u\big((k-\nu)T_0\big) g(\nu T_0)\ . \tag{2.2.9}$$

z-Transfer Function

The sampled and δ-impulse approximated output

$$y^*(t) = \sum_{k=0}^{\infty} y(kT_0)\delta(t - kT_0) \tag{2.2.10}$$

is being subjected to the Laplace transform to obtain

$$y^*(s) = \sum_{\nu=0}^{\infty}\sum_{\mu=0}^{\infty} u(\mu T_0) g\big((\nu-\mu)T_0\big)\mathrm{e}^{-\mu T_0 s}\ . \tag{2.2.11}$$

With the substitution $q = \nu - \mu$ one gets

$$y^*(s) = \sum_{q=0}^{\infty} g(qT_0)\mathrm{e}^{-qT_0 s} \sum_{\mu=0}^{\infty} u(\mu T_0)\mathrm{e}^{-\mu T_0 s} = G^*(s)u^*(s)\ . \tag{2.2.12}$$

Here,

$$G^*(s) = \frac{y^*(s)}{u^*(s)} = \sum_{q=0}^{\infty} g(qT_0)\mathrm{e}^{-qT_0 s} \tag{2.2.13}$$

is called the *impulse transfer function*. The *impulse frequency response* then becomes

$$G^*(\mathrm{i}\omega) = \lim_{s\to \mathrm{i}\omega} G^*(s),\ \omega \le \frac{\pi}{T_0}\ . \tag{2.2.14}$$

It must be kept in mind that for a continuous signal that has been sampled at the angular frequency $\omega_0 = 2\pi/T_0$, only harmonic signals with an angular frequency $\omega < \omega_S$ with

$$\omega_S = \frac{\omega_0}{2} = \frac{\pi}{T_0} \tag{2.2.15}$$

can be detected correctly as harmonic signals with the true angular frequency ω according to *Shannon's theorem*. For signals with $\omega > \omega_S$ one gets phantom output signals with a lower frequency upon sampling of the signal. This is described as the *aliasing effect*.

Upon introduction of the short hand notation $z = e^{T_0 s} = e^{T_0(\delta + i\omega)}$ into (2.2.14), one obtains the z-transfer function (see Fig. 2.4)

$$G(z) = \frac{y(z)}{u(z)} = \sum_{k=0}^{\infty} g(kT_0) z^{-k} = \mathfrak{Z}\{g(kT_0)\} . \tag{2.2.16}$$

For a given s-transfer function $G(s)$ one obtains the z-transfer function by

$$G(z) = \mathfrak{Z}\left\{ \left[\mathfrak{L}^{-1}\{G(s)\} \right]_{t=kT_0} \right\} = \mathcal{Z}\{G(s)\} . \tag{2.2.17}$$

The abbreviation $\mathcal{Z}\{\ldots\}$ means to take the corresponding z-transform for a given s-transform from an s- and z-transform table (e.g. Isermann, 1991).

If the process with the transfer function $G(s)$ is driven by a sample and hold element of order zero, the resulting z-transfer function is given as

$$\begin{aligned} HG(z) &= \mathfrak{Z}\{H(s)G(s)\} = \mathcal{Z}\left\{ \frac{1}{s}\{1 - e^{-T_0 s}\} G(s) \right\} \\ &= \{1 - z^{-1}\} \mathcal{Z}\left\{ \frac{G(s)}{s} \right\} = \frac{z-1}{z} \mathcal{Z}\left\{ \frac{G(s)}{s} \right\} . \end{aligned} \tag{2.2.18}$$

Note that the parameters a_i and b_i of the z-transfer function (2.2.19) are different from those of the s-transfer function in (2.1.22).

z-Transfer Function

If the differential equation (2.1.21) of a linear process is known, one can determine the corresponding s-transfer function by (2.1.22) and subsequently determine the z-transfer function by means of (2.2.17) or (2.2.18) as

$$G(z^{-1}) = \frac{y(z)}{u(z)} = \frac{b_0 + b_1 z^{-1} + \ldots + b_m z^{-m}}{1 + a_1 z^{-1} + \ldots + a_n z^{-n}} = \frac{B(z^{-1})}{A(z^{-1})} . \tag{2.2.19}$$

In many cases, the polynomial order of numerator and denominator will be the same. If the process contains a *dead time* $T_D = dT_0$ with $d = 1, 2, \ldots$, then the z-transfer function is given as

$$G(z^{-1}) = \frac{B(z^{-1})}{A(z^{-1})} z^{-d} . \tag{2.2.20}$$

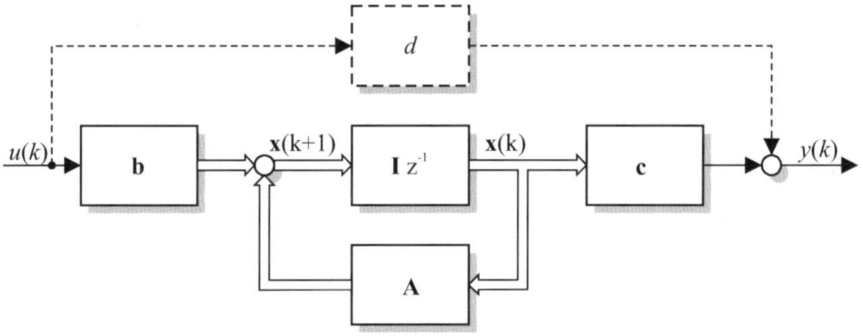

Fig. 2.5. State space representation of a SISO system with sampled input and output

Difference Equations

If (2.2.19) is rewritten in the form

$$y(z)\big(1 + a_1 z^{-1} + \ldots + a_n z^{-n}\big) = u(z)\big(b_0 + b_1 z^{-1} + \ldots + b_m z^{-m}\big)\,, \quad (2.2.21)$$

one can rewrite this in the time domain as

$$y(k) + a_1 y(k-1) + \ldots + a_n y(k-n) = b_0 u(k) + b_1 u(k-1) + \ldots + b_m u(k-m) \quad (2.2.22)$$

with the short hand notation (k) instead of (kT_0). The coefficients of this difference equation will of course be different from the coefficients of the differential equation in (2.1.21). The shape of the impulse response can be derived from the difference equation by choosing the δ-impulse as input, which is equal to

$$u(k) = \begin{cases} 0 \text{ for } k \neq 0 \\ 1 \text{ for } k = 0 \end{cases} \quad (2.2.23)$$

in the discrete time. It follows from (2.2.22) with $y(k) = g(k)$

$$\begin{aligned} g(0) &= b_0 \\ g(1) &= b_1 - a_1 g(0) \\ g(2) &= b_2 - a_1 g(1) - a_2 g(0) \\ &\ \vdots \\ g(k) &= b_k - a_1 g(k-1) - \ldots - a_k g(0) \text{ for } k \leq m \\ g(k) &= -a_1 g(k-1) - \ldots - a_n g(k-n) \text{ for } k > m\,. \end{aligned}$$

State Space Representation

For discrete-time signals, the state space representation is given as

$$x(k+1) = A_d x(k) + b_d u(k) \tag{2.2.24}$$
$$y(k) = c_d^T x(k) + d_d u(k) \tag{2.2.25}$$

with the state vector

$$x(k) = \left(x_1(k)\ x_2(k)\ \ldots\ x_m(k)\right)^T , \tag{2.2.26}$$

see Fig. 2.5. With the relations

$$A_d = \Phi(T_0) = e^{AT_0} \tag{2.2.27}$$
$$b_d = H(T_0) = \int_0^{T_0} e^{A(T_0-\tau)} b \mathrm{d}\tau \tag{2.2.28}$$
$$c_d^T = c^T \tag{2.2.29}$$
$$d_d = d , \tag{2.2.30}$$

compare (2.1.27), one can calculate the discrete-time representation in the state space from a continuous-time model. Furthermore, using

$$G(z^{-1}) = \frac{y(z)}{u(z)} = c_d^T(zI - A_d)^{-1} b_d , \tag{2.2.31}$$

one can derive a transfer function representation in the form given in (2.2.19). The derivation and further properties are in detail discussed e.g. in (Isermann, 1991; Heij et al, 2007). In Sect. 15.2.2, techniques are presented that interpolate the input signal $u(k)$ between the sampling points.

The response of the state space MIMO system to an arbitrary input is given as

$$x(k) = A_d^k x(0) + \sum_{\nu=0}^{k-1} A_d^{k-\nu-1} B_d u(\nu) \tag{2.2.32}$$
$$y(k) = C_d x(k) + D_d u(k) \tag{2.2.33}$$

and will become important for subspace identification methods presented in Chap. 16. For discrete-time state space systems, a few more properties shall be introduced now, also with respect to the later discussed subspace methods.

In the following, the index "d" will not be used for the discrete-time matrices. The term *realization* will be used for a MIMO state space system consisting of the matrices A_d, B, C, and D. There is an infinite number of realizations that describe the same input/output behavior, therefore the term *minimal realization* will be introduced. Such a realization has the least number of state variables necessary to describe a certain input/output behavior.

A realization is called *controllable* if any final state $x(t_1)$ can be reached in the finite time interval $[t_0, t_1]$ from any arbitrary initial state $x(t_0)$ by choosing an appropriate input sequence. A realization with state space dimension n is controllable, if the *controllability matrix* Q_S

$$Q_S = \left(B \; AB \; \dots \; A^{n-1} B \right) \tag{2.2.34}$$

has full rank (i.e. rank n). Other conditions that allow to test for controllability are e.g. given in (Heij et al, 2007) and (Chen, 1999), where also the continuous-time case is treated in detail and controllability indices are introduced.

Observability means that the states at time t_0 and any other time t in the interval $[t_0, t_1]$ can be reconstructed from a measurement of the input $\boldsymbol{u}(k)$ and output $\boldsymbol{y}(k)$ over the time interval $[t_0, t_1]$. A realization with state space dimension n is observable if the *observability matrix* $\boldsymbol{Q}_B$ defined as

$$Q_B = \begin{pmatrix} C \\ CA \\ \vdots \\ C^{n-1} A \end{pmatrix} \tag{2.2.35}$$

has full rank (i.e. rank n), see also (Heij et al, 2007) and (Chen, 1999). Observability does not depend on the measured data, but is a system property (Grewal and Andrews, 2008). In the same citation, it is pointed out that due to the normally inevitable differences between the mathematical model and the real system, the formal measure of observability might fall to short. One should always check the condition number of the observability matrix to see how close this matrix is to being singular. Finally, a realization is *minimal*, if and only if the realization is both *controllable* and *observable*.

2.3 Models for Continuous-Time Stochastic Signals

The course of a *stochastic signal* is random in its nature and can thus not be characterized exactly. However, by means of statistic methods, the calculus of probabilities as well as averaging, properties of these stochastic signals can be described. Measurable stochastic signals are typically not entirely random, but have some internal coherences which can be cast into mathematical signal models. In the following, the most important terms and definitions of stochastic signal models will be presented in brief. The scope is limited to those terms and definitions required for the identification methods described in this book. An extensive treatment of the subject matter can be found e.g. in (Åström, 1970; Hänsler, 2001; Papoulis and Pillai, 2002; Söderström, 2002; Zoubir and Iskander, 2004).

Due to the random behavior, there exists not only one certain realization $x_1(t)$, but rather an entire family (termed *ensemble*) of random time signals

$$\left\{ x_1(t),\; x_2(t), \dots,\; x_n(t) \right\} . \tag{2.3.1}$$

This ensemble of signals is termed a *stochastic process* (signal process). A single realization $x_i(t)$ is termed *sample function*.

Statistical Description

If the signal value of all sample functions $x_i(t)$ is considered at a certain point in time $t = t_\nu$, then the statistical properties of the signal amplitudes of the stochastic process are described by the *probability density function* (PDF), $p(x_i(t_\nu))$ for $i = 1, 2, \ldots, n$.

Internal coherences are described by the joint probability density function at different points in time. For the two points in time t_1 and t_2, the two-dimensional joint PDF is given as

$$p\big(x(t_1), x(t_2)\big) \text{ for } \begin{cases} 0 \leq t_1 < \infty \\ 0 \leq t_2 < \infty \end{cases} , \tag{2.3.2}$$

which is a measure for the probability that the two signals values $x(t_1)$ and $x(t_2)$ appear at t_1 and t_2 respectively. For the appearance of n signal values at the times $t_1, t_2, \ldots, t_n$, one has to consider the n-dimensional joint PDF

$$p\big(x(t_1), x(t_2), \ldots, x(t_n)\big) . \tag{2.3.3}$$

A stochastic process is fully characterized if the PDF and all joint PDFs for all n and all t are known.

So far, it has been assumed that the PDF and all joint PDFs are a function of time. In this case, the stochastic process is termed *non-stationary*. For many areas of application however, it has not proven necessary to use such a broad all-encompassing definition. Therefore, only certain classes of stochastic processes will be considered in the following.

Stationary Processes

A process is *strict sense stationary* (SSS) if all PDFs are independent from a shift in time. By calculation of the *expected value*, denoted by the linear operator E{...},

$$\mathrm{E}\{f(x)\} = \int_{-\infty}^{\infty} f(x)p(x)\mathrm{d}x , \tag{2.3.4}$$

one can derive characteristic values and characteristic curves of stationary processes. With $f(x) = x^n$, one obtains the nth moment of a PDF. The moment of order 1 is the (linear) mean

$$\overline{x} = \mathrm{E}\{x(t)\} = \int_{-\infty}^{\infty} x(t)p(x)\mathrm{d}x \tag{2.3.5}$$

of all sample functions at time t and the central moment of second order is the variance

$$\sigma_x^2 = \mathrm{E}\{(x(t) - \overline{x})^2\} = \int_{-\infty}^{\infty} (x(t) - \overline{x})^2 p(x)\mathrm{d}x . \tag{2.3.6}$$

The two-dimensional joint PDF of a stationary process is according to its definition only dependent on the time difference $\tau = t_2 - t_1$, thus

$$p\big(x(t_1), x(t_2)\big) = p\big(x(t), x(t+\tau)\big) = p\big(x, \tau\big) . \tag{2.3.7}$$

The expected value of the product $x(t)x(t+\tau)$ is then

$$R_{xx}(\tau) = \mathrm{E}\{x(t)x(t+\tau)\} = \int_{-\infty}^{\infty}\int_{-\infty}^{\infty} x(t)x(t+\tau)p(x,\tau)\mathrm{d}x\,\mathrm{d}x\ , \tag{2.3.8}$$

which is also only a function of τ and is termed *auto-correlation function* (ACF).

A process is *wide sense stationary* (WSS) if the expected values

$$\mathrm{E}\{x(t)\} = \overline{x} = \text{const} \tag{2.3.9}$$

$$\mathrm{E}\{x(t)x(t+\tau)\} = R_{xx}(\tau) = \text{const} \tag{2.3.10}$$

are independent of time, i.e. the mean is time independent and the ACF does only depend on the time difference τ. Furthermore, the variance needs to be finite (Verhaegen and Verdult, 2007). The linear combination of stationary processes is also stationary (Box et al, 2008).

Ergodic Processes

The expected values used so far are termed *ensemble averages*, since one averages over multiple similar random signals, which have been generated by statistically identical signal sources at the same time. According to the *ergodic hypothesis*, one can obtain the same statistical information (which one gets from ensemble averaging) also from averaging a single sample function $x(t)$ over time, if infinitely long intervals of time are considered. Thus, the mean of an ergodic process is given as

$$\overline{x} = \mathrm{E}\{x(t)\} = \lim_{T\to\infty}\frac{1}{T}\int_{-\frac{T}{2}}^{\frac{T}{2}} x(t)\mathrm{d}t \tag{2.3.11}$$

and the quadratic mean as

$$\sigma_x^2 = \mathrm{E}\{(x(t)-\overline{x})^2\} = \lim_{T\to\infty}\frac{1}{T}\int_{-\frac{T}{2}}^{\frac{T}{2}} \big(x(t)-\overline{x}\big)^2\mathrm{d}t\ . \tag{2.3.12}$$

Ergodic processes are always stationary. The opposite may not be true.

Correlation Function

Some first information about the internal coherences of stochastic processes can be gathered from the two-dimensional joint PDF as well as from the ACF. For Gaussian processes, this information does already determine all joint PDFs of higher order and thus also all internal coherences. Since many processes can approximately be assumed to be Gaussian, knowledge of the ACF is often sufficient to describe the internal coherences of the signal. By multiplying the signal $x(t)$ with its time shifted counterpart (in negative t-direction) $x(t+\tau)$ and averaging, one gets information about the *internal coherences* respectively *conservation tendency*. If the product is

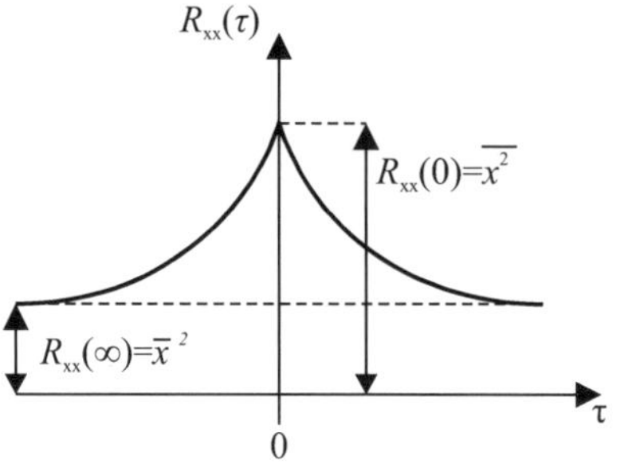

Fig. 2.6. General shape of the auto-correlation function of a stationary stochastic process $x(t)$

large, then there is a strong internal coherence, if it is small, then there is little coherence. By correlating the signals, however, the time information of $x(t)$, i.e. the phase is lost.

The auto-correlation function is given as

$$\begin{aligned} R_{xx}(\tau) &= \mathrm{E}\{x(t)x(t+\tau)\} = \lim_{T\to\infty} \frac{1}{T} \int_{-\frac{T}{2}}^{\frac{T}{2}} x(t)x(t+\tau)\mathrm{d}t \\ &= \lim_{T\to\infty} \frac{1}{T} \int_{-\frac{T}{2}}^{\frac{T}{2}} x(t-\tau)x(t)\mathrm{d}t \;. \end{aligned} \tag{2.3.13}$$

In the past, it had sometimes been termed *correlogram* (Box et al, 2008).

For stationary stochastic signals of infinite length, the ACF has the following properties:

1. The ACF is an even function, $R_{xx}(\tau) = R_{xx}(-\tau)$
2. $R_{xx}(0) = \overline{x^2(t)}$
3. $R_{xx}(\infty) = \overline{x(t)}^2$, which means that for $\tau \to \infty$, the signals can be considered uncorrelated
4. $R_{xx}(\tau) \leq R_{xx}(0)$

With these properties, one gets in principle the curve shown in Fig. 2.6. The faster the ACF decays to both sides, the smaller the *conservation tendency* of the signal, see Fig. 2.7b and Fig. 2.7c. The ACF can also be determined for periodic signals. They show the same periodicity and are ideally suited to separate noise and periodic signals, see Fig. 2.7d and Fig. 2.7e.

The statistical coherence between two different stochastic signals $x(t)$ and $y(t)$ is given by the cross-correlation function CCF,

$$\begin{aligned} R_{xy}(\tau) &= \mathrm{E}\{x(t)y(t+\tau)\} = \lim_{T\to\infty} \frac{1}{T} \int_{-\frac{T}{2}}^{\frac{T}{2}} x(t)y(t+\tau)\mathrm{d}t \\ &= \lim_{T\to\infty} \frac{1}{T} \int_{-\frac{T}{2}}^{\frac{T}{2}} x(t-\tau)y(t)\mathrm{d}t \;. \end{aligned} \tag{2.3.14}$$

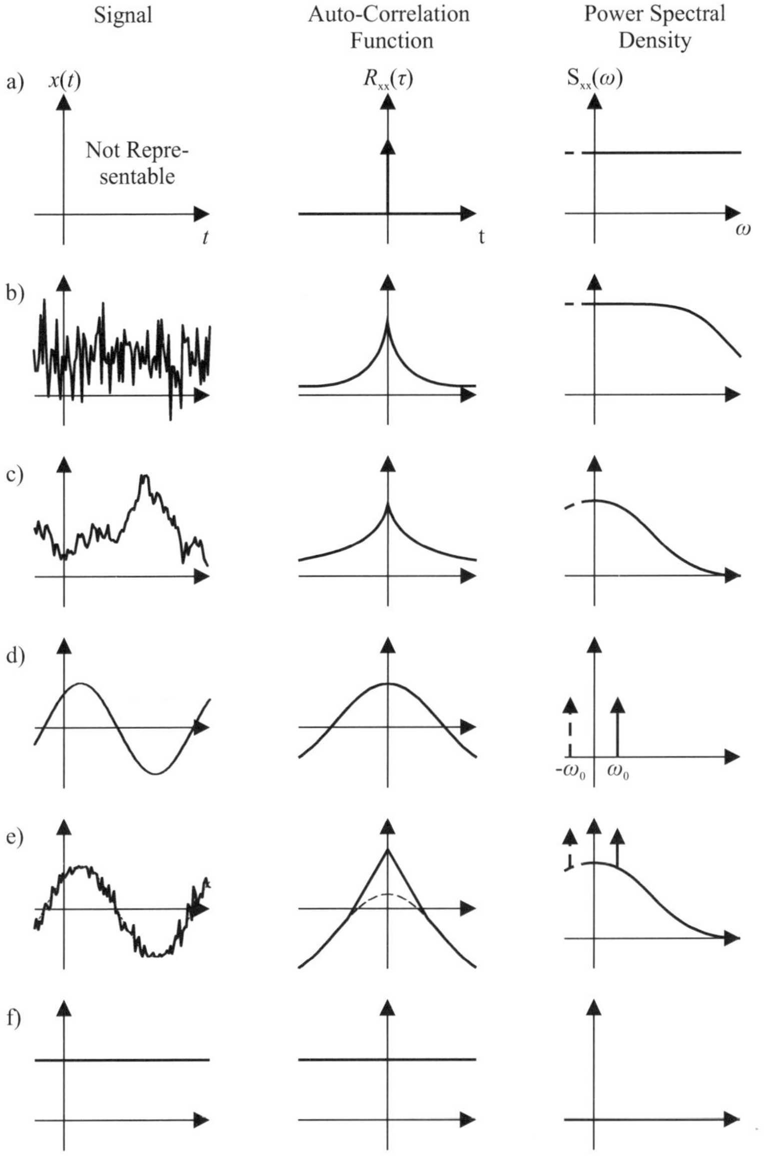

Fig. 2.7. Auto correlation function and power spectral densities of different signals. (**a**) white noise, (**b**) high-frequent noise, (**c**) low-frequent noise, (**d**) harmonic signal, (**e**) harmonic signal and noise, (**f**) constant signal

The CCF is in contrast to the ACF not a symmetric function. The relative phase between the two signals is conserved. The CCF has the following properties:

1. $R_{xy}(\tau) = R_{yx}(-\tau)$
2. $R_{xy}(0) = \overline{x(t)y(t)}$, i.e. mean of the product
3. $R_{xy}(\infty) = \overline{x(t)}\ \overline{y(t)}$, i.e. product of the means
4. $R_{xy}(\tau) \leq 1/2(R_{xx}(0) + R_{yy}(0))$

Covariance Function

The above defined correlation functions depend on the mean of the signals. If one however subtracts the mean of each signal before calculating the correlation functions, one gets the so-called *covariance functions*. For a scalar process $x(t)$, the auto-covariance function is defined as

$$C_{xx}(\tau) = \text{cov}(x, \tau) = \text{E}\big\{(x(t) - \overline{x})(x(t + \tau) - \overline{x})\big\} = \text{E}\big\{x(t)x(t + \tau)\big\} - \overline{x}^2 . \tag{2.3.15}$$

With $\tau = 0$, one gets the variance of the signal process. The cross-covariance function of two scalar processes is defined as

$$C_{xy}(\tau) = \text{cov}(x, y, \tau) = \text{E}\big\{(x(t) - \overline{x})(y(t + \tau) - \overline{y})\big\} = \text{E}\big\{x(t)y(t + \tau)\big\} - \overline{x}\,\overline{y} . \tag{2.3.16}$$

If the mean of the two processes is zero, correlation and covariance function are identical. Vectorial processes will be described by the covariance matrix accordingly.

Power Spectral Density

So far, the stochastic signal processes have been examined in the time domain. By transforming the signals into the frequency domain, one gets a spectral representation. For a non-periodic deterministic function $x(t)$, the complex amplitude density is defined as the Fourier transform of the signal $x(t)$. Accordingly, the power spectral density of a stationary stochastic signal is defined as the Fourier transform of the auto-correlation function, i.e.

$$S_{xx}(\text{i}\omega) = \int_{-\infty}^{\infty} R_{xx}(\tau)\text{e}^{-\text{i}\omega\tau}\text{d}\tau . \tag{2.3.17}$$

The inverse Fourier transform is then given as

$$R_{xx}(\tau) = \frac{1}{2\pi}\int_{-\infty}^{\infty} S_{xx}(\text{i}\omega)\text{e}^{\text{i}\omega\tau}\text{d}\omega . \tag{2.3.18}$$

Since the auto-correlation function is an even function, i.e. $R_{xx}(\tau) = R_{xx}(-\tau)$, the power spectral density is a real-valued function,

$$S_{xx}(\omega) = 2\int_{0}^{\infty} R_{xx}(\tau)\text{e}^{-\text{i}\omega t}\text{d}\tau = 2\int_{0}^{\infty} R_{xx}(\tau)\cos\omega\tau\text{d}\tau . \tag{2.3.19}$$

It is also an even function, since $S_{xx}(\omega) = S_{xx}(-\omega)$. With $\tau = 0$, from (2.3.18) follows

$$R_{xx}(0) = \mathrm{E}\{(x(t)-\overline{x})^2\} = \overline{(x(t)-\overline{x})^2} = \sigma_x^2 = \frac{1}{2\pi}\int_{-\infty}^{\infty} S_{xx}(\mathrm{i}\omega)\mathrm{d}\omega = \frac{1}{\pi}\int_{0}^{\infty} S_{xx}(\mathrm{i}\omega)\mathrm{d}\omega\ . \tag{2.3.20}$$

The quadratic mean respectively the average power of the signal $x(t) - \overline{x}$ is thus proportional to the integral of the power spectral density. Some examples of shapes of the power spectral density are shown in Fig. 2.7.

The *cross power spectral density* of two stochastic signals $x(t)$ and $y(t)$ is defined as the Fourier transform of the cross-correlation function, i.e.

$$S_{xy}(\mathrm{i}\omega) = \int_{-\infty}^{\infty} R_{xy}(\tau)\mathrm{e}^{-\mathrm{i}\omega\tau}\mathrm{d}\tau \tag{2.3.21}$$

and the inverse transform

$$R_{xy}(\tau) = \frac{1}{2\pi}\int_{-\infty}^{\infty} S_{xy}(\mathrm{i}\omega)\mathrm{e}^{\mathrm{i}\omega\tau}\mathrm{d}\omega\ . \tag{2.3.22}$$

Since $R_{xy}(\tau)$ is no symmetric function, $S_{xy}(\mathrm{i}\omega)$ is a complex function with axis-symmetric real part and point-symmetric imaginary part. As a side note, (2.3.17), (2.3.18), (2.3.21), and (2.3.22) are termed *Wiener-Khintchin* relations.

2.3.1 Special Stochastic Signal Processes

Independent, Uncorrelated, and Orthogonal Processes

The stochastic processes $x_1(t), x_2(t), \ldots, x_n(t)$ are termed *statistically independent* if

$$p(x_1, x_2, \ldots, x_n) = p(x_1)p(x_2)\ldots p(x_n)\ , \tag{2.3.23}$$

that is if the joint PDF is equal to the product of the individual PDFs. Pairwise independence

$$\begin{aligned} p(x_1, x_2) &= p(x_1)p(x_2) \\ p(x_1, x_3) &= p(x_1)p(x_3) \end{aligned} \tag{2.3.24}$$

does not imply total statistical independence. It does only indicate that the non-diagonal elements of the covariance matrix will be zero, meaning that the processes are uncorrelated, i.e.

$$\mathrm{cov}(x_i, x_j, \tau) = C_{x_i x_j}(\tau) = 0 \text{ for } i \neq j\ . \tag{2.3.25}$$

Statistical independent processes are always uncorrelated, the opposite may however not be true. Stochastic processes are termed *orthogonal* if they are uncorrelated and their means are zero, so that also the non-diagonal elements of the correlation matrix become zero,

$$R_{x_i x_j}(\tau) = 0 \text{ for } i \neq j . \tag{2.3.26}$$

Zero-mean random variables are orthogonal, if they are uncorrelated, the opposite may not be true.

Gaussian or Normal Distributed Processes

A stochastic process is termed *Gaussian* or *normal distributed*, if it has Gaussian or normal amplitude distribution. Since the Gaussian distribution is entirely determined by the two first moments, i.e. the mean $\overline{x}$ and the variance σ_x^2, the distribution laws of a Gaussian process are entirely defined by the mean and the covariance function. From this follows that a Gaussian process which is wide-sense stationary is also strict-sense stationary. Due to the same reason, uncorrelated Gaussian processes are also statistically independent. Under all linear algebraic operations and also under differentiation and integration, the Gaussian character of the distribution remains. A short hand notation for a Gaussian process is given by $(\overline{x}, \sigma_x)$.

White Noise

A signal process is designated as *white noise* if signal values, which are only infinitesimally small apart in time are still statistically independent, such that the auto-correlation function is given as

$$R_{\text{xx}}(\tau) = S_0 \delta(\tau) . \tag{2.3.27}$$

White noise in continuous-time is thus a signal process with infinitely large amplitudes which has no internal coherences. One can think of this process as a series of δ-impulses with infinitesimally small distances. The power spectral density is given as

$$S_{\text{xx}}(\tau) = \int_{-\infty}^{\infty} S_0 \delta(\tau) \mathrm{e}^{-\mathrm{i}\omega\tau} \mathrm{d}\tau = S_0 . \tag{2.3.28}$$

The power spectral density is thus constant for all angular frequencies. Therefore all angular frequencies from zero to infinity are equally represented (alluding to the frequency spectrum of white light, which contains all vsisble spectral components). The mean power follows from (2.3.28) as

$$\overline{x^2(t)} = \frac{1}{\pi} \int_0^{\infty} S_{\text{xx}}(\omega) \mathrm{d}\omega = \frac{S_0}{\pi} \int_0^{\infty} \mathrm{d}\omega = \infty . \tag{2.3.29}$$

White noise in continuous-time is therefore not realizable. It is a theoretical noise with infinitely large mean power. By applying suitable filters, one can generate broad band-limited "white" noise with finite power or small band-limited colored noise.

Periodic Signals

The correlation functions and power spectral densities are not limited to stochastic signals, but can also be applied to periodic signals. For a harmonic oscillation

$$x(t) = x_0 \sin(\omega_0 t + \alpha) \text{ with } \omega_0 = \frac{2\pi}{T_0}\,, \tag{2.3.30}$$

the auto-correlation function is given as

$$R_{xx}(\tau) = \frac{2x_0^2}{T_0}\int_0^{\frac{T_0}{2}} \sin(\omega_0 t + \alpha)\sin(\omega_0(t+\tau)+\alpha)\mathrm{d}t = \frac{x_0^2}{2}\cos\omega_0\tau\,. \tag{2.3.31}$$

It is sufficient to integrate over half a period. The ACF of a sine oscillation with arbitrary phase α, is thus a cosine oscillation. Frequency ω_0 and amplitude x_0 are conserved, the phase information α is lost. Harmonic signals thus have a harmonic ACF. The harmonic signals are hence treated different than stochastic signals. This is a feature, which makes the correlation function a well suited foundation for many identification methods.

The power spectral density of a harmonic signal follows from (2.3.17) as

$$\begin{aligned} S_{xx}(\omega) &= \frac{x_0^2}{2}\int_{-\infty}^{\infty}\cos\omega_0\tau\cos\omega\tau\mathrm{d}\tau \\ &= \frac{x_0^2}{2}\int_{-\infty}^{\infty}\cos(\omega-\omega_0)\tau\mathrm{d}\tau + \frac{x_0^2}{2}\int_{-\infty}^{\infty}\cos(\omega+\omega_0)\tau\mathrm{d}\tau \\ &= \frac{x_0^2}{2}\big(\delta(\omega-\omega_0)+\delta(\omega+\omega_0)\big)\,. \end{aligned} \tag{2.3.32}$$

As can be seen, the power spectral density of a harmonic oscillation thus consists of two δ-impulses at the frequencies ω_0 and $-\omega_0$. This allows an easy and well performing separation of periodic signals from stochastic signals. The CCF of two periodic signals $x(t) = x_0\sin(n\omega_0 t + \alpha_n)$ with $n = 1, 2, 3, \ldots$ and $y(t) = y_0\sin(m\omega_0 t + \alpha_m)$ with $m = 1, 2, 3, \ldots$ is

$$R_{xy}(\tau) = \frac{x_0 y_0}{T_0}\int_0^{\frac{T}{2}} \sin(n\omega_0 t + \alpha_n)\sin(m\omega_0(t+\tau)+\alpha_m)\mathrm{d}t = 0 \text{ if } n \neq m\,, \tag{2.3.33}$$

which means that only harmonics of the same frequency contribute to the CCF. This is another important property that is exploited in some identification methods, such as the orthogonal correlation, Sect.5.5.2.

Linear Process with Stochastic Signals

A linear process with the impulse response $g(t)$ is driven by a stationary stochastic signal $u(t)$ which evokes the zero-mean output signal $y(t)$. The CCF is then given as

$$R_{\mathrm{uy}}(\tau) = \mathrm{E}\{u(t)y(t+\tau)\} \;. \tag{2.3.34}$$

If one substitutes the convolution integral for $y(t+\tau)$, the CCF becomes

$$\begin{aligned} R_{\mathrm{uy}}(\tau) &= \mathrm{E}\left\{u(t)\int_0^\infty g(t')u(t+\tau-t')\mathrm{d}t'\right\} \\ &= \int_0^\infty g(t')\mathrm{E}\{u(t)u(t+\tau-t')\}\mathrm{d}t' \\ &= \int_0^\infty g(t')R_{\mathrm{uu}}(\tau-t')\mathrm{d}t' \;. \end{aligned} \tag{2.3.35}$$

Similarly to the input $u(t)$ and output $y(t)$ of a linear system, compare (2.1.3), also the ACF and CCF are linked by the convolution integral. The cross power spectral density, i.e. the Fourier transform of the CCF, is given as

$$\begin{aligned} S_{\mathrm{uy}}(\mathrm{i}\omega) &= \int_{-\infty}^{\infty} R_{\mathrm{uy}}(\tau)\mathrm{e}^{-\mathrm{i}\omega\tau}\mathrm{d}\tau \\ &= \int_{-\infty}^{\infty}\int_0^\infty g(t')R_{\mathrm{uu}}(\tau-t')\mathrm{d}t'\mathrm{e}^{-\mathrm{i}\omega\tau}\mathrm{d}\tau \\ &= \int_0^\infty g(t')\mathrm{d}t'\int_{-\infty}^{\infty} R_{\mathrm{uu}}(\tau-t')\mathrm{d}t'\mathrm{e}^{-\mathrm{i}\omega\tau}\mathrm{d}\tau \\ &= \int_0^\infty g(t')e^{-\mathrm{i}\omega t'}\mathrm{d}t'\, S_{uu}(\mathrm{i}\omega) \;. \end{aligned} \tag{2.3.36}$$

Thus, it follows, that

$$S_{\mathrm{uy}}(\mathrm{i}\omega) = G(\mathrm{i}\omega)S_{\mathrm{uu}}(\mathrm{i}\omega) \;, \tag{2.3.37}$$

and furthermore

$$S_{\mathrm{yy}}(\mathrm{i}\omega) = G(\mathrm{i}\omega)S_{\mathrm{yu}}(\mathrm{i}\omega) \tag{2.3.38}$$

$$S_{\mathrm{yu}}(\mathrm{i}\omega) = S_{\mathrm{uy}}(-\mathrm{i}\omega) \tag{2.3.39}$$

$$S_{\mathrm{yy}}(\mathrm{i}\omega) = G(\mathrm{i}\omega)G(-\mathrm{i}\omega)S_{\mathrm{uu}}(\mathrm{i}\omega) = |G(\mathrm{i}\omega)|^2 S_{\mathrm{uu}}(\mathrm{i}\omega) \;. \tag{2.3.40}$$

The term $G(-\mathrm{i}\omega)$ stands for the complex conjugate of the transfer function $G(\mathrm{i}\omega)$, the complex conjugate transfer function is sometimes also denoted as $G^*(\mathrm{i}\omega)$, see e.g. (Hänsler, 2001; Kammeyer and Kroschel, 2009). Using white noise with the power spectral density S_0 as input, one can generate different *colored noises* with the power spectral density

$$S_{\mathrm{yy}}(\mathrm{i}\omega) = \left|G(\mathrm{i}\omega)\right|^2 S_0 \tag{2.3.41}$$

by shaping the frequency response with an appropriate filter.

2.4 Models for Discrete Time Stochastic Signals

Discrete time stochastic signals are typically the result of sampling a continuous-time stochastic signal. The statistical properties are very similar to those just described for continuous-time signals, from the statistic representation to ergodicity up to the calculation of the correlation functions and covariance functions. The main differences are that one uses the discrete-time $k = t/T_0 = 0, 1, 2, \dots$ and that the integrals are replaced by sums. The PDF (probability density function) does not change since the amplitudes remain continuous. A thorough treatment of discrete-time stochastic processes is e.g. presented by Gallager (1996) and Hänsler (2001).

Stationary Processes

The equations are given as follows:

- Mean

$$\overline{x} = \mathrm{E}\{x(k)\} = \lim_{N\to\infty} \frac{1}{N} \sum_{k=1}^{N} x(k) \tag{2.4.1}$$

- Quadratic mean (variance)

$$\sigma_{\mathrm{x}}^2 = \mathrm{E}\{(x(k)-\overline{x})^2\} = \lim_{N\to\infty} \frac{1}{N} \sum_{k=1}^{N} (x(k)-\overline{x})^2 \tag{2.4.2}$$

- Auto-correlation function (ACF)

$$R_{\mathrm{xx}}(\tau) = \mathrm{E}\{x(k)x(k+\tau)\} = \lim_{N\to\infty} \frac{1}{N} \sum_{k=1}^{N} x(k)x(k+\tau) \tag{2.4.3}$$

- Cross-correlation function (CCF)

$$\begin{aligned} R_{\mathrm{xy}}(\tau) &= \mathrm{E}\{x(k)y(k+\tau)\} = \lim_{N\to\infty} \frac{1}{N} \sum_{k=1}^{N} x(k)y(k+\tau) \\ &= \lim_{N\to\infty} \frac{1}{N} \sum_{k=1}^{N} x(k-\tau)y(k) \end{aligned} \tag{2.4.4}$$

- Auto-covariance function

$$\begin{aligned} C_{\mathrm{xx}}(\tau) &= \mathrm{cov}(x,\tau) = \mathrm{E}\{(x(k)-\overline{x})(x(k+\tau)-\overline{x})\} \\ &= \mathrm{E}\{x(k)x(k+\tau)\} - \overline{x}^2 \end{aligned} \tag{2.4.5}$$

- Cross-covariance function

$$\begin{aligned} C_{\mathrm{xy}}(\tau) &= \mathrm{cov}(x,y,\tau) = \mathrm{E}\{(x(k)-\overline{x})(y(k+\tau)-\overline{y})\} \\ &= \mathrm{E}\{x(k)y(k+\tau)\} - \overline{x}\,\overline{y} \end{aligned} \tag{2.4.6}$$

Power Spectral Density

The power spectral density of a stationary signal is defined as the Fourier transform of the auto-correlation function and is given as

$$S^*_{xx}(i\omega) = \mathfrak{F}\{R_{xx}(\tau)\} = \sum_{\tau=-\infty}^{\infty} R_{xx}(\tau)e^{-i\tau\omega T_0} \tag{2.4.7}$$

or by applying the two-sided z-transform

$$S_{xx}(z) = \mathfrak{Z}\{R_{xx}(\tau)\} = \sum_{\tau=-\infty}^{\infty} R_{xx}(\tau)z^{-\tau} \, . \tag{2.4.8}$$

White Noise

A discrete-time signal process is termed *white noise* if the (finitely separated) sampled signal values are statistically independent. Then, the correlation function is given as

$$R_{xx}(\tau) = \sigma_x^2 \, \delta(\tau) \, , \tag{2.4.9}$$

where $\delta(\tau)$ in this context refers to the *Kronecker Delta* function defined as

$$\delta(k) = \begin{cases} 1 \text{ for } k = 0 \\ 0 \text{ for } k \neq 0 \end{cases} \tag{2.4.10}$$

and σ_x^2 refers to the variance. The power spectral density (2.4.8) of a discrete white noise signal is given as

$$S_{xx}(z) = \sigma_x^2 \sum_{\tau=-\infty}^{\infty} \delta(\tau)z^{-\tau} = \sigma_x^2 = S_{xx0} = \text{const} \, . \tag{2.4.11}$$

The power spectral density is thus constant in the interval $0 \leq |\omega| \leq \pi/T_0$. It is noteworthy that the variance of a discrete-time white noise is finite and thus the signal becomes realizable in contrast to a continuous-time white noise signal.

Linear Process with Stochastic Signals

In analogy to the continuous-time case, the ACF $R_{uu}(\tau)$ and the CCF $R_{uy}(\tau)$ are linked by the convolution sum as

$$R_{uy}(\tau) = \sum_{k=0}^{\infty} g(k)R_{uu}(\tau - k) \, . \tag{2.4.12}$$

For the power spectral density, one gets

$$S^*_{uy}(i\omega) = G^*(i\omega)S^*_{uu}(i\omega) \text{ in the interval } |\omega| \leq \frac{\pi}{T_0} \tag{2.4.13}$$

or

$$S_{uy}(z) = G(z)S_{uu}(z) \tag{2.4.14}$$

respectively.

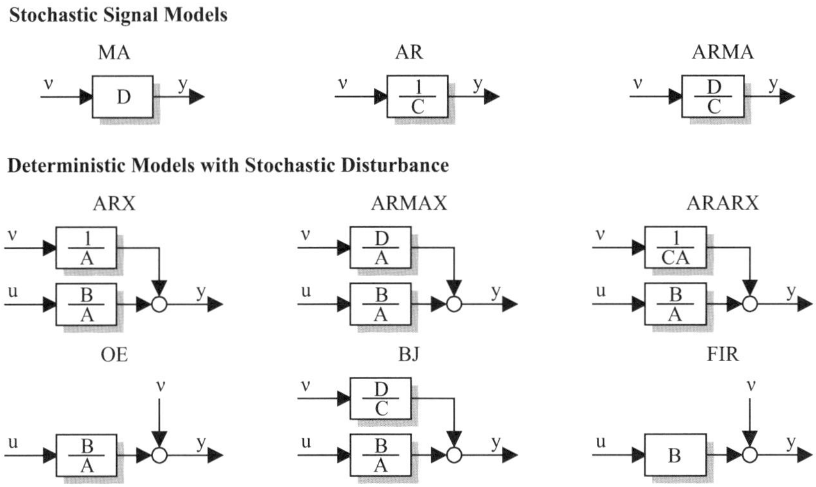

Fig. 2.8. Different models for stochastic signals and deterministic models with stochastic ditsurbance. Model naming as follows: AR=Auto Regressive, MA=Moving Average, X=eXogenous input, OE=Output Error, BJ=Box-Jenkins, FIR=Finite Impulse Response, ν white noise, u process input, y (disturbed) process output, see also (Ljung, 1999; Nelles, 2001)

Stochastic Difference Equation

Scalar stochastic processes can be described by *stochastic difference equations* in the form of a parametric model, which in the linear case is given as

$$y(k)+c_1 y(k-1)+\ldots+c_n y(k-n) = d_0\nu(k)+d_1\nu(k-1)+\ldots+d_m\nu(k-m)\,, \tag{2.4.15}$$

where $y(k)$ is the output of an imaginary filter with the z-transfer function

$$G_{\mathrm{F}}(z^{-1}) = \frac{y(z)}{\nu(z)} = \frac{d_0 + d_1 z^{-1} + \ldots + d_m z^{-m}}{1 + c_1 z^{-1} + \ldots + c_n z^{-n}} = \frac{D(z^{-1})}{C(z^{-1})} \tag{2.4.16}$$

and $\nu(k)$ being a statistically independent signal, i.e. a white noise with $(0,1)$, see Fig. 2.8. Stochastic differential equations thus represent a stochastic process as a function of a discrete-time white noise. For the analysis of stochastic processes, the following typical cases must be considered.

The *auto-regressive process* (AR) of order n is described by the difference equation

$$y(k) + c_1 y(k-1) + \ldots + c_n y(k-n) = d_0\nu(k)\,. \tag{2.4.17}$$

In this case, the signal value $y(k)$ depends on the random value $\nu(k)$ and the weighted past values $y(k-1)$, $y(k-2), \ldots$, thus the term *auto-regression*, see Fig. 2.8. The *moving average process* (MA) in contrast is governed by the difference equation

$$y(k) = d_0 \nu(k) + d_1 \nu(k-1) + \ldots + d_m \nu(k-m) . \tag{2.4.18}$$

It is thus the sum of the weighted random values $\nu(k), \nu(k-1), \ldots, \nu(k-m)$, which is a weighted average and can also be termed as an *accumulating process*. Processes governed by (2.4.17) and (2.4.18) are called *autoregressive moving-average processes* (ARMA), Fig. 2.8. Examples are given e.g. in the book by Box et al (2008). If the output $y(k)$ of such an ARMA process is integrated 1 up to d times over time, then an ARIMA process results, where the I stands for integrating. For the extensive treatment of discrete-time stochastic processes, such as Poisson processes, renewal processes, Markov chains, Random Walks and Martingales, the reader is referred to (Gallager, 1996; Åström, 1970).

Deterministic Models with Stochastic Disturbances

If a deterministic model is combined with a stochastic disturbance, then several different model structures can result, see Fig. 2.8 for a selection of the most typical ones. If the system is also controlled by an exogenous input $u(k)$, then one is faced with an ARX, where the X stands for eXogenous input. The ARX model is given as

$$y(k) + c_1 y(k-1) + \ldots + c_n y(k-n) = d_0 \nu(k) + b_1 u(k-1) + \ldots + b_n u(k-n) . \tag{2.4.19}$$

This model is most often used for identification tasks (Mikleš and Fikar, 2007). Goodwin and Sin (1984) suggested to use a leading "D" to denote a deterministic model. Typically a leading "N" refers to a non-linear model and "C" has been used by some authors to denote a continuous-time model.

2.5 Characteristic Parameter Determination

To get a rough idea about the process to be identified and even find approximate values, it is often advisable to throw a glance at the step response or impulse response. While the step response respectively impulse response is in many cases easy to measure, it can give some rough estimates of important system parameters as e.g. the settling time, the damping coefficient, and such. This section will provide a compilation of characteristic values of special cases of the generic transfer function

$$G(s) = \frac{y(s)}{u(s)} = \frac{B(s)}{A(s)} = \frac{b_0 + b_1 s + \ldots + b_{m-1} s^{m-1} + b_m s^m}{1 + a_1 s + \ldots + a_{n-1} s^{n-1} + a_n s^n} \tag{2.5.1}$$

based on some basic properties of e.g. step responses.

The individual characteristic values can directly be taken from the recorded step responses (or sometimes impulse responses) and can be used to determine coefficients of special transfer functions by means of simple calculations. They are the basis of very simple identification methods. These simple identification methods have been derived in the time around 1950–1965 and allow to generate simple parametric models based on the characteristic values of easy to measure step responses. It is assumed in the following that

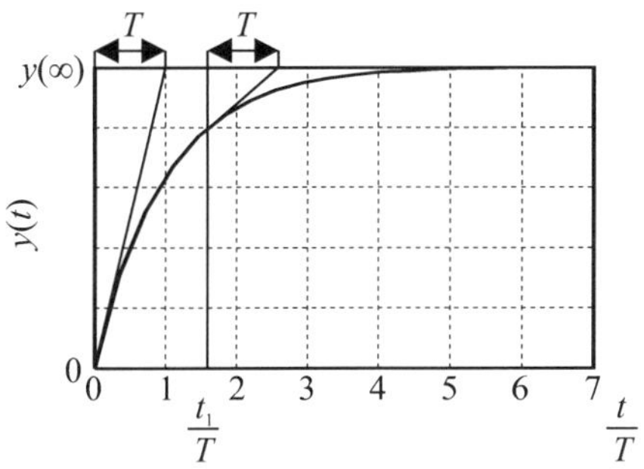

Fig. 2.9. Characteristics of the step response of a first order system

- the recorded step response is almost disturbance-free
- the process can be linearized and approximated by a simple model
- the rough approximation is sufficient for the application at hand.

A detailed derivation of the characteristic values shall not be presented as this can be found in many books on the fundamentals of controls engineering (e.g. Ogata, 2009). In mechanical analysis, one often deals with the impulse response, when a structure is hit e.g. by a hammer and the acceleration of different parts of the structure is recorded.

2.5.1 Approximation by a First Order System

A first order delay is given by the transfer function

$$G(s) = \frac{y(s)}{u(s)} = \frac{b_0}{1 + a_1 s} = \frac{K}{1 + sT} \tag{2.5.2}$$

and the step response

$$y(t) = K u_0 \left(1 - \mathrm{e}^{-\frac{t}{T}}\right) . \tag{2.5.3}$$

For a unit step with $u_0 = 1$, the step response is given as

$$h(t) = K \left(1 - \mathrm{e}^{-\frac{t}{T}}\right) . \tag{2.5.4}$$

The step response is fully described by the gain K and the time-constant T. After the time $t = T, 3T, 5T$, the step response has reached 63%, 95%, 99% of its final value. The *gain* can easily be determined by the ratio between the final value $y(\infty)$ and the step height u_0 as

$$K = \frac{y(\infty)}{u_0} . \tag{2.5.5}$$

In order to obtain the *time constant* T, one can exploit the property that for any arbitrary point in time

$$\frac{\mathrm{d}y(t)}{\mathrm{d}t} = \frac{y(\infty)}{T} \mathrm{e}^{-\frac{t}{T}} . \tag{2.5.6}$$

If one constructs the tangent to the step response at an arbitrary point in time t_1, then

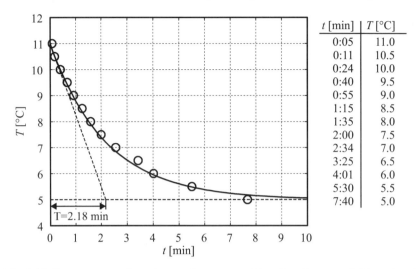

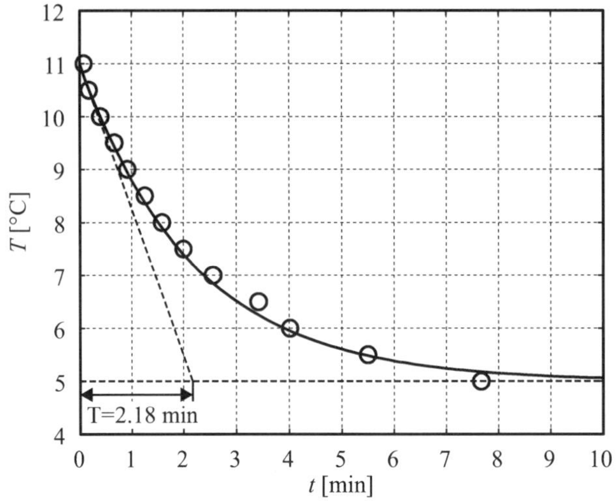

t [min]	T [°C]
0:05	11.0
0:11	10.5
0:24	10.0
0:40	9.5
0:55	9.0
1:15	8.5
1:35	8.0
2:00	7.5
2:34	7.0
3:25	6.5
4:01	6.0
5:30	5.5
7:40	5.0

Fig. 2.10. Step response of a resistance thermometer ($d = 5\,\text{mm}$) surrounded by quiet air

$$\frac{\Delta y(t_1)/y(\infty)}{\Delta t} = \frac{e^{-\frac{t_1}{T}}}{T} , \tag{2.5.7}$$

i.e. it intersects the final value line at a distance T from point t_1, see Fig. 2.9. Especially, for $t_1 = 0$,

$$\frac{\Delta y(0)}{\Delta t} = \frac{y(\infty)}{T} , \tag{2.5.8}$$

so that one can read out the time constant T by constructing the tangent to the step response at the origin and considering the intersection of this tangent with the final value line.

Example 2.1 (Transfer Function of a Resistance Thermometer ($d = 5\,\text{mm}$) Surrounded by Quiet Air).

In this example, the transfer function of a digital thermometer has been identified. For this experiment, the thermometer has first been covered by a shutter and has then been exposed to the outside temperature. The measurements are shown in Fig. 2.10 and the time constant has been identified as $T = 2.18$ min. □

2.5.2 Approximation by a Second Order System

A second order system is governed by the transfer function

$$G(s) = \frac{y(s)}{u(s)} = \frac{b_0}{1 + a_1 s + a_2 s^2} = \frac{K}{1 + T_1 s + T_2^2 s^2} = \frac{K}{1 + \frac{2\zeta}{\omega_n} s + \frac{1}{\omega_n^2} s^2} , \tag{2.5.9}$$

where K is the *gain*, ζ the *damping ratio* and ω_n the *undamped natural frequency*. The two poles are given as

$$s_{1,2} = \omega_{\mathrm{n}}\left(-\zeta \pm \sqrt{\zeta^2 - 1}\right) . \tag{2.5.10}$$

Depending on the size of ζ, the radicant is positive, zero, or negative. One can thus distinguish three cases as discussed in the following. A parameter study of ζ can be seen in Fig. 2.11

Case 1: Overdamped, $\zeta > 1$, Two Real Poles

In this case, the poles are negative real and unequal. The system can hence be realized by a series connection of two first order systems. The step response is given as

$$h(t) = K\left(1 + \frac{1}{s_1 - s_2}\left(s_2 \mathrm{e}^{s_1 t} - s_1 \mathrm{e}^{s_2 t}\right)\right) . \tag{2.5.11}$$

Case 2: Critically Damped, $\zeta = 1$, Double Pole on the Real Axis

Here, the poles are still negative real, but are now equal. This case can be realized by the series connection of two identical first order systems and is characterized by the shortest response time of all second order systems. The step response reads as

$$h(t) = K\left(1 - \mathrm{e}^{-\omega_{\mathrm{n}} t}\left(1 + \omega_{\mathrm{n}}(t)\right)\right) . \tag{2.5.12}$$

Case 3: Underdamped, $0 < \zeta < 1$, Conjugate Complex Pair of Poles

In contrast to the first two cases, the system will show damped oscillations. Two additional characteristic values will now be introduced

$$\omega_{\mathrm{d}} = \omega_{\mathrm{n}}\sqrt{1 - \zeta^2} \quad \text{Damped natural frequency} \tag{2.5.13}$$

$$\gamma = \zeta\omega_{\mathrm{n}} \quad \text{Damping coefficient .} \tag{2.5.14}$$

With these definitions, the step response is given as

$$h(t) = K\left(1 - \frac{1}{\sqrt{1 - \zeta^2}} \mathrm{e}^{-\gamma t} \sin(\omega_{\mathrm{d}} t + \varphi)\right) \tag{2.5.15}$$

with

$$\varphi = \arctan \frac{\omega_{\mathrm{d}}}{\gamma} = \arctan \zeta\sqrt{1 - \zeta^2} . \tag{2.5.16}$$

(2.5.15) describes a phase-shifted damped sine function. The maximum overshoot over the final value line is given by

$$y_{\mathrm{max,K}} = y_{\mathrm{max}} - K = K \exp -\frac{\pi\gamma}{\omega_{\mathrm{d}}} . \tag{2.5.17}$$

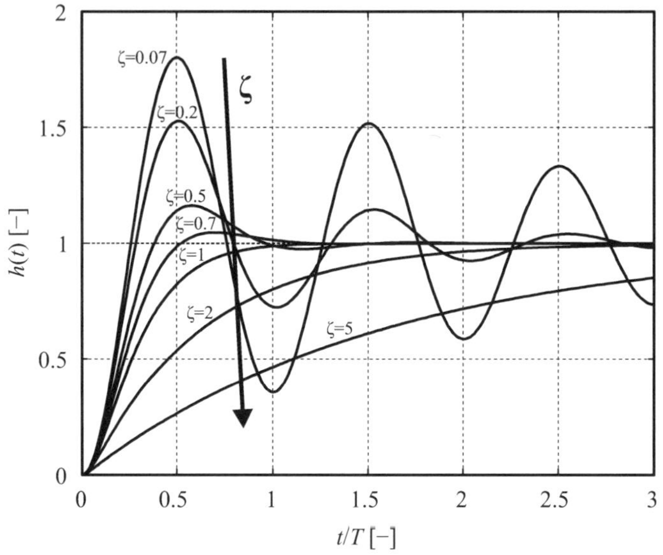

Fig. 2.11. Step response of a second order system for different damping factors ζ

For a given step response, one can first identify the intersection points of the step response with the final value line. The period time of the oscillation T_P leads to the damped natural frequency as

$$\omega_d = \frac{2\pi}{T_P} \,. \tag{2.5.18}$$

From the maximum overshoot, one can then determine the damping coefficient as

$$\gamma = \frac{\omega_d}{\pi} \ln \frac{K}{y_{max,K}} \,. \tag{2.5.19}$$

From (2.5.13) and (2.5.14), one can calculate ω_0 and ζ as

$$\zeta = \frac{1}{\left(\frac{\omega_d}{\gamma}\right)^2 + 1} \tag{2.5.20}$$

$$\omega_n = \frac{\gamma}{\zeta} \,. \tag{2.5.21}$$

The amplitude of the frequency response has a maximum at the *resonant frequency*

$$\omega_r = \omega_n \sqrt{1 - 2\zeta^2} \tag{2.5.22}$$

for $0 < \zeta < 1/\sqrt{2}$ with the maximum magnitude given as

$$|G(\omega_r)| = \frac{K}{\zeta\sqrt{1-\zeta^2}} \,. \tag{2.5.23}$$

2.5.3 Approximation by n^{th} Order Delay with Equal Time Constants

Aperiodic systems of order n are typically realized by a series connection of n mutually independent storages of first order with different time constants, thus

$$\begin{aligned} G(s) = \frac{y(s)}{u(s)} &= \frac{\prod_{k=1}^{n} K_k}{\prod_{k=1}^{n}(1+T_k s)} = \frac{K}{1+a_1 s+\ldots+a_n s^n} \\ &= \frac{K s_1 s_2 \ldots s_n}{(s-s_1)(s-s_2)\ldots(s-s_n)} \end{aligned} \tag{2.5.24}$$

with

$$a_1 = T_1 + T_2 + \ldots + T_n \tag{2.5.25}$$

$$a_n = T_1 T_2 \ldots T_n \tag{2.5.26}$$

$$s_k = \frac{1}{T_k} \,. \tag{2.5.27}$$

Thus, the behavior of the system is fully characterized by the gain K and the n time constants, T_i. The corresponding step response is given as

$$h(t) = K\left(1 + \sum_{\alpha=1}^{n} c_\alpha e^{s_\alpha t}\right) , \tag{2.5.28}$$

where

$$c_\alpha = \lim_{s\to s_\alpha} \frac{1}{s}(s-s_\alpha)G(s) \,. \tag{2.5.29}$$

For passive systems, the energy/mass/momentum stored in the system during the step response is proportional to the individual time constant T_α. Therefore, the total amount of energy/mass/momentum stored in the whole system of order n must be proportional to the sum of all time constants. Thus, the area A in Fig. 2.12 is given as

$$\begin{aligned} A &= K y(\infty) \sum_{\alpha=1}^{n} T_\alpha = K y(\infty)\,(T_1 + T_2 + \ldots + T_n) \\ &= K y(\infty) T_\Sigma = K y(\infty) a_1 \,. \end{aligned} \tag{2.5.30}$$

In the following,

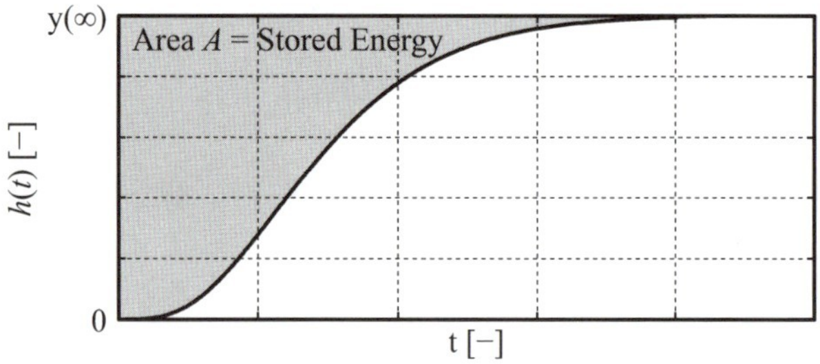

Fig. 2.12. Response of an aperiodic system of order n

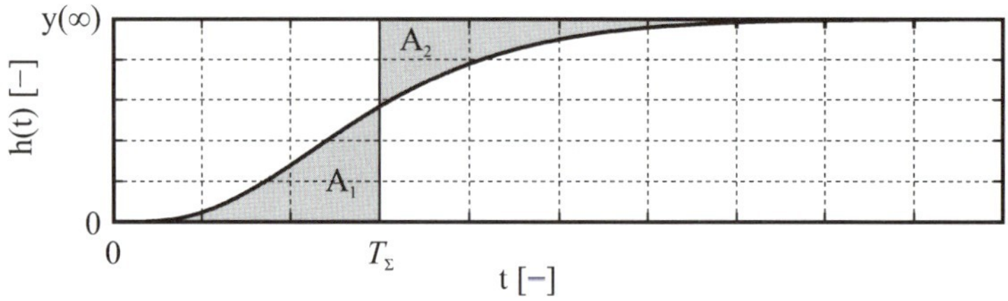

Fig. 2.13. Estimation of the sum of time constants

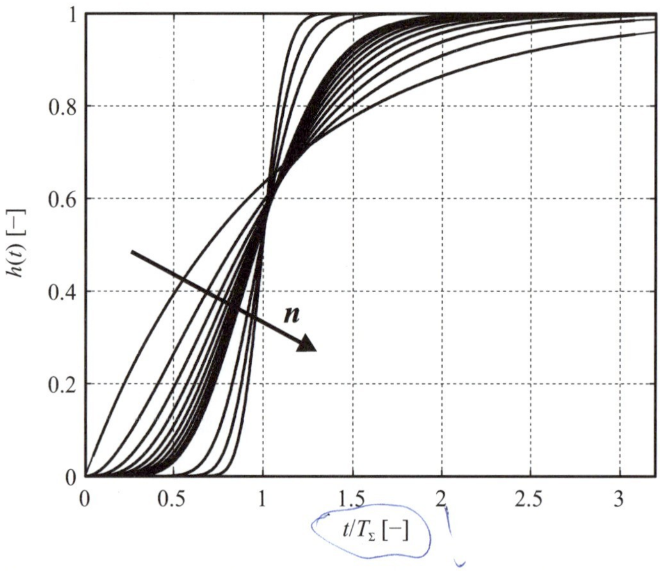

Fig. 2.14. Step responses of an aperiodic system of order n with equal time constants (Radtke, 1966)

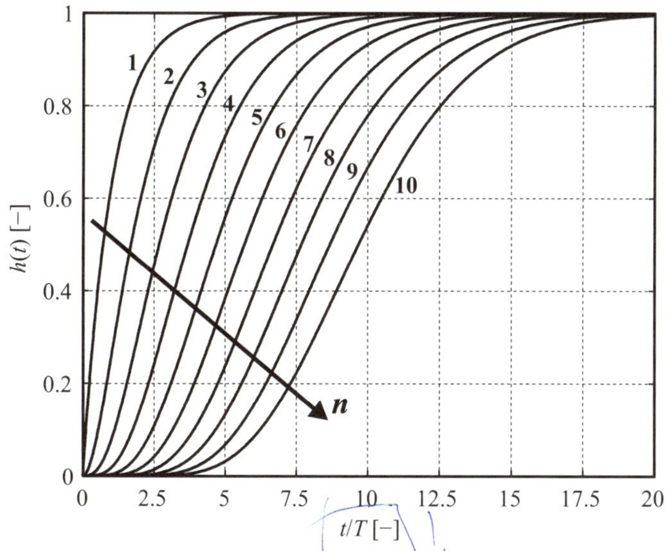

Fig. 2.15. Step responses of an aperiodic system of order n with transfer function $G(s) = 1/(Ts+1)^n$ and $n = 1, 2, \ldots, 10$

$$T_\Sigma = \sum_{\alpha=1}^{n} T_\alpha \tag{2.5.31}$$

will denote the *sum of time constants*. It is an additional characteristic quantity to describe the systems behavior. The sum of time constants can be estimated by drawing a parallel line to the y-axis at $t = T_\Sigma$, such that the area A is divided into two areas A_1 and A_2 of identical size, see Fig. 2.13. Figure 2.14 depicts the step responses of aperiodic systems of order n with equal time constants

$$T = T_1 = T_2 = \ldots = T_n \ . \tag{2.5.32}$$

The step responses are shown in a time scale t which is referred to the sum of time constants T_Σ, which guarantees that all systems store the same amount, i.e. the area A in Fig. 2.12 is the same. Step responses with equal time constants represent the limiting case of a critically damped system. The step responses with $n \geq 2$ intersect each other in one point. The variation of the order n is also shown in Fig. 2.15. It graphs the step response of the system

$$G(s) = \frac{K}{(Ts+1)^n} \tag{2.5.33}$$

on the referred time scale t/T. The step responses are given as

$$h(t) = K\left(1 - e^{-\frac{t}{T}} \sum_{\alpha=0}^{n-1} \frac{1}{\alpha!}\left(\frac{t}{T}\right)^{\alpha}\right) . \tag{2.5.34}$$

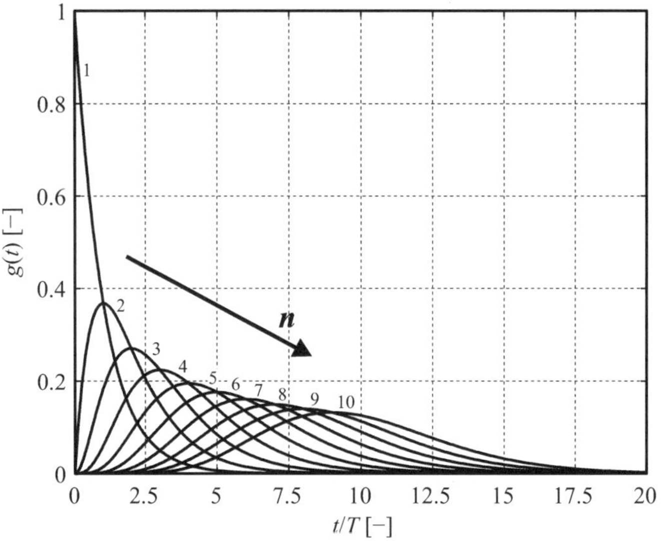

Fig. 2.16. Impulse responses of an aperiodic system of order n with transfer function $G(s) = 1/(Ts+1)^n$ and $n = 1, 2, \ldots, 10$

The impulse responses are governed by (Strejc, 1959)

$$g(t) = \frac{K}{T^n}\frac{t^{n-1}}{(n-1)!}\mathrm{e}^{-\frac{t}{T}}\,. \tag{2.5.35}$$

These impulse responses for $n = 1, 2, \ldots, 10$ are shown in Fig. 2.16. The maximum of the impulse response results as

$$g_{\max}(t_{\max}) = \frac{K(n-1)^{n-1}}{T(n-1)!}\mathrm{e}^{-(n-1)} \tag{2.5.36}$$

at the time

$$t = t_{\max} = (n-1)T \text{ for } n \geq 2\,. \tag{2.5.37}$$

For equal time constants, one gets in the limiting case $n \to \infty$

$$G(s) = \lim_{n\to\infty}(1+Ts)^{-n} = \lim_{n\to\infty}\left(1+\frac{T_\Sigma}{n}s\right)^{-n} = \mathrm{e}^{-T_\Sigma s} \tag{2.5.38}$$

with $T_\Sigma = nT$ and $|T_\Sigma s/n| < 1$. Thus, it follows that the coupling of infinitely many first order systems with infinitesimally small time constants shows the same behavior as a dead time with $T_\mathrm{D} = T_\Sigma$.

A common method to characterize the transfer function of systems with order $n \geq 2$ is by means of the characteristic times T_D and T_S which can be determined by constructing the tangent at the inflection point Q with t_Q and y_Q, see Fig. 2.17. From (2.5.33), one can determine the characteristic quantities t_Q, y_Q, T_D, and T_S,

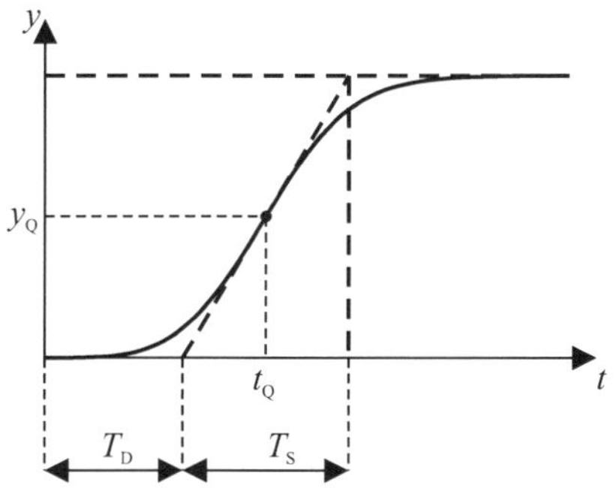

Fig. 2.17. Determination of delay time T_D and settlement time T_S for the step response of system with order $n \geq 2$

$$\frac{t_Q}{T} = n - 1 \tag{2.5.39}$$

$$\frac{y_Q}{y_\infty} = 1 - e^{-(n-1)} \sum_{\nu=0}^{n-1} \frac{(n-1)^\nu}{\nu!} \tag{2.5.40}$$

$$\frac{T_S}{T} = \frac{(n-2)!}{(n-1)^{n-2}} e^{n-1} \tag{2.5.41}$$

$$\frac{T_D}{T} = n - 1 - \frac{(n-2)!}{(n-1)^{n-2}} \left(e^{n-1} - \sum_{\nu=0}^{n-1} \frac{(n-1)^\nu}{\nu!} \right) . \tag{2.5.42}$$

For $n = 1, \ldots, 10$, the values are tabulated in Table 2.1. The characteristic values T_D/T_S and y_Q do not depend on the time constant T but only the oder n. For $1 \leq n \leq 7$, the approximation

$$n \approx 10 \frac{T_D}{T_S} + 1 \tag{2.5.43}$$

is valid. By determining the values T_D, T_S, and y_∞ from the measured step response according to Fig. 2.17, one can then use Table 2.1 to determine the parameters K, T, and n of the approximate continuous-time model according to (2.5.38).

For the approximation by an nth order system with equal time constants one should use the following approach:

1. First of all, onc has to tcst whether the system under investigation can in fact be approximated by the system given in (2.5.38). To determine the feasibility, one has to estimate the total time constant T_Σ according to Fig. 2.13. Then, the measured data can be drawn in a diagram with the axis t referred to the total time constant T_Σ and it can be checked whether the system can in fact be approximated by the model in (2.5.38). If the system under scrutiny contains a dead time, this dead time has to be subtracted from the delay time T_D.
2. Designation of system order n: By means of the ratio between delay time and settlement time, T_D/T_S, the system order can be read out from Table 2.1. The result can be validated by checking the y coordinate of the inflection point. It must be equal to y_Q.

Table 2.1. Characteristic values of a system of order n with equal time constants (Strejc, 1959)

n	$\frac{T_D}{T_S}$	$\frac{t_Q}{T}$	$\frac{T_S}{T}$	$\frac{T_D}{T}$	$\frac{y_Q}{y_\infty}$
1	0	0	1	0	0
2	0.104	1	2.718	0.282	0.264
3	0.218	2	3.695	0.805	0.323
4	0.319	3	4.463	1.425	0.353
5	0.410	4	5.119	2.100	0.371
6	0.493	5	5.699	2.811	0.384
7	0.570	6	6.226	3.549	0.394
8	0.642	7	6.711	4.307	0.401
9	0.709	8	7.164	5.081	0.407
10	0.773	9	7.590	5.869	0.413

3. Specification of the time constant T: The characteristic values t_Q, T_D, T_S allow to determine the time constant T in three different ways based on Table 2.1. Typically, the average of the three (different) estimates for T is taken.
4. Fixing the gain K: The ratio of the height of the step input, u_0, and the final displacement y_∞ of the system yields the gain K as

$$K = \frac{y_\infty}{u_0} \tag{2.5.44}$$

In the case of a non-integer system order n, one can obtain a better approximation by choosing the next lower integer system order n, choosing the corresponding delay time T'_D and assigning the delta $\Delta T_D = T_D - T'_D$ to a newly introduced dead time. The approximation method described in this section requires little effort, but on the other hand is very susceptible to noise and disturbances.

2.5.4 Approximation by First Order System with Dead Time

The step response of an nth order system can be approximated by a first oder system with dead time as

$$\tilde{G}(s) = \frac{K}{1 + T_D s} e^{-T_S s} \tag{2.5.45}$$

with the delay time T_D and the settlement time T_S as defined in Fig. 2.17. The approximation fidelity achieved by this simple model is however in many cases not sufficient.

Other identification methods with the determination of characteristic parameters of step responses, as for second order systems or n^{th} order systems with unequal or staggered time constants are summarized in (Isermann, 1992).

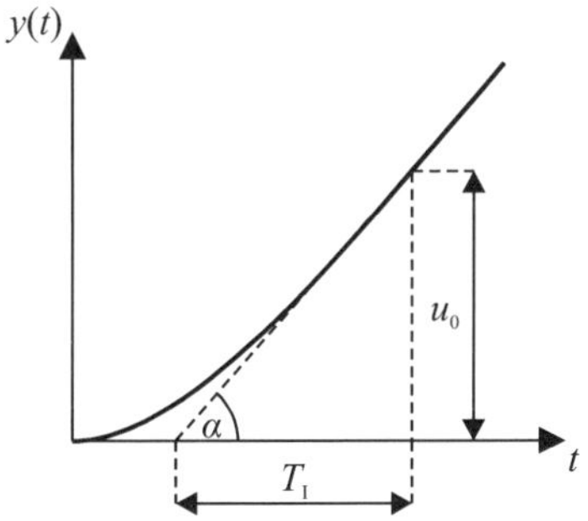

Fig. 2.18. Step response of a system with integral action and additional delays

2.6 Systems with Integral or Derivative Action

The methods presented so far have been targeting proportional acting systems. However, with some easy modifications, also systems with integral or derivative action can be investigated employing the methods derived so far.

2.6.1 Integral Action

An integral element with the transfer function

$$G(s) = \frac{y(s)}{u(s)} = \frac{K_\mathrm{I}}{s} = \frac{1}{T_\mathrm{I} s} \tag{2.6.1}$$

and the integral action coefficient K_I or the integral time T_I shows the response

$$y(t) = \frac{u_0}{T_\mathrm{I}} t \tag{2.6.2}$$

for a step input of height u_0. The slope of the response is thus given as

$$\frac{\mathrm{d}y(t)}{\mathrm{d}t} = \frac{u_0}{T_\mathrm{I}} \, . \tag{2.6.3}$$

By determining the slope $\mathrm{d}y(t)/\mathrm{d}t$, the characteristic value T_I can be determined as

$$T_\mathrm{I} = \frac{u_0}{\frac{\mathrm{d}y(t)}{\mathrm{d}t}} \, . \tag{2.6.4}$$

If the system contains further delay elements according to

$$G(s) = \frac{y(s)}{u(s)} = \frac{1}{T_\mathrm{I} s} \frac{1}{\prod\limits_{k=1}^{n} (1 + T_k s)} \, , \tag{2.6.5}$$

then for the step response with $u(s) = u_0/s$ follows

$$\lim_{t\to\infty} \frac{\mathrm{d}y(t)}{\mathrm{d}t} = \lim_{s\to 0} s^2 y(s) = \lim_{s\to 0} sG(s)u_0 = \frac{u_0}{T_\mathrm{I}} . \tag{2.6.6}$$

The characteristic value T_I can thus be determined from the final value of the slope of the step response by means of (2.6.3) and (2.6.6), see also Fig. 2.18. If the derivative of the step response of the system is created graphically or by means of a computer and therefore

$$\mathfrak{L}\left\{\frac{\mathrm{d}y(t)}{\mathrm{d}t}\right\} = sy(s) \tag{2.6.7}$$

is treated as the system output, then the output can be generated by a proportional acting system with

$$G_\mathrm{P}(s) = \frac{y(s)}{u(s)} = \frac{1}{T_\mathrm{I}} \frac{1}{\prod_{k=1}^{n}(1+T_k s)} , \tag{2.6.8}$$

whose characteristic values T_k can be determined by means of the tools introduced so far.

If the integral acting system with transfer function $G(s)$ is excited by a short *square pulse* of height u_0 and duration T, which can be approximated by a δ-impulse of area $u_0 T$, then the Laplace transform of the output is given as

$$y(s) = G(s)u_0 T = \frac{T}{T_\mathrm{I}} \frac{1}{\prod_{\nu=1}^{n}(1+T_\nu s)} \frac{u_0}{s} . \tag{2.6.9}$$

The response can thus be interpreted as that of a proportional-acting system responding to a step input of height u_0 and can then be examined by the methods learned so far to obtain $K_0 = T/T_\mathrm{I}$ and the T_k.

2.6.2 Derivative Action

Systems with the transfer function

$$G(s) = \frac{y(s)}{u(s)} = \frac{T_\mathrm{D} s}{\prod_{k=1}^{n}(1+T_k s)} , \tag{2.6.10}$$

with the differential action time T_D or differential action coefficient $K_\mathrm{D} = T_\mathrm{D}$ have step responses with the final value $y(\infty) = 0$. If one integrates the recorded step response and interprets the result as the response to a step input, then the input and output are connected by the hypothetical proportional acting system

$$G_\mathrm{P}(s) = \frac{\frac{y(s)}{s}}{u(s)} = \frac{T_\mathrm{D}}{\prod_{\nu=1}^{n}(1+T_\nu s)} , \tag{2.6.11}$$

and its characteristic values T_{D} and T_k can again be identified by the methods presented so far.

Another method is to excite the system by a ramp

$$u(t) = ct \text{ or } u(s) = \frac{c}{s^2} \,. \tag{2.6.12}$$

Then one gets

$$y(s) = \frac{T_{\mathrm{D}}}{\prod_{\nu=1}^{n}(1 + T_\nu s)} \frac{c}{s} \,, \tag{2.6.13}$$

which corresponds to the step response of a proportional acting system. Thus one can reduce the analysis of both systems with integral action as well as systems with differencing action to the analysis of proportional acting systems.

2.7 Summary

After compiling some basic relations for continuous-time and discrete-time processes and stochastic signals in the first section and defining some notations used throughout the sequel chapters, some easy to apply methods for parameter determination of simple linear processes were described. These classic parameter determination methods use characteristic values from measured system responses for *simple models*, and allow a fast and simple evaluation by hand. The methods described in this chapter yield approximate models which allow a rough examination of the system characteristics. They are in general only suitable for measurements with little to no disturbances.

The determination of characteristic values for systems of first and second order can be determined by visual inspection, thus no special methods are needed in this case. For systems of higher order with low-pass characteristics, a bunch of methods has been developed in the past, which allow the determination of characteristic values. While the approximation by a first order system with dead time is seldom accurate enough, an approximation of a higher order system with equal time constants can yield good results in many cases. Furthermore in this chapter, methods have been shown that allow the application of techniques for proportional acting systems (as described mainly in this chapter) to systems with integral and derivative action as well.

Problems

2.1. Fourier Transform

Summarize the conditions for the direct application of the Fourier transform to a signal. Why does the Fourier transform of the step response of a first order system not exist?

2.2. Impulse Response, Step Response, and Frequency Response
How are the impulse response, the step response, and the frequency response related to each other? Calculate these responses for a first order system ($G(s) = K/(Ts + 1)$) with $K = 0.8$ and $T = 1.5$ s.

2.3. First Order Process
A process of first order with a time constant $T = 10$ s is sampled with a sample time of $T_0 = 0.5$ s. What is the largest frequency for determining the frequency response by sinusoidal excitation of the input.

2.4. Sampling
Describe how a signal becomes amplitude- and time-discrete by sampling.

2.5. Stochastic Signals
By which characteristic values and parameters can stationary stochastic signals be described?

2.6. White-Noise
What are the statistical properties of white noise? What is a fundamental difference between continuous-time white noise and discrete-time white noise?

2.7. ARMA Processes
Give the z-transfer function of an auto-regressive and a moving-average process of second order.

2.8. First Order System
Determine the gain and time constant for the thermometer governed in Example 2.1.

2.9. Systems with Integral Action How can the system parameters of systems with integral action be determined?

2.10. Systems with Derivative Action How can the system parameters of systems with derivative action be determined?

References

Åström KJ (1970) Introduction to stochastic control theory. Academic Press, New York

Åström KJ, Murray RM (2008) Feedback systems: An introduction for scientists and engineers. Princeton University Press, Princeton, NJ

Box GEP, Jenkins GM, Reinsel GC (2008) Time series analysis: Forecasting and control, 4th edn. Wiley Series in Probability and Statistics, John Wiley, Hoboken, NJ

Bracewell RN (2000) The Fourier transform and its applications, 3rd edn. McGraw-Hill series in electrical and computer engineering, McGraw Hill, Boston

Bronstein IN, Semendjajew KA, Musiol G, Mühlig H (2008) Taschenbuch der Mathematik. Harri Deutsch, Frankfurt a. M.

Chen CT (1999) Linear system theory and design, 3rd edn. Oxford University Press, New York

Dorf RC, Bishop RH (2008) Modern control systems. Pearson/Prentice Hall, Upper Saddle River, NJ

Föllinger O (2010) Regelungstechnik: Einführung in die Methoden und ihre Anwendung, 10th edn. Hüthig Verlag, Heidelberg

Föllinger O, Kluwe M (2003) Laplace-, Fourier- und z-Transformationen, 8th edn. Hüthig, Heidelberg

Franklin GF, Powell JD, Emami-Naeini A (2009) Feedback control of dynamic systems, 6th edn. Pearson Prentice Hall, Upper Saddle River, NJ

Franklin GG, Powell DJ, Workmann ML (1998) Digital control of dynamic systems, 3rd edn. Addison-Wesley, Menlo Park, CA

Gallager R (1996) Discrete stochastic processes. The Kluwer International Series in Engineering and Computer Science, Kluwer Academic Publishers, Boston

Goodwin GC, Sin KS (1984) Adaptive filtering, prediction and control. Prentice-Hall information and system sciences series, Prentice-Hall, Englewood Cliffs, NJ

Goodwin GC, Graebe SF, Salgado ME (2001) Control system design. Prentice Hall, Upper Saddle River NJ

Grewal MS, Andrews AP (2008) Kalman filtering: Theory and practice using MATLAB, 3rd edn. John Wiley & Sons, Hoboken, NJ

Hänsler E (2001) Statistische Signale: Grundlagen und Anwendungen. Springer, Berlin

Heij C, Ran A, Schagen F (2007) Introduction to mathematical systems theory : linear systems, identification and control. Birkhäuser Verlag, Basel

Isermann R (1991) Digital control systems, 2nd edn. Springer, Berlin

Isermann R (1992) Identifikation dynamischer Systeme: Grundlegende Methoden (Vol. 1). Springer, Berlin

Isermann R (2005) Mechatronic Systems: Fundamentals. Springer, London

Kammeyer KD, Kroschel K (2009) Digitale Signalverarbeitung: Filterung und Spektralanalyse mit MATLAB-Übungen, 7th edn. Teubner, Wiesbaden

Ljung L (1999) System identification: Theory for the user, 2nd edn. Prentice Hall Information and System Sciences Series, Prentice Hall PTR, Upper Saddle River, NJ

Mikleš J, Fikar M (2007) Process modelling, identification, and control. Springer, Berlin

Moler C, van Loan C (2003) Nineteen dubios ways to compute the exponential of a matrix, twenty-five years later. SIAM Rev 45(1):3–49

Nelles O (2001) Nonlinear system identification: From classical approaches to neural networks and fuzzy models. Springer, Berlin

Nise NS (2008) Control systems engineering, 5th edn. Wiley, Hoboken, NJ

Ogata K (2009) Modern control engineering. Prentice Hall, Upper Saddle River, NJ

Papoulis A (1962) The Fourier integral and its applications. McGraw Hill, New York

Papoulis A, Pillai SU (2002) Probability, random variables and stochastic processes, 4th edn. McGraw Hill, Boston

Phillips CL, Nagle HT (1995) Digital control system analysis and design, 3rd edn. Prentice Hall, Englewood Cliffs, NJ

Poularikas AD (1999) The handbook of formulas and tables for signal processing. The Electrical Engineering Handbook Series, CRC Press, Boca Raton, FL

Radtke M (1966) Zur Approximation linearer aperiodischer Übergangsfunktionen. Messen, Steuern, Regeln 9:192–196

Söderström T (2002) Discrete-time stochastic systems: Estimation and control, 2nd edn. Advanced Textbooks in Control and Signal Processing, Springer, London

Strejc V (1959) Näherungsverfahren für aperiodische Übergangscharakteristiken. Regelungstechnik 7:124–128

Verhaegen M, Verdult V (2007) Filtering and system identification: A least squares approach. Cambridge University Press, Cambridge

Zoubir AM, Iskander RM (2004) Bootstrap Techniques for Signal Processing. Cambridge University Press, Cambridge, UK

Part I

Identification of Non-Parametric Models in the Frequency Domain — Continuous Time Signals

3

Spectral Analysis Methods for Periodic and Non-Periodic Signals

Calculating the spectrum of a signal is important for many applications. To be able to automatically calculate the spectrum and also treat signals of arbitrary shape, there is a special interest in methods for numerical determination of the Fourier transform. These methods are typically implemented on digital computers, which makes it necessary to sample and store the signal before it is transformed. This brings along special ramifications that are discussed in later sections of this chapter. As the data sequences can be quite long, one is also especially interested in computationally efficient implementations of the Fourier transform on digital computers.

3.1 Numerical Calculation of the Fourier Transform

Often, one is interested in the frequency content of non-periodic signals to determine the frequency range and magnitude. In the context of identification, it is important to analyze the frequencies, which are excited by a certain test signal or to determine the frequency response function of a system due to non-periodic test signals. In the latter case, the frequency response must be calculated with non-periodic test signals according to (4.1.1), which necessitates knowledge of the Fourier transform of the input and the output. If the input $u(t)$ and/or the output $y(t)$ are provided as sampled signals with measurements taken at the discrete-time points t_k and $k = 0, 1, 2, \dots, N$, then the Fourier transform (4.1.4) must be determined numerically. For the calculation of the Fourier transform of a sampled and time limited signal, one needs the *Discrete Fourier Transform (DFT)*. Of special interest is the *Fast Fourier Transform (FFT)* which is a computationally more time saving realization of the discrete Fourier transform. These aspects will be discussed in the following sections.

3.1.1 Fourier Series for Periodic Signals

Every periodic function $x(t)$ with period time T, i.e. $x(t) = x(t + kT)$ for any integer k can be written as an infinite series

R. Isermann, M. Münchhof, *Identification of Dynamic Systems*,
DOI 10.1007/978-3-540-78879-9_3, © Springer-Verlag Berlin Heidelberg 2011

$$x(t) = \frac{a_0}{2} + \sum_{k=1}^{\infty} a_k \cos(k\omega_0 t) + b_k \sin(k\omega_0 t) \text{ with } \omega_0 = \frac{2\pi}{T} \,. \tag{3.1.1}$$

This series is termed a *Fourier series*. Typically, only a finite number of series elements is considered. The *Fourier coefficients* a_k and b_k can be determined via

$$a_k = \frac{2}{T} \int_0^T x(t) \cos(k\omega_0 t) \mathrm{d}t \tag{3.1.2}$$

$$b_k = \frac{2}{T} \int_0^T x(t) \sin(k\omega_0 t) \mathrm{d}t \,, \tag{3.1.3}$$

where the integration can also be carried out over any other interval of length T. With the complex exponential function, the above Fourier series can also be written as

$$x(t) = \sum_{k=-\infty}^{\infty} c_k \mathrm{e}^{\mathrm{i}k\omega_0 t} \tag{3.1.4}$$

with

$$c_k = \frac{1}{T} \int_0^T x(t) \mathrm{e}^{-\mathrm{i}k\omega_0 t} \mathrm{d}t \,. \tag{3.1.5}$$

A side-note should be made to *Gibbs phenomenon* (Gibbs, 1899). It basically states that a Fourier series cannot approximate a piecewise continuously differentiable periodic functions at jump discontinuities even if the number of series elements goes towards infinity. There will always be an overshoot, whose height in the limit as $N \to \infty$ can be determined to be roughly 18% of the step height. This fact plays an important role in signal processing as it introduces artifacts at stepwise discontinuties e.g. in a signal or a picture that is processed by a $2-D$ Fourier transform. Figure 3.1 illustrates the approximation of a rectangular periodic signal by a Fourier series with an increasing number of elements.

3.1.2 Fourier Transform for Non-Periodic Signals

Now, the interval length can formally be extended to $T \to \infty$ to be able to treat non-periodic signals. The Fourier transform was introduced in (2.1.9) as

$$\mathfrak{F}\{x(t)\} = x(\mathrm{i}\omega) = \int_{-\infty}^{\infty} x(t) \mathrm{e}^{-\mathrm{i}\omega t} \mathrm{d}t \,. \tag{3.1.6}$$

If the non-periodic continuous-time signal $x(t)$ is sampled with a sample time T_0, then the signal can be written as a series of Dirac impulses with the appropriate height

$$x_\delta(k) = \sum_{k=-\infty}^{\infty} x(t)\, \delta(t - kT_0) = \sum_{k=-\infty}^{\infty} x(kT_0)\, \delta(t - kT_0) \,. \tag{3.1.7}$$

Then (2.1.9) respectively (3.1.6) becomes

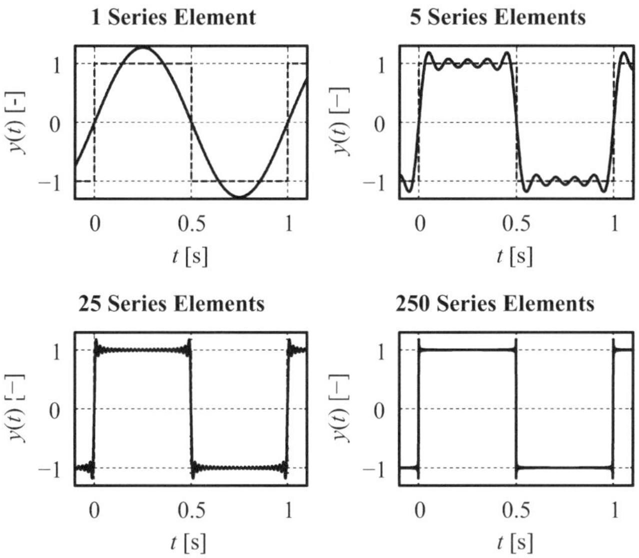

Fig. 3.1. Gibbs phenomenon: Even as the number of series elements is increased, the rectangular wave cannot be fully reconstructed

$$x_\delta(\mathrm{i}\omega) = \int_{-\infty}^{\infty} \sum_{k=-\infty}^{\infty} x(kT_0)\delta(t-kT_0)\mathrm{e}^{-\mathrm{i}k\omega T_0}\mathrm{d}t = \sum_{k=-\infty}^{\infty} x(kT_0)\mathrm{e}^{-\mathrm{i}k\omega T_0} \,. \tag{3.1.8}$$

This transformation is termed the *Discrete Time Fourier Transform (DTFT)*. The inverse transformation of the DTFT is given as

$$x(k) = \frac{T_0}{2\pi} \int_{-\frac{\pi}{T_0}}^{\frac{\pi}{T_0}} x_\delta(\mathrm{i}\omega)\mathrm{e}^{\mathrm{i}k\omega T_0}\mathrm{d}\omega \,. \tag{3.1.9}$$

As the continuous-time signal $x(t)$ is multiplied with a series of Dirac impulses, the resulting Fourier transform $x_\delta(\mathrm{i}\omega)$ is periodic in the frequency domain. This can be explained as follows: The multiplication in the time domain becomes a convolution in the frequency domain. The convolution of a frequency spectrum (i.e. the spectrum of the original, un-sampled signal) and a train of Dirac impulses leads to a periodic continuation. The periodicity can also be derived from the argument of the exponential function, which is periodic with 2π. Since the spectrum is periodic, it must only be evaluated in the range between $0 \leq \omega < 2\pi/T_0$ or $-\pi/T_0 \leq \omega < \pi/T_0$.

The periodicity also motivates *Shannon's theorem*, which states that only frequencies up to

$$f \leq \frac{1}{2T_0} = \frac{1}{2} f_\mathrm{S} \,, \tag{3.1.10}$$

where f_S is the sampling frequency, can be sampled correctly. All other frequencies will be sampled incorrectly due to the periodicity of the frequency spectrum leading to the so-called *aliasing* effect. A fault-free reconstruction of the signal is only possible, if the frequency spectrum is band-limited, i.e. $x(\mathrm{i}\omega)$ must vanish for $|\omega| > \omega_{\max}$, which is only possible for periodic signals. Therefore, all time-limited signals cannot have a band-limited frequency spectrum.

Since the computer has only limited storage capabilities, the summation in (3.1.8) cannot be evaluated in the interval $-\infty \leq k \leq \infty$. The number of datapoints is hence limited to N and sampled between $0 \leq k \leq N-1$. The Discrete Fourier Transform is then given as

$$\begin{aligned} x(\mathrm{i}\omega) &= \sum_{k=0}^{N-1} x(kT_0)\mathrm{e}^{-\mathrm{i}k\omega T_0} \\ &= \sum_{k=0}^{N-1} x(kT_0)\cos(k\omega T_0) - \mathrm{i}\sum_{k=0}^{N-1} x(kT_0)\sin(k\omega T_0) \\ &= \mathrm{Re}\{x(\mathrm{i}\omega)\} + \mathrm{Im}\{x(\mathrm{i}\omega)\} \,. \end{aligned} \tag{3.1.11}$$

The limitation of the number of datapoints that can be processed leads directly to the notion of windowing, see Sect. 3.1.4.

The frequency spectrum is still a continuous, periodic function. However, due to the fact that also in the frequency domain, the computer can only store a limited number of datapoints, the frequency variable ω must be discretized, too. Due to the periodicity, it is sufficient to sample the frequency spectrum in the interval between $0 \leq \omega < 2\pi/T_0$. The continuous spectrum is hence also multiplied with a sampling function, compare (3.1.8),

$$\tilde{x}(\mathrm{i}\nu\Delta\omega) = \sum_{\nu=0}^{M-1} x(\mathrm{i}\omega)\delta(\mathrm{i}\omega - \mathrm{i}\nu\Delta omega) \,, \tag{3.1.12}$$

where $\tilde{x}(\mathrm{i}\nu\Delta\omega)$ denotes the sampled Fourier transform, $\Delta\omega$ is the frequency increment and M is the number of sampling points, which is determined by the frequency increment $\Delta\omega$ as $M = 2\pi/(T_0\Delta\omega)$. This leads to a convolution in the time domain and means that the signal in the time domain is now also periodically continued outside the bounds of the sampling interval by the sampling in the frequency domain, i.e.

$$x(kT_0) = x(kT_0 + \mu T_\mathrm{n}),\ \mu = 0, 1, 2, \ldots, \text{ and } T_\mathrm{n} = \frac{2\pi}{\Delta\omega} \,. \tag{3.1.13}$$

The frequency increment is now chosen such that $T_\mathrm{n} = NT_0$ so that the periodicity is equivalent to the duration of the measurement in the time domain. Hence, also $M = N$ points should be sampled in the frequency domain.

Finally, the pair of transforms for the DFT is given as

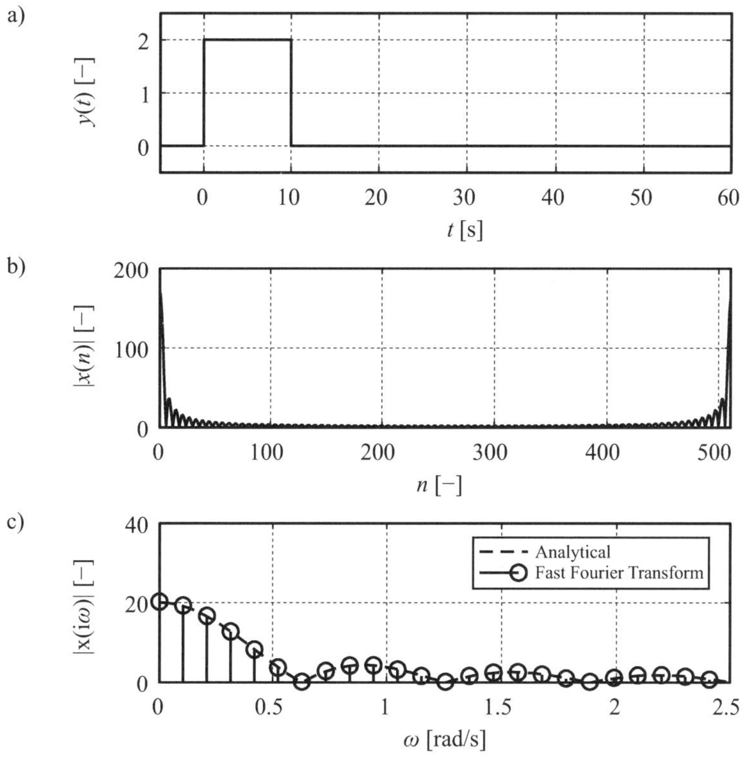

Fig. 3.2. Scaling of the FFT: (**a**) time signal to be converted, (**b**) output of the FFT (unscaled), (**c**) output of the FFT, scaled according to (3.1.16) and the frequency axis calculated by (3.1.17)

$$x(\mathrm{i}n\Delta\omega) = \mathrm{DFT}\{x(kT_0)\} = \sum_{k=0}^{N-1} x(kT_0)\mathrm{e}^{-\mathrm{i}kn\Delta\omega T_0} \tag{3.1.14}$$

$$x(kT_0) = \mathrm{DFT}^{-1}\{x_\mathrm{S}(\mathrm{i}n\Delta\omega)\} = \sum_{k=0}^{N-1} x(\mathrm{i}n\Delta\omega)\mathrm{e}^{\mathrm{i}kn\Delta\omega T_0}\,. \tag{3.1.15}$$

One can conclude that by application of the DFT and the inverse DFT, the signal and its spectrum both become periodic.

For each frequency ω, the DFT needs N multiplications and $(N-1)$ summations. Therefore, for the complete spectrum, one will need N^2 products and $N(N-1)$ summations. The computational expense is obviously quite high. In the next section, more efficient algorithms, the so-called *Fast Fourier Transforms* will be introduced. A detailed discussion of the Discrete Fourier Transform can be found e.g. in (Brigham, 1988; Stearns, 2003).

As the DFT (and subsequently also the FFT) processes only a vector of numbers, it shall finally shortly be discussed, how the output of the FFT must be interpreted.

Figure 3.2 shows the FFT and the analytically determined Fourier transform of a rectangular pulse. In order to determine the correct height of the amplitudes, the output of the FFT must be scaled by the sample time T_0 as

$$\mathfrak{F}\{x(kT_0)\} = T_0 x(\mathrm{i}k\Delta\omega) \,, \tag{3.1.16}$$

see (Isermann, 1991). The frequency vector belonging to the data is given as

$$\omega = \big(0, \Delta\omega, 2\Delta\omega, \ldots, (N-1)\Delta\omega\big) \text{ with } \Delta\omega = \frac{2\pi}{T_\mathrm{M}} \,, \tag{3.1.17}$$

where T_M is the measurement time and N the number of samples.

3.1.3 Numerical Calculation of the Fourier Transform

The calculation of the DFT (3.1.14) necessitates the multiplication of the sampled signal with the complex rotary operator

$$\mathrm{e}^{-\mathrm{i}kn\Delta\omega T_0} = W_N^{nk} \,. \tag{3.1.18}$$

Thus, the discrete Fourier transform can be rewritten as

$$x(n) = \sum_{k=0}^{N-1} x(k) W_N^{nk} \,, \tag{3.1.19}$$

where the sample time T_0 and the frequency increment $\Delta\omega$ will no longer be written down.

For the derivation of algorithms for the Fast Fourier Transform, typically the fact is exploited that the rotary operator W_N^{nk} has *cyclic* and *symmetric* properties. The Fourier transform can e.g. be split into two sums, one for the odd and one for the even numbered elements

$$\begin{aligned} x(n) &= \sum_{k=0}^{N-1} x(k) W_N^{nk} \\ &= \sum_{k=0}^{\frac{N}{2}-1} x(2k) W_N^{2nk} + \sum_{k=0}^{\frac{N}{2}-1} x(2k+1) W_N^{n(2k+1)} \\ &= \sum_{k=0}^{\frac{N}{2}-1} x(2k) W_N^{2nk} + W_N^n \sum_{k=0}^{\frac{N}{2}-1} x(2k+1)) W_N^{2kn} \\ &= x_\mathrm{e}(n) + W_N^n x_\mathrm{o}(n) \,, \end{aligned} \tag{3.1.20}$$

which is the basis for the radix-2 decimation-in-time (DIT) FFT, a very easy to motivate and possibly the most commonly used form of the Cooley-Tuckey algorithm (Cooley and Tukey, 1965). Both sums now have a rotary operator that is periodic

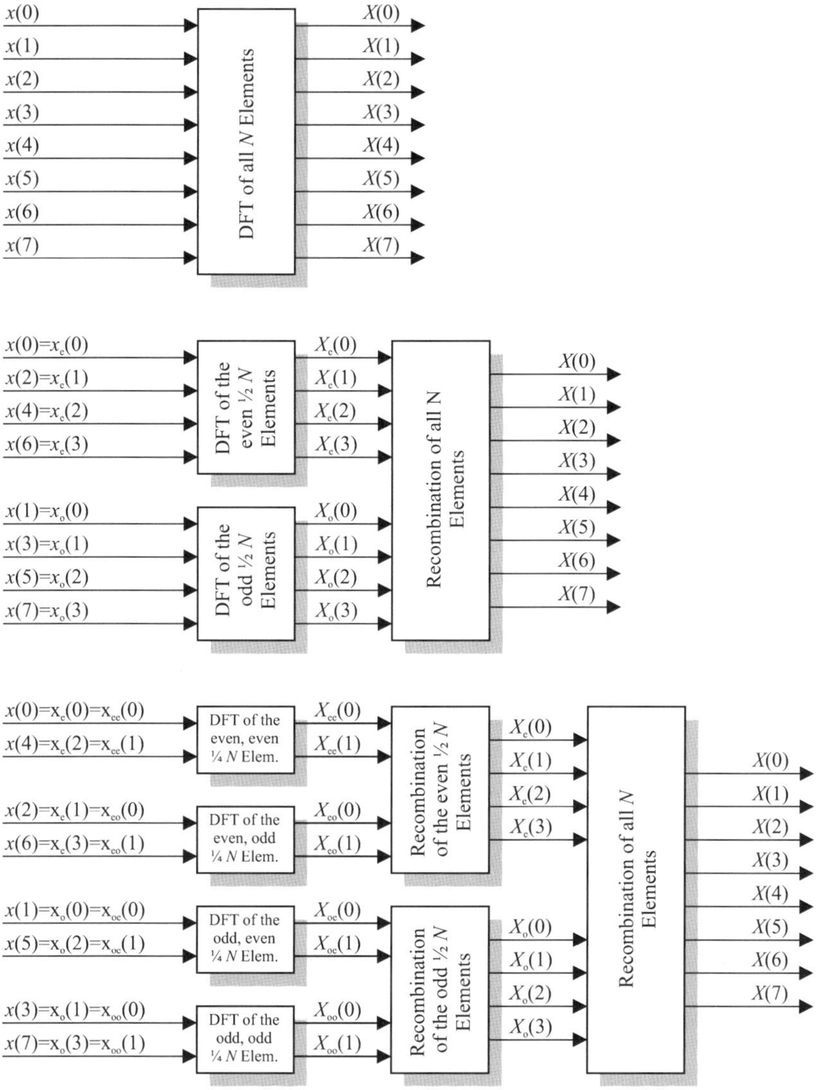

Fig. 3.3. Application of the FFT to $N = 8$ data points

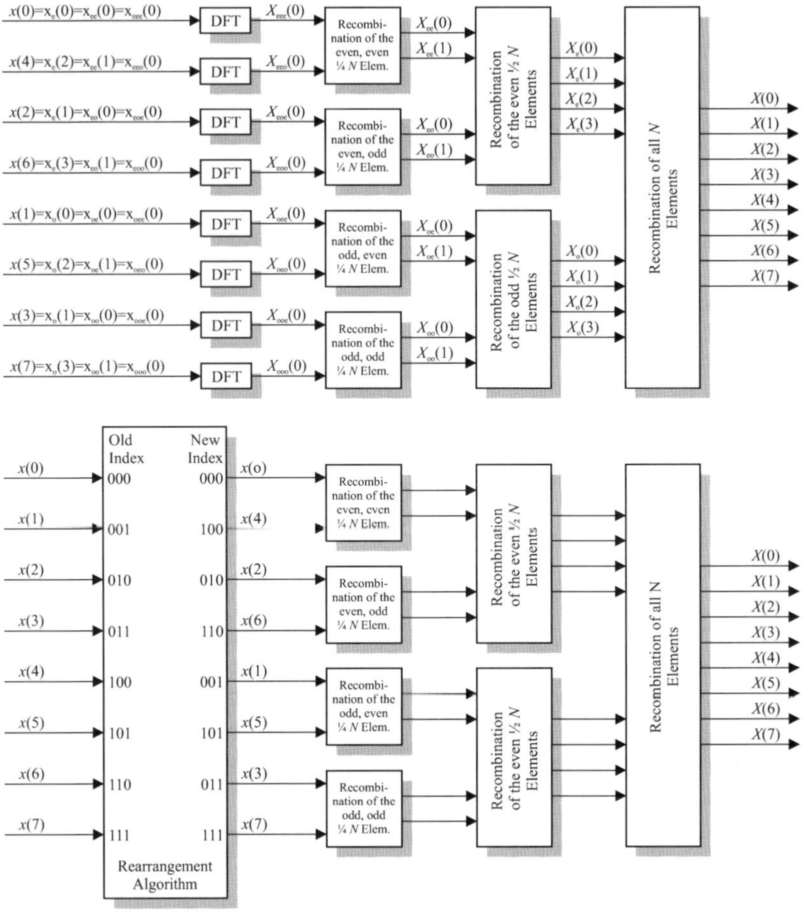

Fig. 3.3. Application of the FFT to $N = 8$ data points *(continued)*

with $N/2$. The two sums have been denoted as "e" for even and "o" for odd. These two sums can now again be split into two sums each. For each split, an additional letter is added to the index, therefore "ee" denotes then even, even elements. The principle of this divide et impera algorithm is illustrated for $N = 8$ in Fig. 3.3.

In the subsequent steps, the sums for the FFT are iteratively split up into two sums, each processing half of the elements of the initial sum. The algorithm stops, when the Fourier transform is applied to single numbers only, since the Fourier transform of a single element is just that very element itself. The entire decomposition can be seen in Fig. 3.3. One can also see from the figures that the recombination algorithm works on adjacent elements throughout the entire algorithm. Before the first recombination however, the samples have to be rearranged.

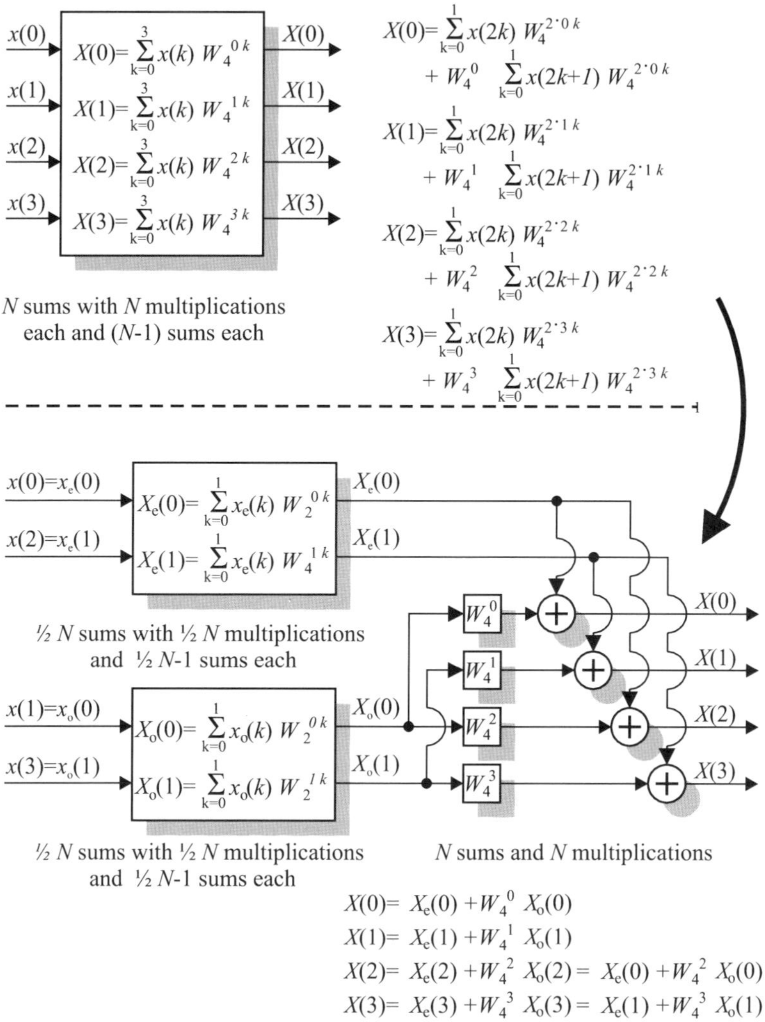

Fig. 3.4. First step of the recursion in the FFT's underlying algorithm in (3.1.20)

For this regrouping, the elements must be ordered according to their inverse binary index. For example, the element with index $n = 100_2$ becomes element $n = 001_2$ after this regrouping. This can be done by a simple, fast, and efficient algorithm. In this algorithm, two pointers are maintained. One pointer `k` is walking through the elements from 0 to $N-1$. The other pointer `l` always maintains the corresponding inverse bit order of `k`. The variable `n` will hold the number of elements in this algorithm.

The algorithm works in two stages: The first part of the loop adds 1 to the pointer `k` and also the inverse pointer `l` utilizing the following rules:

- if the LSB (Left-Most Bit) is 0, it is set to 1
- if the LSB (Left-Most Bit) is 1, it is set to zero and the second Left-Most Bit is increased by 1. Once again there are two case:
  - If the second Left-Most Bit is 0, it is set to 1
  - If the second Left-Most Bit is 1, it is set to zero and the third Left-most Bit is increased by one and so forth

One other result that can be exploited is the fact the while the inverse bit order of `k` is `l`, at the same time, the inverse bit order of `l` is `k`. Thus, one can always swap two elements. One only swaps elements if `k` is less than `l` to avoid back-swapping. The second part of the algorithm covers the recombination, here one repeatedly multiplies with the *twiddle-factors*, a name coined by Gentleman and Sande (1966). These factors should be computed ahead of time and stored in a table. Finally, the results of these multiplications are added up as illustrated in Fig. 3.4.

Today, there is a multitude of algorithms for the Fast Fourier Transform available. There are radix-4, radix-8 or radix-2^n algorithms available, which, while processing on larger data blocks, can bring down the total number of multiplications between 25% and 40% for the radix-4 and radix-8 algorithms respectively. Other authors that have contributed to the FFT are Bruun (1978), Rader (1968), Bluestein (1970), and Goertzel (1958). The Goertzel algorithm is of special interest whenever it is only necessary to compute a few spectral lines and not the entire spectrum. Also, split-radix techniques have been proposed (e.g. Sorensen et al, 1986).

Good implementations of the FFT and other algorithms can be found in the book by Press et al (2007) and the `FFTW` library (Frigo and Johnson, 2005). The latter is a software library for computing the discrete Fourier transform in one or more dimensions with arbitrary input size and both of real and complex data.

While the DFT needed a total of N^2 complex multiplications and $N(N-1)$ complex additions, where $O(N)$ operations could be saved, because they are trivial (multiplication with 1), the FFT as implemented above can bring this number down to $N \log_2 N$ complex multiplications and $N \log_2 N$ complex additions, where once again, trivial operations are included in the count and could be saved.

The FFT accepts both real-valued and complex data as it is formulated for complex data also in the time domain. If measurements, consisting of real-valued data only, are supplied to the FFT, then unnecessary operations, e.g. multiplications with zero, are carried out. One remedy would be to modify the calculation to accept real-valued data only. However, after the first stage of the FFT, the data to be processed will inevitably become complex. Therefore, one would need two realizations of the same algorithm, one for real and one for complex data. A much better idea is to combine two real-valued data points into one complex data point such that the results can be separated again after the Fourier transform has been carried out (Kammeyer and Kroschel, 2009; Chu and George, 2000).

One can transform two-real valued signals $y(k)$ and $z(k)$ at the same time by first combining the signals,

$$\tilde{x}(k) = y(k) + \mathrm{i}z(k) , \tag{3.1.21}$$

Table 3.1. Comparison of FLOPS for DFT and FFT

Complex-Valued Function in Time Domain, Complex Operations

N	DFT		FFT	
	Add	Mul	Add	Mul
128	16 256	16 384	896	896
1 024	1 047 552	1 048 576	10 240	10 240
4 096	16 773 120	16 777 216	49 152	49 152

Complex-Valued Function in Time Domain, Real Operations

N	DFT		FFT	
	Add	Mul	Add	Mul
128	65 280	65 536	3 584	3 584
1 024	4 192 256	4 194 304	40 960	40 960
4 096	67 100 672	67 108 864	196 608	196 608

Real-Valued Functions in Time Domain, Real Operations

N	DFT		FFT	
	Add	Mul	Add	Mul
128	16 256	16 384	768	512
1 024	1 047 552	1 048 576	7 680	5 632
4 096	16 773 120	16 777 216	34 816	26 624

Note: For the calculation of the FFT, a radix-2 algorithm was assumed, because of its wide spread.

then transforming the combined signal,

$$\tilde{x}(n) = \mathrm{DFT}\{\tilde{x}(k)\} , \tag{3.1.22}$$

and then separating the results

$$y(n) = \frac{1}{2}\big(\tilde{x}(n) + \tilde{x}^*(N-n)\big) \tag{3.1.23}$$

$$z(n) = \frac{1}{2\mathrm{i}}\big(\tilde{x}(n) - \tilde{x}^*(N-n)\big) . \tag{3.1.24}$$

If only one sequence is to be transformed, one can divide it into two sequences of half the length,

$$\left.\begin{aligned} y(k) &= x(2k) \\ z(k) &= x(2k+1) \end{aligned}\right\} \text{ with } k = 0, \dots, \frac{N-1}{2} . \tag{3.1.25}$$

Then, these sequences of real-valued data are merged into

$$\tilde{x}(k) = y(k) + \mathrm{i}z(k), \tag{3.1.26}$$

and transformed into the frequency domain by

$$\tilde{x}(n) = \mathrm{DFT}\{\tilde{x}(k)\} . \tag{3.1.27}$$

Finally, the Fourier transform of $x(k)$ is given as

$$x(n) = \frac{1}{2}\big(\tilde{x}(n) + \tilde{x}^*(N-n)\big) + \mathrm{e}^{-\mathrm{i}\frac{\pi n}{N}} \frac{1}{2\mathrm{i}}\big(\tilde{x}(n) - \tilde{x}^*(N-n)\big) . \tag{3.1.28}$$

This combination of two real-valued data points into one complex data point further speeds up the calculation of the FFT by a factor of almost 2. Table 3.1 illustrates the computational expense for different applications.

Every time-limited signal of length N can be augmented with an arbitrary number of zeros to the total length L,

$$x(k) = \begin{cases} x(k) \text{ for } 0 \le k \le N-1 \\ 0 \text{ for } N \le k \le L-1 \text{ and } L > N \end{cases} \tag{3.1.29}$$

with the effect of an increase in the resolution of the spectrum. This technique is termed *zero padding* and is often used to bring the length of a signal to an optimal length for the use of different FFT algorithms (Kammeyer and Kroschel, 2009). However, zero padding will inevitably give cause to the leakage effect for periodic signals.

3.1.4 Windowing

The limitation of the number of data points in the time domain can be understood as a multiplication of the time function $x(t)$ with a so-called *window function* to obtain the time-limited function $x_\mathrm{w}(t)$. The data that are sampled hence do not stem from $x(t)$, but rather from the product

$$x_\mathrm{w}(t) = x(t)\, w(t) . \tag{3.1.30}$$

When applying the Fourier transform, one does not obtain the Fourier transform of $x(\mathrm{i}\omega) = \mathfrak{F}\{x(t)\}$, where $x(t)$ is a signal of long or infinite time duration, but rather

$$x_\mathrm{w}(\mathrm{i}\omega) = \mathfrak{F}\{x_\mathrm{w}(t)\} = \mathfrak{F}\{x(t)\, w(t)\} = x(\mathrm{i}\omega) * f(\mathrm{i}\omega) , \tag{3.1.31}$$

i.e. the convolution of the frequency spectrum of the original signal and the frequency spectrum of the window function.

The effect of the window functions can best be illustrated by considering a single, distinct spectral line of a periodic sine at the moment. The considerations can however easily be applied to arbitrary non-periodic signals as well. If considering a single spectral line convoluted with the Fourier transform of the window function, one will see that the spectral line has "spilled over" to adjacent frequencies, thus this effect is also called *spill over*. The spill over effect can be controlled by means of so-called *window functions*, where one always has to find a compromise between a narrow and high main maximum and the suppression of side maxima.

For example, the *Bartlett Window* achieves a strong suppression of the side maxima at the cost of a low and broad main maximum. An even stronger suppression of the side maxima can be achieved by the *Hamming Window*. which in fact minimizes the main maximum in the restricted band. The *Hann window* and the *Blackmann window* are other typical window functions and are all shown in Table 3.2. An overview and detailed comparison of these along with many other window functions can be found in the books by Poularikas (1999) and Hamming (2007). The function of windowing is explained by Schoukens et al (2009) by showing that the Hann window does nothing else than take the second derivative of the frequency spectrum, thereby reducing the leakage effects. At the same time however, a smoothing of the spectrum is carried out, which introduces a smoothing error. A different tradeoff between the two types of errors is presented as the Diff window, which takes the difference between two adjacent spectral lines. For a thorough treatment of windowing, see also (Harris, 1978).

3.1.5 Short Time Fourier Transform

The Fourier transform, which was introduced at the beginning of this chapter, has an infinite time scale. However, in applications such as the identification of time-varying systems or fault detection, one wants to know how the frequency content varies as a function of the time. The *Short Time Fourier Transform* (STFT) has been introduced as a tool which has been tailored to the specific task of a joint time-frequency analysis and will be presented in the following, see also (Qian and Chen, 1996).

The Fourier transform as introduced in (2.1.9),

$$\mathfrak{F}\{x(t)\} = x(\mathrm{i}\omega) = \int_{-\infty}^{\infty} x(t)\mathrm{e}^{-\mathrm{i}\omega t}\,\mathrm{d}t \,. \tag{3.1.32}$$

is again rewritten as a sum of sampled values. The summation is once again carried out over a finite interval and the factor T_0 neglected. Thus, one obtains

$$x(\mathrm{i}\omega) = \sum_{k=0}^{N-1} x(kT_0)\mathrm{e}^{-\mathrm{i}\omega kT_0} \,. \tag{3.1.33}$$

A window function is now introduced along with the time shift parameter τ. The resulting calculation rule for the STFT is then given as

$$x(\omega,\tau) = \sum_{k=0}^{R-1} x((k-\tau)T_0)w(k)\mathrm{e}^{-\mathrm{i}\omega kT_0} \tag{3.1.34}$$

or

$$x(\omega,\tau) = \sum_{k=0}^{R-1} x(kT_0)w(k+\tau)\mathrm{e}^{-\mathrm{i}\omega kT_0} \,. \tag{3.1.35}$$

Table 3.2. Typical window functions (left: function over time; right: Fourier transform) (Harris, 1978)

Name, Shape and Equation
Bartlett Window
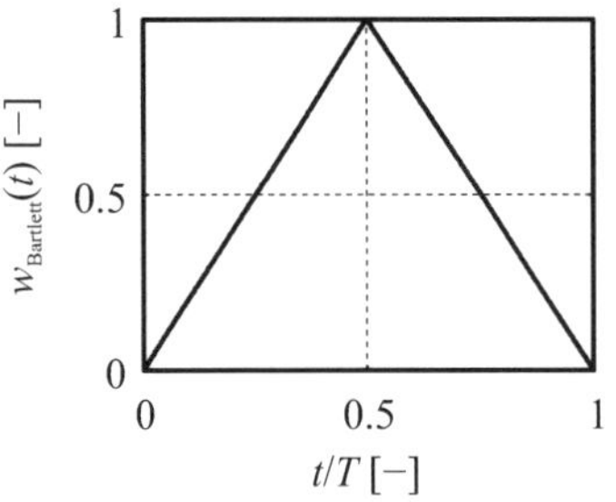 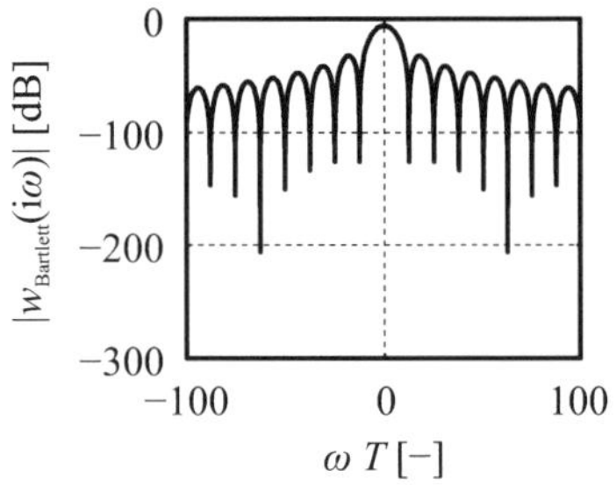
$w_{\text{Bartlett}}(t) = \begin{cases} 1 - \frac{2\left\lvert t - \frac{T}{2} \right\rvert}{T} & \text{if } 0 \le t \le T \\ 0 \text{ otherwise} \end{cases}$
Hamming Window
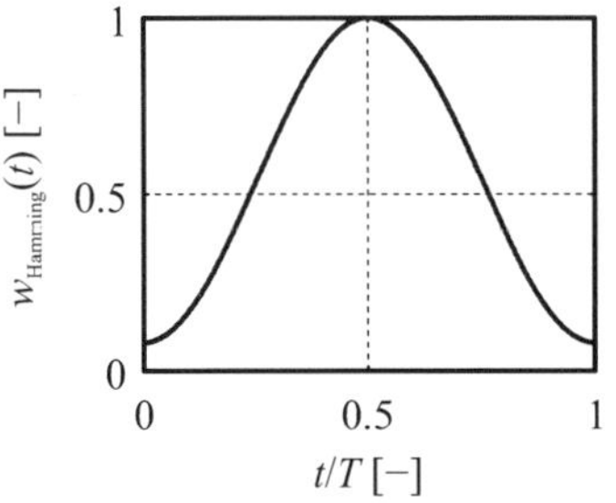 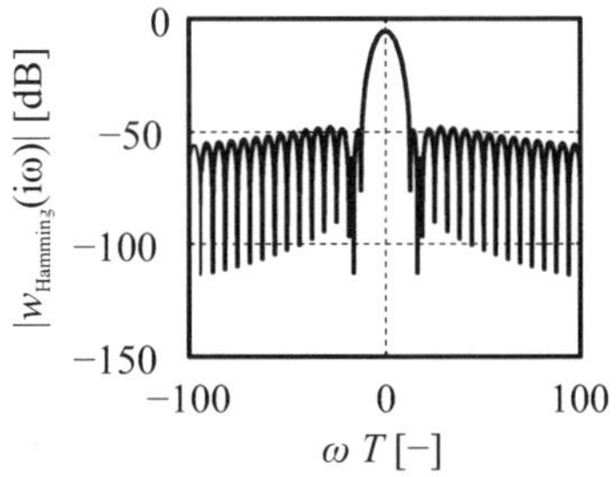
$w_{\text{Hamming}}(t) = \begin{cases} 0.54 - 0.46 \cos\left(2\pi \frac{t}{T}\right) & \text{if } 0 \le t \le T \\ 0 \text{ otherwise} \end{cases}$

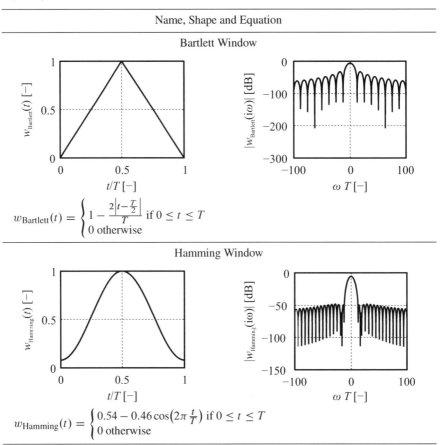

To see the changes of a signal over time, the signal is divided into small blocks and the STFT is calculated for each block separately as defined in (3.1.34) or (3.1.35), see Fig. 3.5. Therefore, the STFT depends on both the time and the frequency.

The two-dimensional plot of the STFT is termed *spectrogram*. The tuning parameters of the algorithm are the *block length* R and the *overlap*. A long block length R provides a higher frequency solution and a coarser resolution in the time domain. It is termed *narrowband spectrogram*. A short block length R on the contrary provides a higher time resolution and is termed *wideband spectrogram*. The overlap of the individual blocks allows to use longer blocks and thus increase the frequency domain resolution. It also allows to detect changes in the frequency spectrum earlier.

The spectrogram of a time-varying signal is shown in Fig. 3.6, where the STFT has been applied to a chirp signal.

Table 3.2. Typical window functions (left: function over time; right: Fourier transform (Harris, 1978)) *(continued)*

Name, Shape and Equation

Hann Window

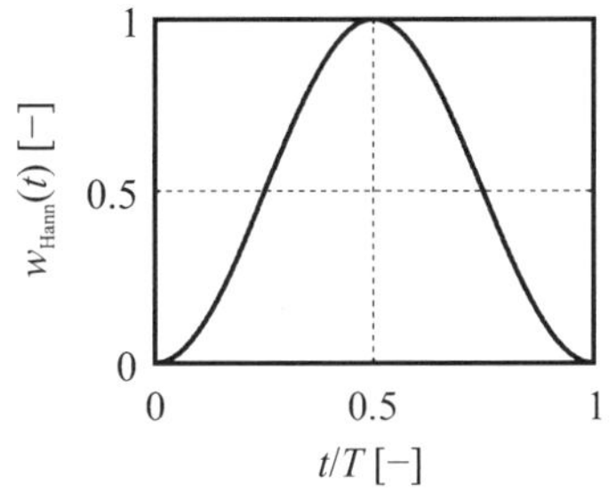

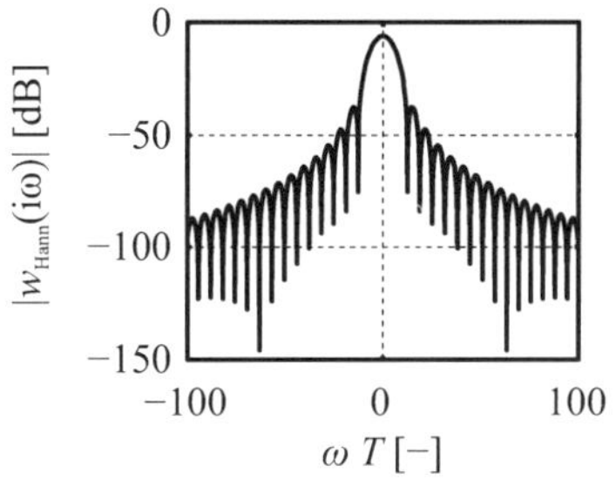

$$w_{\text{Hann}}(t) = \begin{cases} 0.5\left(1 - \cos\left(2\pi \frac{t}{T}\right)\right) \text{ if } 0 \leq t \leq T \\ 0 \text{ otherwise} \end{cases}$$

Blackmann Window

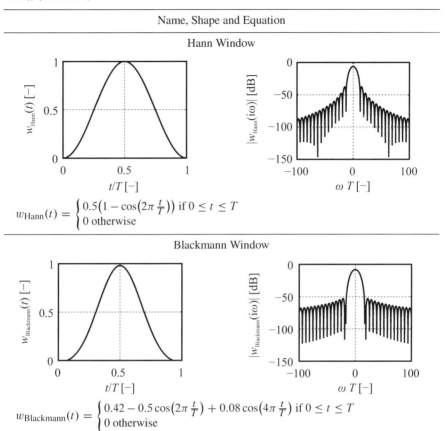

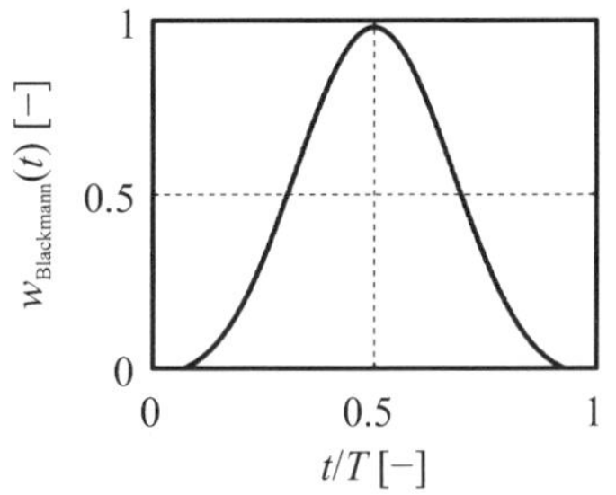

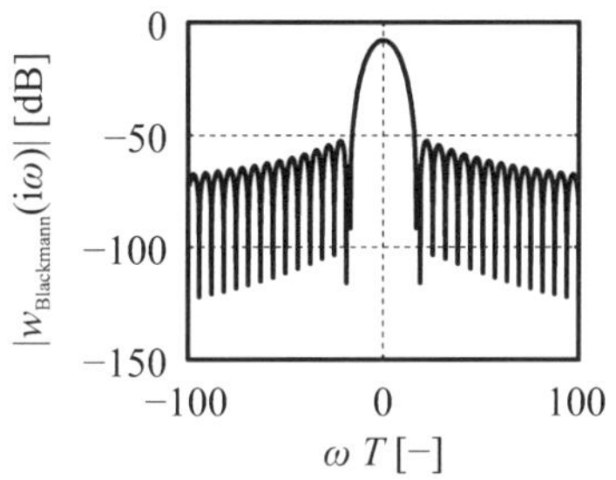

$$w_{\text{Blackmann}}(t) = \begin{cases} 0.42 - 0.5\cos\left(2\pi \frac{t}{T}\right) + 0.08\cos\left(4\pi \frac{t}{T}\right) \text{ if } 0 \leq t \leq T \\ 0 \text{ otherwise} \end{cases}$$

3.2 Wavelet Transform

The STFT determines the similarity between the investigated signal and a windowed harmonic signal. In order to obtain a better approximation of short time signal changes with sharp transients, the similarity with a short time *prototype function* of finite duration can be calculated. Such prototype or basis functions which show some damped oscillating behavior are *wavelets* that origin from a mother wavelet $\Psi(t)$, see (Qian and Chen, 1996; Best, 2000). Table 3.3 shows some typical mother wavelets, which can now be time-scaled (*dilatation*) by a factor a and time shifted (*translation*) by τ, leading to

$$\Psi^*(t, a, \tau) = \frac{1}{\sqrt{a}} \Psi\left(\frac{t - \tau}{a}\right) . \tag{3.2.1}$$

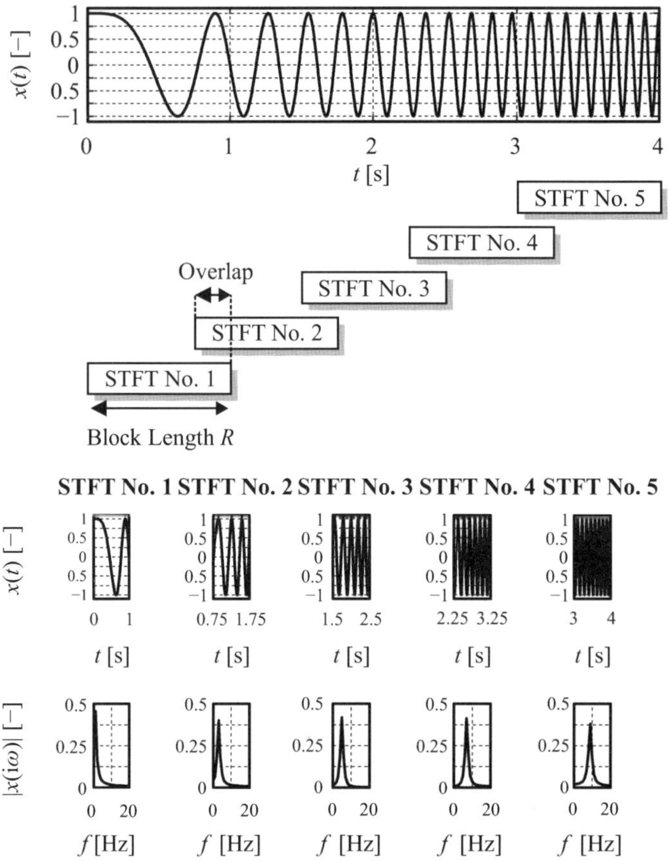

Fig. 3.5. Application of the STFT: Calculation of five Fourier transforms for overlapping regions of the time signal applied to a chirp signal starting at $f = 0\,\mathrm{Hz}$ for $t = 0\,\mathrm{s}$ and ending at $f = 10\,\mathrm{Hz}$ for $t = 4\,\mathrm{s}$

The factor $1/\sqrt{a}$ is introduced in order to reach a correct scaling of the power-density spectrum. If the mean frequency of the wavelet is ω_0, the scaling of the wavelet by t/a results in the scaled mean frequency ω_0/a.

The continuous-time wavelet transform (CWT) then is given as

$$\mathrm{CWT}(y, a, \tau) = \frac{1}{\sqrt{a}} \int_{-\infty}^{\infty} y(t) \Psi\left(\frac{t-\tau}{a}\right) \mathrm{d}t \,, \qquad (3.2.2)$$

which is real-valued for real-valued $y(t)$ and $\Psi(t)$. Note that in contrast, the STFT is typically a complex-valued function. Some sample wavelet functions have been tabulated in Table 3.3. The advantages of the wavelet transform stem from the signal adapted basis function and the better resolution in time and frequency. The signal

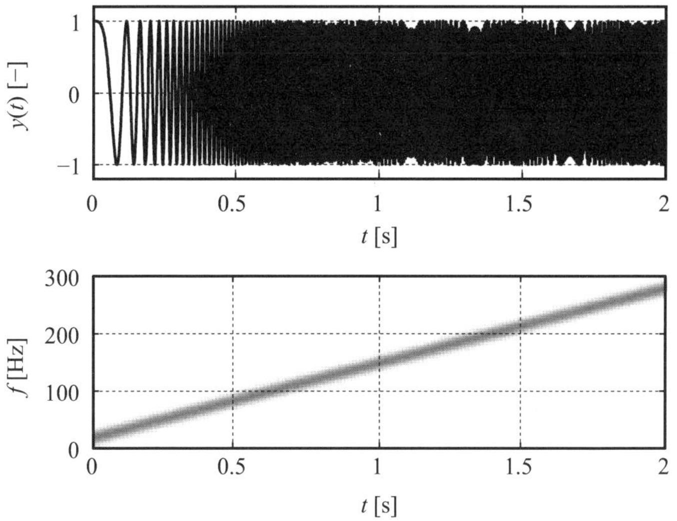

Fig. 3.6. Spectrogram of a chirp signal starting at $f = 0\,\text{Hz}$ for $t = 0\,\text{s}$ and ending at $f = 300\,\text{Hz}$ for $t = 2\,\text{s}$

adaptation for example is illustrated by the Haar wavelet which does not give raise to Gibbs phenomenon.

The wavelet functions correspond to certain band-pass filters where, for example, by a reduction of the mean frequency through the scale factor also a reduction of the bandwidth is achieved, compared to the STFT, where the bandwidth stays constant.

3.3 Periodogram

The periodogram is often also mentioned as a tool to determine the spectrum of a signal. It is defined as

$$\hat{S}_{\text{xx}}(\text{i}\omega) = \frac{1}{N}\left|x(\text{i}\omega)\right|^2 = \frac{1}{N}x(\text{i}\omega)x^*(\text{i}\omega) = \frac{1}{N}\sum_{\nu=0}^{N-1}\sum_{\mu=0}^{N-1} x(\nu)x(\mu)\text{e}^{-\text{i}\omega(\nu+\mu)T_0}\,. \tag{3.3.1}$$

It can be shown (e.g. Kammeyer and Kroschel, 2009) that the expected value of the estimate is given as

$$\text{E}\{\hat{S}_{\text{xx}}(\text{i}\omega)\} = \sum_{\nu=-(N-1)}^{N-1} w_{\text{Bartlett}}(\nu)R_{\text{xx}}(\nu)\text{e}^{-\text{i}\omega\nu T_0}\,, \tag{3.3.2}$$

where $R_{\text{xx}}(\nu)$ denotes the auto-correlation function of the signal $x(t)$. Hence the estimate of the spectrum is given as the true power spectral density $S_{\text{xx}}(\text{i}\omega)$ convoluted with the Fourier transform of the Bartlett window. So, the periodogram is only

Table 3.3. Typical Wavelets Functions

Name, Shape and Equation	Name, Shape and Equation
Haar Wavelet 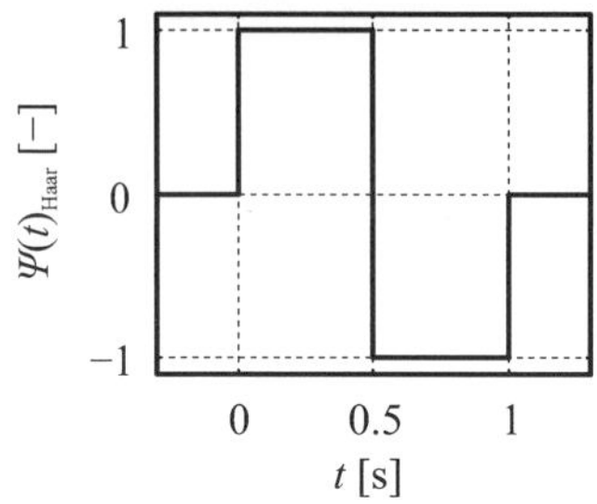 $\Psi_{\mathrm{Haar}}(t) = \begin{cases} 1 \text{ if } 0 \leq t \leq 0.5 \\ -1 \text{ if } 0.5 < t \leq 1 \\ 0 \text{ otherwise} \end{cases}$	Mexican Hat Wavelet 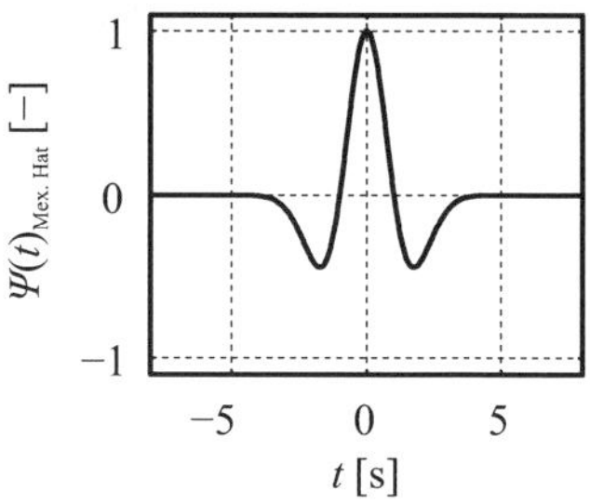 $\Psi_{\mathrm{Mexican\ Hat}}(t)(1-t^2)\mathrm{e}^{-\frac{t^2}{2}}$
One-Cycle-Sine Wavelet 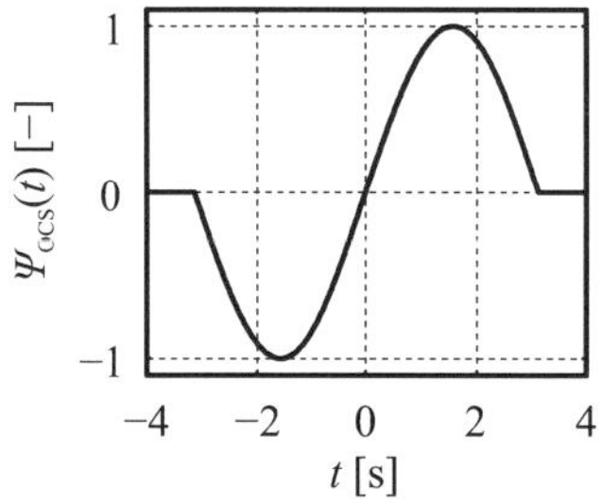 $\Psi_{\mathrm{OCS}}(t) = \begin{cases} \sin(t) \text{ if } \lvert t \rvert \leq \pi \\ 0 \text{ otherwise} \end{cases}$	

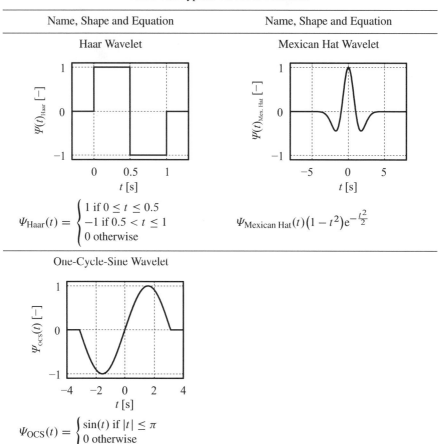

asymptotically bias-free at the frequency points ω_{n}, it is *not a consistent estimate* as the variance of the periodogram does not go to zero for $k \to \infty$ (Verhaegen and Verdult, 2007; Heij et al, 2007). Due to this negative property, the Periodogram should not be used per se, but only with certain modifications.

Bartlett proposed to divide the measurement into several data sets, then calculate the periodogram for each data set separately and finally average over the individual periodograms (e.g. Proakis and Manolakis, 2007). It can be shown that the variance can effectively be reduced by a factor $1/M$ if M individual periodograms are calculated. The expected value is still given by (3.3.2), hence the estimate still is biased for a finite number of data points. Also, the number of data points for each individual periodogram is reduced, which reduces the spectral resolution. However, by the averaging, the estimate becomes consistent in the mean square.

Welch (1977) also divides the data into shorter sequences that are processed individually. However, a window function is applied to the individual time sequences before they are processed. Furthermore, Welch suggested to use overlapping data segments hence allowing to form more data segment. The overlap can be up to 50%, consequently doubling the number of available data sets for the averaging and hence reducing the variance by 50%. Different window functions can be used, Welch (1977) for example suggested a Hann window.

3.4 Summary

In this chapter, methods for the spectral analysis of non-periodic signals have been presented. The Fourier transform has been introduced as a tool to determine the frequency content of signals. While the Fourier transform is applied to continuous-time signals and has an infinite time and frequency support, the signals that are processed in experimental applications are typically sampled and furthermore only recorded over a finite measurement interval.

Sampled data in the time domain can be processed by means of the Discrete Time Fourier Transform. It was shown that by the sampling in the time domain, the frequency spectrum becomes periodic. Also the frequency spectrum will only be determined at a finite number of discrete frequencies, which leads to the discrete Fourier transform.

As the DFT is very computationally demanding, different algorithms have been developed that allow a much faster calculation of the Fourier transform and are called the Fast Fourier Transform. The idea behind many of these algorithms is to divide the original sequence into a number of shorter subsequences that are transformed separately and then recombined appropriately. It has also been shown, how the output of an FFT algorithm can be interpreted correctly.

Since the signals are evaluated over a finite interval by the DFT/FFT, the frequency spectrum might get corrupted, which is the so-called leakage effect respectively spill-over. The time signal can be multiplied with a window function to mitigate the leakage effect. In windowing, there is always a trade-off between a good suppression of the side maxima and a narrow main lobe.

To analyze changes in the spectral properties as a function of time, joint time frequency representation methods have been developed. Two examples, the short time Fourier transform and the wavelet transform have been presented.

The periodogram has been introduced as an estimator for the power spectrum of a signal. It was shown that this estimator is only asymptotically bias-free and that furthermore the variance does not go to zero as $N \to \infty$. Methods proposed by Bartlett and Welsh, which are based on averaging multiple periodograms determined from different intervals of the measured signal avoid this disadvantage.

Problems

3.1. Fourier Transform
How is the Fourier transform defined for analytical signals? Determine the Fourier transform of the sawtooth:

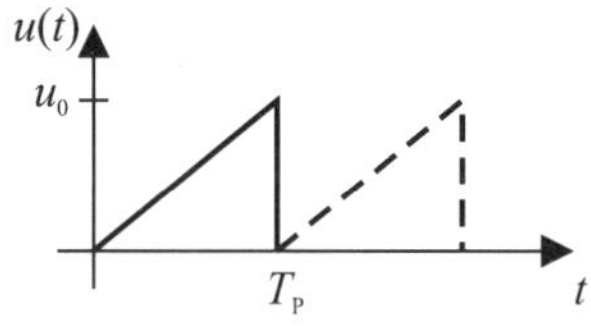

How does the frequency spectrum change, if $2, 3, \ldots$ sawtooth pulses are combined in series? What calculation rule must be used for an infinite number of pulses? What effects does this have on the resulting spectrum.

3.2. Fast Fourier Transform
Use the FFT algorithm as implemented in a numerical software package to determine the Fourier transform of the time signal $x(t) = \sin(2\pi t)$. Compare it to the theoretically expected result and try to understand the scaling of the results as well as the frequency resolution.

3.3. Fast Fourier Transform 1
In your own words, describe the algorithms involved in the Fast Fourier Transform.

3.4. Fast Fourier Transform 2
How can the resolution in the frequency domain be increased if the number of datapoints, that have been measured, is fixed?

3.5. Windowing 1
Describe the effect of the windowing. What is the typical trade-off in windowing? Try to find more window functions.

3.6. Windowing 2
Why is the spectrum obtained by the DFT not falsified, if a periodic signal is sampled over an integer number of periods?

3.7. Short Time Fourier Transform
Generate a chirp signal and analyze it using the short time Fourier transform. Comment on overlapping and on the choice of the parameter R.

3.8. Short Time Fourier Transform and Wavelet Transform
What are differences between the wavelet transform and the short time Fourier transform?

3.9. Periodogram
How is the periodogram defined? What is critical about the application of the periodogram?

References

Best R (2000) Wavelets: Eine praxisorientierte Einführung mit Beispielen: Teile 2 & 8. Tech Mess 67(4 & 11):182–187; 491–505

Bluestein L (1970) A linear filtering approach to the computation of discrete Fourier transform. IEEE Trans Audio Electroacoust 18(4):451–455

Brigham EO (1988) The fast Fourier transform and its applications. Prentice-Hall Signal Processing Series, Prentice Hall, Englewood Cliffs, NJ

Bruun G (1978) z-transform DFT filters and FFT's. IEEE Trans Acoust Speech Signal Process 26(1):56–63

Chu E, George A (2000) Inside the FFT black box: Serial and parallel fast Fourier transform algorithms. Computational mathematics series, CRC Press, Boca Raton, FL

Cooley JW, Tukey JW (1965) An algorithm for the machine calculation of complex fourier series. Math Comput 19(90):297–301

Frigo M, Johnson SG (2005) The design and implementation of FFTW3. Proc IEEE 93(2):216–231

Gentleman WM, Sande G (1966) Fast Fourier Transforms: for fun and profit. In: AFIPS '66 (Fall): Proceedings of the fall joint computer conference, San Francisco, CA, pp 563–578

Gibbs JW (1899) Fourier series. Nature 59:606

Goertzel G (1958) An algorithm for the evaluation of finite trigonometric series. Am Math Mon 65(1):34–35

Hamming RW (2007) Digital filters, 3rd edn. Dover books on engineering, Dover Publications, Mineola, NY

Harris FJ (1978) On the use of windows for harmonic analysis with the discrete Fourier transform. Proceedings of the IEEE 66(1):51–83

Heij C, Ran A, Schagen F (2007) Introduction to mathematical systems theory : linear systems, identification and control. Birkhäuser Verlag, Basel

Isermann R (1991) Digital control systems, 2nd edn. Springer, Berlin

Kammeyer KD, Kroschel K (2009) Digitale Signalverarbeitung: Filterung und Spektralanalyse mit MATLAB-Übungen, 7th edn. Teubner, Wiesbaden

Poularikas AD (1999) The handbook of formulas and tables for signal processing. The Electrical Engineering Handbook Series, CRC Press, Boca Raton, FL

Press WH, Teukolsky SA, Vetterling WT, Flannery BP (2007) Numerical recipes: The art of scientific computing, 3rd edn. Cambridge University Press, Cambridge, UK

Proakis JG, Manolakis DG (2007) Digital signal processing, 4th edn. Pearson Prentice Hall, Upper Saddle River, NJ

Qian S, Chen D (1996) Joint-time frequency analysis: Methods and applications. PTR Prentice Hall, Upper Saddle River, NJ

Rader CM (1968) Discrete Fourier transforms when the number of data samples is prime. Proc IEEE 56(6):1107–1108

Schoukens J, Vandersteen K, Barbé, Pintelon R (2009) Nonparametric preprocessing in system identification: A powerful tool. In: Proceedings of the European Control Conference 2009 - ECC 09, Budapest, Hungary, pp 1–14

Sorensen H, Heideman M, Burrus C (1986) On computing the split-radix FFT. Speech Signal Proc Acoust 34(1):152–156

Stearns SD (2003) Digital signal processing with examples in MATLAB. CRC Press, Boca Raton, FL

Verhaegen M, Verdult V (2007) Filtering and system identification: A least squares approach. Cambridge University Press, Cambridge

Welch P (1977) On the variance of time and frequency averages over modified periodograms. In: Acoustics, Speech, and Signal Processing, IEEE International Conference on ICASSP '77, vol 2, pp 58–62

4

Frequency Response Measurement with Non-Periodic Signals

The Fourier analysis with non-periodic test signals can be applied to determine the non-parametric frequency response function of linear processes by first bringing the input and the output signal to the frequency domain and then determining the transfer function by an element-wise division of the former two.

4.1 Fundamental Equations

The *frequency response function* in non-parametric form can be determined from non-periodic test signals by means of the relation

$$G(\mathrm{i}\omega) = \frac{y(\mathrm{i}\omega)}{u(\mathrm{i}\omega)} = \frac{\mathfrak{F}\{y(t)\}}{\mathfrak{F}\{u(t)\}} = \frac{\int_0^\infty y(t)\mathrm{e}^{-\mathrm{i}\omega t}\,\mathrm{d}t}{\int_0^\infty u(t)\mathrm{e}^{-\mathrm{i}\omega t}\,\mathrm{d}t}\,, \tag{4.1.1}$$

where the integral can furthermore be split up into real and imaginary part as

$$y(\mathrm{i}\omega) = \lim_{T\to\infty}\left(\int_0^T y(t)\cos\omega t\,\mathrm{d}t - \mathrm{i}\int_0^T y(t)\sin\omega t\,\mathrm{d}t\right)\,. \tag{4.1.2}$$

Here, the Fourier transform of the input as well as the output must be determined, i.e. the (typically noisy) signals must be subjected to a *Fourier transform*. Since the Fourier transform of many typical test signals, such as e.g. the step or the ramp function, does not converge, one typically uses the Laplace transforms with the limit $s \to \mathrm{i}\omega$ instead of (4.1.1). For the step and the ramp response, for example, there exists with $\lim_{s\to\mathrm{i}\omega} u(s)$, $(\omega \neq 0)$ a representation, which is similar to the Fourier transform, see Sec. 4.2.3. Hence, if the Fourier transform does not converge, one can use the ratio of the Laplace transforms with the limiting value $s \to \mathrm{i}\omega$ instead of (4.1.1) as

$$G(\mathrm{i}\omega) = \lim_{s\to\mathrm{i}\omega}\frac{y(s)}{u(s)} = \lim_{s\to\mathrm{i}\omega}\frac{\int_0^\infty y(t)\mathrm{e}^{-st}\,\mathrm{d}t}{\int_0^\infty u(t)\mathrm{e}^{-st}\,\mathrm{d}t} = \frac{y(\mathrm{i}\omega)}{u(\mathrm{i}\omega)}\,. \tag{4.1.3}$$

R. Isermann, M. Münchhof, *Identification of Dynamic Systems*,
DOI 10.1007/978-3-540-78879-9_4, © Springer-Verlag Berlin Heidelberg 2011

The integral can furthermore be split up into real and imaginary part as

$$y(\mathrm{i}\omega) = \lim_{\substack{\delta \to 0 \\ T \to \infty}} \left(\int_0^T y(t)\mathrm{e}^{-\delta t} \cos \omega t \,\mathrm{d}t - \mathrm{i} \int_0^T y(t)\mathrm{e}^{-\delta t} \sin \omega t \,\mathrm{d}t \right) . \tag{4.1.4}$$

The transform $u(\mathrm{i}\omega)$ can be written correspondingly.

As far as the signals are concerned, one has to use the small signal quantities, i.e. the deviation from their steady-state levels. If $U(t)$ and $Y(t)$ denote the large signal values and U_{00}, Y_{00} denote their respective steady-state values before the measurement, then

$$y(t) = Y(t) - Y_{00} \tag{4.1.5}$$

$$u(t) = U(t) - U_{00} \,. \tag{4.1.6}$$

In order to simplify the generation and subsequent evaluation of the results, one typically chooses test signals with a simple shape. Figure 4.1 shows some examples. For these simple signals, the Fourier transform can be determined a priori, see Sect. 4.2. The simple analytical expressions allow for example to optimize the test signal with respect to the identification task at hand, the plant under investigation and so on. Furthermore, one must only determine the Fourier transform of the output $y(t)$ (Isermann, 1967, 1982).

The frequency response function determination from non-parametric test signal excitation is typically applied to get a first quick system model, which gives hints on how to design the subsequent, more time-consuming experiments, which then yield the final system model. This method is often used in the analysis of mechanical structures, when e.g. the object is hit by a special hammer and the accelerations of different parts of the structure are measured. The Fourier transform of an impulse is just the height of the impulse or the contact force during the hammer hit respectively. Therefore, the input signal does not need to be Fourier transformed at all.

4.2 Fourier Transform of Non-Periodic Signals

To be able to determine the frequency response according to (4.1.3) and to attenuate the influence of noise in the frequency band of interest, the Fourier transform and amplitude density of various test signals in analytical form should be known. Therefore, this section will provide the Fourier transforms of the test signals shown in Fig. 4.1 and will analyze their amplitude density (Bux and Isermann, 1967). For simplicity in notation, the pulse width T_{P} is replaced by T.

4.2.1 Simple Pulses

First, the test signals a) through d) of Fig. 4.1 will be analyzed. They have in common that the test signal is always positive. Their major disadvantage is that they are not perfectly well suited for systems with integral action as the integrator will not come back to zero at the end of the test signal. Further discussion of this topic can be found later in following subsections.

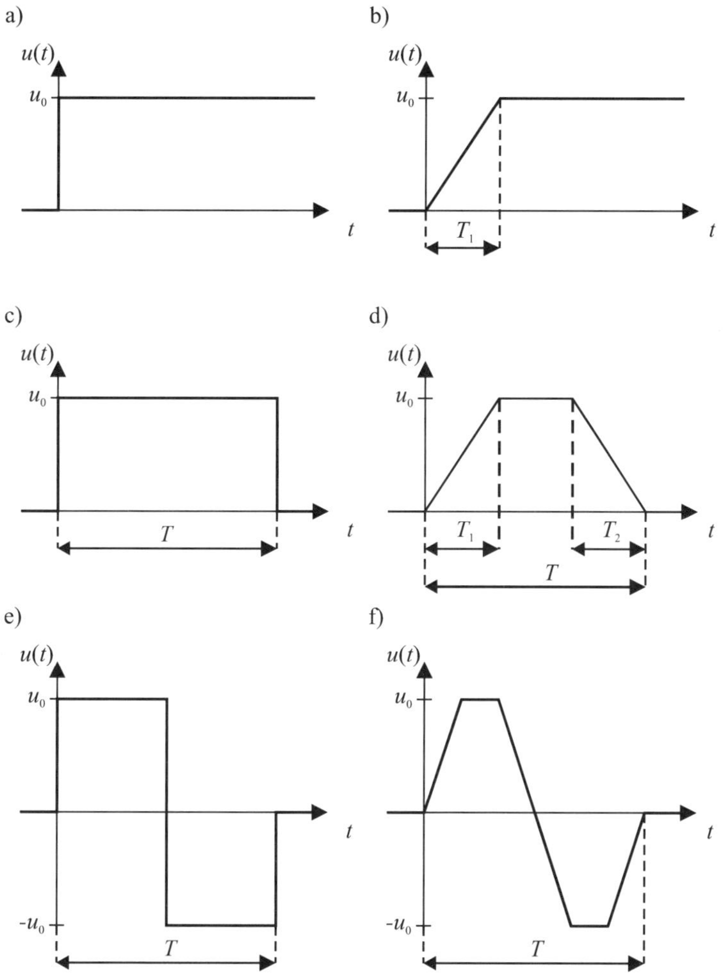

Fig. 4.1. Simple non-periodic test signals: (**a**) step function (**b**) ramp function (**c**) rectangular pulse (**d**) trapezoidal pulse (**e**) double rectangular pulse (**f**) double trapezoidal pulse

Trapezoidal Pulse

The Fourier transform for the case $T_2 = T_1$ can be determined. By transforming the three sections (ramp, constant, ramp) separately and adding the results together, one gets from using (4.1.4)

$$u_{\mathrm{tr}}(\mathrm{i}\omega) = u_0(T - T_1)\left(\frac{\sin\frac{\omega T_1}{2}}{\frac{\omega T_1}{2}}\right)\left(\frac{\sin\frac{\omega(T-T_1)}{2}}{\frac{\omega(T-T_1)}{2}}\right)\mathrm{e}^{-\mathrm{i}\frac{\omega T}{2}}\ . \tag{4.2.1}$$

Rectangular Pulse

The rectangular pulse can easily be determined from (4.2.1) by considering the limit $T_1 \rightarrow 0$. This limiting case is given as

$$u_{\mathrm{sq}}(\mathrm{i}\omega) = u_0 T \left(\frac{\sin \frac{\omega T}{2}}{\frac{\omega T}{2}} \right) \mathrm{e}^{-\mathrm{i}\frac{\omega T}{2}} \, . \tag{4.2.2}$$

Pintelon and Schoukens (2001) suggest to choose the pulse length T as

$$T = \frac{1}{2.5 f_{\max}} \, , \tag{4.2.3}$$

where $f_{\max}$ is the highest interesting frequency to be identified.

Triangular Pulse

With $T_1 = T_2 = T/2$ the Fourier transform follows from (4.2.1) as

$$u_{\mathrm{tri}}(\mathrm{i}\omega) = u_0 \frac{T}{2} \left(\frac{\sin \frac{\omega T}{4}}{\frac{\omega T}{4}} \right)^2 \mathrm{e}^{-\mathrm{i}\frac{\omega T}{2}} \, . \tag{4.2.4}$$

Dimensionless Representation

To be able to compare the Fourier transforms of the individual test signals in an easy way, the related (starred) quantities

$$u^*(t) = \frac{u(t)}{u_0} \tag{4.2.5}$$

$$t^* = \frac{t}{T} \tag{4.2.6}$$

$$\omega^* = \frac{\omega T}{2\pi} \tag{4.2.7}$$

are introduced. The Fourier transforms are furthermore normalized with respect to the maximum possible amplitude of the Fourier transform of a rectangular pulse,

$$u_{\mathrm{sq}}(\mathrm{i}\omega)|_{\omega=0} = \int_0^T u_0 \mathrm{d}t = u_0 T \, . \tag{4.2.8}$$

By the use of the referred quantities, test signals which are similar in their shape, but differ in their height u_0 and their pulse length T have the same amplitude density $|u^*(\mathrm{i}\omega^*)|$ and the same phase $\angle u^*(\mathrm{i}\omega^*)$. Thus, only the pulse shape determines the Fourier transform. For the three pulse shapes introduced so far, one gets

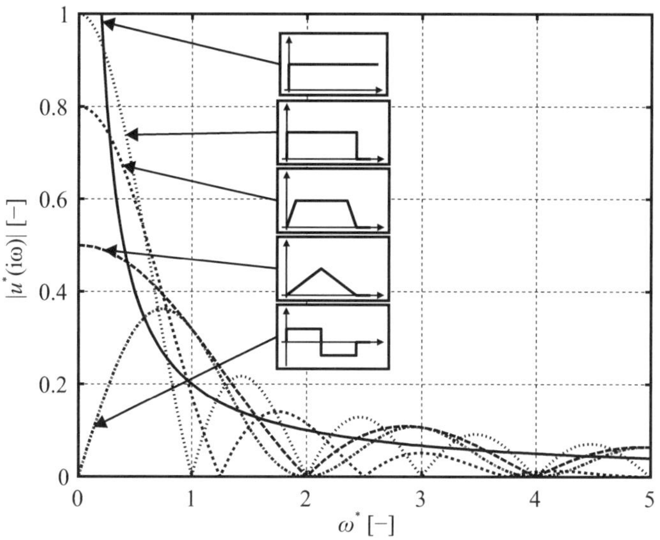

Fig. 4.2. Referred amplitude density of various non-periodic test signals

$$u^*_{\mathrm{tr}}(\mathrm{i}\omega^*) = (T^* - T_1^*)\left(\frac{\sin \pi\omega^* T_1^*}{\pi\omega^* T_1^*}\right)\left(\frac{\sin \pi\omega^*(T^* - T_1^*)}{\pi\omega^*(T^* - T_1^*)}\right)\mathrm{e}^{-\mathrm{i}\pi\omega^*} \quad (4.2.9)$$

$$u^*_{\mathrm{sq}}(\mathrm{i}\omega^*) = \left(\frac{\sin \pi\omega^*}{\pi\omega^*}\right)\mathrm{e}^{-\mathrm{i}\pi\omega^*} \quad (4.2.10)$$

$$u^*_{\mathrm{tri}}(\mathrm{i}\omega^*) = \frac{1}{2}\left(\frac{\sin \frac{\pi\omega^*}{2}}{\frac{\pi\omega^*}{2}}\right)^2 \mathrm{e}^{-\mathrm{i}\pi\omega^*}\,. \quad (4.2.11)$$

The amplitude density is plotted as a function of the referred angular frequency in Fig. 4.2. The largest value of each amplitude density is at $\omega^* = 0$ and is given as the area under the pulse. With increasing frequency, the amplitude spectra of the pulses decrease until the first zero is reached. This first zero is followed by additional zeros and intermediate local maxima of the amplitude spectra. These individual zeros are given as

- $\omega_1 = 2\pi n / T_1$ or $\omega_1^* = n / T_1^* \Rightarrow$ first row of zeros for trapezoidal pulse
- $\omega_2 = 2\pi n / (T - T_1)$ or $\omega_2^* = n / (T^* - T_1^*) \Rightarrow$ second row of zeros for trapezoidal pulse
- $\omega = 2\pi n / T$ or $\omega^* = n \Rightarrow$ only row of zeros for rectangular pulse
- $\omega = 4\pi n / T$ or $\omega^* = 2n \Rightarrow$ only row of zeros for triangular pulse

with $n = 1, 2, \ldots$. Trapezoidal and rectangular pulses have single zeros, triangular pulses double zeros. In the former case, the amplitude density curve intersects the ω-axis, in the latter case, it is a tangent to the ω-axis.

Variation of the Pulse Width

If the duration T of a pulse is increased, then the amplitude density increases at small frequencies, as the area underneath the pulse curve increases, see Fig. 4.3. The decay at higher frequencies at the same time gets steeper, since the zeros move toward smaller frequencies. One can construct an envelope which displays the highest possible amplitude at any given angular frequency ω. This envelope is given as

$$|u^*_{\mathrm{sq}}(\mathrm{i}\omega^*)|_{\max} = \frac{1}{\pi\omega^*} = \frac{0.3183}{\omega^*} \tag{4.2.12}$$

for the rectangular pulse and

$$|u^*_{\mathrm{tri}}(\mathrm{i}\omega^*)|_{\max} = \frac{0.2302}{\omega^*} \tag{4.2.13}$$

for the triangular pulse. For the trapezoidal test signal, one gets various envelopes depending on the shape. They are all bounded by the envelopes for the rectangular and triangular pulse respectively. Rectangular pulses have the largest amplitude at low frequencies compared with all other single-sided pulses with the same maximum height u_0. One can state the following reasons for this:

- At low frequencies, the area underneath the pulse curve determines the amplitude. Rectangular pulses have the maximum possible area for any given pulse width T.
- For medium frequencies, the envelope determines the amplitude density. Rectangular pulses have the highest envelope and thus in the area at $\omega^* = 1/2$ the highest amplitude density. In Fig. 4.2, one can see that the rectangular pulse has the highest amplitude density in the entire range of low to medium frequencies, $0 \leq \omega^* \leq 1/2$.
- For higher frequencies, rectangular pulses also have the highest amplitude density in certain areas left and right of the second, third, etc. extremum. This is for most applications however not of interest as the excitation is too low.

4.2.2 Double Pulse

Point-Symmetric Rectangular Pulse

Next, the double rectangular pulse as depicted in Fig. 4.1e with height u_0 and pulse width T will be considered. The Fourier transform is in this case given as

$$u(\mathrm{i}\omega) = u_0 T \left(\frac{\sin^2 \frac{\omega T}{4}}{\frac{\omega T}{4}} \right) \mathrm{e}^{-\mathrm{i}\frac{\omega T - \pi}{2}} \tag{4.2.14}$$

and with referred quantities as

$$u^*(\mathrm{i}\omega^*) = \left(\frac{\sin^2 \frac{\pi\omega^*}{2}}{\frac{\pi\omega^*}{2}} \right) \mathrm{e}^{-\mathrm{i}\pi\frac{2\omega^* - 1}{2}} \,. \tag{4.2.15}$$

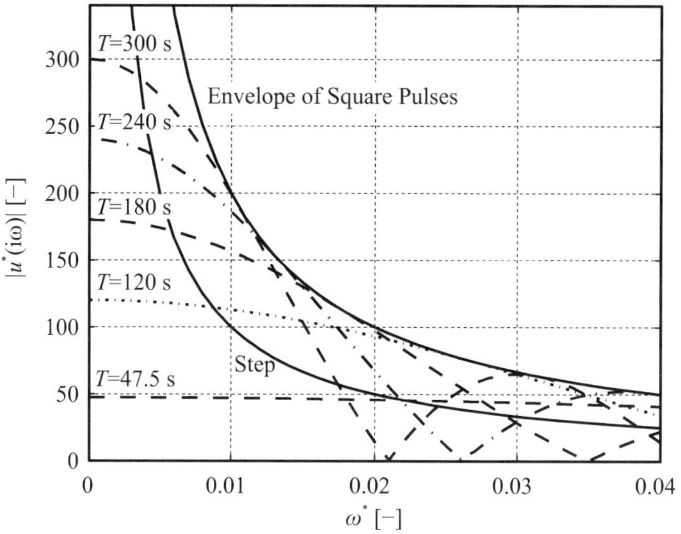

Fig. 4.3. Amplitude density of rectangular pulses with different pulse width T

The zeros are located at

$$\omega = \frac{4\pi}{T}n \text{ or } \omega^* = 2n \text{ with } n = 0, 1, 2, \dots \,. \tag{4.2.16}$$

With exception of $n = 0$, all roots are double roots. The amplitude density thus touches the ω^* axis for $n = 1, 2, \dots$. In contrast to the simple rectangular pulse, the amplitude density is zero at $\omega^* = 0$ and has a maximum at the finite frequency

$$|u^*(\mathrm{i}\omega^*)|_{\max} = 0.362 \text{ at } \omega^* = 0.762 \,, \tag{4.2.17}$$

see also Fig. 4.2.

Axis-Symmetric Rectangular Pulse

For an axis-symmetric double rectangular pulse, one gets

$$u(\mathrm{i}\omega) = u_0 T \frac{\sin \frac{\omega T}{2}}{\frac{\omega T}{2}} 2 \cos \omega T \,. \tag{4.2.18}$$

Figure 4.4 shows (for the time shifted double pulse) that the amplitude density at $\omega^* = 0$ and $\omega^* = 0.5$ is twice as high as that of the single rectangular pulse. The frequency interval of interest, $\omega_1^* < \omega^* < \omega_2^*$, is quite small. Outside of this area of interest, the amplitude density is smaller than that of the single rectangular pulse. The concatenation of two rectangular pulses yields an increase in the level of excitation in the area around $\omega^* = 0.5$ and all mutiples, i.e. $\omega^* = 1.5$, $\omega^* = 2.5$, etc. at the

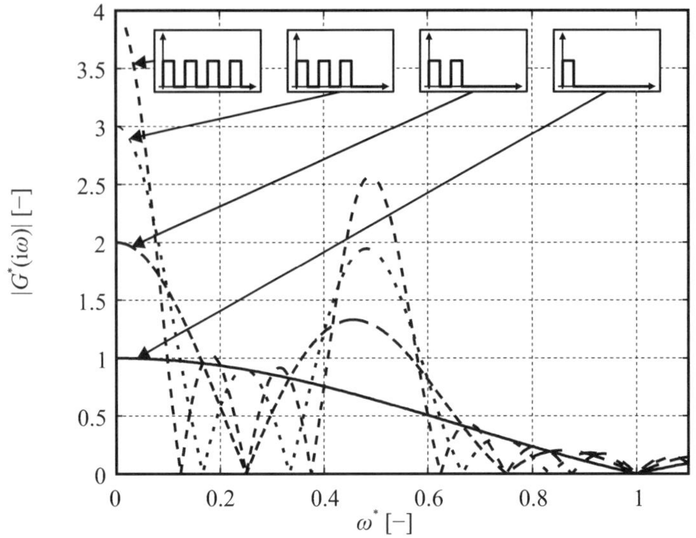

Fig. 4.4. Amplitude spectra of rectangular pulses with different number of rectangles

expense of an attenuation in the lower intermediate and all upper frequency ranges except for integer multiples.

If one concatenates not only two, but even more rectangular pulses at a distance of $2T$, the amplitude density in the vicinity of $\omega^* = 0.5$ (and $\omega^* = 0$) grows ever larger. At the same time the interval $\omega_1^* < \omega^* < \omega_2^*$ gets smaller. In the limiting case of an infinite number of rectangular pulses, the Fourier transform develops to a δ-impulse at $\omega^* = 0.5$ (and $\omega^* = 0$ as well as $\omega^* = 1.5$, $\omega^* = 2.5$ and so forth).

4.2.3 Step and Ramp Function

Step and ramp functions do not satisfy the convergence criteria (2.1.11) so that a Fourier transform cannot directly be determined by (2.1.9) or (4.1.4) respectively. However, there are still means to determine a frequency domain representation.

By calculating the limiting value $s \to i\omega$, one can obtain the Fourier transform in a strictly formal way from the Laplace transform

$$u_{st}(i\omega) = \lim_{s\to i\omega} \mathfrak{L}\{u_{st}(t)\} = \lim_{s\to i\omega} \frac{u_0}{s} = \frac{u_0}{i\omega} = \frac{u_0}{\omega} e^{-i\frac{\pi}{2}} \,. \tag{4.2.19}$$

Rewriting this equation with referred quantities yields

$$u_{st}^*(i\omega) = \frac{1}{2\pi\omega^*} e^{-\frac{\pi}{2}} \,. \tag{4.2.20}$$

Similarly, one can now cope with the ramp function. For a ramp with rise time T_1, see Fig. 4.1b), one gets

$$u_{\mathrm{r}}(\mathrm{i}\omega) = \lim_{s\to\mathrm{i}\omega} \mathfrak{L}\{u_{\mathrm{r}}(t)\} = \lim_{s\to\mathrm{i}\omega} \frac{u_0}{T_1 s^2}\left(1-\mathrm{e}^{-T_1 s}\right) = \frac{u_0}{\omega}\left(\frac{\sin\frac{\omega T_1}{2}}{\frac{\omega T_1}{2}}\right) \tag{4.2.21}$$

or

$$u_{\mathrm{r}}^{*}(\mathrm{i}\omega^{*}) = \frac{1}{2\pi\omega^{*}}\left(\frac{\sin\pi\omega^{*}T_1^{*}}{\pi\omega^{*}T_1^{*}}\right)\exp\left(\frac{\pi}{2}-\pi\omega^{*}\right) \tag{4.2.22}$$

respectively.

The amplitude density of a step function follows as

$$|u_{\mathrm{st}}(\mathrm{i}\omega)| = \frac{u_0}{\omega} \text{ or } |u_{\mathrm{st}}^{*}(\mathrm{i}\omega^{*})| = \frac{1}{2\pi\omega^{*}} \text{ for } \omega \neq 0,\ \omega^{*} \neq 0 \tag{4.2.23}$$

and thus represents a hyperbolic function. Since there are no zeros, all frequencies in the range $0 < \omega < \infty$ will be excited. Figure 4.2 as well as (4.2.12) exhibit that the amplitude spectrum of the step function is always half as high as the envelope of the amplitude spectrum of the rectangular pulses. The amplitude density of a step function and any rectangular pulse are equally high at

$$\omega_{\mathrm{sr}} = \frac{\pi}{3T} \text{ or } \omega_{\mathrm{sr}}^{*} = \frac{1}{6} = 0.1667\,. \tag{4.2.24}$$

In the area of small frequencies, $0 < \omega < \omega_{\mathrm{sr}}$, the step function has the larger amplitude density compared to the rectangular pulse and thus the largest amplitude density of all non-periodic test signals of height u_0.

The amplitude density of a ramp function is by a factor

$$\kappa = \frac{|u_{\mathrm{r}}(\mathrm{i}\omega)|}{|u_{\mathrm{st}}(\mathrm{i}\omega)|} = \frac{\sin\frac{\omega T_1}{2}}{\frac{\omega T_1}{2}} \tag{4.2.25}$$

smaller. This factor is equal to the shape of the amplitude density function of the rectangular pulses. At $\omega = 2\pi n/T_1$, with $n = 1, 2, \ldots$ one can find zeros. This is in contrast to the step function, which has no zeros. The first zero moves to higher frequencies as the rise time T_1 of the ramp gets smaller, which means that the edge of the signal gets steeper. This points to a general property of all test signals: *The steeper the edges, the stronger the excitation at high frequencies.* P4.5

In many cases, one is interested in whether a step function can be assumed for a ramp-wise excitation with rise time T_1. The factor κ in (4.2.25) provides an answer to this question. If one accepts an error of $\leq 1\%$ or $\leq 5\%$ up to the largest angular frequency $\omega_{\max}$, then the factor κ can be $\kappa \geq 0.95$ or $\kappa \geq 0.99$ respectively. Thus, the rise time is limited to

$$T_{1,\max} \leq \frac{1.1}{\omega_{\max}} \text{ or } T_{1,\max} \leq \frac{0.5}{\omega_{\max}}\,. \tag{4.2.26}$$

In conclusion, the analysis of the amplitude density of different non-periodic test signals shows that for a given test signal height u_0, the highest amplitude density of all possible test signals can be achieved with

- step functions for small frequencies
- rectangular pulses for medium to high frequencies

Thus, according to (4.3.6), these signals provide the smallest error in the identification of the frequency responses from noisy measurements of the response (Isermann, 1967). Non-periodic test signals in contrast to periodic test signals excite all frequencies in the range $0 \leq \omega < \infty$ at once with the exception of the zeros which show up for pulse responses and ramp responses.

4.3 Frequency Response Determination

Now, the properties of the frequency response determination by means of (4.1.1), i.e.

$$\hat{G}(\mathrm{i}\omega) = \frac{y(\mathrm{i}\omega)}{u(\mathrm{i}\omega)} = \frac{\mathfrak{F}\{y(t)\}}{\mathfrak{F}\{u(t)\}}, \tag{4.3.1}$$

will be analyzed. Here, special attention must be paid to the influence of noise on the output: The systems response $y_{\mathrm{u}}(t)$ evoked by the test signal $u(t)$ is usually affected by noise which is superimposed as

$$y(t) = y_{\mathrm{u}}(t) + n(t). \tag{4.3.2}$$

By substituting into (4.3.1), one gets

$$\hat{G}(\mathrm{i}\omega) = \frac{1}{u(\mathrm{i}\omega)} \lim_{s \to \mathrm{i}\omega} \left(\int_0^\infty y_u(t) \mathrm{e}^{-st} \mathrm{d}t + \int_0^\infty n(t) \mathrm{e}^{-st} \mathrm{d}t \right) \tag{4.3.3}$$

and

$$\hat{G}(\mathrm{i}\omega) = G_0(\mathrm{i}\omega) + \Delta G_{\mathrm{n}}(\mathrm{i}\omega). \tag{4.3.4}$$

The estimated frequency response $\hat{G}(\mathrm{i}\omega)$ thus consists not only of the exact frequency response $G_0(\mathrm{i}\omega)$, but also of the frequency response error $\Delta G_{\mathrm{n}}(\mathrm{i}\omega)$ which is evoked by the noise $n(t)$ and is given as

$$\Delta G_{\mathrm{n}}(\mathrm{i}\omega) = \lim_{s \to \mathrm{i}\omega} \frac{n(s)}{u(s)} = \frac{n(\mathrm{i}\omega)}{u(\mathrm{i}\omega)}. \tag{4.3.5}$$

Hence, the magnitude of the error results as

$$|\Delta G_{\mathrm{n}}(\mathrm{i}\omega)| = \frac{|n(\mathrm{i}\omega)|}{|u(\mathrm{i}\omega)|}. \tag{4.3.6}$$

The frequency response error gets smaller as $|u(\mathrm{i}\omega)|$ becomes larger in relation to $|n(\mathrm{i}\omega)|$. Thus, for a given noise $|n(\mathrm{i}\omega)|$, one must try to make $|u(\mathrm{i}\omega)|$, i.e. the amplitude density at ω of the test signal $u(t)$, as large as possible. This can be achieved by

- Choosing the height u_0 of the test signal as large as possible. However, limitations on u_0 are often imposed by the process itself, the operating range in which the process can be linearized, limitations of the actuator and so on, see Sect. 1.2.
- Selecting an appropriate *shape* of the test signal so that the amplitude density is concentrated in the frequency area of interest.

The influence of noise and the properties of the frequency response function estimate will be discussed again in the next section.

Example 4.1 (Frequency Response Function). An example of a frequency response function estimate of the Three-Mass Oscillator, see Fig. B.1, using a non-periodic test signals is shown in Fig. 4.5. The test signal employed was a rectangular pulse with the length $T = 0.15$ s. One can see the relative good match between the frequency response determined by means of the orthogonal correlation described in Sect. 5.5.2 (which will serve as a reference) and the frequency response determined by the aid of the Fourier transform for $\omega < 25\text{rad/s}$. In Fig. 4.6 on the contrary, the excitation was chosen as a triangular pulse, where the first zero coincides with the maximum magnitude of the transfer function $G(\mathrm{i}\omega)$. Hence, the frequency response can only be determined for $\omega < 13.5\,\text{Hz}$. □

4.4 Influence of Noise

The output signals of many processes do not only contain the response to the test signal, but also some noise, see Fig. 1.5. This noise can have manifold reasons. Noise can be caused by external disturbances acting on the process or by internal disturbances located within the processes boundaries. As has been outlined in Sect. 1.2, one can differentiate between higher-frequent quasi-stochastic disturbances (Fig. 1.6a), low-frequent non-stationary stochastic disturbances, e.g. drift, (Fig. 1.6b), and disturbances of unknown character, e.g. outliers, (Fig. 1.6c).

Identifying a process with a single non-stationary test signal as described in this chapter is typically only possible if the noise has a *small amplitude* compared to the test signal and if the noise has a *constant mean*. If non-stationary noise or noise with an unknown type is acting on the system, it is in most cases impossible to obtain any useful identification results from the relative short time interval in which the response $y(t)$ is recorded. One rather has to wait for a time period, where the noise has constant mean or take resort to other identification methods that can better cope with non-stationary noises.

In the following, the influence of *stationary, stochastic noise* $n(t)$ with $\mathrm{E}\{n(t)\} = 0$ on the fidelity of the identified frequency response will be investigated. The investigation will assume that the noise $n(t)$ is additively superimposed onto the undisturbed output $y_{\mathrm{u}}(t)$ evoked by the test signal (4.3.2). The noise can be created by a form filter with transfer function $G_{\mathrm{n}}(\mathrm{i}\omega)$ from white noise with the power spectral density $S_{\nu 0}$, see Fig. 4.7.

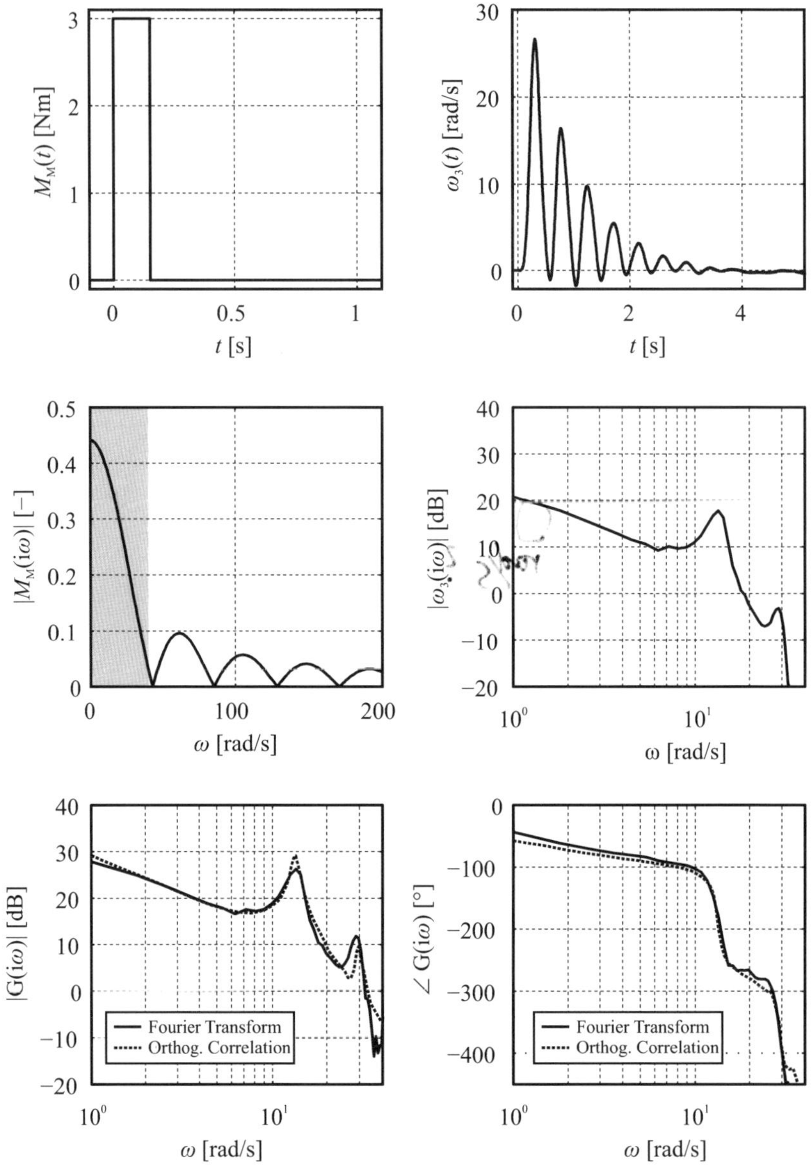

Fig. 4.5. Frequency response measurement of the Three-Mass Oscillator with a rectangular pulse of length $T = 0.15$ s. Measurement by Fourier transform (*solid line*), orthogonal correlation as reference (*dashed line*), frequency range of the Bode diagram (*gray shaded area*). The input signal is the torque $M_M(t)$ applied by the electric motor, the output the rotational speed $\omega_3(t)$ of the third mass.

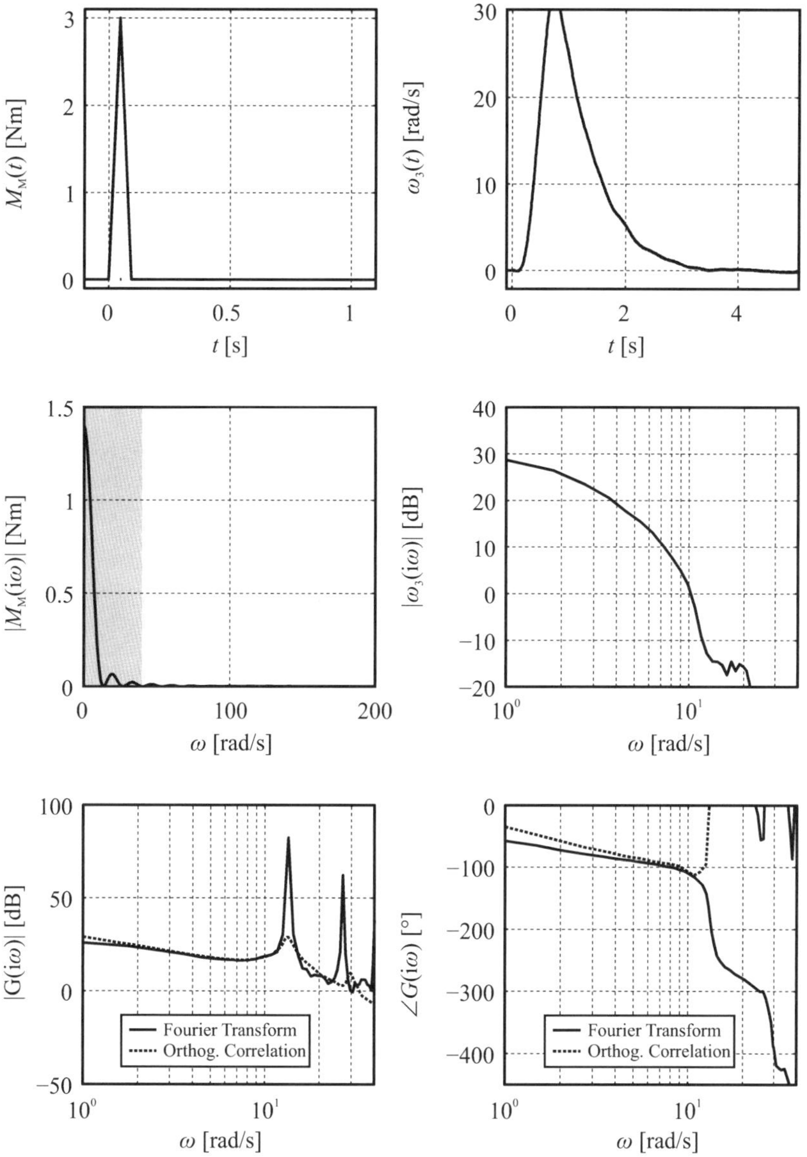

Fig. 4.6. Frequency response measurement of the Three-Mass Oscillator with a triangular pulse of length $T = 0.09\,\mathrm{s}$. First zero of the input frequency spectrum located at the first resonant frequency at $\omega \approx 15\,\mathrm{rad/s}$ of the process. Erroneous results are obtained for angular frequencies around $\omega \approx 13.5\,\mathrm{rad/s}$ and at all integer multiples k. Also for higher frequencies, the excitation by the test signal is too small. Measurement by Fourier transform (*solid line*), orthogonal correlation as reference (*dashed line*), frequency range of the Bode diagram (*gray shaded area*)

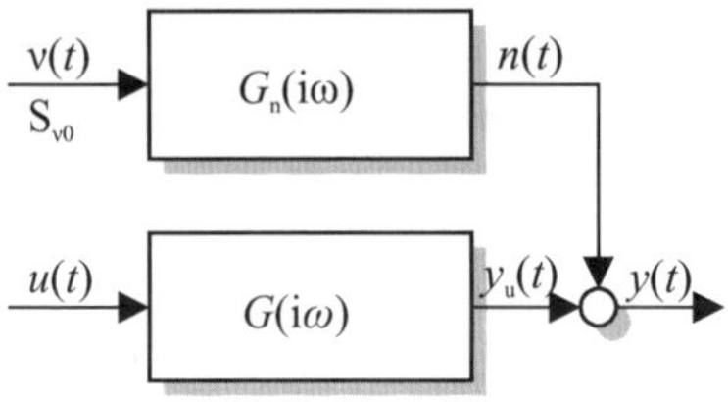

Fig. 4.7. Block diagram of a linear process disturbed by a stochastic noise $n(t)$

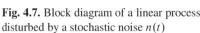

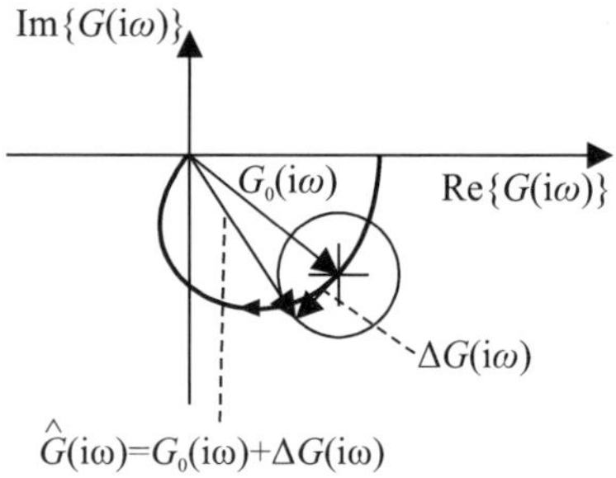

Fig. 4.8. Error $\Delta G_{\mathrm{n}}(\mathrm{i}\omega)$ of a frequency response

In the following, the error caused by a disturbed response will be investigated. A stochastic noise acting on the process in the time period $0 \leq t \leq T_{\mathrm{E}}$ causes the error

$$\Delta G_{\mathrm{n}}(\mathrm{i}\omega) = \frac{n_{\mathrm{T}}(\mathrm{i}\omega)}{u(\mathrm{i}\omega)} \tag{4.4.1}$$

in the frequency response, see Fig. 4.8. Its magnitude is given by

$$|\Delta G_{\mathrm{n}}(\mathrm{i}\omega)| = \frac{|n_{\mathrm{T}}(\mathrm{i}\omega)|}{|u(\mathrm{i}\omega)|} \; . \tag{4.4.2}$$

The auto-correlation function of noise of finite duration can be estimated by

$$\hat{\Phi}_{\mathrm{nn}}(\tau) = \frac{1}{T_{\mathrm{E}}} \int_0^{T_{\mathrm{E}}} n_{\mathrm{T}}(t) n_{\mathrm{T}}(t+\tau) \mathrm{d}\tau \; , \tag{4.4.3}$$

see Sect. 6.1. One can then determine the power spectral density and determine the expected value. Knowledge of the power spectral densities of the noise and the input (4.4.2) allows to determine the expected value of the magnitude of the error

$$\mathrm{E}\big\{|\Delta G_{\mathrm{n}}(\mathrm{i}\omega)|^2\big\} = \frac{\mathrm{E}\big\{S_{\mathrm{nn}}(\mathrm{i}\omega)\big\}}{S_{\mathrm{uu}}(\omega)} \; . \tag{4.4.4}$$

Since the test signal is deterministic,

$$S_{\mathrm{uu}}(\omega) = \frac{|u(\mathrm{i}\omega)|^2}{T_{\mathrm{E}}} \; . \tag{4.4.5}$$

The variance of the relative frequency response error upon evaluation of a response of length T_{E} is thus given as

$$\sigma^2_{\mathrm{G1}} = \mathrm{E}\left\{\frac{|\Delta G_{\mathrm{n}}(\mathrm{i}\omega)|^2}{|G(\mathrm{i}\omega)|^2}\right\} = \frac{S_{\mathrm{nn}}(\omega) T_{\mathrm{E}}}{|G(\mathrm{i}\omega)|^2 |u(\mathrm{i}\omega)|^2} \; . \tag{4.4.6}$$

If N responses are evaluated, then

$$S_{\mathrm{uu}}(\omega) = \frac{|Nu(\mathrm{i}\omega)|^2}{NT_{\mathrm{E}}} = N\frac{|u(\mathrm{i}\omega)|^2}{T_{\mathrm{E}}} . \tag{4.4.7}$$

The standard deviation is in this case given as

$$\sigma_{\mathrm{Gn}} = \frac{\sqrt{S_{\mathrm{nn}}(\omega)T_{\mathrm{E}}}}{|G(\mathrm{i}\omega)||u(\mathrm{i}\omega)|\sqrt{N}} . \tag{4.4.8}$$

The error in the frequency response is thus inversely proportional to the signal-to-noise ratio and inversely proportional to the square root $\sqrt{N}$ of the number of responses recorded. Hence, in order to decrease the influence of a stochastic noise $n(t)$, one can record more than one responses evoked by the same test signal and determine the averaged response by

$$\bar{y}(t) = \frac{1}{N}\sum_{k=0}^{N-1} y_k(t) . \tag{4.4.9}$$

Especially in the case of different test signals, one can also determine the mean frequency response as

$$\bar{G}(\mathrm{i}\omega) = \frac{1}{N}\sum_{k=1}^{N} G_k(\mathrm{i}\omega) = \frac{1}{N}\sum_{k=1}^{N} \mathrm{Re}\big\{G_k(\mathrm{i}\omega)\big\} + \mathrm{i}\frac{1}{N}\sum_{k=1}^{N} \mathrm{Im}\big\{G_k(\mathrm{i}\omega)\big\} . \tag{4.4.10}$$

As can be seen in (4.4.8), the standard deviation decays with a factor $1/\sqrt{N}$ such that

$$\sigma_{\mathrm{GN}} = \frac{1}{\sqrt{N}}\sigma_{\mathrm{G1}} . \tag{4.4.11}$$

It is important however, that one may only take the mean of the real and imaginary part and never the mean of the amplitude and phase.

For the form filter in Fig. 4.7, one can write

$$S_{\mathrm{nn}}(\omega) = |G_n(s)|^2 S_{\nu 0} , \tag{4.4.12}$$

so that finally

$$\sigma_{\mathrm{GN}} = \frac{|G_n(\mathrm{i}\omega)|\sqrt{S_{\nu 0}T_{\mathrm{E}}}}{|G(\mathrm{i}\omega)||u(\mathrm{i}\omega)|\sqrt{N}} . \tag{4.4.13}$$

Example 4.2 (Influence of Noise on the Frequency Response Function Estimate). The accuracy of the estimation in the presence of noise shall now be illustrated for the Three-Mass Oscillator excited by rectangular pulses. The standard deviation of the noise $n(t)$ given as σ_{n} shall be 1. The noise-to-signal ratio is given as

$$\eta = \frac{\sigma_{\mathrm{n}}}{y_{\max}} \approx 4\% \mathrel{\hat{=}} 1 : 25 , \tag{4.4.14}$$

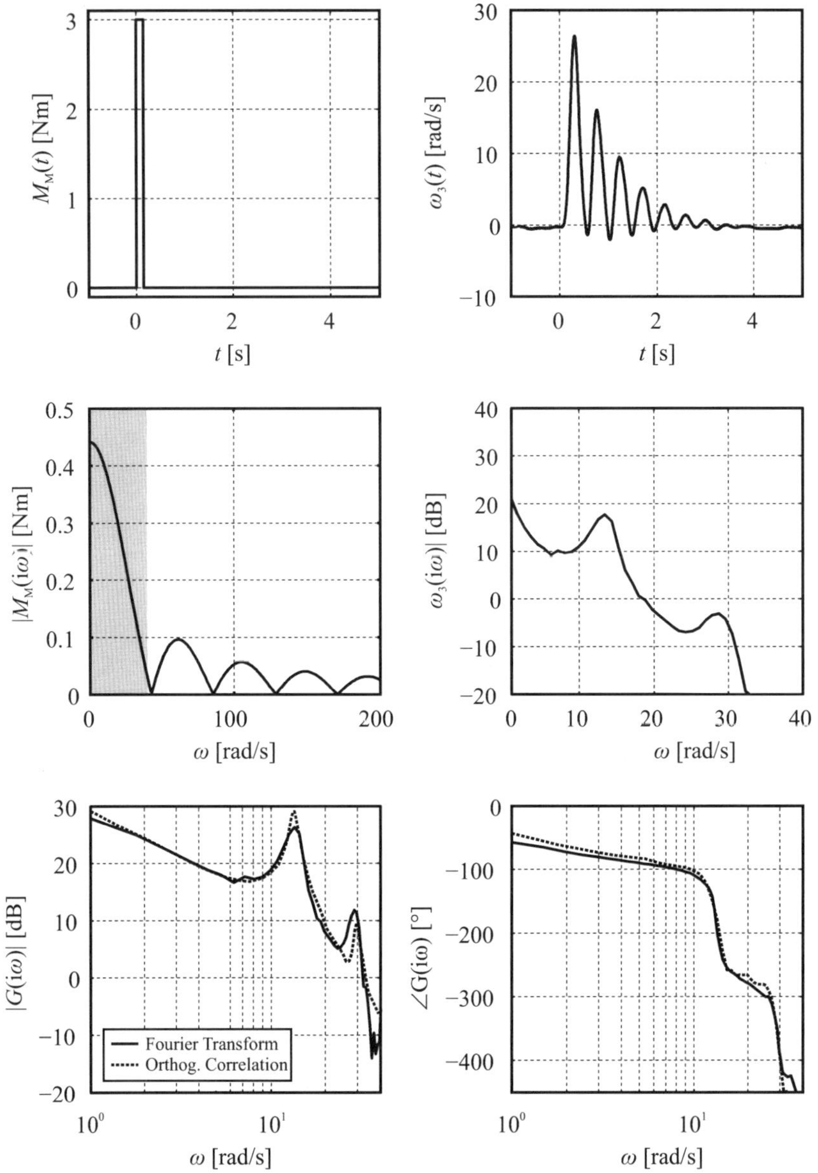

Fig. 4.9. Rectangular pulse response with step height $u_0 = 3\,\text{Nm}$ and duration $\Delta T = 0.15\,\text{s}$ and frequency response without noise

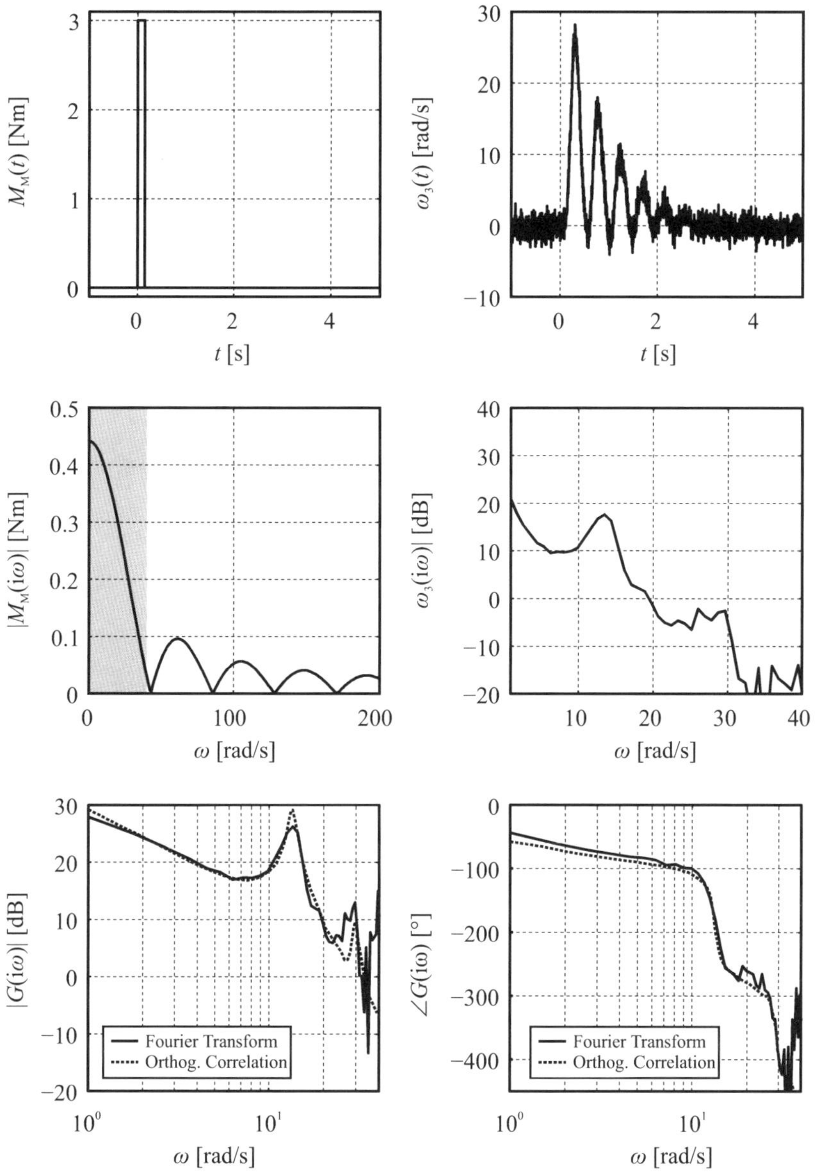

Fig. 4.10. Rectangular pulse response with step height $u_0 = 3\,\mathrm{Nm}$ and duration $\Delta T = 0.15\,\mathrm{s}$ and frequency response with noise $\sigma = 1$

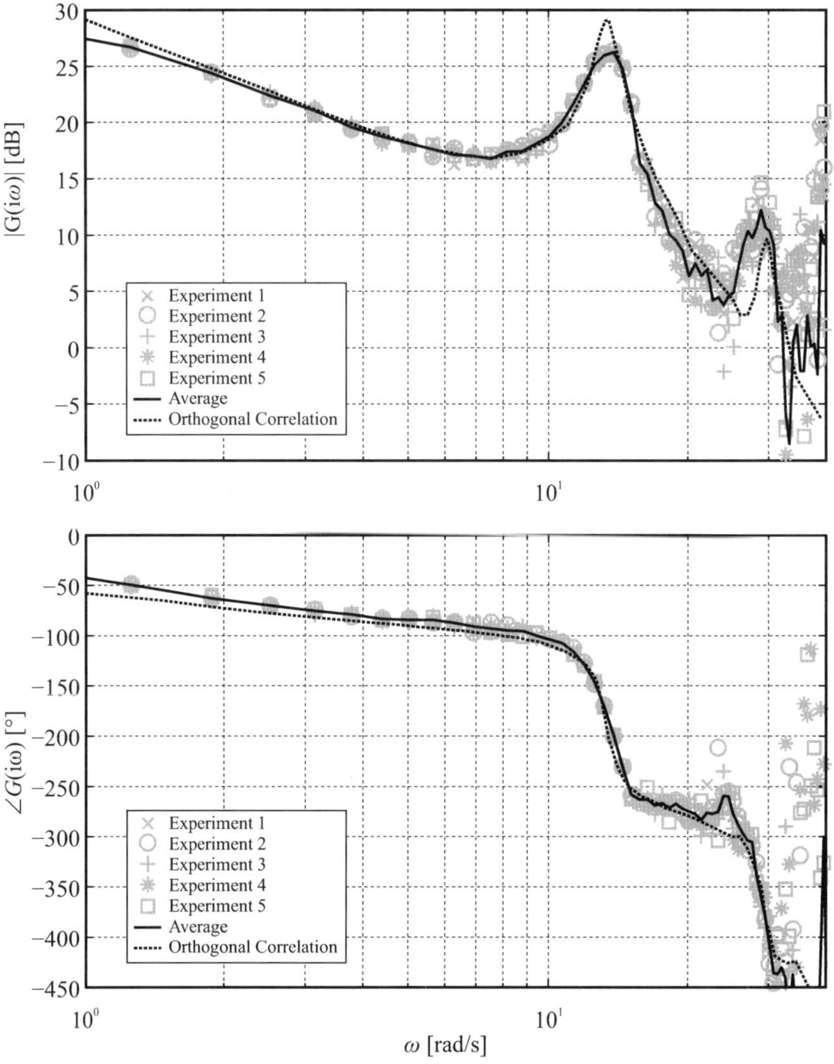

Fig. 4.11. Frequency responses determined from averaging multiple measurements with a rectangular pulse with noise $\sigma = 1$. (*solid line*): Average of 5 frequency responses, (*dashed line*): Direct frequency response measurement with sinusoidal excitation and evaluation with orthogonal correlation (see Sect. 5.5.2)

which corresponds to a signal-to-noise ratio of $1/\eta = 25 : 1$. Such a disturbance is quite small, the peak-to-peak value for a Gaussian distribution of the signal amplitudes is given as

$$b \approx 4\sigma_{\mathrm{n}} \,. \tag{4.4.15}$$

Figure 4.9 shows the noise-free case, whereas Fig. 4.10 depicts the case of a noisy measurement. One can see that the results in the range of medium frequencies is still relatively good, but the peak of the second resonance in the amplitude plot cannot be determined any longer. Figure 4.11 shows how the fidelity of the estimated frequency response can be increased by averaging multiple noisy measurements.

The agreement with the discrete measured frequency response is very good in the range of medium frequencies, i.e. $10\mathrm{rad/s} \leq \omega \leq 25\mathrm{rad/s}$ which validates the analysis of the excitation signals in Sect. 4.2. □

Summarizing, the spectral estimation of the transfer function as

$$\hat{G}(\mathrm{i}\omega) = \frac{y(\mathrm{i}\omega)}{u(\mathrm{i}\omega)} \tag{4.4.16}$$

has the following properties

$$\lim_{N\to\infty} \mathrm{E}\big\{\hat{G}(\mathrm{i}\omega)\big\} = G(\mathrm{i}\omega) \tag{4.4.17}$$

$$\lim_{N\to\infty} \mathrm{var}\big(\hat{G}(\mathrm{i}\omega)\big) = \frac{S_{\mathrm{nn}}(\mathrm{i}\omega)}{S_{\mathrm{uu}}(\mathrm{i}\omega)} \,, \tag{4.4.18}$$

see (Ljung, 1999; Heij et al, 2007; Verhaegen and Verdult, 2007). As one can see, the variance does not diminish as $N \to \infty$. Also, the estimate is only unbiased if there are no transients (Broersen, 1995), and if there is no noise acting on the input $u(t)$. Transients can only be avoided if there are no responses due to $u(t) \neq 0$ for $t < 0$ and also that the input signal is of finite duration and also the system response has died out before the end of the measurement period. For signals that have not died out at the end of the measurement period, the issue of windowing and the effect of windowing on the estimates comes up. For this topic, see (Schoukens et al, 2006). Windowing has already been discussed in detail in Sect. 3.1.4.

4.5 Summary

In this chapter, the estimation of the frequency response function by means of dividing the Fourier transform of the output $y(t)$ by the Fourier transform of the input $u(t)$ was presented. As the quality of the estimate strongly depends on the excitation of the dominant process dynamics, the amplitude density of different test signals has been derived analytically and then compared with each other. Based on this analysis, suggestions on the design of advantageous test signals can now be given. Windowing can also have a detrimental effect on the identification results. The interested reader is referred to the studies by Schoukens et al (2006) and Antoni and Schoukens (2007).

The use of this method as an initial tool for quick system analysis is also suggested by Verhaegen and Verdult (2007).

The term *advantageous* shall denote those realizable test signals, which have the highest amplitude density in a certain frequency range. These signals will yield the smallest error in the estimate of the frequency response in that particular frequency range. As has already been shown in this chapter, the most advantageous test signals for small frequencies are the *step signal* and the *rectangular pulse* for medium frequencies. (4.4.8) can be used to determine the required amplitude for the identification of a process with stochastic disturbances as

$$|u(\mathrm{i}\omega)|_{\mathrm{req}} = \frac{\sqrt{S_{\mathrm{nn}}(\mathrm{i}\omega)T_{\mathrm{E}}}}{|G(\mathrm{i}\omega)|\sigma_G(\mathrm{i}\omega)\sqrt{N}} \,. \tag{4.5.1}$$

In this equation, $\sigma_{\mathrm{G}}(\mathrm{i}\omega)$ denotes the maximum allowable standard deviation for the frequency response error. From (4.5.1), one can see that the required amplitude density of the test signal depends on the power spectral density of the noise and the intended application, which determines $\sigma_{\mathrm{G}}(\mathrm{i}\omega)$. Generic requirements on the amplitude density can hardly be formulated without knowledge of the process. The experience gained in controller synthesis (Isermann, 1991) however shows that the relative frequency response error must be small in the medium frequency range, which is for example fulfilled for short rectangular pulses.

It is now self-evident that one should use not only one test signal, but rather a test sequence combined of different test signals, where each test signal is advantageous for the identification of a certain frequency range. For processes, which may be displaced permanently, one should use,

- a sequence of a few step responses to determine the frequency response at low frequencies
- a sequence of rectangular pulses to determine the frequency response at medium and high frequencies

A guiding value for the distribution can be $20\% - 30\%$ of the measurement time for step responses and $80\%-70\%$ rectangular pulses. The length T of the rectangular pulses is determined so that the highest amplitude density is approximately at the highest interesting process frequency $\omega_{\max}$, i.e.

$$T = \frac{\pi}{\omega_{\max}} \,. \tag{4.5.2}$$

If possible, one should evaluate the response for both directions to attenuate certain non-linear effects by the subsequent averaging. At this point, the close resemblance between these test sequences and the binary test signals, which are treated later in Sect. 6.3 shall already be pointed out.

After discussing the design of ideal test signals, the properties of the frequency response function estimate have been discussed in detail. It has been shown that if the system is excited with the same test signal $u(t)$ in each experiment, one can calculate the average of the system response $y(t)$ and then determine the frequency response function based on this average. If different test signals are used, one can also estimate a frequency response function for each measurement individually and then calculate the average of all frequency response functions.

Problems

4.1. Frequency Response Measurement with Non-Periodic Test Signals
How can the frequency response of a linear system be determined with non-periodic test signals? What are the advantages/disadvantages compared to periodic test signals?

4.2. Fourier Transform of Test Signals
Which test signals yield the highest amplitude density for very low, low, medium, or high frequencies respectively? Assume that all test signals are constrained to the same maximum height u_0.

4.3. Trapezoidal Pulse
Determine the Fourier transform of the trapezoidal pulse.

4.4. Rectangular Pulse
Determine the Fourier transform of the rectangular pulse for $T = 20\,\mathrm{s}$.

4.5. Test Signals
How do the steepness of the edges and the excitation of high frequencies relate to each other?

4.6. Noise
How can one improve the identification result, if the process is
(a) excited multiple times with the same test signals
(b) excited multiple times with different test signals

4.7. Advantageous Test Signals
Describe an advantageous test signal sequence.

References

Antoni J, Schoukens J (2007) A comprehensive study of the bias and variance of frequency-response-function measurements: Optimal window selection and overlapping strategies. Automatica 43(10):1723–1736

Broersen PMT (1995) A comparison of transfer function estimators. IEEE Trans Instrum Meas 44(3):657–661

Bux D, Isermann R (1967) Vergleich nichtperiodischer Testsignale zur Messung des dynamischen Verhaltens von Regelstrecken. Fortschr.-Ber. VDI Reihe 8 Nr. 9. VDI Verlag, Düsseldorf

Heij C, Ran A, Schagen F (2007) Introduction to mathematical systems theory : linear systems, identification and control. Birkhäuser Verlag, Basel

Isermann R (1967) Zur Messung des dynamischen Verhaltens verfahrenstechnischer Regelstrecken mit determinierten Testsignalen (On the masurement of dynamic behavior of processes with deterministic test signals). Regelungstechnik 15:249–257

Isermann R (1982) Parameter-adaptive control algorithms: A tutorial. Automatica 18(5):513–528

Isermann R (1991) Digital control systems, 2nd edn. Springer, Berlin

Ljung L (1999) System identification: Theory for the user, 2nd edn. Prentice Hall Information and System Sciences Series, Prentice Hall PTR, Upper Saddle River, NJ

Pintelon R, Schoukens J (2001) System identification: A frequency domain approach. IEEE Press, Piscataway, NJ

Schoukens J, Rolain Y, Pintelon R (2006) Analysis of windowing/leakage effects in frequency response function measurements. Automatica 42(1):27–38

Verhaegen M, Verdult V (2007) Filtering and system identification: A least squares approach. Cambridge University Press, Cambridge

5

Frequency Response Measurement for Periodic Test Signals

The frequency response measurement with periodic test signals allows the determination of the relevant frequency range for linear systems for certain, discrete points in the frequency spectrum. Typically, one uses sinusoidal signals at fixed frequencies, see Sect. 5.1. However, one can also use other periodic signals such as e.g. rectangular, trapezoidal, or triangular signals as shown in Sect. 5.2. The analysis can be carried out manually or with the aid of digital computers, where the Fourier analysis or special correlation methods come into play.

Based on the determination of *correlation functions*, special frequency response measurement techniques have been developed, which work well even in the presence of larger disturbances and noise, Sect. 5.5. Here, Sect. 5.5.1 describes the general approach in determining the frequency response from correlation functions. An especially well suited approach for the determination of the frequency response is governed in Sect. 5.5.2, which describes the *orthogonal correlation*, a very powerful technique. It is remarkably well suited for disturbance rejection and performs very reliably in the presence of large noise levels.

Special attention must be paid to the characteristics of the actuator, as was already stressed in Sect. 1.5. The use of sinusoidal test signals typically requires that the static and dynamic behavior of the actuator is linear within the input signal interval that is used for the experiments. If the *static behavior* is linear, then one can realize a sinusoidal input to the plant by connecting the actuator to a signal generator. This applies to actuators which are proportional acting or exhibit integral action with variable actuation speed. For actuators with integral action, it is often advisable to use an underlying position controller and supply a sinusoidal setpoint signal to this controller to maintain a constant mean and avoid drifting of the actuator. For actuators with integral action and constant speed operation (e.g. AC motor drives), one can use a three point controller for the displacement of the actuated variable to generate an approximate sinusoidal oscillation at lower frequencies. For higher frequencies however, such an actuator can only generate trapezoidal or triangular signals.

Frequently, the static characteristics of the actuator is non-linear, so that the actuator generates distorted sinusoidal or trapezoidal signals, which show a frequency spectrum that is different from the original test signal. One can try to determine the

R. Isermann, M. Münchhof, *Identification of Dynamic Systems*,
DOI 10.1007/978-3-540-78879-9_5, © Springer-Verlag Berlin Heidelberg 2011

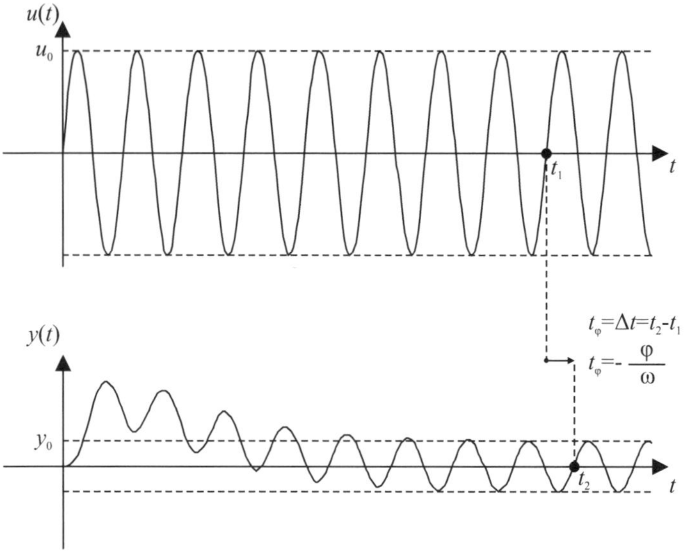

Fig. 5.1. Direct determination of the frequency response by analyzing the measured input and output. Evaluation after the stationary oscillation has fully developed

frequency response of the subsequent process by analyzing the first harmonic of the response only. The problem of the non-linear distortion can however be avoided by employing rectangular signals (or trapezoidal signals with steep edges) as an input signal. Here, one does only switch between two discrete points of the non-linearity at the input of the actuator. Therefore the non-linear behavior between the two single operating points must not be considered. If the actuator can be operated manually, it is also possible to apply rectangular or trapezoidal signals by hand.

5.1 Frequency Response Measurement with Sinusoidal Test Signals

The easiest and probably most well-known identification method for the determination of a discrete point of the frequency response is the direct determination of the amplitude ratio and phase angle of the recorded input and output oscillation, see Fig. 5.1. For this identification technique, one needs only a two channel oscilloscope or two channel plotter. The experiment has to be repeated for each frequency ω_k that is of interest. The gain and phase can then be determined from

$$|G(\mathrm{i}\omega_\nu)| = \frac{y_0(\omega_\nu)}{u_0(\omega_\nu)} \tag{5.1.1}$$

$$\angle G(\mathrm{i}\omega_\nu) = -t_\varphi \omega_\nu \,, \tag{5.1.2}$$

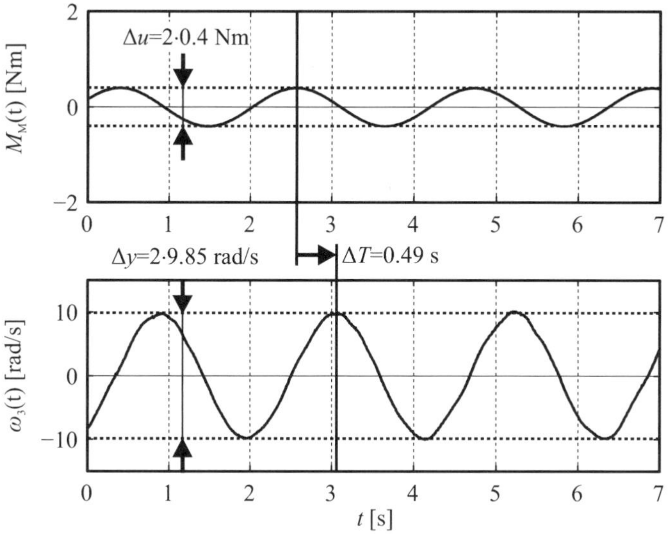

Fig. 5.2. Direct determination of the frequency response by analyzing the measured input and output of the Three-Mass Oscillator

where t_φ denotes the time of the phase lag and is positive if the output $y(t)$ is "later" than the input $u(t)$ or has a lag compared to the input. The phase $\angle G(\mathrm{i}\omega)$ is in this case negative. If disturbances are present, the gain and phase can be determined by averaging the results from (5.1.1) and (5.1.2) for multiple points of the recorded input and output signals.

Example 5.1 (Direct Determination of the Frequency Response).

The direct determination of the frequency response is shown for the Three-Mass Oscillator in Fig. 5.2. The amplitude of the input signal is $u_0 = 0.4\,\mathrm{Nm}$. The amplitude of the output oscillation is $y_0 = 9.85\,\mathrm{rad/s}$. Thus, the gain is given as

$$|G(\mathrm{i}\omega)|_{\omega=2.89\,\mathrm{rad/s}} = \frac{y_0}{u_0} = \frac{9.85\,\mathrm{rad/s}}{0.4\,\mathrm{Nm}} = 24.63\,\frac{\mathrm{rad}}{\mathrm{Nm\,s}}\,. \tag{5.1.3}$$

The phase can be determined by two subsequent zero crossings of the input and output signal for stationary oscillation of the output. The input signal has a zero crossing at $t_1 = 2.57\,\mathrm{s}$ and the output has its corresponding zero crossing at $t_2 = 3.06\,\mathrm{s}$. The phase is thus given as

$$\begin{aligned}\varphi(\mathrm{i}\omega)|_{\omega=2.89\,\mathrm{rad/s}} &= -t_\varphi\omega = (3.06\,\mathrm{s} - 2.57\,\mathrm{s})\,2.89\,\frac{\mathrm{rad}}{\mathrm{s}}\\ &= -1.41\,\mathrm{rad} = -81.36^\circ\,.\end{aligned} \tag{5.1.4}$$

This result can later be compared with the frequency response as determined by the orthogonal correlation. □

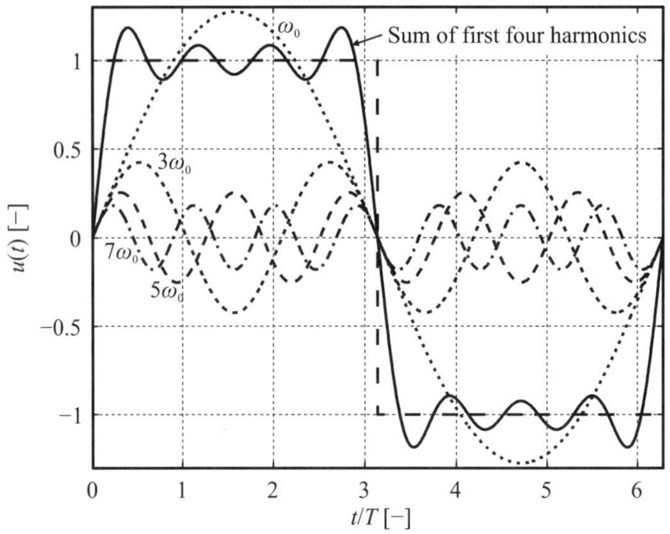

Fig. 5.3. Harmonic decomposition of a rectangular wave

5.2 Frequency Response Measurement with Rectangular and Trapezoidal Test Signals

In some cases, it is more convenient to apply rectangular signals or trapezoidal signals instead of the typical sinusoidal signals. In the following, a simple identification method shall be described, which allows the recording of the frequency response with rectangular waves and is especially well suited for slow processes with order $n \geq 3$ (Isermann, 1963).

A rectangular wave of the amplitude u_0 and the frequency $\omega_0 = 2\pi/T$ can be written as a Fourier series,

$$u(t) = \frac{4}{\pi} u_0 \left(\sin \omega_0 t + \frac{1}{3} \sin 3\omega_0 t + \frac{1}{5} \sin 5\omega_0 + \ldots \right) . \tag{5.2.1}$$

Figure 5.3 shows the first four harmonics and their superposition, which resembles a rectangular wave already quite well. The response to this input is given as

$$\begin{aligned} y(t) = \frac{4}{\pi} u_0 \Big(& |G(\mathrm{i}\omega_0)| \sin\big(\omega_0 t + \varphi(\omega_0)\big) \\ & + \frac{1}{3} |G(\mathrm{i}3\omega_0)| \sin\big(3\omega_0 t + \varphi(3\omega_0)\big) \\ & + \frac{1}{5} |G(\mathrm{i}5\omega_0)| \sin\big(5\omega_0 t + \varphi(5\omega_0)\big) + \ldots \Big) . \end{aligned} \tag{5.2.2}$$

One starts off with the identification of the frequency response for *high frequencies*. In this frequency range, the amplitude of the second harmonic with the frequency

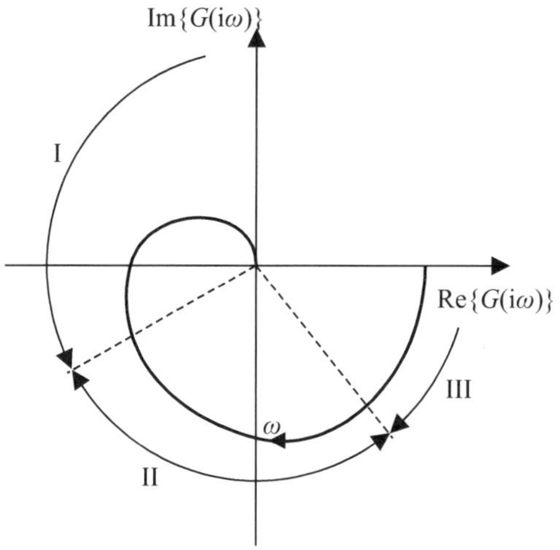

Fig. 5.4. Evaluation sections for the measurement of the frequency response with rectangular waves

$3\omega_0$ is by a factor of $\gamma = 1/3^n$ (n is the order of a delay with identical time constants) smaller than the amplitude of the fundamental. For $n \geq 3$ follows that $\gamma \leq 0.04$. The higher harmonics are thus damped so strongly that the resulting output closely resembles a pure sinusoidal oscillation whose amplitude and phase with respect to the input signal can easily be determined. In this manner, one can determine part I of the Nyquist plot shown in Fig. 5.4 (Isermann, 1963).

For *medium frequencies*, the amplitude of the second harmonic at $3\omega_0$ grows to a value where it can no longer be neglected, thus

$$y(t) \approx \frac{4}{\pi} u_0 \Big(|G(\mathrm{i}\omega_0)| \sin\big(\omega_0 t + \varphi(\omega_0)\big) + \frac{1}{3} |G(\mathrm{i}3\omega_0)| \sin\big(3\omega_0 t + \varphi(3\omega_0)\big)\Big) . \tag{5.2.3}$$

The third harmonic with the frequency $5\omega_0$ can still be neglected as can be all higher-frequent harmonics. One can obtain the response that belongs to the fundamental frequency by subtracting the response evoked by the second harmonic

$$y_{3\omega_0}(t) = \frac{4}{\pi}\frac{1}{3} u_0 |G(\mathrm{i}3\omega_0)| \sin\big(3\omega_0 t + \varphi(3\omega_0)\big) \tag{5.2.4}$$

from the measured system output $y(t)$. The amplitude and phase of the component $y_{3\omega_0}$ are known from the frequency response identification for high frequencies (part I of the Nyquist plot). One can thus obtain part II of the Nyquist plot shown in Fig. 5.4

For *lower frequencies*, part III of the Nyquist plot (see Fig. 5.4) can be determined by subtracting the response due to as many harmonics as necessary.

The response due to the sinusoidal fundamental is given as

$$\begin{aligned}&\frac{4}{\pi}u_0|G(\mathrm{i}\omega_0)|\sin\big(\omega_0 t+\varphi(\omega_0)\big)\\&=y-u_0\frac{1}{3}|G(\mathrm{i}3\omega_0)|\sin\big(3\omega_0 t+\varphi(3\omega_0)\big)\\&-u_0\frac{1}{5}|G(\mathrm{i}5\omega_0)|\sin\big(5\omega_0 t+\varphi(5\omega_0)\big)-\dots\,.\end{aligned}\tag{5.2.5}$$

One will however apply this method typically only for the identification of the higher-frequent part of the frequency response where the evaluation work is small.

However, it is more efficient to determine the lower-frequent part of the frequency response from recorded step responses, see Sect. 4.2.3 and 4.3. The evaluation can also be carried out by means of a Fourier analysis.

The advantages of this identification method can be summarized as follows:

- The rectangular wave test signal can often easier be realized than the sinusoidal test signal.
- The static transfer behavior of the actuator does not need to be linear.
- For a given amplitude u_0, the rectangular wave has the highest amplitude of the fundamental sine wave compared to all other periodic input signals (e.g. sinusoidal, trapezoidal or triangular oscillation). Thus the ratio for a given disturbance with respect to the wanted output is the smallest.

The jump from $+u_0$ to $-u_0$ must not be carried out in an infinitely small time interval, but can take a certain time T_1^*. As can be seen by a comparison of the coefficients of the Fourier transformed trapezoidal and rectangular pulse, the coefficients of the Fourier transform of the trapezoidal pulse are by a factor

$$\kappa=\frac{\sin\frac{\omega T_1}{2}}{\frac{\omega T_1}{2}}\tag{5.2.6}$$

smaller. If one accepts an error of 5% (respectively 1%), which means $\kappa = 0.95$ (respectively $\kappa = 0.99$), then the switching time from $+u_0$ to $-u_0$ and vice versa may be as large as

$$T_1^*<\frac{1.1}{\omega_{\max}}\ \text{resp.}\ T_1^*<\frac{0.5}{\omega_{\max}}\,,\tag{5.2.7}$$

where $\omega_{\max}$ is the highest frequency of interest. If the actuation time T_1^* gets larger and the error resulting by the approximation with a rectangular pulse should be avoided, then one has to determine the Fourier coefficients of the trapezoidal oscillation.

5.3 Frequency Response Measurement with Multi-Frequency Test Signals

The periodic test signals treated in the last section use only the basic frequency. This requires several evaluation runs with each having an unused settling phase before

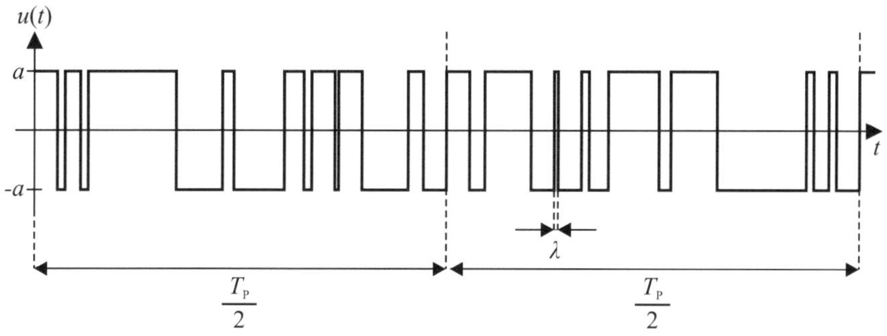

Fig. 5.5. Example for a binary multi-frequency signal for 6 frequencies $\omega_0,\ 2\omega_0,\ 4\omega_0, 8\omega_0,\ 16\omega_0,\ 32\omega_0$ and $N = 256$ intervals

reaching stationary oscillations. This drawback can, however, be avoided if test signals are designed which contain several frequency components at once with relatively low amplitudes.

Levin (1960) superimposed several sinusoidal oscillations at the frequencies $\omega_0,\ 2\omega_0,\ 4\omega_0,\ 8\omega_0, \ldots$. Using binary signals, Jensen (1959) designed a multifrequency signal with 7 frequencies at $\omega_0,\ 2\omega_0,\ 4\omega_0, \ldots$ with $u(t) = u(t)/|u(t)|$. Another proposal was made by Werner (1965) with rectangular oscillations of frequencies $\omega_0,\ 3\omega_0,\ 9\omega_0, \ldots$ and amplitudes $u_0,\ 2/3u_0, 2/3u_0, \ldots$. However, these signals did not result in best efficiency or only in small amplitudes with regard to the usable frequencies. Therefore, van den Bos (1967) then optimized binary signals with regard to the largest amplitudes for 6 frequencies $\omega_0,\ 2\omega_0,\ 4\omega_0, 8\omega_0,\ 16\omega_0,\ 32\omega_0$, with periods T_0 of the lowest frequency as $N = 512,\ 256,$ or 128 intervals. The size of the amplitudes is about $u_0 = 0.585a$.

Figure 5.5 shows a binary multifrequency signal for $N = 256$. The discrete-time instants for switches are for half a period $12 + 2 - 4 + 2 - 23 + 12 - 3 + 13 - 5 + 2 - 6 + 1 - 6 + 12 - 4 + 6-$.

The evaluation of the frequency response due to the multisine signal follows from the Fourier coefficients as

$$\left.\begin{aligned} a_{y\nu} &= \frac{2}{nT_\mathrm{P}} \int_0^{nT_\mathrm{P}} y(t) \cos \omega_\nu t \,\mathrm{d}t \\ b_{y\nu} &= \frac{2}{nT_\mathrm{P}} \int_0^{nT_\mathrm{P}} y(t) \sin \omega_\nu t \,\mathrm{d}t \end{aligned}\right\} \tag{5.3.1}$$

with integer values for n representing the total measurement time $T_\mathrm{M} = nT_\mathrm{P}$ and

$$\left.\begin{aligned} |G(\mathrm{i}\omega_\nu)| &= \tfrac{1}{u_{0\nu}} \sqrt{a_{y\nu}^2 + b_{y\nu}^2} \\ \varphi(\omega_\nu) &= \arctan \tfrac{a_{y\nu}}{b_{y\nu}} \end{aligned}\right\} . \tag{5.3.2}$$

Finally, a Schroeder multisine (Schroeder, 1970) is given as

$$u(t) = \sum_{k=1}^{N} A \cos(2\pi f_k t + \varphi_k) \tag{5.3.3}$$

with

$$f_k = l_k f_0 \text{ with } l_k \in N \tag{5.3.4}$$

$$\varphi_k = -\frac{k(k+1)\pi}{N} \, . \tag{5.3.5}$$

The goal in designing this signal was to reduce the maximum amplitude of the combined signal as much as possible.

5.4 Frequency Response Measurement with Continuously Varying Frequency Test Signals

In telecommunication engineering, analysis of electronic circuits and audio engineering, one often uses a sweep sine test signal, which is also referred to as a *chirp* signal. Here, the frequency of the signal varies as a function of time. This brings up the question, how the current frequency of a signal can be measured. The Fourier transform is only defined for an infinite time interval and also the short time Fourier transform requires at least an interval of finite length and hence does not allow to determine the frequency of a signal at a single point in time. Here, the notion of the *instantaneous frequency* (Cohen, 1995) of a signal comes into play. The instantaneous frequency is defined as the time derivative of the phase of a complex valued signal,

$$\omega = \frac{\mathrm{d}}{\mathrm{d}t}\big(\angle x(t)\big) \, . \tag{5.4.1}$$

This notion can easily be applied to the sweep sine signal.

A sweep sine is given as

$$x(t) = \sin\big(2\pi f(t)\, t\big) \, . \tag{5.4.2}$$

The phase is hence given as the argument of the sine function and the instantaneous frequency can therefore be determined as

$$\omega = \frac{\mathrm{d}}{\mathrm{d}t}\big(2\pi f(t)\, t\big) \, . \tag{5.4.3}$$

Now, a function $f(t)$ shall be defined for the case of a linear transition from frequency f_0 to f_1 in time T and for a logarithmic transition. For the linear transition, the frequency function $f(t)$ will be given as

$$f(t) = at + b \, . \tag{5.4.4}$$

Hence, the instantaneous frequency is given as

$$\omega = \frac{\mathrm{d}}{\mathrm{d}t}\big(2\pi(at + b)\, t\big) = 2\pi(2at + b) \; . \tag{5.4.5}$$

To obtain the instantaneous frequency $\omega(t = 0) = 2\pi f_0$ at $t = 0$ and $\omega(t = T) = 2\pi f_1$ at T, one has to select the frequency function $f(t)$ as

$$f(t) = f_0 + \frac{f_1 - f_0}{2T} t \; . \tag{5.4.6}$$

With a similar derivation, one can determine the frequency function $f(t)$ for an exponential transition as

$$f(t) = f_0 \left(\frac{f_1}{f_0}\right)^{\frac{1}{T}} \; . \tag{5.4.7}$$

These two frequency sweeps have been shown in Table 5.1. Sweep sines are often used in circuit and network analysis. Here, a so-called wobble generator produces a sweep sine in linear or exponential form which is then used as an input to a circuit. The output is analyzed, the amplitude and phase are determined and are displayed on a screen or saved for later reference.

5.5 Frequency Response Measurement with Correlation Functions

The frequency response measurement methods presented so far have mainly only been suitable for small disturbances. For larger disturbances, techniques are required, which automatically separate the wanted, useful signal from the noise. Especially well suited for this task are correlation methods which correlate the test signal and the disturbed output. In Sects. 5.5.1 and 5.5.2, identification techniques based on the determination of correlation functions are presented. The methods are basically exploiting the fact that the correlation of a periodic function is again periodic and thus is easily separable from the correlation functions of stochastic disturbances, as was already illustrated in Sect. 2.3.

5.5.1 Measurement with Correlation Functions

For a linear system, the auto-correlation function (ACF) of the input signal is given as

$$R_{\mathrm{uu}}(\tau) = \lim_{T\to\infty} \frac{1}{T} \int_{-\frac{T}{2}}^{\frac{T}{2}} u(t) u(t + \tau) \mathrm{d}t \; , \tag{5.5.1}$$

see (2.3.8). For the cross-correlation function (CCF) follows from (2.3.14)

$$\begin{aligned} R_{\mathrm{uy}}(\tau) = \mathrm{E}\big\{u(t) y(t + \tau)\big\} &= \lim_{T\to\infty} \frac{1}{T} \int_{-\frac{T}{2}}^{\frac{1}{T}} u(t) y(t + \tau) \mathrm{d}t \\ &= \lim_{T\to\infty} \frac{1}{T} \int_{-\frac{T}{2}}^{\frac{1}{T}} u(t - \tau) y(t) \mathrm{d}t \; . \end{aligned} \tag{5.5.2}$$

Table 5.1. Linear and Exponential Sine Sweep

Name, Shape and Equation

Linear Sweep

Equation

$x(t) = \sin(2\pi f(t)\, t)$
with
$f(t) = f_0 + \frac{f_1 - f_0}{2T} t$

Time Response

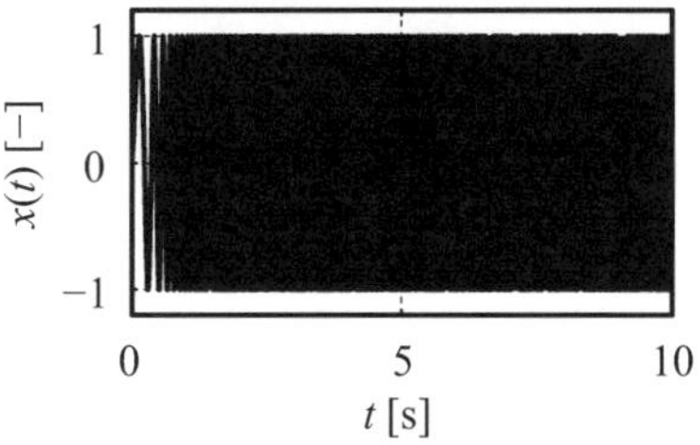

Fourier Transform

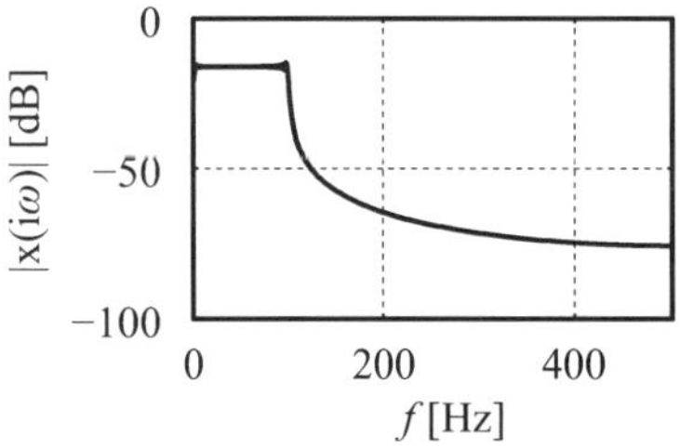

Spectrogram

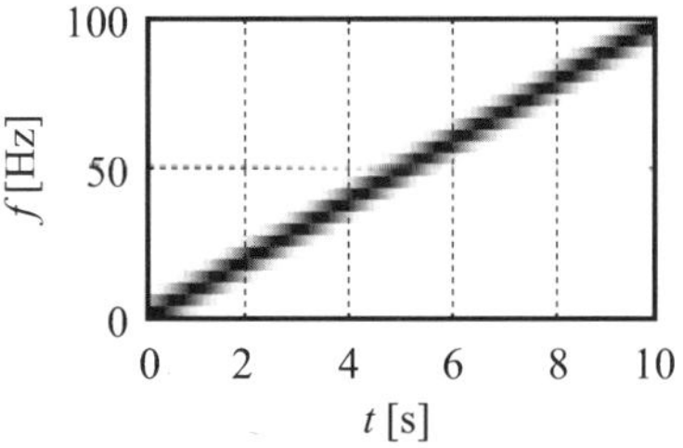

Exponential Sweep

Equation

$x(t) = \sin(2\pi f(t)\, t)$
with
$f(t) = f_0 \left(\frac{f_1}{f_0}\right)^{\frac{1}{T}}$

Time Response

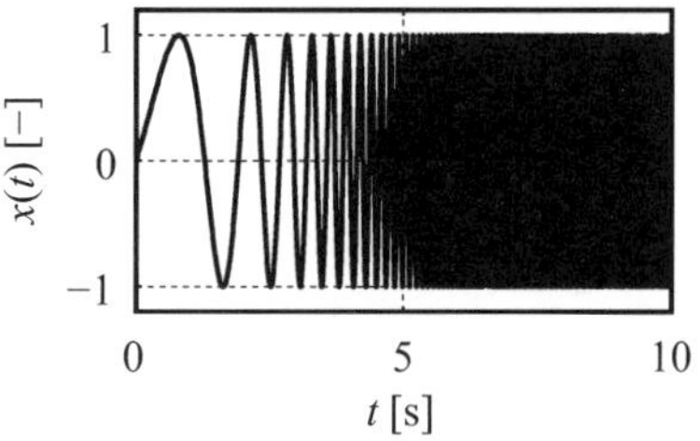

Fourier Transform

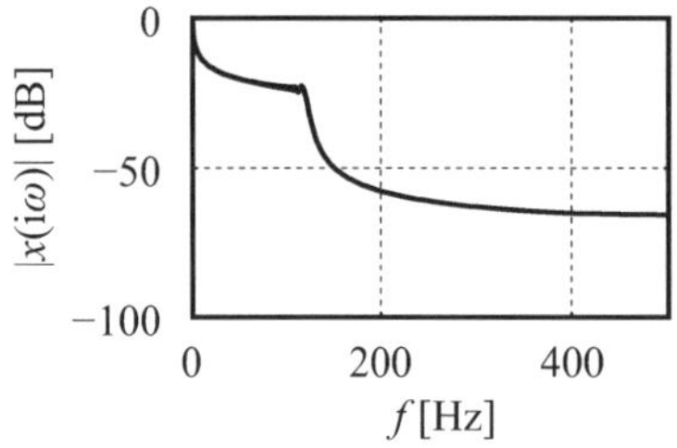

Spectrogram

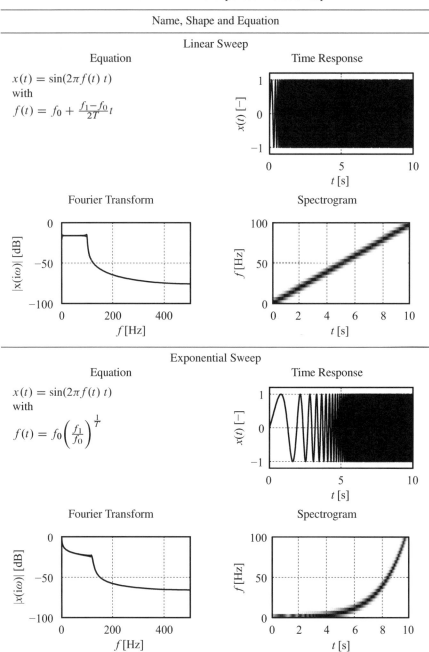

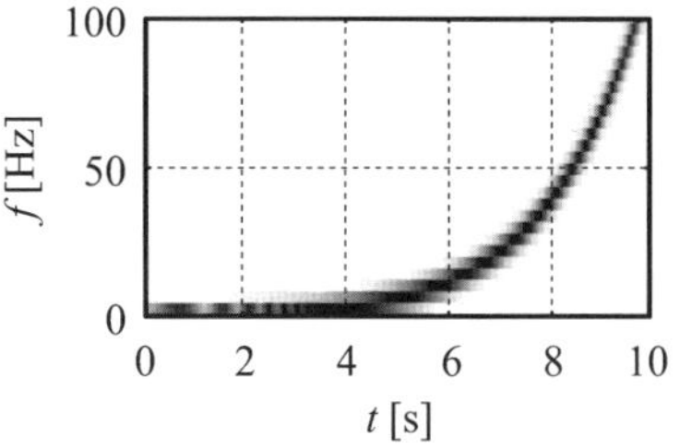

They are both connected by means of the convolution integral, (2.3.35),

$$R_{uy}(\tau) = \int_0^\infty g(t')R_{uu}(\tau - t')\mathrm{d}t' \ . \tag{5.5.3}$$

These relations have been developed for stochastic signals in Sect. 2.3, but are also valid for periodic signals. In order to determine the frequency response, one could determine the impulse response $g(t')$ from (5.5.3) and then calculate the Fourier transform of the impulse response to determine the frequency response $G(\mathrm{i}\omega)$. However, due to some special features of the correlation function, a direct determination of the amplitude and phase of the frequency response can be derived as is shown in the following. For a sinusoidal test signal

$$u(t) = u_0 \sin \omega_0 t \tag{5.5.4}$$

with the frequency

$$\omega_0 = \frac{2\pi}{T_P} \ , \tag{5.5.5}$$

the ACF is given as (2.3.31)

$$R_{uu}(\tau) = \frac{2u_0^2}{T_0}\int_0^{\frac{T_0}{2}} \sin(\omega_0 t + \alpha)\sin\big(\omega_0(t+\tau)+\alpha\big)\mathrm{d}t = \frac{u_0^2}{2}\cos\omega_0\tau \ . \tag{5.5.6}$$

The CCF of the test signal (5.5.4) and the test signals response

$$y(t) = u_0|G(\mathrm{i}\omega_0)|\sin\big(\omega_0 t - \varphi(\omega_0)\big) \tag{5.5.7}$$

yield with (5.5.3)

$$\begin{aligned} R_{uy}(\tau) &= |G(\mathrm{i}\omega_0)|\frac{2u_0^2}{T_P}\int_0^{\frac{T_P}{2}} \sin\omega_0(t-\tau)\sin\big(\omega_0 t - \varphi(\omega_0)\big)\mathrm{d}t \\ &= |G(\mathrm{i}\omega_0)|\frac{u_0^2}{2}\cos\big(\omega_0\tau - \varphi(\omega_0)\big) \ . \end{aligned} \tag{5.5.8}$$

Due to the periodicity of the CCF, one can confine the integration to half a period.

By considering (5.5.6), one obtains

$$R_{uy}(\tau) = |G(\mathrm{i}\omega_0)|R_{uu}\left(\tau - \frac{\varphi(\omega_0)}{\omega_0}\right) \ . \tag{5.5.9}$$

If the ACF and CCF are graphed over time, see Fig. 5.6, then the amplitude of the frequency response is always the ratio of the CCF at the point τ to the ACF at the point $\tau - \varphi(\omega_0)/\omega_0$ (Welfonder, 1966),

$$|G(\mathrm{i}\omega_0)| = \frac{R_{uy}(\tau)}{R_{uu}\left(\tau - \frac{\varphi(\omega_0)}{\omega_0}\right)} = \frac{R_{uy,\max}}{R_{uu}(0)} = \frac{R_{uy}\left(\frac{\varphi(\omega_0)}{\omega_0}\right)}{R_{uu}(0)} \ . \tag{5.5.10}$$

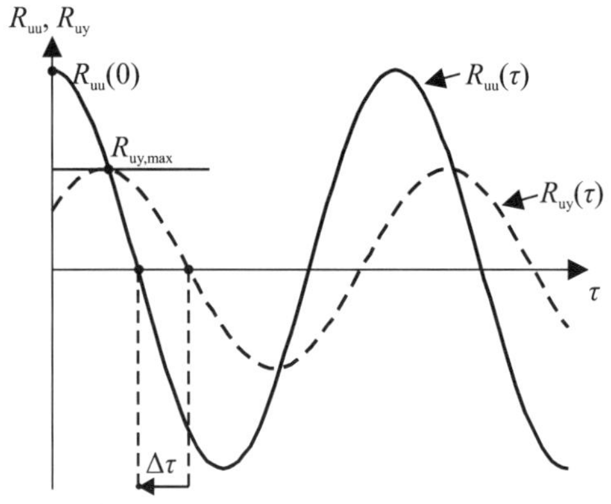

Fig. 5.6. ACF and CCF for sinusoidal input signal

The phase can be determined from the *time lag* $\Delta\tau$ of the two correlation functions

$$\varphi(\omega_0) = -\omega_0 \Delta\tau \; . \tag{5.5.11}$$

$\Delta\tau$ can best be determined from the *zero crossings* of the two functions. Amplitude and phase can thus be determined from four discrete points of the two correlation functions only. However, for averaging, one can process more points of the periodic correlation functions as necessary.

The application of the method is not limited to sinusoidal signals. One can employ any arbitrary periodic signal since the higher harmonics of the test signal do not influence the result as long as the input and output are correlated with sinusoidal reference signals (Welfonder, 1966).

If stochastic disturbances $n(t)$ are superimposed onto the output, then one will use larger measurement periods for the determination of the CCF according to (5.5.2). The influence of stochastic signals on the determination of the ACF is covered in Chap. 6. It will be shown there that the error vanishes if the stochastic disturbance $n(t)$ is not correlated with the test signal $u(t)$ and either $\overline{u(t)} = 0$ or $\overline{n(t)} = 0$. This is also valid for arbitrary periodic signals as long as their frequency is different from the measurement frequency ω_0.

Example 5.2 (Determination of the Frequency Response Using Correlation Functions).

An example of the determination of the frequency response function using correlation functions, again applied to the Three-Mass Oscillator, is shown in Fig. 5.7. Here, noise has been added to the output signal with $\sigma_{n_y} = 4\,\text{rad/s}$. As can clearly be seen from the input signal and the output signal graphed in Fig. 5.7, the direct determination of the frequency response is impossible due to the noise superimposed onto the systems measured output. Thus, the ACF and CCF of the input and output

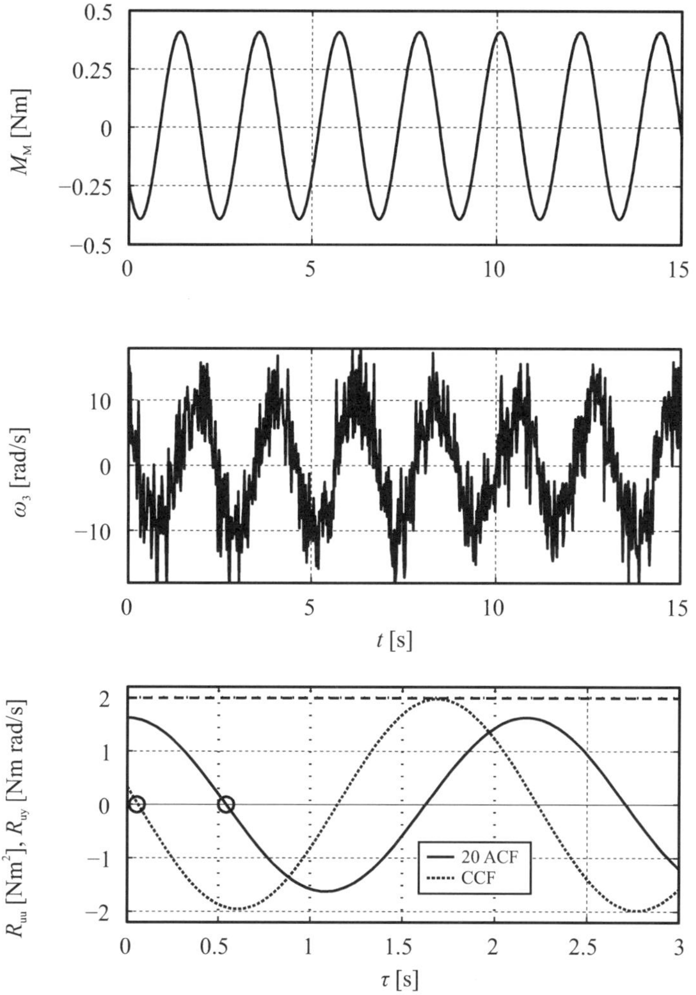

Fig. 5.7. Frequency response measurement with correlation functions for a noisy output signal of the Three-Mass Oscillator

have to be determined. As can be seen from the bottom diagram in Fig. 5.7, the CCF shows a smooth course and is obviously not much affected by the noise. The zero crossings and the maximum amplitude of the CCF can easily be determined.

The amplitude of the ACF can be read out at the time difference $\tau = 0$ as $R_{\mathrm{uu}}(0) = 0.1152\,\mathrm{Nm}^2$ (note the scaling of the ACF with a factor of 20 in Fig. 5.7), the maximum amplitude of the CCF is given as $\max(R_{\mathrm{uy}}(\tau)) = R_{\mathrm{uy,max}} = 3.217\,\mathrm{Nm\,rad/s}$. Thus, the gain is given as

$$|G(\mathrm{i}\omega_0)|_{\omega=2.8947\,\mathrm{rad/s}} = \frac{R_{\mathrm{uy,max}}}{R_{\mathrm{uu}}(0)} = \frac{1.99\,\mathrm{Nm}\frac{\mathrm{rad}}{\mathrm{s}}}{0.081\,\mathrm{Nm}^2} = 24.49\,\frac{\frac{\mathrm{rad}}{\mathrm{s}}}{\mathrm{Nm}}\,. \tag{5.5.12}$$

The phase can be determined by two subsequent zero crossings of the input and the output signal. The auto-correlation function has a zero crossing at $\tau = 0.053$ s and the cross-correlation function has its corresponding zero crossing at $\tau = 0.542$ s. The phase is thus given as

$$\begin{aligned}\varphi(\mathrm{i}\omega)|_{\omega=2.8947\,\mathrm{rad/s}} &= -\Delta\tau\omega = (0.542\,\mathrm{s} - 0.053\,\mathrm{s})\,2.8947\,\frac{\mathrm{rad}}{\mathrm{s}}\\ &= -1.41\,\mathrm{rad} = -81.1^\circ\,.\end{aligned} \tag{5.5.13}$$

One can see that the amplitude matches relatively well with the value derived by the direct evaluation, see Example 5.1.

5.5.2 Measurement with Orthogonal Correlation

The following section will cover the *most important frequency response measurement technique* for linear systems that allow the injection of special test signals and offline identification.

The Principle

The characteristics of the frequency response for a certain frequency ω_0 can be determined from *two points* of the CCF of the test signal and the system output. Real and imaginary part can both be estimated from the CCF, (5.5.8),

$$|G(\mathrm{i}\omega_0)|\cos\bigl(\omega_0\tau - \varphi(\omega_0)\bigr) = \frac{R_{\mathrm{uy}}(\tau)}{\frac{u_0^2}{2}}\,. \tag{5.5.14}$$

For $\tau = 0$, one obtains the real part of the frequency response as

$$\mathrm{Re}\bigl\{G(\mathrm{i}\omega_0)\bigr\} = |G(\mathrm{i}\omega_0)|\cos\bigl(\varphi(\omega_0)\bigr) = \frac{R_{\mathrm{uy}}(0)}{\frac{u_0^2}{2}} \tag{5.5.15}$$

and for $\tau = T_{\mathrm{P}}/4 = \pi/2\omega_0$ or $\omega_0\tau = \pi/2$, the imaginary part of the frequency response can be determined as

$$\mathrm{Im}\bigl\{G(\mathrm{i}\omega_0)\bigr\} = |G(\mathrm{i}\omega_0)|\sin\bigl(\varphi(\omega_0)\bigr) = \frac{R_{\mathrm{uy}}\Bigl(\frac{\pi}{2\omega_0}\Bigr)}{\frac{u_0^2}{2}}\,. \tag{5.5.16}$$

Thus, one has to determine the CCF merely for two points and not its entire course as a function of τ. The CCF for $\tau = 0$ can according to (5.5.2) be calculated by multiplying the test signal with the system output as

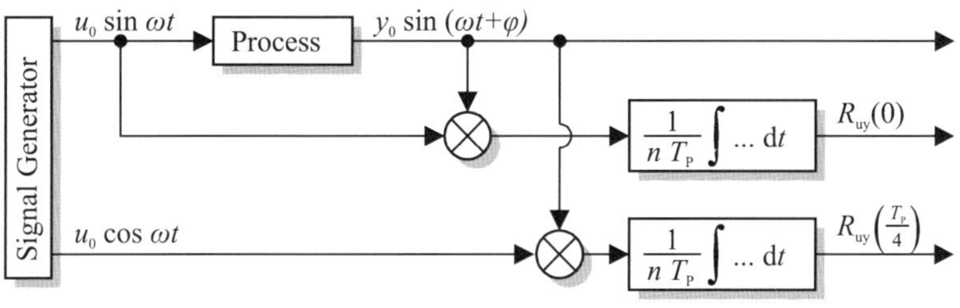

Fig. 5.8. Setup for the orthogonal correlation

$$R_{\mathrm{uy}}(0) = \frac{u_0^2}{2}\mathrm{Re}\{G(\mathrm{i}\omega_0)\} = \frac{u_0}{nT_\mathrm{P}}\int_0^{nT_\mathrm{P}} y(t)\sin\omega_0 t\,\mathrm{d}t \tag{5.5.17}$$

and the CCF for $\tau = T_\mathrm{P}/4$ can similarly be determined by a multiplication of the phase shifted test signal with the output as

$$R_{\mathrm{uy}}\left(\frac{T_\mathrm{P}}{4}\right) = \frac{u_0^2}{2}\mathrm{Im}\{G(\mathrm{i}\omega_0)\} = -\frac{u_0}{nT_\mathrm{P}}\int_0^{nT_\mathrm{P}} y(t)\cos\omega_0 t\,\mathrm{d}t\ , \tag{5.5.18}$$

where the phase shift by $\pi/2$ transforms the sine into a cosine of the same frequency. The multiplied signals are subsequently integrated over n full periods.

This measurement principle exploits the orthogonality relations of the trigonometric functions. Signal components which are (integer multiple) harmonics of the fundamental frequency ω_0 as well as signal components which have the same fundamental frequency ω_0, but are orthogonal to $\sin\omega_0 t$ or $\cos\omega_0 t$ do not contribute to the identification of the real and imaginary part respectively.

Figure 5.8 shows the corresponding experimental setup (Schäfer and Feissel, 1955; Balchen, 1962; Elsden and Ley, 1969). Before starting the integration, one must wait for transient effects to have settled. In contrast to the technique presented in the preceding section, the orthogonal correlation allows to directly display the real and imaginary part immediately after each integration over n periods.

Despite the fact that the relations for the real and imaginary part have already been shown in (5.5.15) and (5.5.16), the relations will be derived from scratch again in the following. This time the CCFs will be considered as they are shown in Fig. 5.8.

At the output of the integrators, the following values can be tapped according to (5.5.17) and (5.5.18)

$$\begin{aligned} R_{\mathrm{uy}}(0) &= \frac{1}{nT_\mathrm{P}}\int_0^{nT_\mathrm{P}} u_0\sin\omega_0 t y_0\sin(\omega_0 t+\varphi)\mathrm{d}t \\ &= \frac{1}{nT_\mathrm{P}}u_0 y_0\int_0^{nT_\mathrm{P}}(\sin\omega_0 t\cos\varphi+\cos\omega_0 t\sin\varphi)\sin\omega_0 t\,\mathrm{d}t \end{aligned} \tag{5.5.19}$$

Application of the orthogonality relation then yields

$$R_{uy}(0) = \frac{1}{nT_P} u_0 y_0 \Bigg(\int_0^{nT_P} \sin^2 \omega_0 t \cos \varphi \mathrm{d}t + \underbrace{\int_0^{nT_P} \sin \omega_0 t \cos \omega_0 t \sin \varphi \mathrm{d}t}_{=0} \Bigg)$$
$$= \frac{y_0}{u_0} \frac{u_0^2}{2} \cos \varphi = |G(\mathrm{i}\omega_0)| \cos \varphi \frac{u_0^2}{2} = \mathrm{Re}\big\{G(\mathrm{i}\omega_0)\big\} \frac{u_0^2}{2} \,. \tag{5.5.20}$$

Similarly, one obtains for $R_{uy}(T_P/4)$

$$R_{uy}\left(\frac{T_P}{4}\right) = \frac{1}{nT_P} \int_0^{nT_P} u_0 \cos \omega_0 t y_0 \sin(\omega_0 t + \varphi) \mathrm{d}t$$
$$= \mathrm{Im}\big\{G(\mathrm{i}\omega_0)\big\} \frac{u_0^2}{2} \,. \tag{5.5.21}$$

Amplitude and phase of the frequency response can then be determined by the relations

$$|G(\mathrm{i}\omega_0)| = \sqrt{\mathrm{Re}^2\{G(\mathrm{i}\omega_0)\} + \mathrm{Im}^2\{G(\mathrm{i}\omega_0)\}} \tag{5.5.22}$$

$$\varphi(\omega_0) = \arctan \frac{\mathrm{Im}\{G(\mathrm{i}\omega_0)\}}{\mathrm{Re}\{G(\mathrm{i}\omega_0)\}} \,. \tag{5.5.23}$$

This measurement principle has found widespread distribution and is also part of commercially available frequency responses measurement systems (Seifert, 1962; Elsden and Ley, 1969). Due to its easy application, it is not only used in the presence of large disturbances, but often also in the case of little or no disturbances. Frequency response measurement systems that are based on this working principle are termed *correlation frequency response analyzers*. The competing measurement principle is the *sweep frequency response analysis*, which is based on a sweep sine generator and a spectrum analyzer carrying out an FFT, see Sect. 5.4.

Example 5.3 (Orthogonal Correlation).

The orthogonal correlation has been applied to the Three-Mass-Oscillator and the measured frequency response has been compared with the theoretically derived frequency response, see Fig. 5.9. Experiment and theory match very well. □

Stochastic signals and periodic signals with $\omega \neq \omega_0$ do not influence the result in the case of infinite measurement time as was the case for the technique presented in the preceding section. However, in practical applications, the measurement time is always limited and in many cases quite short. Due to this, attention will be paid to the resulting errors in the case of a finite measurement time nT_P in the next section.

Influence of Noise

The disturbances $y_z(t)$, which are superimposed onto the response $y_u(t)$ (see Fig. 1.5) cause the following error in the determination of the real and imaginary part of the frequency response according to (5.5.17) and (5.5.18)

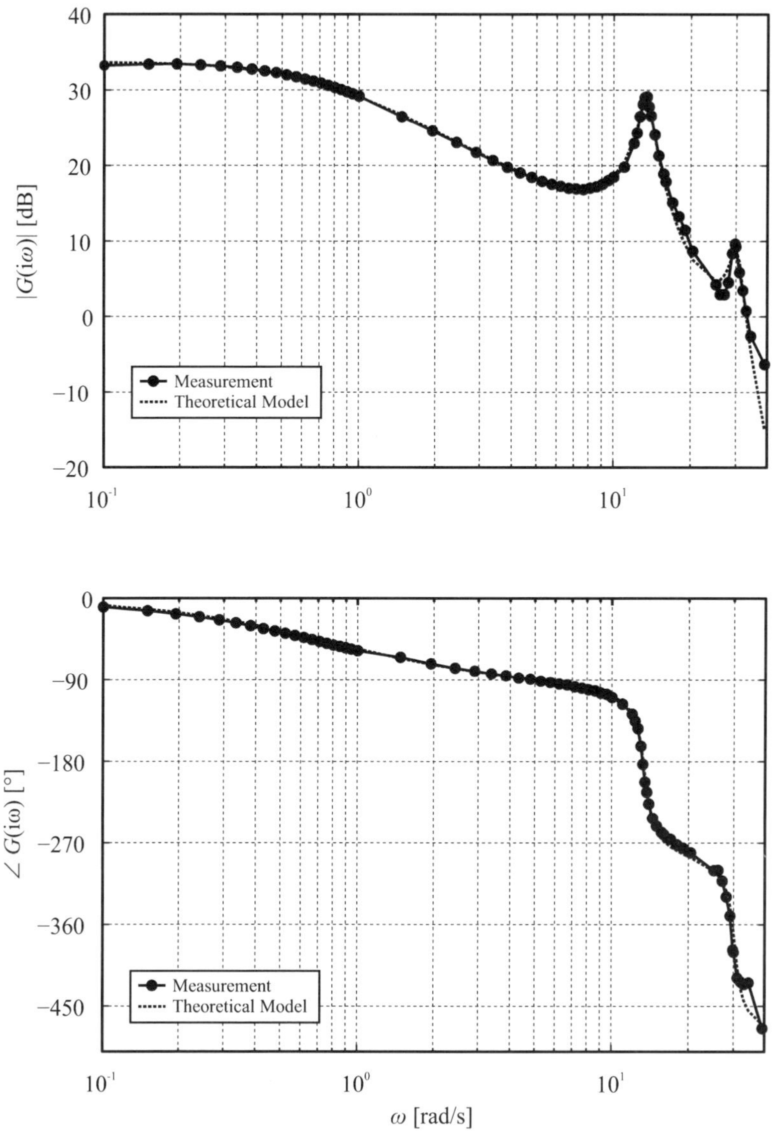

Fig. 5.9. Frequency response measurement of the Three-Mass Oscillator by means of the orthogonal correlation

$$\Delta\mathrm{Re}\{G(\mathrm{i}\omega_0)\} = \frac{2}{u_0 n T_\mathrm{P}} \int_0^{nT_\mathrm{P}} y_\mathrm{z}(t)\sin\omega_0 t\,\mathrm{d}t \tag{5.5.24}$$

$$\Delta\mathrm{Im}\{G(\mathrm{i}\omega_0)\} = -\frac{2}{u_0 n T_\mathrm{P}} \int_0^{nT_\mathrm{P}} y_\mathrm{z}(t)\cos\omega_0 t\,\mathrm{d}t\ . \tag{5.5.25}$$

The magnitude of the resulting error in the frequency response is then given as

$$|\Delta G(\mathrm{i}\omega_0)|^2 = \Delta\mathrm{Re}^2\{G(\mathrm{i}\omega_0)\} + \Delta\mathrm{Im}^2\{G(\mathrm{i}\omega_0)\}\ . \tag{5.5.26}$$

Now, the influence of a stochastic noise $n(t)$, a periodic disturbance $p(t)$, and a drift $d(t)$ will be investigated.

For a stationary *stochastic disturbance* $n(t)$, the expected value of the squared error of the real part is given as

$$\begin{aligned}\mathrm{E}\{\Delta\mathrm{Re}^2(\omega_\mathrm{z})\} &= \frac{4}{u_0^2 n^2 T_\mathrm{P}^2}\mathrm{E}\left\{\int_0^{nT_\mathrm{P}} n(t')\sin\omega_0(t')\mathrm{d}t' \int_0^{nT_\mathrm{P}} n(t'')\sin\omega_0(t'')\mathrm{d}t''\right\}\\ &= \frac{4}{u_0^2 n^2 T_\mathrm{P}^2}\int_0^{nT_\mathrm{P}}\int_0^{nT_\mathrm{P}} \mathrm{E}\{n(t')n(t'')\}\sin\omega_0 t'\sin\omega_0 t''\mathrm{d}t'\mathrm{d}t''\ .\end{aligned} \tag{5.5.27}$$

With

$$R_\mathrm{nn}(\tau) = R_\mathrm{nn}(t'-t'') = \mathrm{E}\{n(t')n(t'')\} \tag{5.5.28}$$

and the substitution $\tau = t' - t''$ follows

$$\mathrm{E}\{\Delta\mathrm{Re}^2(\omega_0)\} = \frac{4}{u_0^2 n T_\mathrm{P}}\int_0^{nT_\mathrm{P}} R_\mathrm{nn}(\tau)\left(\left(1-\frac{\tau}{nT_\mathrm{P}}\right)\cos\omega_0\tau + \frac{\sin\omega_0\tau}{\omega_0 n T_\mathrm{P}}\right)\mathrm{d}\tau\ . \tag{5.5.29}$$

The derivation is shown in (Eykhoff, 1974) and (Papoulis and Pillai, 2002). For $\mathrm{E}\{\Delta\mathrm{Im}^2(\omega_0)\}$, one can derive a similar equation with a minus sign in front of the last addend. Plugging these terms into (5.5.26) yields

$$\begin{aligned}\mathrm{E}\{|\Delta G(\mathrm{i}\omega_0)|^2\} &= \frac{8}{u_0^2 n T_\mathrm{P}}\int_0^{nT_\mathrm{P}} R_\mathrm{nn}(\tau)\left(1-\frac{|\tau|}{nT_\mathrm{P}}\right)\cos\omega_0\tau\,\mathrm{d}\tau\\ &= \frac{4}{u_0^2 n T_\mathrm{P}}\int_{-nT_\mathrm{P}}^{nT_\mathrm{P}} R_\mathrm{nn}(\tau)\left(1-\frac{|\tau|}{nT_\mathrm{P}}\right)\mathrm{e}^{-\mathrm{i}\omega_0\tau}\mathrm{d}\tau\ .\end{aligned} \tag{5.5.30}$$

One has to take into account that $\mathrm{E}\{\Delta\mathrm{Re}(\omega_0)\Delta\mathrm{Im}(\omega_0)\} = 0$ (Sins, 1967; Eykhoff, 1974). If $n(t)$ is a *white noise* with the power spectral density S_0 and thus

$$R_\mathrm{nn}(\tau) = S_0\delta(\tau)\ , \tag{5.5.31}$$

then (5.5.30) simplifies to

$$\mathrm{E}\{|\Delta G(\mathrm{i}\omega_0)|^2\} = \frac{4S_0}{u_0^2 n T_\mathrm{P}}\ . \tag{5.5.32}$$

The standard deviation of the relative frequency response error is then given as

$$\sigma_G = \sqrt{\mathrm{E}\left\{\frac{|\Delta G(\mathrm{i}\omega_0)|^2}{|G(\mathrm{i}\omega_0)|^2}\right\}} = \frac{2\sqrt{S_0}}{|G(\mathrm{i}\omega_0)|u_0\sqrt{nT_\mathrm{P}}} \,. \tag{5.5.33}$$

Now, it is assumed that $n(t)$ is a *colored noise* that has been derived by filtering the white noise $\nu(t)$ with the power spectral density $S_{\nu 0}$. The filter can e.g. be a first order low-pass filter with the corner frequency $\omega_\mathrm{C} = 1/T_\mathrm{C}$,

$$G_\nu(\mathrm{i}\omega) = \frac{n(\mathrm{i}\omega)}{\nu(\mathrm{i}\omega)} = \frac{1}{1+\mathrm{i}\omega T_\mathrm{C}} \,. \tag{5.5.34}$$

The ACF is then given as

$$R_\mathrm{nn}(\tau) = \frac{S_{\nu 0}}{2T_\mathrm{C}} \mathrm{e}^{-\frac{|\tau|}{T_\mathrm{C}}} \,. \tag{5.5.35}$$

and $R_\mathrm{nn}(\tau) \approx 0$ for $|\tau_\mathrm{max}| < kT_\mathrm{C}$ where e.g. $k > 3$. Then, from (5.5.30) follows for large measurement times $nT_\mathrm{P} \gg |\tau_\mathrm{max}|$

$$\mathrm{E}\{|\Delta G(\mathrm{i}\omega_0)|^2\} \approx \frac{4}{u_0^2 nT_\mathrm{P}} S_\mathrm{nn}(\omega_0) \,. \tag{5.5.36}$$

Therefore, for a colored noise $n(t)$ with the power spectral density $S_\mathrm{nn}(\omega)$ for large measurement periods follows

$$\sigma_G \approx \frac{2\sqrt{S_\mathrm{nn}(\omega_0)}}{|G(\mathrm{i}\omega_0)|u_0\sqrt{nT_\mathrm{P}}} \tag{5.5.37}$$

with

$$S_\mathrm{nn}(\omega) = |G_\nu(\mathrm{i}\omega)|^2 S_{\nu 0} \tag{5.5.38}$$

and with (5.5.34)

$$\sigma_\mathrm{G} \approx \frac{\sqrt{2S_{\nu 0}\omega_\mathrm{C}}}{|G(\mathrm{i}\omega_0)|u_0} \underbrace{\frac{\sqrt{\dfrac{\omega_0}{\omega_\mathrm{C}}}}{\sqrt{\pi\left(1+\left(\dfrac{\omega_0}{\omega_\mathrm{C}}\right)^2\dfrac{1}{\sqrt{n}}\right)}}}_{Q} \,. \tag{5.5.39}$$

The factor Q is shown in Fig. 5.11. For a given colored noise created by a filter with the corner frequency ω_C, the absolute frequency response error is largest for the measurement frequency $\omega_0 = \omega_\mathrm{C}$. *The error diminishes proportionally to the square root of the number of full periods measured.*

Example 5.4 (Disturbance Rejection of the Orthogonal Correlation).

The good rejection of disturbances can be seen in Fig. 5.10. Here, noise has been added to the output of the Three-Mass Oscillator. The topmost plot shows the noisy measurement. The lower two plots illustrate that the frequency response despite the large noise is still measured and that the first resonance can be detected relatively precisely for $\omega < 20\,\mathrm{rad/s}$. □

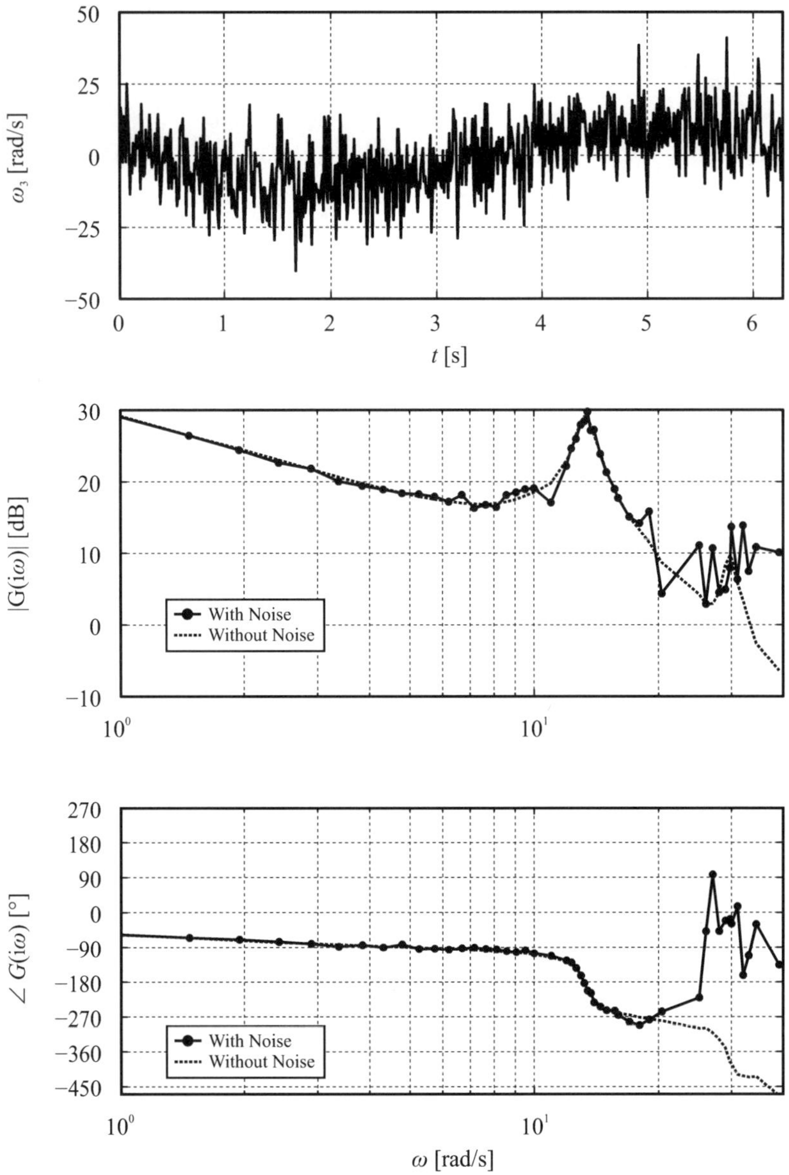

Fig. 5.10. Frequency response as determined by the orthogonal correlation in the presence of noise

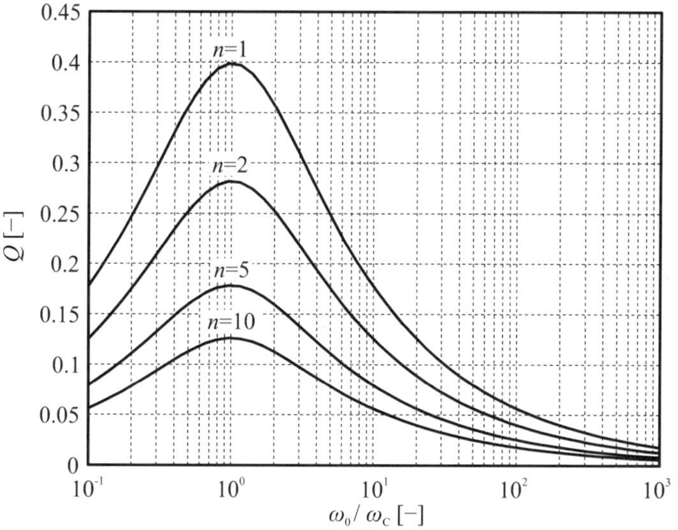

Fig. 5.11. Factor Q of the frequency response error caused by a stochastic disturbance with corner frequency ω_C and measurement frequency ω_0 (Balchen, 1962)

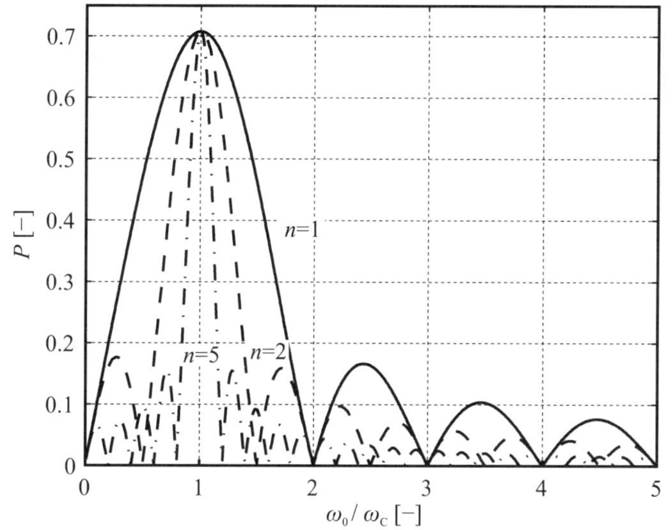

Fig. 5.12. Factor P of the frequency response error caused by a periodic disturbance $p_0 \cos \omega t$ with measurement frequency ω_0 (Balchen, 1962)

For a *periodic disturbance* $p(t)$

$$p(t) = p_0 \cos \omega t \tag{5.5.40}$$

the error of the real and the imaginary part of the frequency response can be calculated according to (5.5.24) and (5.5.25). Calculating the integral yields

$$\Delta\mathrm{Re}\left(\frac{\omega}{\omega_0}\right) = \frac{p_0}{u_0}\pi n\left(1-\left(\frac{\omega}{\omega_0}\right)^2\right)\left(1-\cos 2\pi\frac{\omega}{\omega_0}n\right) \tag{5.5.41}$$

$$\Delta\mathrm{Im}\left(\frac{\omega}{\omega_0}\right) = -\frac{p_0\left(\frac{\omega}{\omega_0}\right)}{u_0\pi n\left(1-\left(\frac{\omega}{\omega_0}\right)^2\right)\sin 2\pi\frac{\omega}{\omega_0}}n\,. \tag{5.5.42}$$

The magnitude of the relative frequency response error is then given as

$$\begin{aligned}\delta_{\mathrm{G}} = \frac{|\Delta G(\mathrm{i}\omega_0)|}{|G(\mathrm{i}\omega_0)|} &= \frac{2p_0\left(\frac{\omega}{\omega_0}\right)\sqrt{1-\left(1-\left(\frac{\omega}{\omega_0}\right)^2\right)\cos^2\pi\frac{\omega}{\omega_0}n}}{u_0|G(\mathrm{i}\omega_0)|\left|\left(1-\left(\frac{\omega}{\omega_0}\right)^2\right)\right|}\frac{\left|\sin\pi\frac{\omega}{\omega_0}n\right|}{\pi\left(\frac{\omega}{\omega_0}\right)n}\\ &\approx \frac{p_0\sqrt{2}}{u_0|G(\mathrm{i}\omega_0)|}\underbrace{\frac{\left(\frac{\omega}{\omega_0}\right)\sqrt{1+\left(\frac{\omega}{\omega_0}\right)^2}}{\left|1-\left(\frac{\omega}{\omega_0}\right)^2\right|}\frac{\left|\sin\pi\frac{\omega}{\omega_0}n\right|}{\pi\left(\frac{\omega}{\omega_0}\right)n}}_{P}\quad\text{for } \omega\neq\omega_0\,.\end{aligned} \tag{5.5.43}$$

The approximation can be derived by taking the average of $\cos^2(\ldots) = 0.5$. The factor P which is decisive for the frequency dependency of the frequency response error is graphed in Fig. 5.12, see (Balchen, 1962; Elsden and Ley, 1969). The factor has zeros at $\omega/\omega_0 = j/n$ with $j = 0, 2, 3, 4, 5, \ldots$. Periodic disturbances which have a frequency ω that is an integer multiple of the measurement frequency ω_0 do not contribute to the error of the frequency response measurement. Periodic disturbances with any other frequency ω cause an error in the frequency response measurement for finite measuring periods nT_{P}, which are proportional to the factor P. The most severe error is caused by disturbances whose frequency ω is quite close to measuring frequency ω_0. If one interprets $P(\omega/\omega_0)$ as a filter, then the "pass-band" gets smaller as the measuring time increases. For $n \to \infty$, the identified frequency response can only be falsified by periodic disturbances with the very same frequency ω_0 as the measuring frequency. The envelope of $P(\omega/\omega_0)$ is given as

$$\delta_{\mathrm{G}}\left(\frac{\omega}{\omega_0}\right)\bigg|_{\max} = \frac{p_0\sqrt{2}}{u_0|G(\mathrm{i}\omega_0)|}\frac{\sqrt{1+\left(\frac{\omega}{\omega_0}\right)^2}}{\left|1-\left(\frac{\omega}{\omega_0}\right)^2\right|}\frac{1}{n\pi}\,. \tag{5.5.44}$$

For $\omega/\omega_0 \neq j/n$ the *error diminishes proportionally to the number of full periods* n, which is faster than for stochastic disturbances.

Finally, the influence of *very low frequent* disturbances shall be investigated. Over the measurement period, these can be seen approximately as a non-periodic disturbance $d(t)$. From (5.5.24), (5.5.25), and (5.5.26) follows

$$\begin{aligned}|\Delta G(\mathrm{i}\omega)|^2 &= \frac{2}{u_0^2 n^2 T_{\mathrm{P}}^2}\int_0^{nT_{\mathrm{P}}} d(t')\mathrm{e}^{-\mathrm{i}\omega_0 t'}\mathrm{d}t' \int_0^{nT_{\mathrm{P}}} d(t'')\mathrm{e}^{\mathrm{i}\omega_0 t''}\mathrm{d}t'' \\ &= \frac{2}{u_0^2 n^2 T_{\mathrm{P}}^2} d_{\mathrm{T}}(-\mathrm{i}\omega) d_{\mathrm{T}}(\mathrm{i}\omega) \\ &= \frac{2}{u_0^2 n^2 T_{\mathrm{P}}^2}|d_{\mathrm{T}}(\mathrm{i}\omega)|^2\end{aligned} \tag{5.5.45}$$

with $d_{\mathrm{T}}(\mathrm{i}\omega)$ being the Fourier transform of the disturbance of length $T = nT_{\mathrm{P}}$. For a drift

$$d(t) = at \tag{5.5.46}$$

of duration $T = nT_{\mathrm{P}}$ follows

$$d_{\mathrm{T}}(\mathrm{i}\omega) = \int_0^{nT_{\mathrm{P}}} at\mathrm{e}^{-\mathrm{i}\omega_0 t}\mathrm{d}t = -\frac{2\pi n}{\omega_0^2}\mathrm{i} \tag{5.5.47}$$

and the frequency response error becomes

$$|\Delta G(\mathrm{i}\omega)| = \frac{\sqrt{2}a}{u_0\omega_0}\,. \tag{5.5.48}$$

The error in the frequency response caused by a drift does not diminish as the measurement time increases and is proportional to the drift factor a. Thus, one has to employ special means to suppress the disturbance of low frequent disturbances. One example is to filter the signal by means of a high-pass filter with the transfer function

$$G_{\mathrm{HP}}(s) = \frac{T_{\mathrm{D}}s}{1+T_1 s}\,, \tag{5.5.49}$$

where the time constants have to be adapted to the measurement frequency ω_0.

Another remedy is to approximate the drift-wise disturbance by a polynomial

$$d(t) = a_0 + a_1 t + a_2 t^2 + \ldots \tag{5.5.50}$$

and subsequently eliminate $d(t)$ by subtraction of the polynomial drift model from the measured signal. A method for drift elimination that is based on this approach has been presented by Liewers (1964).

5.6 Summary

The direct methods for the determination of the frequency response allow a pointwise determination of the frequency response with little effort and quite good results as long as the disturbances acting on the process are small. It is however time-consuming for processes with slow dynamics as the transitional phases between the measurements cannot be exploited. For linear processes with larger disturbances, the frequency response measurement with correlation functions has proven to be a powerful tool. The therefrom derived method of orthogonal correlation is employed in many commercial frequency response measurement devices and software tools.

Due to its long measuring time, it is mainly used for processes with small settling times. A reduction of the total time can be achieved if the frequency response for small frequencies is determined by means of the Fourier analysis of recorded step responses as was shown in Chap. 4 and only the frequency response for the higher frequencies is determined by means of correlation methods. Thus, one can combine non-periodic and periodic test signals into "advantageous" test signal sequences (Isermann, 1971).

Problems

5.1. Frequency Response Measurement with Monofrequent Signals
What are the advantages and disadvantages of determining the frequency response with monofrequent signals as illustrated in this chapter?

5.2. Rectangular Wave Test Signal
How can the frequency response be determined with rectangular waves?

5.3. Orthogonal Correlation
Derive a method to employ the orthogonal correlation method with rectangular waves as an input.

5.4. Orthogonal Correlation
How does the frequency response measurement error decay in the presence of stochastic or periodic disturbances as the number of measurement periods increases? By which factor must the measurement time be increased to reduce the influence of these disturbances by half.

References

Balchen JG (1962) Ein einfaches Gerät zur experimentellen Bestimmung des Frequenzganges von Regelungsanordnungen. Regelungstechnik 10:200–205

van den Bos A (1967) Construction of binary multifrequency testsignals. In: Preprints of the IFAC Symposium on Identification, Prag

Cohen L (1995) Time frequency analysis. Prentice Hall signal processing series, Prentice Hall PTR, Englewood Cliffs, NJ

Elsden CS, Ley AJ (1969) A digital transfer function analyser based on pulse rate techniques. Automatica 5(1):51–60

Eykhoff P (1974) System identification: Parameter and state estimation. Wiley-Interscience, London

Isermann R (1963) Frequenzgangmessung an Regelstrecken durch Eingabe von Rechteckschwingungen. Regelungstechnik 11:404–407

Isermann R (1971) Experimentelle Analyse der Dynamik von Regelsystemen. BI-Hochschultaschenbücher, Bibliographisches Institut, Mannheim

Jensen JR (1959) Notes on measurement of dynamic characteristics of linear systems, Part III. Servoteknisk forsksingslaboratorium, Copenhagen

Levin MJ (1960) Optimum estimation of impulse response in the presence of noise. IRE Trans Circuit Theory 7(1):50–56

Liewers P (1964) Einfache Methode zur Drifteliminierung bei der Messung von Frequenzgängen. Messen, Steuern, Regeln 7:384–388

Papoulis A, Pillai SU (2002) Probability, random variables and stochastic processes, 4th edn. McGraw Hill, Boston

Schäfer O, Feissel W (1955) Ein verbessertes Verfahren zur Frequenzgang-Analyse industrieller Regelstrecken. Regelungstechnik 3:225–229

Schroeder M (1970) Synthesis of low-peak-factor signals and binary sequences with low autocorrelation. IEEE Trans Inf Theory 16(1):85–89

Seifert W (1962) Kommerzielle Frequenzgangmeßeinrichtungen. Regelungstechnik 10:350–353

Sins AW (1967) The determination of a system transfer function in presence of output noise. Ph. D. thesis. Electrical Engineering, University of Eindhoven, Eindhoven

Welfonder E (1966) Kennwertermittlung an gestörten Regelstrecken mittels Korrelation und periodischen Testsignalen. Fortschritt Berichte VDI-Z 8(4)

Werner GW (1965) Entwicklung einfacher Verfahren zur Kennwertermittlung an lienaren, industriellen Regelstrecken mit Testsignalen. Dissertation. TH Ilmenau, Ilmenau

Part II

Identification of Non-Parametric Models with Correlation Analysis — Continuous and Discrete Time

6

Correlation Analysis with Continuous Time Models

The correlation methods for single periodic test signals, which have been described in Chap. 5 provide only one discrete point of the frequency response at each measurement with one measurement frequency. At the start of each experiment, one must wait for the decay of the transients. Due to these circumstances, the methods are not suitable for online identification in real time. Thus, it is interesting to employ test signals which have a broad frequency spectrum and thus excite more frequencies at once as did the non-periodic deterministic test signals. This requirement is fulfilled by the properties of stochastic signals and the therefrom derived pseudo-stochastic signals. The stochastic signals can be generated artificially or one can use the signals which appear during normal operation of the plant, if they are suitable. By the correlation of the test signal and the output signal, the response evoked by the test signal is weighted differently than the noise. This results in an automatic separation of the wanted signal from the noise and thus a suppression of the noise.

This chapter covers correlation methods for the identification of non-periodic models for *continuous-time signals*. Since nowadays, the correlation functions are typically evaluated by digital computers, the use of correlation functions will also be presented in Chap. 7 for the discrete-time case. Section 6.1 covers the *estimation of correlation functions* in finite time and formulates conditions for the convergence of the estimate. Next, the identification of processes which are excited by stochastic signals by means of the ACF and CCF will be presented in Sect. 6.2. The correlation analysis with binary test signals, especially with pseudo-random binary signals and generalized random binary signals, is covered in Sect. 6.3. Issues of the identification by the aid of the correlation analysis in *closed-loop* are discussed in Sect. 6.4.

6.1 Estimation of Correlation Functions

In this section, the estimation of the CCF and the ACF of stationary stochastic signals for limited measurement periods is covered.

R. Isermann, M. Münchhof, *Identification of Dynamic Systems*,
DOI 10.1007/978-3-540-78879-9_6, © Springer-Verlag Berlin Heidelberg 2011

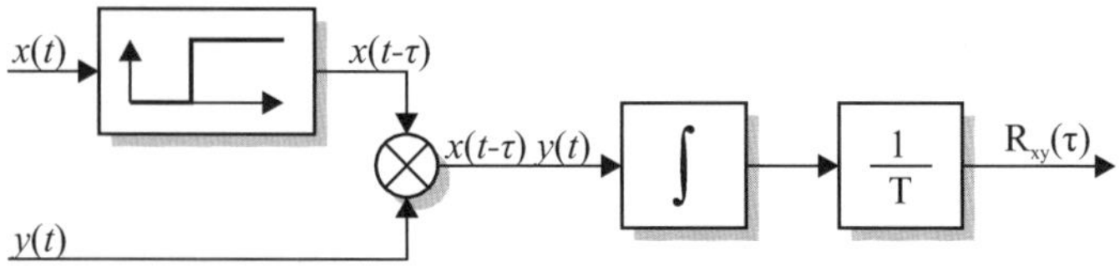

τ time delay; T measurement time

Fig. 6.1. Block diagram for the estimation of the cross-correlation function

6.1.1 Cross-Correlation Function

The cross-correlation function (CCF) (Hänsler, 2001; Papoulis, 1962) in the case of two continuous-time stationary random signals $x(t)$ and $y(t)$ is according to (2.3.14) defined as

$$\begin{aligned} R_{\mathrm{xy}}(\tau) &= \mathrm{E}\{x(t)y(t+\tau)\} = \lim_{T\to\infty}\frac{1}{T}\int_{-\frac{T}{2}}^{\frac{T}{2}} x(t)y(t+\tau)\mathrm{d}t \\ &= \lim_{T\to\infty}\frac{1}{T}\int_{-\frac{T}{2}}^{\frac{T}{2}} x(t-\tau)y(t)\mathrm{d}t \end{aligned} \quad (6.1.1)$$

and

$$\begin{aligned} R_{\mathrm{yx}}(\tau) &= \mathrm{E}\{y(t)x(t+\tau)\} = \lim_{T\to\infty}\frac{1}{T}\int_{-\frac{T}{2}}^{\frac{T}{2}} y(t)x(t+\tau)\mathrm{d}t \\ &= \lim_{T\to\infty}\frac{1}{T}\int_{-\frac{T}{2}}^{\frac{T}{2}} y(t-\tau)x(t)\mathrm{d}t \end{aligned} \quad . \quad (6.1.2)$$

Therefore,

$$R_{\mathrm{xy}}(\tau) = -R_{\mathrm{yx}}(\tau)\ . \quad (6.1.3)$$

In most applications however, the measurement period is quite limited and only of the (short) finite duration T. Thus, the influence of the measurement period T on the estimation of the correlation function must be taken into account and will now be investigated.

It is assumed that the signals $x(t)$ and $y(t)$ are known in the time interval $0 \leq t \leq T + \tau$ and that $\mathrm{E}\{x(t)\} = 0$ and $\mathrm{E}\{y(t)\} = 0$. (The case of a time interval $0 \leq t \leq T$ is covered in Chap. 7). The CCF can then be estimated by

$$\begin{aligned} \hat{R}_{\mathrm{xy}}(\tau) &= \frac{1}{T}\int_0^T x(t)y(t+\tau)\mathrm{d}t \\ &= \frac{1}{T}\int_0^T x(t-\tau)y(t)\mathrm{d}t\ . \end{aligned} \quad (6.1.4)$$

Figure 6.1 shows the block diagram of this estimator. First, one signal must be delayed by the time τ and then the two signals must be multiplied with each other.

Finally, the mean of the product has to be determined. The expected value of this estimation is given as

$$\begin{aligned} \mathrm{E}\{\hat{R}_{\mathrm{xy}}(\tau)\} &= \frac{1}{T}\int_0^T \mathrm{E}\{x(t)y(t+\tau)\}\mathrm{d}t \\ &= \frac{1}{T}\int_0^T R_{\mathrm{xy}}(\tau)\mathrm{d}t = R_{\mathrm{xy}}(\tau) \,. \end{aligned} \tag{6.1.5}$$

Thus, the estimate is unbiased. The variance of this estimate is given as

$$\begin{aligned} \operatorname{var}\hat{R}_{\mathrm{xy}}(\tau) &= \mathrm{E}\{\hat{R}_{\mathrm{xy}}(\tau) - R_{\mathrm{xy}}(\tau)\}^2 = \mathrm{E}\{\hat{R}^2_{\mathrm{xy}}(\tau)\} - R^2_{\mathrm{xy}}(\tau) \\ &= \frac{1}{T^2}\int_0^T\int_0^T \big(x(t)y(t+\tau)x(t')y(t'+\tau)\big)\mathrm{d}t'\mathrm{d}t - R^2_{\mathrm{xy}}(\tau) \,. \end{aligned} \tag{6.1.6}$$

Under the assumption that $x(t)$ and $y(t)$ are normally distributed, one obtains

$$\begin{aligned} \operatorname{var}\hat{R}_{\mathrm{xy}}(\tau) = \frac{1}{T^2}\int_0^T\int_0^T &\big(R_{\mathrm{xx}}(t'-t)R_{\mathrm{yy}}(t'-t) \\ &+ R_{\mathrm{xy}}(t'-t+\tau)R_{\mathrm{yx}}(t'-t-\tau)\big)\mathrm{d}t'\mathrm{d}t \,. \end{aligned} \tag{6.1.7}$$

By substituting $t'-t = \xi$, and $\mathrm{d}t' = \mathrm{d}\xi$ and exchanging the order of the integrals (Bendat and Piersol, 2010), it follows that

$$\begin{aligned} \operatorname{var}\hat{R}_{\mathrm{xy}}(\tau) = \frac{1}{T}\int_0^T \Big(1 - \frac{|\xi|}{T}\Big)&\big(R_{\mathrm{xx}}(\xi)R_{\mathrm{yy}}(\xi) \\ &R_{\mathrm{xy}}(\xi+\tau)R_{\mathrm{yx}}(\xi-\tau)\big)\mathrm{d}\xi = \sigma^2_{R1} \,. \end{aligned} \tag{6.1.8}$$

If the correlation functions are absolutely integrable, which necessitates $\mathrm{E}\{x(t)\} = 0$ or $\mathrm{E}\{y(t)\} = 0$, it follows that

$$\lim_{T\to\infty} \operatorname{var}\hat{R}_{\mathrm{xy}}(\tau) = 0 \,, \tag{6.1.9}$$

which means that (6.1.4) is consistent in the mean square.

For $T \gg \tau$, the variance of the estimate is given as

$$\begin{aligned} \operatorname{var}\hat{R}_{\mathrm{xy}}(\tau) &\approx \frac{1}{T}\int_{-T}^T \big(R_{\mathrm{xx}}(\xi)R_{\mathrm{yy}}(\xi) + R_{\mathrm{xy}}(\xi+\tau)R_{\mathrm{yx}}(\xi-\tau)\big)\mathrm{d}\xi \\ &= \frac{1}{T}\int_{-T}^T \big(R_{\mathrm{xx}}(\xi)R_{\mathrm{yy}}(\xi) + R_{\mathrm{xy}}(\tau+\xi)R_{\mathrm{xy}}(\tau-\xi)\big)\mathrm{d}\xi \,. \end{aligned} \tag{6.1.10}$$

The variance of the estimate of the CCF is only determined by the stochastic nature of the two signals. In a finite time horizon T, it is not possible to determine the stochastic correlation between two random signals without a certain uncertainty. This is termed the *intrinsic statistic uncertainty* (Eykhoff, 1964).

If one can assume $R_{\mathrm{xy}}(\tau) \approx 0$ for large τ and additionally $T \gg \tau$, then (6.1.10) can be simplified as

$$\operatorname{var} \hat{R}_{xy}(\tau) \approx \frac{2}{T} \int_0^T R_{xx}(\xi) R_{yy}(\xi) \mathrm{d}\xi \ . \tag{6.1.11}$$

Often, one must use correlation functions because one signal is disturbed by a *stochastic disturbance* $n(t)$, as e.g.

$$y(t) = y_0(t) + n(t) \ . \tag{6.1.12}$$

This additive noise $n(t)$ shall be zero-mean, $\mathrm{E}\{n(t)\} = 0$, and statistically independent from the useful signals $y_0(t)$ and $x(t)$. Then, it holds for the correlation functions

$$R_{yy}(\xi) = R_{y_0y_0}(\xi) + R_{nn}(\xi) \tag{6.1.13}$$

$$R_{xy}(\xi) = R_{xy_0}(\xi) \ . \tag{6.1.14}$$

According to (6.1.5) follows that the estimation is unbiased, i.e.

$$\mathrm{E}\{\hat{R}_{xy}(\tau)\} = R_{xy_0}(\tau) \ . \tag{6.1.15}$$

The variance of the estimate, (6.1.8) is augmented by another term as

$$\operatorname{var}\big(\hat{R}_{xy}(\tau)\big)_n = \frac{1}{T} \int_{-T}^{T} \left(1 - \frac{|\xi|}{T}\right) R_{xx}(\xi) R_{nn}(\xi) \mathrm{d}\xi = \sigma_{R_2}^2 \tag{6.1.16}$$

with

$$\lim_{T \to \infty} \operatorname{var}\big(\hat{R}_{xy}(\tau)\big)_n = 0 \ , \tag{6.1.17}$$

such that the estimate is still consistent in the mean square. The influence of the disturbance is eliminated as the measurement period T is increased, so that the variance of the estimate of the CCF decays inversely proportional to the measurement time T. If the disturbance is superimposed onto the other signal $x(t)$,

$$x(t) = x_0(t) + n(t) \ , \tag{6.1.18}$$

one can derive analogous results, which means that with respect to the convergence it does not matter which signal is disturbed. Now, it is assumed that both signals are similarly disturbed, i.e.

$$y(t) = y_0(t) + n_1(t) \tag{6.1.19}$$

$$x(t) = x_0(t) + n_2(t) \ . \tag{6.1.20}$$

With $\mathrm{E}\{n_1(t)\} = 0$ and $\mathrm{E}\{n_2(t)\} = 0$ follows that

$$R_{yy}(\xi) = R_{y_0y_0}(\xi) + R_{n_1n_1}(\xi) \tag{6.1.21}$$

$$R_{xx}(\xi) = R_{x_0x_0}(\xi) + R_{n_2n_2}(\xi) \ , \tag{6.1.22}$$

and, if the two disturbances are statistically independent from the respective useful signals, then

$$R_{\mathrm{xy}}(\xi) = R_{\mathrm{x_0y_0}}(\xi) + R_{\mathrm{n_1n_2}}(\xi) \,. \tag{6.1.23}$$

The estimation of the CCF is in this case only unbiased if $n_1(t)$ and $n_2(t)$ are uncorrelated. Under this prerequisite, the additional term for (6.1.16) is given as

$$\begin{aligned}\mathrm{var}\big(\hat{R}_{\mathrm{xy}}(\tau)\big)_{\mathrm{n_1n_2}} = \frac{1}{T}\int_{-T}^{T}\Big(1-\frac{|\xi|}{T}\Big)\Big(&R_{\mathrm{x_0x_0}}(\xi)R_{\mathrm{n_1n_1}}(\xi)\\ &+ R_{\mathrm{y_0y_0}}(\xi)R_{\mathrm{n_2n_2}}(\xi) + R_{\mathrm{n_1n_1}}(\xi)R_{\mathrm{n_2n_2}}(\xi)\Big)\mathrm{d}\xi \,.\end{aligned} \tag{6.1.24}$$

For $T \to \infty$, this variance also approaches zero. However, for finite T, its magnitude is larger than for the case of only one disturbance acting on the system.

Theorem 6.1 (Convergence of the Cross Correlation Function).

For the estimation of the cross-correlation function of two stationary stochastic signals according to (6.1.5), errors are caused by

- *the intrinsic statistical uncertainty according to (6.1.8)*
- *the uncertainty due to disturbances $n(t)$ according to (6.1.16)*

The estimate of the CCF for a finite time horizon T is unbiased, if the disturbance $n(t)$ is statistically independent from the respective wanted signal $x_0(t)$ and $y_0(t)$ and $\mathrm{E}\{n(t)\} = 0$. For the variance of the estimate in the presence of a disturbance $n(t)$ follows, see (6.1.8) and (6.1.16),

$$\mathrm{var}\,\hat{R}_{\mathrm{xy}}(\tau) = \sigma_{R1}^2 + \sigma_{R2}^2 \,. \tag{6.1.25}$$

If both signals are affected by disturbances, then the estimate is only unbiased if both disturbances are uncorrelated with each other. □

6.1.2 Auto-Correlation Function

As an estimate for the auto-correlation function (ACF) of a continuous-time stationary random signal $x(t)$, which exists in the time interval $0 \le t \le T + \tau$, it is suggested to use

$$\hat{R}_{\mathrm{xx}}(\tau) = \frac{1}{T}\int_0^T x(t)x(t+\tau)\mathrm{d}t \,. \tag{6.1.26}$$

The expected value of this estimate is

$$\mathrm{E}\big\{\hat{R}_{\mathrm{xx}}(\tau)\big\} = R_{\mathrm{xx}}(\tau) \,. \tag{6.1.27}$$

The estimate thus is unbiased. For a normally distributed signal $x(t)$ follows from (6.1.8) that

$$\begin{aligned}\mathrm{var}\,\hat{R}_{\mathrm{xx}}(\tau) &= \mathrm{E}\big\{\big(\hat{R}_{\mathrm{xx}}(\tau) - R_{\mathrm{xx}}(\tau)\big)^2\big\}\\ &= \frac{1}{T}\int_{-T}^{T}\Big(1-\frac{|\xi|}{T}\Big)\big(R_{\mathrm{xx}}^2(\xi) + R_{\mathrm{xx}}(\xi+\tau)R_{\mathrm{xx}}(\xi-\tau)\big)\mathrm{d}\xi = \sigma_{R1}^2 \,.\end{aligned} \tag{6.1.28}$$

If the ACF is absolutely integrable, then

$$\lim_{T\to\infty} \operatorname{var} \hat{R}_{\mathrm{xx}}(\tau) = 0 \,, \tag{6.1.29}$$

which means that (6.1.26) is consistent in the mean square. The variance σ_{R1} is caused by the intrinsic uncertainty.

For $T \gg \tau$, it follows that

$$\operatorname{var} \hat{R}_{\mathrm{xx}}(\tau) \approx \frac{1}{T} \int_{-T}^{T} \big(R_{\mathrm{xx}}^2(\xi) + R_{\mathrm{xx}}(\xi+\tau)R_{\mathrm{xx}}(\xi-\tau)\big)\mathrm{d}\xi \,. \tag{6.1.30}$$

Under the assumption of large measurement times T, the following special cases can be discussed:

1. $\tau = 0$:
$$\operatorname{var} \hat{R}_{\mathrm{xx}}(0) \approx \frac{2}{T} \int_{-T}^{T} R_{\mathrm{xx}}^2(\xi)\mathrm{d}\xi \,. \tag{6.1.31}$$
2. τ large and thus $R_{\mathrm{xx}}(\tau) \approx 0$: Due to $R_{\mathrm{xx}}^2(\xi) \gg R_{\mathrm{xx}}(\xi+\tau)R_{\mathrm{xx}}(\xi-\tau)$ it follows that
$$\operatorname{var} \hat{R}_{\mathrm{xx}}(0) \approx \frac{1}{T} \int_{-T}^{T} R_{\mathrm{xx}}^2(\xi)\mathrm{d}\xi \,. \tag{6.1.32}$$

The variance for large τ is thus only half as large as for $\tau = 0$.

If the signal $x(t)$ is disturbed by $n(t)$, such that

$$x(t) = x_0(t) + n(t) \,, \tag{6.1.33}$$

then the ACF is given as

$$R_{\mathrm{xx}}(\tau) = R_{\mathrm{x_0x_0}}(\tau) + R_{\mathrm{nn}}(\tau) \,, \tag{6.1.34}$$

provided that the useful signal $x_0(t)$ and the noise $n(t)$ are uncorrelated and furthermore $\mathrm{E}\{n(t)\} = 0$. The auto-correlation function of the disturbed signal is thus the sum of the two auto-correlation functions for the noise free signal $x(t)$ and the noise $n(t)$.

6.2 Correlation Analysis of Dynamic Processes with Stationary Stochastic Signals

6.2.1 Determination of Impulse Response by Deconvolution

According to (5.5.3) or (2.3.35) respectively, the auto-correlation function and the cross-correlation function are linked by the convolution integral, i.e.

$$R_{\mathrm{uy}}(\tau) = \int_0^\infty g(t')R_{\mathrm{uu}}(\tau - t')\mathrm{d}t' \,, \tag{6.2.1}$$

where $g(t)$ is the impulse response of the process with input $u(t)$ and output $y(t)$. As estimates of the correlation functions for a finite time horizon T, one uses

$$\hat{R}_{\mathrm{uu}}(\tau) = \frac{1}{T}\int_0^T u(t-\tau)u(t)\mathrm{d}t \tag{6.2.2}$$

$$\hat{R}_{\mathrm{uy}}(\tau) = \frac{1}{T}\int_0^T u(t-\tau)y(t)\mathrm{d}t \tag{6.2.3}$$

according to (6.1.4) and (6.1.26). The required impulse response $g(t')$ can be determined by *de-convolution* of (6.2.1). First however, the equation must be discretized with the sample time T_0 as

$$\hat{R}_{\mathrm{uy}}(\nu T_0) \approx T_0 \sum_{\mu=0}^{M} g(\mu T_0)\hat{R}_{\mathrm{uu}}\big((\nu-\mu)T_0\big) \,. \tag{6.2.4}$$

To determine the impulse response for $k = 0, \ldots, N$, one must formulate $N+1$ equations of the form (6.2.4). This is covered in Sect. 7.2.1. The direct convolution of the input $u(t)$ and the output $y(t)$ is used by Sage and Melsa (1971). While this will result in the inversion of a lower triangular matrix, it is not advisable to do so as the calculation of the correlation functions beforehand will reduce the influence of noise already.

Since the correlation functions are estimated according to (6.1.26) and (6.1.5) and thus are only approximately known for finite measurement times T, the estimated impulse response will be counterfeit to some degree.

As had been shown in Sect. 6.1, the ACF and the CCF are estimated bias-free for stationary signals $u(t)$ and $y(t)$ in the absence of noise. More important for the application however is the case of a disturbed output $y(t)$, (6.1.12) through (6.1.15), which will be reviewed in the following.

For a stochastically disturbed output

$$y(t) = y_{\mathrm{u}}(t) + n(t) \,. \tag{6.2.5}$$

follows according to (6.1.12) and (6.1.14)

$$\mathrm{E}\{\hat{R}_{\mathrm{uy}}(\tau)\} = R^0_{\mathrm{uy}}(\tau) + \mathrm{E}\{\Delta R_{\mathrm{uy}}(\tau)\} \tag{6.2.6}$$

with

$$R^0_{\mathrm{uy}}(\tau) = \frac{1}{T}\int_0^T \mathrm{E}\{u(t-\tau)y_{\mathrm{u}}(t)\}\mathrm{d}t \tag{6.2.7}$$

$$\mathrm{E}\{\Delta R_{\mathrm{uy}}(\tau)\} = \frac{1}{T}\int_0^T \mathrm{E}\{u(t-\tau)n(t)\}\mathrm{d}t = R_{\mathrm{un}}(\tau) \,. \tag{6.2.8}$$

If the input and the disturbance are uncorrelated, then it follows

$$\mathrm{E}\{u(t-\tau)n(t)\} = \mathrm{E}\{u(t-\tau)\}\mathrm{E}\{n(t)\} \,, \tag{6.2.9}$$

so that if either $\mathrm{E}\{u(t)\} = 0$ or $\mathrm{E}\{n(t)\} = 0$, then

$$\mathrm{E}\big\{\Delta R_{\mathrm{uy}}(\tau)\big\} = 0 \,. \tag{6.2.10}$$

The CCF according to (6.2.6) is henceforth unbiased even for estimation over a finite time horizon T. The variances of the estimated correlation functions can be determined as follows: Due to the stochastic nature of the input signal, the ACF has an intrinsic statistical uncertainty according to (6.1.28)

$$\mathrm{var}\big\{\hat{R}_{\mathrm{uu}}(\tau)\big\} = \frac{1}{T}\int_{-T}^{T}\left(1-\frac{|\xi|}{T}\right)\big(R_{\mathrm{uu}}^2(\xi) + R_{\mathrm{uu}}(\xi+\tau)R_{\mathrm{uu}}(\xi-\tau)\big)\mathrm{d}\xi \,. \tag{6.2.11}$$

The CCF also has an intrinsic statistical uncertainty, which can be determined from (6.1.8) as

$$\mathrm{var}\big(\hat{R}_{\mathrm{uy}}(\tau)\big) = \frac{1}{T}\int_{0}^{T}\left(1-\frac{|\xi|}{T}\right)\big(R_{\mathrm{uu}}(\xi)R_{\mathrm{yy}}(\xi)R_{\mathrm{uy}}(\xi+\tau)R_{\mathrm{yu}}(\xi-\tau)\big)\mathrm{d}\xi \tag{6.2.12}$$

and an additional uncertainty if a noise $n(t)$ is superimposed onto the output, see (6.1.16),

$$\mathrm{var}\big(\hat{R}_{\mathrm{uy}}(\tau)\big)_{\mathrm{n}} = \frac{1}{T}\int_{-T}^{T}\left(1-\frac{|\xi|}{T}\right)R_{\mathrm{uu}}(\xi)R_{\mathrm{nn}}(\xi)\mathrm{d}\xi \,. \tag{6.2.13}$$

All these variances vanish for $T \to \infty$, if the individual correlation functions respectively their products are absolutely integrable, which means that at least $\mathrm{E}\{u(t)\} = 0$. Then, all correlation function estimates are consistent in the mean square.

Theorem 6.2 (Convergence of the Correlation Functions for a Linear Process).
The auto-correlation function $R_{\mathrm{uu}}(\tau)$ and the cross-correlation function $R_{\mathrm{uy}}(\tau)$ for a linear process with the impulse response $g(t)$ are estimated consistently in the mean square according to (6.2.2) and (6.2.3) under the following conditions:

- *The useful signals $u(t)$ and $y_{\mathrm{u}}(t)$ are stationary*
- $\mathrm{E}\{u(t)\} = 0$
- *The disturbance $n(t)$ is stationary and uncorrelated with $u(t)$*

□

As has been shown in Sect. 6.1, the above theorem also holds true if the input $u(t)$ is disturbed by $n(t)$ or if both the input $u(t)$ and the output $y(t)$ are disturbed by $n_1(t)$ and $n_2(t)$ respectively, where $n_1(t)$ and $n_2(t)$ may not be correlated. If the theorem is valid for a given application, then the impulse response can also be estimated consistently in the mean square according to (6.2.4), see Sect. 7.2.1 An example for the assessment of the resulting error in the estimation of the frequency response is shown in the following section (Sect. 6.2.2).

6.2.2 White Noise as Input Signal

Ideal White Noise

If the input signal is a *white noise*, then its ACF is given as

$$R_{\mathrm{uu}}(\tau) = S_{\mathrm{u0}}\,\delta(\tau) \tag{6.2.14}$$

and from (6.2.1) with the masking property of the δ-function

$$R_{\mathrm{uy}}(\tau) = S_{\mathrm{u0}} \int_0^\infty g(t')\delta(\tau - t')\mathrm{d}t' = S_{\mathrm{u0}}\, g(\tau)\,. \tag{6.2.15}$$

The required impulse response is thus proportional to the CCF as

$$g(\tau) = \frac{1}{S_{\mathrm{u0}}} R_{\mathrm{uy}}(\tau) \tag{6.2.16}$$

and the de-convolution of the correlation functions is hence unnecessary. This idealized white noise with constant, frequency independent power spectral density S_{u0} is however not realizable. Therefore, this investigation shall thus be carried out again using *broadband noise* which has an approximately constant power spectral density in the interesting frequency range.

Broad-Band Noise

A broadband noise can be generated hypothetically by filtering a white noise. It then has the power spectral density

$$S_{\mathrm{uu}}(\omega) = |G_{\mathrm{F}}(\mathrm{i}\omega)|^2 S_{\mathrm{u0}}\,. \tag{6.2.17}$$

For a filter of first order with the corner frequency $\omega_{\mathrm{C}} = 1/T_{\mathrm{C}}$, one obtains by using (2.3.22) and the tables for Fourier transform of simple linear dynamic systems

$$\begin{aligned} R_{\mathrm{uu}}(\tau) &= \frac{1}{2\pi}\int_{-\infty}^{\infty} |G_{\mathrm{F}}(\mathrm{i}\omega)|^2 S_{\mathrm{u0}} \mathrm{e}^{\mathrm{i}\omega t}\mathrm{d}\omega \\ &= \frac{1}{\pi}\int_0^\infty \frac{S_{\mathrm{u0}}}{1 + T_{\mathrm{C}}^2\omega^2}\cos\omega\tau\mathrm{d}\omega \\ &= \frac{1}{2} S_{\mathrm{u0}}\omega_{\mathrm{C}}\mathrm{e}^{-\omega_{\mathrm{C}}|\tau|}\,. \end{aligned} \tag{6.2.18}$$

The shape of the ACF and the corresponding power spectral density S_{uu} is shown in Fig. 6.2. For a sufficiently large bandwidth, i.e. corner frequency ω_{C}, the ACF approaches the shape of a δ-function, so that the conditions for the application of (6.2.16) are approximately staisfied.

The error which stems from the limited bandwidth of the excitation and the subsequently "wrong" application of (6.2.16) has been investigated e.g. by Hughes and

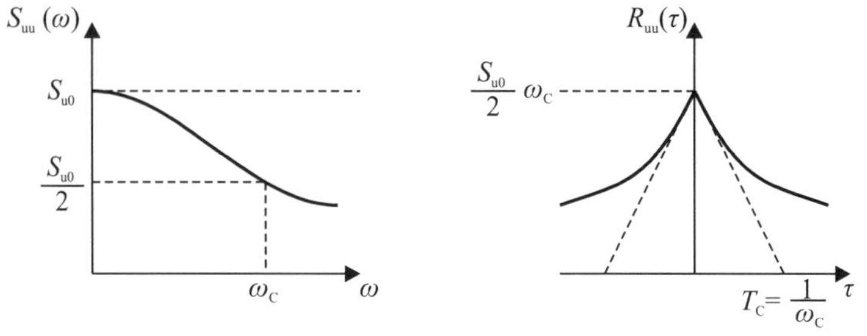

Fig. 6.2. Power spectral density and auto-correlation function of a broadband noise of first order

Norton (1962) and Cummins (1964). For this investigation, the ACF according to (6.2.18) has been approximated by a triangular pulse of width $T_C = 1/\omega_C$. The largest error of the impulse response estimate turns up at $\tau = 0$ and amounts to

$$\frac{\Delta g(0)}{g(0)} \approx \frac{1}{3\omega_C} \frac{\dot{g}(0)}{g(0)} . \tag{6.2.19}$$

For a filter of first order with the time constant T_1, one gets

$$\frac{\Delta g(0)}{g(0)} \approx -\frac{1}{3T_1\omega_C} . \tag{6.2.20}$$

If one chooses $\omega_C = 5/T_1$, then one gets for $\Delta g(0)/g(0) \approx 0.07$. The error which is caused by the finite bandwidth of the test signal gets smaller as the bandwidth is increased. However, the error which is caused by the disturbances gets larger. Thus, the bandwidth $\omega_C = 1/T_C$ may not be chosen to large.

6.2.3 Error Estimation

For the case of a *white noise*, the variance of the estimated impulse response $g(\tau)$, a non-parametric process model, shall be calculated.

The intrinsic statistical uncertainty of the CCF causes according to (6.2.12), (6.2.14), and (6.2.16) for large measurement periods $T \gg \tau$ a variance of the impulse response which is given as

$$\begin{aligned}\sigma_{g,1}^2 &= \operatorname{var} g(\tau) = \mathrm{E}\{\Delta g^2(\tau)\} \\ &\approx \frac{1}{S_{u0}^2 T}\int_{-T}^{T}\big(R_{uu}(\xi)R_{yy}(\xi) + R_{uy}(\tau+\xi)R_{uy}(\tau-\xi)\big)\mathrm{d}\xi \\ &= \frac{1}{S_{u0}^2 T}\left(R_{yy}(0) + S_{u0}\int_{-T}^{T} g(\tau+\xi)g(\tau-\xi)\mathrm{d}\xi\right) .\end{aligned} \tag{6.2.21}$$

For $\tau = 0$ and processes without a direct feedthrough (i.e. $g(0) = 0$) or for large τ with $g(\tau) \approx 0$ follows

$$\sigma_{g,1}^2 \approx \frac{1}{S_{u0}^2 T} R_{yy}(0) = \frac{1}{S_{u0}^2 T} \overline{y^2(t)} = \frac{1}{S_{u0}^2 T} \sigma_y^2 \,. \tag{6.2.22}$$

$R_{yy}(0)$ is in this case given as

$$R_{yy}(\tau) = \int_0^\infty g(t') R_{uy}(\tau + t') \mathrm{d}t' \,, \tag{6.2.23}$$

which follows from (2.3.14) in analogy to (2.3.35). With (6.2.15), one obtains

$$R_{yy}(\tau) = S_{u0} \int_0^\infty g(t') g(\tau + t') \mathrm{d}t' \tag{6.2.24}$$

and

$$R_{yy}(0) = S_{u0} \int_0^\infty g^2(t') \mathrm{d}t' \,. \tag{6.2.25}$$

Finally,

$$\sigma_{g,1}^2 \approx \frac{1}{T} \int_0^\infty g^2(t') \mathrm{d}t' \tag{6.2.26}$$

follows. The variance of the impulse response estimate caused by the variance from the intrinsic uncertainty of the CCF is thus independent from the amplitude of the test signal and depends only on the measurement time T and the quadratic area of the impulse response.

The uncertainty caused by the noise $n(t)$ follows from (6.2.13) for large measurement times as

$$\begin{aligned} \sigma_{g,2}^2 &= \operatorname{var}\big(g(\tau)\big)_{\mathrm{n}} \approx \frac{1}{S_{u0}^2 T} \int_{-T}^{T} R_{uu}(\xi) R_{nn}(\xi) \mathrm{d}\xi \\ &= \frac{1}{S_{u0}^2 T} R_{nn}(0) = \frac{1}{S_{u0}^2 T} \overline{n^2(t)} = \frac{1}{S_{u0}^2 T} \sigma_n^2 \,. \end{aligned} \tag{6.2.27}$$

If $n(t)$ is a white noise with power spectral density N_0, then

$$\sigma_{g,2}^2 = \frac{N_0}{S_{u0}} \frac{1}{T} \,. \tag{6.2.28}$$

The variance gets smaller as the signal-to-noise ratio σ_n^2 / S_{u0} or N_0 / S_{u0} decreases and as the measurement period T increases. The variance of the impulse response estimate is then given as

$$\sigma_g^2 = \sigma_{g,1}^2 + \sigma_{g,2}^2 \,. \tag{6.2.29}$$

To get a better insight for the magnitude of the two components contributing to the variance of the impulse response estimation error, the terms shall be calculated for a first order system with the transfer function

$$G(s) = \frac{y(s)}{u(s)} = \frac{K}{1 + T_1 s} \tag{6.2.30}$$

Table 6.1. Standard deviations of the impulse response, identifying a first order system with the CCF and white noise excitation as a function of the measurement time

$\frac{T}{T_1}$	50	250	1000
$\frac{\sigma_{g1}}{g_{max}}$	0.100	0.044	0.022
$\frac{\sigma_{g2}}{g_{max}}$	0.063	0.028	0.014
$\frac{\sigma_{g}}{g_{max}}$	0.118	0.052	0.026

and the impulse response

$$g(t) = \frac{K}{T_1} e^{-\frac{t}{T_1}} , \tag{6.2.31}$$

which will be excited by a white noise with power spectral density S_{u0}. The intrinsic statistical uncertainty of the CCF contributes as

$$\sigma_{g1}^2 \approx \frac{1}{T} \int_0^\infty g^2(t')\mathrm{d}t' = \frac{K^2}{2T_1 T} , \tag{6.2.32}$$

and the uncertainty due to the disturbance $n(t)$ contributes as

$$\sigma_{g2}^2 \approx \frac{\sigma_n^2}{S_{u0} T} \tag{6.2.33}$$

If the variances are normalized with respect to $g_{max} = g(0) = K/T_1$, then one obtains for the standard deviations of the relative impulse response error

$$\frac{\sigma_{g1}}{g_{max}} = \sqrt{\frac{T_1}{2T}} \tag{6.2.34}$$

$$\frac{\sigma_{g2}}{g_{max}} = \frac{\sqrt{T_1}}{K} \frac{\sigma_n}{\sqrt{S_{u0}}} \sqrt{\frac{T_1}{T}} . \tag{6.2.35}$$

Consequently, if the input to the system is now a discrete binary noise with the amplitude a and a small cycle time λ and thus with the power spectral density

$$S_{u0} \approx a^2 \lambda , \tag{6.2.36}$$

see Sect. 6.3, it follows that

$$\frac{\sigma_{g2}}{g_{max}} = \frac{1}{K} \frac{\sigma_n}{a} \sqrt{\frac{T_1}{\lambda} \frac{T_1}{T}} . \tag{6.2.37}$$

For $K = 1$, $\sigma_n/a = 0.2$, $\lambda/T_1 = 0.2$, one obtains the standard deviations of the impulse response estimate listed in Table 6.1.

This example illustrates that the contributions from the intrinsic statistical uncertainty of the CCF and the uncertainty caused by the disturbance are roughly of the same magnitude. Only for very unfavorable (small) signal-to-noise ratios σ_y/σ_u does the latter dominate. In Chap. 7 an example for an application of the identification with correlation functions (de-convolution) will be shown.

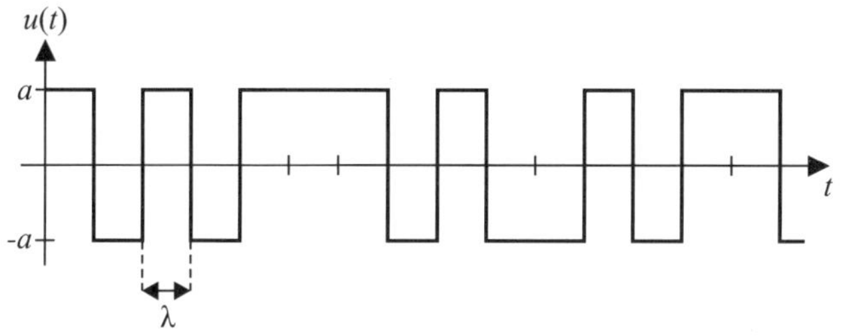

Fig. 6.3. Discrete random binary signal (DRBS)

6.2.4 Real Natural Noise as Input Signal

For some applications, it may be necessary to determine the dynamic behavior of a process without disrupting its operation by injecting artificial test signals. Then, one can only try to use the disturbances which appear during normal operation of the plant as test signals. This natural input signal must however fulfill the following properties:

- Stationarity
- The bandwidth must exceed the highest interesting frequency of the process
- The power spectral density must be larger than the disturbances acting on the output of the process to avoid extremely long measurement times
- It may not be correlated with other disturbances
- No closed-loop control, also no manual control

However, the requirements can only be satisfied in few rare cases. Thus, it is in general *advisable to inject an artificial test signal.* One can try to work with very small amplitudes as not to unnecessarily disturb the process.

6.3 Correlation Analysis of Dynamic Processes with Binary Stochastic Signals

The detailed discussion of deterministic non-periodic and periodic test signals has shown that for given constraints on the amplitude of the test signal, square signals, i.e. binary signals, have delivered the largest amplitude density (or oscillation amplitudes) and thus utilized the given amplitude range in the best way.

Continuous-Time Random Binary Signals (RBS)

A binary stochastic signal, which is also termed *random binary signal* (RBS) is characterized by the following two properties: First, the signal $u(t)$ has two states, $+a$ and $-a$, and second, the change from one state to the other can occur at any arbitrary time. Compared to other random signals with a continuous amplitude distribution, these signals have the following advantages:

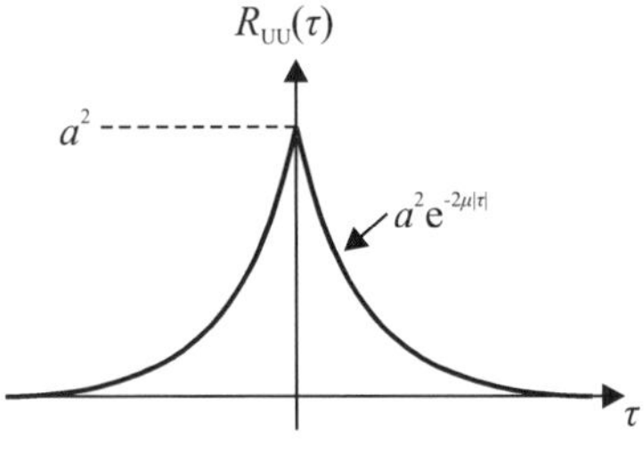

Fig. 6.4. Auto correlation function of a random binary signal (RBS)

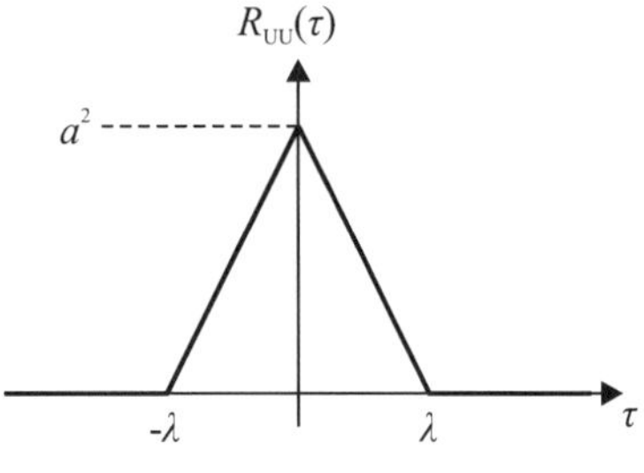

Fig. 6.5. Auto correlation function of a discrete random binary signal (DRBS)

- Simple generation
- Simple calculation of the cross-correlation function as the output of the system under investigation must only be multiplied with $+a$ or $-a$ respectively
- Largest amplitude density under constraints on the signal amplitude

While the first two points do not have much weight given the nowadays typically available computational possibilities, the third point still represents a strong argument for the use of binary test signals.

The ACF of an RBS can be determined as follows (Solodownikow, 1964): The probability of n sign changes in a given period of time Δt is Poisson-distributed with

$$P(n) = \frac{(\mu \Delta t)^n}{n!} \mathrm{e}^{-\mu \Delta t} \,, \tag{6.3.1}$$

where μ is the average number of sign changes over a given period of time.

The product $u(t)u(t+\tau)$ of an RBS signal assumes the values $+a^2$ or $-a^2$ depending on whether both values $u(t)$ and $u(t+\tau)$ have the same sign or opposite signs. Consequently, the expected value $\mathrm{E}\{u(t)u(t+\tau)\}$ is $+a^2$ for $\tau = 0$. For $\tau > 0$, the product becomes $-a^2$ if, compared to $\tau = 0$, a total of $1, 3, \ldots$ (i.e. an odd number) of sign changes took place. On the contrary, the product is $+a^2$ if $0, 2, 4, \ldots$ (i.e. an even number) of sign changes took place. Since the sign changes are random, one obtains with $\Delta t = |\tau|$,

$$\begin{aligned} \mathrm{E}\big\{u(t)u(t+\tau)\big\} &= a^2\big(P(0) + P(2) + \ldots\big) - a^2\big(P(1) + P(3) + \ldots\big) \\ &= a^2 \mathrm{e}^{-\mu|\tau|}\left(1 - \frac{\mu\tau}{1!} + \frac{(\mu\tau)^2}{2!} \pm \ldots\right) \\ &= a^2 \mathrm{e}^{-2\mu|\tau|} \,. \end{aligned} \tag{6.3.2}$$

The shape of the ACF for an RBS is shown in Fig. 6.4. Basically, it has the same shape as a broadband noise of first oder. The ACFs are identical for

$$a^2 = \frac{S_{u0}\omega_C}{2} \text{ and } \mu = \frac{\omega_C}{2} \,, \tag{6.3.3}$$

which means that μ, i.e. the average number of sign changes in a given time period, is equal to half the corner frequency.

Discrete Random Binary Signals

Due to its easy generation by means of shift registers and digital computers, the *discrete random binary signal* (DRBS) is much more widespread used in practice. Here, the sign changes take place at discrete points in time $k\lambda$ with $k = 1, 2, 3, \ldots$ where λ is the length of the time interval and is also termed cycle time, see Fig. 6.3.

The ACF,

$$R_{\mathrm{uu}}(\tau) = \lim_{T\to\infty} \int_{-T}^{T} u(t)u(t-\tau)\mathrm{d}\tau \;, \tag{6.3.4}$$

of the DRBS can be calculated as follows. For $\tau = 0$, there will only be positive products and the integral thus covers the area $2a^2T$, so that $R_{\mathrm{uu}}(0) = a^2$. For small shifts in time $|\tau| < \lambda$, there will also be negative products so that $R_{\mathrm{uu}}(\tau) < a^2$. The areas that have to be counted negatively under the integration are proportional to τ. For $|\tau| \geq \lambda$, there are as many positive as negative products, such that $R_{\mathrm{uu}} = 0$. Thus, in total,

$$R_{\mathrm{uu}}(\tau) = \begin{cases} a^2\left(1 - \frac{|\tau|}{\lambda}\right) \text{ for } |\tau| < \lambda \\ 0 \text{ for } |\tau| \geq \lambda \;. \end{cases} \tag{6.3.5}$$

The power spectral density of a DRBS follows from the Fourier transform according to (2.3.17) as the Fourier transform of a triangular pulse of the width 2λ, see (4.2.4), as

$$S_{\mathrm{uu}} = a^2\lambda\left(\frac{\sin\frac{\omega\lambda}{2}}{\frac{\omega\lambda}{2}}\right)^2 \;. \tag{6.3.6}$$

The discrete-time power spectral density is given as

$$S_{\mathrm{uu}}(z) = \sum_{\tau=-\infty}^{\infty} R_{\mathrm{uu}}(z)z^{-\tau} = R_{\mathrm{uu}}(0) = S^*_{\mathrm{uu}}(\omega) = a^2 \text{ for } 0 \leq |\omega| \leq \frac{\pi}{T_0} \;. \tag{6.3.7}$$

The ACF of a discrete random binary signal is shown in Fig. 6.5

If one equates the magnitude of this power spectral density for $\omega = \omega_{\mathrm{C}}$ with the power spectral density of the band limited noise, $S_{\mathrm{uu}}(\omega_{\mathrm{C}}) = S_{\mathrm{u0}}/2$, (6.2.18),

$$S_{\mathrm{u0}} = a^2\lambda \text{ and } \lambda \approx \frac{2.77}{\omega_{\mathrm{C}}} \tag{6.3.8}$$

follows. Thus, band limited noise and a DRBS have approximately the same power spectral density for $\omega < \omega_{\mathrm{C}}$.

As the cycle time gets smaller, the ACF approaches a small impulse with the area $a^2\lambda$. If λ is small compared to the total time constant of the subsequent plant, then one can approximate the triangular ACF by a δ-function with the same area, i.e.

$$R_{\mathrm{uu}}(\tau) = a^2\lambda\delta(\tau) \;, \tag{6.3.9}$$

and the power spectral density becomes

$$S_{u0} \approx a^2 \lambda \, . \tag{6.3.10}$$

The estimation of the impulse response can be performed according to Sect. 6.2.1 in analogy to the determination of the impulse response by white noise excitation. In this case

$$\begin{aligned} g(\tau) &= \frac{1}{a^2 \lambda} R_{uy}(\tau) \text{ for } \tau \geq \lambda \\ g(0) &= \frac{2}{a^2 \lambda} R_{uy}(0) \, . \end{aligned} \tag{6.3.11}$$

For $\tau = 0$, one has to use twice the value of the CCF, since in this case only one half of the triangular ACF ($\tau \leq 0$) is in effect. For this simplified evaluation, the error estimation as presented in Sect. 6.2.3 remains valid. For a given amplitude a, the cycle time λ may not be chosen too small, because otherwise the variance of the estimate of the impulse response might grow too large.

All of the above considerations are only valid for infinite measurement times. For finite measurement times, the correlation function and power spectral density has to be calculated for each experiment individually.

The use of a discrete random binary signal has the big advantage that the amplitude a and cycle time λ can better be matched with the process under investigation than the parameters of a stochastic signal with a continuous amplitude distribution. However, the intrinsic uncertainty in the determination of the ACF and CCF is still cumbersome. Furthermore, the experiments cannot be reproduced due to the stochastic nature of the test signal. These disadvantages can however be eliminated by the use of periodic binary test signals, which have almost the same ACF as the DRBS.

Pseudo-Random Binary Signals (PRBS) for Continuous-Time

Periodic binary signals can for example be generated by clipping N samples from a discrete random binary signal and repeating it one or multiple times. The problematic aspects of this admittedly simple approach are manifold: First of all, the random sequence cannot be parameterized easily. Secondly, the properties shown in (6.3.5) and (6.3.6) are only valid for sequences of infinite length. For sequences of finite length, the ACF and the power spectral density must be determined for each sequence individually.

Due to these impracticalities, one prefers periodic binary sequences, which have almost the same ACF as a stochastic DRBS. They are typically generated by means of shift registers with n stages whose outputs are fed back. For a shift register with n stages, the binary information 0 or 1 is passed on to the next stage as the clock input is activated. The shift register is augmented with a feedback to allow the generation of periodic sequences with a length $N > n$. Typically, two or more stages are fed-back to an XOR gate, see Fig. 6.6.

The XOR gate is a non-equal element, which outputs a zero if both input gates have equal states (i.e. $0/0$ or $1/1$) and outputs a one, if both input gates have unequal states (i.e. $0/1$ or $1/0$). If one excludes the case that all states of the shift register

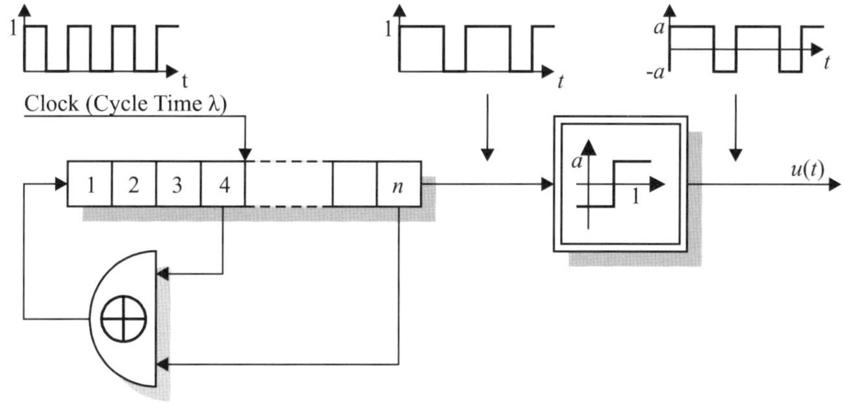

Fig. 6.6. Pseudo random binary signal generator

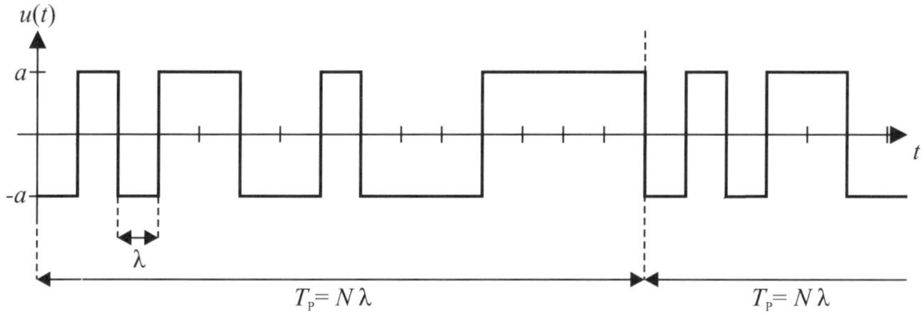

Fig. 6.7. Pseudo random binary signal generated by a shift register with 4 stages and $N = 15$

Table 6.2. Feed-back structures of shift registers for PRBS signals of maximum possible length N

No. of Stages	Feedback Law	Length
2	1 XOR 2	3
3	1 XOR 3 or 2 XOR 3	7
4	1 XOR 4 or 3 XOR 4	15
5	2 XOR 5 or 3 XOR 5	31
6	1 XOR 6 or 5 XOR 6	63
7	1 XOR 7 or 3 XOR 7 or 4 XOR 7 or 6 XOR 7	127
8	1 XOR 2 XOR 7 XOR 8	255
9	4 XOR 9 or 5 XOR 9	511
10	3 XOR 10 or 7 XOR 10	1023
11	2 XOR 11 or 9 XOR 11	2047

Remark: "XOR" denotes the XOR gate, "or" denotes different possible feedback laws resulting in the same maximum possible sample length

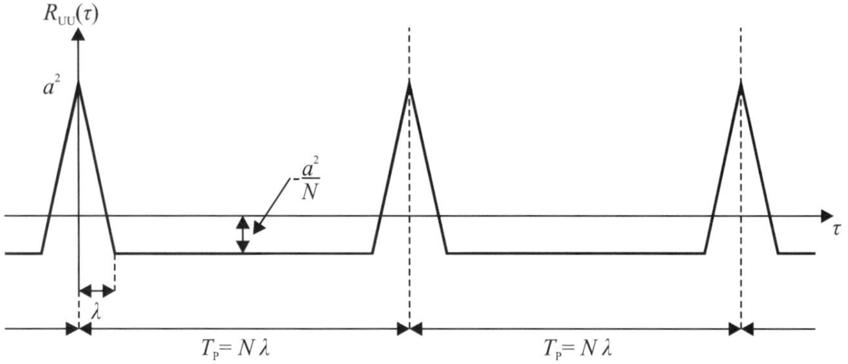

Fig. 6.8. Auto correlation function of a pseudo-random binary signal for continuous time

are zero, then one obtains a periodic signal for any arbitrary initialization of the shift register. Since for a shift register with n stages, the maximum number of different states is 2^n and since the case of zeros in all states is excluded, the maximum possible length of the signal (maximum period) sequence generated by the shift register is

$$N = 2^n - 1 \, , \tag{6.3.12}$$

because after each clock pulse, there will be a new composition of the states of the shift register. A signal sequence with the maximum possible sequence length can however only be obtained for certain feedback set-ups (Chow and Davies, 1964; Davies, 1970), see Table 6.2. If one maps the output 0 to $-a$ and the output 1 to $+a$, then one obtains the desired pseudo-random binary signal. Figure 6.7 shows the signal generated by a shift register with 4 stages.

In the following, the properties of a *continuous-time PRBS* random signal will be investigated (Davies, 1970). The clock time or cycle time is denoted with λ. Due to its periodicity, the PRBS becomes a deterministic signal. It is reproducible and can be tuned to suit individual processes. Since the ACF for this signal is exactly known, there will be no intrinsic statistical uncertainty in the determination of the ACF and CCF. The discrete ACF of a PRBS signal is given as

$$R_{uu}(\tau) = \begin{cases} a^2 \text{ for } \tau = 0 \\ -\frac{a^2}{N} \text{ for } \lambda \leq |\tau| < (N-1)\lambda \end{cases} . \tag{6.3.13}$$

Due to the uneven number N, there is an offset of $-a^2/N$ which can be neglected for large N. Reconsidering the calculation of the ACF of a DRBS, one obtains for the continuous-time ACF

$$R_{uu}(\tau) = a^2\left(1 - \frac{|\tau|}{\lambda}\left(1 + \frac{1}{N}\right)\right) \text{ for } 0 < |\tau| \leq \lambda \, . \tag{6.3.14}$$

Thus, the signal has the same triangular shape of the ACF as the DRBS. This explains the denomination as *pseudo-random*. By the periodicity, the ACF however also becomes periodic, see Fig. 6.8,

$$R_{\mathrm{uu}}(\tau) = \begin{cases} a^2\left(1 - \left(1 + \dfrac{1}{N}\right)\dfrac{|\tau - \nu N\lambda|}{\lambda}\right) & \text{for } |\tau - \nu N\lambda| \leq \lambda \\ -\dfrac{a^2}{N} & \text{for } (\lambda + \nu N\lambda) < |\tau| < (N-1)\lambda + \nu N\lambda\,. \end{cases} \tag{6.3.15}$$

If one looks at the distribution of the amplitudes, one can note the following:

- A PRBS signal contains $(N+1)/2$ times the amplitude $+a$ and $(N-1)/2$ times the amplitude $-a$. The mean is thus given as

$$\overline{u(k)} = \frac{a}{N}\,. \tag{6.3.16}$$

- If one regards the PRBS signal as a concatenation of square pulses of amplitude $+a$ and $-a$ respectively, then the frequencies of occurrence for the individual pulse lengths are given as

$$\alpha = \begin{cases} \left.\begin{array}{l} \frac{1}{2}\frac{N+1}{2} \text{ impulses of length } \lambda \\ \frac{1}{4}\frac{N+1}{2} \text{ impulses of length } 2\lambda \\ \frac{1}{8}\frac{N+1}{2} \text{ impulses of length } 3\lambda \\ \vdots \end{array}\right\} \alpha > 1 \\ \\ \left.\begin{array}{l} 1 \text{ impulse of length } (n-1)\lambda \\ 1 \text{ impulse of length } n\lambda \end{array}\right\} \alpha = 1 \end{cases}\,. \tag{6.3.17}$$

The number of pulses with amplitude $+a$ and $-a$ is always equal except that there is only one pulse of length $(n-1)\lambda$ for the amplitude $+a$ and one pulse of length $n\lambda$ for the amplitude $-a$.

The power spectral density does not have a continuous spectrum, but rather discrete spectral lines because of the periodicity. These discrete spectral lines can be calculated from the Fourier transform of the ACF as

$$S_{\mathrm{uu}}(\omega) = \int_{-\infty}^{\infty} R_{\mathrm{uu}}(\tau)\mathrm{e}^{-\mathrm{i}\omega\tau}\mathrm{d}\tau\,. \tag{6.3.18}$$

The ACF will first be developed into a Fourier series (Davies, 1970)

$$R_{\mathrm{uu}}(\tau) = \sum_{\nu=-\infty}^{\infty} c_\nu \mathrm{e}^{-\mathrm{i}\nu\omega_0\tau} \tag{6.3.19}$$

with the Fourier coefficients

$$\begin{aligned} c_\nu(\mathrm{i}\nu\omega_0) &= \frac{1}{T_{\mathrm{P}}}\int_{-\frac{T_{\mathrm{P}}}{2}}^{\frac{T_{\mathrm{P}}}{2}} R_{\mathrm{uu}}(\tau)\mathrm{e}^{-\mathrm{i}\nu\omega_0\tau}\mathrm{d}\tau \\ &= \frac{2}{T_{\mathrm{P}}}\int_{0}^{\frac{T_{\mathrm{P}}}{2}} R_{\mathrm{uu}}(\tau)\cos\nu\omega_0\tau\,\mathrm{d}\tau\,. \end{aligned} \tag{6.3.20}$$

Using (6.3.13) and (6.3.14) yields

$$\begin{aligned}c_\nu(\mathrm{i}\nu\omega_0) &= \frac{2}{T_\mathrm{P}}\int_0^\lambda a^2\left(1-\frac{\tau}{\lambda}\left(\frac{N+1}{N}\right)\right)\cos\nu\omega_0\tau\mathrm{d}\tau \\ &+ \frac{2}{T_\mathrm{P}}\int_\lambda^{\frac{T_\mathrm{P}}{2}} -\frac{a^2}{N}\cos\nu\omega_0\tau\mathrm{d}\tau \\ &= \frac{2a^2}{N\lambda}\left(\frac{1}{\nu\omega_0}\sin\nu\omega_0\lambda - \frac{N+1}{N\lambda(\nu\omega_0)^2}(\cos\nu\omega_0\lambda-1)\right. \\ &\left.-\frac{N+1}{N\nu\omega_0}\sin\nu\omega_0\lambda + \frac{1}{N\nu\omega_0}\sin\nu\omega_0\lambda\right) \\ &= \frac{2a^2(N+1)}{(N\lambda\nu\omega_0)^2}(1-\cos\nu\omega_0\lambda) \\ &= \frac{a^2(N+1)}{N^2}\left(\frac{\sin\frac{\nu\omega_0\lambda}{2}}{\frac{\nu\omega_0\lambda}{2}}\right)^2 .\end{aligned} \tag{6.3.21}$$

The Fourier coefficients are thus real-valued and the Fourier series for the ACF is given as

$$R_\mathrm{uu}(\tau) = \sum_{\nu=-\infty}^{\infty}\frac{a^2(N+1)}{N^2}\left(\frac{\sin\frac{\nu\omega_0\lambda}{2}}{\frac{\nu\omega_0\lambda}{2}}\right)^2\cos\nu\omega_0\tau . \tag{6.3.22}$$

Inserting the above term in (6.3.18) yields

$$S_\mathrm{uu}(\omega_0) = \frac{a^2(N+1)}{N^2}\sum_{\nu=-\infty}^{\infty}\left(\frac{\sin\frac{\nu\omega_0\lambda}{2}}{\frac{\nu\omega_0\lambda}{2}}\right)^2\delta(\omega-\nu\omega_0) \tag{6.3.23}$$

with

$$S_\mathrm{uu}(0) = \frac{a^2(N+1)}{N^2}\delta(\omega) . \tag{6.3.24}$$

The form factor in (6.3.23) has been denoted as Q and is given as

$$Q(\nu\omega_0) = \frac{a^2}{N}\left(1+\frac{1}{N}\right)\left(\frac{\sin\frac{\nu\omega_0\lambda}{2}}{\frac{\nu\omega_0\lambda}{2}}\right)^2 = \frac{a^2}{N}\left(1+\frac{1}{N}\right)\left(\frac{\sin\frac{\nu}{N}\pi}{\frac{\nu}{n}\pi}\right)^2 \tag{6.3.25}$$

has been graphed in Fig. 6.9 for different values $\nu = 0, 1, 2, \ldots$. The resulting discrete spectrum has the following properties:

- The spectral lines have the distance $\Delta\omega = \omega_0 = 2\pi/N\lambda$
- The lines diminish as the frequency increases with zeros at $\nu\omega_0 = 2\pi j/\lambda$ with $j = 1, 2, \ldots$
- The bandwidth of the signal can be defined by taking the first zero (Fig. 6.10) into account as

$$\omega_\mathrm{B} = \frac{2\pi}{\lambda} . \tag{6.3.26}$$

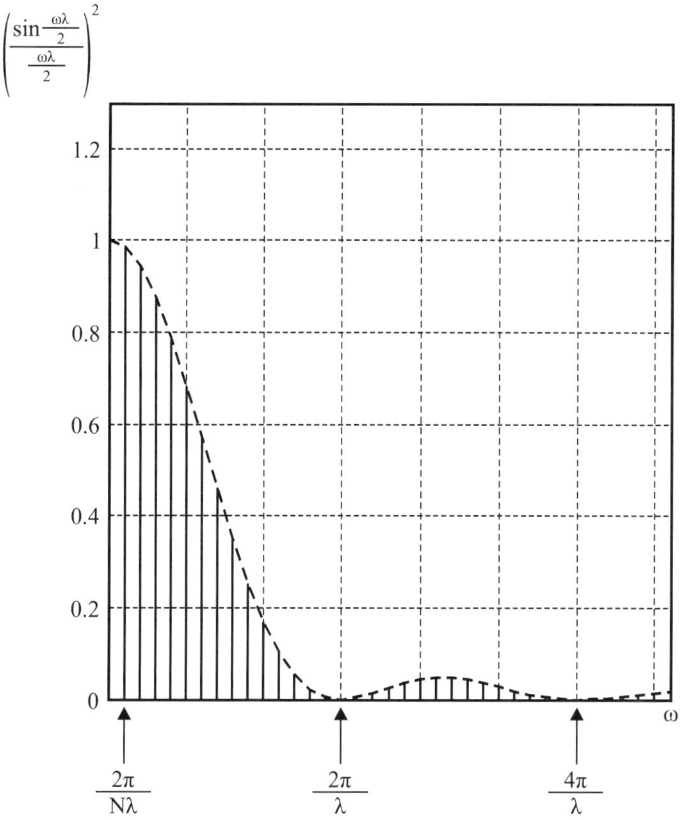

Fig. 6.9. Factor Q of the discrete power spectral density of a PRBS with period length $T_P = N\lambda$ for $N = 15$

- The cut-off frequency is with $S_{uu}(\omega_c) = S_{uu}/2$ in accordance to (6.3.8) defined as

$$\omega_c \approx \frac{2.77}{\lambda} . \qquad (6.3.27)$$

- For $\nu \approx N/3$, the factor $Q(\nu\omega_0)$ has decreased by a factor of 3 dB compared to $Q(0)$. Which means that one can assume a constant power spectral density up to the frequency

$$\omega_{3dB} = \frac{\omega_B}{3} = \frac{2\pi}{3\lambda} . \qquad (6.3.28)$$

Figure 6.10 shows the factor Q of the discrete power spectral density for changes in the cycle time λ. Figure 6.10a shows Q for the original PRBS with a cycle time λ_1 and $T_P = N_1\lambda_1$. Now the cycle time is increased under different assumptions:

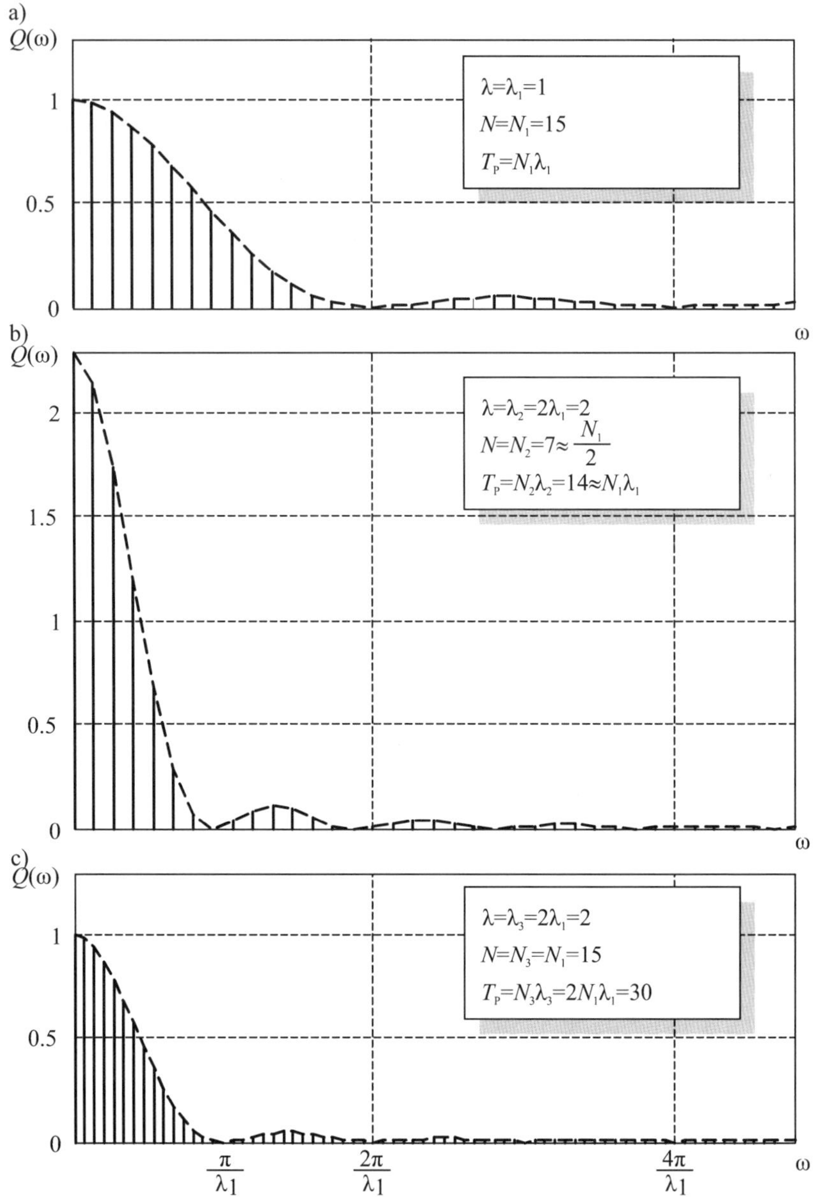

Fig. 6.10. Factor Q of the discrete power spectral density of a PRBS for different values of the period length N and cycle time λ

- Cycle time λ is increased while the period time T_P remains constant. Figure 6.10b shows Q for the case $\lambda = 2\lambda_1$, i.e.

$$Q(\nu\omega_0) = 2\frac{a^2}{N_1}\left(1 + \frac{2}{N_1}\right)\left(\frac{\sin k\omega_0\lambda_1}{k\omega_0\lambda_1}\right)^2 . \tag{6.3.29}$$

  - The distance of the spectral lines $\Delta\omega = 2\pi/T_P$ remains the same
  - The first zero is located at $\omega = 2\pi/\lambda_2 = \pi/\lambda_1$, i.e. is reached at lower frequencies
  - There are less, but higher spectral lines (The total power remains approximately constant)
- The cycle time λ is increased at a constant cycle length N: Figure 6.10c shows for $\lambda = \lambda_1$

$$Q(\nu\omega_0) = \frac{a^2}{N_1}\left(1 + \frac{1}{N_1}\right)\left(\frac{\sin\frac{\nu\omega_0\lambda_1}{2}}{\frac{\nu\omega_0\lambda_1}{2}}\right)^2 . \tag{6.3.30}$$

  - The distance of the spectral lines $\Delta\omega = 2\pi/2N_1\lambda_1 = \pi/N_1\lambda_1$ gets smaller
  - The first zero is located at $\omega = 2\pi/\lambda_2 = \pi/\lambda_1$, i.e. is reached at lower frequencies
  - There are more, but equally high spectral lines

This parameter study illustrates in both cases that a stronger excitation of the lower frequent dynamics can be obtained by increasing the cycle time λ. For a large period time T_P with respect to the transient time and thus for a $\lambda \ll N$ and a large N, the ACF of the PRBS approaches the ACF of the DRBS (6.3.5), where also the DC value $-a^2/N$ gets negligibly small. Then the impulse response can be determined according to the simplifications derived for the DRBS. If in addition the cycle time λ is small compared to the total time constant of the process, then

$$\Phi_{uu}(\tau) \approx a^2\lambda\delta(\tau) \tag{6.3.31}$$

with the power spectral density

$$S_{u0} \approx a^2\lambda \tag{6.3.32}$$

$$g(\tau) = \frac{1}{a^2\lambda}R_{uy}(\tau) , \tag{6.3.33}$$

and the evaluation can be carried out in analogy to the case with a white noise input driving the system, see (6.2.16). The de-convolution with discrete-time signals is covered in Sect. 7.2.1.

For this case, the error estimation for a disturbance $n(t)$ can be carried out according to Sect. 6.2.3. One has to bear in mind however that due to its deterministic nature, the CCF does not have an intrinsic uncertainty (6.1.8) and thus the term σ_{g1}, (6.2.32) gets zero.

For the choice of the *free parameters* a, λ, and N of a PRBS, the following rules of thumb can be helpful:

- The amplitude a shall always be chosen as large as possible so that the corruption of the output signal by a given disturbance $n(t)$ gets as small as possible. One however has to take the process limits for the input $u(t)$ and output $y(t)$ into account
- The cycle time λ should be chosen as large as possible, so that for a given amplitude a the power spectral density $S_{\mathrm{uu}}(\omega)$ gets as large as possible. If the impulse response is determined by the simplified approach in (6.2.16), then the evaluation becomes erroneous and an error according to (6.2.20) is introduced. Thus the cut-off frequency of the test signal $\omega_{\mathrm{c}} = 1/\lambda$ may not be too small and consequently λ may not be chosen too large. It is thus suggested to chose $\lambda \leq T_i/5$ where T_i denotes the smallest interesting time constant of the process
- The period time $T_{\mathrm{P}} = N\lambda$ may not be smaller than the transient time T_{95} of the system under investigation so that there is no overlap of the impulse responses. A guiding value is $T_{\mathrm{P}} \approx 1.5 T_{95}$.

The number M of the periods of the PRBS signal is determined by the total required measurement time $T = MT_{\mathrm{P}} = MN\lambda$ which for given signal parameters a, λ and, N depends mainly on the signal-to-noise ratio, see Sect. 6.2.3.

For a *discrete-time PRBS*, the choice of the cycle time λ is coupled to the choice of the sample time T_0 as λ can only be an integer multiple of the sample time, i.e.

$$\lambda = \mu T_0 \text{ with } \mu = 1, 2, \ldots \, . \tag{6.3.34}$$

For $\mu = 1$ and large N, the properties of the discrete-time PRBS approach those of a discrete white noise. If one increases λ by a choice of $\mu = 2, 3, \ldots$, then the excitation of the lower frequencies is increased for both $N = \text{const}$ as well as $T_{\mathrm{P}} = \text{const}$. Pintelon and Schoukens (2001) pointed out that a PRBS signal is not ideally suited for the determination of the frequency response function as it never has a period length of 2^n, which would be ideally suited for the Fast Fourier Transform. A further side-note should be made on the applicability of PRBS signals for non-linear systems. A PRBS signal can in general not be used to detect non-linearities at the input, i.e. is unsuitable for e.g. a Hammerstein model, see Chap. 18.

A different way of generating RBS signals is described by Ljung (1999) as follows: A zero-mean white noise Gaussian signal is first filtered by a form filter to generate a test signal with the appropriate frequency content and then, just the sign of resulting signal is retained and scaled accordingly to the requirements on the test signal amplitude. This non-linear operation however changes the frequency content, so that the spectrum of the resulting RBS signal must be analyzed to ensure the suitability of the signal.

Generalized Random Binary Signals (GRBS) for Discrete Time

The *generalized random binary signal (GRBS)* (Tulleken, 1990) is a generalized form of the random binary signal. For a discrete random binary signal, one assumes that the change of the amplitude is random, i.e. at each time step k the probability that the signal keeps the same amplitude is 50% as is the probability that the signal

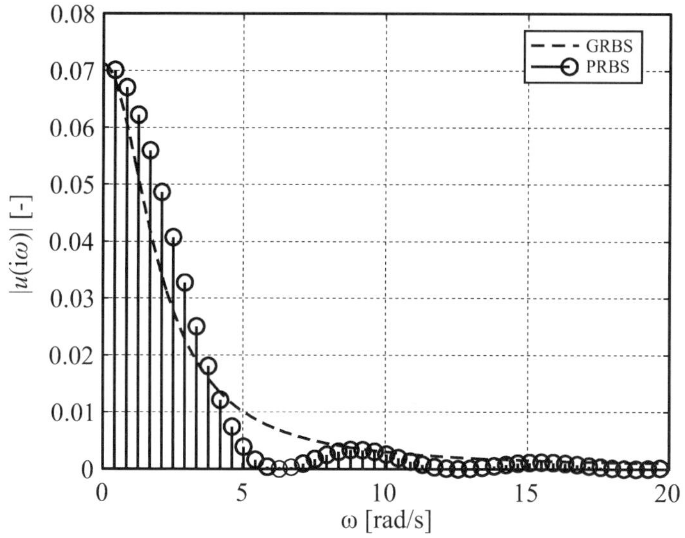

Fig. 6.11. Comparison of the frequency spectra of an PRBS signal ($\mu T_0 = 1\,\mathrm{s}$, $a = 1$, $N = 15$) and a GRBS signal ($T_0 = 0.1\mathrm{s}$, $p \approx 0.9$) (Zimmerschied, 2002)

changes its amplitude. For such a signal, it is not very likely that a signal level is held for a long time.

Therefore, the generalized random binary signal introduces a different value for the probability of a change and the probability that the signal value is held respectively. In the following, the probability that the signal value is held is denoted as p, such that

$$P\big(u(k) = u(k-1)\big) = p \tag{6.3.35}$$

$$P\big(u(k) \neq u(k-1)\big) = (1-p)\,. \tag{6.3.36}$$

The *expected impulse length* is then given as

$$\mathrm{E}\{T_\mathrm{P}\} = \sum_{k=1}^{\infty}(kT_0)p^{k-1}(1-p) = \frac{T_0}{1-p}\,. \tag{6.3.37}$$

One can see that longer impulse lengths appear more often if p is increased. An important difference to the PRBS signal is however that also an impulse of length T_0 can always appear. For a PRBS, the minimum impulse length was given by μT_0. The auto-correlation function of a GRBS of infinite length is given as

$$R_\mathrm{uu}(\tau) = a^2(2p-1)^{|\tau|}\,. \tag{6.3.38}$$

The power spectral density is then given as

$$S_\mathrm{uu} = \frac{(1-\beta)^2 T_0}{1 - 2\beta\cos\omega T_0 + \beta^2} \text{ with } \beta = 2p-1\,. \tag{6.3.39}$$

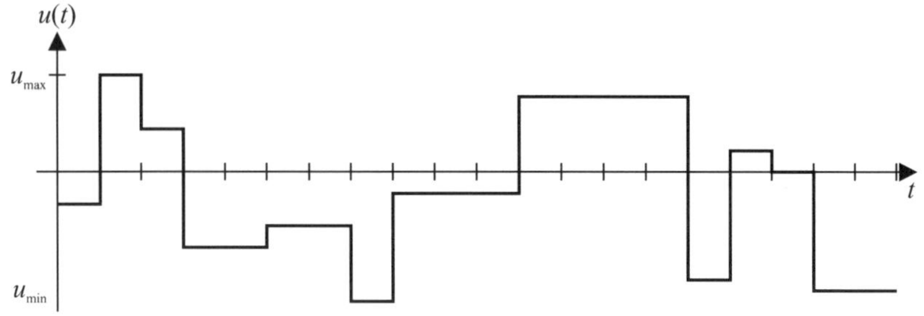

Fig. 6.12. Amplitude-modulated pseudo-random binary signal (APRBS)

A comparison between the frequency spectra of a PRBS signal and a GRBS signal is shown in Fig. 6.11. One can see that the excitation using the GRBS signal covers a wider frequency range than the PRBS signal, which has zeros in the frequency spectrum at comparably low frequencies due to its periodicity (Zimmerschied, 2002).

Amplitude-modulated PRBS and GRBS

For the identification of non-linear systems, the PRBS (see Sect. 6.3) and GRBS (see Sect. 6.3) are not well suited as they do only have two different values of $u(k)$ and hence do not excite non-linear systems over their full input range $u(k) \in (u_{\min} \dots u_{\max})$. Therefore, one must use test signals that do not only vary the frequency of the excitation, but also the amplitude. This means that now there are much more design parameters that have to be taken into account, such as (Doyle et al, 2002):

- length of the input sequence N
- range of input amplitudes $u(k) \in (u^*_{\min} \dots u^*_{\max})$
- distribution of input amplitudes $u(k)$
- frequency spectrum or shape of the resulting signal respectively

As a basis for the development of non-linear excitation signals, one can use the PRBS signal or GRBS signal. They have proven well in many applications and their properties are well known. A simple and straightforward extension to the design of input sequences suitable for non-linear systems is as follows:

One uses the PRBS or GRBS signal to determine the length of each impulse. The length of each impulse is then taken from a set of predefined amplitudes. Here, one can either split the input range from $u_{\min}$ to $u_{\max}$ equidistantly and use each of these amplitude levels exactly one time. An alternative is to use a random number generator to randomly chose values of $u(k)$ from the interval $u_{\min}$ to $u_{\max}$. Although the distribution of amplitudes and frequencies of the resulting test signal is not equally distributed over the entire operating range, this does not present a severe drawback for sufficiently long test signal sequences, see Fig. 6.12.

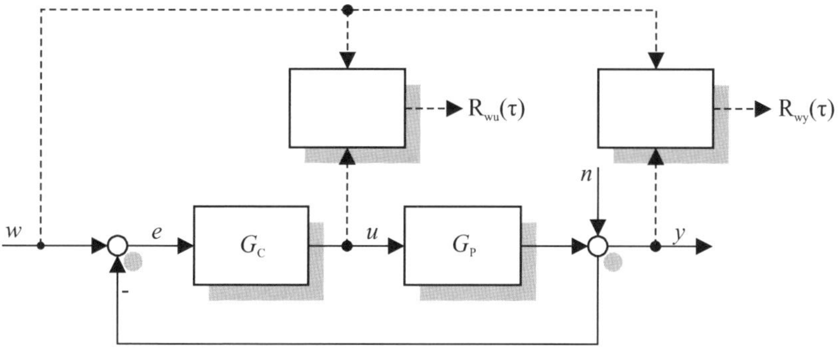

Fig. 6.13. Block diagram for the correlation analysis in closed-loop

The auto-correlation functions of an APRBS or AGRBS signal are very similar to the auto-correlation functions of the PRBS and GRBS signal. According to (Pearson, 1999), the auto-correlation function of an AGRBS is given as

$$R_{\mathrm{uu}}(\tau) = \mu^2 + \sigma^2 p(\tau) \,. \tag{6.3.40}$$

Here μ is the mean and σ^2 the variance of the signal. The second term correlates to the auto-correlation function of the GRBS signal, compare (6.3.38) (Zimmerschied, 2002). An example of an amplitude modulated pseudo-random binary signal is shown in Fig. 6.12.

6.4 Correlation Analysis in Closed-Loop

If a disturbed process G_{P} as shown in Fig. 6.13 is operated in closed loop, then the input $u(t)$ to the process is correlated with the disturbance $n(t)$ by means of the feedback loop and the controller G_{C}. In this case, it is typically impossible to determine the dynamic behavior of the process from the cross-correlation function $R_{\mathrm{uy}}(\tau)$. This is independent of the location where the test signal is injected into the control loop. If one nevertheless tries to identify a model in the previously described way, one will typically have non-zero values for negative times in the impulse response (Godman and Reswick, 1956; Rödder, 1973, 1974).

The dynamic behavior of the process can be identified, if its input and output are correlated with an external test signal, e.g. $w(t)$. Then, one obtains

$$R_{\mathrm{wu}}(\tau) = \mathrm{E}\{w(t-\tau)u_0(t)\} + \underbrace{\mathrm{E}\{w(t-\tau)u_{\mathrm{n}}(t)\}}_{=0} \,, \tag{6.4.1}$$

where $u_0(t)$ denotes the part of $u(t)$ that can be attributed to $w(t)$ and $u_{\mathrm{n}}(t)$ denotes the part evoked by the reaction of the controller to $n(t)$. Since the disturbance is not correlated with the set-point, one obtains

$$R_{\mathrm{wy}}(\tau) = \mathrm{E}\{w(t-\tau)y_0(t)\} + \underbrace{\mathrm{E}\{w(t-\tau)n(t)\}}_{=0} \; . \tag{6.4.2}$$

From $R_{\mathrm{wu}}(\tau)$ one obtains the impulse response $g_{\mathrm{wu}}(\tau)$ and, by application of the Fourier transformation, the transfer function

$$G_{\mathrm{uw}} = \frac{u(s)}{w(s)} = \frac{G_{\mathrm{C}}}{1 + G_{\mathrm{C}}G_{\mathrm{P}}} \; , \tag{6.4.3}$$

and in analogy, one obtains the impulse response $g_{\mathrm{wy}}(\tau)$ from the correlation function $R_{\mathrm{wy}}(\tau)$ and from there

$$G_{\mathrm{wy}} = \frac{G_{\mathrm{C}}G_{\mathrm{P}}}{1 + G_{\mathrm{C}}G_{\mathrm{P}}} \; . \tag{6.4.4}$$

From these two equations, one can determine G_{P} by

$$G_{\mathrm{P}} = \frac{G_{\mathrm{wy}}}{G_{\mathrm{wu}}} \; . \tag{6.4.5}$$

If the process however is undisturbed between $u(t)$ and $y(t)$, it can identified error-free if the test signal is injected in an appropriate place into the control loop (Rödder, 1974).

6.5 Summary

The correlation analysis with stochastic or pseudo-stochastic test signals allows the estimation of non-parametric models for linear processes. They can be used for on-line identification in real-time and deliver the impulse response of the process if the process is driven by a colored or white input signal. For a white noise input signal, the impulse response is directly proportional to the cross-correlation function. Since the cross-correlation of stationary signals automatically separates the wanted signal from the noise, one can apply these methods even in the presence of large disturbances and unfavorable signal-to-noise ratios. The only requirement is that a sufficiently long measurement time is allowed. While the use of real natural noise as a test signal is possible under certain conditions, it is seldom advisable to do so. In practice, it is in general better to use an artificial test signal. Pseudo-random binary signals (PRBS) have found wide-spread use, since they can easily be generated, have an easy to determine and favorable auto-correlation function, and allow to identify impulse responses directly. In addition to pseudo-random binary signals, also generalized binary random signals have been introduced in this chapter. They have a wider frequency range that is excited compared to the pseudo-random binary signals, which have zeros in the amplitude spectrum at comparably low frequencies. For the excitation of non-linear systems, binary signals are not well suited if also the non-linearities shall be identified, because a binary signal does not cover the full input range $u \in (u_{\min}, u_{\max})$.

Problems

6.1. Auto-Correlation Function
Describe the shape of the auto-correlation function of a white noise and a broadband noise of first order with the corner frequency $f_C = 1\,\text{Hz}$.

6.2. Cross-Correlation Function
Determine the cross-correlation function of input and output for a first order process with the transfer function $G(s) = K/(1 + Ts)$ with $K = 1$ and $T = 0.2\,\text{s}$ and the input signals from Problem 6.1

6.3. Discrete Random Binary Signal
Determine the auto-correlation function of a discrete random binary signal with $a = 2\,\text{V}$ and $\lambda = 2\,\text{s}$.

6.4. Correlation Analysis
Describe, how one could measure the frequency response of a process G_P if the control loop may not be opened, i.e. the system must always be operated in closed-loop. However, it is possible to choose the setpoint $w(t)$ freely and to measure the actuated variable $u(t)$ as well as the (disturbed) output $y(t)$. Software is available to determine correlation functions and to calculate the Fourier transform. Describe the overall approach, which correlation functions have to be determined and how the frequency response in continuous-time can be determined.

6.5. Discrete Random Binary Signal and Pseudo-Random Binary Signal
Discuss the differences between a discrete random binary signal and a pseudo-random binary signal under the following aspects: Reproducible generation, mean, auto-correlation function, power spectral density, and periodicity.

References

Bendat JS, Piersol AG (2010) Random data: Analysis and measurement procedures, 4th edn. Wiley-Interscience, New York

Chow P, Davies AC (1964) The synthesis of cyclic code generators. Electron Eng 36:253–259

Cummins JC (1964) A note on errors and signal to noise ratio of binary cross-correlation measurements of system impulse response. Atom Energy Establishment, Winfrith (AEEW), Dorset

Davies WDT (1970) System identification for self-adaptive control. Wiley-Interscience, London

Doyle FJ, Pearson RK, Ogunnaike BA (2002) Identification and control using Volterra models. Communications and Control Engineering, Springer, London

Eykhoff P (1964) Process parameter estimation. Progress in Control Engineering 2:162–206

Godman TP, Reswick JB (1956) Determination of th system characteristics from normal operation modes. Trans ASME 78:259–271

Hänsler E (2001) Statistische Signale: Grundlagen und Anwendungen. Springer, Berlin

Hughes M, Norton A (1962) The measurement of control system characteristics by means of cross-correlator. Proc IEE Part B 109(43):77–83

Ljung L (1999) System identification: Theory for the user, 2nd edn. Prentice Hall Information and System Sciences Series, Prentice Hall PTR, Upper Saddle River, NJ

Papoulis A (1962) The Fourier integral and its applications. McGraw Hill, New York

Pearson RK (1999) Discrete-time dynamic models. Topics in chemical engineering, Oxford University Press, New York

Pintelon R, Schoukens J (2001) System identification: A frequency domain approach. IEEE Press, Piscataway, NJ

Rödder P (1973) Systemidentifikation mit stochastischen Signalen im geschlossenen Regelkreis – Verfahren der Fehlerabschätzung. Dissertation. RWTH Aachen, Aachen

Rödder P (1974) Nichtbeachtung der Rückkopplung bei der Systemanalyse mit stochastischen Signalen. Regelungstechnik 22:154–156

Sage AP, Melsa JL (1971) System identification. Academic Press, New York

Solodownikow WW (1964) Einführung in die statistische Dynamik linearer Regelsysteme. Oldenbourg Verlag, München

Tulleken HJAF (1990) Generalized binary noise test-signal concept for improved identification-experiment design. Automatica 26(1):37–49

Zimmerschied R (2002) Entwurf von Anregungssignalen für die Identifikation nichtlinearer dynamischer Prozesse. Diplomarbeit. Institut für Regelungstechnik, TU Darmstadt, Darmstadt

7

Correlation Analysis with Discrete Time Models

Based on the fundamentals of the correlation analysis as outlined in Chap. 6 for the continuous-time case, the discrete-time case will now be examined more closely in this chapter. This case is required for the implementation on digital computers. The difference in the treatment of continuous-time and discrete-time signals is rather small as it only affects the calculation of the correlation functions, where basically the continuous-time integration must be replaced by the summation of discrete values. In Sect. 7.1, the estimation of the correlation function is treated again. This time however, it is closely analyzed for the case of signal samples of finite length and the subsequently appearing intrinsic estimation uncertainty. Also, a fast implementation of the calculation of the correlation function is presented in this section. An attractive feature for online applications is to estimate the correlation functions recursively. Section 7.2 covers the correlation analysis of sampled linear dynamic systems in the discrete-time case. *Binary test signals*, which are well suited as test signals for the de-convolution have already been treated in the preceding chapter in Sect. 6.3 and will only shortly be discussed.

7.1 Estimation of the Correlation Function

7.1.1 Auto-Correlation Function

The auto-correlation function of a discrete-time stationary stochastic process $x(k)$ with the discrete-time $k = t/T_0 = 0, 1, 2, \ldots$ and T_0 being the sample time, is according to (2.4.3) given as

$$R_{xx}(\tau) = \mathrm{E}\{x(k)x(k+\tau)\} = \lim_{N\to\infty} \frac{1}{N} \sum_{k=1}^{N} x(k)x(k+\tau)\,. \qquad (7.1.1)$$

In this simple case, it has been assumed that the measurement period is infinitely long. Recorded signals however, are always of limited length. It is henceforth interesting to determine the possible accuracy of the estimate of the auto-correlation

R. Isermann, M. Münchhof, *Identification of Dynamic Systems*,
DOI 10.1007/978-3-540-78879-9_7, © Springer-Verlag Berlin Heidelberg 2011

function of a signal $x(k)$ of an individual sample function $\{x(k)\}$ based on a series of datapoints of finite length N and constant sample time T_0. From (7.1.1), the estimate can first be written as

$$\hat{R}_{xx}(\tau) \approx R_{xx}^N(\tau) = \frac{1}{N}\sum_{k=0}^{N-1} x(k)x(k+\tau)\;. \tag{7.1.2}$$

If $x(k)$ has however only been sampled in the finite interval $0 \le k \le N-1$, then

$$\hat{R}_{xx}(\tau) = \frac{1}{N}\sum_{k=0}^{N-1-|\tau|} x(k)x(k+|\tau|),\ \text{for } 0 \le |\tau| \le N-1 \tag{7.1.3}$$

since $x(k) = 0$ for $k < 0$ and $k > N-1$ or $x(k+|\tau|)$ for $k > N-1-|\tau|$ respectively. In this case, only $N - |\tau|$ product terms exist. Thus, one could use the alternative estimate

$$\hat{R}'_{xx}(\tau) = \frac{1}{N-|\tau|}\sum_{k=0}^{N-1-|\tau|} x(k)x(k+|\tau|),\ \text{for } 0 \le |\tau| \le N-1\;, \tag{7.1.4}$$

where one divides by the effective number of terms $N - |\tau|$.

Now, the question arises, which of the two estimates is more favorable. For this investigation, it will be assumed that $\mathrm{E}\{x(k)\} = 0$. The expected values of the two estimates can then be determined for the interval $0 \le |\tau| \le N-1$

$$\begin{aligned}\mathrm{E}\big\{\hat{R}_{xx}(\tau)\big\} &= \frac{1}{N}\sum_{k=0}^{N-1-|\tau|} \mathrm{E}\big\{x(k)x(k+|\tau|)\big\} = \frac{1}{N}\sum_{k=0}^{N-1-|\tau|} R_{xx}(\tau) \\ &= \left(1-\frac{|\tau|}{N}\right)R_{xx}(\tau) = R_{xx}(\tau) + b(\tau)\end{aligned} \tag{7.1.5}$$

and

$$\mathrm{E}\big\{\hat{R}'_{xx}(\tau)\big\} = R_{xx}(\tau)\;. \tag{7.1.6}$$

It can be seen from (7.1.2) that the estimate has a systematic error $b(\tau)$ (bias) for a finite sample length N, which however vanishes for $N \to \infty$ and $|\tau| \ll N$,

$$\lim_{N\to\infty} \mathrm{E}\big\{\hat{R}_{xx}(\tau)\big\} = R_{xx}(\tau) \text{ for } |\tau| \ll N\;. \tag{7.1.7}$$

Hence, the estimate is consistent. (7.1.4) however is also unbiased for finite measurement periods N.

For a signal with Gaussian distribution, the variance of the estimate (7.1.2) follows from the variance of the cross-correlation function, which is covered in the following section, as

$$\begin{aligned}\lim_{N\to\infty} \operatorname{var} \hat{R}_{xx}(\tau) &= \lim_{N\to\infty} \mathrm{E}\Big\{\big(\hat{R}_{xx}(\tau) - R_{xx}(\tau)\big)^2\Big\} \\ &= \lim_{N\to\infty} \sum_{\nu=-(N-1)}^{N-1} \big(R_{xx}^2(\nu) + R_{xx}(\nu+\tau)R_{xx}(\nu-\tau)\big)\;.\end{aligned} \tag{7.1.8}$$

This represents the intrinsic uncertainty of the estimated auto-correlation function, see Chap. 6. If the auto-correlation function is finite and $\mathrm{E}\{x(k)\} = 0$, then the variance diminishes as $N \to \infty$. The estimation of the ACF according to (7.1.2) is thus consistent in the mean square. From (7.1.8), one can derive the following special cases for large N:

- $\tau = 0$:

$$\operatorname{var} \hat{R}_{xx}(\tau) \approx \frac{2}{N} \sum_{\xi=-(N-1)}^{N-1} R_{xx}^2(\xi) \,. \tag{7.1.9}$$

If $x(k)$ is a white noise, then

$$\operatorname{var} \hat{R}_{xx}(0) \approx \frac{2}{N} R_{xx}^2(0) = \left(\overline{x^2(k)}\right)^2 \,. \tag{7.1.10}$$

- Large τ: It holds that

$$R_{xx}^2(\nu) \gg R_{xx}(\nu+\tau) R_{xx}(\nu-\tau) \text{ since } R_{xx}(\tau) \approx 0 \,. \tag{7.1.11}$$

Thus, one obtains

$$\operatorname{var} \hat{R}_{xx}(\tau) \approx \frac{1}{N} \sum_{\nu=-(N-1)}^{N-1} R_{xx}^2(\nu) \,. \tag{7.1.12}$$

From (7.1.10) and (7.1.11), it can furthermore be shown that

$$\operatorname{var} \hat{R}_{xx}(0) \approx 2 \operatorname{var} \hat{R}_{xx}(\tau) \,. \tag{7.1.13}$$

The variance for large τ is thus only half as big as the one for $\tau = 0$.

For the biased estimate, (7.1.2) one has to replace the term N by $N - |\tau|$ in (7.1.9) and thus it follows for finite N that

$$\operatorname{var} \hat{R}'_{xx}(\tau) = \frac{N}{N - |\tau|} \operatorname{var}\big(\hat{R}_{xx}(\tau)\big) \,. \tag{7.1.14}$$

The unbiased estimate therefore always delivers estimates with a larger variance for $|\tau| > 0$. For $|\tau| \to N$, the variance approaches infinity. Thus, one typically uses the biased estimate in (7.1.2). Table 7.1 summarizes the main features of the two estimates.

Since $\mathrm{E}\{x(k)\} = 0$ has been assumed, all equations can similarly be applied to the estimation of the *auto-covariance function* $C_{xx}(\tau)$. For additionally superimposed disturbances $n(t)$, the considerations from Sect. 6.1 are equally applicable.

7.1.2 Cross-Correlation Function

The cross-correlation function of two discrete-time stationary processes is according to (2.4.4) given as

Table 7.1. Properties of the estimates of the auto-correlation function

Estimate	Bias for Finite N	Variance for Finite N	Bias for $N \to \infty$
$\hat{R}_{\mathrm{xx}}(\tau)$	$-\frac{\lvert\tau\rvert}{N} R_{\mathrm{xx}}(\tau)$	$\mathrm{var}\big(\hat{R}_{\mathrm{xx}}(\tau)\big)$	0
$\hat{R}'_{\mathrm{xx}}(\tau)$	0	$\frac{N}{N-\lvert\tau\rvert} \mathrm{var}\big(\hat{R}_{\mathrm{xx}}(\tau)\big)$	0

$$R_{\mathrm{xy}}(\tau) = \mathrm{E}\big\{x(k)y(k+\tau)\big\} = \lim_{N\to\infty} \frac{1}{N} \sum_{k=0}^{N-1} x(k)y(k+\tau) = \mathrm{E}\big\{x(k-\tau)y(k)\big\} \,. \tag{7.1.15}$$

As an estimate for the cross-correlation,

$$\hat{R}_{\mathrm{xy}}(\tau) \approx R^N_{\mathrm{xy}}(\tau) = \frac{1}{N} \sum_{k=0}^{N-1} x(k)y(k+\tau) \tag{7.1.16}$$

will be introduced according to (7.1.2). For $-(N-1) \leq \tau \leq (N-1)$ follows

$$\hat{R}_{\mathrm{xy}}(\tau) = \begin{cases} \dfrac{1}{N} \displaystyle\sum_{k=0}^{N-1-\tau} x(k)y(k+\tau) \text{ for } 0 \leq \tau \leq N-1 \\ \dfrac{1}{N} \displaystyle\sum_{k=-\tau}^{N-1} x(k)y(k+\tau) \text{ for } -(N-1) \leq \tau < 0 \end{cases} \tag{7.1.17}$$

since $y(k) = 0$ and $x(k) = 0$ for $k < 0$ and $k > N-1$. The expected value of this estimate is given as

$$\mathrm{E}\big\{\hat{R}_{\mathrm{xy}}(\tau)\big\} = \left(1 - \frac{|\tau|}{N}\right) R_{\mathrm{xy}}(\tau) \,, \tag{7.1.18}$$

compare (7.1.5).

For finite measurement times N, the estimate is thus biased, the bias vanishes only for $N \to \infty$,

$$\lim_{N\to\infty} \mathrm{E}\big\{\hat{R}_{\mathrm{xy}}(\tau)\big\} = R_{\mathrm{xy}}(\tau) \,. \tag{7.1.19}$$

If one would divide by $N - |\tau|$ instead of N in (7.1.17), then the cross-correlation function estimate would also be bias-free for finite measurement periods, but the variance would increase as was the case for the auto-correlation function.

Now, the variance of (7.1.17) will be determined. The first calculation of the variance (but for the auto-correlation function) can already be found in (Bartlett, 1946).

According to the definition of the cross-correlation function,

$$\mathrm{var}\, \hat{R}_{\mathrm{xy}}(\tau) = \mathrm{E}\Big\{\big(\hat{R}_{\mathrm{xy}}(\tau) - R_{\mathrm{xy}}(\tau)\big)^2\Big\} = \mathrm{E}\Big\{\big(\hat{R}_{\mathrm{xy}}(\tau)\big)^2\Big\} - R^2_{\mathrm{xy}}(\tau) \,, \tag{7.1.20}$$

where the result in (7.1.19) has been exploited. Furthermore, one can rewrite the expected value of the estimate as

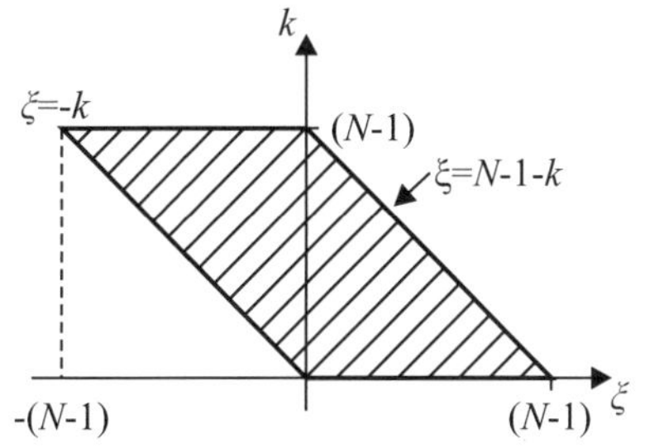

Fig. 7.1. Summation area of (7.1.25)

$$\mathrm{E}\left\{\left(\hat{R}_{\mathrm{xy}}(\tau)\right)^2\right\} = \frac{1}{N^2}\sum_{k=0}^{N-1}\sum_{k'=0}^{N-1}\mathrm{E}\left\{x(k)y(k+\tau)x(k')y(k'+\tau)\right\}\,. \tag{7.1.21}$$

To simplify the notation, the boundaries of (7.1.15) will be used instead of those in (7.1.17). It is now assumed that both $x(k)$ and $y(k)$ have a Gaussian distribution. In this case, (7.1.21) contains four random variables z_1, z_2, z_3, and z_4, for which, according to Bendat and Piersol (2010), one can write

$$\begin{aligned}\mathrm{E}\left\{z_1, z_2, z_3, z_4\right\} &= \mathrm{E}\left\{z_1, z_2\right\}\mathrm{E}\left\{z_3, z_4\right\} + \mathrm{E}\left\{z_1, z_3\right\}\mathrm{E}\left\{z_2, z_4\right\}\\ &\quad - \mathrm{E}\left\{z_1, z_4\right\}\mathrm{E}\left\{z_2, z_3\right\} - 2\overline{z_1}\,\overline{z_2}\,\overline{z_3}\,\overline{z_4}\,.\end{aligned} \tag{7.1.22}$$

If $\mathrm{E}\{x(k)\} = 0$ or $\mathrm{E}\{y(k)\} = 0$, then

$$\begin{aligned}&\mathrm{E}\left\{x(k)y(k+\tau)x(k')y(k'+\tau)\right\}\\ &= R_{\mathrm{xy}}^2(\tau) + R_{\mathrm{xx}}(k'-k)R_{\mathrm{yy}}(k'-k) + R_{\mathrm{xy}}(k'-k+\tau)R_{\mathrm{yx}}(k'-k-\tau)\,.\end{aligned} \tag{7.1.23}$$

Thus, by inserting (7.1.23) into (7.1.21) follows that

$$\begin{aligned}\mathrm{var}\,\hat{R}_{\mathrm{xy}}(\tau) = \frac{1}{N^2}\sum_{k=0}^{N-1}\sum_{k'=0}^{N-1}\Big(&R_{\mathrm{xx}}(k'-k)R_{\mathrm{yy}}(k'-k)\\ &+ R_{\mathrm{xy}}(k'-k+\tau)R_{\mathrm{yx}}(k'-k-\tau)\Big)\,.\end{aligned} \tag{7.1.24}$$

Now, $k' - k = \xi$, which leads to

$$\mathrm{var}\,\hat{R}_{\mathrm{xy}}(\tau) = \frac{1}{N^2}\sum_{k=0}^{N-1}\sum_{\xi=-k}^{N-1-k}\Big(R_{\mathrm{xx}}(\xi)R_{\mathrm{yy}}(\xi) + R_{\mathrm{xy}}(\xi+\tau)R_{\mathrm{yx}}(\xi-\tau)\Big)\,. \tag{7.1.25}$$

The addend shall be denoted with $F(\xi)$. Its summation area is shown in Fig. 7.1. After exchanging the order of the sums, one obtains

$$
\begin{aligned}
\sum_{k=0}^{N-1}\sum_{\xi=-k}^{N-1-k} F(\xi) &= \underbrace{\sum_{\xi=0}^{N-1} F(\xi) \sum_{k=0}^{N-1-\xi} 1}_{\text{Right Triangle}} + \underbrace{\sum_{\xi=-(N-1)}^{0} F(\xi) \sum_{k=-\xi}^{N-1} 1}_{\text{Left Triangle}} \\
&= \sum_{\xi=0}^{N-1} (N-\xi)F(\xi) + \sum_{\xi=-(N-1)}^{0} (N+\xi)F(\xi) \\
&= \sum_{\xi=-(N-1)}^{N-1} (N-|\xi|)F(\xi) \,.
\end{aligned}
\tag{7.1.26}
$$

With these considerations, (7.1.25) can be written as

$$
\operatorname{var}\hat{R}_{\mathrm{xy}}(\tau) = \frac{1}{N}\sum_{\xi=-(N-1)}^{N-1}\left(1-\frac{|\xi|}{N}\right)\big(R_{\mathrm{xx}}(\xi)R_{\mathrm{yy}}(\xi) + R_{\mathrm{xy}}(\xi+\tau)R_{\mathrm{yx}}(\xi-\tau)\big)
\tag{7.1.27}
$$

and finally

$$
\lim_{N\to\infty}\operatorname{var}\hat{R}_{\mathrm{xy}}(\tau) = \lim_{N\to\infty}\frac{1}{N}\sum_{\xi=-(N-1)}^{N-1}\big(R_{\mathrm{xx}}(\xi)R_{\mathrm{yy}}(\xi) + R_{\mathrm{xy}}(\xi+\tau)R_{\mathrm{yx}}(\xi-\tau)\big)\,.
\tag{7.1.28}
$$

This variance is only determined by the stochastic nature of the two random signals. It expresses the intrinsic uncertainty of the CCF, see Sect. 6.1. For $N \to \infty$, the variance becomes zero if the correlation functions are finite and either $\mathrm{E}\{x(k)\} = 0$ or $\mathrm{E}\{y(k)\} = 0$. Henceforth, the estimation of the correlation function according to (7.1.19) is consistent in the mean square for Gaussian distributed signals. In the case of additionally superimposed disturbances, the considerations in Sect. 6.1 can be applied accordingly.

7.1.3 Fast Calculation of the Correlation Functions

As the correlation functions often have to be calculated for large numbers of data points N, computationally efficient algorithms shall now be discussed. An algorithm which is based on the Fast Fourier Transform (Sect. 3.1.3) shall be outlined (Kammeyer and Kroschel, 2009). This algorithm makes use of the fact that the biased estimate of the correlation function,

$$
\hat{R}_{\mathrm{xx}}(\tau) = \frac{1}{N}\sum_{k=0}^{N-1-|\tau|} x(k)x(k+|\tau|),\ \text{for } 0 \le |\tau| \le N-1\,,
\tag{7.1.29}
$$

which was presented in (7.1.4), can be expressed as a convolution in the time domain and hence a multiplication in the frequency domain.

The signal $x(k)$ is augmented with zeros to bring it to the total length L as

$$x_L(k) = \begin{cases} x(k) \text{ for } 0 \le k \le N-1 \\ 0 \text{ for } N-1 < k \le L-1 \end{cases} . \tag{7.1.30}$$

The estimate of the auto-correlation function can be rewritten as

$$\hat{R}_{xx}(\tau) = \frac{1}{N} \sum_{k=0}^{N-1-|\tau|} x(k)x(k+|\tau|) = \frac{1}{N} \sum_{k=0}^{L-1-|\tau|} x_L(k)x_L(k+|\tau|) . \tag{7.1.31}$$

Due to the symmetry of the auto-correlation function, only the values for $\tau \ge 0$ must be determined, since $R_{xx}(-\tau) = R_{xx}(\tau)$. Hence, it can be assumed in the following that $\tau > 0$ and therefore, the absolute value operator can be disposed.

$$\begin{aligned} \hat{R}_{xx}(\tau)\big|_{\tau\ge 0} &= \frac{1}{N} \sum_{k=0}^{L-1-\tau} x_L(k)x_L(k+\tau) \\ &= \frac{1}{N} \sum_{\nu=\tau}^{L-1} x_L(\nu-\tau)x_L(\nu) = \frac{1}{N} \sum_{\nu=\tau}^{L-1} x_L(-(\tau-\nu))x_L(\nu) . \end{aligned} \tag{7.1.32}$$

Since $x_L(-(\tau-\nu)) = 0$ for $-(\tau-\nu) < 0$ and hence $\nu < \tau$, the index can also start from $\nu = 0$ instead of $\nu = \tau$. The addends in the range $0 \le \nu \le \tau$ are zero, because $x_L(-(\tau-\nu)) = 0$ in this interval and therefore, these addends do not contribute to the sum.

$$\begin{aligned} \hat{R}_{xx}(\tau)\big|_{\tau\ge 0} &= \frac{1}{N} \sum_{\nu=0}^{L-1} x_L\big(-(\tau-\nu)\big)x_L(\nu) \\ &= \frac{1}{N} x_L(-k) * x_L(k) \end{aligned} \tag{7.1.33}$$

for $0 \le |\tau| \le N-1$, which represents the convolution of the signals $x_L(k)$ and $x_L(-k)$. This convolution can in the frequency domain be determined as a simple multiplication. Therefore, a solution to calculate the correlation function is to first transform the series $x_L(k)$ and $x_L(-k)$ into the frequency domain, then multiply the two Fourier transforms and finally bring this product back into the time domain. Although this may look like a cumbersome and slow process at first, this approach greatly benefits from the many efficient implementations of the Fast Fourier Transform, see Chap. 3.

While the extension of the signals with zeros at first seemed arbitrarily, its reason will become clear by looking at Fig. 7.2 and noting that the discrete pair of Fourier transforms will be used to calculate the correlation function.

Figure 7.2a shows the normal calculation of the correlation function. The sequence and its time-shifted time counterpart are multiplied in the interval $\tau \le k \le N-1$. Here, two problems of using the Fourier transform to compute the convolution of the two sequences immediately become apparent, see also Fig. 7.2b: First, the summation index for the convolution always runs from 0 to $N-1$, if the sequence has N elements. Secondly, the discrete Fourier transform causes a periodic repetition of the signal, i.e. $x(k+iN) = x(k) \ne 0$ for $i = \ldots, -3, -2, -1, 1, 2, 3, \ldots$.

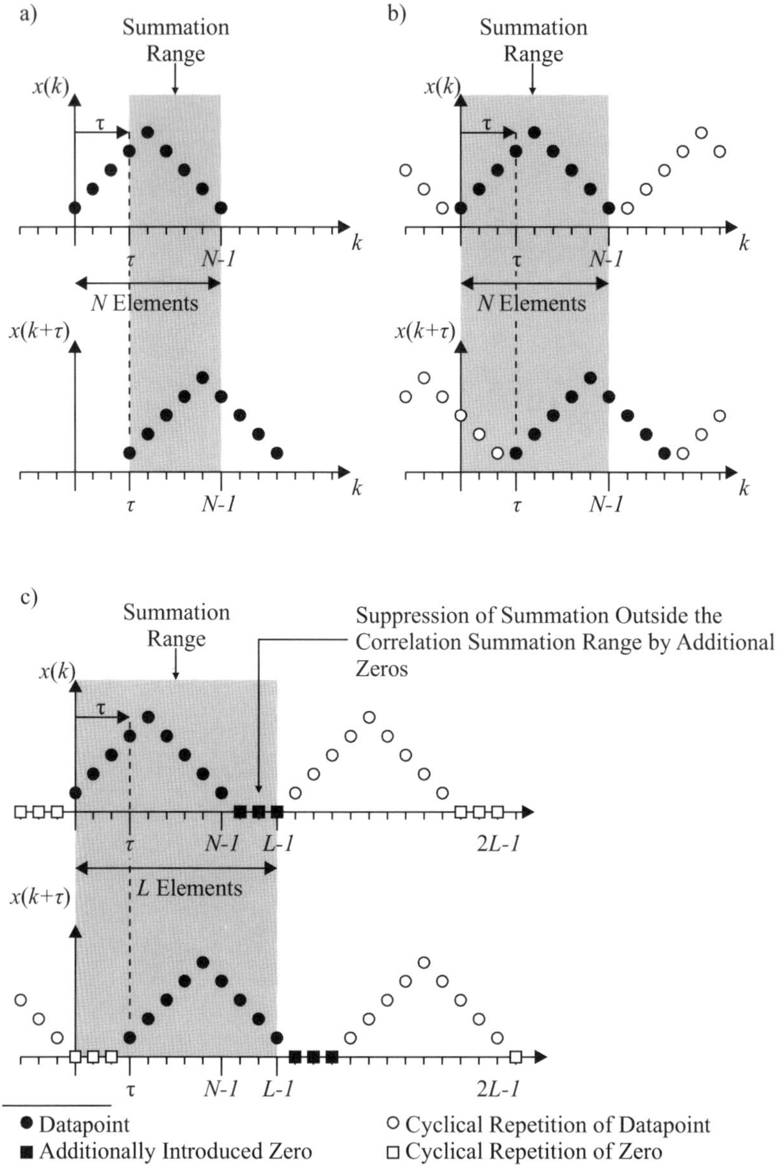

Fig. 7.2. Error in summation by (**b**) cyclical convolution instead of (**a**) normal convolution and (**c**) correction by inserting zeros into the data sequences

While the original data points have been marked as black circles in Fig. 7.2b, one can also see white circles that symbolize this periodic repetition. A remedy to avoid the adverse effects of this periodic repetition is to introduce additional zeros into the signal, see Fig. 7.2c. The summation is now carried out over $L \geq N + \tau_{\max}$ elements, where the additionally introduced zero elements lead to a suppression of the adverse effects of the repetition of the signal outside the measurement interval.

To summarize the above statement: If the FFT is used to determine the frequency domain representations of the signal $x(k)$, the signal will be repeated periodically outside the measurement interval and furthermore, the summation is always carried out over the full N respectively L elements. By introducing additional zeros and bringing the signal to any length

$$L \geq N + \max \tau \ , \tag{7.1.34}$$

an error by this cyclical repetition introduced by the Fourier transform of the discrete-time signals can be avoided. L can be chosen arbitrarily large and can conveniently be chosen as a power of 2 or 4 to apply computationally efficient realizations of the Fast Fourier Transform. Note also that all results for $\tau > N-1$ must be disposed as they are invalid.

Now, the auto-correlation function can be calculated as

$$\begin{aligned} \hat{R}_{xx}(\tau) &= \frac{1}{N} \mathrm{DFT}^{-1}\Big\{\mathrm{DFT}\{x_L(-k)\}\,\mathrm{DFT}\{x_L(k)\}\Big\} \\ &= \frac{1}{N} \mathrm{DFT}^{-1}\Big\{\big|\,\mathrm{DFT}\{x_L(k)\}\big|^2\Big\} \ . \end{aligned} \tag{7.1.35}$$

Hence, the calculation of the auto-correlation function can be formulated as a problem in the frequency domain (Kammeyer and Kroschel, 2009; Press et al, 2007). Further tricks can be employed to speed up the calculation of the correlation function by applying the Fourier transform repeatedly to smaller blocks of data (Rader, 1970).

In the following, the calculation will be developed for the cross-correlation of two signals $x(k)$ and $y(k)$ as the auto-correlation is automatically included in this more general case. In this setting, one is interested in providing an estimate

$$\hat{R}_{xy}(\tau) = \frac{1}{N} \sum_{k=0}^{N-1-|\tau|} x(k)y(k+|\tau|) = \frac{1}{N} x_L(-k) * y_L(k) \tag{7.1.36}$$

for $0 \leq |\tau| \leq N-1$ of the cross-correlation function for $x(k)$ and $y(k)$. A reflection in the time domain results also in a reflection in the frequency domain

$$x_L(-k) \circ\!\!-\!\!\bullet\ x_L(-\mathrm{i}\omega) \ . \tag{7.1.37}$$

Typically, the length N of the dataset is much larger than the maximum value of τ that is of interest. This can be exploited to speed up the calculation of the correlation function. The same technique is also used for the fast convolution. The time sequences $x(k)$ and $y(k)$ are split up into sequences

$$x_i = \begin{cases} x(n+iM) \text{ for } 0 \le n \le M-1,\ i=0,1,2,\ldots \\ 0 \text{ for} M \ \le n \le 2M-1 \end{cases} \tag{7.1.38}$$

and

$$y_i = y(n+iM) \text{ for } 0 \le n \le 2M-1,\ i=0,1,2,\ldots\ , \tag{7.1.39}$$

where $M = \max|\tau|$ and a total of I blocks have been formed. It is now assumed that the last block has been padded with zeros to the length $2M$. Furthermore $\tau > 0$ from now on.

Then (7.1.36) can be rewritten

$$\begin{aligned}
\hat{R}_{xy}(\tau) &= \frac{1}{N}\sum_{k=0}^{N-1-|\tau|} x(k)y(k+|\tau|) \\
&= \frac{1}{N}\sum_{k=0}^{N-1-|\tau|} x(k)y(k+|\tau|) \\
&- \frac{1}{N}\sum_{i=0}^{I-1}\sum_{k=0}^{M-1} x(k+iM)y(k+iM+\tau) \\
&= \frac{1}{N}\sum_{i=0}^{I-1}\sum_{k=0}^{M-1} x_i(k)y_i(k+\tau) \\
&= \frac{1}{N}\sum_{i=0}^{I-1}\mathrm{DFT}^{-1}\Big\{\mathrm{DFT}\{x_i(-k)\}\,\mathrm{DFT}\{y_i(k)\}\Big\} \\
&= \frac{1}{NL}\sum_{i=0}^{I-1}\sum_{n=0}^{N-1}\mathrm{DFT}\{x_i(-k)\}\,\mathrm{DFT}\{y_i(k)\}W_N^{nk} \\
&= \frac{1}{NL}\sum_{n=0}^{N-1}\sum_{i=0}^{I-1}\mathrm{DFT}\{x_i(-k)\}\,\mathrm{DFT}\{y_i(k)\}W_N^{nk} \\
&= \frac{1}{N}\mathrm{DFT}^{-1}\left\{\sum_{i=0}^{I-1}\mathrm{DFT}\{x_i(-k)\}\,\mathrm{DFT}\{y_i(k)\}\right\}\ .
\end{aligned} \tag{7.1.40}$$

For the de-convolution, only values of $R_{uy}(\tau)$ for $\tau \ge 0$ are of interest. If one also wants to obtain values of $R_{uy}(\tau)$ for $\tau < 0$ one can either exploit the fact that $R_{xy}(-\tau) = R_{yx}(\tau)$ and $R_{yx}(\tau)$ for $\tau \ge 0$ can be determined with the algorithm that was just presented. Fransaer and Fransaer (1991) presented a method, which is based on a change of the vectors $x_i(k)$ by $x_{i+1}(k)$.

As was already discussed above, it may at first not seem likely that this approach is computationally more efficient than the direct evaluation of the convolution sum. A short look at the numerical expense may however give an idea, why the method is more efficient.

Provided that the convolution of a series with N elements shall be determined in the range $0 \le \tau \le M-1$, the direct evaluation will require additions and multiplications in an amount of the order of NM multiplications and NM additions, provided that $M \ll N$. By using the FFT and the segmentation as described above, the effort can be brought down to the order of $N \log_2 M$ for both additions and multiplications. Obviously, for large values of M, the saving can become quite tremendous.

7.1.4 Recursive Correlation

The correlation functions can also be determined recursively. This will now be explained for the cross-correlation function as the transfer to the auto-correlation function is again straightforward. For the time-step $k-1$, the non-recursive estimation is given as (7.1.16)

$$\hat{R}_{xy}(\tau, k-1) = \frac{1}{k}\sum_{l=0}^{k-1} x(l-\tau)y(l)\,. \tag{7.1.41}$$

For the time k, the estimate can then be written as

$$\begin{aligned}\hat{R}_{xy}(\tau, k) &= \frac{1}{k+1}\sum_{l=0}^{k} x(l-\tau)y(l)\\ &= \frac{1}{k+1}\Bigg(\underbrace{\sum_{l=0}^{k-1} x(l-\tau)y(\tau)}_{k\hat{R}_{xy}(\tau,k-1)} + x(k-\tau)y(k)\Bigg)\,.\end{aligned} \tag{7.1.42}$$

Thus,

$$\underset{\substack{\text{New}\\ \text{Estimate}}}{\hat{R}_{xy}(\tau, k)} = \underset{\substack{\text{Old}\\ \text{Estimate}}}{\hat{R}_{xy}(\tau, k)} + \underset{\substack{\text{Correction}\\ \text{Factor}}}{\frac{1}{k+1}} \Big(\underset{\substack{\text{New}\\ \text{Product}}}{x(k-\tau)y(k)} - \underset{\substack{\text{Old}\\ \text{Estimate}}}{\hat{R}_{xy}(\tau, k-1)}\Big)\,. \tag{7.1.43}$$

If the last addend is interpreted as an error or *innovation*

$$e(k) = x(k-\tau)y(k) - \hat{R}_{xy}(\tau, k-1)\,, \tag{7.1.44}$$

then one can also write

$$\hat{R}_{xy}(\tau, k) = \hat{R}_{xy}(\tau, k-1) + \gamma(k)e(k)\,. \tag{7.1.45}$$

The correction factor is given as

$$\gamma(k) = \frac{1}{k+1} \tag{7.1.46}$$

and weights the new contribution with less weight as the measurement period k increases, which is in line with the normal averaging, where all terms $0 \leq l \leq k$ have the same weight.

If the correction factor is fixed to a certain k_1, then all new contributions are weighted with the same weight $\gamma(k_1)$. The recursive estimation algorithm then corresponds to a discrete low-pass filter. With these modifications, it is also possible to analyze slowly time-varying processes.

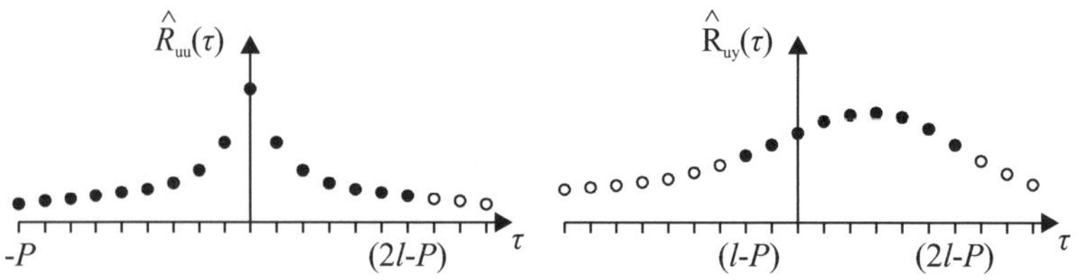

Fig. 7.3. Values of the correlation functions needed for de-convolution

7.2 Correlation Analysis of Linear Dynamic Systems

By means of the correlation functions, an easy to apply identification technique for the time-domain can be derived, which is termed *de-convolution* and will be developed in the following.

7.2.1 Determination of Impulse Response by De-Convolution

If a linear, stable, and time-invariant process is excited by a stationary colored stochastic input signal $u(k)$, then the output $y(k)$ will also be a stationary stochastic signal once the transients have vanished. Therefore, one can estimate the auto-correlation function $\hat{R}_{uu}(\tau)$ and the cross-correlation function $\hat{R}_{uy}(\tau)$.

It is for now assumed that both $\mathrm{E}\{u(k)\} = 0$ and $\mathrm{E}\{y(k)\} = 0$. Then, both correlation functions are linked by the convolution sum (2.4.12)

$$R_{uy}(\tau) = \sum_{\nu=0}^{\infty} R_{uu}(\tau - \nu) g(\nu) , \tag{7.2.1}$$

where $g(k)$ denotes the discrete-time impulse response. It is now assumed that both $R_{uu}(\tau)$ and $R_{uy}(\tau)$ have been determined for different τ, as e.g. shown in Fig. 7.3.

Now, the impulse response $g(\nu)$ shall be determined. According to (7.2.1), one obtains for each τ an equation with a different number of elements. In order to determine the individual values of the impulse response, $g(0)$, $g(1)$, up to $g(l)$, these individual equations will be written as a system of $l + 1$ equations as

$$\underbrace{\begin{pmatrix} R_{uy}(-P+l) \\ \vdots \\ R_{uy}(-1) \\ R_{uy}(0) \\ R_{uy}(1) \\ \vdots \\ R_{uy}(M) \end{pmatrix}}_{\hat{\boldsymbol{R}}_{uy}} \approx \underbrace{\begin{pmatrix} R_{uu}(-P+l) & \dots & R_{uu}(-P) \\ \vdots & & \vdots \\ R_{uu}(-1) & \dots & R_{uu}(-1-l) \\ R_{uu}(0) & \dots & R_{uu}(-l) \\ R_{uu}(1) & \dots & R_{uu}(1-l) \\ \vdots & & \vdots \\ R_{uu}(M) & \dots & R_{uu}(M-l) \end{pmatrix}}_{\hat{\boldsymbol{R}}_{uu}} \underbrace{\begin{pmatrix} g(0) \\ \vdots \\ g(l) \end{pmatrix}}_{\boldsymbol{g}} . \tag{7.2.2}$$

The largest negative time shift of $R_{uu}(\tau)$ is $\tau_{min} = -P$ and the largest positive time shift is $\tau_{max} = M$. The system of equations then consists of $P - l + M + 1$ equations. If one chooses $M = P + 2l$, then there are $l + 1$ equations such that $\hat{\boldsymbol{\Phi}}_{uu}$ becomes a square matrix and it follows that

$$\boldsymbol{g} \approx \hat{\boldsymbol{R}}_{uu}^{-1} \hat{\boldsymbol{R}}_{uy} . \tag{7.2.3}$$

If one chooses $P = l$ in addition, then for positive and negative values τ of the ACF R_{uu}, the same number of elements is used (a symmetric ACF, since $\tau_{min} = -P = -l$ and $\tau_{max} = M = l$). Considering the impulse response $g(\nu)$ only up to the finite value $\nu = l$, instead of $\nu \to \infty$ causes a round-off error. The estimate in (7.2.3) typically gets more accurate as l increases.

A condition for the existence of the inverse of $\hat{\boldsymbol{R}}_{uu}$ in (7.2.3) is that

$$\det \hat{\boldsymbol{R}}_{uu} \neq 0 , \tag{7.2.4}$$

which means that the system of equations may not contain linearly dependent rows or columns. At least one value of $R_{uu}(\tau)$ must change from one line to the next, which is guaranteed if the process is driven by a dynamically exciting input $u(k)$.

As can be seen from Fig. 7.3, not all available values of $R_{uy}(\tau)$ and $R_{uu}(\tau)$ are employed to determine $g(\nu)$. If one wants to use also the other values of the correlation functions, which are different from zero and thus employ more available information about the process, then one can shift P further to the left and M further to the right. One now obtains $(P + M + 1) > l + 1$ equations to determine the $l + 1$ unknown values of $g(\nu)$. By means of the pseudo inverse, one can determine a typically more accurate estimate of $\boldsymbol{g}$ as

$$\hat{\boldsymbol{g}} = \left(\hat{\boldsymbol{R}}_{uu}^{T} \hat{\boldsymbol{R}}_{uu}\right)^{-1} \hat{\boldsymbol{R}}_{uu}^{T} \hat{\boldsymbol{R}}_{uy} . \tag{7.2.5}$$

The estimation of the impulse response can drastically be simplified by exciting the process using a *white noise* with the auto-correlation function

$$R_{uu}(\tau) = \sigma_u^2 \delta(\tau) = R_{uu}(0)\delta(\tau) \tag{7.2.6}$$

with

$$\delta(\tau) = \begin{cases} 1 \text{ for } \tau = 0 \\ 0 \text{ for } \tau \neq 0 \end{cases} . \tag{7.2.7}$$

Then follows from (7.2.1) that

$$R_{uy}(\tau) = R_{uu}(0) g(\tau) \tag{7.2.8}$$

and thus

$$\hat{g}(\tau) = \frac{1}{\hat{R}_{uu}(0)} \hat{R}_{uy}(\tau) . \tag{7.2.9}$$

The impulse response is in this case proportional to the cross-correlation function.

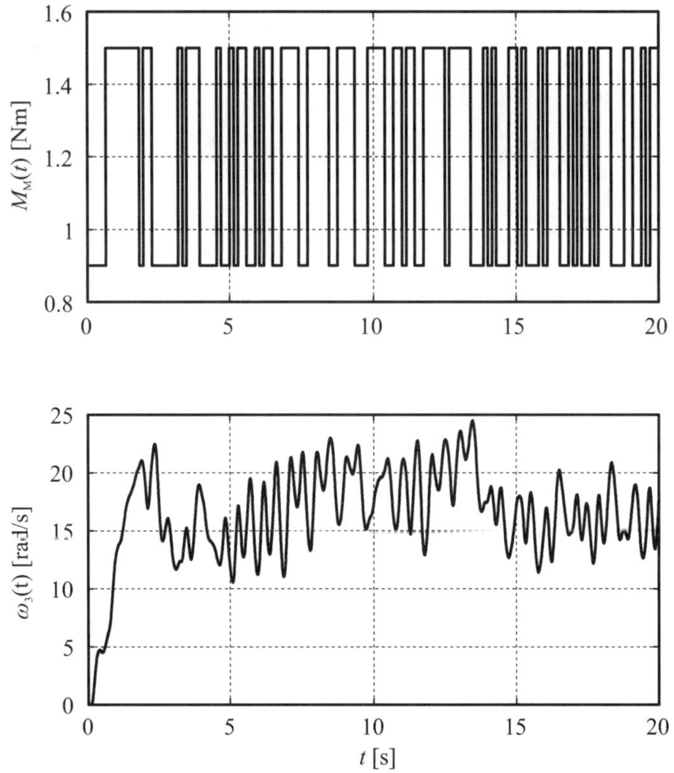

Fig. 7.4. Detail of the excitation of the Three-Mass Oscillator with a PRBS signal. The excitation has the parameters: $\mu = 50$, $n = 11$, $T_0 = 0.003\,\mathrm{s}$ and hence a period time of $T_P = 307.5\,\mathrm{s}$

Example 7.1 (De-Convolution Applied to the Three-Mass Oscillator).

In the following example, the Three-Mass Oscillator is excited with a PRBS signal. To avoid the negative side effects of the break-away torque due to adhesive and Coulomb friction, the oscillator is operated around a certain mean rotational velocity. This can be interpreted as the operating point, around which the system is linearized. Figure 7.4 shows the PRBS signal $u(t)$ that has been used for the excitation of the system as well as the rotational velocity of the third mass, $\omega_3(t)$. The PRBS generator had a cycle time of $\lambda = 0.15\,\mathrm{s}$. The measurements have been sampled at $T_0 = 0.003\,\mathrm{s}$, therefore the output of the PRBS is always held for $\mu = 50$ samples.

The ACF and CCF for a PRBS excitation are shown in Fig. 7.5. One can see that the calculated ACF converges to the exact course after $t = 80\,\mathrm{s}$ and that the CCF then approximates the direct measured impulse response according to (7.2.9). Despite the good results, it should be kept in mind that the ACF of a PRBS satisfies (6.3.15) only for full periods. Also, in Fig. 7.6, the de-convolution is calculated based on the

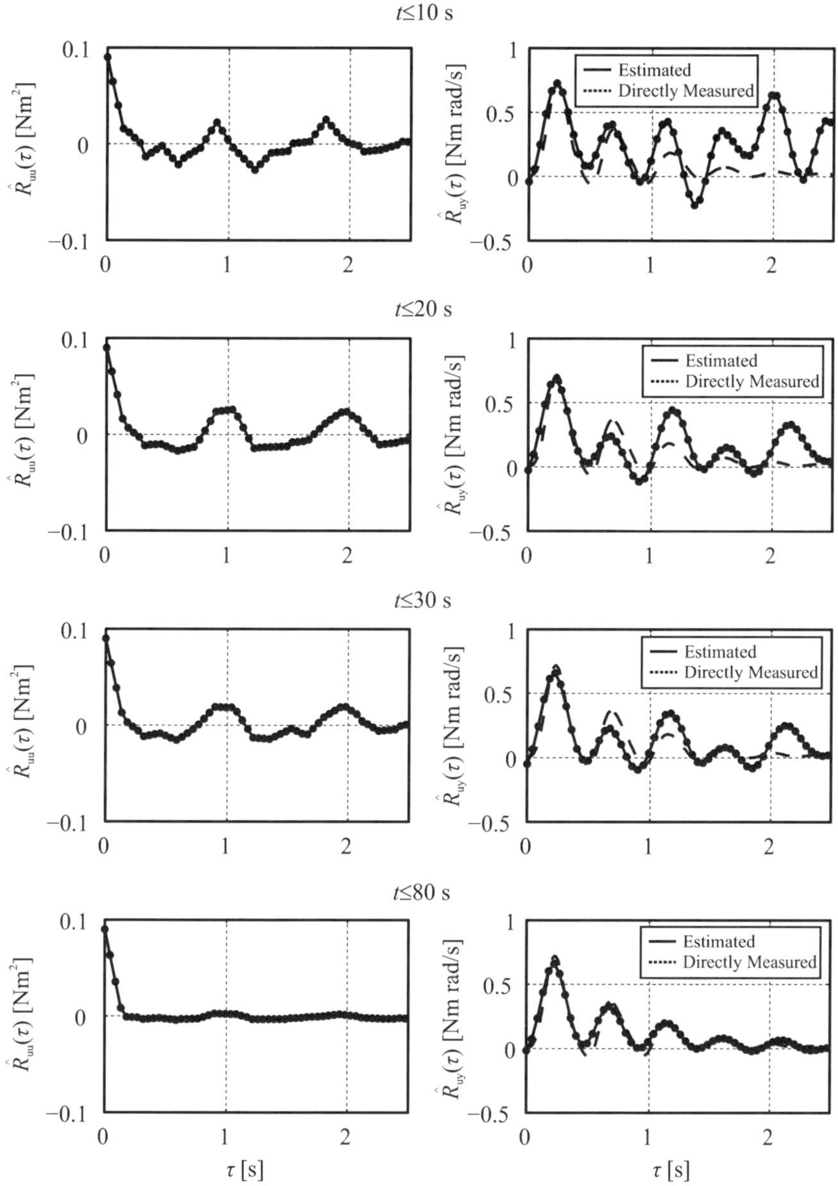

Fig. 7.5. Estimated ACF and CCF for the signals in Fig. 7.4 for measurement intervals $0 \leq t \leq T$ of different length. Estimated (*solid line*) and directly recorded (*dashed line*) impulse response

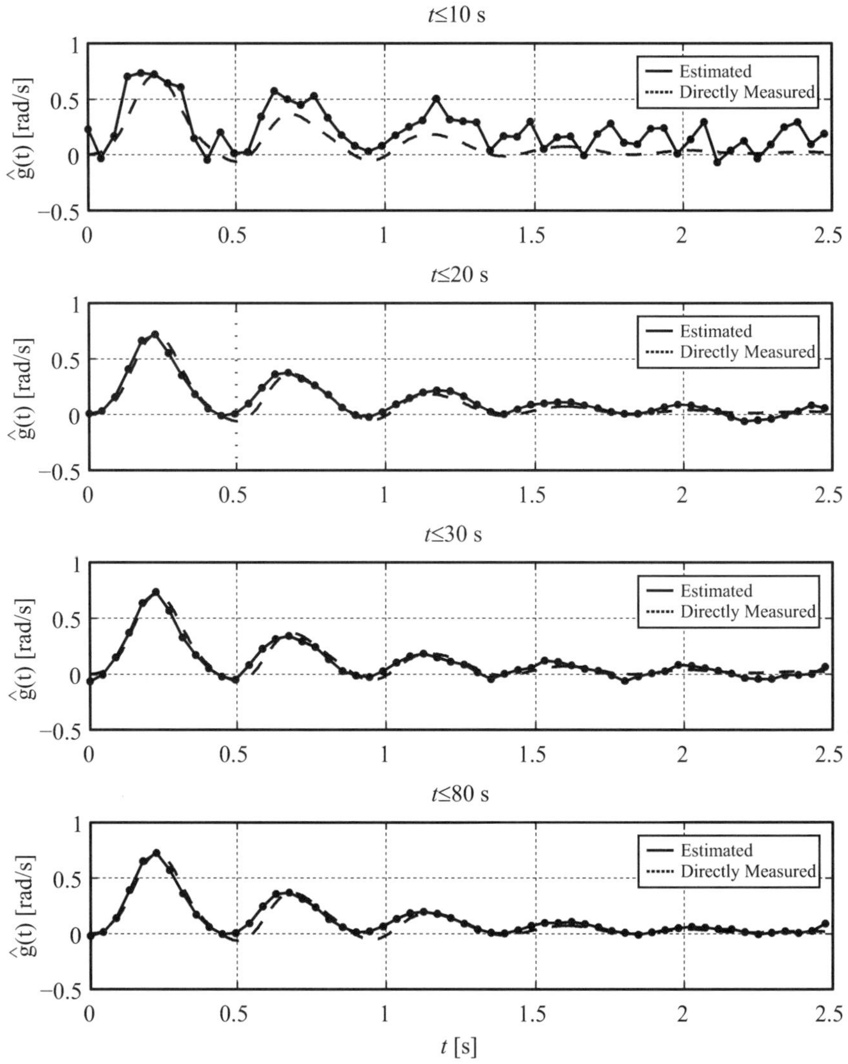

Fig. 7.6. Estimated impulse response due to (7.2.5) based on the correlation function estimates in Fig. 7.5 for measurement intervals $0 \leq t \leq T$ of different length. Estimated (*solid line*) and directly recorded (*dashed line*) impulse response

matrix inversion of $R_{\mathrm{uu}}(\tau)$ governed by (7.2.5). Here, one can see that the estimate of the impulse response converges to the true impulse response much earlier, a good match is already obtained for $t = 20\,\mathrm{s}$. □

7.2.2 Influence of Stochastic Disturbances

Now, the influence of stochastic disturbances in the output signal on the determination of the cross-correlation function $R_{\mathrm{uy}}(\tau)$ shall be determined. For this examination, it is again assumed that the exact output signal $y_{\mathrm{u}}(k)$ shall be affected by a superimposed stochastic disturbance $n(k)$, such that

$$y(k) = y_{\mathrm{u}}(k) + n(k) \tag{7.2.10}$$

and the input signal $u(k)$ and its auto-correlation function $R_{\mathrm{uu}}(\tau)$ shall be known exactly. Then, the cross-correlation follows from

$$\hat{R}_{\mathrm{uy}}(\tau) = \frac{1}{N}\sum_{\nu=0}^{N-1} u(\nu)y(\nu+\tau)\,. \tag{7.2.11}$$

With (7.2.10), the error is given as

$$\Delta R_{\mathrm{uy}}(\tau) = \frac{1}{N}\sum_{\nu=0}^{N-1} u(\nu)n(\nu+\tau)\,. \tag{7.2.12}$$

If the disturbance $n(k)$ is not correlated with the input signal and either $\mathrm{E}\{n(k)\} = 0$ or $\mathrm{E}\{u(k)\} = 0$, then follows

$$\mathrm{E}\{\Delta R_{\mathrm{uy}}(\tau)\} = \frac{1}{N}\sum_{\nu=0}^{N-1} \mathrm{E}\{u(k)\}\mathrm{E}\{n(k+\tau)\} = \mathrm{E}\{u(k)\}\mathrm{E}\{n(k)\} = 0\,. \tag{7.2.13}$$

The variance of the error is given as

$$\begin{aligned}\mathrm{E}\Big\{\big(\Delta\hat{R}_{\mathrm{uy}}(\tau)\big)^2\Big\} &= \frac{1}{N^2}\mathrm{E}\left\{\sum_{\nu=0}^{N-1}\sum_{\nu'=0}^{N-1} u(\nu)u(\nu')n(\nu+\tau)n(\nu'+\tau)\right\}\\ &= \frac{1}{N^2}\sum_{\nu=0}^{N-1}\sum_{\nu'=0}^{N-1} R_{\mathrm{uu}}(\nu'-\nu)R_{\mathrm{nn}}(\nu'-\nu)\end{aligned} \tag{7.2.14}$$

if $u(k)$ and $y(k)$ are statistically independent. If the input is a white noise with the auto-correlation function according to (7.2.6), then (7.2.14) can be simplified to

$$\mathrm{E}\Big\{\big(\Delta\hat{R}_{\mathrm{uy}}(\tau)\big)^2\Big\} = \frac{1}{N}R_{\mathrm{uu}}(0)R_{\mathrm{nn}}(0) = \frac{1}{N}S_{\mathrm{uu}0}\,\overline{n^2(k)}\,. \tag{7.2.15}$$

The standard deviation of the impulse response estimation error,

$$\Delta g(\tau) = \frac{1}{S_{\mathrm{uu0}}} \Delta \hat{R}_{\mathrm{uy}}(\tau) , \tag{7.2.16}$$

is then given as

$$\sigma_{\mathrm{g}} = \sqrt{\mathrm{E}\{\Delta g^2(\tau)\}} = \sqrt{\frac{\overline{n^2(k)}}{S_{\mathrm{uu0}} N}} = \frac{\sqrt{\overline{n^2(k)}}}{\sigma_{\mathrm{u}}} \frac{1}{\sqrt{N}} = \frac{\sigma_{\mathrm{n}}}{\sigma_{\mathrm{u}}} \frac{1}{\sqrt{N}} . \tag{7.2.17}$$

The standard deviation of the impulse response is proportional to the noise-to-signal ratio $\sigma_{\mathrm{n}}/\sigma_{\mathrm{u}}$ and inversely proportional to the square root of the measurement time N.

Thus, in the presence of disturbances $n(k)$, it follows from (7.2.13) and (7.2.17) that

$$\mathrm{E}\{\hat{g}(\tau)\} = g_0(\tau) \tag{7.2.18}$$

and

$$\lim_{N\to\infty} \operatorname{var} \hat{g}(\tau) = 0 . \tag{7.2.19}$$

The impulse response according to (7.2.9) is determined consistent in the mean square. A corresponding examination of the convergence can also be carried out for the more general estimation given in (7.2.3) and (7.2.5). Under the prerequisite that the estimates of the correlation functions are both consistent, then

$$\begin{aligned} \lim_{N\to\infty} \mathrm{E}\{\hat{g}\} &\approx \lim_{N\to\infty} \mathrm{E}\{\hat{\boldsymbol{R}}_{\mathrm{uu}}^{-1} \hat{\boldsymbol{R}}_{\mathrm{uy}}\} \\ &\approx \lim_{N\to\infty} \mathrm{E}\{\hat{\boldsymbol{R}}_{\mathrm{uu}}^{-1}\} \lim_{N\to\infty} \mathrm{E}\{\hat{\boldsymbol{R}}_{\mathrm{uy}}\} = \boldsymbol{R}_{\mathrm{uu}}^{-1} \boldsymbol{R}_{\mathrm{uy}} \\ &\approx \boldsymbol{g}_0 . \end{aligned} \tag{7.2.20}$$

Since also the variances of the correlation function estimates go to 0 as $N \to \infty$ (as shown in Sect. 7.1), it follows (disregarding the effects of the truncation of the impulse response):

Theorem 7.1 (Convergence of the Impulse Response Estimate Based on Correlation Functions).

The impulse response $g(k)$ of a linear time-invariant process can be estimated consistent in the mean square by de-convolution according to (7.2.3), (7.2.5), or (7.2.9) under the following necessary conditions:

- *The signals $u(k)$ and $y_{\mathrm{u}}(k)$ are stationary*
- $\mathrm{E}\{u(k)\} = 0$
- *The input signal is persistently exciting so that* $\det \hat{\boldsymbol{R}}_{\mathrm{uu}} \neq 0$
- *The disturbance $n(k)$ is stationary and uncorrelated with $u(k)$*

□

If the auto-correlation function is known exactly, as e.g. for a PRBS, then the intrinsic uncertainty of the ACF according to (7.1.8) will vanish. Furthermore, from (7.1.28) and (7.2.13) follows that $\mathrm{E}\{u(k)\}$ may be non-zero if $\mathrm{E}\{y(k)\} = 0$ and $\mathrm{E}\{n(k)\} = 0$.

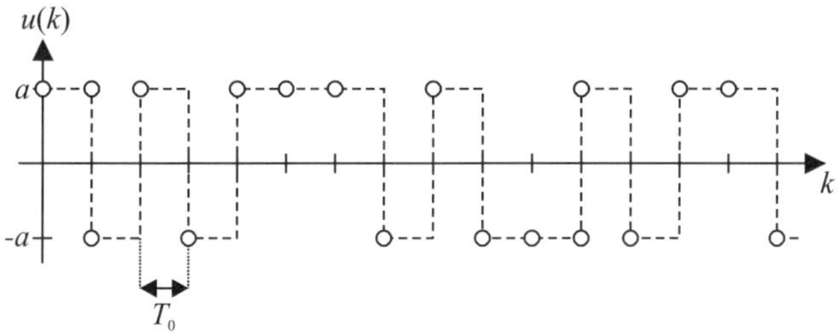

Fig. 7.7. Discrete random binary signal (DRBS)

This can also be illustrated with the following reflection. If the input or output has a non-zero mean, then it follows

$$u(k) = U(k) - U_{00} \tag{7.2.21}$$
$$y(k) = Y(k) - Y_{00} \tag{7.2.22}$$

with $U_{00} = \overline{U(k)}$ and $Y_{00} = \overline{Y(k)}$ and after plugging the large signals values into (7.2.11) follows

$$\hat{R}_{\mathrm{uy}}(\tau) = \frac{1}{N}\sum_{\nu=0}^{N-1}\big(U(\nu)Y(\nu+\tau)\big) - U_{00}Y_{00}\,. \tag{7.2.23}$$

Thus, the mean values of $U(k)$ and $Y(k)$ have to be determined separately during the measurement and their product must be subtracted. If however $U_{00} = 0$ or $Y_{00} = 0$ and $\mathrm{E}\{n(k)\} = 0$, then one does not have to carry out this separate averaging since in this case (7.2.23) and (7.2.13) yield the same results. However, due to the finite word length and the resulting computational errors, it is usually recommended to program the deviation from the signal due to (7.2.21) and (7.2.22) and determine the operating point U_{00} and Y_{00} separately, if U_{00} and Y_{00} are constant during the dynamic measurement.

7.3 Binary Test Signals for Discrete Time

The identification of linear processes for sampled signals via correlation functions is preferably performed with binary test signals. A discrete binary random signal (DRBS) is generated by random changes of the binary values at discrete-time instants kT_0, see Fig. 7.7.

The discrete-time auto-correlation function of such a discrete random binary signal is given as

$$R_{\mathrm{uu}}(\tau) = \begin{cases} a^2 \text{ for } \tau = 0 \\ 0 \text{ for } \tau \neq 0 \end{cases}, \tag{7.3.1}$$

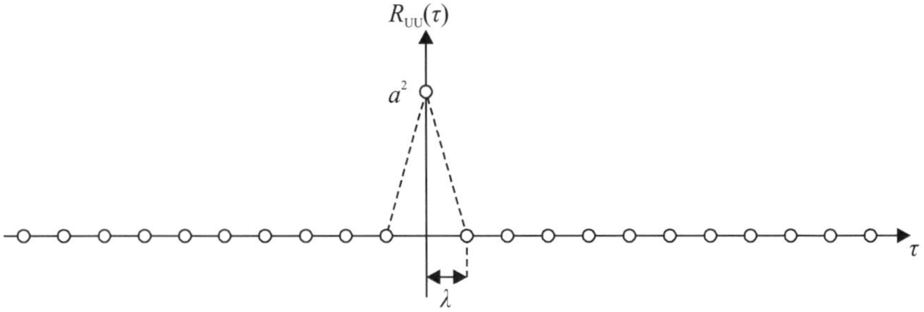

Fig. 7.8. Auto correlation function of a discrete random binary signal (DRBS)

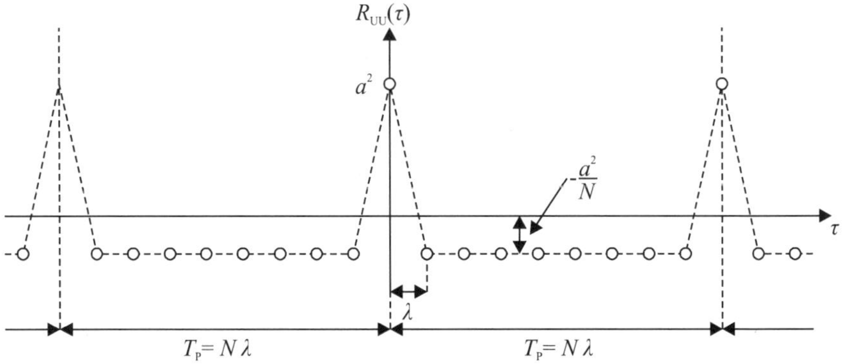

Fig. 7.9. Auto correlation function of a pseudo-random binary signal for discrete time

see Fig. 7.7. For $\tau \neq 0$, positive and negative values appear equally often, hence the auto-correlation function becomes zero, see Fig. 7.8. The power spectral density follows as

$$S_{\mathrm{uu}}(\tau) = \sum_{\tau=-\infty}^{\infty} R_{\mathrm{uu}}(\tau) z^{-\tau} = R_{\mathrm{uu}}(0) = S^*_{\mathrm{uu}}(\omega) = a^2 \text{ for } 0 \leq |\omega| \leq \frac{\pi}{T_0} \,. \quad (7.3.2)$$

The discrete random binary signal hence has the same auto-correlation function and power spectral density as a discrete white noise with an arbitrary amplitude density.

(7.3.1) and (7.3.2) are only valid for infinitely long measurement times. For finite measurement times, the auto-correlation and the power spectral density can deviate drastically from the values in (7.3.1) and (7.3.2) and hence have to be determined individually for each measurement. Due to this, one typically prefers periodic binary signals, which are deterministic signals, but have almost the same auto-correlation function already for finite measurement times as the stochastic signals have for $t \to \infty$. The auto-correlation function for the discrete pseudo-random binary signal is shown in Fig. 7.9. Such a signal is generated from a shift register as was presented in Sect. 6.3, see (Chow and Davies, 1964; Pittermann and Schweizer, 1966; Davies, 1970).

7.4 Summary

Correlation functions are defined for an infinitely long measurement interval. In practice, however, the measurement interval is always confined to a maximum of N data points. Two estimates for the discrete-time auto-correlation function have been presented that individually cope with the finite measurement time for sampled data. One estimate is bias-free, whereas the other one has a smaller variance. The advantages and disadvantages of the two estimates have been discussed. The results have then been generalized to include the cross-correlation as well. An approach for the fast calculation of the correlation functions has been presented that interprets the calculation of the correlation function as a convolution of the two signals. The convolution is then carried out in the frequency domain. By means of the discrete Fourier transform, the two signals are transformed into the frequency domain, then multiplied with each other and then transformed back into the time domain. This method can reduce the computational effort for large data sets and a large number of different time lags to be calculated. By dividing the time sequence into smaller blocks that can be processed separately, the calculation of the correlation function can be accelerated even more. Also, a recursive formulation of the correlation function estimation is presented. The estimates of the correlation function can then be used to determine the impulse response of a system by means of the de-convolution.

The presented correlation analysis with stochastic as well as pseudo-stochastic signals is well suited for the identification of non-parametric models of linearizable processes with discrete-time signals. The method can be implemented easily on digital signal processors or micro-controllers. In its recursive form, it is also well suited for online application in real-time.

Problems

7.1. Estimation of the Correlation Functions
Describe the two ways to estimate the auto-correlation function for finite time measurements and discuss their bias and variance in dependence of the measurement time.

7.2. Fast Calculation of Correlation Functions
Program the fast calculation of the correlation functions using the built-in Fourier transform routines of a mathematical software package.

7.3. De-Convolution I
How can you determine the impulse response of a linear system by means of the de-convolution. How does the problem simplify for a white noise input signal.

7.4. De-Convolution II
Given is the process

$$G(z) = \frac{y(z)}{u(z)} = \frac{0.5z^{-1}}{1 - 0.5z^{-1}}$$

As input signal $u(k)$ use a PRBS signal with $N = 4$ and the initial values $(1, 0, 0, 1)$. The following questions can either be answered by manual calculation or by use of a mathematical program.

a) Determine the values of $y(k)$ for $k = 1, 2, \ldots, 25$.

b) Determine the auto-correlation and cross-correlation functions.

c) Determine the impulse response by de-convolution.

References

Bartlett MS (1946) On the theoretical specification and sampling properties of autocorrelated time series. J Roy Statistical Society B 8(1):27–41

Bendat JS, Piersol AG (2010) Random data: Analysis and measurement procedures, 4th edn. Wiley-Interscience, New York

Chow P, Davies AC (1964) The synthesis of cyclic code generators. Electron Eng 36:253–259

Davies WDT (1970) System identification for self-adaptive control. Wiley-Interscience, London

Fransaer J, Fransaer D (1991) Fast cross-correlation algorithm with application to spectral analysis. IEEE Trans Signal Process 39(9):2089–2092

Kammeyer KD, Kroschel K (2009) Digitale Signalverarbeitung: Filterung und Spektralanalyse mit MATLAB-Übungen, 7th edn. Teubner, Wiesbaden

Pittermann F, Schweizer G (1966) Erzeugung und Verwendung von binärem Rauschen bei Flugversuchen. Regelungstechnik 14:63–70

Press WH, Teukolsky SA, Vetterling WT, Flannery BP (2007) Numerical recipes: The art of scientific computing, 3rd edn. Cambridge University Press, Cambridge, UK

Rader C (1970) An improved algorithm for high speed autocorrelation with applications to spectral estimation. IEEE Trans Audio Electroacoust 18(4):439–441

Part III

Identification with Parametric Models — Discrete Time Signals

8

Least Squares Parameter Estimation for Static Processes

This chapter lays the foundation of the least squares parameter estimation, which allows to determine model parameters from (noisy) measurements. The fundamental method described in this chapter for static non-linear systems will be applied to linear dynamic discrete-time systems in Chap. 9. In Chap. 9, also a recursive formulation will be presented. This allows to identify processes in real time. Several modifications to this basic approach for linear dynamic processes will then be presented in Chap. 10. The method of least squares will also be applied to linear dynamic continuous time processes in Chap. 15. Furthermore, the method will be employed for the identification of processes from frequency response data (Chap. 14), for processes in closed-loop (Chap. 13), for non-linear systems (Chap. 18), and for MIMO systems (Chap. 17).

8.1 Introduction

The fundamental task of the parameter estimation can be formulated as follows: Given is a real process with the parameters

$$\boldsymbol{\theta}_0^{\mathrm{T}} = \begin{pmatrix} \theta_{10} \; \theta_{20} \; \ldots \; \theta_{m0} \end{pmatrix} \tag{8.1.1}$$

and the output $y_{\mathrm{u}}(k)$. It is assumed that this process follows physical laws with the parameters $\boldsymbol{\theta}_0$, like a planetary system, where only outputs can be observed,

$$y_{\mathrm{u}}(k) = f(\boldsymbol{\theta}_0) \; . \tag{8.1.2}$$

The output can however not be measured directly. One can only measure $y_{\mathrm{P}}(k)$ which is the true process output falsified by a superimposed disturbance $n(k)$, see Fig. 8.1.

Furthermore, a model of the process shall be known

$$y_{\mathrm{M}} = f(\boldsymbol{\theta}) \; , \tag{8.1.3}$$

where

R. Isermann, M. Münchhof, *Identification of Dynamic Systems*,
DOI 10.1007/978-3-540-78879-9_8, © Springer-Verlag Berlin Heidelberg 2011

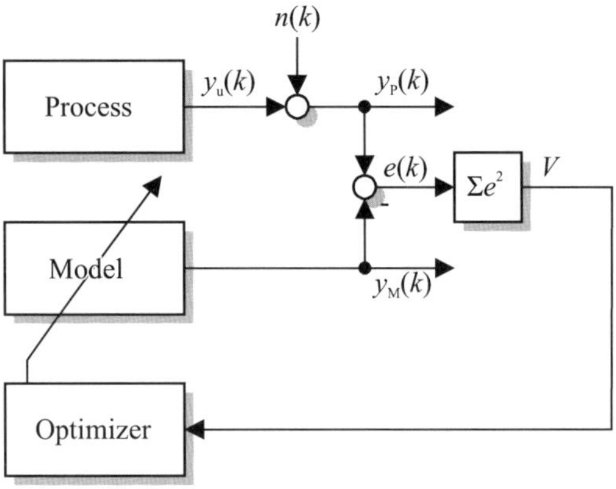

Fig. 8.1. Schematic diagram of the general arrangement for the method of least squares

$$\boldsymbol{\theta}^{\mathrm{T}} = \begin{pmatrix} \theta_1 & \theta_2 & \dots & \theta_m \end{pmatrix} \tag{8.1.4}$$

are the unknown model parameters. The task now is to find the model parameters $\boldsymbol{\theta}$ that result in a model which best fits with N observations $y_{\mathrm{P}}(k)$.

This task has first been solved by Gauss in the year 1795 (at the age of 18 years). Gauss later published the papers *Theoria combinatoris observationum erroribus minimis obnoxiae I* and *II* in the years 1821 and 1823, where he motivated and formally derived the method of least squares. In this original problem formulation, the parameters θ_i were the orbit parameters of planets, the model $\boldsymbol{y}_M = f(\boldsymbol{\theta}$ were Kepler's laws of planetary motion, the model output $\boldsymbol{y}_M$ were the coordinates of planets at different times and the measurements $\boldsymbol{y}_{\mathrm{P}}$ their observed, i.e. "measured" positions.

There, the *best fit* had been defined by first introducing the observation error

$$e(k) = y_{\mathrm{P}}(k) - y_{\mathrm{M}}(k) \tag{8.1.5}$$

and determining the minimum of the sum of the squared errors,

$$V = e^2(1) + e^2(2) + \dots + e^2(N) = \sum_{k=1}^{N} \big(e(k)\big)^2 \,. \tag{8.1.6}$$

The arrangement can be seen in Fig. 8.1.

There are several reasons which promote the choice of a quadratic cost function. First of all, it is easier to minimize than many other cost functions, as e.g. the absolute error $|e(k)|$. The main reason is however that for a normally distributed noise, it yields asymptotically the best unbiased estimates in terms of the parameter error variance , as will be shown later in Sect. 8.5.

The quadratic criterion however overemphasizes the effect of single, large outliers compared to small, but steadily occurring errors due to model impurities. Therefore, other criteria have been promoted as well as e.g. the least absolute value cost

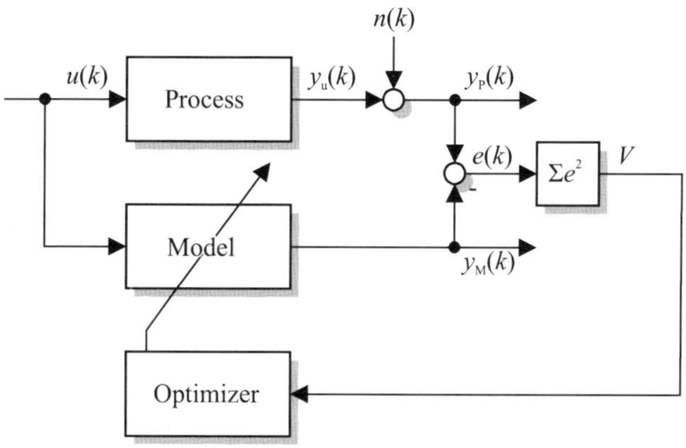

Fig. 8.2. Schematic diagram of the general arrangement for parameter estimation of a process with measured input and output signals with the method of least squares

function or a mixed linear/quadratic cost function, see Sect. 19.1 and (e.g. Pintelon and Schoukens, 2001).

The above formulated problem is the starting point for this chapter as well as for the following chapters on parameter estimation. The method of least squares will be introduced in this chapter for the simple case of static processes now with measurable input and output signals, see Fig. 8.2. This is made in an easy to understand tutorial style, beginning with scalar calculations and then transferring the procedure to vectorial notation. Two different derivations will be presented. One is based on the differential calculus and the other on a geometrical interpretation. In other contexts, the method of least squares is one of several *regression methods*. The following chapters will deal with the more difficult case of dynamic processes as well as recursive formulations of the parameter estimation problem, modifications for special applications, and, finally, computationally efficient methods.

8.2 Linear Static Processes

The static behavior of a simple linear process shall be given

$$y = Ku \; . \tag{8.2.1}$$

In general, it must be assumed that at least the sampled output $y_u(k)$ of the process (so-called wanted or useful signal) is affected by disturbances $n(k)$, so that the measured output is given as

$$y_P(k) = y_u(k) + n(k) \; , \tag{8.2.2}$$

where $n(k)$ is a discrete-time stationary random signal with $\mathrm{E}\{n(k)\} = 0$. Then, the *disturbed process* is given as

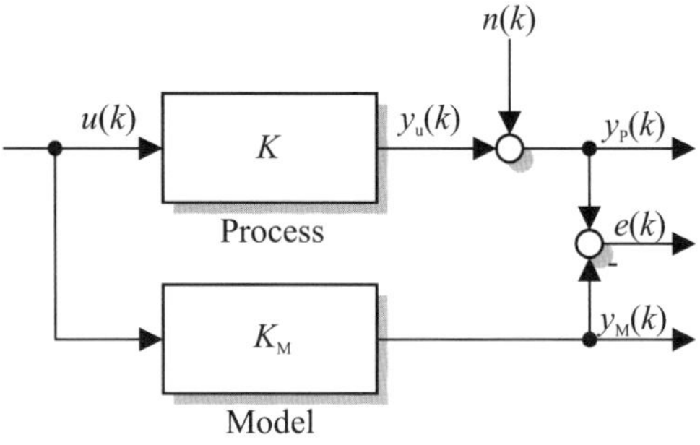

Fig. 8.3. Linear static process with one parameter, arrangement of process and model for calculation of error $e(k)$

$$y_P(k) = Ku(k) + n(k) \,, \tag{8.2.3}$$

see the topmost part of Fig. 8.3. The task is now to determine the parameter K from N measurements given as pairs of $u(k)$ and $y_P(k)$ as $(u(1), y_P(1))$ up to $(u(N), y_P(N))$.

Since the structure of the process is known, one can now place a *model* of the form

$$y_M(k) = K_M u(k) \tag{8.2.4}$$

in parallel to the process (see Fig. 8.3), so that the error between the process and the model is given as the difference between the corresponding output signals, i.e.

$$e(k) = y_P(k) - y_M(k) \,. \tag{8.2.5}$$

With (8.1.5) and (8.2.4), one obtains

$$\underset{\text{Error}}{e(k)} = \underset{\text{Observation}}{y_P(k)} - \underset{\text{Model Prediction}}{K_M u(k)} \,. \tag{8.2.6}$$

For the method of least squares, the cost function

$$V = \sum_{k=1}^{N} e^2(k) = \sum_{k=1}^{N} \big(y_P(k) - K_M u(k)\big)^2 \tag{8.2.7}$$

has to be minimized for the parameter K_M. To find the minimum, one first determines the first derivative with regard to the model parameter K_M

$$\frac{\mathrm{d}V}{\mathrm{d}K_M} = -2 \sum_{k=1}^{N} \big(y_P(k) - K_M u(k)\big) u(k) \,. \tag{8.2.8}$$

The first derivative of (8.2.7) with respect to K_M is set to zero to find the optimum value of K_M that minimizes (8.2.7). This optimal choice shall be denoted as the parameter estimate $\hat{K}$, i.e.

$$\left.\frac{\mathrm{d}V}{\mathrm{d}K_\mathrm{M}}\right|_{K_\mathrm{M}=\hat{K}} \stackrel{!}{=} 0 \Rightarrow -2\sum_{k=1}^{N}\big(y_\mathrm{P}(k) - \hat{K}u(k)\big)u(k) = 0\,. \tag{8.2.9}$$

This equation can be solved to provide an estimate for the model coefficient as

$$\hat{K} = \frac{\sum_{k=1}^{N} y_\mathrm{P}(k)u(k)}{\sum_{k=1}^{N} u^2(k)} \tag{8.2.10}$$

and, after multiplying $1/N$ into the numerator and denominator,

$$\hat{K} = \frac{\hat{R}_\mathrm{uy}(0)}{\hat{R}_\mathrm{uu}(0)}\,. \tag{8.2.11}$$

The best estimate $\hat{K}$ is thus the ratio of the estimates of the cross-correlation function and the auto-correlation function for $\tau = 0$. A condition for the existence of the estimate is that

$$\sum_{k=1}^{N} u^2(k) \neq 0 \text{ or } \hat{R}_\mathrm{uu}(0) \neq 0\,. \tag{8.2.12}$$

This equation demands that the input signal $u(k)$ must be non-zero or to rephrase it: The input signal $u(k)$ must "excite" the process with its parameter K. K is also termed the *regression coefficient*, since in mathematical settings, the parameter estimation problem is also known under the name regression problem.

In the following, the convergence of the estimation shall be investigated. For a review of the definition of notions for the convergence, see also App. A. Considering (8.2.2) and (8.2.10), the expected value of $\hat{K}$ is given as

$$\begin{aligned}\mathrm{E}\{\hat{K}\} &= \mathrm{E}\left\{\frac{\sum_{k=1}^{N} y_\mathrm{P}(k)u(k)}{\sum_{k=1}^{N} u^2(k)}\right\} = \mathrm{E}\left\{\frac{\sum_{k=1}^{N}\big(y_\mathrm{u}(k)+n(k)\big)u(k)}{\sum_{k=1}^{N} u^2(k)}\right\} \\ &= \frac{1}{\sum_{k=1}^{N} u^2(k)}\left(\sum_{k=1}^{N} y_\mathrm{u}(k)u(k) + \sum_{k=1}^{N}\mathrm{E}\{n(k)u(k)\}\right) = K\,,\end{aligned} \tag{8.2.13}$$

if the input signal $u(k)$ is uncorrelated with the noise $n(k)$ and thus

$$\mathrm{E}\{u(k)n(k)\} = \mathrm{E}\{u(k)\}\mathrm{E}\{n(k)\} \tag{8.2.14}$$

and $\mathrm{E}\{n(k)\} = 0$ and/or $\mathrm{E}\{u(k)\} = 0$. The estimate according to (8.2.10) is thus bias-free.

Given (8.2.2), (8.2.10), and (8.2.13), the variance of the parameter estimate $\hat{K}$ can be calculated as

$$\sigma_{\mathrm{K}}^2 = \mathrm{E}\{(\hat{K} - K)^2\} = \frac{1}{\left(\sum_{k=1}^{N} u^2(k)\right)^2} \mathrm{E}\left\{\left(\sum_{k=1}^{N} n(k)u(k)\right)^2\right\} . \tag{8.2.15}$$

If $n(k)$ and $u(k)$ are uncorrelated, then

$$\mathrm{E}\left\{\sum_{k=1}^{N} n(k)u(k) \cdot \sum_{k'=1}^{N} n(k')u(k')\right\} = \sum_{k=1}^{N}\sum_{k'=1}^{N} R_{\mathrm{nn}}(k-k')R_{\mathrm{uu}}(k-k') = Q . \tag{8.2.16}$$

This equation can be simplified if either $n(k)$ or $u(k)$ is a white noise.

In the first case, it is assumed that $n(k)$ is a white noise. In this case

$$R_{\mathrm{nn}}(\tau) = \sigma_{\mathrm{n}}^2\delta(\tau) = \overline{n^2(k)}\delta(\tau) \tag{8.2.17}$$

$$Q = NR_{\mathrm{uu}}(0)\,\overline{n^2(k)} = \overline{n^2(k)}\sum_{k=1}^{N} u^2(k) \tag{8.2.18}$$

$$\sigma_{\mathrm{K}}^2 = \frac{\overline{n^2(k)}}{\sum_{k=1}^{N} u^2(k)} . \tag{8.2.19}$$

If the other case is investigated, i.e. if $u(k)$ is a white-noise, identical results can be obtained. The standard deviation of the estimated parameter is thus given as

$$\sigma_{\mathrm{K}} = \sqrt{\mathrm{E}\{(\hat{K} - K)^2\}} = \sqrt{\frac{\overline{n^2(k)}}{\overline{u^2(k)}}\frac{1}{\sqrt{N}}} = \left(\frac{\sigma_{\mathrm{n}}}{\sigma_{\mathrm{u}}}\right)\frac{1}{\sqrt{N}} , \tag{8.2.20}$$

if $u(k)$ and/or $n(k)$ is a white noise. The standard deviation thus gets smaller as the signal-to-noise ratio $(\sigma_{\mathrm{u}}/\sigma_{\mathrm{n}})$ gets better (i.e. bigger) and is furthermore inversely proportional to the square root of the number of measured data points N. The true but unknown process parameter shall be denoted as K_0 in the following. Since both

$$\mathrm{E}\{\hat{K}\} = K_0 \tag{8.2.21}$$

and

$$\lim_{N\to\infty} \mathrm{E}\{(\hat{K} - K_0)^2\} = 0 \tag{8.2.22}$$

are fulfilled, the the estimation in (8.2.13) is consistent in the mean square (App. A). Now, the *vectorial notation* will be introduced for this parameter estimation problem. Upon introduction of the vectors

$$\boldsymbol{u} = \begin{pmatrix} u(1) \\ u(2) \\ \vdots \\ u(N) \end{pmatrix}, \; \boldsymbol{y}_\mathrm{P} = \begin{pmatrix} y_\mathrm{P}(1) \\ y_\mathrm{P}(2) \\ \vdots \\ y_\mathrm{P}(N) \end{pmatrix}, \; \boldsymbol{e} = \begin{pmatrix} e(1) \\ e(2) \\ \vdots \\ e(N) \end{pmatrix}, \tag{8.2.23}$$

the equation for the error can be written as

$$\boldsymbol{e} = \boldsymbol{y}_\mathrm{P} - \boldsymbol{u}K\,, \tag{8.2.24}$$

and the cost function is given as

$$V = \boldsymbol{e}^\mathrm{T}\boldsymbol{e} = (\boldsymbol{y}_\mathrm{P} - \boldsymbol{u}K)^\mathrm{T}(\boldsymbol{y}_\mathrm{P} - \boldsymbol{u}K)\,. \tag{8.2.25}$$

Taking the first derivative of the cost function yields

$$\frac{\mathrm{d}V}{\mathrm{d}K} = \frac{\mathrm{d}\boldsymbol{e}^\mathrm{T}}{\mathrm{d}K}\boldsymbol{e} + \boldsymbol{e}^\mathrm{T}\frac{\mathrm{d}\boldsymbol{e}}{\mathrm{d}K}\,. \tag{8.2.26}$$

The derivatives can be determined as

$$\frac{\mathrm{d}\boldsymbol{e}}{\mathrm{d}K} = -\boldsymbol{u} \text{ and } \frac{\mathrm{d}\boldsymbol{e}^\mathrm{T}}{\mathrm{d}K} = -\boldsymbol{u}^\mathrm{T}\,. \tag{8.2.27}$$

Equating the first derivative (8.2.26) to zero and considering (8.2.27) yields

$$\left.\frac{\mathrm{d}V}{\mathrm{d}K}\right|_{K=\hat{K}} = -2\boldsymbol{u}^\mathrm{T}(\boldsymbol{y}_\mathrm{P} - \boldsymbol{u}K) \stackrel{!}{=} 0\,. \tag{8.2.28}$$

The solution is given as

$$\boldsymbol{u}^\mathrm{T}\boldsymbol{u}\hat{K} = \boldsymbol{u}^\mathrm{T}\boldsymbol{y}_\mathrm{P} \Leftrightarrow \hat{K} = \left(\boldsymbol{u}^\mathrm{T}\boldsymbol{u}\right)^{-1}\boldsymbol{u}^\mathrm{T}\boldsymbol{y}_\mathrm{P}\,. \tag{8.2.29}$$

This solution is identical to (8.2.10).

If one wants to work with the large signal quantities $U(k)$ and $Y(k)$, then one obtains

$$Y_\mathrm{P}(k) - Y_{00} = K\big(U(k) - U_{00}\big) + n(k) \tag{8.2.30}$$

$$Y_\mathrm{M}(k) - Y_{00} = K_\mathrm{M}\big(U(k) - U_{00}\big)\,. \tag{8.2.31}$$

The error is then given as

$$e(k) = Y_\mathrm{P}(k) - Y_\mathrm{M}(k) = y_\mathrm{P}(k) - y_\mathrm{M}(k) = y_\mathrm{P}(k) - K_\mathrm{M}\big(U(k) - U_{00}\big) \tag{8.2.32}$$

and is identical to (8.2.5).

The DC quantity Y_{00} cancels out and thus does not have to be known exactly (or can be chosen arbitrarily). The DC value U_{00} must however be known exactly. This will lead to

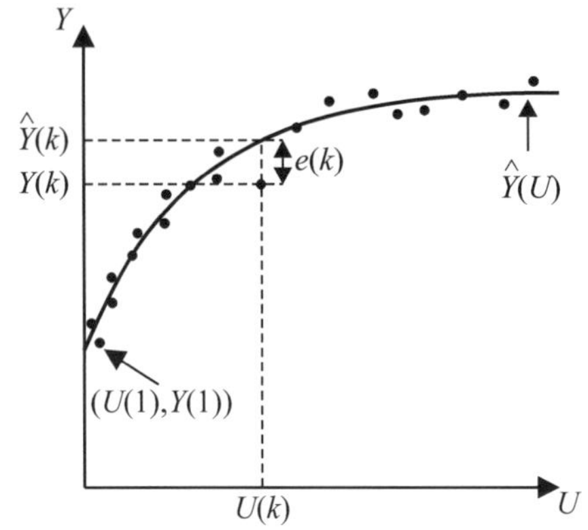

Fig. 8.4. Parameter estimation for a non-linear (polynomial) static process

Theorem 8.1 (Convergence of the Parameter Estimation of a Linear Static Process).

The parameter K of a linear static process is estimated consistently in the mean square by the methods of least squares if the following necessary conditions are satisfied:

- *The input signal $u(k) = U(k) - U_{00}$ is exactly measurable and U_{00} is known exactly.*
- $\sum_{k=1}^{N} u^2(k) \neq 0$*, i.e. the process is sufficiently excited*
- *The disturbance $n(k)$ is stationary and thus* $\mathrm{E}\{n(k)\} = const$
- *The input signal $u(k)$ is uncorrelated with the disturbance $n(k)$*
- *Either* $\mathrm{E}\{n(k)\} = 0$ *or* $\mathrm{E}\{u(k)\} = 0$

□

8.3 Non-Linear Static Processes

Now, a static process shall be considered, where the output depends non-linearly on the input quantity $U(k)$, but linearly on the process parameters K_i

$$Y_\mathrm{u}(k) = K_0 + U(k)K_1 + U^2(k)K_2 + \ldots + U^q(k)K_q = K_0 + \sum_{\nu=1}^{q} U^\nu K_\nu \,, \quad (8.3.1)$$

see Fig. 8.4. This process is called to be linear in parameters.

It is again assumed that $U(k)$ shall be known exactly. If the measured output signal is affected by random disturbances $n(k)$, then the output is given as

$$Y(k) = Y_\mathrm{u}(k) + n(k) \,. \quad (8.3.2)$$

The following matrices and vectors shall be agreed upon,

$$\boldsymbol{\Psi} = \begin{pmatrix} 1 & U(1) & U^2(1) & \dots & U^q(1) \\ 1 & U(2) & U^2(2) & \dots & U^q(2) \\ \vdots & \vdots & \vdots & & \vdots \\ 1 & U(N) & U^2(N) & \dots & U^q(N) \end{pmatrix},$$

as well as

$$\boldsymbol{y} = \begin{pmatrix} Y_{\mathrm{P}}(1) \\ Y_{\mathrm{P}}(2) \\ \vdots \\ Y_{\mathrm{P}}(N) \end{pmatrix} \boldsymbol{e} = \begin{pmatrix} e(1) \\ e(2) \\ \vdots \\ e(N) \end{pmatrix} \boldsymbol{n} = \begin{pmatrix} n(1) \\ n(2) \\ \vdots \\ n(N) \end{pmatrix} \boldsymbol{\theta} = \begin{pmatrix} K_0 \\ K_1 \\ \vdots \\ K_q \end{pmatrix},$$

where at this point, the notation is changed to the well-established notation for the regression problem, with $\boldsymbol{\Psi}$ being the data matrix, $\boldsymbol{\theta}$ the parameter vector, and $\boldsymbol{y}$ the output vector. The process equation is then given as

$$\boldsymbol{y} = \boldsymbol{\Psi}\boldsymbol{\theta}_0 + \boldsymbol{n}\,, \tag{8.3.3}$$

where $\boldsymbol{\theta}_0$ denotes the true (but unknown) parameters and the model equation can be written as

$$\hat{\boldsymbol{y}} = \boldsymbol{\Psi}\boldsymbol{\theta}\,, \tag{8.3.4}$$

where $\hat{\boldsymbol{y}}$ shall denote the model output from now on. Process and model are once again placed in parallel so that the error between process and model amounts to

$$\boldsymbol{e} = \boldsymbol{y} - \boldsymbol{\Psi}\boldsymbol{\theta}\,. \tag{8.3.5}$$

The cost function is then given as

$$\begin{aligned} V = \boldsymbol{e}^{\mathrm{T}}\boldsymbol{e} &= \big(\boldsymbol{y}^{\mathrm{T}} - \boldsymbol{\theta}^{\mathrm{T}}\boldsymbol{\Psi}^{\mathrm{T}}\big)\big(\boldsymbol{y} - \boldsymbol{\Psi}\boldsymbol{\theta}\big) \\ &= \boldsymbol{y}^{\mathrm{T}}\boldsymbol{y} - \boldsymbol{\theta}^{\mathrm{T}}\boldsymbol{\Psi}^{\mathrm{T}}\boldsymbol{y} - \boldsymbol{y}^{\mathrm{T}}\boldsymbol{\Psi}\boldsymbol{\theta} + \boldsymbol{\theta}^{\mathrm{T}}\boldsymbol{\Psi}^{\mathrm{T}}\boldsymbol{\Psi}\boldsymbol{\theta} \end{aligned} \tag{8.3.6}$$

and

$$V = \boldsymbol{y}^{\mathrm{T}}\boldsymbol{y} - \boldsymbol{\theta}^{\mathrm{T}}\boldsymbol{\Psi}^{\mathrm{T}}\boldsymbol{y} - \big(\boldsymbol{\Psi}^{\mathrm{T}}\boldsymbol{y}\big)^{\mathrm{T}}\boldsymbol{\theta} + \boldsymbol{\theta}^{\mathrm{T}}\boldsymbol{\Psi}^{\mathrm{T}}\boldsymbol{\Psi}\boldsymbol{\theta}\,. \tag{8.3.7}$$

With the calculus for vectors and matrices, see App. A.3, the derivative of the above term with respect to the parameter vector $\boldsymbol{\theta}$ can be determined as

$$\frac{\mathrm{d}}{\mathrm{d}\boldsymbol{\theta}}\big(\boldsymbol{\theta}^{\mathrm{T}}\boldsymbol{\Psi}^{\mathrm{T}}\boldsymbol{y}\big) = \boldsymbol{\Psi}^{\mathrm{T}}\boldsymbol{y} \tag{8.3.8}$$

$$\frac{\mathrm{d}}{\mathrm{d}\boldsymbol{\theta}}\Big(\big(\boldsymbol{\Psi}^{\mathrm{T}}\boldsymbol{y}\big)^{\mathrm{T}}\boldsymbol{\theta}\Big) = \boldsymbol{\Psi}^{\mathrm{T}}\boldsymbol{y} \tag{8.3.9}$$

$$\frac{\mathrm{d}}{\mathrm{d}\boldsymbol{\theta}}\big(\boldsymbol{\theta}^{\mathrm{T}}\boldsymbol{\Psi}^{\mathrm{T}}\boldsymbol{\Psi}\boldsymbol{\theta}\big) = 2\boldsymbol{\Psi}^{\mathrm{T}}\boldsymbol{\Psi}\boldsymbol{\theta}\,, \tag{8.3.10}$$

and thus

$$\frac{\mathrm{d}V}{\mathrm{d}\boldsymbol{\theta}} = -2\boldsymbol{\Psi}^{\mathrm{T}}\boldsymbol{y} + 2\boldsymbol{\Psi}^{\mathrm{T}}\boldsymbol{\Psi}\boldsymbol{\theta} = -2\boldsymbol{\Psi}^{\mathrm{T}}\big(\boldsymbol{y} - \boldsymbol{\Psi}\boldsymbol{\theta}\big)\,. \tag{8.3.11}$$

From the optimality condition

$$\left.\frac{\mathrm{d}V}{\mathrm{d}\boldsymbol{\theta}}\right|_{\boldsymbol{\theta}=\hat{\boldsymbol{\theta}}} \stackrel{!}{=} 0\,, \tag{8.3.12}$$

follows the estimation as

$$\hat{\boldsymbol{\theta}} = \left(\boldsymbol{\Psi}^{\mathrm{T}}\boldsymbol{\Psi}\right)^{-1}\boldsymbol{\Psi}^{\mathrm{T}}\boldsymbol{y}\,. \tag{8.3.13}$$

For the existence of the solution, $\boldsymbol{\Psi}^{\mathrm{T}}\boldsymbol{\Psi}$ may not be singular, thus the condition for a sufficient excitation of the process is given as

$$\det\left(\boldsymbol{\Psi}^{\mathrm{T}}\boldsymbol{\Psi}\right) \neq 0\,. \tag{8.3.14}$$

The expected value of this estimation is given as

$$\mathrm{E}\{\hat{\boldsymbol{\theta}}\} = \boldsymbol{\theta} + \mathrm{E}\{(\boldsymbol{\Psi}^{\mathrm{T}}\boldsymbol{\Psi})^{-1}\boldsymbol{\Psi}^{\mathrm{T}}\boldsymbol{n}\} = \boldsymbol{\theta}\,, \tag{8.3.15}$$

if the elements from $\boldsymbol{\Psi}$ and $\boldsymbol{n}$, i.e. input and noise, are not correlated and $\mathrm{E}\{n(k)\} = 0$. $\hat{\boldsymbol{\theta}}$ thus is a bias-free estimate. The variance can be determined similarly to the approach presented in Chap. 9, see also (Ljung, 1999).

Theorem 8.2 (Convergence of the Parameter Estimation of a Non-Linear Static Process). *The parameters $\boldsymbol{\theta}$ of a non-linear static process according to (8.3.1) are estimated consistently in the mean square by the method of least squares, (8.3.13), if the following necessary conditions are satisfied:*

- *The input signal $U(k)$ is exactly measurable*
- $\det(\boldsymbol{\Psi}^{\mathrm{T}}\boldsymbol{\Psi}) \neq 0$
- *The disturbance $n(k)$ is stationary and zero-mean, i.e.* $\mathrm{E}\{n(k)\} = 0$
- *The input signal $U(k)$ is uncorrelated with the disturbance $n(k)$*

□

8.4 Geometrical Interpretation

The method of least squares can also be interpreted geometrically (Himmelblau, 1970; van der Waerden, 1969; Björck, 1996; Golub and van Loan, 1996; Ljung, 1999; Verhaegen and Verdult, 2007) by means of the *orthogonality relation*. In this section, the problem is thus revisited under geometrical aspects.

The error $\boldsymbol{e}$ has been defined as the difference between the model output $\hat{\boldsymbol{y}}$ and the process output $\boldsymbol{y}$,

$$\boldsymbol{e} = \boldsymbol{y} - \hat{\boldsymbol{y}} = \boldsymbol{y} - \boldsymbol{\Psi}\hat{\boldsymbol{\theta}}\,. \tag{8.4.1}$$

The cost function then is given as

$$V = \boldsymbol{e}^{\mathrm{T}}\boldsymbol{e}\,. \tag{8.4.2}$$

The vector product $\boldsymbol{e}^{\mathrm{T}}\boldsymbol{e}$ can be rewritten as the squared Euclidian distance, therefore

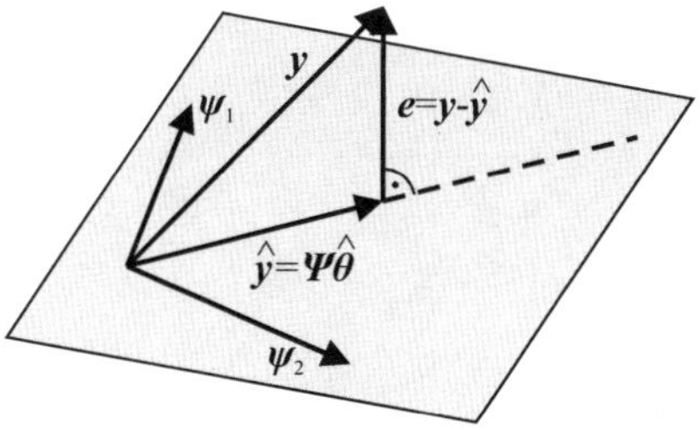

Fig. 8.5. Geometric interpretation of the method of least squares

$$V = \boldsymbol{e}^{\mathrm{T}}\boldsymbol{e} = \|\boldsymbol{e}\|_2^2 \,. \tag{8.4.3}$$

To find the optimal parameter set, the minimum Euclidian distance must be determined, i.e.

$$\min_{\hat{\theta}} \|\boldsymbol{e}\|_2 = \min_{\hat{\theta}} \|\boldsymbol{y} - \hat{\boldsymbol{y}}\|_2 = \min_{\hat{\theta}} \|\boldsymbol{y} - \boldsymbol{\Psi}\hat{\boldsymbol{\theta}}\|_2 \,. \tag{8.4.4}$$

The Euclidian distance of $\boldsymbol{e}$ becomes the smallest, if the vector $\boldsymbol{e}$ is orthogonal to the plane spanned by the regressors $\boldsymbol{\psi}_1$ and $\boldsymbol{\psi}_2$, which are the columns of the data matrix $\boldsymbol{\Psi}$. This is obvious from the geometric point-of-view, but can also be proven mathematically. As was already shown before, the optimality criterion

$$\left.\frac{\mathrm{d}V}{\mathrm{d}\boldsymbol{\theta}}\right|_{\boldsymbol{\theta}=\hat{\boldsymbol{\theta}}} \overset{!}{=} 0 \tag{8.4.5}$$

requires that

$$\boldsymbol{\Psi}^{\mathrm{T}}\big(\boldsymbol{y} - \boldsymbol{\Psi}\hat{\boldsymbol{\theta}}\big) = 0 \,. \tag{8.4.6}$$

As (8.4.6) can be rewritten in terms of the error $\boldsymbol{e}$, one can see that

$$\boldsymbol{\Psi}^{\mathrm{T}}\big(\boldsymbol{y} - \boldsymbol{\Psi}\hat{\boldsymbol{\theta}}\big) = \boldsymbol{\Psi}^{\mathrm{T}}\boldsymbol{e} = 0 \,. \tag{8.4.7}$$

To satisfy (8.4.7), the *orthogonality relation* demands that the error $\boldsymbol{e}$ must be orthogonal to the regressors, i.e. the columns of $\boldsymbol{\Psi}$.

This shall now be illustrated for an experiment with three measurements, see Fig. 8.5: These three measurements are made at different times $k = 1, 2, 3$ as $(\psi_1(1), \psi_2(1), y(1))$, $(\psi_1(2), \psi_2(2), y(2))$, and $(\psi_1(3), \psi_2(3), y(3))$ or in vector matrix notation,

$$\boldsymbol{\Psi} = \begin{pmatrix} \psi_1(1) & \psi_2(1) \\ \psi_1(2) & \psi_2(2) \\ \psi_1(3) & \psi_2(3) \end{pmatrix} \tag{8.4.8}$$

and

$$\boldsymbol{y} = \begin{pmatrix} y_1 \\ y_2 \\ y_3 \end{pmatrix} \,. \tag{8.4.9}$$

The model output $\hat{\boldsymbol{y}}$ is now given as

$$\underbrace{\begin{pmatrix} \hat{y}_1 \\ \hat{y}_2 \\ \hat{y}_3 \end{pmatrix}}_{\hat{\boldsymbol{y}}} = \theta_1 \underbrace{\begin{pmatrix} \psi_1(1) \\ \psi_1(2) \\ \psi_1(3) \end{pmatrix}}_{\boldsymbol{\psi}_1} + \theta_2 \underbrace{\begin{pmatrix} \psi_2(1) \\ \psi_2(2) \\ \psi_2(3) \end{pmatrix}}_{\boldsymbol{\psi}_2} . \quad (8.4.10)$$

Hence, one now tries to represent the vector $\boldsymbol{y}$ of the measurements by the vector $\hat{\boldsymbol{y}}$ of the model output. The model output $\hat{\boldsymbol{y}}$ is given as a linear combination of the column vectors $\boldsymbol{\psi}_1$ and $\boldsymbol{\psi}_2$ of $\boldsymbol{\Psi}$. This means that the vector $\boldsymbol{y}$ must be projected onto the plane spanned by the vector $\boldsymbol{\psi}_1$ and $\boldsymbol{\psi}_2$.

The error $\boldsymbol{e}$ between the model output $\hat{\boldsymbol{y}}$ and the process output $\boldsymbol{y}$ has the smallest norm, which can be seen as the shortest length, if it stands orthogonally on the plane spanned by $\boldsymbol{\psi}_1$ and $\boldsymbol{\psi}_2$. This is what the orthogonality relation states. This orthogonality relation will be revisited in Chap. 16 for the derivation of the subspace methods and Chap. 22 in the derivation of numerically better suited methods for the parameter estimation by the method of least squares. Numerically improved methods typically avoid the direct inversion as used e.g. in (8.2.10) and (8.3.13), but rather construct an orthonormal basis of $\boldsymbol{\Psi}$. The QR-decomposition, described in Chap. 22, also presents an attractive way of solving the method of least squares by decomposing the cost function V, which is written as the squared 2-norm. Such a derivation is also shown by Verhaegen and Verdult (2007).

For linear static processes according to Sect. 8.2, the output signal y and the error e have been linearly dependent on the input signal u and the single parameter K. The non-linear static processes that have been treated in depth in Sect. 8.3 also had a linear dependency between y and the parameters $\boldsymbol{\theta}$ as well as e and $\boldsymbol{\theta}$, but have been non-linear in $\boldsymbol{\psi}$. The parameter estimation that has been described in this chapter is thus also suited for non-linear processes as long as the error e is *linear in the parameters* $\boldsymbol{\theta}$, i.e. the error e depends linearly on the parameters $\boldsymbol{\theta}$ to be estimated.

Although this seems quite confining at first, the limitation is rather small for many practical applications. One can often apply transformations to come up with a problem that is linear in its parameters. For example,

$$Y_{\mathrm{u}}(k) = K_1 \mathrm{e}^{-K_2 U(k)}$$

can be transformed into

$$\log Y_{\mathrm{u}}(k) = \log K_1 - K_2 U(k) ,$$

which is linear in its parameters. Another often needed case is the estimation of amplitude and phase of an oscillation

$$Y_{\mathrm{u}}(k) = a \sin(\omega k T_0 + \varphi) .$$

To identify a and φ, one can use

$$Y_{\mathrm{u}}(k) = b \cos(\omega U(k)) + c \sin(\omega U(k)) \text{ with } U(k) = kT_0 ,$$

where the equation is linear in the parameters b and c. From there, one can obtain estimates of φ and a by

$$\hat{a} = \sqrt{\hat{b}^2 + \hat{c}^2} \text{ and } \hat{\varphi} = \arctan \frac{\hat{b}}{\hat{c}} .$$

Such non-linear expressions, which can be transformed so that they become linear in parameters, are called *intrinsically linear* (Åström and Eykhoff, 1971).

Furthermore, many functions can sufficiently precisely be approximated by polynomials of low order (e.g. 2 or 3) as

$$Y_{\mathrm{u}}(k) = f\big(U(k)\big) \approx K_0 + K_1 U(k) + K_2 U^2(k) . \tag{8.4.11}$$

One can also use piecewise linear approximations, splines and many other approaches.

8.5 Maximum Likelihood and the Cramér-Rao Bound

Initially in this chapter, it has been claimed that the least squares cost function is a natural cost function to choose in the case of disturbances at the output which follow a Gaussian distribution in their probability density function. In the following, a proof of this claim will be presented and also the *quality* of the estimator will be investigated. The question here is what is the *best* estimate that one can obtain.

Before however the estimation quality is discussed, the terms *likelihood* shall be introduced and the *maximum likelihood estimator* shall be derived.

The maximum likelihood estimate is based on the conditional probability of the measurement. This function is given as

$$p_{\boldsymbol{y}}(\boldsymbol{y}|\boldsymbol{u}, \boldsymbol{\theta}) \tag{8.5.1}$$

and is termed *likelihood* function. One can clearly see that the probability that a certain series of values $\boldsymbol{y}$ is measured ("observed") depends on the input $\boldsymbol{u}$ and the parameters $\boldsymbol{\theta}$ that are to be estimated. The input $\boldsymbol{u}$ is now neglected in the argument to come to a more compact notation.

The idea is now to select the parameter estimate $\hat{\boldsymbol{\theta}}$ such that it maximizes the likelihood function as estimates for the true parameters $\boldsymbol{\theta}$, hence

$$p_{\boldsymbol{y}}(\boldsymbol{y}|\boldsymbol{\theta})\big|_{\boldsymbol{\theta}=\hat{\boldsymbol{\theta}}} \rightarrow \max . \tag{8.5.2}$$

Thus, those parameters are taken as the estimates, which make the measurement *most likely* to occur. The maximum can be determined by the classical way in calculus, i.e. by taking the first derivative with respect to the unknown parameters $\boldsymbol{\theta}$ and equate it to zero, i.e.

$$\left.\frac{\partial p_{\boldsymbol{y}}(\boldsymbol{y}|\boldsymbol{\theta})}{\partial \boldsymbol{\theta}}\right|_{\boldsymbol{\theta}=\hat{\boldsymbol{\theta}}} \stackrel{!}{=} \boldsymbol{0} . \tag{8.5.3}$$

This technique shall now be applied to the estimation of a *static non-linearity*. The measured output is given as

$$y = \boldsymbol{\Psi}\boldsymbol{\theta} + \boldsymbol{n}\,. \tag{8.5.4}$$

Each noise sample $n(k)$ is Gaussian distributed with the probability density function

$$p\big(n(k)\big) = \frac{1}{\sqrt{2\pi\sigma_{\mathrm{n}}^2}} \exp\left(-\frac{(n(k)-\mu)^2}{2\sigma_{\mathrm{n}}^2}\right), \tag{8.5.5}$$

where $\mu = 0$ because the noise was assumed to be zero-mean. For white noise, the individual elements are uncorrelated, the probability density function for the entire noise sequence $\boldsymbol{n}$ with N samples is therefore given as the product of the individual probability density functions of each sample,

$$p(\boldsymbol{n}) = \prod_{k=0}^{N-1} \frac{1}{\sqrt{2\pi\sigma_{\mathrm{n}}^2}} \exp\left(-\frac{(n(k))^2}{2\sigma_{\mathrm{n}}^2}\right) = \frac{1}{(2\pi)^{\frac{N}{2}}\sqrt{\det \boldsymbol{\Sigma}}} \mathrm{e}^{-\frac{1}{2}\boldsymbol{n}^{\mathrm{T}}\boldsymbol{\Sigma}^{-1}\boldsymbol{n}}\,, \tag{8.5.6}$$

which is an N-dimensional Gaussian distribution with $\boldsymbol{\Sigma}$ being the covariance matrix. For uncorrelated elements, $\boldsymbol{\Sigma} = \sigma_{\mathrm{n}}^2\boldsymbol{I}$ where $\boldsymbol{I}$ is the identity matrix and $\det \boldsymbol{\Sigma} = N\sigma_{\mathrm{n}}^2$.

Now, the probability density function of the measurements $\boldsymbol{y}$ with $\boldsymbol{y} = \boldsymbol{\Psi}\boldsymbol{\theta} + \boldsymbol{n}$ and hence $\boldsymbol{n} = \boldsymbol{y} - \boldsymbol{\Psi}\boldsymbol{\theta}$ is given as

$$p(\boldsymbol{y}|\boldsymbol{\theta}) = \frac{1}{(2\pi)^{\frac{N}{2}}\sqrt{N}\sigma_{\mathrm{n}}} \exp\left(-\frac{1}{2\sigma_{\mathrm{n}}^2}(\boldsymbol{y}-\boldsymbol{\Psi}\boldsymbol{\theta})^{\mathrm{T}}(\boldsymbol{y}-\boldsymbol{\Psi}\boldsymbol{\theta})\right), \tag{8.5.7}$$

For this function, the maximum must now be determined. A rather common approach in the calculation of maximum likelihood estimates involving Gaussian distributions is to take the logarithm first as $p(x)$ and $\log p(x)$ have their maximum at the same value of x, but the resulting terms are much easier to work with. Hence,

$$\left.\frac{\partial \log f(\boldsymbol{y}|\boldsymbol{\theta})}{\partial \boldsymbol{\theta}}\right|_{\boldsymbol{\theta}=\hat{\boldsymbol{\theta}}} = \frac{1}{2\sigma_{\mathrm{N}}^2}\Big(\big(\boldsymbol{y}-\boldsymbol{\Psi}\hat{\boldsymbol{\theta}}\big)^{\mathrm{T}}\big(\boldsymbol{y}-\boldsymbol{\Psi}\hat{\boldsymbol{\theta}}\big)\Big) \overset{!}{=} 0\,. \tag{8.5.8}$$

Solving this equation for $\hat{\boldsymbol{\theta}}$ leads to

$$\hat{\boldsymbol{\theta}} = \big(\boldsymbol{\Psi}^{\mathrm{T}}\boldsymbol{\Psi}\big)^{-1}\boldsymbol{\Psi}^{\mathrm{T}}\boldsymbol{y}\,, \tag{8.5.9}$$

which is identical to (8.3.13). Hence the *least squares estimator* and the *maximum likelihood estimator* yield identical solutions.

Now, the quality of the estimate shall be discussed. The question is whether there exists a lower bound on the variance of the estimate. If this is the case, then the estimator that delivers estimates with the minimum attainable variance would be the best estimator that is available. This measure of quality will be called *efficiency* of an estimator (see App. A), and a lower bound on the minimum attainable variance will be derived as the *Cramér-Rao bound*.

For the derivation of this bound, an arbitrary bias-free estimate $\hat{\boldsymbol{\theta}}$ shall be considered, hence $\mathrm{E}\{\hat{\boldsymbol{\theta}}|\boldsymbol{\theta}\} = \boldsymbol{\theta}$. With this prerequisite, one can now try to determine the variance of the estimate $\mathrm{E}\{(\boldsymbol{\theta} - \hat{\boldsymbol{\theta}}|\boldsymbol{\theta})^2\}$.

The following derivation of the Cramér-Rao bound has been described by Hänsler (2001) and is in more depth presented there. The idea behind the derivation is that for a bias-free estimate, the expected value of $\mathrm{E}\{\hat{\boldsymbol{\theta}} - \boldsymbol{\theta}\}$ is zero, i.e.

$$\mathrm{E}\{\hat{\boldsymbol{\theta}} - \boldsymbol{\theta}\} = \int_{-\infty}^{\infty} (\hat{\boldsymbol{\theta}} - \boldsymbol{\theta}) p_{\mathrm{y}}(\boldsymbol{y}|\boldsymbol{\theta}) \mathrm{d}\boldsymbol{y} \,. \tag{8.5.10}$$

One can take the derivative with respect to the parameter vector $\boldsymbol{\theta}$ and then one obtains

$$\int_{-\infty}^{\infty} \frac{\partial}{\partial \boldsymbol{\theta}} \Big((\hat{\boldsymbol{\theta}} - \boldsymbol{\theta}) p_{\boldsymbol{y}}(\boldsymbol{y}|\boldsymbol{\theta}) \mathrm{d}\boldsymbol{y} \Big) = 0 \,. \tag{8.5.11}$$

With the Cauchy-Schwartz inequality $(\mathrm{E}\{xy\})^2 \leq \mathrm{E}\{x^2\}\mathrm{E}\{y^2\}$, one can state a lower bound for the variance of the estimate as

$$\mathrm{E}\big\{(\hat{\boldsymbol{\theta}} - \boldsymbol{\theta})^2\big\} \geq \frac{1}{\mathrm{E}\left\{\left(\dfrac{\partial}{\partial \boldsymbol{\theta}} \log p_{\boldsymbol{y}}(\boldsymbol{y}|\boldsymbol{\theta})\right)^2\right\}} \tag{8.5.12}$$

or

$$\mathrm{E}\big\{(\hat{\boldsymbol{\theta}} - \boldsymbol{\theta})^2\big) \geq \frac{-1}{\mathrm{E}\left\{\dfrac{\partial^2}{\partial \boldsymbol{\theta}^2} \log p_{\boldsymbol{y}}(\boldsymbol{y}|\boldsymbol{\theta})\right\}} \,. \tag{8.5.13}$$

A detailed derivation can also be found in (e.g. Raol et al, 2004).

An estimator is termed *efficient* if it attains this lower bound of the variance. The term *BLUE* stands for *best linear unbiased estimator* and hence denotes the estimator that attains the minimal variance of all unbiased estimators. According to the Gauss-Markov theorem, the least squares estimator is the best linear unbiased estimator, it will in the following be shown that this estimator also reaches the Cramér-Rao bound in the case of Gaussian distributed noise $n(k)$ respectively errors $e(k)$. The Cramér-Rao bound is not always attainable, because it is too conservative as pointed out by Pintelon and Schoukens (2001). Relaxed bounds exist, but are very difficult to calculate and are hence seldom used.

The Cramér-Rao bound can now also be applied to the least squares estimator, which in this case is equivalent to the maximum likelihood estimator. The minimum attainable variance is hence given as

$$\mathrm{E}\big\{(\hat{\boldsymbol{\theta}} - \boldsymbol{\theta})^2\big\} \geq \frac{-1}{\mathrm{E}\left\{\dfrac{\partial^2}{\partial \boldsymbol{\theta}^2} \log p_{\boldsymbol{y}}(\boldsymbol{y}|\boldsymbol{\theta})\right\}} = \sigma_{\mathrm{e}}^2 \mathrm{E}\big\{(\boldsymbol{\Psi}^{\mathrm{T}}\boldsymbol{\Psi})^{-1}\big\} \,. \tag{8.5.14}$$

If the error $e(k)$ / noise $n(k)$ is Gaussian distributed, then the estimator attains the Cramér-Rao bound and hence the minimum variance of all estimators. The denominator is termed *Fisher information matrix* (Fisher, 1922, 1950). An extension of the Cramér-Rao bound to biased estimates is discussed in (van den Bos, 2007).

8.6 Constraints

Constraints are additional conditions that the solution must satisfy. One can discern *equality* and *inequality constraints*. Linear inequality constraints require the parameters to satisfy the set of inequalities

$$\boldsymbol{A}\boldsymbol{\theta} \leq \boldsymbol{b}\,, \tag{8.6.1}$$

whereas equality constraints require that the parameters satisfy a set of equations given as

$$\boldsymbol{C}\boldsymbol{\theta} = \boldsymbol{d}\,. \tag{8.6.2}$$

The inclusion of equality constraints shall be introduced first as it can easily be solved directly (e.g. Björck, 1996). It is required that $\boldsymbol{C}$ has linearly independent rows so that the system of equations in (8.6.2) is consistent. Then, one can first solve the unrestricted problem of least squares as usual as

$$\hat{\boldsymbol{\theta}} = \left(\boldsymbol{\Psi}^{\mathrm{T}}\boldsymbol{\Psi}\right)^{-1}\boldsymbol{\Psi}^{\mathrm{T}}\boldsymbol{y}\,. \tag{8.6.3}$$

Then, the solution to the restricted problem, $\tilde{\boldsymbol{\theta}}$ can be determined as

$$\tilde{\boldsymbol{\theta}} = \hat{\boldsymbol{\theta}} - \left(\boldsymbol{\Psi}^{\mathrm{T}}\boldsymbol{\Psi}\right)^{-1}\boldsymbol{C}\Big(\boldsymbol{C}\left(\boldsymbol{\Psi}^{\mathrm{T}}\boldsymbol{\Psi}\right)^{-1}\boldsymbol{C}^{\mathrm{T}}\Big)^{-1}\left(\boldsymbol{C}\hat{\boldsymbol{\theta}} - \boldsymbol{d}\right), \tag{8.6.4}$$

see e.g. (Doyle et al, 2002).

Inequality constraints can in theory be solved with *active set methods*, although in numerical implementations prefer alternatives, such as interior point methods (Nocedal and Wright, 2006). The basic idea is that inequality constraints can either be inactive, then they do not need to be regarded as part of the solution of the optimization problem. Or, if they are active, then they can be treated as equality constraints as the design variable is fixed to the boundary of the *feasible space*. Active in this context means a constraint that actively constrains the solution, whereas inactive constraints are currently not influencing the solution. The critical part in this algorithm is the determination of the active set, which can be of exponential complexity. The recursive method of least squares (RLS) also allows the inclusion of constraints in a very elegant way, see Sect. 9.6.1.

8.7 Summary

In this chapter, the method of least squares was derived for linear and non-linear processes. It is well suited for static processes described by linear and non-linear algebraic equations. Important for the direct, i.e. non-iterative solution is the condition that the error between the process output and the model output is linear in the parameters. However, also many functions which are at first non-linear in parameters can either be transformed so that they become linear in parameters or can be approximated by polynomials or piecewise linear models.

With the Gauss-Markov theorem, it can be shown that for these applications, the method of least squares provides the best linear unbiased estimate. It has furthermore been shown that the method of least squares for Gaussian noise at the output is equivalent to the maximum likelihood estimator and that the variance of the estimate asymptotically attains the Cramér-Rao bound, which is a lower bound for the variance of an estimate. This makes it an asymptotically efficient estimator.

Problems

8.1. Non-Linear Static SISO Process

The process shall be modeled with a non-linear static model of second order,

$$y(k) = K_0 + K_1 u(k) + K_2 u^2(k)$$

The parameter K_0 is zero, so it does not have to be considered in the following. Determine the parameters K_1 and K_2 based on the measured data points

Data point k	1	2	3	4	5
Input signal u	-1.5	-0.5	4.5	7	8
Output signal y	5.5	1.5	-3.5	4.5	8.5

using the method of least squares.

8.2. Non-Linear Static MISO Process

A MISO process is described by a non-linear second order model

$$y(k) = K_0 - K_1 u_1(k) u_2(k) + K_2 u_1^2(k)$$

The process shall be identified by the method of least aquares. Set up the data matrix $\boldsymbol{\Psi}$, data vector $\boldsymbol{y}$, and the parameter vector $\boldsymbol{\theta}$. Given the measurements

Data point k	1	2	3	4	5
Input signal u_1	-1	-0.5	0	1	2
Input signal u_2	2	2	2	2	2
Output signal y	3.5	1.875	0	-4.5	-10

determine the parameters K_1 and K_2 under the assumption that $K_0 = 0$.

8.3. Non-Linear Static SISO Process

A static non-linear process with the structure

$$y(k) = \sqrt{au(k)} + (b+1)u^2(k)$$

shall be identified for the measurements

Data point k	1	2	3	4	5
Input signal u	0.5	1	1.5	2	2.5
Output signal y	2.2247	5.7321	11.1213	18.4495	27.7386

Set up the data matrix $\boldsymbol{\Psi}$, data vector $\boldsymbol{y}$ and the parameter vector $\boldsymbol{\theta}$. Then, determine the parameters a and b.

8.4. Sinusoidal Oscillation
In the following, the methods of least square shall be utilized to determine the phase φ and amplitude A of an oscillation with known frequency ω_0, i.e.

$$y(t) = A\sin(\omega_0 t + \varphi)$$

The signal has been sampled with the sample time $T_0 = 0.1$ s. The frequency of the oscillation is known to be $\omega_0 = 10\,\text{rad/sec}$. The following measurements have been determined:

Data point k	0	1	2	3	4
Output signal $y(k)$	0.52	1.91	1.54	-0.24	-1.80

Determine the parameters A and φ.

8.5. Consistent Estimate and BLUE
What is a consistent estimate? What does the term BLUE stand for?

8.6. Bias
What is a bias? How is it defined mathematically?

References

Åström KJ, Eykhoff P (1971) System identification – a survey. Automatica 7(2):123–162

Björck Å (1996) Numerical methods for least squares problems. SIAM, Philadelphia

van den Bos A (2007) Parameter estimation for scientists and engineers. Wiley-Interscience, Hoboken, NJ

Doyle FJ, Pearson RK, Ogunnaike BA (2002) Identification and control using Volterra models. Communications and Control Engineering, Springer, London

Fisher RA (1922) On the mathematical foundation of theoretical statistics. Philos Trans R Soc London, Ser A 222:309–368

Fisher RA (1950) Contributions to mathematical statistics. J. Wiley, New York, NY

Golub GH, van Loan CF (1996) Matrix computations, 3rd edn. Johns Hopkins studies in the mathematical sciences, Johns Hopkins Univ. Press, Baltimore

Hänsler E (2001) Statistische Signale: Grundlagen und Anwendungen. Springer, Berlin

Himmelblau DM (1970) Process analysis by statistical methods. Wiley & Sons, New York, NY

Ljung L (1999) System identification: Theory for the user, 2nd edn. Prentice Hall Information and System Sciences Series, Prentice Hall PTR, Upper Saddle River, NJ

Nocedal J, Wright SJ (2006) Numerical optimization, 2nd edn. Springer series in operations research, Springer, New York

Pintelon R, Schoukens J (2001) System identification: A frequency domain approach. IEEE Press, Piscataway, NJ

Raol JR, Girija G, Singh J (2004) Modelling and parameter estimation of dynamic systems, IEE control engineering series, vol 65. Institution of Electrical Engineers, London

Verhaegen M, Verdult V (2007) Filtering and system identification: A least squares approach. Cambridge University Press, Cambridge

van der Waerden BL (1969) Mathematical statistics. Springer, Berlin

9

Least Squares Parameter Estimation for Dynamic Processes

The application of the method of least squares to static models has been described in the previous chapter and is well known to scientists for a long time already. The application of the method of least squares to the identification of dynamic processes has been tackled with much later in time. First works on the parameter estimation of AR models have been reported in the analysis of time series of economic data (Koopmans, 1937; Mann and Wald, 1943) and for the difference equations of linear dynamic processes (Kalman, 1958; Durbin, 1960; Levin, 1960; Lee, 1964).

The application of the method of least squares to dynamic processes is dealt with in this chapter for the discrete-time case and later in Chap. 15 for the continuous-time case, as well. In the chapter at hand, first the original non-recursive setting is derived, then the recursive form is presented in detail. Also, the weighted method of least squares is presented and the highly important case of least squares with exponential forgetting.

In order not to conceal the train of thoughts for the application of the method of least squares to dynamic systems, only one mathematical solution of the problem will be presented. In Chap. 22, different ways of solving the least squares problem are presented and compared in terms of accuracy and speed. Modifications to the method of least squares, which e.g. allow better estimation results for noise acting on the input and other cases can be found in Chap. 10.

9.1 Non-Recursive Method of Least Squares (LS)

In the following, the classical method of least squares for discrete-time linear processes will be derived.

9.1.1 Fundamental Equations

The transfer function of a discrete-time linear process is given as

$$G(z^{-1}) = \frac{y(z)}{u(z)} = \frac{b_0 + b_1 z^{-1} + \ldots + b_m z^{-m}}{1 + a_1 z^{-1} + \ldots + a_m z^{-m}} = \frac{B(z^{-1})}{A(z^{-1})}\,, \tag{9.1.1}$$

R. Isermann, M. Münchhof, *Identification of Dynamic Systems*,
DOI 10.1007/978-3-540-78879-9_9, © Springer-Verlag Berlin Heidelberg 2011

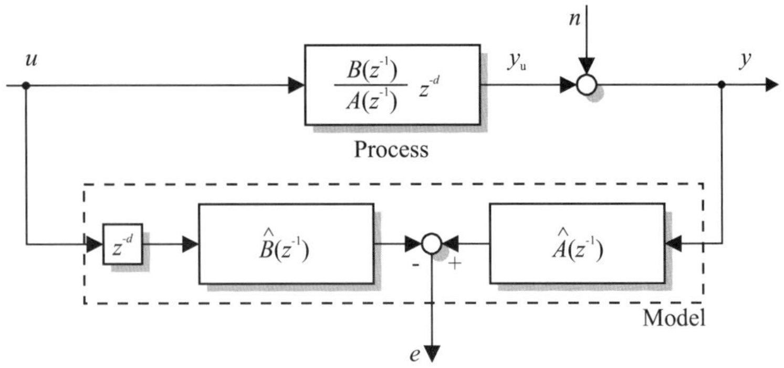

Fig. 9.1. Block diagram for the non-recursive parameter estimation according to the method of least squares

with

$$\begin{aligned} u(k) &= U(k) - U_{00} \\ y(k) &= Y(k) - Y_{00} \end{aligned} \tag{9.1.2}$$

as deviations of the signals from the steady-state values U_{00} and Y_{00}, see also (2.2.19). The parameter b_0 will be disregarded in the following as biproper systems, i.e. systems that can directly follow a step input, are hardly ever encountered in nature and the dimensionality of the subsequent parameter estimation problem is reduced by one dimension by neglecting b_0. The process (9.1.1) will now be extended by introducing a dead time T_{d} with integer values $d = T_{\mathrm{d}}/T_0 = 0, 1, \ldots$. It is further assumed that the model order and dead time are known exactly. If this is not the case, one can apply the methods described in Chap. 23 to determine an appropriate model order and dead time.

The transfer function is then given as

$$G_{\mathrm{P}}(z) = \frac{y_{\mathrm{u}}(z)}{u(z)} = \frac{B(z^{-1})}{A(z^{-1})} z^{-d} = \frac{b_1 z^{-1} + \ldots + b_m z^{-m}}{1 + a_1 z^{-1} + \ldots + a_m z^{-m}} z^{-d} \,. \tag{9.1.3}$$

The measurable signal $y(k)$ is assumed to be a superposition of the real process output $y_{\mathrm{u}}(k)$ (useful signal) and a stochastic disturbance $n(k)$ as

$$y(k) = y_{\mathrm{u}}(k) + n(k) \,, \tag{9.1.4}$$

see Fig. 9.1

Now, the task is to determine the unknown parameters a_i and b_i of the process transfer function (9.1.3) from N pairs of the input signal $u(k)$ and the measured output $y(k)$. First, some assumptions have to be made:

- The process is for $k < 0$ in its steady state
- The model order m and dead time d are known exactly (see Chap. 23 on how to determine the model order m and dead time d if they are not known a priori)

- The input $u(k)$ and its DC value U_{00} are known exactly
- The disturbance $n(k)$ must be stationary with $\mathrm{E}\{n(k)\} = 0$
- The DC value Y_{00} must be known exactly and must belong to U_{00}

(9.1.3) will be transformed to the time domain, resulting in the difference equation

$$y_{\mathrm{u}}(k)+a_1 y_{\mathrm{u}}(k-1)+\ldots+a_m y_{\mathrm{u}}(k-m) = b_1 u(k-d-1)+\ldots+b_m u(k-d-m)\,. \quad (9.1.5)$$

Now, the measured values $y(k)$ are used instead of the model output $y_{\mathrm{u}}(k)$ and furthermore the estimated parameters are plugged into the equation, leading to

$$\begin{aligned} &y(k) + \hat{a}_1(k-1)y(k-1) + \ldots + \hat{a}_m(k-1)y(k-m) \\ &\quad - \hat{b}_1(k-1)u(k-d-1) - \ldots - \hat{b}_m(k-1)u(k-d-m) = e(k)\,. \end{aligned} \quad (9.1.6)$$

The equation error $e(k)$ (residual) is introduced which is caused by the use of the measured values $y(k)$ instead of the "true" output $y_{\mathrm{u}}(k)$ and the use of parameter estimates instead of the "true" parameters. Figure 9.1 shows the overall setup. According to Chap. 1 (see Fig. 1.8), this setup is called the *generalized equation error*. This error definition is *linear in parameters*, which is a requirement for the application of *direct methods* of parameter estimation. As was already stated in Sect. 1.3, direct methods allow the determination of the parameter estimates in one pass.

(9.1.6) can be interpreted as the prediction $\hat{y}(k|k-1)$ of the output signal $y(k)$ for one time step ahead into the future (*one-step prediction*) based on the measurements that have been available up to the time step $k-1$. In this context, (9.1.6) can be written as

$$\begin{aligned} \hat{y}(k|k-1) &= -\hat{a}_1(k-1)y(k-1) - \ldots - \hat{a}_m(k-1)y(k-m) \\ &\quad + \hat{b}_1(k-1)u(k-d-1) + \ldots - \hat{b}_m(k-1)u(k-d-m) \\ &= \boldsymbol{\psi}^{\mathrm{T}}(k)\hat{\boldsymbol{\theta}}(k-1) \end{aligned} \quad (9.1.7)$$

with the data vector

$$\boldsymbol{\psi}^{\mathrm{T}}(k) = \big(-y(k-1)\,\ldots\,-y(k-m)\big|u(k-d-1)\,\ldots\,u(k-d-m)\big) \quad (9.1.8)$$

and the parameter vector

$$\hat{\boldsymbol{\theta}}^{\mathrm{T}}(k) = \big(\hat{a}_1\,\ldots\,\hat{a}_m\big|\hat{b}_1\,\ldots\,\hat{b}_m\big)\,. \quad (9.1.9)$$

As one can see, (9.1.7) corresponds to the regression model in the non-linear static case, (8.3.3). The equation error in (9.1.6) can thus be interpreted as

$$\underset{\substack{\text{Equation}\\ \text{Error}}}{e(k)} = \underset{\substack{\text{New}\\ \text{Observation}}}{y(k)} - \underset{\substack{\text{One Step Prediction}\\ \text{of the Model}}}{\hat{y}(k|k-1)}\,. \quad (9.1.10)$$

Now, the input and output are sampled for the data points $k = 0, 1, 2, \ldots, m+d+N$. For the time step $k = m + d$, the data vector $\boldsymbol{\psi}$ (9.1.8) can be filled up for the first time. For the time steps $k = m+d, m+d+1, \ldots, m+d+N$, a system of $N+1$ equations of the form

$$y(k) = \boldsymbol{\psi}^{\mathrm{T}}(k)\hat{\boldsymbol{\theta}}(k-1) + e(k) \tag{9.1.11}$$

can be set up. In order to determine the $2m$ parameters, one needs at least $2m$ equations, thus $N \geq 2m-1$. In order to suppress the influence of the disturbance $n(k)$, one will typically use much more equations, so that $N \gg 2m-1$.

These $N+1$ equations can be written in matrix form as

$$\boldsymbol{y}(m+d+N) = \boldsymbol{\Psi}(m+d+N)\hat{\boldsymbol{\theta}}(m+d+N-1) + \boldsymbol{e}(m+d+N) \tag{9.1.12}$$

with

$$\boldsymbol{y}^{\mathrm{T}}(m+d+N) = \big(y(m+d)\; y(m+d+1)\; \ldots\; y(m+d+N)\big) \tag{9.1.13}$$

and the data matrix

$$\boldsymbol{\Psi}(m+d+N) = \\ \left(\begin{array}{ccc|ccc} -y(m+d-1) & \ldots & -y(d) & u(m-1) & \ldots & u(0) \\ -y(m+d) & \ldots & -y(d+1) & u(m) & \ldots & u(1) \\ \vdots & & \vdots & \vdots & & \vdots \\ -y(m+d+N-1) & \ldots & -y(d+N) & u(m+N-1) & \ldots & u(N) \end{array}\right). \tag{9.1.14}$$

For the minimization of the cost function

$$V = \boldsymbol{e}^{\mathrm{T}}(m+d+N)\boldsymbol{e}(m+d+N) = \sum_{k=m+d}^{m+d+N} e^2(k) \tag{9.1.15}$$

with

$$\boldsymbol{e}^{\mathrm{T}}(m+d+N) = \big(e(m+d)\; e(m+d+1)\; \ldots\; e(m+d+N)\big), \tag{9.1.16}$$

one must in analogy to Sect. 8.3 determine the first derivative of the cost function, see (8.3.11), and equate it to zero,

$$\left.\frac{\mathrm{d}V}{\mathrm{d}\boldsymbol{\theta}}\right|_{\boldsymbol{\theta}=\hat{\boldsymbol{\theta}}} = -2\boldsymbol{\Psi}^{\mathrm{T}}\big(\boldsymbol{y} - \boldsymbol{\Psi}\hat{\boldsymbol{\theta}}\big) \stackrel{!}{=} \boldsymbol{0}. \tag{9.1.17}$$

For the time step $k = m+d+N$, the solution of the over-determined system of equations is given as

$$\hat{\boldsymbol{\theta}} = \big(\boldsymbol{\Psi}^{\mathrm{T}}\boldsymbol{\Psi}\big)^{-1}\boldsymbol{\Psi}^{\mathrm{T}}\boldsymbol{y}, \tag{9.1.18}$$

compare (8.3.13). With the abbreviation

$$\boldsymbol{P} = \big(\boldsymbol{\Psi}^{\mathrm{T}}\boldsymbol{\Psi}\big)^{-1}, \tag{9.1.19}$$

the equation is given as

$$\hat{\boldsymbol{\theta}} = \boldsymbol{P}\boldsymbol{\Psi}^{\mathrm{T}}\boldsymbol{y} \,. \tag{9.1.20}$$

To calculate the parameter estimate $\hat{\boldsymbol{\theta}}$, one must thus invert the matrix

$$\boldsymbol{\Psi}^{\mathrm{T}}\boldsymbol{\Psi} = \boldsymbol{P}^{-1} \text{ see next page} \tag{9.1.21}$$

and multiply it with the vector

$$\boldsymbol{\Psi}^{\mathrm{T}}\boldsymbol{y} \text{ see next page} \,, \tag{9.1.22}$$

see e.g. (Åström and Eykhoff, 1971).

The matrix $\boldsymbol{\Psi}$ has the dimension $(N+1)\times 2m$ and thus grows quite large for large measurement periods. The matrix $\boldsymbol{\Psi}^{\mathrm{T}}\boldsymbol{\Psi}$ is for stationary input and output signals symmetric and has independent of the measurement period the dimension $2m \times 2m$. In order for the inverse to exist, the matrix $\boldsymbol{\Psi}^{\mathrm{T}}\boldsymbol{\Psi}$ must have the (full) rank $2m$ or

$$\det\bigl(\boldsymbol{\Psi}^{\mathrm{T}}\boldsymbol{\Psi}\bigr) = \det\bigl(\boldsymbol{P}^{-1}\bigr) \neq 0 \,. \tag{9.1.23}$$

This necessary condition means among others that the process under investigation must be excited sufficiently by the input signal, see Sect. 9.1.4. According to (9.1.17), the second derivative is given as

$$\frac{\mathrm{d}V}{\mathrm{d}\boldsymbol{\theta}\,\mathrm{d}\boldsymbol{\theta}^{\mathrm{T}}} = \boldsymbol{\Psi}^{\mathrm{T}}\boldsymbol{\Psi} \,. \tag{9.1.24}$$

In order for the cost function to be (locally) minimal at the optimum determined from (9.1.17) and for $\hat{\boldsymbol{\theta}}$ to have a unique solution, the matrix $\boldsymbol{\Psi}^{\mathrm{T}}\boldsymbol{\Psi}$ must be positive definite, i.e.

$$\det \boldsymbol{\Psi}^{\mathrm{T}}\boldsymbol{\Psi} > 0 \,. \tag{9.1.25}$$

If both $\boldsymbol{\Psi}^{\mathrm{T}}\boldsymbol{\Psi}$ and $\boldsymbol{\Psi}^{\mathrm{T}}\boldsymbol{y}$ are divided by N, then the individual elements of the matrix and the vector are correlation functions with different starting and ending points as (9.1.21) and (9.1.22) show. For large N however, one can neglect these different starting and ending times and build a "correlation matrix"

$$(N+1)^{-1}\boldsymbol{\Psi}^{\mathrm{T}}\boldsymbol{\Psi} =$$

$$\left(\begin{array}{cccc|ccc}
\hat{R}_{\mathrm{yy}}(0) & \hat{R}_{\mathrm{yy}}(1) & \ldots & \hat{R}_{\mathrm{yy}}(m-1) & -\hat{R}_{\mathrm{uy}}(d) & \ldots & -\hat{R}_{\mathrm{uy}}(d+m-1) \\
\hat{R}_{\mathrm{yy}}(1) & \hat{R}_{\mathrm{yy}}(0) & \ldots & \hat{R}_{\mathrm{yy}}(m-2) & -\hat{R}_{\mathrm{uy}}(d-1) & \ldots & -\hat{R}_{\mathrm{uy}}(d+m-2) \\
\vdots & \vdots & & \vdots & \vdots & & \vdots \\
 & & \ldots & \hat{R}_{\mathrm{yy}}(0) & -\hat{R}_{\mathrm{uy}}(d-m+1) & \ldots & -\hat{R}_{\mathrm{uy}}(d) \\
\hline
 & & & & \hat{R}_{\mathrm{uu}}(0) & \ldots & \hat{R}_{\mathrm{uu}}(m-1) \\
\vdots & \vdots & & \vdots & \vdots & & \vdots \\
 & & & & & & \hat{R}_{\mathrm{uu}}(0)
\end{array}\right) \tag{9.1.26}$$

and a "correlation vector"

$$\boldsymbol{\Psi}^{\mathrm{T}}\boldsymbol{\Psi} = \left(\begin{array}{ccc|ccc}
\sum\limits_{k=m+d-1}^{m+d+N-1} y^2(k) & \dots & \sum\limits_{k=m+d-1}^{m+d+N-1} y(k)y(k-m+1) & -\sum\limits_{k=m+d-1}^{m+d+N-1} y(k)u(k-d) & \dots & \sum\limits_{k=m+d-1}^{m+d+N-1} y(k)u(k-d-m+1) \\
\sum\limits_{k=m+d-2}^{m+d+N-2} y(k)y(k+1) & \dots & \sum\limits_{k=m+d-2}^{m+d+N-2} y(k)y(k-m+2) & -\sum\limits_{k=m+d-2}^{m+d+N-2} y(k)u(k-d+1) & \dots & \sum\limits_{k=m+d-2}^{m+d+N-2} y(k)u(k-d-m+2) \\
\vdots & & \vdots & \vdots & & \vdots \\
 & & \sum\limits_{k=d}^{d+N} y^2(k) & -\sum\limits_{k=d}^{d+N} y(k)u(k-d+m-1) & \dots & \sum\limits_{k=d}^{d+N} y(k)u(k-d) \\
\hline
 & & & \sum\limits_{k=m-1}^{m+N-1} u^2(k) & \dots & \sum\limits_{k=m-1}^{m+N-1} u(k)u(k-m+1) \\
 & & & & & \sum\limits_{k=m-2}^{m+N-2} u(k)u(k-m+2) \\
 & & & & & \vdots \\
 & & & & & \sum\limits_{k=0}^{N} u^2(k)
\end{array}\right) \tag{9.1.21}$$

$$\boldsymbol{\Psi}^{\mathrm{T}}\boldsymbol{y} = \left(\begin{array}{c}
-\sum\limits_{k=m+d}^{m+d+N} y(k)y(k-1) \\
-\sum\limits_{k=m+d}^{m+d+N} y(k)y(k-2) \\
\vdots \\
-\sum\limits_{k=m+d}^{m+d+N} y(k)y(k-m) \\
\hline
-\sum\limits_{k=m+d}^{m+d+N} y(k)u(k-d-1) \\
\vdots \\
-\sum\limits_{k=m+d}^{m+d+N} y(k)u(k-d-m)
\end{array}\right) \tag{9.1.22}$$

$$(N+1)^{-1}\boldsymbol{\Psi}^{\mathrm{T}}\boldsymbol{y} = \begin{pmatrix} -\hat{R}_{\mathrm{yy}}(1) \\ -\hat{R}_{\mathrm{yy}}(2) \\ \vdots \\ -\hat{R}_{\mathrm{yy}}(m) \\ \hline \hat{R}_{\mathrm{uy}}(d+1) \\ \vdots \\ \hat{R}_{\mathrm{uy}}(d+m) \end{pmatrix} . \tag{9.1.27}$$

The method of least squares can thus for the dynamic case also be expressed by correlation functions. If one determines $\hat{\boldsymbol{\theta}}$ according to

$$\hat{\boldsymbol{\theta}} = \left(\frac{1}{N+1}\boldsymbol{\Psi}^{\mathrm{T}}\boldsymbol{\Psi}\right)^{-1} \frac{1}{N+1}\boldsymbol{\Psi}^{\mathrm{T}}\boldsymbol{y} , \tag{9.1.28}$$

then the elements of the matrix and the vector will approach constant values of the correlation function in case of convergence. Thus, the entries of the matrix and the vector are very well suited as non-parametric and easy to interpret intermediate results to check the progress of the parameter estimation.

One should keep in mind that only the following correlation functions are employed,

$$\begin{aligned} \hat{R}_{\mathrm{yy}}(0), \hat{R}_{\mathrm{yy}}(1), \ldots, \hat{R}_{\mathrm{yy}}(m-1) \\ \hat{R}_{\mathrm{uu}}(0), \hat{R}_{\mathrm{uu}}(1), \ldots, \hat{R}_{\mathrm{uu}}(m-1) \\ \hat{R}_{\mathrm{uy}}(d), \hat{R}_{\mathrm{uy}}(d+1), \ldots, \hat{R}_{\mathrm{uy}}(d+m-1) . \end{aligned}$$

Thus, always m values will be used for the calculation of the correlation functions. If the correlation functions are also considerably different from zero for other time shifts τ, i.e. $\tau < 0$ and $\tau > m-1$ or $\tau < d$ and $\tau > d+m-1$ respectively, the technique does not employ the entire available information about the process dynamics. This topic will again be discussed in Sect. 9.3.

In order to calculate the estimates for the parameters, one has the following options:

- Set up $\boldsymbol{\Psi}$ and $\boldsymbol{y}$. Calculate $\boldsymbol{\Psi}^{\mathrm{T}}\boldsymbol{\Psi}$ and $\boldsymbol{\Psi}^{\mathrm{T}}\boldsymbol{y}$. Then solve the parameter estimation problem using (9.1.18)
- Determine the elements of $\boldsymbol{\Psi}^{\mathrm{T}}\boldsymbol{\Psi}$ and $\boldsymbol{\Psi}^{\mathrm{T}}\boldsymbol{y}$ in form of the sums given by (9.1.21), (9.1.22). Then use (9.1.18)
- Determine the elements of $(N+1)^{-1}\boldsymbol{\Psi}^{\mathrm{T}}\boldsymbol{\Psi}$ and $(N+1)^{-1}\boldsymbol{\Psi}^{\mathrm{T}}\boldsymbol{y}$ in the form of the correlation functions according to (9.1.26), (9.1.27). Then use (9.1.28)

9.1.2 Convergence

In order to examine the convergence, the expected values and the convergence of the parameter estimates will be analyzed for the case assumed in (9.1.4), where the output has been affected by a stochastic noise $n(k)$.

For the expected value of the estimate, one obtains by inserting (9.1.12) into (9.1.18) under the assumption that the estimated parameters $\hat{\boldsymbol{\theta}}$ of the model (9.1.12) already agree with the true process parameters $\boldsymbol{\theta}_0$,

$$\mathrm{E}\{\hat{\boldsymbol{\theta}}\} = \mathrm{E}\big\{(\boldsymbol{\Psi}^{\mathrm{T}}\boldsymbol{\Psi})^{-1}\boldsymbol{\Psi}^{\mathrm{T}}\boldsymbol{\Psi}\boldsymbol{\theta}_0 + (\boldsymbol{\Psi}^{\mathrm{T}}\boldsymbol{\Psi})^{-1}\boldsymbol{\Psi}^{\mathrm{T}}\boldsymbol{e}\big\} = \boldsymbol{\theta}_0 + \mathrm{E}\big\{(\boldsymbol{\Psi}^{\mathrm{T}}\boldsymbol{\Psi})^{-1}\boldsymbol{\Psi}^{\mathrm{T}}\boldsymbol{e}\big\}\,, \tag{9.1.29}$$

where

$$\boldsymbol{b} = \mathrm{E}\big\{(\boldsymbol{\Psi}^{\mathrm{T}}\boldsymbol{\Psi})^{-1}\boldsymbol{\Psi}^{\mathrm{T}}\boldsymbol{e}\big\} \tag{9.1.30}$$

is a bias. The afore mentioned assumption that $\hat{\boldsymbol{\theta}} = \boldsymbol{\theta}_0$ is satisfied, if the bias vanishes. This leads to

Theorem 9.1 (First Property of the Bias-Free Parameter Estimation).

If the parameters of the dynamic process governed by (9.1.5) are estimated bias-free by the method of least squares, then $\boldsymbol{\Psi}^{\mathrm{T}}$ and $\boldsymbol{e}$ are uncorrelated and furthermore $\mathrm{E}\{\boldsymbol{e}\} = 0$. *Then*

$$\boldsymbol{b} = \mathrm{E}\big\{(\boldsymbol{\Psi}^{\mathrm{T}}\boldsymbol{\Psi})^{-1}\boldsymbol{\Psi}^{\mathrm{T}}\big\}\mathrm{E}\{\boldsymbol{e}\} = \boldsymbol{0} \tag{9.1.31}$$

for an arbitrary, also finite measurement time N. ⊓

This means that according to (9.1.27),

$$(N+1)^{-1}\mathrm{E}\{\boldsymbol{\Psi}^{\mathrm{T}}\boldsymbol{e}\} = \mathrm{E}\left\{\begin{pmatrix} -\hat{R}_{\mathrm{ye}}(1) \\ \vdots \\ -\hat{R}_{\mathrm{ye}}(m) \\ \hline -\hat{R}_{\mathrm{ue}}(d+1) \\ \vdots \\ -\hat{R}_{\mathrm{ue}}(d+m) \end{pmatrix}\right\} = \boldsymbol{0}\,. \tag{9.1.32}$$

For $\hat{\boldsymbol{\theta}} = \boldsymbol{\theta}_0$, the input signal $u(k)$ is not correlated with the error signal $e(k)$, so that $R_{\mathrm{ue}}(\tau) = 0$. (9.1.32) will be revisited later, see (9.1.54).

It will now be investigated, which conditions must be fulfilled so that a bias-free parameter estimate can be obtained. For this investigation, it is assumed that the signals are stationary processes, so that the estimates of the correlation functions are consistent and furthermore

$$\lim_{N\to\infty} \mathrm{E}\big\{\hat{R}_{\mathrm{uu}}(\tau)\big\} = R_{\mathrm{uu}}(\tau)$$
$$\lim_{N\to\infty} \mathrm{E}\big\{\hat{R}_{\mathrm{yy}}(\tau)\big\} = R_{\mathrm{yy}}(\tau)$$
$$\lim_{N\to\infty} \mathrm{E}\big\{\hat{R}_{\mathrm{uy}}(\tau)\big\} = R_{\mathrm{uy}}(\tau)\,.$$

From the theorem of Slutsky, see App. A.1, it follows with (9.1.28) for the convergence of the parameters in probability

$$\operatorname*{plim}_{N\to\infty} \hat{\boldsymbol{\theta}} = \left(\operatorname*{plim}_{N\to\infty} \frac{1}{N+1}\boldsymbol{\Psi}^{\mathrm{T}}\boldsymbol{\Psi}\right)^{-1}\left(\operatorname*{plim}_{N\to\infty} \frac{1}{N+1}\boldsymbol{\Psi}^{\mathrm{T}}\boldsymbol{y}\right). \tag{9.1.33}$$

This includes

$$\lim_{N\to\infty} \mathrm{E}\{\hat{\boldsymbol{\theta}}\} = \left(\lim_{N\to\infty} \mathrm{E}\left\{\frac{1}{N+1}\boldsymbol{\Psi}^{\mathrm{T}}\boldsymbol{\Psi}\right\}\right)^{-1}\left(\lim_{N\to\infty} \mathrm{E}\left\{\frac{1}{N+1}\boldsymbol{\Psi}^{\mathrm{T}}\boldsymbol{y}\right\}\right) . \quad (9.1.34)$$

This means that the terms in brackets each individually converge to steady values and are then statistically independent. Now (9.1.34) is separated for the useful signal and the disturbance. With (9.1.4), (9.1.8) becomes

$$\begin{aligned}\boldsymbol{\psi}^{\mathrm{T}}(k) &= \big(-y_{\mathrm{u}}(k-1)\,\ldots\,-y_{\mathrm{u}}(k-m)\big|u(k-d-1)\,\ldots\,u(k-d-m)\big)\\ &\quad+\big(-n(k-1)\,\ldots\,-n(k-m)\big|0\,\ldots\,0\big)\\ &= \boldsymbol{\psi}_{\mathrm{u}}^{\mathrm{T}}(k)+\boldsymbol{\psi}_{\mathrm{n}}^{\mathrm{T}}(k)\end{aligned} \quad (9.1.35)$$

and consequently

$$\boldsymbol{\Psi}^{\mathrm{T}} = \boldsymbol{\Psi}_{\mathrm{u}}^{\mathrm{T}} + \boldsymbol{\Psi}_{\mathrm{n}}^{\mathrm{T}} . \quad (9.1.36)$$

Furthermore, according to (9.1.4)

$$y(k) = y_{\mathrm{u}}(k) + n(k) = \boldsymbol{\psi}_{\mathrm{u}}^{\mathrm{T}}(k)\boldsymbol{\theta}_0 + n(k) , \quad (9.1.37)$$

where the $\boldsymbol{\theta}_0$ are the true process parameters and thus

$$\boldsymbol{y} = \boldsymbol{\Psi}_{\mathrm{u}}\boldsymbol{\theta}_0 + \boldsymbol{n} = \big(\boldsymbol{\Psi} - \boldsymbol{\Psi}_{\mathrm{n}}\big)\boldsymbol{\theta}_0 + \boldsymbol{n} . \quad (9.1.38)$$

If (9.1.38) is inserted into (9.1.34)

$$\begin{aligned}\lim_{N\to\infty} \mathrm{E}\{\hat{\boldsymbol{\theta}}\} &= \left(\lim_{N\to\infty} \mathrm{E}\left\{\frac{1}{N+1}\boldsymbol{\Psi}^{\mathrm{T}}\boldsymbol{\Psi}\right\}\right)^{-1}\\ &\quad\left(\lim_{N\to\infty} \mathrm{E}\left\{\frac{1}{N+1}\boldsymbol{\Psi}^{\mathrm{T}}\big(\boldsymbol{\Psi}-\boldsymbol{\Psi}_{\mathrm{n}}\big)\boldsymbol{\theta}_0 + \frac{1}{N+1}\boldsymbol{\Psi}^{\mathrm{T}}\boldsymbol{n}\right\}\right)\\ &= \boldsymbol{\theta}_0 + \boldsymbol{b} ,\end{aligned} \quad (9.1.39)$$

where

$$\begin{aligned}\lim_{N\to\infty} \boldsymbol{b} &= \left(\lim_{N\to\infty} \mathrm{E}\left\{\frac{1}{N+1}\boldsymbol{\Psi}^{\mathrm{T}}\boldsymbol{\Psi}\right\}\right)^{-1}\\ &\quad\left(\lim_{N\to\infty} \mathrm{E}\left\{\frac{1}{N+1}\boldsymbol{\Psi}^{\mathrm{T}}\boldsymbol{n} - \frac{1}{N+1}\boldsymbol{\Psi}^{\mathrm{T}}\boldsymbol{\Psi}_{\mathrm{n}}\boldsymbol{\theta}_0\right\}\right)\end{aligned} \quad (9.1.40)$$

represents an asymptotic bias. As an abbreviation, a "correlation matrix" is introduced as

$$\hat{\boldsymbol{R}}(N+1) = \frac{1}{N+1}\boldsymbol{\Psi}^{\mathrm{T}}\boldsymbol{\Psi} \quad (9.1.41)$$

$$\boldsymbol{R} = \lim_{N\to\infty} \mathrm{E}\left\{\frac{1}{N+1}\boldsymbol{\Psi}^{\mathrm{T}}\boldsymbol{\Psi}\right\} , \quad (9.1.42)$$

and it follows on the basis of (9.1.26) and (9.1.27)

$$\lim_{N\to\infty} \boldsymbol{b} = \boldsymbol{R}^{-1} \lim_{N\to\infty} \mathrm{E}\left\{ \begin{pmatrix} -\hat{R}_{\mathrm{yn}}(1) \\ \vdots \\ -\hat{R}_{\mathrm{yn}}(m) \\ \hline 0 \\ \vdots \\ 0 \end{pmatrix} - \begin{pmatrix} a_1 \hat{R}_{\mathrm{yn}}(0) + \ldots + a_m \hat{R}_{\mathrm{yn}}(1-m) \\ \vdots \\ a_1 \hat{R}_{\mathrm{yn}}(m-1) + \ldots + a_m \hat{R}_{\mathrm{yn}}(0) \\ \hline 0 \\ \vdots \\ 0 \end{pmatrix} \right\} , \tag{9.1.43}$$

where $\hat{R}_{\mathrm{un}}(\tau) = 0$, i.e. it is assumed that input signal $u(k)$ and the noise $n(k)$ are uncorrelated. For the CCF, one obtains with $y(k) = y_{\mathrm{u}}(k) + n(k)$

$$\begin{aligned} \mathrm{E}\{\hat{R}_{\mathrm{yn}}(\tau)\} &= \mathrm{E}\left\{\frac{1}{N+1}\sum_{k=0}^{N} y(k)n(k+\tau)\right\} \\ &= \underbrace{\mathrm{E}\left\{\frac{1}{N+1}\sum_{k=0}^{N} y_{\mathrm{u}}(k)n(k+\tau)\right\}}_{=0} + \mathrm{E}\left\{\frac{1}{N+1}\sum_{k=0}^{N} n(k)n(k+\tau)\right\} \\ &= R_{\mathrm{nn}}(\tau) , \end{aligned} \tag{9.1.44}$$

and thus

$$\lim_{N\to\infty} \boldsymbol{b} = -\boldsymbol{R}^{-1} \lim_{N\to\infty} \mathrm{E}\left\{ \begin{pmatrix} \hat{R}_{\mathrm{nn}}(1) + a_1 \hat{R}_{\mathrm{nn}}(0) + \ldots + a_m \hat{R}_{\mathrm{nn}}(1-m) \\ \vdots \\ \hat{R}_{\mathrm{nn}}(m) + a_1 \hat{R}_{\mathrm{nn}}(m-1) + \ldots + a_m \hat{R}_{\mathrm{nn}}(0) \\ \hline 0 \\ \vdots \\ 0 \end{pmatrix} \right\} . \tag{9.1.45}$$

The bias vanishes if for $N \to \infty$,

$$\sum_{j=0}^{m} a_j R_{\mathrm{nn}}(\tau - j) = 0 \text{ for } 1 \le \tau \le m \text{ and } a_0 = 1 . \tag{9.1.46}$$

This is the *Yule-Walker equation* of the auto-regressive signal process (2.4.17)

$$\begin{aligned} n(k) + a_1 n(k-1) + \ldots + a_m n(k-m) &= \nu(k) \\ A(z^{-1}) n(z) &= \nu(z) \end{aligned} , \tag{9.1.47}$$

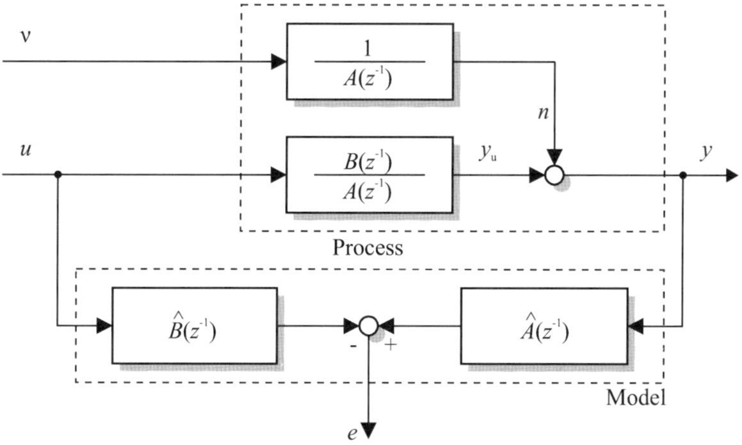

Fig. 9.2. Required structure of a process for the bias-free parameter estimation with the method of least squares, ν is a white noise

where $\nu(k)$ is a statistically independent Gaussian random signal with $(\overline{\nu}, \sigma_\nu) = (0, 1)$. (9.1.47) means that the noise $n(k)$ must be generated from white noise $\nu(k)$ by a filter with the transfer function $1/A(z^{-1})$ so that a bias-free estimate can be obtained with $\boldsymbol{b} = \boldsymbol{0}$. Therefore,

$$G_\nu(z) = \frac{n(z)}{\nu(z)} = \frac{1}{A(z^{-1})}\,, \tag{9.1.48}$$

see Fig. 9.2. The output is then given as

$$y(z) = \frac{1}{A}\nu(z) + \frac{B}{A}u(z) \tag{9.1.49}$$

and the error signal as

$$e(z) = -\hat{B}u(z) + \hat{A}y(z) = -\hat{B}u(z) + \frac{\hat{A}}{A}\nu(z) + \hat{A}\frac{B}{A}u(z) \tag{9.1.50}$$

If the process and model parameters match exactly, i.e. $\hat{\boldsymbol{\theta}} = \boldsymbol{\theta}_0$ or $\hat{A} = A$ and $\hat{B} = B$ respectively and thus the bias $\boldsymbol{b}$ vanishes, then

$$e(z) = \nu(z)\,. \tag{9.1.51}$$

Theorem 9.2 (Conditions for a Consistent Parameter Estimation).

The parameters of a dynamic process governed by (9.1.5) are estimated consistent (asymptotically bias-free) by the method of least squares, if the error $e(k)$ is uncorrelated, *that is if*

$$R_{\mathrm{ee}}(\tau) = \sigma_{\mathrm{e}}^2 \delta(\tau) \textit{ with } \delta(\tau) = \begin{cases} 1 \textit{ for } \tau = 0 \\ 0 \textit{ for } \tau \neq 0 \end{cases} \tag{9.1.52}$$

is valid and $e(k)$ *is furthermore zero-mean, i.e.*

$$\mathrm{E}\{e(k)\} = 0\,. \tag{9.1.53}$$

□

An annotation to the above theorem: If the conditions of the above theorem are satisfied, then the parameter estimates are also bias-free for finite measurement times N.

From (9.1.32) follows for a bias-free estimate in finite measurement time N

$$\begin{aligned} \hat{R}_{\mathrm{ye}}(\tau) &= \frac{1}{N+1} \sum_{k=m+d}^{m+d+N} e(k) y(k-\tau) \\ &= \frac{1}{N+1} \sum_{k=m+d}^{m+d+N} e(k+\tau) y(k) - 0 \text{ for } \tau = 1, 2, \ldots, m \end{aligned}\,. \tag{9.1.54}$$

For $e(k)$, one obtains by considering (9.1.4), (9.1.7), (9.1.10), (9.1.35), and $\hat{\boldsymbol{\theta}} = \boldsymbol{\theta}_0$

$$\begin{aligned} e(k) &= y(k) - \boldsymbol{\psi}^{\mathrm{T}}(k)\boldsymbol{\theta}_0 = y_{\mathrm{u}}(k) + n(k) - \boldsymbol{\psi}_{\mathrm{u}}^{\mathrm{T}}(k)\boldsymbol{\theta}_0 - \boldsymbol{\psi}_{\mathrm{n}}^{\mathrm{T}}(k)\boldsymbol{\theta}_0 \\ &= n(k) - \boldsymbol{\psi}_{\mathrm{n}}^{\mathrm{T}}(k)\boldsymbol{\theta}_0 \end{aligned} \tag{9.1.55}$$

and

$$\boldsymbol{\psi}_{\mathrm{n}}^{\mathrm{T}}(k) = \big(-n(k-1) \ldots -n(k-m) \big| 0 \ldots 0\big)\,. \tag{9.1.56}$$

The equation error then only depends on $n(k)$.

If one introduces (9.1.4) into (9.1.54) and bears in mind that upon convergence with $\hat{\boldsymbol{\theta}} = \boldsymbol{\theta}_0$ the wanted signal $y_{\mathrm{u}}(k)$ does not correlate with $e(k)$, then follows

$$\mathrm{E}\{\hat{R}_{\mathrm{ye}}(\tau)\} = \mathrm{E}\left\{\frac{1}{N+1} \sum_{k=m+d}^{m+d+N} e(k) n(k-\tau)\right\} = R_{\mathrm{ne}}(\tau) \text{ for } \tau = 1, 2, \ldots, m\,. \tag{9.1.57}$$

If the disturbance $n(k)$ can be governed by an auto-regressive signal process according to (9.1.47), then it follows from multiplication of this equation with $n(k-\tau)$ and determination of the expected value

$$R_{\mathrm{nn}}(\tau) + a_1 R_{\mathrm{nn}}(\tau-1) + \ldots + a_m R_{\mathrm{nn}}(\tau-m) = R_{\mathrm{ne}}(\tau)\,. \tag{9.1.58}$$

According to the Yule-Walker equation and for $\tau > 0$, the term $R_{\mathrm{ne}}(\tau)$ vanishes,

$$R_{\mathrm{nn}}(\tau) + a_1 R_{\mathrm{nn}}(\tau-1) + \ldots + a_m R_{\mathrm{nn}}(\tau-m) = 0 \tag{9.1.59}$$

with

$$R_{\mathrm{ne}}(\tau) = 0 \text{ for } \tau = 1, 2, \ldots, m\,, \tag{9.1.60}$$

and thus, according to (9.1.32) and (9.1.57), the bias $\boldsymbol{b}$ vanishes, i.e. $\boldsymbol{b} = \boldsymbol{0}$.

The theorems that have been presented in this chapter so far are valid assuming a special form filter for the noise acting on the systems output, (9.1.48). Then, the parameter estimates are also unbiased for finite measurement periods.

The necessary form filter

$$G_\nu(z^{-1}) = \frac{D(z^{-1})}{C(z^{-1})} = \frac{1}{A(z^{-1})} \tag{9.1.61}$$

is quite particular. For dynamic systems of an order greater than 1, the numerator of the disturbance transfer function $D(z^{-1})$ is typically not equal to 1, but rather has the form

$$D(z^{-1}) = d_0 + d_1 z^{-1} + d_2 z^{-2} + \ldots \,. \tag{9.1.62}$$

Therefore, the parameter estimation will typically yield biased results for dynamic processes affected by disturbances. As (9.1.40) shows, the bias grows larger as the amplitude of the disturbance $n(k)$ increases compared to the wanted signal. The model structure as depicted in Fig. 9.2 is termed ARX (Ljung, 1999).

If the conditions in Theorem 9.2 cannot be satisfied, biased parameter estimates will result. The *magnitude* of the bias is given by the results from (9.1.43) and the annotations to Theorem 9.2 as

$$\begin{aligned} \mathrm{E}\{\boldsymbol{b}(N+1)\} &= -\mathrm{E}\{\boldsymbol{R}^{-1}(N+1)\} \\ &= \mathrm{E}\left\{\begin{pmatrix} \hat{R}_{\mathrm{nn}}(1) + a_1 \hat{R}_{\mathrm{nn}}(0) + \ldots + a_m \hat{R}_{\mathrm{nn}}(1-m) \\ \vdots \\ \hat{R}_{\mathrm{nn}}(m) + a_1 \hat{R}_{\mathrm{nn}}(m-1) + \ldots + a_m \hat{R}_{\mathrm{nn}}(0) \\ \hline 0 \\ \vdots \\ 0 \end{pmatrix}\right\} . \end{aligned} \tag{9.1.63}$$

For the special case of the noise $n(k)$ being a white noise, this equation can be simplified using

$$\mathrm{E}\{\hat{R}_{\mathrm{nn}}(0)\} = R_{\mathrm{nn}}(0) = \mathrm{E}\{n^2(k)\} = \sigma_{\mathrm{n}}^2 \,, \tag{9.1.64}$$

resulting in

$$\begin{aligned} \mathrm{E}\{\boldsymbol{b}(N+1)\} &= -\mathrm{E}\{\boldsymbol{R}^{-1}(N+1)\} \begin{pmatrix} a_1 \\ \vdots \\ a_m \\ \hline 0 \\ \vdots \\ 0 \end{pmatrix} \sigma_{\mathrm{n}}^2 \\ &= -\mathrm{E}\{\boldsymbol{R}^{-1}(N+1)\} \left(\begin{array}{c|c} \boldsymbol{I} & \boldsymbol{0} \\ \hline \boldsymbol{0} & \boldsymbol{0} \end{array}\right) \boldsymbol{\theta}_0 \sigma_{\mathrm{n}}^2 \,. \end{aligned} \tag{9.1.65}$$

Further studies on the bias can be found e.g. in (Sagara et al, 1979).

9.1.3 Covariance of the Parameter Estimates and Model Uncertainty

Considering (9.1.29) and assuming $\hat{\boldsymbol{\theta}} = \boldsymbol{\theta}_0$, the *covariance matrix* of the parameter estimates is given as

$$\begin{aligned}\operatorname{cov} \Delta\boldsymbol{\theta} &= \mathrm{E}\big\{(\hat{\boldsymbol{\theta}} - \boldsymbol{\theta}_0)(\hat{\boldsymbol{\theta}} - \boldsymbol{\theta}_0)^{\mathrm{T}}\big\} \\ &= \mathrm{E}\Big\{\big((\boldsymbol{\Psi}^{\mathrm{T}}\boldsymbol{\Psi})^{-1}\boldsymbol{\Psi}^{\mathrm{T}}\boldsymbol{e}\big)\big((\boldsymbol{\Psi}^{\mathrm{T}}\boldsymbol{\Psi})^{-1}\boldsymbol{\Psi}^{\mathrm{T}}\boldsymbol{e}\big)^{\mathrm{T}}\Big\} \\ &= \mathrm{E}\big\{(\boldsymbol{\Psi}^{\mathrm{T}}\boldsymbol{\Psi})^{-1}\boldsymbol{\Psi}^{\mathrm{T}}\boldsymbol{e}\boldsymbol{e}^{\mathrm{T}}\boldsymbol{\Psi}(\boldsymbol{\Psi}^{\mathrm{T}}\boldsymbol{\Psi})^{-1}\big\}\,.\end{aligned} \tag{9.1.66}$$

One has to consider that $((\boldsymbol{\Psi}^{\mathrm{T}}\boldsymbol{\Psi})^{-1})^{\mathrm{T}} = (\boldsymbol{\Psi}^{\mathrm{T}}\boldsymbol{\Psi})^{-1}$, since $(\boldsymbol{\Psi}^{\mathrm{T}}\boldsymbol{\Psi})$ is a symmetric matrix. If $\boldsymbol{\Psi}$ and $\boldsymbol{e}$ are statistically independent, then

$$\operatorname{cov} \Delta\boldsymbol{\theta} = \mathrm{E}\big\{(\boldsymbol{\Psi}^{\mathrm{T}}\boldsymbol{\Psi})\boldsymbol{\Psi}^{\mathrm{T}}\big\}\mathrm{E}\big\{\boldsymbol{e}\boldsymbol{e}^{\mathrm{T}}\big\}\mathrm{E}\big\{\boldsymbol{\Psi}(\boldsymbol{\Psi}^{\mathrm{T}}\boldsymbol{\Psi})^{-1}\big\} \tag{9.1.67}$$

and if furthermore $\boldsymbol{e}$ is uncorrelated,

$$\mathrm{E}\big\{\boldsymbol{e}\boldsymbol{e}^{\mathrm{T}}\big\} = \sigma_{\mathrm{e}}^2\boldsymbol{I}\,. \tag{9.1.68}$$

Under these conditions and satisfying the requirements of Theorem 9.2, i.e. for a bias-free parameter estimate, the covariance matrix becomes

$$\begin{aligned}\operatorname{cov} \Delta\boldsymbol{\theta} &= \sigma_{\mathrm{e}}^2\mathrm{E}\big\{(\boldsymbol{\Psi}^{\mathrm{T}}\boldsymbol{\Psi})^{-1}\big\} = \sigma_{\mathrm{e}}^2\mathrm{E}\{\boldsymbol{P}\} \\ &= \sigma_{\mathrm{e}}^2\mathrm{E}\big\{((N+1)^{-1}\boldsymbol{\Psi}^{\mathrm{T}}\boldsymbol{\Psi})^{-1}\big\}\frac{1}{N+1} = \sigma_{\mathrm{e}}^2\frac{1}{N+1}\mathrm{E}\big\{\hat{\boldsymbol{R}}^{-1}(N+1)\big\}\,.\end{aligned} \tag{9.1.69}$$

For $N \to \infty$, one obtains

$$\lim_{N\to\infty} \operatorname{cov} \Delta\boldsymbol{\theta} = \boldsymbol{R}^{-1} \lim_{N\to\infty} \frac{\sigma_{\mathrm{e}}^2}{N+1} = 0\,. \tag{9.1.70}$$

The parameter estimates are thus consistent in the mean square if Theorem 9.2 is satisfied.

In general, σ_{e}^2 is unknown. It can be estimated bias-free by (Stuart et al, 1987; Kendall and Stuart, 1977b,a; Johnston and DiNardo, 1997; Mendel, 1973; Eykhoff, 1974)

$$\sigma_{\mathrm{e}}^2 \approx \hat{\sigma}_{\mathrm{e}}^2(m+d+N) = \frac{1}{N+1-2m}\boldsymbol{e}^{\mathrm{T}}(m+d+N)\boldsymbol{e}(m+d+N)\,, \tag{9.1.71}$$

where

$$\boldsymbol{e} = \boldsymbol{y} - \boldsymbol{\Psi}\hat{\boldsymbol{\theta}}\,. \tag{9.1.72}$$

Thus, one cannot only determine the parameter estimates according to (9.1.18) or (9.1.28), but at the same time also estimates for the variances and covariances employing (9.1.69) and (9.1.71).

Besides expressions for the covariance of the parameter estimates, it is also interesting to find metrics for the model uncertainty. There is no unique way to do so.

Hence, in the following, some methods that are presented in literature are summarized, providing a tool-set to judge the model uncertainty.

The first approach is based on the covariance matrix of the parameter estimates. It is assumed that the parameter error $\hat{\boldsymbol{\theta}} - \boldsymbol{\theta}_0$ is Gaussian distributed around zero with the covariance matrix $\boldsymbol{P}_{\boldsymbol{\theta}}$. Then, each single parameter error is Gaussian distributed with the probability distribution function

$$p(\hat{\theta}_k) = \frac{1}{\sqrt{2\pi P_{\boldsymbol{\theta},kk}}} \exp\left(-\frac{\hat{\theta}_k - \theta_{0,k}}{2P_{\boldsymbol{\theta},kk}} \right) , \tag{9.1.73}$$

where $\hat{\theta}_k$ is the estimate of the kth parameter, $\theta_{0,k}$ the true parameter and $P_{\boldsymbol{\theta},kk}$ the corresponding element on the diagonal of $\boldsymbol{P}_{\boldsymbol{\theta}}$. One can use this equation to determine the probability that the estimate $\hat{\theta}_k$ is more than a distance a away from the true value $\theta_{0,k}$ by evaluating the integral

$$P(|\hat{\theta}_k - \theta_{0,k}| > a) = 1 - \int_{-a}^{a} \frac{1}{\sqrt{2\pi P_{\boldsymbol{\theta},kk}}} \exp\left(-\frac{x}{2P_{\boldsymbol{\theta},kk}} \right) \mathrm{d}x , \tag{9.1.74}$$

see also (Ljung, 1999; Box et al, 2008) for a similar derivation.

Next, a confidence interval for the parameter vector shall be determined. Here, the χ^2 distribution will be used. The sum of k independent Gaussian distributed random variables has a χ^2 distribution with k degrees of freedom. Consequently, the quantity

$$r^2 = \sum_k \frac{\left(\hat{\theta}_k - \theta_{0,k}\right)^2}{P_{\boldsymbol{\theta},kk}} \tag{9.1.75}$$

is χ^2 distributed with $d = \dim \boldsymbol{\theta}$ degrees of freedom. Confidence intervals which state that r does not exceed $r_{\max}$ with a certain probability can be taken from any table for the χ^2 distribution as found in textbooks on statistics and can be used to calculate confidence ellipsoids, see also (Ljung, 1999).

By means of the rules of error propagation, one can now deduce the uncertainty of the resulting model as well. The model, which shall be denoted as M in the following, is basically a non-linear function of the estimated parameters, i.e.

$$M = f(\hat{\boldsymbol{\theta}}) . \tag{9.1.76}$$

Then, the covariance can be transformed as

$$\operatorname{cov} f(\hat{\boldsymbol{\theta}}) = \left(\left. \frac{\partial f(\boldsymbol{\theta})}{\partial \boldsymbol{\theta}} \right|_{\boldsymbol{\theta}=\hat{\boldsymbol{\theta}}} \right) \boldsymbol{P}_{\boldsymbol{\theta}} \left(\left. \frac{\partial f(\boldsymbol{\theta})}{\partial \boldsymbol{\theta}} \right|_{\boldsymbol{\theta}=\hat{\boldsymbol{\theta}}} \right)^{\mathrm{T}} \tag{9.1.77}$$

see e.g. (Vuerinckx et al, 2001). Consequently, the covariance of the output of a system with the model $\boldsymbol{y} = \boldsymbol{\Psi}\boldsymbol{\theta}$ is given as

$$\operatorname{cov} \hat{\boldsymbol{y}} = \boldsymbol{\Psi} \operatorname{cov} \Delta\hat{\boldsymbol{\theta}} \boldsymbol{\Psi}^{\mathrm{T}} . \tag{9.1.78}$$

The above derivations are all based on the assumptions that the estimated parameters are random variables with Gaussian distribution and are estimated bias-free. As this does not have to be the case, especially for finite, short sample lengths N, alternatives will be mentioned in the following.

An extension of the confidence intervals to the case of finite sample lengths has been described in (Campi and Weyer, 2002; Weyer and Campi, 2002).

For the location of poles and zeros in transfer functions, it is suggested to determine the confidence regions by perturbing each estimated zero and pole respectively and check whether the resulting model still represents the system with a sufficient fidelity, (Vuerinckx et al, 2001). It has been shown that the shape of the confidence intervals can differ drastically from the typically assumed ellipsoids, see also (Pintelon and Schoukens, 2001). The uncertainty ellipsoids have also been discussed in (Gevers, 2005) and (Bombois et al, 2005), where it has been pointed out that the calculation of uncertainty ellipsoids in the frequency domain is often based on wrong assumptions about the underlying distribution as the uncertainty is analyzed at each point separately. A survey on further methods to determine the model quality can be found in (Ninness and Goodwin, 1995).

Example 9.1 (First Order System for LS Parameter Estimation).

The method of least squares for dynamic systems shall now be illustrated for two examples. First, a simple difference equation of first order will be analyzed, then the Three-Mass Oscillator will be treated.

The simple difference equation of first order is given as

$$y_{\mathrm{u}}(k) + a_1 y_{\mathrm{u}}(k-1) = b_1 u(k-1) \tag{9.1.79}$$

$$y(k) = y_{\mathrm{u}}(k) + n(k)\,. \tag{9.1.80}$$

This difference equation e.g. represents a first order continuous-time systems with a ZOH (Zero Order Hold). As a process model for the parameter estimation, one will use in analogy to (9.1.6) the model

$$y(k) + \hat{a}_1 y(k-1) - \hat{b}_1 u(k-1) = e(k)\,. \tag{9.1.81}$$

A total of $N+1$ values of $u(k)$ and $y(k)$ will be measured,

$$\boldsymbol{\Psi} = \left(\begin{array}{c|c} -y(0) & u(0) \\ -y(1) & u(1) \\ \vdots & \\ -y(N-1) & u(N-1) \end{array}\right) \tag{9.1.82}$$

and

$$\boldsymbol{y} = \begin{pmatrix} y(1) \\ y(2) \\ \vdots \\ y(N) \end{pmatrix}. \tag{9.1.83}$$

Then,

$$(N+1)^{-1}\boldsymbol{\Psi}^{\mathrm{T}}\boldsymbol{\Psi} = \begin{pmatrix} \hat{R}_{\mathrm{yy}}(0) & -\hat{R}_{\mathrm{uy}}(0) \\ -\hat{R}_{\mathrm{uy}}(0) & \hat{R}_{\mathrm{uu}}(0) \end{pmatrix} \tag{9.1.84}$$

$$(N+1)^{-1}\boldsymbol{\Psi}^{\mathrm{T}}\boldsymbol{y} = \begin{pmatrix} -\hat{R}_{\mathrm{yy}}(1) \\ -\hat{R}_{\mathrm{uy}}(1) \end{pmatrix}. \tag{9.1.85}$$

The inverse is then given as

$$(N+1)\big(\boldsymbol{\Psi}^{\mathrm{T}}\boldsymbol{\Psi}\big)^{-1} = \frac{1}{\hat{R}_{\mathrm{uu}}(0)\hat{R}_{\mathrm{yy}}(0) - \big(\hat{R}_{\mathrm{uy}}(0)\big)^2} \begin{pmatrix} \hat{R}_{\mathrm{uu}}(0) & \hat{R}_{\mathrm{uy}}(0) \\ \hat{R}_{\mathrm{uy}}(0) & \hat{R}_{\mathrm{yy}}(0) \end{pmatrix} \tag{9.1.86}$$

and the parameter estimates finally become

$$\begin{pmatrix} \hat{a}_1 \\ \hat{b}_1 \end{pmatrix} = \frac{1}{\hat{R}_{\mathrm{uu}}(0)\hat{R}_{\mathrm{yy}}(0) - \big(\hat{R}_{\mathrm{uy}}(0)\big)^2} \begin{pmatrix} -\hat{R}_{\mathrm{uu}}(0)\hat{R}_{\mathrm{yy}}(1) + \hat{R}_{\mathrm{uy}}(0)\hat{R}_{\mathrm{uy}}(1) \\ -\hat{R}_{\mathrm{uy}}(0)\hat{R}_{\mathrm{yy}}(1) + \hat{R}_{\mathrm{yy}}(0)\hat{R}_{\mathrm{uy}}(1) \end{pmatrix}. \tag{9.1.87}$$

If the requirements in Theorem 9.2 are not satisfied, then the resulting bias can according to (9.1.45) be estimated as

$$\begin{aligned} \boldsymbol{b} &= -\mathrm{E}\Big\{(N+1)\big(\boldsymbol{\Psi}^{\mathrm{T}}\boldsymbol{\Psi}\big)^{-1}\Big\}\mathrm{E}\begin{pmatrix} \hat{R}_{\mathrm{nn}}(1) + \hat{a}_1\hat{R}_{\mathrm{nn}}(0) \\ 0 \end{pmatrix} \\ &= -\frac{1}{\hat{R}_{\mathrm{uu}}(0)\hat{R}_{\mathrm{yy}}(0) - \big(\hat{R}_{\mathrm{uy}}(0)\big)^2} \begin{pmatrix} -\hat{R}_{\mathrm{uu}}(0)\hat{R}_{\mathrm{nn}}(1) + \hat{a}_1\hat{R}_{\mathrm{uu}}(0)\hat{R}_{\mathrm{nn}}(0) \\ -\hat{R}_{\mathrm{uy}}(0)\hat{R}_{\mathrm{nn}}(1) + \hat{a}_1\hat{R}_{\mathrm{uy}}(0)\hat{R}_{\mathrm{nn}}(0) \end{pmatrix}. \end{aligned} \tag{9.1.88}$$

This expression becomes much easier to read if for both $u(k)$ and $n(k)$, one assumes a white noise such that $R_{\mathrm{uy}}(0) = g(0) = 0$. Then

$$\mathrm{E}\big\{\Delta\hat{a}_1\big\} = -a_1\frac{R_{\mathrm{nn}}(0)}{R_{\mathrm{yy}}(0)} = -a_1\frac{\overline{n^2(k)}}{\overline{y^2(k)}} = -a_1\frac{1}{1+\frac{\overline{y_{\mathrm{u}}^2(k)}}{\overline{n^2(k)}}} \tag{9.1.89}$$

$$\mathrm{E}\big\{\Delta\hat{b}_1\big\} = 0\,. \tag{9.1.90}$$

The bias of $\hat{a}_1$ gets larger as the noise amplitude increases. The parameter $\hat{b}_1$ is in this example estimated bias-free.

The covariance matrix of the parameter error is according to (9.1.69) given as

$$\mathrm{cov}\begin{pmatrix} \Delta\hat{a}_1 \\ \Delta\hat{b}_1 \end{pmatrix} = \mathrm{E}\left\{\frac{\sigma_{\mathrm{e}}^2}{\hat{R}_{\mathrm{uu}}(0)\hat{R}_{\mathrm{yy}}(0) - \big(\hat{R}_{\mathrm{uy}}(0)\big)^2}\begin{pmatrix} \hat{R}_{\mathrm{uu}}(0) & \hat{R}_{\mathrm{uy}}(0) \\ \hat{R}_{\mathrm{uy}}(0) & \hat{R}_{\mathrm{yy}}(0) \end{pmatrix}\right\}\frac{1}{N+1}. \tag{9.1.91}$$

If $u(k)$ is a white noise, then the variances can be obtained as

$$\mathrm{var}\,\Delta\hat{a}_1 = \frac{\overline{e^2(k)}}{\overline{y^2(k)}}\frac{1}{N+1} \tag{9.1.92}$$

and

$$\operatorname{var} \Delta \hat{b}_1 = \frac{\overline{e^2(k)}}{\overline{u^2(k)}} \frac{1}{N+1} . \tag{9.1.93}$$

If furthermore $n(k)$ is a white noise, then one obtains for the bias-free estimation $\hat{\boldsymbol{\theta}} = \boldsymbol{\theta}_0$ after squaring (9.1.55)

$$\operatorname{var} \Delta \hat{a}_1 = \left(1 + a_1^2\right) \frac{\overline{n^2(k)}}{\overline{y^2(k)}} \frac{1}{N+1} \tag{9.1.94}$$

$$\operatorname{var} \Delta \hat{b}_1 = \left(1 + a_1^2\right) \frac{\overline{n^2(k)}}{\overline{u^2(k)}} \frac{1}{N+1} . \tag{9.1.95}$$

The standard deviation of the parameter estimates thus diminishes proportionally to the square root of the measurement time. An interesting aspect is the fact that

$$\frac{\operatorname{var}\left(\Delta \hat{b}_1\right)}{\operatorname{var}\left(\Delta \hat{a}_1\right)} = \frac{\overline{y^2(k)}}{\overline{u^2(k)}} = \frac{\overline{y_{\mathrm{u}}^2(k)}}{\overline{u^2(k)}} + \frac{\overline{n^2(k)}}{\overline{u^2(k)}} . \tag{9.1.96}$$

The variance of the parameter $\hat{b}_1$ gets smaller in relation to the variance of the parameter $\hat{a}_1$ the smaller $n(k)$ and the smaller $y_{\mathrm{u}}(k)$, i.e. as the input signal $u(k)$ has a higher-frequent content. □

Example 9.2 (First Order System in Continuous Time for LS Estimation).

Now, a first order system with the transfer function

$$G(s) = \frac{K}{1 + sT_1} \tag{9.1.97}$$

will be studied. The system parameters have been chosen as $K = 2$ and $T_1 = 0.5$ s.

As the parameter estimation has so far only been introduced for discrete-time dynamic systems, the system must first be subjected to the z-transform. For the treatment of continuous-time systems, the reader is referred to Chap. 15. The discrete-time model is given as

$$G\left(z^{-1}\right) = \frac{b_1 z^{-1}}{1 + a_1 z^{-1}} = \frac{0.09754 z^{-1}}{1 - 0.9512 z^{-1}} \tag{9.1.98}$$

with the coefficients being given as

$$b_1 = K\left(1 - \mathrm{e}^{-\frac{T_0}{T_1}}\right) = 0.09754 \tag{9.1.99}$$

and

$$a_1 = -\mathrm{e}^{-\frac{T_0}{T_1}} = 0.9512 \tag{9.1.100}$$

for a sample time of $T_0 = 0.025$ s.

The matrix $\boldsymbol{\Psi}$ is set up according to (9.1.82) and the vector $\boldsymbol{y}$ analog to (9.1.83).

Table 9.1. Parameter estimation results for the first order process

σ_n	$\hat{a}_1$	$\hat{b}_1$	$\hat{K}$	$\hat{T}_1$	$\Delta\hat{K}[\%]$	$\Delta\hat{T}_1[\%]$	Remark
≈ 0	−0.9510	0.09802	2	0.4975	≈ 0	−0.50	
0.0002	−0.9500	0.09807	1.9617	0.4875	−1.92	−2.51	see Fig. 9.3
0.002	−0.9411	0.09851	1.6735	0.4121	−16.32	−17.59	
0.02	−0.8607	0.10241	0.7354	0.1667	−63.23	−66.66	see Fig. 9.4
0.2	−0.4652	0.12148	0.2272	0.0327	−88.64	−93.47	
2.0	−0.0828	0.1399	0.1525	0.0100	−92.37	−97.99	see Fig. 9.5

A white noise has been generated and has been superimposed onto the output. Three different noise levels have been added to the output and the results have been graphed in Figs. 9.3 through 9.5. The first case, depicted in Fig. 9.3, represents a very small noise level ($\sigma_n = 0.0002$). One can see that the parameter estimates match very well with the theoretical values that have been marked by the dash-dotted lines. Moderate noise ($\sigma_n = 0.02$) has been added in Fig. 9.4 and one can already witness a bias in the parameter estimates. Finally, Fig. 9.5 illustrates the effect of even larger noise ($\sigma_n = 2$). Despite the long time-base, the parameter estimates do not converge to the real values, but rather settle to the biased values. The diagrams have been generated with the DSFI algorithm, see Chap. 22, which is numerically more robust. The results however are comparable to those that would have been obtained with the direct calculation of the pseudo-inverse of the matrix $\boldsymbol{\Psi}$.

When using the method of least squares to determine a discrete-time model of a physical process, which is governed by ODEs, one has to convert the parameters of the discrete-time model back to the parameters of the corresponding model in continuous-time to obtain physical parameters such as e.g. inductances, spring constants, and such. The two physical parameters of the first order system are given as

$$K = \frac{b_1}{1 + a_1} \tag{9.1.101}$$

and

$$T_1 = -\frac{T_0}{\ln -a_1} \, . \tag{9.1.102}$$

The estimated results, the errors and such are tabulated in Table 9.1. One can see that the bias of the estimated system parameters can become quite large, rendering the estimated parameter values useless. As can be seen, the bias mainly affects the parameter estimate $\hat{a}_1$, while the estimate $\hat{b}_1$ still converges to the true value (see Example 9.1). Regarding the noise, one should however keep in mind that the noise levels chosen for the illustration of the bias are really large. Secondly, in many cases one can resort to other methods, which are better suited for noisy signals, such as e.g. the orthogonal correlation, see Sect. 5.5.2.

Example 9.3 (Discrete Time Model of the Three-Mass Oscillator). To apply the method of least squares to a real process, another example will now be presented. The

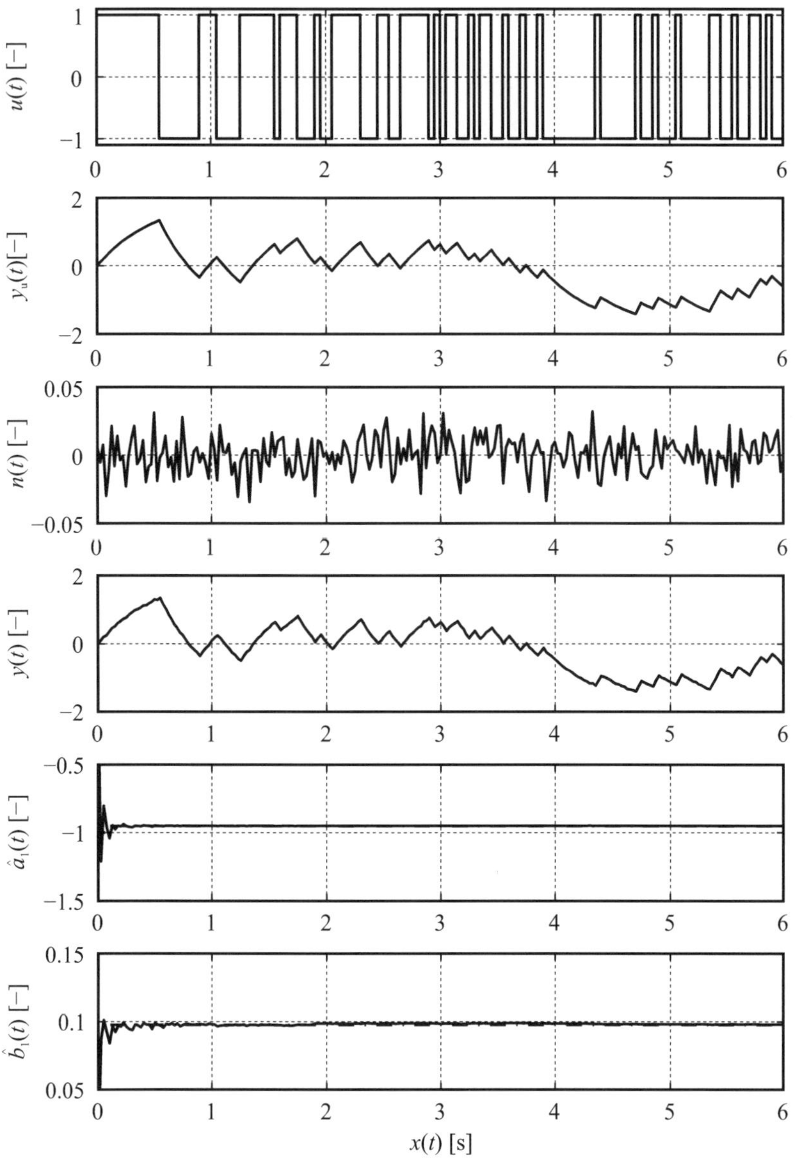

Fig. 9.3. Parameter estimation for a first order system. True parameter values (*dash-dotted line*), $\sigma_n = 0.0002$, $\sigma_n/\sigma_y = 0.0004$

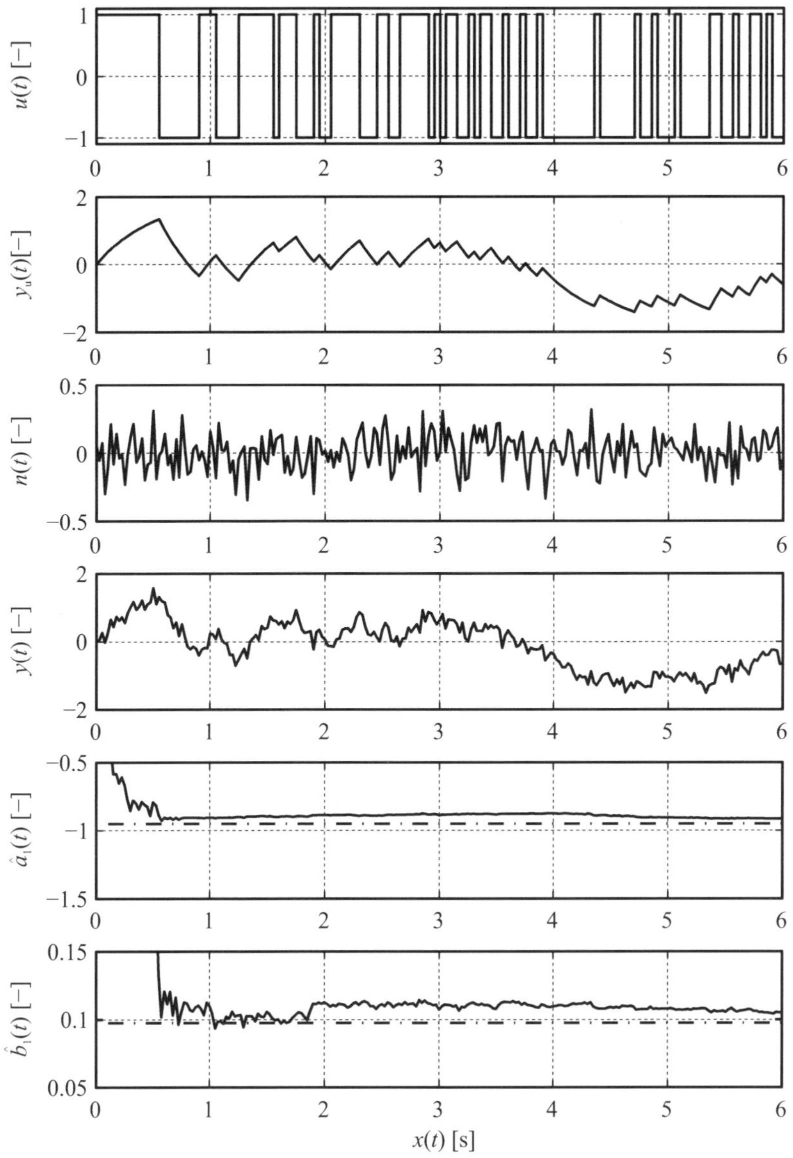

Fig. 9.4. Parameter estimation for a first order system. True parameter values (*dash-dotted line*), $\sigma_n = 0.02$, $\sigma_n/\sigma_y = 0.04$

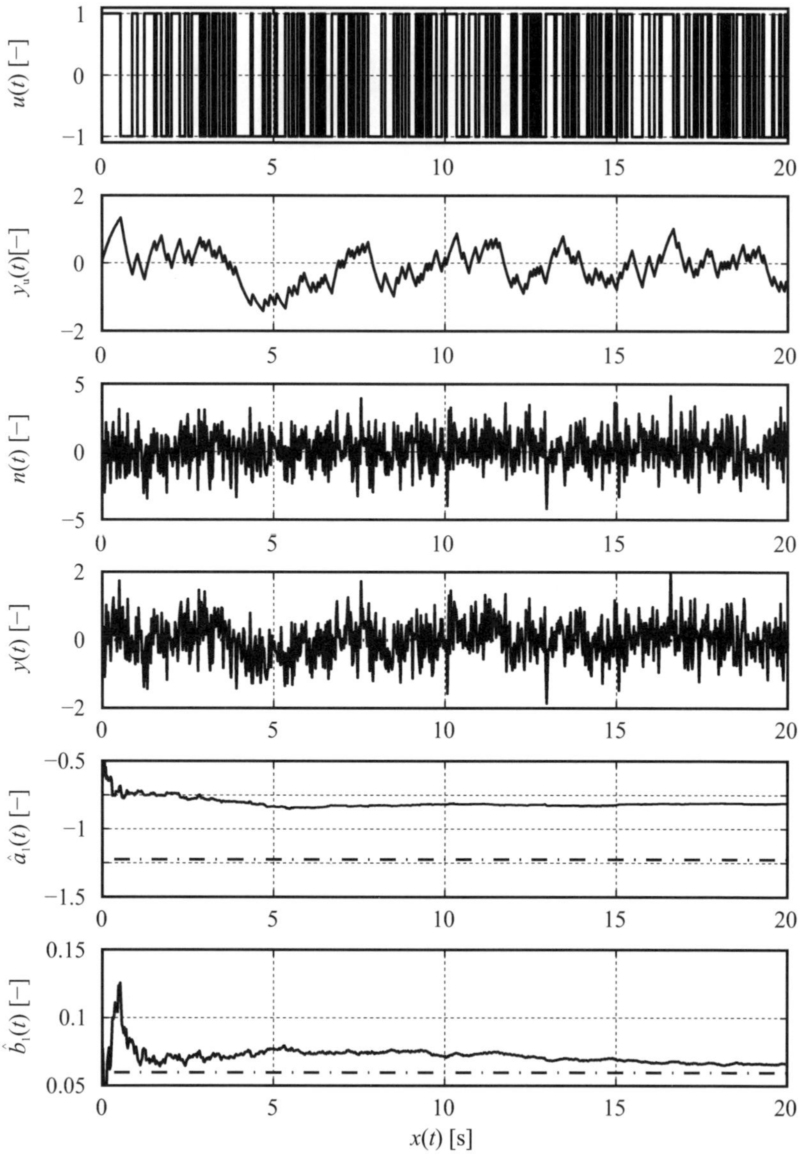

Fig. 9.5. Parameter estimation for a first order system. True parameter values (*dash-dotted line*), $\sigma_n = 2$, $\sigma_n/\sigma_y = 4$

Three-Mass Oscillator will be covered now. In this chapter, a discrete-time model will be identified, an estimation of the physical parameters of a continuous-time model will be presented in Chap. 15.

The continuous-time transfer function from the torque of the motor to the position of the last mass is given as a state space model with six states. In transfer function notation, the system has six poles, thus the discrete-time transfer function will theoretically also be of order six and is thus given as

$$G(z^{-1}) = \frac{b_1 z^{-1} + b_2 z^{-2} + b_3 z^{-3} + b_4 z^{-4} + b_5 z^{-5} + b_6 z^{-6}}{1 + a_1 z^{-1} + a_2 z^{-2} + a_3 z^{-3} + a_4 z^{-4} + a_5 z^{-5} + a_6 z^{-6}} \,. \tag{9.1.103}$$

The setup of the matrix $\boldsymbol{\Psi}$, see (9.1.82), and the vector $\boldsymbol{y}$, see (9.1.83), is shown here to clarify the case of a higher order system. $\boldsymbol{\Psi}$ is written as

$$\boldsymbol{\Psi} = \left(\begin{array}{cccc|ccc} -y(5) & -y(4) & \dots & -y(0) & u(5) & \dots & u(0) \\ -y(6) & -y(5) & \dots & -y(1) & u(6) & \dots & u(1) \\ \vdots & \vdots & & \vdots & \vdots & & \vdots \\ -y(N-1) & -y(N-2) & \dots & -y(N-6) & u(N-1) & \dots & u(N-6) \end{array}\right) \tag{9.1.104}$$

and $\boldsymbol{y}$ as

$$\boldsymbol{y} = \begin{pmatrix} y(6) \\ y(7) \\ \vdots \\ y(N) \end{pmatrix} . \tag{9.1.105}$$

The parameter vector $\boldsymbol{\theta}$ then consists of the elements

$$\boldsymbol{\theta}^{\mathrm{T}} = \big(a(1)\; a(2) \dots a(6) \big| b(1) \dots b(6)\big) \,. \tag{9.1.106}$$

The process has been excited with a PRBS signal (see Sect. 6.3). The process input is the torque of the motor M_{M} acting on the first mass. The output is the rotational speed of the third mass, $\omega_3 = \dot{\varphi}_3$, as shown in Fig. 9.6. The parameters can reliably be estimated in 20 seconds of the excitation as can be witnessed in Fig. 9.7. An important issue in the estimation of discrete-time models is the sample rate. The data for the Three-Mass Oscillator have been sampled with a sample time of $T_0 = 0.003\,\mathrm{s}$. This sample rate was too high to obtain reasonable results, thus the data have been downsampled by a factor of $N = 16$, leading to $T_0 = 48\,\mathrm{ms}$. Section 23.2 discusses the optimal choice of the sample rate.

In order to judge the quality of the estimated model, the frequency response of the discrete-time model has been graphed against the frequency response determined by direct measurement with the orthogonal correlation (see Sect. 5.5.2). This comparison in Fig. 9.8 demonstrates the good fidelity of the estimated model. □

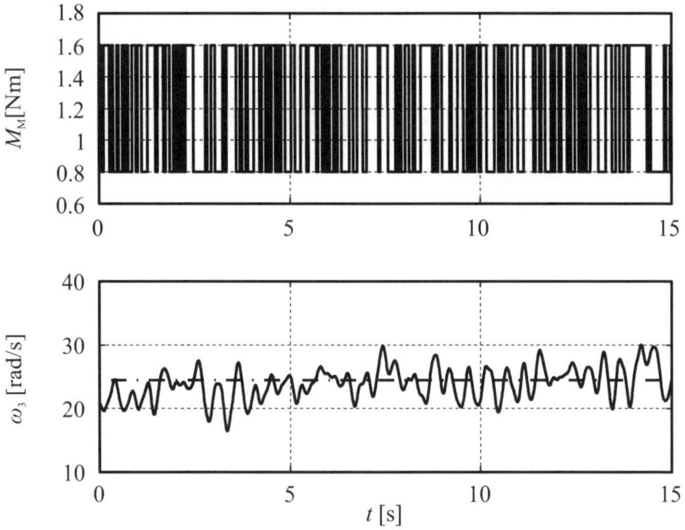

Fig. 9.6. Input and output signals for the parameter estimation of a discrete-time model of the Three-Mass Oscillator

9.1.4 Parameter Identifiability

Before applying any parameter identification method, one has to check the identifiability of the parameters. Identifiability in general relates to the issue whether the true system can be described by means of a model that is identified using a certain identification method. This property hence depends on the

- System S
- Experimental setup X
- Model structure M
- Identification method I

Many different definitions have been introduced. In (Bellmann and Åström, 1970), identifiability is defined to be satisfied if the identification criterion, i.e. the cost function, has an unambiguous minimum. In most cases however, identifiability is linked to the consistency of the estimation. For a parametric system model, the model parameters $\boldsymbol{\theta}$ are identifiable, if their estimates $\hat{\boldsymbol{\theta}}(N)$ converge to the true values $\boldsymbol{\theta}_0$ for $N \rightarrow \infty$. The convergence criterion is however chosen differently by different authors. Åström and Bohlin (1965) as well as Tse and Anton (1972) tested for convergence in probability, (Staley and Yue, 1970) used the convergence in the mean square. In the lines of Staley and Yue (1970) and Young (1984), the following concepts shall be used.

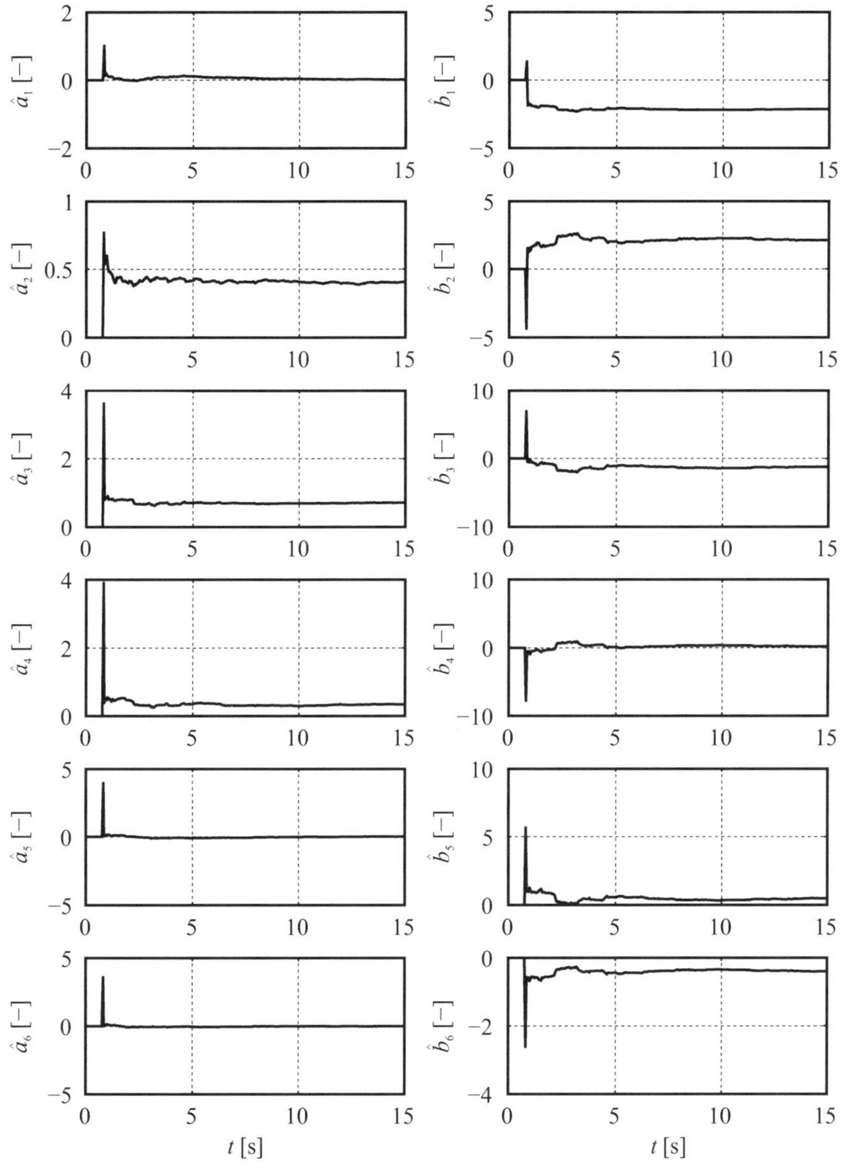

Fig. 9.7. Parameter estimation for a discrete-time model of the Three-Mass Oscillator, parameter estimates in dependence of time

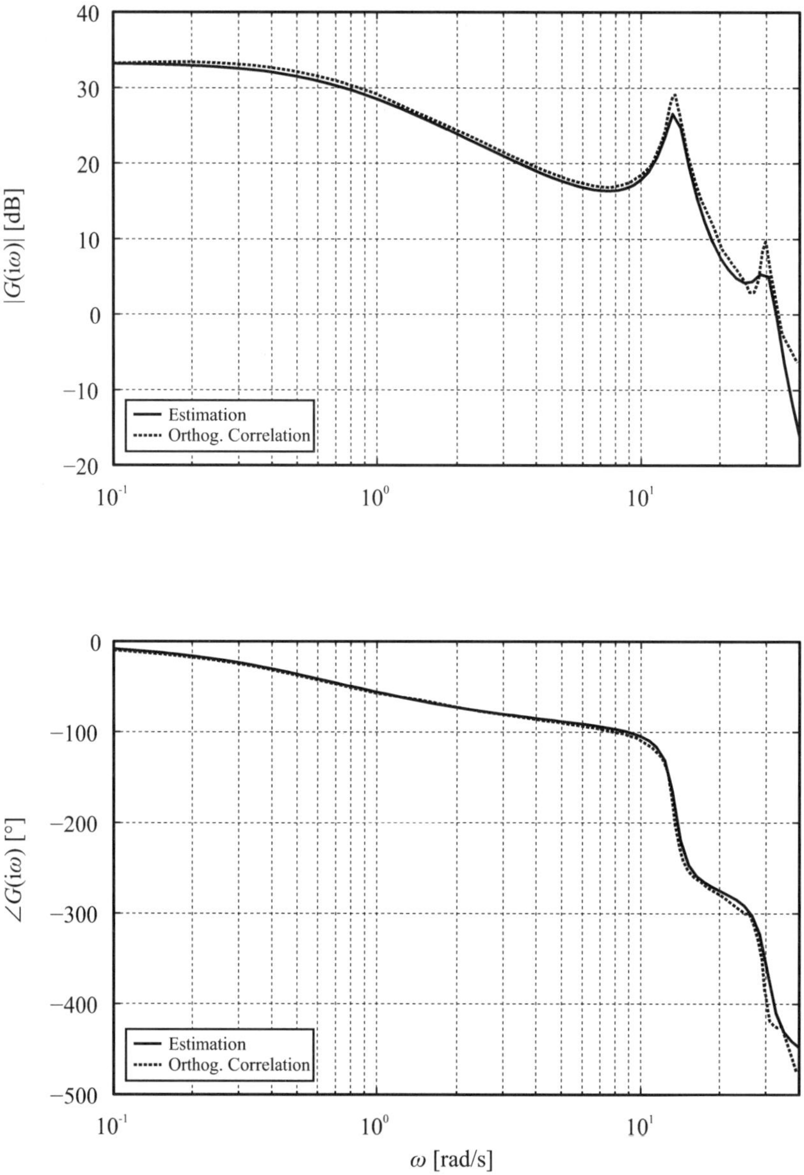

Fig. 9.8. Calculation of the frequency response based on the parameter estimation for a discrete-time model of the Three-Mass Oscillator, comparison with direct measurement of the frequency responses

Definition of Parameter Identifiability

The parameter vector $\boldsymbol{\theta}$ of a model is identifiable, if the estimated values $\hat{\boldsymbol{\theta}}$ converge to the true parameters $\boldsymbol{\theta}_0$ in the mean square. This means that

$$\lim_{N\to\infty} \mathrm{E}\{\hat{\boldsymbol{\theta}}(N)\} = \boldsymbol{\theta}_0$$

$$\lim_{N\to\infty} \operatorname{cov} \Delta\hat{\boldsymbol{\theta}}(N) = \mathbf{0}$$

and hence requires an estimator that is consistent in the mean square.

In the following, the conditions that have to be fulfilled by the system S, the experiment X, the model structure M, and the identification method I to guarantee parameter identifiability shall be analyzed and shall be tested for the method of least squares.

It is first assumed that the model structure M and the system structure S agree and that the model structure has been chosen such that Theorem 9.2 is fulfilled, i.e. that a consistent estimate can be obtained. Now, it shall be analyzed, which additional requirements have to be fulfilled by the system S and the experiment X.

To be able to obtain the parameter estimates $\hat{\boldsymbol{\theta}}$ according to (9.1.18), one must ensure that $\det(\boldsymbol{\Psi}^{\mathrm{T}}\boldsymbol{\Psi}) \neq 0$, (9.1.23), and (9.1.25) to ensure that the cost function reaches its global minimum and hence $\hat{\boldsymbol{\theta}}$ becomes the optimal parameter set. Both conditions are satisfied if

$$\det \boldsymbol{\Psi}^{\mathrm{T}}\boldsymbol{\Psi} = \det \boldsymbol{P}^{-1} > 0 \, . \tag{9.1.107}$$

With the correlation matrix, one can write

$$\det \frac{1}{N}\boldsymbol{\Psi}^{\mathrm{T}}\boldsymbol{\Psi} = \det \hat{\boldsymbol{R}}(N) > 0 \, . \tag{9.1.108}$$

With (9.1.70), also $\operatorname{cov} \Delta\boldsymbol{\theta}$ will convergence to zero as $N \to \infty$, so that the estimate is consistent in the mean square. The correlation matrix is now divided and analyzed in its limit $N \to \infty$,

$$\boldsymbol{R} = \left(\begin{array}{c|c} \boldsymbol{R}_{11} & \boldsymbol{R}_{12} \\ \hline \boldsymbol{R}_{21} & \boldsymbol{R}_{22} \end{array}\right) \tag{9.1.109}$$

such that e.g.

$$\boldsymbol{R}_{22} = \begin{pmatrix} R_{\mathrm{uu}}(0) & R_{\mathrm{uu}}(1) & \dots & R_{\mathrm{uu}}(m-1) \\ R_{\mathrm{uu}}(-1) & R_{\mathrm{uu}}(0) & \dots & R_{\mathrm{uu}}(m-2) \\ \vdots & \vdots & & \vdots \\ R_{\mathrm{uu}}(-m+1) & R_{\mathrm{uu}}(-m+2) & \dots & R_{\mathrm{uu}}(0) \end{pmatrix} \tag{9.1.110}$$

or due to the symmetry of R_{uu},

$$\boldsymbol{R}_{22} = \begin{pmatrix} R_{\mathrm{uu}}(0) & R_{\mathrm{uu}}(1) & \dots & R_{\mathrm{uu}}(m-1) \\ R_{\mathrm{uu}}(1) & R_{\mathrm{uu}}(0) & \dots & R_{\mathrm{uu}}(m-2) \\ \vdots & \vdots & & \vdots \\ R_{\mathrm{uu}}(m-1) & R_{\mathrm{uu}}(m-2) & \dots & R_{\mathrm{uu}}(0) \end{pmatrix} . \tag{9.1.111}$$

For the determinant in (9.1.108) with the decomposition in (9.1.109), one can now write (Young, 1984)

$$|\boldsymbol{R}| = |\boldsymbol{R}_{11}|\,|\boldsymbol{R}_{22} - \boldsymbol{R}_{21}\boldsymbol{R}_{11}^{-1}\boldsymbol{R}_{12}| \tag{9.1.112}$$

or

$$|\boldsymbol{R}| = |\boldsymbol{R}_{22}|\,|\boldsymbol{R}_{11} - \boldsymbol{R}_{12}\boldsymbol{R}_{22}^{-1}\boldsymbol{R}_{21}|\,. \tag{9.1.113}$$

Necessary conditions are hence that

$$\det \boldsymbol{R}_{22} > 0 \tag{9.1.114}$$

and

$$\det \boldsymbol{R}_{11} > 0\,. \tag{9.1.115}$$

These conditions lead to requirements on both the input signal and the system as will be discussed in the following. One can in this context discern structural identifiability, which means that the system is in general identifiable and identifiability, which means that the chosen input allows indeed to identify the system.

Conditions on the Input Signal

In order to satisfy the condition in (9.1.114), one must check that (9.1.110) fulfills the requisite that its determinant is greater than zero. According to the Sylvester criterion for positive definite matrices, one must ensure that all northeastern sub-determinants are also positive, i.e.

$$\det \boldsymbol{R}_i > 0 \text{ for } i = 1, 2, \ldots, m\,. \tag{9.1.116}$$

This means that

$$\begin{aligned}
\det \boldsymbol{R}_1 &= R_{\mathrm{uu}}(0) > 0 \\
\det \boldsymbol{R}_2 &= \begin{vmatrix} R_{\mathrm{uu}}(0) & R_{\mathrm{uu}}(1) \\ R_{\mathrm{uu}}(-1) & R_{\mathrm{uu}}(0) \end{vmatrix} > 0 \\
&\;\vdots
\end{aligned}$$

and finally

$$\det \boldsymbol{R}_{22} > 0\,. \tag{9.1.117}$$

Here, $\boldsymbol{R}_{22} > 0$ does only depend on the input signal $u(k)$, hence (9.1.114) can always be fulfilled by an appropriate choice of $u(k)$.

Theorem 9.3 (Condition for Persistent Excitation).

A necessary condition for the parameter estimation by means of the method of least squares is that the input signal $u(k) = U(k) - \overline{U}$ *fulfills the conditions*

$$\overline{U} = \lim_{N\to\infty} \frac{1}{N} \sum_{k=m+d}^{m+d+N-1} U(k) \tag{9.1.118}$$

and

$$R_{\mathrm{uu}}(\tau) = \lim_{N \to \infty} \big(U(k) - \overline{U}\big)\big(U(k+\tau) - \overline{U}\big) \qquad (9.1.119)$$

exist and that the matrix R_{22} is positive definite. □

These conditions have been stated by Åström and Bohlin (1965) for the maximum likelihood method and have been termed *persistently exciting of order m*. One will note that the condition (9.1.114) is the same as for the correlation analysis, (7.2.4) with the only difference being the order of the persistent excitation. Some examples of persistently exciting input signals are

- $R_{\mathrm{uu}}(0) > R_{\mathrm{uu}}(1) > \ldots > R_{\mathrm{uu}}(m)$, a moving average signal process of order m
- $R_{\mathrm{uu}}(0) \neq 0$, $R_{\mathrm{uu}}(1) = \ldots = R_{\mathrm{uu}}(m) = 0$, white noise for $m \to \infty$
- $R_{\mathrm{uu}}(0) = a^2$ for $\tau = 0, N\lambda, 2N\lambda, \ldots$, $R_{\mathrm{uu}}(\tau) = -a^2/N$ for $\lambda(1 + \nu N) < \tau < \lambda(N - 1 + \nu N)$, $\nu = 0, 1, 2, \ldots$, PRBS with amplitude a, cycle time $\lambda = T_0$ and period length N, persistently exciting of order m if $N = m + 1$

The condition stated in Theorem 9.3 can easily be examined by evaluating the auto-correlation function of deterministic or stochastic signals.

The conditions for a persistent excitation of order m can also be interpreted in the frequency domain. From the Fourier analysis, one knows that a necessary condition for the existence of the power spectral density of a signal process in discrete-time,

$$S^*_{\mathrm{uu}}(\omega) = \sum_{n=-\infty}^{\infty} R_{\mathrm{uu}}(n) \mathrm{e}^{-\mathrm{i}\omega T_0 n} = R_{\mathrm{uu}}(0) + 2 \sum_{n=1}^{\infty} R_{\mathrm{uu}}(n) \cos \omega T_0 n \ , \qquad (9.1.120)$$

in the range $0 < \omega < \pi/T_0$ is that the auto-correlation function $R_{\mathrm{uu}}(\tau) > 0$ for all τ. Then, the signal is persistently exciting of arbitrary order. Therefore, if $S^*_{\mathrm{uu}}(\omega) > 0$ for all ω, then the signal is persistently exciting of any order (Åström and Eykhoff, 1971). Persistent excitation of finite order means that $S^*_{\mathrm{uu}}(\omega) = 0$ for certain frequencies (as e.g. the Fourier transform of pulses, Sect. 4.2, or of the PRBS, Sect. 6.3).

Ljung (1999) requires that for the identification of a transfer function of order m, the signal should be persistently exciting of order $2m$. It is hence sufficient to use m sinusoids, see also Chap. 4 for multi-sine signals.

Conditions on the Process

In order to satisfy (9.1.115), one must ensure

$$\det R_1 = R_{\mathrm{yy}}(0) > 0$$

$$\det \boldsymbol{R}_2 = \begin{vmatrix} R_{\mathrm{yy}}(0) & R_{\mathrm{yy}}(1) \\ R_{\mathrm{yy}}(-1) & R_{\mathrm{yy}}(0) \end{vmatrix} = R^2_{\mathrm{yy}}(0) - R_{\mathrm{yy}}(1) > 0$$

$$\vdots$$

and finally

$$\det \boldsymbol{R}_{11} > 0\,. \tag{9.1.121}$$

While (9.1.117) had to be satisfied by choosing an appropriate input signal, the condition in (9.1.121) depends on the system. If $\boldsymbol{R}_{22}$ is positive definite, then it follows:

Theorem 9.4 (Condition on the Process).

A necessary condition for the parameter estimation by means of the method of least squares is that for the output $y(k) = Y(k) - \overline{Y}$ *with*

$$\overline{Y} = \lim_{N\to\infty} \frac{1}{N+1} \sum_{k=m+d}^{m+d+N} Y(k) \tag{9.1.122}$$

and

$$R_{yy}(\tau) = \lim_{N\to\infty} \sum_{k=m+d}^{m+d+N} \big(Y(k) - \overline{Y}\big)\big(Y(k+\tau) - \overline{Y}\big) \tag{9.1.123}$$

the matrix

$$\boldsymbol{R}_{11} = \big(R_{ij} = R_{yy}(i \quad j)\big)\; i, j = 1, \ldots, m \tag{9.1.124}$$

is positive definite. □

In order to satisfy these requirements, one must ensure the following:

- The system must be stable. All poles of $A(z)$ must lie within the unit circle.
- Not all coefficients $b_i,\ i = 1, 2, \ldots, m$ may be zero. To ensure that for a persistently exciting input of order m, the output signal is persistently excited of the same order m and hence the matrix R_{11} is positive definite, one must ensure that ...
- $A(z)$ and $B(z)$ may not have common roots.

This also means that the correct order m of the model must be chosen. If the order of the system model is chosen too high, then poles and zeros can cancel. The above results can be combined as follows (Tse and Anton, 1972):

- If the minimal dimension m is known, then stability, controllability, and observability also ensure identifiability.

If the conditions in Theorem 9.3 and 9.4 are satisfied and hence (9.1.114) and (9.1.115) and satisfied, it is still not sure that (9.1.108) is satisfied since according to (9.1.112) and (9.1.113), also the right factors must be positive definite. This shall be illustrated by an example.

Example 9.4 (Parameter Identifiability for a Harmonic Excitation).

A linear, discrete-time process shall be excited by a sinusoidal excitation

$$u(kT_0) = u_0 \sin \omega_1 k T_0\,.$$

It shall now be investigated up to which order m, the parameters of the processes

$$G_{\mathrm{P,A}} = \frac{b_0 + b_1 z^{-1} + \ldots + b_m z^{-m}}{1 + a_1 z^{-1} + \ldots + a_m z^{-m}}$$

$$G_{\mathrm{P,B}} = \frac{b_1 z^{-1} + \ldots + b_m z^{-m}}{1 + a_1 z^{-1} + \ldots + a_m z^{-m}}$$

are identifiable if the transients in the output have settled. In both cases, the output will be given as

$$y(kT_0) = y_0 \sin\bigl(\omega_1 k T_0 + \varphi\bigr)$$

with different y_0 and φ. The correlation functions are given as

$$R_{\mathrm{uu}}(\tau) = \frac{u_0^2}{2} \cos \omega_1 \tau T_0$$

$$R_{\mathrm{yy}}(\tau) = \frac{y_0^2}{2} \cos \omega_1 \tau T_0 \ .$$

First process A ($b_0 \neq 0$) will be examined. One has to set up the matrices

$$\boldsymbol{R}_{22} = \begin{pmatrix} R_{\mathrm{uu}}(0) & \ldots & R_{\mathrm{uu}}(m) \\ \vdots & & \vdots \\ R_{\mathrm{uu}}(m) & \ldots & R_{\mathrm{uu}}(0) \end{pmatrix}$$

and

$$\boldsymbol{R}_{11} = \begin{pmatrix} R_{\mathrm{yy}}(0) & \ldots & R_{\mathrm{yy}}(m-1) \\ \vdots & & \vdots \\ R_{\mathrm{yy}}(m-1) & \ldots & R_{\mathrm{yy}}(0) \end{pmatrix} .$$

For the process order $m = 1$,

$$\begin{aligned} \det \boldsymbol{R}_{22} &= R_{\mathrm{uu}}^2(0) - R_{\mathrm{uu}}^2(1) = \frac{u_0^2}{2}\bigl(1 - \cos^2 \omega_1 T_0\bigr) \\ &= \frac{u_0^2}{2} \sin^2 \omega_1 T_0 > 0 \text{ if } \omega_1 T_0 \neq 0, \pi, 2\pi, \ldots \\ \det \boldsymbol{R}_{11} &= R_{\mathrm{yy}}(0) = \frac{y_0^2}{2} > 0 \\ \det \boldsymbol{R} &> 0 \text{ according to (9.1.112)} . \end{aligned}$$

The process is hence identifiable.

Now the case of a process order $m = 2$ is investigated,

$$\begin{aligned} \det \boldsymbol{R}_{22} &= R_{\mathrm{uu}}^3(0) + 2R_{\mathrm{uu}}^2(1) R_{\mathrm{uu}}(2) - R_{\mathrm{uu}}^2(2) R_{\mathrm{uu}}(0) - 2R_{\mathrm{uu}}^2(1) R_{\mathrm{uu}}(0) \\ &= \left(\frac{u_0^2}{2}\right)^3 \bigl(1 - \cos^4 \omega_1 T_0 - \sin^4 \omega_1 T_0 - 2\cos^2 \omega_1 T_0 \sin^2 \omega_1 T_0\bigr) \\ &= 0 \end{aligned}$$

$$\begin{aligned} \det \boldsymbol{R}_{11} &= R_{\mathrm{yy}}^2(0) - R_{\mathrm{yy}}^2(1) = \frac{y_0^2}{2} \sin^2 \omega_1 T_0 > 0 \\ \det \boldsymbol{R} &= 0 \text{ according to (9.1.113)} . \end{aligned}$$

The process is in this case not identifiable.

Now, process B ($b_0 = 0$) will be examined. The matrices

$$
\boldsymbol{R}_{22} = \begin{pmatrix} R_{\mathrm{uu}}(0) & \dots & R_{\mathrm{uu}}(m-1) \\ \vdots & & \vdots \\ R_{\mathrm{uu}}(m-1) & \dots & R_{\mathrm{uu}}(0) \end{pmatrix}
$$

and

$$
\boldsymbol{R}_{11} = \begin{pmatrix} R_{\mathrm{yy}}(0) & \dots & R_{\mathrm{yy}}(m-1) \\ \vdots & & \vdots \\ R_{\mathrm{yy}}(m-1) & \dots & R_{\mathrm{yy}}(0) \end{pmatrix}
$$

have to be analyzed. For $m = 1$, the analysis yields

$$
\begin{aligned}
\det \boldsymbol{R}_{22} &= R_{\mathrm{uu}}(0) = \frac{u_0^2}{2} > 0 \\
\det \boldsymbol{R}_{11} &- R_{yy}(0) - \frac{y_0^2}{2} > 0 \\
\det \boldsymbol{R} &> 0 \, .
\end{aligned}
$$

The process is identifiable. For $m = 2$,

$$
\begin{aligned}
\det \boldsymbol{R}_{22} &= R_{\mathrm{uu}}^2(0) - R_{\mathrm{uu}}^2(1) = \frac{u_0^2}{2} \sin^2 \omega_1 T_0 > 0 \\
\det \boldsymbol{R}_{11} &= R_{\mathrm{yy}}^2(0) - R_{\mathrm{yy}}^2(1) = \frac{y_0^2}{2} \sin^2 \omega_1 T_0 > 0 \text{ if } \omega_1 T_0 \neq 0, \pi, 2\pi, \dots \, .
\end{aligned}
$$

However, even though $\boldsymbol{R}_{22}$ and $\boldsymbol{R}_{11}$ are positive definite, $\boldsymbol{R}$ is not, as can be shown by e.g. choosing $\varphi = \pi/2$ and then evaluating the determinants.

This example shown that for b_0, the conditions in (9.1.114) and (9.1.115) already suffice to ensure that $\det \boldsymbol{R} > 0$, but not for $b_0 \neq 0$. (The conditions that have been stated in (Åström and Bohlin, 1965) only cover the case $b_0 \neq 0$). The common result is that with a single sinusoidal oscillation, one can only identify a process of a maximum order of 1. One should note however that for process A, one can identify three parameters b_0, b_1, a_1 and for process B the two parameters b_1, a_1. □

All important conditions for the method of least squares can now be summarized in the following theorem.

Theorem 9.5 (Conditions for a Consistent Parameter Estimation by the Method of Least Squares).

The parameters of a linear, time-invariant difference equation can be estimated consistent in the mean square by the method of least squares if the following necessary conditions are satisfied:

- *Order m and dead time d are known.*

- *The input signal $u(k) = U(k) - U_{00}$ must be exactly measurable and the DC value U_{00} must be known.*
- *The matrix*

$$\boldsymbol{R} = \frac{1}{N+1}\boldsymbol{\Psi}^{\mathrm{T}}\boldsymbol{\Psi}$$

  must be positive definite. This requires that
  - *The input signal $u(k)$ must be persistently exciting of at least order m, see Theorem 9.3.*
  - *The process must be stable, controllable and observable, see Theorem 9.4.*
- *The stochastic disturbance $n(k)$ which is superimposed onto the output signal $y(k) = Y(k) - Y_{00}$ must be stationary. The DC value Y_{00} must be known exactly and must correspond to U_{00}.*
- *The error $e(k)$ may not be correlated and* $\mathrm{E}\{e(k)\} = 0$.

□

From this conditions follows for $\hat{\boldsymbol{\theta}} = \boldsymbol{\theta}_0$.

$$1)\mathrm{E}\big\{n(k)\big\} = 0 \text{ (which follows from (9.1.47), (9.1.51) and Theorem 9.5)} \tag{9.1.125}$$

$$2)R_{\mathrm{ue}}(\tau) = 0 \text{ (which follows from (9.1.55))} \; . \tag{9.1.126}$$

These equations can be used in addition to validate the parameter estimates. Extensions of the above notions to non-linear systems are e.g. shown in (van Doren et al, 2009).

9.1.5 Unknown DC Values

As for process parameter estimation the variations of $u(k)$ and $y(k)$ of the measured signals $U(k)$ and $Y(k)$ have to be used, the DC (direct current or steady-state) values U_{00} and Y_{00} either have also to be estimated or have to be removed. The following methods are available for dealing with unknown DC values U_{00} and Y_{00}.

Differencing

The easiest way to obtain the variations without knowing the DC values is just to take the differences

$$\begin{aligned} U(k) - U(k-1) &= u(k) - u(k-1) = \Delta u(k) \\ Y(k) - Y(k-1) &= y(k) - y(k-1) = \Delta y(k) \; . \end{aligned} \tag{9.1.127}$$

Instead of $u(z)$ and $y(z)$, the signals $\Delta u(z) = u(z)(1 - z^{-1})$ and $\Delta y(z) = y(z)(1 - z^{-1})$ are then used for the parameter estimation. As this special high-pass filtering is applied to both the process input and output, the process parameters can be estimated in the same way as in the case of measuring $u(k)$ and $y(k)$. In the parameter estimation algorithms $u(k)$ and $y(k)$ have to be replaced by $\Delta u(k)$ and $\Delta y(k)$. However, the signal-to-noise ratio may become worse. If the DC values are required explicitly, other methods have to be used.

Averaging

The DC values can be estimated simply by averaging from steady-state measurement

$$\hat{Y}_{00} = \frac{1}{N} \sum_{k=0}^{N-1} Y(k) \tag{9.1.128}$$

before starting the dynamic excitation. The recursive version of this is

$$\hat{Y}_{00} = \hat{Y}_{00}(k-1) + \frac{1}{k}\big(Y(k) - \hat{Y}_{00}(k-1)\big) \,. \tag{9.1.129}$$

For slowly time varying DC values, recursive averaging with exponential forgetting leads to

$$\hat{Y}_{00} = \lambda \hat{Y}_{00}(k-1) + (1-\lambda) Y(k) \tag{9.1.130}$$

with $\lambda < 1$. A similar argument applies for U_{00}. The variations $u(k)$ and $y(k)$ can be determined by

$$u(k) = U(k) - U_{00} \tag{9.1.131}$$

$$y(k) = Y(k) - Y_{00} \,. \tag{9.1.132}$$

Implicit Estimation of a Constant

The estimation of the DC values U_{00} and Y_{00} can also be included into the parameter estimation problem. Substituting (9.1.132) and (9.1.131) into (9.1.5) results in

$$\begin{aligned} Y(k) = &- a_1 Y(k-1) - \ldots - a_m Y(k-m) + b_1 U(k-d-1) \\ &+ \ldots + b_m U(k-d-m) + C \,, \end{aligned} \tag{9.1.133}$$

where

$$C = (1 + a_1 + \ldots + a_m) Y_{00} - (b_1 + \ldots + b_m) U_{00} \,. \tag{9.1.134}$$

Extending the parameter vector $\hat{\boldsymbol{\theta}}$ by including the element C and the data vector $\boldsymbol{\psi}^{\mathrm{T}}(k)$ by adding the number 1, the measured $Y(k)$ and $U(k)$ can directly be used for the estimation and C can also be estimated. Then, for one given DC value the other can be calculated, using (9.1.134). For closed-loop identification, it is convenient to use

$$Y_{00} = W(k) \,. \tag{9.1.135}$$

Explicit Estimation of a Constant

The parameters $\hat{a}_i$ and $\hat{b}_i$ for the dynamic behavior and the DC constant C can also be estimated separately. First the dynamic parameters are estimated using the differencing method above. Then with

$$L(k) = Y(k) + \hat{a}_1 Y(k-1) + \ldots + \hat{a}_m Y(k-m) \\ -\hat{b}_1 U(k-d-1) - \ldots - \hat{b}_m U(k-d-m) \,, \tag{9.1.136}$$

the equation error becomes

$$e(k) = L(k) - C \tag{9.1.137}$$

and, after applying the LS method,

$$C(m+d+N) = \frac{1}{N+1} \sum_{k=m+d}^{m+d+N} L(k) \,. \tag{9.1.138}$$

For large N, one obtains

$$\hat{C} \approx \left(1 + \sum_{i=1}^{m} \hat{a}_i\right) \hat{Y}_{00} - \left(\sum_{i=1}^{m} \hat{b}_i\right) \hat{U}_{00} \,. \tag{9.1.139}$$

If the $\hat{Y}_{00}$ is of interest and U_{00} is known, it can be calculated from (9.1.139) using the estimate $\hat{C}$. In this case $\hat{\boldsymbol{\theta}}$ and $\hat{C}$ are only coupled in one direction, as $\hat{\boldsymbol{\theta}}$ does not depend on $\hat{C}$. A disadvantage can be the worse noise-to-signal ratio caused by the differencing. The final selection of the DC method depends on the particular application.

9.2 Spectral Analysis with Periodic Parametric Signal Models

Many problems associated with the determination of the Fourier transformation (see Sect. 3.1) would vanish if the course of the transformed signal would also be known outside the measurement interval. For this reason, Burg (1968) was looking for techniques to predict the unknown signal course from the measured data points without making any a priori assumptions about the signal course. This estimation of the values with maximum uncertainty concerning the signal course led to the term *maximum entropy* and to a substantially improved spectral estimation.

9.2.1 Parametric Signal Models in the Time Domain

A method to obtain the phase and angle of oscillations with known frequency is suggested by Heij et al (2007) as follows: In the least squares setting (see Chap. 8), the data matrix is set up as

$$\boldsymbol{\Psi} = \begin{pmatrix} \cos\omega_1 & \sin\omega_1 & \ldots & \cos\omega_n & \sin\omega_n \\ \cos 2\omega_1 & \sin 2\omega_1 & \ldots & \cos 2\omega_n & \sin 2\omega_n \\ \vdots & \vdots & & \vdots & \vdots \\ \cos N\omega_1 & \sin N\omega_1 & \ldots & \cos N\omega_n & \sin N\omega_n \end{pmatrix} \tag{9.2.1}$$

and the output vector chosen as the signal $x(k)$ to be analyzed, i.e.

$$\boldsymbol{y}^{\mathrm{T}} = \big(x(1)\; x(2)\; \ldots\; x(N)\big)\,. \tag{9.2.2}$$

The vector of estimated parameters then has the form

$$\hat{\boldsymbol{\theta}}^{\mathrm{T}} = \big(\hat{b}_1\; \hat{c}_1\; \ldots\; \hat{b}_n\; \hat{c}_n\big) \tag{9.2.3}$$

The phase and amplitude of the oscillation

$$y_i(k) = a_i \sin\big(\omega_i k + \varphi_i\big) \tag{9.2.4}$$

can then be estimated as

$$\hat{a}_i = \sqrt{\hat{b}_i^2 + \hat{c}_i^2} \tag{9.2.5}$$

and

$$\hat{\varphi}_i = \arctan \frac{\hat{b}_i}{\hat{c}_i}\,. \tag{9.2.6}$$

Nice features of the estimation technique are the fact that neither the omission of relevant frequencies nor the inclusion of irrelevant frequencies influences the result of the parameter estimation.

9.2.2 Parametric Signal Models in the Frequency Domain

The periodic signal is treated as if it was generated by a fictitious form filter $F(z)$ or $F(\mathrm{i}\omega)$ respectively. The form filter is driven by a δ- impulse $\delta(k)$ (2.4.10) to generate a steady state oscillation $y(k)$. The aim now is to match the frequency response of the form filter $F(z)$ with the amplitude spectrum of the measured signal $y(z)$. This is equivalent to match the power spectral densities,

$$S_{yy}(z) \stackrel{!}{=} |F(z)|^2 S_{\delta\delta}(z) = |F(z)|^2\,. \tag{9.2.7}$$

In general, a parametric signal model can have three different profile structures as has been discussed in Sect. 2.4. For the *moving average model*, which would be

$$F_{\mathrm{MA}}(z) = \beta_0 + \beta_1 z^{-1} + \ldots + \beta_n z^{-m}\,, \tag{9.2.8}$$

the signal spectrum is approximated by a polynomial of the (limited) order m and would in general be more appropriate for the approximation of smooth spectra. On the contrary, this model is extremely unsuited for the modeling of oscillations, which manifest themselves as distinct peaks in the frequency spectrum.

This leads to the *auto-regressive model* as the next possible candidate structure. The purely auto-regressive model, given as

$$F_{\mathrm{AR}}(z) = \frac{\beta_0}{1 + \alpha_1 z^{-1} + \ldots + \alpha_m z^{-n}}\,, \tag{9.2.9}$$

is able to approximate the sharp spectral lines of periodic signals according to the poles of the denominator polynomial. This makes it an appealing choice for the estimation of the spectra of harmonic oscillations (Tong, 1975, 1977; Pandit and Wu, 1983).

The third possible setup is the combined *auto-regressive moving average model* given by

$$F_{\text{ARMA}}(z) = \frac{\beta_0 + \beta_1 z^{-1} + \ldots + \beta_n z^{-m}}{1 + \alpha_1 z^{-1} + \ldots + \alpha_m z^{-n}}\,, \tag{9.2.10}$$

which combines the possibilities of both the AR and the MA model. The biggest disadvantage is the increase in the number of parameters, which doubles compared to the AR and the MA model respectively. This can also lead to convergence problems. More complex and more special estimators have been described in literature (Makhoul, 1975, 1976).

Coming back to (9.2.7), one obtains for the power spectral density of the AR-model,

$$S_{\text{yy}}(z) = F(z)F^*(z)S_{\delta\delta}(z) = \frac{\beta_0^2}{\left|1 + \sum\limits_{\nu=0}^{n} \alpha_\nu z^{-\nu}\right|^2}\,. \tag{9.2.11}$$

Estimation of the coefficients b_0 and a_k from the measured signal $y(k)$ leads to a *parametric, auto-regressive model in the frequency domain* for the power spectral density $S_{\text{yy}}(\mathrm{i}\omega)$, characterized by its $n+1$ parameters, with n typically being in the range $4\ldots30$. This technique, where the maximum entropy for the power spectral density $S_{\text{yy}}(\mathrm{i}\omega)$ is determined for a pure AR filter, has e.g. been described by Edward and Fitelson (1973) as well as Ulrych and Bishop (1975).

9.2.3 Determination of the Coefficients

In order to suppress stochastic disturbances, the maximum entropy spectral estimation will also employ correlation functions instead of the measured signal $y(t)$. The measured signal shall for now be composed of a number of damped oscillations as

$$y(t) = \sum_{\nu=1}^{m} y_{0\nu}\mathrm{e}^{-d_\nu t}\sin\bigl(\omega_\nu t + \varphi_\nu\bigr)\,. \tag{9.2.12}$$

Its auto-correlation function is then given as

$$R_{\text{yy}}(\tau) = \mathrm{E}\bigl\{y(t)y(t+\tau)\bigr\} = \sum_{\nu=1}^{m} \frac{y_{0\nu}^2}{2}\mathrm{e}^{-d_\nu\tau}\cos\omega_\nu\tau\,. \tag{9.2.13}$$

As is discussed in Sect. 5.5, the ACF $R_{\text{yy}}(\tau)$ of a periodic signal $y(t)$ is periodic in τ. Taking the above considerations into account, one can surely approximate the ACF by a form filter as well, i.e.

$$R_{\text{yy}}(z) = F(z)\delta(z)\,. \tag{9.2.14}$$

By the calculation of the ACF, the phase information gets lost and the amplitude is changed by a constant factor

$$R_{0\nu} = \frac{1}{2} y_{0\nu}^2 \,. \tag{9.2.15}$$

To capture m frequencies in the model of the ACF, the model must have an order of $2m$. As one is only interested in the stationary steady-state oscillations, one can use exclusively an AR model which for a signal of order $2m$ is given as

$$R_{\mathrm{nn}}(\tau) = R_{\mathrm{yy}}(\tau) + \alpha_1 R_{\mathrm{yy}}(\tau-1) + \alpha_2 R_{\mathrm{yy}}(\tau-2) + \ldots + \alpha_{2m} R_{\mathrm{yy}}(\tau-2m)\,, \tag{9.2.16}$$

where also an additive, zero-mean, uncorrelated disturbance $n(t)$ has been taken into account with its ACF

$$R_{\mathrm{nn}}(\tau) = \begin{cases} n_0 \text{ for } \tau = 0 \\ 0 \text{ else} \end{cases} \,. \tag{9.2.17}$$

For different time lags τ, a system of equations can be constructed as

$$\begin{pmatrix} R_{\mathrm{yy}}(0) & R_{\mathrm{yy}}(1) & \ldots & R_{\mathrm{yy}}(2m) \\ R_{\mathrm{yy}}(1) & R_{\mathrm{yy}}(0) & \ldots & R_{\mathrm{yy}}(2m-1) \\ \vdots & \vdots & & \vdots \\ R_{\mathrm{yy}}(2m) & R_{\mathrm{yy}}(2m-1) & \ldots & R_{\mathrm{yy}}(0) \end{pmatrix} \begin{pmatrix} 1 \\ \alpha_1 \\ \vdots \\ \alpha_{2m} \end{pmatrix} = \begin{pmatrix} n_0 \\ 0 \\ \vdots \\ 0 \end{pmatrix}, \tag{9.2.18}$$

where the fact has been exploited that the ACF is axis-symmetric, i.e. $R_{\mathrm{yy}}(k) = R_{\mathrm{yy}}(-k)$. The coefficient n_0 with

$$R_{\mathrm{nn}}(0) = n_0 = \mathrm{E}\{n^2(k)\} = \mathrm{E}\Big\{\big(y(k) - \hat{y}(k)\big)^2\Big\} \tag{9.2.19}$$

is a measure for the mean square model error with $\hat{y}(k)$ being the model prediction for $y(k)$.

To set up this system of equations, estimates of the ACF $R_{\mathrm{yy}}(\tau)$ have to be supplied for $\tau = 0, \ldots, 2m$ from the measured signal sequence $y(k)$ at the time steps $k = 0, \ldots, N-1$. They are determined according to (7.1.4) given as

$$\hat{R}_{\mathrm{yy}}(\tau) = \frac{1}{N - |\tau|} \sum_{\nu=0}^{N-1-|\tau|} y(\nu) y(\nu + |\tau|) \text{ for } 0 \leq |\tau| \leq N-1 \,. \tag{9.2.20}$$

The system of equations in (9.2.18) can efficiently be solved by the Burg algorithm (Press et al, 2007). The frequencies of the significant discrete oscillations in $y(t)$ can then be determined by a pole decomposition of the denominator polynomial of the AR signal model,

$$z^{2m}\big(1 + a_1 z^{-1} + a_2 z^{-2} + \ldots + a_{2m} z^{-2m}\big) \stackrel{!}{=} \prod_{\nu=1}^{m} \big(1 + a_{1\nu} z^{-1} + a_{2\nu} z^{-2}\big) \,. \tag{9.2.21}$$

From a corresponding table for the z-transform (Isermann, 1991), one can obtain for each conjugate complex pair of poles the angular frequency ω_k of the appropriate sinusoidal oscillation in $y(t)$ as

$$\omega_k = \frac{1}{T_0} \arccos\left(\frac{-a_{1\nu}}{2\sqrt{a_{2\nu}}}\right) . \tag{9.2.22}$$

Thus, one can determine all significant oscillation frequencies of the signal $y(t)$.

9.2.4 Estimation of the Amplitudes

The amplitudes y_{0k} of each contributing oscillation could theoretically be determined from the AR signal model. Unfortunately, this proves to be a very inaccurate method as the result depends on the denominator coefficients a_k and the constant numerator coefficient b_0. The slightest errors in the estimation of the coefficients could result in large errors for the amplitude estimate. To avoid these undesired effects, a second estimation is carried out to determine the amplitudes (Neumann and Janik, 1990; Neumann, 1991).

The damping term in the ACF of a periodic oscillation,

$$R_{yy}(\tau) = \mathrm{E}\{y(t)y(t+\tau)\} = \sum_{\nu=1}^{m} \frac{y_{0\nu}^2}{2} \mathrm{e}^{-d_\nu \tau} \cos\omega_\nu \tau \tag{9.2.23}$$

can be neglected for small damping values. In this case, one obtains

$$R_{yy}(\tau) = \mathrm{E}\{y(t)y(t+\tau)\} = \sum_{\nu=1}^{m} \frac{y_{0\nu}^2}{2} \cos\omega_\nu \tau , \tag{9.2.24}$$

which has been used widespread in Chap. 5. Provided that the frequencies of the oscillations contributing to $y(t)$ are known from the antecedent estimation problem, one can now set up a second system of equations to determine the amplitudes of the ACF, i.e. the values of R_{0k} as

$$\begin{pmatrix} R_{yy}(1) \\ R_{yy}(2) \\ \vdots \\ R_{yy}(m) \end{pmatrix} = \begin{pmatrix} \cos(\omega_1 T_0) & \cos(\omega_2 T_0) & \dots & \cos(\omega_m T_0) \\ \cos(\omega_1 2T_0) & \cos(\omega_2 2T_0) & \dots & \cos(\omega_m 2T_0) \\ \vdots & \vdots & & \vdots \\ \cos(\omega_1 m T_0) & \cos(\omega_2 m T_0) & \dots & \cos(\omega_m m T_0) \end{pmatrix} \begin{pmatrix} R_{01} \\ R_{02} \\ \vdots \\ R_{0m} \end{pmatrix} . \tag{9.2.25}$$

The signal amplitudes can finally be determined from the amplitudes of the ACF by

$$y_{0\nu} = \sqrt{2R_{0\nu}} . \tag{9.2.26}$$

Thus, a parametric model representation for the power spectral density S_{yy} of the measured signal $y(kT_0)$ was found, representing the spectrum by a parametric AR model respectively by frequencies of significant sinusoidal components. The order m has to be selected or searched for (Isermann, 2005, 2006).

9.3 Parameter Estimation with Non-Parametric Intermediate Model

If the structure of the process model is not known a priori, it can be useful to first identify a non-parametric model and then determine a parametric model based on the previously identified intermediate model. In the first step, no assumptions about the model structure have to be made. The determination of an appropriate model order and dead time then takes place during the second step of the identification. Since the non-parametric model already condenses the measured data, much less data have to be operated upon in the second step. The non-parametric model also allows to already make statements about the quality of the identification results. In the following, two different approaches shall be presented.

9.3.1 Response to Non-Periodic Excitation and Method of Least Squares

For good signal-to-noise ratios, one can identify a process by first recording several responses to the *same deterministic input signal* and then determine the average of these signals to eliminate the influence of stochastic disturbances. The input signals must excite the interesting process dynamics, but apart from that can have any arbitrary form. One often prefers *steps* and *ramps* or *rectangular* and *trapezoidal pulses* respectively.

Now, the deterministic input signal $u_j(k)$ and the corresponding output $y_j(k)$ shall be considered. Here, $u_j(k)$ and $y_j(k)$ are once again small signal values. If the identical input signal $u_j(k)$ is applied M times, then the average of the system response is given as

$$\overline{y}(k) = \frac{1}{M}\sum_{j=1}^{M} y_j(k)\,. \tag{9.3.1}$$

If the output from the process $y_j(k)$ is superpositioned with a stochastic noise $n(k)$,

$$y_j(k) = y_{\mathrm{u}j}(k) + n_j(k)\,, \tag{9.3.2}$$

the expected value is given as

$$\mathrm{E}\{\,\overline{y}(k)\,\} = y_{\mathrm{u}}(k) + \mathrm{E}\{\,\overline{n}(k)\,\}\,. \tag{9.3.3}$$

The expected value of the averaged output signal hence is identical with the useful signal if $\mathrm{E}\{\,\overline{n}(k)\} = 0$. For the parameter estimation, now $y(k) = \overline{y}(k)$ will be written. A parametric model in the form of a difference equation is assumed as

$$\begin{aligned} y(k) = &-a_1 y(k-1) - a_2 y(k-2) - \ldots - a_m y(k-m) \\ &+ b_1 u(k-d-1) + b_2 u(k-d-2) + \ldots + b_m u(k-d-m)\,. \end{aligned} \tag{9.3.4}$$

(9.3.4) can now be written in vector form for the interval $1 \le k \le l$ as

$$\begin{pmatrix} y(1) \\ y(2) \\ y(3) \\ \vdots \\ y(l) \end{pmatrix} =$$

$$\left(\begin{array}{cccc|ccc} 0 & 0 & \dots & 0 & u(-d) & \dots & 0 \\ -y(1) & 0 & \dots & 0 & u(1-d) & \dots & 0 \\ -y(2) & -y(1) & \dots & 0 & u(2-d) & \dots & 0 \\ \vdots & \vdots & & \vdots & \vdots & & \vdots \\ -y(l-1) & -y(l-2) & \dots & -y(l-m) & u(l-d-1) & \dots & u(l-d-m) \end{array}\right) \begin{pmatrix} a_1 \\ a_2 \\ \vdots \\ a_m \\ b_1 \\ \vdots \\ b_m \end{pmatrix}, \tag{9.3.5}$$

which can then be written as

$$\boldsymbol{y} = \boldsymbol{R}\boldsymbol{\theta} . \tag{9.3.6}$$

Introducing the error $\boldsymbol{e}$ as

$$\boldsymbol{e} = \boldsymbol{y} - \boldsymbol{R}\boldsymbol{\theta} \tag{9.3.7}$$

and using a quadratic cost function V as

$$V = \boldsymbol{e}^{\mathrm{T}}\boldsymbol{e} \tag{9.3.8}$$

will lead to the parameter estimates as

$$\hat{\boldsymbol{\theta}} = \left(\boldsymbol{R}^{\mathrm{T}}\boldsymbol{R}\right)^{-1}\boldsymbol{R}^{\mathrm{T}}\boldsymbol{y} . \tag{9.3.9}$$

The parameter estimates are consistent in the mean square, since for the limiting value of the error, one obtains

$$\lim_{M\to\infty} \mathrm{E}\{e(k)\}\big|_{\hat{\boldsymbol{\theta}}=\boldsymbol{\theta}_0} = \lim_{M\to\infty} \mathrm{E}\{\, \overline{n}(k) + a_1\overline{n}(k-1) + \ldots + a_m\overline{n}(k-m) \,\} = 0 , \tag{9.3.10}$$

if $\mathrm{E}\{\, \overline{n}(k) \,\} = 0$ and hence

$$\lim_{M\to\infty} \mathrm{E}\{\hat{\boldsymbol{\theta}} - \boldsymbol{\theta}_0\} = \lim_{M\to\infty} \mathrm{E}\Big\{\left(\boldsymbol{R}^{\mathrm{T}}\boldsymbol{R}\right)^{-1}\boldsymbol{R}^{\mathrm{T}}\boldsymbol{e}\Big\} = \boldsymbol{0} \tag{9.3.11}$$

$$\lim_{M\to\infty} \mathrm{E}\Big\{(\hat{\boldsymbol{\theta}} - \boldsymbol{\theta}_0)(\hat{\boldsymbol{\theta}} - \boldsymbol{\theta}_0)^{\mathrm{T}}\Big\} = \lim_{M\to\infty} \mathrm{E}\Big\{\left(\boldsymbol{R}^{\mathrm{T}}\boldsymbol{R}\right)^{-1}\boldsymbol{R}^{\mathrm{T}}\boldsymbol{e}^{\mathrm{T}}\boldsymbol{e}\boldsymbol{R}\left(\boldsymbol{R}^{\mathrm{T}}\boldsymbol{R}\right)^{-1}\Big\} = \boldsymbol{0} . \tag{9.3.12}$$

l must be chosen such that the entire transient is covered. A lower bound is given by the number of parameters to be estimated as $l \geq 2m$, an upper bound by the condition

$$\det\left(\boldsymbol{R}^{\mathrm{T}}\boldsymbol{R}\right) \neq 0 . \tag{9.3.13}$$

The matrix $\boldsymbol{R}^{\mathrm{T}}\boldsymbol{R}$ becomes approximately singular if l is chosen too large and too many datapoints stem from the steady state. Then, the individual rows are roughly linearly dependent. The difference of this method compared to the normal method of least squares is that the datapoints are first averaged which reduces the influence of disturbances and at the same time avoids the bias problem for correlated error signals.

9.3.2 Correlation Analysis and Method of Least Squares (COR-LS)

If a stochastic or pseudo-stochastic signal is used as an input, then the auto-correlation function of the input is given as

$$R_{\mathrm{uu}}(\tau) = \lim_{N\to\infty} \frac{1}{N+1} \sum_{k=0}^{N} u(k-\tau)u(k) \tag{9.3.14}$$

and the cross-correlation between the input and the output as

$$R_{\mathrm{uy}}(\tau) = \lim_{N\to\infty} \frac{1}{N+1} \sum_{k=0}^{N} u(k-\tau)y(k)\,. \tag{9.3.15}$$

The correlation functions can also be determined recursively as

$$\hat{R}_{\mathrm{uy}}(\tau,k) = \hat{R}_{\mathrm{uy}}(\tau,k-1) + \frac{1}{k+1}\big(u(k-\tau)y(k) - \hat{R}_{\mathrm{uy}}(\tau,k-1)\big)\,. \tag{9.3.16}$$

The process model is once again given by the difference equation

$$\begin{aligned} y(k) = &-a_1 y(k-1) - a_2 y(k-2) - \ldots - a_m y(k-m) \\ &+ b_1 u(k-d-1) + b_2 u(k-d-2) + \ldots + b_m u(k-d-m)\,. \end{aligned} \tag{9.3.17}$$

Upon multiplying with $u(k-\tau)$ and calculation of the expected values, one obtains

$$\begin{aligned} R_{\mathrm{uy}}(\tau) = &-a_1 R_{\mathrm{uy}}(\tau-1) - a_2 R_{\mathrm{uy}}(\tau-2) - \ldots - a_m R_{\mathrm{uy}}(\tau-m) \\ &+ b_1 R_{\mathrm{uu}}(\tau-d-1) + b_2 R_{\mathrm{uy}}(\tau-d-2) + \ldots + b_m R_{\mathrm{uu}}(\tau-d-m)\,. \end{aligned} \tag{9.3.18}$$

This relation is the basis for the following identification technique (Isermann and Baur, 1974). A similar method has been presented by Scheuer (1973). (9.3.18) does also result, if only a finite number N of measurements is used as a basis for the determination of the correlation functions. In this case, the correlation functions shall be replaced by their estimates

$$\hat{R}_{\mathrm{uy}}(\tau) = \frac{1}{N+1} \sum_{k=0}^{N} u(k-\tau)y(k)\,. \tag{9.3.19}$$

The values of the cross-correlation function that are used for the parameter estimation shall be $R_{\mathrm{uy}}(\tau) \neq 0$ in the interval $-P \leq \tau \leq M$ and $R_{\mathrm{uy}}(\tau) \approx 0$ for $\tau < -P$ and $\tau > M$, see Fig. 9.9.

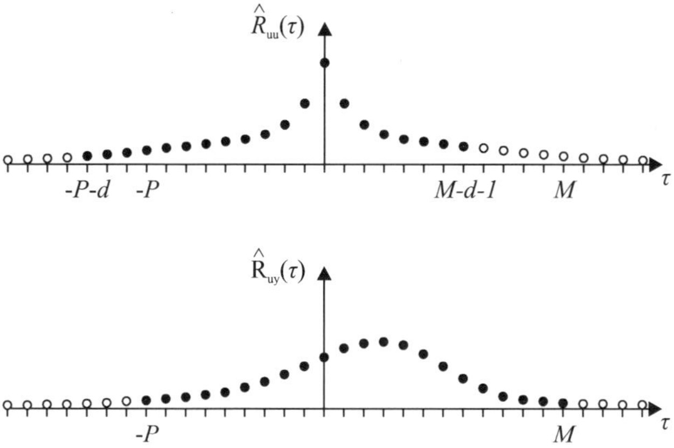

Fig. 9.9. Correlation function values employed for parameter estimation (colored noise input)

Then, one obtains the system of equations given as

$$\begin{pmatrix} R_{uy}(-P+m) \\ \vdots \\ R_{uy}(-1) \\ R_{uy}(0) \\ R_{uy}(1) \\ \vdots \\ R_{uy}(M) \end{pmatrix}$$

$$= \left(\begin{array}{ccc|cc} -R_{uy}(-P+m+1) & \ldots & -R_{uy}(-P) & R_{uu}(-P+m-d-1) & \ldots \\ \vdots & & \vdots & \vdots & \\ -R_{uy}(-2) & \ldots & -R_{uy}(-1-m) & R_{uu}(-2-d) & \ldots \\ -R_{uy}(-1) & \ldots & -R_{uy}(-m) & R_{uu}(-d-1) & \ldots \\ -R_{uy}(0) & \ldots & -R_{uy}(1-m) & R_{uu}(-d) & \ldots \\ \vdots & & \vdots & \vdots & \\ -R_{uy}(M-1) & \ldots & -R_{uy}(M-m) & R_{uu}(M-d-1) & \ldots \end{array}\right) \begin{pmatrix} a_1 \\ \vdots \\ a_m \\ b_1 \\ \vdots \end{pmatrix} , \tag{9.3.20}$$

which can then be written as

$$\boldsymbol{R}_{uy} = \boldsymbol{S}\boldsymbol{\theta} , \tag{9.3.21}$$

and application of the method of least squares leads to the parameter estimates

$$\hat{\boldsymbol{\theta}} = \left(\boldsymbol{S}^{T}\boldsymbol{S}\right)^{-1}\boldsymbol{S}^{T}\boldsymbol{R}_{uy} . \tag{9.3.22}$$

Example 9.5 (Parameter Estimation by Means of the COR-LS Method).

The parameters of the discrete-time transfer function of the Three-Mass Oscillator shall be estimated by means of the COR-LS method. Hence, first the correlation function estimates $\hat{R}_{\mathrm{uu}}(\tau)$ and $\hat{R}_{\mathrm{uy}}(\tau)$ have been determined, see Fig. 9.10. Here, the input $u(k)$ was a PRBS signal, hence the auto-correlation function in the evaluated interval shows a close resemblance to the auto-correlation function of a white noise. The input $u(k)$ and output $y(k) = \omega_3(k)$ have already been shown in Fig. 9.6.

Once again, the parameters of a transfer function of order $m = 6$ between the input and the output shall be determined. The setup of the matrix $\boldsymbol{\Psi}$ and the vector $\boldsymbol{y}$, see (9.3.20), is as follows (with $P = 0$ according to the PRBS input),

$$\boldsymbol{\Psi} = \left(\begin{array}{ccc|ccc} -\hat{R}_{\mathrm{uy}}(5) & \ldots & -\hat{R}_{\mathrm{uy}}(0) & \hat{R}_{\mathrm{uu}}(5) & \ldots & \hat{R}_{\mathrm{uu}}(0) \\ -\hat{R}_{\mathrm{uy}}(6) & \ldots & -\hat{R}_{\mathrm{uy}}(1) & \hat{R}_{\mathrm{uu}}(6) & \ldots & \hat{R}_{\mathrm{uu}}(1) \\ \vdots & & \vdots & \vdots & & \vdots \\ -\hat{R}_{\mathrm{uy}}(M-1) & \ldots & -\hat{R}_{\mathrm{uy}}(M-6) & \hat{R}_{\mathrm{uu}}(M-1) & \ldots & \hat{R}_{\mathrm{uu}}(M-6) \end{array}\right) \tag{9.3.23}$$

and $\boldsymbol{y}$ as

$$\boldsymbol{y} = \begin{pmatrix} \hat{R}_{\mathrm{uy}}(6) \\ \hat{R}_{\mathrm{uy}}(7) \\ \vdots \\ \hat{R}_{\mathrm{uy}}(M) \end{pmatrix} . \tag{9.3.24}$$

The parameter vector $\boldsymbol{\theta}$ is constructed as

$$\boldsymbol{\theta}^{\mathrm{T}} = \left(a_1 \; a_2 \; \ldots \; a_6 \middle| b_1 \; b_2 \; \ldots \; b_6 \right) . \tag{9.3.25}$$

In order to judge the quality of the estimated model, the frequency response of the estimated discrete-time model has been shown together with the frequency response determined by the orthogonal correlation (see Sect. 5.5.2) in Fig. 9.11. One can see the good fidelity of the estimated model. □

The *convergence* of the estimate shall now be investigated. From Chap. 7, it is known that the estimates of the correlation function converge for $N \to \infty$ as

$$\lim_{N\to\infty} \mathrm{E}\{\hat{R}_{\mathrm{uu}}(\tau)\} = R_{\mathrm{uu},0}(\tau) \tag{9.3.26}$$

$$\lim_{N\to\infty} \mathrm{E}\{\hat{R}_{\mathrm{uy}}(\tau)\} = R_{\mathrm{uy},0}(\tau) \, , \tag{9.3.27}$$

if $\mathrm{E}\{n(k)\} = 0$ and $\mathrm{E}\{u(k-\tau)n(k)\} = 0$. Hence the estimates which were determined over a finite time horizon will converge towards the true values of the correlation functions and thus, it follows that

$$\lim_{N\to\infty} \mathrm{E}\{\boldsymbol{e}\} = \boldsymbol{0} \tag{9.3.28}$$

provided that the model matches in structure and model order with the process. Then, it can be shown that this method provides estimates that are *consistent in the mean square*. The technique can either be applied recursively or non-recursively.

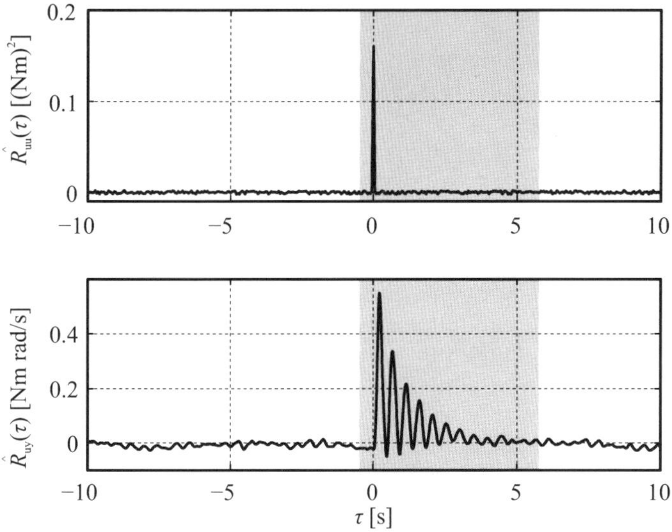

Fig. 9.10. Parameter estimation using COR-LS for a discrete-time model of the Three-Mass Oscillator, correlation functions. Values in *gray-shaded area* used for the parameter estimation

Non-Recursive Method (COR-LS)

For the non-recursive method, the following steps need to be taken:

1. $u(k)$ and $y(k)$ are stored
2. $R_{uu}(\tau)$ and $R_{uy}(\tau)$ are determined according to (9.3.19), $\hat{\boldsymbol{\theta}}$ is determined according to (9.3.22)

Recursive Method (RCOR-LS)

The recursive method requires the following steps to be taken:

1. $R_{uy}(\tau, k)$ and if necessary $R_{uu}(\tau, k)$ are determined recursively according to (9.3.16) at each time step k
2. $\hat{\boldsymbol{\theta}}$ is determined according to (9.3.22) either after every time step or in larger intervals

The method of correlation analysis and least squares differs in the following aspects from the normal method of least squares:

1. Instead of the $N \times 2m$ matrix $\boldsymbol{\Psi}$, one processes the matrix $\boldsymbol{S}$ of size $P + M - m + 1) \times 2m$ which normally has a smaller size. The matrices $\boldsymbol{\Psi}^{\mathrm{T}}\boldsymbol{\Psi}$ and $\boldsymbol{S}^{\mathrm{T}}\boldsymbol{S}$ however both have the same dimension $2m \times 2m$
2. The COR-LS method uses $P + M + 1$ values of the CCF, the normal LS method only $2m - 1$. If P and M are chosen accordingly, one can consider more values of the CCF

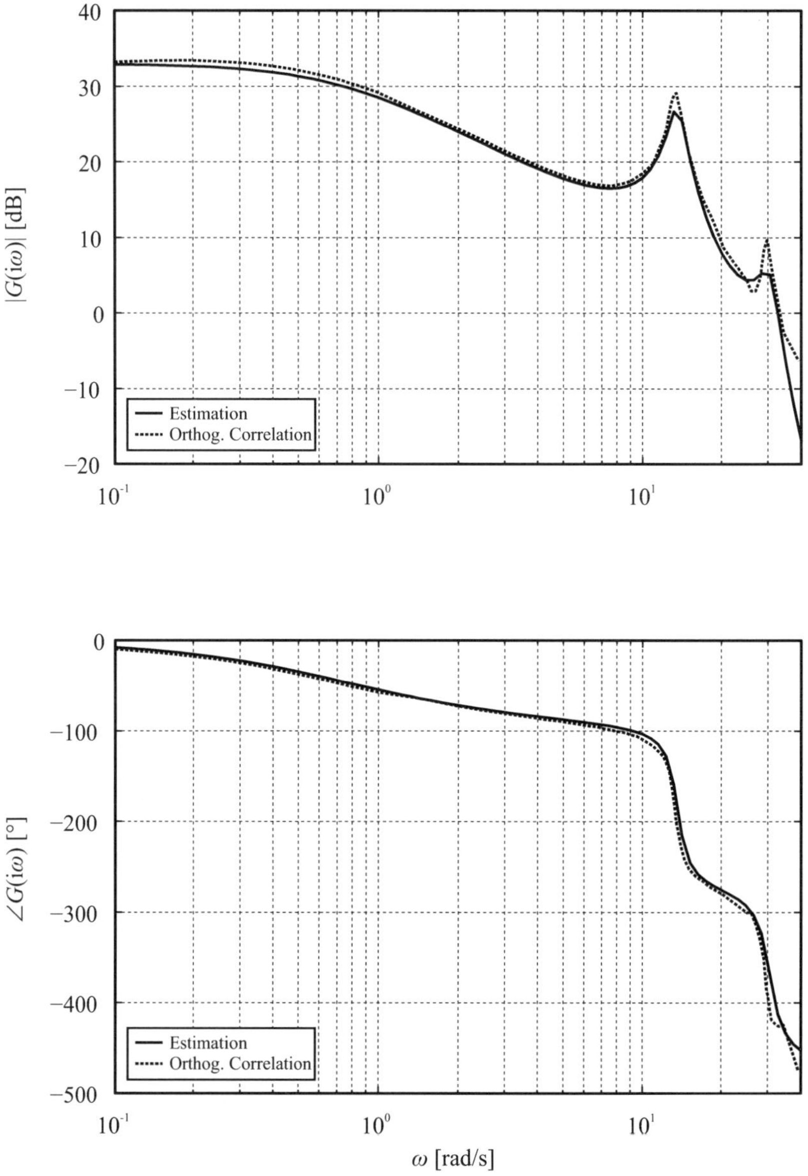

Fig. 9.11. Calculated frequency response based on the parameter estimation using COR-LS for a discrete-time model of the Three-Mass Oscillator. Comparison with directly measured frequency response

3. One obtains consistent parameter estimates for arbitrary stationary disturbances

In (9.3.18), which was the basis for the formulation of the parameter estimation problem, only the small signal quantities had been evaluated. For large signal quantities $U(k)$ and $Y(k)$, it can first be shown that the results are independent of Y_{00} if

$$U_{00} = \mathrm{E}\{u(k)\} = 0 \, .$$

If however, the DC values are unknown, one can use the techniques that had been outlined in Sect. 9.1.5. For the implicit estimation of a constant, one has to calculate the estimates of the correlation functions as follows

$$\hat{R}_{\mathrm{uu}}(\tau) = \frac{1}{N+1} \sum_{k=0}^{N} U(k-\tau)U(k) \tag{9.3.29}$$

$$\hat{R}_{\mathrm{uy}}(\tau) = \frac{1}{N+1} \sum_{k=0}^{N} U(k-\tau)Y(k) \, . \tag{9.3.30}$$

The matrix $\boldsymbol{S}$ is augmented with a column of 1 as

$$\boldsymbol{R}_{\mathrm{uy}} = \underbrace{\left(\boldsymbol{S} \left| \begin{matrix} 1 \\ \vdots \\ 1 \end{matrix} \right. \right)}_{\boldsymbol{S}^*} \boldsymbol{\theta}^* \, . \tag{9.3.31}$$

Then, the new parameter vector $\boldsymbol{\theta}^*$ contains as the last element the constant C as described in Sect. 9.1.5.

9.4 Recursive Methods of Least Squares (RLS)

The method of least squares as presented until now assumed that all parameters had first been stored and had then been processed in one pass (batch processing). This also means that the parameter estimates are only available after the end of the measurement. The non-recursive method of least squares is hence more suitable for offline identification.

If the process however shall be identified online and in real-time, then new parameter estimates should be available during the measurement, e.g. after each sample step. If one would apply the non-recursive method of least squares, which was already introduced, one would augment the data matrix $\boldsymbol{\Psi}$ with one row after each sample step and would then process all available data (even from the previous sample steps). Such an approach would require a lot of computations and is hence inappropriate. Recursive methods reduce the computational effort and provide an update of the parameter estimates after each sample step. Previous measurements do not have to be stored. With appropriate modifications that will be presented later in this chapter, it is also possible to identify time varying processes.

The recursive method of least squares was also described by Gauss (1809), see (Genin, 1968). First applications of this technique to dynamic systems have been presented by Lee (1964) and Albert and Sittler (1965). In Sect. 9.4.1, the fundamental equations will be derived. Then, the recursive parameter estimation is extended to stochastic signals in Sect. 9.4.2. Methods for the treatment of unknown DC values are presented in Sect. 9.4.3. The convergence is analyzed in a later chapter (Sect. 12.4).

9.4.1 Fundamental Equations

The parameter estimate of the non-recursive method of least squares for the sample step k is given as

$$\hat{\boldsymbol{\theta}}(k) = \boldsymbol{P}(k)\boldsymbol{\Psi}^{\mathrm{T}}(k)\boldsymbol{y}(k) \tag{9.4.1}$$

with

$$\boldsymbol{P}(k) = \big(\boldsymbol{\Psi}^{\mathrm{T}}(k)\boldsymbol{\Psi}(k)\big)^{-1} \tag{9.4.2}$$

$$\boldsymbol{y}(k) = \begin{pmatrix} y(1) \\ y(2) \\ \vdots \\ y(k) \end{pmatrix} \tag{9.4.3}$$

$$\boldsymbol{\Psi}(k) = \begin{pmatrix} \boldsymbol{\psi}^{\mathrm{T}}(1) \\ \boldsymbol{\psi}^{\mathrm{T}}(2) \\ \vdots \\ \boldsymbol{\psi}^{\mathrm{T}}(k) \end{pmatrix} \tag{9.4.4}$$

and

$$\boldsymbol{\psi}^{\mathrm{T}} = \big(-y(k-1)\; -y(k-2)\; \ldots\; -y(k-m)\big|u(k-d-1)\; \ldots\; u(k-d-m)\big)\,. \tag{9.4.5}$$

Accordingly, the parameter estimate for the sample step $k+1$ is given as

$$\hat{\boldsymbol{\theta}}(k+1) = \boldsymbol{P}(k+1)\boldsymbol{\Psi}^{\mathrm{T}}(k+1)\boldsymbol{y}(k+1)\,. \tag{9.4.6}$$

This equation can be split up as

$$\begin{aligned} \hat{\boldsymbol{\theta}}(k+1) &= \boldsymbol{P}(k+1)\begin{pmatrix} \boldsymbol{\Psi}(k) \\ \boldsymbol{\psi}^{\mathrm{T}}(k+1) \end{pmatrix}^{\mathrm{T}}\begin{pmatrix} \boldsymbol{y}(k) \\ y(k+1) \end{pmatrix} \\ &= \boldsymbol{P}(k+1)\big(\boldsymbol{\Psi}(k)\boldsymbol{y}(k) + \boldsymbol{\psi}^{\mathrm{T}}(k+1)y(k+1)\big)\,. \end{aligned} \tag{9.4.7}$$

Based on (9.4.1), one can substitute $\boldsymbol{\Psi}(k)\boldsymbol{y}(k) = \boldsymbol{P}^{-1}(k)\hat{\boldsymbol{\theta}}(k)$ and obtains

$$\hat{\boldsymbol{\theta}}(k+1) = \hat{\boldsymbol{\theta}}(k) + \big(\boldsymbol{P}(k+1)\boldsymbol{P}^{-1}(k) - \boldsymbol{I}\big)\hat{\boldsymbol{\theta}}(k) + \boldsymbol{P}(k+1)\boldsymbol{\psi}(k+1)y(k+1)\,. \tag{9.4.8}$$

According to (9.4.2),

$$P(k+1) = \left(\begin{pmatrix} \boldsymbol{\Psi}(k) \\ \boldsymbol{\psi}^{\mathrm{T}}(k+1) \end{pmatrix}^{\mathrm{T}} \begin{pmatrix} \boldsymbol{\Psi}(k) \\ \boldsymbol{\psi}^{\mathrm{T}}(k+1) \end{pmatrix} \right)^{-1} , \\ = \left(\boldsymbol{P}^{-1}(k) + \boldsymbol{\psi}(k+1)\boldsymbol{\psi}^{\mathrm{T}}(k+1) \right)^{-1} \tag{9.4.9}$$

and hence

$$\boldsymbol{P}^{-1}(k) = \boldsymbol{P}^{-1}(k+1) - \boldsymbol{\psi}(k+1)\boldsymbol{\psi}^{\mathrm{T}}(k+1) \, . \tag{9.4.10}$$

Together with (9.4.8), one then obtains

$$\underbrace{\hat{\boldsymbol{\theta}}(k+1)}_{\substack{\text{New Parameter} \\ \text{Estimate}}} = \underbrace{\hat{\boldsymbol{\theta}}(k)}_{\substack{\text{Old Parameter} \\ \text{Estimate}}} + \underbrace{\boldsymbol{P}(k+1)\boldsymbol{\psi}(k+1)}_{\substack{\text{Correction} \\ \text{Vector}}} \Big(\underbrace{y(k+1)}_{\substack{\text{New} \\ \text{Measurement}}} - \underbrace{\boldsymbol{\psi}^{\mathrm{T}}(k+1)\hat{\boldsymbol{\theta}}(k)}_{\substack{\text{Predicted Measurement based} \\ \text{on Last Parameter Estimate}}} \Big) \, . \tag{9.4.11}$$

In this way, a recursive formulation of the estimation problem has been found. According to (9.1.7), the term

$$\boldsymbol{\psi}^{\mathrm{T}}(k+1)\hat{\boldsymbol{\theta}}(k) = \hat{y}(k+1|k) \tag{9.4.12}$$

can be interpreted as a one-step prediction of the model with the parameters and the measurements up to sample step k. The factor in brackets in (9.4.11) is according to (9.1.10) the equation error

$$\left(y(k+1) - \boldsymbol{\psi}^{\mathrm{T}}(k+1)\hat{\boldsymbol{\theta}}(k) \right) = e(k+1) \, , \tag{9.4.13}$$

so that (9.4.11) can finally be written as

$$\hat{\boldsymbol{\theta}}(k+1) = \hat{\boldsymbol{\theta}}(k) + \boldsymbol{P}(k+1)\boldsymbol{\psi}(k+1)e(k+1) \, . \tag{9.4.14}$$

One has to determine $\boldsymbol{P}(k+1)$ according to (9.4.9) respectively $\boldsymbol{P}^{-1}(k+1)$ according to (9.4.10) recursively. This requires one matrix inversion per update step. The inversion of the matrix can be avoided by exploiting the matrix inversion theorem presented in App. A.4. Then, instead of (9.4.9), one can write

$$\boldsymbol{P}(k+1) = \boldsymbol{P}(k) - \boldsymbol{P}(k)\boldsymbol{\psi}(k+1) \\ \left(\boldsymbol{\psi}^{\mathrm{T}}(k+1)\boldsymbol{P}(k)\boldsymbol{\psi}(k+1) + 1 \right)^{-1} \boldsymbol{\psi}^{\mathrm{T}}(k+1)\boldsymbol{P}(k) \, . \tag{9.4.15}$$

Since the term in brackets is a scalar quantity only, one does not have to invert a full matrix any longer. Upon multiplication with $\boldsymbol{\psi}(k+1)$, one obtains

$$\boldsymbol{P}(k+1)\boldsymbol{\psi}(k+1) = \frac{1}{\boldsymbol{\psi}^{\mathrm{T}}(k+1)\boldsymbol{P}(k)\boldsymbol{\psi}(k+1) + 1} \boldsymbol{P}(k)\boldsymbol{\psi}(k+1) \, , \tag{9.4.16}$$

which combined with (9.4.11) yields the recursive method of least squares as

$$\hat{\boldsymbol{\theta}}(k+1) = \hat{\boldsymbol{\theta}}(k) + \boldsymbol{\gamma}(k)\left(y(k+1) - \boldsymbol{\psi}^{\mathrm{T}}(k+1)\hat{\boldsymbol{\theta}}(k) \right) . \tag{9.4.17}$$

The correction vector $\boldsymbol{\gamma}(k)$ is given as

$$\boldsymbol{\gamma}(k) = \boldsymbol{P}(k+1)\boldsymbol{\psi}(k+1) = \frac{1}{\boldsymbol{\psi}^{\mathrm{T}}(k+1)\boldsymbol{P}(k)\boldsymbol{\psi}(k+1)+1}\boldsymbol{P}(k)\boldsymbol{\psi}(k+1)\,. \tag{9.4.18}$$

From (9.4.15) follows

$$\boldsymbol{P}(k+1) = \big(\boldsymbol{I} - \boldsymbol{\gamma}(k)\boldsymbol{\psi}^{\mathrm{T}}(k+1)\big)\boldsymbol{P}(k)\,. \tag{9.4.19}$$

The recursive method of least squares is hence given by the three equations above, which have to be evaluated in the sequence (9.4.18), (9.4.17), (9.4.19), see also (Goodwin and Sin, 1984). The matrix $\boldsymbol{P}(k+1)$ is a scaled estimate of the covariance matrix of the estimation error, since according to (9.1.69)

$$\mathrm{E}\big\{\boldsymbol{P}(k+1)\big\} = \frac{1}{\sigma_{\mathrm{e}}^2}\operatorname{cov}\Delta\boldsymbol{\theta}(k+1) \tag{9.4.20}$$

holds true for bias-free estimates.

In order to start the recursive method of least squares, initial values for $\hat{\boldsymbol{\theta}}(k)$ and $\boldsymbol{P}(k)$ must be known. For an appropriate choice of these values, the following methods have proven successful (Klinger, 1968):

- *Start with non-recursive method of least squares*: One uses the non-recursive method of least squares on at least $2m$ equations, e.g. from $k = d+1$ up to $k = d + 2m = k'$ and then uses the values $\hat{\boldsymbol{\theta}}(k')$ and $\boldsymbol{P}(k')$ as initial values for the recursive method of least squares which will start at time step $k = k'$.
- *Use of a priori estimates*: If one knows a priori approximate values of the parameters, their covariance and the variance of the equation error, then these values can be used to initialize $\hat{\boldsymbol{\theta}}(0)$ and $\boldsymbol{P}(0)$, see (9.4.20).
- *Assumption of appropriate initial values*: The easiest solution however is to assume appropriate initial values for $\hat{\boldsymbol{\theta}}(0)$ and $\boldsymbol{P}(0)$ (Lee, 1964).

An appropriate choice of $\boldsymbol{P}(0)$ can be derived as follows: From (9.4.10) follows

$$\begin{aligned}
\boldsymbol{P}^{-1}(k+1) &= \boldsymbol{P}^{-1}(k) + \boldsymbol{\psi}(k+1)\boldsymbol{\psi}^{\mathrm{T}}(k+1)\\
\boldsymbol{P}^{-1}(1) &= \boldsymbol{P}^{-1}(0) + \boldsymbol{\psi}(1)\boldsymbol{\psi}^{\mathrm{T}}(1)\\
\boldsymbol{P}^{-1}(2) &= \boldsymbol{P}^{-1}(1) + \boldsymbol{\psi}(2)\boldsymbol{\psi}^{\mathrm{T}}(2)\\
&= \boldsymbol{P}^{-1}(0) + \boldsymbol{\psi}(1)\boldsymbol{\psi}^{\mathrm{T}}(1) + \boldsymbol{\psi}(2)\boldsymbol{\psi}^{\mathrm{T}}(2)\\
&\ \ \vdots\\
\boldsymbol{P}^{-1}(k) &= \boldsymbol{P}^{-1}(0) + \boldsymbol{\Psi}^{\mathrm{T}}(k)\boldsymbol{\Psi}(k)\,.
\end{aligned} \tag{9.4.21}$$

If one chooses

$$\boldsymbol{P}(0) = \alpha\boldsymbol{I} \tag{9.4.22}$$

with a large value α, then one obtains

$$\lim_{\alpha \to \infty} \boldsymbol{P}^{-1}(0) = \frac{1}{\alpha}\boldsymbol{I} = \boldsymbol{0} \tag{9.4.23}$$

and (9.4.21) matches with (9.4.2), which was how $\boldsymbol{P}(k)$ was defined for the non-recursive case.

With large values of α, $\boldsymbol{P}(0)$ has a negligibly small influence on the recursively calculated $\boldsymbol{P}(k)$. Furthermore, it follows from (9.4.11) that

$$\begin{aligned}\hat{\boldsymbol{\theta}}(1) &= \hat{\boldsymbol{\theta}}(0) + \boldsymbol{P}(1)\boldsymbol{\psi}(1)\big(y(1) - \boldsymbol{\psi}^{\mathrm{T}}(1)\hat{\boldsymbol{\theta}}(0)\big) \\ &= \boldsymbol{P}(1)\Big(\boldsymbol{\psi}(1)y(1) + \big(-\boldsymbol{\psi}(1)\boldsymbol{\psi}^{\mathrm{T}}(1) + \boldsymbol{P}^{-1}(1)\big)\hat{\boldsymbol{\theta}}(0)\Big)\end{aligned}$$

with (9.4.21) follows

$$\hat{\boldsymbol{\theta}}(1) = \boldsymbol{P}(1)\big(\boldsymbol{\psi}(1)y(1) + \boldsymbol{P}^{-1}(0)\hat{\boldsymbol{\theta}}(0)\big) \tag{9.4.24}$$

and correspondingly

$$\begin{aligned}\hat{\boldsymbol{\theta}}(2) &= \boldsymbol{P}(2)\big(\boldsymbol{\psi}(2)y(2) + \boldsymbol{P}^{-1}(1)\hat{\boldsymbol{\theta}}(1)\big) \\ &= \boldsymbol{P}(2)\big(\boldsymbol{\psi}(2)y(2) + \boldsymbol{\psi}(1)y(1) + \boldsymbol{P}^{-1}(0)\hat{\boldsymbol{\theta}}(0)\big)\,,\end{aligned}$$

so that one finally obtains

$$\hat{\boldsymbol{\theta}}(k) = \boldsymbol{P}(k)\big(\boldsymbol{\Psi}(k)\boldsymbol{y}(k) + \boldsymbol{P}^{-1}(0)\hat{\boldsymbol{\theta}}(0)\big)\,. \tag{9.4.25}$$

Because of (9.4.23), (9.4.25) will for large α and arbitrary $\hat{\boldsymbol{\theta}}(0)$ match with the non-recursive estimation in (9.4.1). The choice of large values of α can be interpreted such that at the beginning, a large variance of the error of the estimates $\hat{\boldsymbol{\theta}}(0)$ is assumed, (9.4.20). To start the recursive method, one can thus choose $\boldsymbol{P}(0)$ according to (9.4.22) and an arbitrary $\hat{\boldsymbol{\theta}}(0)$ or for reasons of simplicity $\hat{\boldsymbol{\theta}}(0) = \boldsymbol{0}$.

Now, it has to be investigated how large α should at least be chosen. From (9.4.18), one can see that $\boldsymbol{P}(0) = \alpha\boldsymbol{I}$ has no substantial influence on the correction vector $\boldsymbol{\gamma}(0)$ (Isermann, 1974), if

$$\boldsymbol{\psi}^{\mathrm{T}}(1)\boldsymbol{P}(0)\boldsymbol{\psi}(1) = \alpha\boldsymbol{\psi}^{\mathrm{T}}(1)\boldsymbol{\psi}(1) \gg 1\,, \tag{9.4.26}$$

because in this case

$$\lim_{\alpha \to \infty} \gamma(0) = \lim_{\alpha \to \infty} \boldsymbol{P}(0)\boldsymbol{\psi}(1)\big(\boldsymbol{\psi}^{\mathrm{T}}(1)\boldsymbol{P}(0)\boldsymbol{\psi}(1)\big)^{-1} = \boldsymbol{\psi}(1)\big(\boldsymbol{\psi}^{\mathrm{T}}(1)\boldsymbol{\psi}(1)\big)^{-1}\,. \tag{9.4.27}$$

If the process has been in its steady state for $k < 0$, i.e. $u(k) = 0$ and $y(k) = 0$ before the test signals started at $k = 0$, then it follows for e.g. $d = 0$,

$$\boldsymbol{\psi}^{\mathrm{T}}(1) = \big(0 \ldots 0 \big| u(0) \ldots\big) \tag{9.4.28}$$

and from (9.4.26) follows $\alpha u^2(0) \gg 1$ or

$$\alpha \gg \frac{1}{u^2(0)}\,. \tag{9.4.29}$$

If the process was not in its steady state, so one can derive the relation

$$\alpha \gg \frac{1}{\sum_{k=0}^{m-1} y^2(k) + \sum_{k=0}^{m-1} u^2(k)} \tag{9.4.30}$$

for the correct choice of α. One can see that α depends on the square of the signal changes. The larger the signal changes, the smaller α can be chosen. For $u(0) = 1$, a value of $\alpha = 10$ is sufficient (Lee, 1964). Baur (1976) has shown by means of simulation that for $\alpha = 10$ or $\alpha = 10^5$, only small differences appear for sufficiently long measurement times. In practice, one will choose values of α in the range $\alpha = 100, \ldots, 10\,000$.

Example 9.6 (Recursive Identification of a First Order Difference Equation with 2 Parameters).

The same process that was already used in Example 9.2 shall be used again. The process is governed by

$$\begin{aligned} y_{\mathrm{u}}(k) + a_1 y_{\mathrm{u}}(k-1) &= b_1 u(k-1) \\ y(k) &= y_{\mathrm{u}}(k) + n(k)\,. \end{aligned}$$

The process model according to (9.1.6) is given as

$$y(k) + \hat{a}_1 y(k-1) - \hat{b}_1 u(k-1) = e(k)$$

or

$$y(k) = \boldsymbol{\psi}^{\mathrm{T}}(k)\hat{\boldsymbol{\theta}}(k-1) + e(k)$$

with

$$\begin{aligned} \boldsymbol{\psi}^{\mathrm{T}}(k) &= \big(-y(k-1)\;u(k-1)\big) \\ \hat{\boldsymbol{\theta}}(k-1) &= \big(\hat{a}_1(k-1)\;\hat{b}_1(k-1)\big)^{\mathrm{T}}\,. \end{aligned}$$

The estimation will now be programmed in the following form

1. The new data $u(k)$, $y(k)$ are measured at time step k
2. $e(k) = y(k) - \big(-y(k-1)\;u(k-1)\big)\begin{pmatrix}\hat{a}_1(k-1)\\ \hat{b}_1(k-1)\end{pmatrix}$
3. The new parameter estimates are then given as
$$\begin{pmatrix}\hat{a}_1(k)\\ \hat{b}_1(k)\end{pmatrix} = \begin{pmatrix}\hat{a}_1(k-1)\\ \hat{b}_1(k-1)\end{pmatrix} + \underbrace{\begin{pmatrix}\hat{\gamma}_1(k-1)\\ \hat{\gamma}_2(k-1)\end{pmatrix}}_{\text{from step 7}} e(k)$$
4. The data $y(k)$ and $u(k)$ are plugged into $\boldsymbol{\psi}^{\mathrm{T}}(k+1) = \big(-y(k)\;u(k)\big)$
5. $$\underbrace{\boldsymbol{P}(k)}_{\text{from step 8}}\;\boldsymbol{\psi}(k+1) = \begin{pmatrix}p_{11}(k) & p_{12}(k)\\ p_{21}(k) & p_{22}(k)\end{pmatrix}\begin{pmatrix}-y(k)\\ u(k)\end{pmatrix}$$
$$= \begin{pmatrix}-p_{11}(k)y(k) + p_{12}(k)u(k)\\ -p_{21}(k)y(k) + p_{22}(k)u(k)\end{pmatrix} = \begin{pmatrix}i_1\\ i_2\end{pmatrix} = \boldsymbol{i}$$

6. $\boldsymbol{\psi}^{\mathrm{T}}(k+1)\ \underbrace{\boldsymbol{P}(k)\boldsymbol{\psi}(k+1)}_{\text{from step 5}} = \big(-y(k)\ u(k)\big)\begin{pmatrix} i_1 \\ i_2 \end{pmatrix} = -i_1 y(k) + i_2 u(k) = j$

7. $\begin{pmatrix} \hat{\gamma}_1(k) \\ \hat{\gamma}_2(k) \end{pmatrix} = \frac{1}{j+\lambda}\begin{pmatrix} i_1 \\ i_2 \end{pmatrix}$

8. Now, one can determine $\boldsymbol{P}(k+1)$

$$\begin{aligned}
\boldsymbol{P}(k+1) &= \frac{1}{\lambda}\big(\boldsymbol{P}(k) - \boldsymbol{\gamma}(k)\boldsymbol{\psi}^{\mathrm{T}}(k+1)\boldsymbol{P}(k)\big) \\
&= \frac{1}{\lambda}\Big(\boldsymbol{P}(k) - \gamma(k)\ \underbrace{\big(\boldsymbol{P}(k)\boldsymbol{\psi}(k+1^{)}\big)^{\mathrm{T}}}_{\text{from step 5}}\Big) \\
&= \frac{1}{\lambda}\big(\boldsymbol{P}(k) - \boldsymbol{\gamma}(k)\boldsymbol{i}^{\mathrm{T}}\big) \\
&= \frac{1}{\lambda}\begin{pmatrix} p_{11}(k) - \gamma_1 i_1 & p_{12}(k) - \gamma_1 i_2 \\ p_{21}(k) - \gamma_2 i_1 & p_{22}(k) - \gamma_2 i_2 \end{pmatrix}
\end{aligned}$$

9. For the next time step $(k+1)$ is now replaced by k and one starts over with step 1.

   For starting at time $k = 0$, one chooses

$$\hat{\boldsymbol{\theta}}(0) = \begin{pmatrix} 0 \\ 0 \end{pmatrix} \text{ and } \boldsymbol{P}(0) = \begin{pmatrix} \alpha & 0 \\ 0 & \alpha \end{pmatrix}$$

where α is a large number. □

The recursive method of least squares can now be represented as a block diagram, see Fig. 9.12.

In order to develop the block diagram, (9.4.17) will be rewritten as

$$\begin{aligned}
\hat{\boldsymbol{\theta}}(k+1) &= \hat{\boldsymbol{\theta}}(k) + \Delta\hat{\boldsymbol{\theta}}(k+1) \\
\Delta\hat{\boldsymbol{\theta}}(k+1) &= \boldsymbol{\gamma}(k)\big(y(k+1) - \boldsymbol{\psi}^{\mathrm{T}}(k+1)\hat{\boldsymbol{\theta}}(k)\big) = \boldsymbol{\gamma}(k)e(k+1)\ .
\end{aligned}$$

This yields a closed loop with the controlled quantity $\Delta\hat{\boldsymbol{\theta}}(k+1)$, the setpoint $\boldsymbol{w} = \boldsymbol{0}$, the integral-acting discrete-time "controller"

$$\frac{\hat{\boldsymbol{\theta}}(z)}{\Delta\hat{\boldsymbol{\theta}}(z)} = \frac{1}{1-z^{-1}}$$

and the manipulated variable $\hat{\boldsymbol{\theta}}(k)$. The "plant" consists of the process model and the correction vector $\boldsymbol{\gamma}$. Since the model as well as the factor $\boldsymbol{\gamma}$ are adjusted according to the signals $u(k)$ and $y(k)$, the "plant" shows a time-variant behavior. This consideration was used by Becker (1990) to design improved recursive algorithms with "better control behavior".

The underlying idea is that one replaces the integral-acting discrete-time "controller" by other control algorithms leading to the RLS-IF (Recursive Least Squares

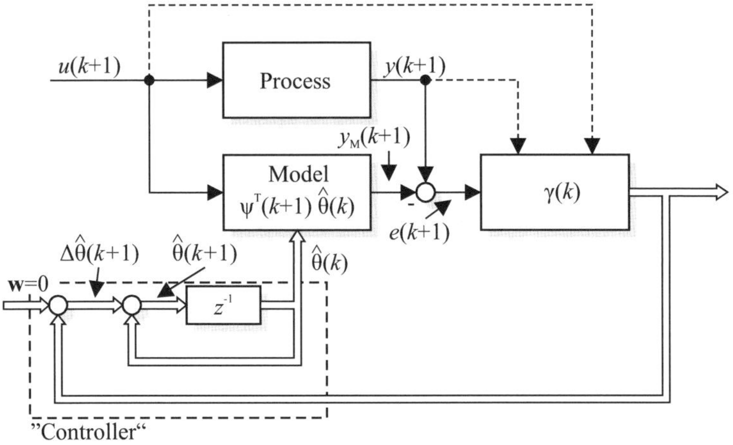

Fig. 9.12. Block diagram of the recursive parameter estimation by the method of least squares

with Improved Feedback). The requirement of bias-free estimates then correlates to the requirement that the control shows no steady-state error. An analysis further shows that the controller matrix only contains elements on the diagonal, which means that each parameter can be controlled by a SISO controller, thus simplifying the controller design enormously. Furthermore, one can use methods such as a "disturbance feedforward control". If one knows, how the estimated parameters change in dependence of measured signals (e.g. operating-point dependent changes, temperature-dependent changes), then one can use this knowledge to speed up the parameter adaptation by feedforward control. By using higher order polynomials for the controller transfer function, it was for example possible to avoid a lag error on the parameter estimates for monotonous/non-monotonous parameter changes. As a controller architecture, the transfer function

$$G(z) = \mu \frac{z - a}{(z - 1)(z - b)} \tag{9.4.31}$$

was suggested, i.e. one additional pole and zero were introduced. The tuning of the parameters of the controller however depends strongly on the process to be identified and hence no general tuning rules can be presented here.

9.4.2 Recursive Parameter Estimation for Stochastic Signals

The method of least squares can also be used for the parameter estimation of stochastic signal models. As a model, a stationary auto-regressive moving-average (ARMA) process is chosen, i.e.

$$\begin{aligned} &y(k) + c_1 y(k-1) + \ldots + c_p y(k-p) \\ &= \nu(k) + d_1 \nu(k-1) + \ldots + d_p \nu(k-p) \,, \end{aligned} \tag{9.4.32}$$

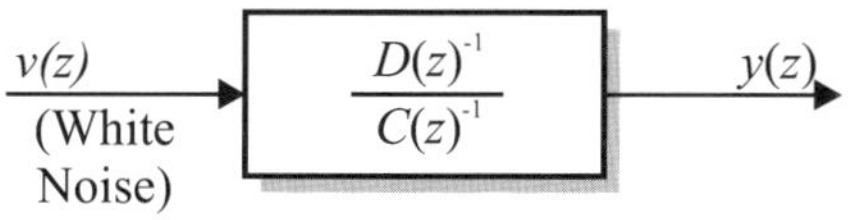

Fig. 9.13. Auto-regressive moving-average (ARMA) stochastic signal model

see Fig. 9.13. Here, $y(k)$ is a measurable signal and $\nu(k)$ a virtual white noise with $\mathrm{E}\{\nu(k)\} = 0$ and variance σ_ν^2. According to (9.1.7), one can write

$$y(k) = \boldsymbol{\psi}^{\mathrm{T}}(k)\hat{\boldsymbol{\theta}}(k-1) + \nu(k) \tag{9.4.33}$$

with

$$\boldsymbol{\psi}^{\mathrm{T}}(k) = \big(-y(k-1) \ldots -y(k-p)\big|\nu(k-1) \ldots \nu(k-p)\big) \tag{9.4.34}$$

and

$$\hat{\boldsymbol{\theta}}^{\mathrm{T}}(k) = \big(c_1 \ldots c_p\big|d_1 \ldots d_p\big) \tag{9.4.35}$$

If the values $\nu(k-1), \ldots, \nu(k-p)$ would be known, one could use the recursive method of least squares as it was applied to the estimation of dynamic process models, since $\nu(k)$ can be interpreted as an equation error and per definition is statistically independent.

Now, the time after the measurement of $y(k)$ will be discussed. At this point in time, $y(k-1), \ldots, y(k-p)$ are known. If one assumes that the estimates $\hat{\nu}(k-1), \ldots, \hat{\nu}(k-p)$ and $\hat{\boldsymbol{\theta}}(k-1)$ are known, then one can estimate the latest input $\hat{\nu}(k)$ by using (9.4.33) (Panuska, 1969)

$$\hat{\nu}(k) = y(k) - \hat{\boldsymbol{\psi}}^{\mathrm{T}}(k)\hat{\boldsymbol{\theta}}(k-1) \tag{9.4.36}$$

with

$$\hat{\boldsymbol{\psi}}^{\mathrm{T}}(k) = \big(-y(k-1) \ldots -y(k-p)\big|\hat{\nu}(k-1) \ldots \hat{\nu}(k-p)\big) \,. \tag{9.4.37}$$

Then, also

$$\hat{\boldsymbol{\psi}}^{\mathrm{T}}(k+1) = \big(-y(k) \ldots -y(k-p+1)\big|\hat{\nu}(k) \ldots \hat{\nu}(k-p+1)\big) \tag{9.4.38}$$

is available, so that now the recursive estimation algorithm in (9.4.17) through (9.4.19) can be used to estimate $\hat{\boldsymbol{\theta}}(k+1)$ if $\boldsymbol{\psi}^{\mathrm{T}}(k+1)$ is replaced by $\hat{\boldsymbol{\psi}}^{\mathrm{T}}(k+1)$. Then, $\hat{\nu}(k+1)$ and $\hat{\boldsymbol{\theta}}(k+2)$ can be estimated and so on. For the initialization, one uses

$$\hat{\nu}(0) = \hat{y}(0),\quad \hat{\boldsymbol{\theta}}(0) = \boldsymbol{0},\quad \boldsymbol{P}(0) = \alpha\boldsymbol{I} \tag{9.4.39}$$

with α being appropriately large. Since $\nu(k)$ is statistically independent and also $\nu(k)$ and $\hat{\boldsymbol{\psi}}^{\mathrm{T}}(k)$ are uncorrelated, one will according to Theorems 9.1 and 9.2 and (9.1.70) obtain estimates that are *consistent in the mean square*.

Additional requirements for parameter identifiability are given as follows:

1. The poles of $C(z) = 0$ must be asymptotically stable, i.e. inside the unit circle to ensure that (9.4.32) is stationary and that the correlation functions in $\boldsymbol{R}$ converge towards fixed values. This corresponds to the requirement of stability of the dynamic process that is excited by the input $u(k)$.
2. The zeros $D(z) = 0$ must lie inside the unit circle as well to ensure that the estimate of the white noise according to (9.4.36) or

$$\hat{\nu}(z) = \frac{\hat{C}(z^{-1})}{\hat{D}(z^{-1})} y(z)$$

does not diverge.

The variance of $\nu(k)$ can be estimated according to (9.1.71) by

$$\hat{\sigma}_\nu^2(k) = \frac{1}{k+1-2p} \sum_{i=0}^{k} \hat{\nu}^2(k) \tag{9.4.40}$$

or in the recursive form

$$\hat{\sigma}_\nu^2(k+1) = \hat{\sigma}_\nu^2(k) + \frac{1}{k+2-2p}\big(\hat{\nu}^2(k+1) - \hat{\sigma}_\nu^2(k)\big)\,. \tag{9.4.41}$$

In general, the estimates of a stochastic signal process converge significantly slower than those of a dynamic process, since the input signal $\nu(k)$ is unknown and must be estimated as well.

9.4.3 Unknown DC values

If the DC values U_{00} and Y_{00} are unknown, one can use in principle the same methods that were introduced in Sect. 9.1.5. These techniques must only be reformulated for the recursive application.

Averaging

The recursive equation for providing an estimate of the average is given as

$$\hat{Y}_{00}(k) = \hat{Y}_{00}(k-1) + \frac{1}{k}\big(Y(k) - \hat{Y}_{00}(k-1)\big)\,. \tag{9.4.42}$$

For slowly time-varying DC values, one should use an averaging with exponential forgetting,

$$\hat{Y}_{00}(k) = \lambda \hat{Y}_{00}(k-1) + (1-\lambda)Y(k) \tag{9.4.43}$$

with $\lambda < 1$ (Isermann, 1987).

Differencing

The differencing is done as described in (9.1.127).

Implicit Estimation of a Constant

Also for the recursive estimation, one can determine a constant by extending the parameter vector $\hat{\boldsymbol{\theta}}$ by including the element C and extending the data vector $\boldsymbol{\psi}^{\mathrm{T}}(k)$ by adding the number 1. The measured $Y(k)$ and $U(k)$ can directly be used for the estimation and C can also be estimated.

Explicit Estimation of a Constant

For the explicit estimation of a constant, one must use a recursive estimation for $\hat{K}_0$ similarly to (9.4.42) or (9.4.43).

9.5 Method of weighted least squares (WLS)

9.5.1 Markov Estimation

For the method of least squares presented so far, all values $e(k)$ of the equation error have been weighted with the same weight. In the following, the weighting for the individual values of $e(k)$ shall be chosen differently to obtain a more general method of least squares. The cost function (9.1.15) is then given as

$$\begin{aligned} V =& w(m+d)e^2(m+d) + w(m+d+1)e^2(m+d+1) + \ldots \\ &+ w(m+d+N)e^2(m+d+N) \end{aligned} \tag{9.5.1}$$

or in a more general form

$$V = \boldsymbol{e}^{\mathrm{T}}\boldsymbol{W}\boldsymbol{e} \ , \tag{9.5.2}$$

where $\boldsymbol{W}$ must be a symmetric positive-definite matrix, since only the symmetric part of $\boldsymbol{W}$ contributes to V and only a positive definite matrix ensures the existence of a unique solution. For the weighting as in (9.5.1), the matrix $\boldsymbol{W}$ is a diagonal matrix,

$$\boldsymbol{W} = \begin{pmatrix} w(m+d) & 0 & \ldots & 0 \\ 0 & w(m+d+1) & \ldots & 0 \\ \vdots & \vdots & & \vdots \\ 0 & 0 & \ldots & w(m+d+N) \end{pmatrix} . \tag{9.5.3}$$

Analog to (9.1.18) and starting with (9.5.2), one obtains the non-recursive parameter estimation by means of the weighted least squares as

$$\hat{\boldsymbol{\theta}} = \left(\boldsymbol{\Psi}^{\mathrm{T}}\boldsymbol{W}\boldsymbol{\Psi}\right)^{-1}\boldsymbol{\Psi}^{\mathrm{T}}\boldsymbol{W}\boldsymbol{y} \ . \tag{9.5.4}$$

The conditions for a consistent estimation are the same as those stated in Theorem 9.5. For the covariance of the estimate, one obtains in analogy

$$\operatorname{cov}\Delta\boldsymbol{\theta} = \mathrm{E}\left\{(\boldsymbol{\Psi}^{\mathrm{T}}\boldsymbol{W}\boldsymbol{\Psi})^{-1}\boldsymbol{\Psi}^{\mathrm{T}}\right\}\boldsymbol{W}\mathrm{E}\left\{\boldsymbol{e}\boldsymbol{e}^{\mathrm{T}}\right\}\boldsymbol{W}\mathrm{E}\left\{\boldsymbol{\Psi}(\boldsymbol{\Psi}^{\mathrm{T}}\boldsymbol{W}\boldsymbol{\Psi})^{-1}\right\} . \tag{9.5.5}$$

If the weighting matrix $\boldsymbol{W}$ is chosen as

$$\boldsymbol{W} = \left(\mathrm{E}\{\boldsymbol{e}\boldsymbol{e}^{\mathrm{T}}\}\right)^{-1} , \tag{9.5.6}$$

then (9.5.5), reduces to

$$\operatorname{cov} \Delta\boldsymbol{\theta}_{\mathrm{MV}} = \left(\boldsymbol{\Psi}^{\mathrm{T}}\left(\mathrm{E}\{\boldsymbol{e}\boldsymbol{e}^{\mathrm{T}}\}\right)^{-1}\boldsymbol{\Psi}\right)^{-1} , \tag{9.5.7}$$

and therefore

$$\operatorname{cov} \Delta\boldsymbol{\theta}_{\mathrm{MV}} \leq \operatorname{cov} \Delta\boldsymbol{\theta} , \tag{9.5.8}$$

which means that the choice of (9.5.6) as the weighting matrix yields parameter estimates with the smallest possible variance (Deutsch, 1965; Eykhoff, 1974). Estimates with minimum variance are also termed *Markov estimates*. One should however note here that the covariance matrix of the equation error is in general not known a priori. If the error $\boldsymbol{e}$ is uncorrelated, then its covariance matrix is a diagonal matrix and from (9.5.4) and (9.5.6) yield the estimate with minimum variance as

$$\hat{\boldsymbol{\theta}} = \left(\boldsymbol{\Psi}^{\mathrm{T}}\boldsymbol{\Psi}\right)^{-1}\boldsymbol{\Psi}^{\mathrm{T}}\boldsymbol{y} , \tag{9.5.9}$$

which is the method of least squares.

The recursive method of weighted least squares can be derived as follows: According to (9.4.2), one introduces

$$\boldsymbol{P}_{\mathrm{W}}(k) = \left(\boldsymbol{\Psi}^{\mathrm{T}}(k)\boldsymbol{W}(k)\boldsymbol{\Psi}(k)\right)^{-1} \tag{9.5.10}$$

with the symbols

$$\boldsymbol{\Psi}_{\mathrm{W}} = \boldsymbol{W}(k)\boldsymbol{\Psi}(k) \text{ and } \boldsymbol{y}_{\mathrm{W}}(k) = \boldsymbol{W}(k)\boldsymbol{y}(k) , \tag{9.5.11}$$

one obtains

$$\begin{aligned}\boldsymbol{\psi}_{\mathrm{W}}^{\mathrm{T}}(k) &= \boldsymbol{\psi}^{\mathrm{T}}(k)w(k)\\ \boldsymbol{P}_{\mathrm{W}}(k) &= \left(\boldsymbol{\Psi}^{\mathrm{T}}(k)\boldsymbol{\Psi}_{\mathrm{W}}(k)\right)^{-1} .\end{aligned} \tag{9.5.12}$$

Then, the estimates at the time k and $k+1$ are given as

$$\hat{\boldsymbol{\theta}}(k) = \boldsymbol{P}_{\mathrm{W}}(k)\boldsymbol{\Psi}^{\mathrm{T}}(k)\boldsymbol{y}_{\mathrm{W}}(k) \tag{9.5.13}$$

$$\begin{aligned}\hat{\boldsymbol{\theta}}(k+1) &= \boldsymbol{P}_{\mathrm{W}}(k+1)\boldsymbol{\Psi}^{\mathrm{T}}(k+1)\boldsymbol{y}_{\mathrm{W}}(k+1)\\ &= \boldsymbol{P}_{\mathrm{W}}(k+1)\begin{pmatrix}\boldsymbol{\Psi}(k)\\ \boldsymbol{\psi}^{\mathrm{T}}(k+1)\end{pmatrix}^{\mathrm{T}}\begin{pmatrix}\boldsymbol{y}_{\mathrm{W}}(k)\\ y_{\mathrm{W}}(k+1)\end{pmatrix}\\ &= \boldsymbol{P}_{\mathrm{W}}(k+1)\left(\boldsymbol{\Psi}^{\mathrm{T}}(k)\boldsymbol{y}_{\mathrm{W}}(k) + \boldsymbol{\psi}(k+1)y_{\mathrm{W}}(k+1)\right) .\end{aligned} \tag{9.5.14}$$

The further calculations can be carried out in analogy to (9.4.8 ff). One obtains the following equations for the estimation by the method of the recursive weighted least squares as

$$\hat{\boldsymbol{\theta}}(k+1) = \hat{\boldsymbol{\theta}}(k) + \boldsymbol{\gamma}_{\mathrm{W}}(k)\big(y_{\mathrm{W}}(k+1) - \boldsymbol{\psi}_{\mathrm{W}}^{\mathrm{T}}(k+1)\hat{\boldsymbol{\theta}}(k)\big) \tag{9.5.15}$$

$$\boldsymbol{\gamma}_{\mathrm{W}}(k+1) = \frac{1}{\boldsymbol{\psi}_{\mathrm{W}}^{\mathrm{T}}(k+1)\boldsymbol{P}_{\mathrm{W}}(k)\boldsymbol{\psi}(k+1)+1}\boldsymbol{P}_{\mathrm{W}}(k)\boldsymbol{\psi}(k+1) \tag{9.5.16}$$

$$\boldsymbol{P}_{\mathrm{W}}(k+1) = \big(\boldsymbol{I} - \boldsymbol{\gamma}_{\mathrm{W}}(k)\boldsymbol{\psi}_{\mathrm{W}}^{\mathrm{T}}(k+1)\big)\boldsymbol{P}_{\mathrm{W}}(k)\,. \tag{9.5.17}$$

If one assumes a diagonal matrix as a weighting matrix

$$\boldsymbol{W}(k) = \begin{pmatrix} w(0) & 0 & \dots & 0 \\ 0 & w(1) & \dots & 0 \\ \vdots & \vdots & & \vdots \\ 0 & 0 & \dots & w(k) \end{pmatrix}, \tag{9.5.18}$$

then according to (9.5.11), one obtains

$$\boldsymbol{\psi}_{\mathrm{W}}^{\mathrm{T}}(k) = \boldsymbol{\psi}^{\mathrm{T}}(k)w(k) \text{ and } y_{\mathrm{W}}(k) = y(k)w(k)\,, \tag{9.5.19}$$

and by using (9.5.15) through (9.5.17), one finally obtains

$$\hat{\boldsymbol{\theta}}(k+1) = \hat{\boldsymbol{\theta}}(k) + \boldsymbol{\gamma}(k)\big(y(k+1) - \boldsymbol{\psi}^{\mathrm{T}}(k+1)\hat{\boldsymbol{\theta}}(k)\big) \tag{9.5.20}$$

$$\boldsymbol{\gamma}(k) = \frac{1}{\boldsymbol{\psi}^{\mathrm{T}}(k+1)\boldsymbol{P}_{\mathrm{W}}(k)\boldsymbol{\psi}(k+1)+\frac{1}{w(k+1)}}\boldsymbol{P}_{\mathrm{W}}(k)\boldsymbol{\psi}(k+1) \tag{9.5.21}$$

$$\boldsymbol{P}_{\mathrm{W}}(k+1) = \big(\boldsymbol{I} - \boldsymbol{\gamma}(k)\boldsymbol{\psi}^{\mathrm{T}}(k+1)\big)\boldsymbol{P}_{\mathrm{W}}(k)\,. \tag{9.5.22}$$

Compared with the standard recursive least squares formulation, the calculation of the correction vector $\boldsymbol{\gamma}$ changes, as in the denominator, the term 1 is replaced by $1/w(k+1)$. This means that also the values of $\boldsymbol{P}_{\mathrm{W}}(k+1)$ change.

9.6 Recursive Parameter Estimation with Exponential Forgetting

If the recursive estimation algorithms should be able to follow slowly time-varying process parameters, more recent measurements must be weighted more strongly than old measurements. Therefore the estimation algorithms should have a *fading memory*. This can be incorporated into the least squares method by time-depending weighting of the squared errors, as was introduced in the previous section.

By choice of

$$w(k) = \lambda^{(m+d+N)-k} = \lambda^{N'-k},\ 0 < \lambda < 1\,, \tag{9.6.1}$$

the errors $e(k)$ are weighted as shown in Table 9.2 for $N' = 50$. The weighting then increases exponentially to 1 for time step N'. Hence, one talks about the exponential forgetting memory and λ is termed *forgetting factor*.

The weighting matrix (9.5.3) for the non-recursive estimation is then given as

Table 9.2. Weighting factors due to (9.6.1) for $N' = 50$

k	1	10	20	30	40	47	48	49	50
$\lambda = 0.99$	0.61	0.67	0.73	0.82	0.90	0.97	0.98	0.99	1
$\lambda = 0.95$	0.08	0.13	0.21	0.35	0.60	0.85	0.90	0.95	1

$$\boldsymbol{W}(m+d+n) = \begin{pmatrix} \lambda^N & & & & & \\ & \lambda^{N-1} & & & & \\ & & \ddots & & & \\ & & & \lambda^2 & & \\ & & & & \lambda & \\ & & & & & 1 \end{pmatrix} . \tag{9.6.2}$$

Upon the arrival of a new measurement, the weighting matrix is updated as

$$\boldsymbol{W}(k+1) = \begin{pmatrix} \lambda \boldsymbol{W}(k) & \boldsymbol{0} \\ \boldsymbol{0}^{\mathrm{T}} & 1 \end{pmatrix} . \tag{9.6.3}$$

Therefore, the parameter estimates are now updated as

$$\begin{aligned} \hat{\boldsymbol{\theta}}(k+1) &= \boldsymbol{P}_{\mathrm{W}}(k+1) \begin{pmatrix} \boldsymbol{\Psi}(k) \\ \boldsymbol{\psi}^{\mathrm{T}}(k+1) \end{pmatrix} \begin{pmatrix} \lambda \boldsymbol{W}(k) & \boldsymbol{0} \\ \boldsymbol{0}^{\mathrm{T}} & 1 \end{pmatrix} \begin{pmatrix} \boldsymbol{y}(k) \\ y(k+1) \end{pmatrix} \\ &= \boldsymbol{P}_{\mathrm{W}}(k+1)\big(\lambda \boldsymbol{\Psi}^{\mathrm{T}}(k) \boldsymbol{W}(k) \boldsymbol{y}(k) + \boldsymbol{\psi}(k+1) y(k+1)\big) \\ &= \boldsymbol{P}_{\mathrm{W}}(k+1)\big(\lambda \boldsymbol{P}_{\mathrm{W}}(k)^{-1} \hat{\boldsymbol{\theta}}(k) + \boldsymbol{\psi}(k+1) y(k+1)\big) , \end{aligned} \tag{9.6.4}$$

see also (9.5.4) and (9.5.13). Furthermore,

$$\begin{aligned} \boldsymbol{P}_{\mathrm{W}}(k+1) &= \left(\begin{pmatrix} \boldsymbol{\Psi}(k) \\ \boldsymbol{\psi}^{\mathrm{T}}(k+1) \end{pmatrix}^{\mathrm{T}} \begin{pmatrix} \lambda \boldsymbol{W}(k) & \boldsymbol{0} \\ \boldsymbol{0}^{\mathrm{T}} & 1 \end{pmatrix} \begin{pmatrix} \boldsymbol{\Psi}(k) \\ \boldsymbol{\psi}^{\mathrm{T}}(k+1) \end{pmatrix} \right)^{-1} \\ &= \big(\lambda \boldsymbol{\Psi}^{\mathrm{T}}(k) \boldsymbol{W}(k) \boldsymbol{\Psi}(k) + \boldsymbol{\psi}(k+1) \boldsymbol{\psi}^{\mathrm{T}}(k+1)\big)^{-1} \\ &= \big(\lambda \boldsymbol{P}_{\mathrm{W}}(k)^{-1} + \boldsymbol{\psi}(k) \boldsymbol{\psi}^{\mathrm{T}}(k+1)\big)^{-1} . \end{aligned} \tag{9.6.5}$$

Therefore,

$$\boldsymbol{P}_{\mathrm{W}}^{-1}(k+1) = \lambda \boldsymbol{P}_{\mathrm{W}}^{-1}(k) + \boldsymbol{\psi}(k+1) \boldsymbol{\psi}^{\mathrm{T}}(k+1) . \tag{9.6.6}$$

Then follows according to (9.4.8) from (9.6.4)

$$\begin{aligned} \hat{\boldsymbol{\theta}}(k+1) =& \hat{\boldsymbol{\theta}}(k) + \big(\lambda \boldsymbol{P}_{\mathrm{W}}(k+1) \boldsymbol{P}_{\mathrm{W}}^{-1}(k) - \boldsymbol{I}\big) \hat{\boldsymbol{\theta}}(k) \\ &+ \boldsymbol{P}_{\mathrm{W}}(k+1) \boldsymbol{\Psi}(k+1) y(k+1) \end{aligned} \tag{9.6.7}$$

and, after inserting (9.6.5),

$$\hat{\boldsymbol{\theta}}(k+1) = \hat{\boldsymbol{\theta}}(k) + \boldsymbol{P}_{\mathrm{W}}(k+1) \boldsymbol{\psi}(k+1) \big(y(k+1) - \boldsymbol{\psi}^{\mathrm{T}}(k+1) \hat{\boldsymbol{\theta}}(k)\big) . \tag{9.6.8}$$

Application of the matrix inversion lemma similarly to (9.4.15) then yields

$$\boldsymbol{P}_{\mathrm{W}}(k+1) = \frac{1}{\lambda}\boldsymbol{P}_{\mathrm{W}}(k) - \frac{1}{\lambda}\boldsymbol{P}_{\mathrm{W}}(k)\boldsymbol{\psi}(k+1)\left(\boldsymbol{\psi}^{\mathrm{T}}(k+1)\frac{1}{\lambda}\boldsymbol{P}_{\mathrm{W}}(k)\boldsymbol{\psi}(k+1)+1\right)^{-1}\boldsymbol{\psi}^{\mathrm{T}}(k+1)\frac{1}{\lambda}\boldsymbol{P}_{\mathrm{W}}(k) \tag{9.6.9}$$

and

$$\boldsymbol{P}_{\mathrm{W}}(k+1)\boldsymbol{\psi}(k+1) = \frac{\boldsymbol{P}_{\mathrm{W}}(k)\boldsymbol{\psi}(k+1)}{\boldsymbol{\psi}^{\mathrm{T}}(k+1)\boldsymbol{P}_{\mathrm{W}}(k)\boldsymbol{\psi}(k+1)+\lambda} = \boldsymbol{\gamma}_{\mathrm{W}}(k)\,. \tag{9.6.10}$$

Finally, the recursive estimation algorithms are now given as

$$\hat{\boldsymbol{\theta}}(k+1) = \hat{\boldsymbol{\theta}}(k) + \boldsymbol{\gamma}_{\mathrm{W}}(k)\big(y(k+1) - \boldsymbol{\psi}^{\mathrm{T}}(k+1)\hat{\boldsymbol{\theta}}(k)\big) \tag{9.6.11}$$

$$\boldsymbol{\gamma}_{\mathrm{W}}(k) = \frac{1}{\boldsymbol{\psi}^{\mathrm{T}}(k+1)\boldsymbol{P}_{\mathrm{W}}(k)\boldsymbol{\psi}(k+1)+\lambda}\boldsymbol{P}_{\mathrm{W}}(k)\boldsymbol{\psi}(k+1) \tag{9.6.12}$$

$$\boldsymbol{P}_{\mathrm{W}}(k+1) = \big(\boldsymbol{I} - \boldsymbol{\gamma}_{\mathrm{W}}(k)\boldsymbol{\psi}^{\mathrm{T}}(k+1)\big)\boldsymbol{P}_{\mathrm{W}}(k)\frac{1}{\lambda} \tag{9.6.13}$$

and will be evaluated in the order (9.6.12), (9.6.11), and finally (9.6.13).

The effect of λ on the parameter estimates can easily be seen from (9.6.6) and (9.6.8). $\boldsymbol{P}_{\mathrm{W}}^{-1}(k)$ is for $\lambda = 1$ proportional to the covariance matrix of the parameter estimates. $\boldsymbol{P}_{\mathrm{W}}^{-1}(k+1)$ now is constructed in such a way that the new measured data $\boldsymbol{\psi}(k+1)\boldsymbol{\psi}^{\mathrm{T}}(k+1)$ are weighted with 1, the old data $\boldsymbol{P}_{\mathrm{W}}^{-1}(k)$ are however weighted with the smaller weight $\lambda < 1$. This is tantamount to increasing the covariance values of the last step or equivalently increasing the *uncertainty* of the old parameter estimates.

The choice of λ presents a compromise between better suppression of disturbances ($\lambda \to 1$) or a better tracking of time-varying systems ($\lambda < 1$). In practical applications, values between $0.9 < \lambda < 0.995$ have been proven well. The detailed choice of λ either as a constant or as a time-varying parameter is discussed in detail in Chap. 12.

9.6.1 Constraints and the Recursive Method of Least Squares

The recursive method of least squares allows to incorporate constraints in a very elegant way (Goodwin and Sin, 1984), which will shortly be outlined in the following. Equality constraints should be taken care of a priori by transforming the set of parameters accordingly. Inequality constraints, e.g. bounds for the individual parameters to ensure stability etc., can be incorporated in the following way:

After each iteration of the RLS algorithm, check whether the estimated parameters $\hat{\boldsymbol{\theta}}$, lie in the *feasible* area, i.e. are within the set of admissible values, which shall be denoted as $\mathcal{C}$. If so, proceed with the next iteration as usual. If not, the new parameter vector is projected onto the boundary of $\mathcal{C}$ and then proceed with the next step.

The projection onto the boundary of $\mathcal{C}$ has to be done such that the value of the cost function remains as small as possible under the constraint. This is done as follows, (Goodwin and Sin, 1984):

1. Transform the parameter vector to a new coordinate basis by
$$\boldsymbol{\rho} = \boldsymbol{P}^{-\frac{1}{2}}\hat{\boldsymbol{\theta}} \tag{9.6.14}$$
2. Orthogonally project $\boldsymbol{\rho}$ onto the boundary of the transformed feasible area $\overline{\mathcal{C}}$
3. Back transform the result to obtain $\hat{\boldsymbol{\theta}}'$

9.6.2 Tikhonov Regularization

The Tikhonov regularization (Tikhonov and Arsenin, 1977; Tikhonov, 1995) adds a penalty term to the quadratic cost function as

$$V(\boldsymbol{\theta}) = \sum_{k=1}^{N} e^2(k, \boldsymbol{\theta}) + \gamma \Omega(\boldsymbol{\theta}) \,. \tag{9.6.15}$$

In this equation, $\gamma > 0$ is a scalar parameter that determines the degree of regularization and $\Omega(\boldsymbol{\theta})$ is the regularization term that depends on the parameters to be estimated. Often, $\Omega(\boldsymbol{\theta})$ is calculated by means of the weighted vector 2-norm of the parameters as

$$\Omega(\boldsymbol{\theta}) = \boldsymbol{\theta}^{\mathrm{T}} \boldsymbol{K} \boldsymbol{\theta} \tag{9.6.16}$$

For this choice, the problem of least squares can still be solved directly as

$$\hat{\boldsymbol{\theta}} = \left(\boldsymbol{\Psi}^{\mathrm{T}}\boldsymbol{\Psi} + \gamma \boldsymbol{K}\right)^{-1} \boldsymbol{\Psi}^{\mathrm{T}} \boldsymbol{y} \,. \tag{9.6.17}$$

One can choose the matrix $\boldsymbol{K}$ as the identity matrix, which will cause unnecessary parameters to go to zero. In the more general case, one can choose $\Omega(\boldsymbol{\theta}) = (\boldsymbol{\theta} - \boldsymbol{\theta}_0)^{\mathrm{T}}(\boldsymbol{\theta} - \boldsymbol{\theta}_0)$ to draw the parameters towards $\boldsymbol{\theta}_0$. This is also known under the term *ridge regression* (Hoerl and Kennard, 1970a,b).

9.7 Summary

The method of least squares is very suitable for the identification of linear dynamic discrete-time processes and for non-linear (static) processes which are linear in the parameters. It has been shown in this chapter that the estimation is then based on the generalized equation error. However, the disturbance must be generated by a very special filter from white noise, in order to obtain unbiased estimates. For more general disturbances, the method of least squares will result in biased parameter estimates. The next chapter will show further strategies to avoid this bias or limit its impact on the estimates.

Besides process models, also the estimation of signal model parameters in the time and frequency domain has been discussed. Here, the estimation of the spectral density by means of parametric signal models has been treated. The biggest advantage is the elimination of the leakage effect as it is no longer assumed that the signal is periodically repeated outside the measurement interval. Several different techniques have been presented. The first is based on the formulation of a pure LS parameter estimation problem in the time domain. Another approach tries to match the power spectral densities of the measured signal and a colored noise, which is realized from white noise by means of a form filter.

Furthermore, conditions have been provided which allow to judge whether the parameters are identifiable. In particular, it has been found that the input must be persistently exciting of a certain order. Tests to find out whether an input is persistently exciting or not have been formulated along with examples for often applicable persistently exciting inputs. The method of least squares has also been formulated in a recursive form, which allows a computationally efficient online parameter estimation in real-time. By introducing the method of weighted least squares and subsequently the exponential forgetting, it is also possible to identify time-varying processes.

The use of an intermediate non-parametric model has allowed to obtain unbiased estimates and also allowed to condense the experimental data before the parameter estimation takes place.

Also in this chapter, the inclusion of constraints has been discussed and the Tikhonov regularization, which is also termed ridge regression, had been introduced to "pull" unused parameters towards zero.

Problems

9.1. Discrete-Time Process 1

Given is the discrete-time process

$$G(z) = \frac{0.5z^{-1}}{(1-0.5z^{-1})(1-0.1z^{-1})}$$

Determine the step response for $u(k) = \sigma(k)$.
Determine the response for the input signal $u(k) = \sin \pi k/5$.
Given is a simplified model of the process as

$$G_{\mathrm{m}}(z) = \frac{b_1 z^{-1}}{1 + a_1 z^{-1}}$$

Determine the parameters a_1 and b_1 by the method of least squares for the step input and the sinusoidal input.

9.2. Discrete-Time Moving Average Process

Given is the second order process

$$y(k) = b_0 u(k) + b_1 u(k-1)$$

and the following measurements

Data point k	0	1	2	3	4	5	6	7	8	9	10
Input $u(k)$	0	1	−1	1	1	1	−1	−1	0	0	0
Output $y(k)$	0	1.1	−0.2	0.1	0.9	1	0.1	−1.1	−0.8	−0.1	0

Estimate the parameters b_0 and b_1 by means of the method of least squares. Determine the disturbance $n(k)$, its mean and variance.

9.3. Discrete-Time Process 2

The PRBS signal with

$$u(k) = 1, -1, 1, 1, 1, -1, -1, 1, -1, 1, 1, 1, -1, -1, \ldots$$

which is periodic with $N = 7$ is used as an input signal for the process

$$G(z) = \frac{0.7z^{-1}}{1 - 0.3z^{-1}}$$

Determine the output $y(k)$, the auto-correlation function $R_{\mathrm{uu}}(\tau)$ and the cross-correlation function $R_{\mathrm{uy}}(\tau)$.
Use $R_{\mathrm{uu}}(\tau)$ as an input for the process and compare it with $R_{\mathrm{uy}}(\tau)$ as calculated before.
Estimate the impulse response of the system and compare it with $R_{\mathrm{uy}}(\tau)$.
Estimate the parameters b_0 and b_1 by the method of least squares.

9.4. Discrete-Time Process 3

Given is the process

$$y(k) + a_1 y(k-1) = b_1 u(k-1) \, .$$

State the equation of recursive least squares. What changes if the process has a dead time with $d = 2$?

9.5. Discrete-Time Process 4

A process of first order shall be identified by the method of least squares. The process cannot respond directly to a step. For the identification $N = 18$ pairs of input-/output data have been recorded.
Draw the block diagram for the calculation of the input error, output error, and equation error between process and model. Which setup must be chosen to ensure that the error is linear in the parameters?
What is the non-recursive estimation equation for the method of least squares if $\boldsymbol{\theta}$ is the parameter vector, $\boldsymbol{\Psi}$ the data matrix and $\boldsymbol{y}$ the output vector? What dimension do the individual vectors and matrices have?

As an input signal, a PRBS signal of amplitude 1 is used. One obtains the following data for the auto-correlation and cross-correlation:

$$R_{\mathrm{uy}}(0) = -0.0662 \quad R_{\mathrm{uy}}(1) = 0.4666$$
$$R_{\mathrm{yy}}(0) = 0.278 \quad R_{\mathrm{yy}}(1) = 0.112$$

Determine the parameters a_1 and b_1 of the model.

9.6. Bias-free Estimation
What are the conditions for a bias-free estimation of the parameters of a first order model by means of the method of least squares? Which estimates show a bias for white noise as a disturbance $n(k)$?

9.7. DC Value
Compare the advantages and disadvantages of the different methods of working with large signal values $U(k)$ and $Y(k)$.

9.8. Exponential Forgetting
What happens for the method of least squares with exponential forgetting ($\lambda < 1$) if the input signal does not change?

9.9. Sinusoidal Excitation
If a process is excited with a single sinusoidal oscillation, what is the maximum model order that can be handled?

9.10. Stochastic Disturbances
According to which relation diminishes the error of the parameter estimates as a function of the measurement time if stochastic disturbances are acting on the process.

References

Albert R, Sittler RW (1965) A method for computing least squares estimators that keep up with the data. SIAM J Control Optim 3(3):384–417

Åström KJ, Bohlin T (1965) Numerical identification of linear dynamic systems from normal operating records. In: Proceedings of the IFAC Symposium Theory of Self-Adaptive Control Systems, Teddington

Åström KJ, Eykhoff P (1971) System identification – a survey. Automatica 7(2):123–162

Baur U (1976) On-Line Parameterschätzverfahren zur Identifikation linearer, dynamischer Prozesse mit Prozeßrechnern: Entwicklung, Vergleich, Erprobung: KfK-PDV-Bericht Nr. 65. Kernforschungszentrum Karlsruhe, Karlsruhe

Becker HP (1990) Beiträge zur rekursiven Parameterschätzung zeitvarianter Prozesse. Fortschr.-Ber. VDI Reihe 8 Nr. 203. VDI Verlag, Düsseldorf

Bellmann R, Åström KJ (1970) On structural identifiability. Math Biosci 7(3–4):329–339

Bombois X, Anderson BDO, Gevers M (2005) Quantification of frequency domain error bounds with guaranteed confidence level in prediction error identification. Syst Control Lett 54(11):471–482

Box GEP, Jenkins GM, Reinsel GC (2008) Time series analysis: Forecasting and control, 4th edn. Wiley Series in Probability and Statistics, John Wiley, Hoboken, NJ

Burg JP (1968) A new analysis technique for time series data. In: Proceedings of NATO Advanced Study Institute on Signal Processing, Enschede

Campi MC, Weyer E (2002) Finite sample properties of system identification methods. IEEE Trans Autom Control 47(8):1329–1334

Deutsch R (1965) Estimation theory. Prentice-Hall, Englewood Cliffs, NJ

van Doren JFM, Douma SG, van den Hof PMJ, Jansen JD, Bosgra OH (2009) Identifiability: From qualitative analysis to model structure approximation. In: Proceedings of the 15th IFAC Symposium on System Identification, Saint-Malo, France

Durbin J (1960) Estimation of parameters in time-series regression models. J Roy Statistical Society B 22(1):139–153

Edward J, Fitelson M (1973) Notes on maximum-entropy processing (Corresp.). IEEE Trans Inf Theory 19(2):232–234

Eykhoff P (1974) System identification: Parameter and state estimation. Wiley-Interscience, London

Gauss KF (1809) Theory of the motion of the heavenly bodies moving about the sun in conic sections: Reprint 2004. Dover phoenix editions, Dover, Mineola, NY

Genin Y (1968) A note on linear minimum variance estimation problems. IEEE Trans Autom Control 13(1):103–103

Gevers M (2005) Identification for control: From early achievements to the revival of experimental design. Eur J Cont 2005(11):1–18

Goodwin GC, Sin KS (1984) Adaptive filtering, prediction and control. Prentice-Hall information and system sciences series, Prentice-Hall, Englewood Cliffs, NJ

Heij C, Ran A, Schagen F (2007) Introduction to mathematical systems theory : linear systems, identification and control. Birkhäuser Verlag, Basel

Hoerl AE, Kennard RW (1970a) Ridge regression: Application to nonorthogonal problems. Technometrics 12(1):69–82

Hoerl AE, Kennard RW (1970b) Ridge regression: Biased estimation for nonorthogonal problems. Technometrics 12(1):55–67

Isermann R (1974) Prozessidentifikation: Identifikation und Parameterschätzung dynamischer Prozesse mit diskreten Signalen. Springer, Heidelberg

Isermann R (1987) Digitale Regelsysteme Band 1 und 2. Springer, Heidelberg

Isermann R (1991) Digital control systems, 2nd edn. Springer, Berlin

Isermann R (2005) Mechatronic Systems: Fundamentals. Springer, London

Isermann R (2006) Fault-diagnosis systems: An introduction from fault detection to fault tolerance. Springer, Berlin

Isermann R, Baur U (1974) Two-step process identification with correlation analysis and least-squares parameter estimation. J Dyn Syst Meas Contr 96:426–432

Johnston J, DiNardo J (1997) Econometric Methods: Economics Series, 4th edn. McGraw-Hill, New York, NY

Kalman RE (1958) Design of a self-optimizing control system. Trans ASME 80:468–478

Kendall MG, Stuart A (1977a) The advanced theory of statistics: Design and analysis, and time-series (vol. 3). Charles Griffin, London

Kendall MG, Stuart A (1977b) The advanced theory of statistics: Inference and relationship (vol. 2). Charles Griffin, London

Klinger A (1968) Prior information and bias in sequential estimation. IEEE Trans Autom Control 13(1):102–103

Koopmans TC (1937) Linear regression analysis of economic time series. Netherlands Economic Institute, Haarlem

Lee KI (1964) Optimal estimation, identification, and control, Massachusetts Institute of Technology research monographs, vol 28. MIT Press, Cambridge, MA

Levin MJ (1960) Optimum estimation of impulse response in the presence of noise. IRE Trans Circuit Theory 7(1):50–56

Ljung L (1999) System identification: Theory for the user, 2nd edn. Prentice Hall Information and System Sciences Series, Prentice Hall PTR, Upper Saddle River, NJ

Makhoul J (1975) Linear prediction: A tutorial review. Proc IEEE 63(4):561–580

Makhoul J (1976) Correction to "Linear prediction : A tutorial review"'. Proc IEEE 64(2):285

Mann HB, Wald W (1943) On the statistical treatment of linear stochastic difference equations. Econometrica 11(3/4):173–220

Mendel JM (1973) Discrete techniques of parameter estimation: The equation error formulation, Control Theory, vol 1. Marcel Dekker, New York

Neumann D (1991) Fault diagnosis of machine-tools by estimation of signal spectra. In: Proceedings of the IFAC/IMACS Sympsoium on Fault Detection, Supervision, and Safety for Technical Processes SAFEPROCESS'91, Baden-Baden, Germany

Neumann D, Janik W (1990) Fehlerdiagnose an spanenden Werkzeugmaschinen mit parametrischen Signalmodellen von Schwingungen. In: VDI-Schwingungstagung Mannheim, VDI-Verlag, Düsseldorf, Germany

Ninness B, Goodwin GC (1995) Estimation of model quality. Automatica 31(12):32–74

Pandit SM, Wu SM (1983) Time series and system analysis with applications. Wiley, New York

Panuska V (1969) An adaptive recursive least squares identification algorithm. In: Proceedings of the IEEE Symposium in Adaptive Processes, Decision and Control

Pintelon R, Schoukens J (2001) System identification: A frequency domain approach. IEEE Press, Piscataway, NJ

Press WH, Teukolsky SA, Vetterling WT, Flannery BP (2007) Numerical recipes: The art of scientific computing, 3rd edn. Cambridge University Press, Cambridge, UK

Sagara S, Wada K, Gotanda H (1979) On asymptotic bias of linear least squares estimator. In: Proceedings of the 5th IFAC Symposium on Identification and System Parameter Estimation Darmstadt, Pergamon Press, Darmstadt, Germany

Scheuer HG (1973) Ein für den Prozessrechnereinsatz geeignetes Identifikationsverfahren auf der Grundlage von Korrelationsfunktionen. Dissertation. Universität Trier, Trier

Staley RM, Yue PC (1970) On system parameter identifability. Inf Sci 2(2):127–138

Stuart TA, Ord JK, Kendall MG (1987) Kendalls advanced theory of statistics: Distribution theory (vol. 1). Charles Griffin Book

Tikhonov AN (1995) Numerical methods for the solution of ill-posed problems, Mathematics and its applications, vol 328. Kluwer Academic Publishers, Dordrecht

Tikhonov AN, Arsenin VY (1977) Solutions of ill-posed problems. Scripta series in mathematics, Winston, Washington, D.C.

Tong H (1975) Autoregressive model fitting with noisy data by Akaike's information criterion. IEEE Trans Inf Theory 21(4):476–480

Tong H (1977) More on autoregressive model fitting with noisy data by Akaike's information criterion. IEEE Trans Inf Theory 23(3):409–410

Tse E, Anton J (1972) On the identifiability of parameters. IEEE Trans Autom Control 17(5):637–646

Ulrych TJ, Bishop TN (1975) Maximum entropy spectral analysis and autoregressive decomposition. Rev Geophys 13(1):183–200

Vuerinckx R, Pintelon R, Schoukens J, Rolain Y (2001) Obtaining accurate confidence regions for the estimated zeros and poles in system identification problems. IEEE Trans Autom Control 46(4):656–659

Weyer E, Campi MC (2002) Non-asymptotic confidence ellipsoids for the least-squares estimate. Automatica 38(9):1539–1547

Young P (1984) Recursive estimation and time-series analysis: An introduction. Communications and control engineering series, Springer, Berlin

10

Modifications of the Least Squares Parameter Estimation

In order to obtain bias-free estimates of linear dynamic processes by the method of least squares, the error signal $e(k)$ may not be correlated. This requirement is only satisfied if the disturbance $n(k)$ that is acting on the system is a colored noise that is generated from a white noise $\nu(k)$ filtered by a form filter with the transfer function $1/A(z^{-1})$. Since this prerequisite is hardly ever met in practice, the method of least squares typically works on a correlated error signal and hence yields biased estimates. The bias can be so high for larger noise levels that the results are unusable. To avoid this problem, in the following, methods are presented which yield bias-free estimates for larger classes of dynamic processes. Furthermore, methods of stochastic approximation are introduced, namely the Robbins-Monro algorithm, the Kiefer-Wolfowitz algorithm, the least mean squares and the normalized least mean squares algorithms, all of which allow to approximate the solution of the Least Squares method with much less computational effort.

10.1 Method of Generalized Least Squares (GLS)

In the following, methods are introduced, which give greater flexibility to the noise model by introducing additional degrees of freedom into the transfer function of the form filter. The most flexible model is the Box-Jenkins model, which allows to freely parameterize numerator and denominator (Ljung, 1999) of the noise. However, more degrees of freedom also raise the question on how these additional parameters can be identified. Therefore, limiting the degrees of freedom can be attractive and/or necessary to e.g. be able to use a direct least squares parameter estimation method.

10.1.1 Non-Recursive Method of Generalized Least Squares (GLS)

The fundamental idea behind the method of generalized least squares is that the error signal in the model for the normal method of least squares

$$A(z^{-1})y(z) - B(z^{-1})z^{-d}u(z) = e(z) \tag{10.1.1}$$

R. Isermann, M. Münchhof, *Identification of Dynamic Systems*,
DOI 10.1007/978-3-540-78879-9_10, © Springer-Verlag Berlin Heidelberg 2011

with the uncorrelated error signal $e(k)$ is being replaced by a correlated signal, i.e. a colored noise $\xi(k)$, which is generated by means of a form filter

$$\xi(z) = \frac{1}{F(z^{-1})} e'(z) , \tag{10.1.2}$$

where $e'(z)$ is uncorrelated. $\xi(k)$ is assumed to be an auto regressive signal process. Since the filter polynomial $F(z^{-1})$ is unknown, Clarke (1967) proposed the following iterative procedure:

Step 1

The method of least squares is applied to the measurements in the interval $m + d \leq k \leq m + d + N$ based on the model

$$A(z^{-1})y(z) - B(z^{-1})z^{-d}u(z) = \xi(z) , \tag{10.1.3}$$

where the estimates $\hat{\boldsymbol{\theta}}_1$ are biased and $\xi(z)$ is a correlated signal.

Step 2

The error signal $\xi(k)$ is determined for the estimated parameters $\hat{\boldsymbol{\theta}}_1$ according to (10.1.3). Using the AR model

$$\xi(k) = \boldsymbol{\psi}_\xi^{\mathrm{T}}(k)\boldsymbol{f} + e'(k) \tag{10.1.4}$$

yields

$$\boldsymbol{\psi}_\xi^{\mathrm{T}}(k) = \big(-\xi(k-1) \; -\xi(k-2) \; \ldots \; -\xi(k-\nu)\big) \tag{10.1.5}$$

$$\boldsymbol{f}^{\mathrm{T}} = \big(f_1 \; f_2 \; \ldots \; f_\nu\big) . \tag{10.1.6}$$

The order ν has to be assumed appropriately, e.g. $\nu = m$. Then, the parameters are estimated according to the method of least squares as

$$\hat{\boldsymbol{f}} = \big(\boldsymbol{\Xi}^{\mathrm{T}}\boldsymbol{\Xi}\big)^{-1}\boldsymbol{\Xi}^{\mathrm{T}}\boldsymbol{\xi} , \tag{10.1.7}$$

where $\boldsymbol{\Xi}$ is made up of the row vectors of $\boldsymbol{\psi}_\xi^{\mathrm{T}}(k)$.

Step 3

The measured input and output signal $u(k)$ and $y(k)$ are processed by the filter

$$G_{\mathrm{F}}(z^{-1}) = \hat{F}(z^{-1}) , \tag{10.1.8}$$

so that

$$\tilde{u}(z) = G_{\mathrm{F}}(z^{-1})u(z) \text{ and } \tilde{y}(z) = G_{\mathrm{F}}(z^{-1})y(z) . \tag{10.1.9}$$

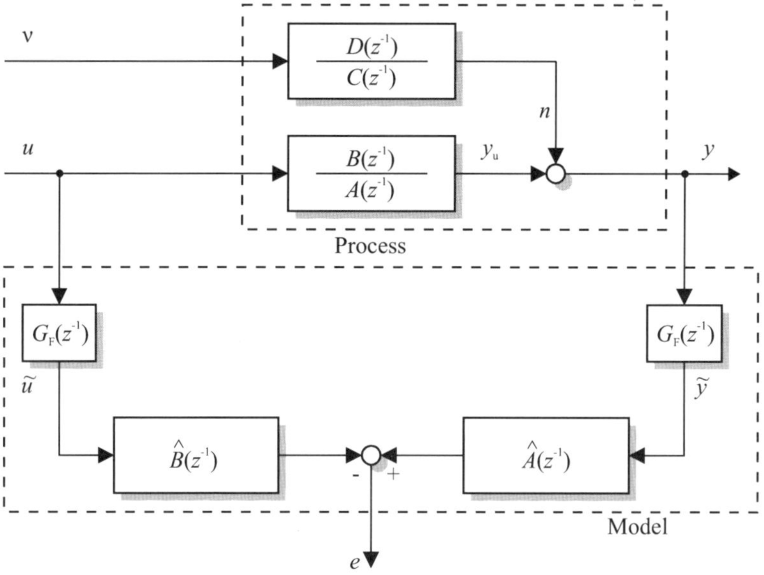

Fig. 10.1. Block diagram for the method of generalized least squares (GLS)

Step 4

The method of least squares is now applied to the filtered signals $\tilde{u}$ and $\tilde{y}$, i.e. to the model

$$A(z^{-1})\tilde{y}(z) - B(z^{-1})z^{-d}\tilde{u}(z) = \xi'(z) \,, \tag{10.1.10}$$

see Fig. 10.1. One obtains the parameters $\hat{\boldsymbol{\theta}}_2$.

Step 5

Steps 2 through 4 are repeated until $\hat{\boldsymbol{\theta}}_j$ does not change significantly from one iteration to the next.

In order to obtain bias-free parameter estimates, the error signal $\xi'(k)$ must be uncorrelated. This is the case, if $\hat{F}(z^{-1})$ matches with $F(z^{-1})$ according to (10.1.2). The method of generalized least squares provides bias-free estimates of the noise generating form filter of the form

$$G_\nu(z) = \frac{n(z)}{\nu(z)} = \frac{D(z^{-1})}{C(z^{-1})} = \frac{1}{A(z^{-1})F(z^{-1})} \,, \tag{10.1.11}$$

what can be proven by inserting (10.1.2) into (10.1.3) and $\nu(z) = e'(z)$. If this form of the noise filter does not apply to the disturbance, then the method of generalized least squares provides biased estimates or does not converge at all.

A simple form of the GLS method has been suggested by Steiglitz and McBride (1965): In the ith iteration, it is suggested to set $\hat{F}_j(z^{-1}) = \hat{A}_{(j-1)}(z^{-1})$. This however leads to a very specific filter $G_\nu(z^{-1})$.

Stoica and Söderström (1977) proposed a further GLS method which replaced $\xi(k)$ in (10.1.3) by a moving-average process,

$$\xi(z) = H(z^{-1})e'(z) \,, \tag{10.1.12}$$

see also (Isermann, 1974). This method however is similar to the ELS method, Sect. 10.2, which does not require an iterative approach. Compared to the normal method of least squares, the method of generalized least squares requires a much larger effort, but as a remuneration for that also delivers a model of the noise generating signal process.

10.1.2 Recursive Method of Generalized Least Squares (RGLS)

In a similar manner to the method of least squares, one can also derive a recursive formulation for the method of generalized least squares. Only the resulting set of equations shall be presented here, the derivation is e.g. presented by Hastings-James and Sage (1969). The recursive equations are given as

$$\begin{aligned}\hat{\boldsymbol{\theta}}(k+1) = \hat{\boldsymbol{\theta}}(k) + \big(\boldsymbol{\psi}^{\mathrm{T}}(k+1)\tilde{\boldsymbol{P}}(k)\tilde{\boldsymbol{\psi}}(k+1)+1\big)^{-1}\\ \tilde{\boldsymbol{P}}(k)\tilde{\boldsymbol{\psi}}(k+1)\big(\tilde{y}(k+1) - \boldsymbol{\psi}^{\mathrm{T}}(k+1)\hat{\boldsymbol{\theta}}(k)\big)\end{aligned} \tag{10.1.13}$$

$$\begin{aligned}\tilde{\boldsymbol{P}}(k+1) = \tilde{\boldsymbol{P}}(k)\Big(\boldsymbol{I} - \tilde{\boldsymbol{\psi}}^{\mathrm{T}}(k+1)\tilde{\boldsymbol{\psi}}(k+1)\tilde{\boldsymbol{P}}(k)\\ \big(\tilde{\boldsymbol{\psi}}^{\mathrm{T}}(k+1)\tilde{\boldsymbol{P}}(k)\tilde{\boldsymbol{\psi}}(k+1)+1\big)^{-1}\Big)\end{aligned} \tag{10.1.14}$$

$$\begin{aligned}\tilde{\boldsymbol{f}}(k+1) = \tilde{\boldsymbol{f}}(k) + \big(\boldsymbol{\psi}_{\xi}^{\mathrm{T}}(k+1)\boldsymbol{Q}(k)\boldsymbol{\psi}_{\xi}(k+1)+1\big)^{-1}\\ \boldsymbol{Q}(k)\boldsymbol{\psi}_{\xi}(k+1)\big(\xi(k+1) - \boldsymbol{\psi}_{\xi}^{\mathrm{T}}(k+1)\hat{\boldsymbol{f}}(k)\big)\end{aligned} \tag{10.1.15}$$

$$\begin{aligned}\boldsymbol{Q}(k+1) = \boldsymbol{Q}(k)\Big(\boldsymbol{I} - \boldsymbol{\psi}_{\xi}^{\mathrm{T}}(k+1)\boldsymbol{\psi}_{\xi}(k+1)\boldsymbol{Q}(k)\\ \big(\boldsymbol{\psi}_{\xi}^{\mathrm{T}}(k+1)\boldsymbol{Q}(k)\boldsymbol{\psi}_{\xi}(k+1)+1\big)^{-1}\Big)\,.\end{aligned} \tag{10.1.16}$$

The elements of $\tilde{\boldsymbol{\psi}}$ are the filtered signals from (10.1.10). The initial values of $\boldsymbol{P}(0)$ and $\boldsymbol{Q}(0)$ are chosen as diagonal matrices with large elements according to (9.4.22). α may however not be chosen too large as the estimate may diverge in this case. As initial values for the parameters, one can choose $\hat{\boldsymbol{\theta}}(0) = \boldsymbol{0}$.

An exponential weighting of the past data with λ can be achieved by replacing the terms

$$\big(\boldsymbol{\psi}^{\mathrm{T}}(k+1)\tilde{\boldsymbol{P}}(k)\tilde{\boldsymbol{\psi}}(k+1)+\lambda\big) \text{ in (10.1.13) and (10.1.14)} \tag{10.1.17}$$

$$\tilde{\boldsymbol{P}}(k+1) = \frac{1}{\lambda}\tilde{\boldsymbol{P}}(k)(\boldsymbol{i} - \ldots) \text{ in (10.1.14)} \tag{10.1.18}$$

and also the respective terms in (10.1.15) and (10.1.16). This exponential weighting can also lead to better convergence if the first 100 to 200 values are weighted with $\lambda = 0.99$ (Isermann and Baur, 1973). The initial values then have less influence on the results in subsequent recursions.

10.2 Method of Extended Least Squares (ELS)

If instead of the LS method

$$A(z^{-1})y(z) - B(z^{-1})\, z^{-d}\, u(z) = \varepsilon(z) \tag{10.2.1}$$

with a correlated error signal $\varepsilon(z)$ the ARMAX model

$$A(z^{-1})y(z) - B(z^{-1})z^{-d}u(z) = D(z^{-1})e'(z) \tag{10.2.2}$$

with a correlated signal $\varepsilon(z) = D(z^{-1})e'(z)$ is used, the recursive methods for dynamic processes and for stochastic signals can be combined to form the extended least squares method (Young, 1968; Panuska, 1969). Based on

$$y(k) = \boldsymbol{\psi}^{\mathrm{T}}(k)\hat{\boldsymbol{\theta}}(k-1) + e(k)\,, \tag{10.2.3}$$

the following extended vectors are introduced:

$$\begin{aligned}\boldsymbol{\psi}^{\mathrm{T}}(k) = \big(&-y(k-1)\,\dots\,-y(k-m)\big|u(k-d-1)\,\dots\\ &u(k-d-m)\big|\hat{\nu}(k-1)\,\dots\,\hat{\nu}(k-p)\big)\end{aligned} \tag{10.2.4}$$

$$\hat{\boldsymbol{\theta}}^{\mathrm{T}} = \big(\hat{a}_1\,\dots\,a_m\big|\hat{b}_1\,\dots\,\hat{b}_m\big|\hat{d}_1\,\dots\,\hat{d}_p\big)\,. \tag{10.2.5}$$

Herewith, as in the case of the ARMA signal process (9.4.35), the virtual and hence unknown white noise $\nu(k)$ is taken as an estimate $\hat{\boldsymbol{k}} = e'(k)$, which can be determined recursively by

$$\hat{\nu}(k) = y(k) - \boldsymbol{\psi}^{\mathrm{T}}(k)\hat{\boldsymbol{\theta}}(k-1)\,. \tag{10.2.6}$$

Then, the recursive version is applied, i.e.

$$\hat{\boldsymbol{\theta}}(k+1) = \hat{\boldsymbol{\theta}}(k) + \gamma(k)\big(y(k+1) - \hat{\boldsymbol{\psi}}^{\mathrm{T}}(k+1)\hat{\boldsymbol{\theta}}(k)\big)\,, \tag{10.2.7}$$

and the equations corresponding to (9.4.17) through (9.4.19). Instead of (10.2.6), also

$$\hat{\nu}(k) = y(k) - \boldsymbol{\psi}^{\mathrm{T}}(k)\hat{\boldsymbol{\theta}}(k) \tag{10.2.8}$$

can be used. This implies that

$$H(z) = \frac{1}{D(z)} - \frac{1}{2} \tag{10.2.9}$$

must be positive real.

The signal values $\hat{\nu}(k) = e(k)$ in $\hat{\boldsymbol{\psi}}^{\mathrm{T}}(k+1)$ are calculated recursively. Therefore the roots of $D(z) = 0$ must lie within the unit circle of the z-plane. The parameter estimation is unbiased and consistent in the mean square if the convergence conditions of the LS method are transferred to the model equation (10.2.3). That means that (10.2.2) has to be valid.

The noise form filter now is given as

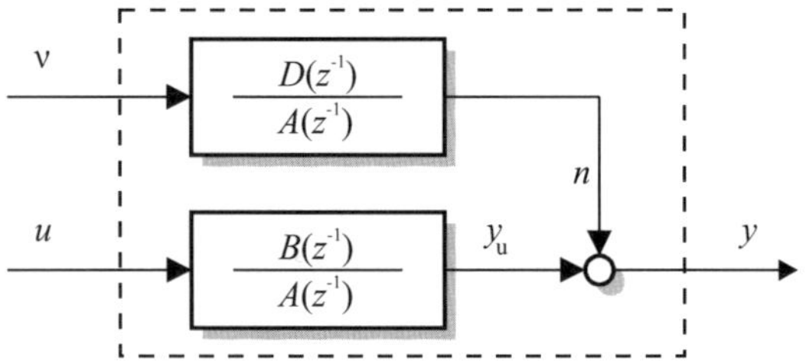

Fig. 10.2. Block diagram for the method of extended least squares (ELS)

$$G_\nu(z) = \frac{n(z)}{\nu(z)} = \frac{D(z^{-1})}{A(z^{-1})} , \tag{10.2.10}$$

see also Fig. 10.2. Although the fixed denominator polynomial limits the general applicability, there are still enough degrees of freedom by the variable order of the numerator polynomial to be able to approximate a given disturbance $n(k)$ precisely enough. The parameters of $D(z^{-1})$ converge slower than those of the process. The method however requires little additional effort and has proven itself in many applications.

10.3 Method of Bias Correction (CLS)

The methods presented so far tried to avoid biased estimates by making special assumptions about the signal process for the disturbance and hence could accommodate a correlated error $e(k)$ in the original underlying model of the method of least squares. A different solution to this problem is to determine the resulting bias and then use this result to correct the biased estimates of the method of least squares. This however requires that the bias can be determined with reasonable effort, which is only possible in very special cases, first and foremost for white noise as a disturbance. An overview over the different methods is given by Stoica and Söderström (1982). Basis can be a model of the form

$$y(z) = \frac{B(z^{-1})}{A(z^{-1})} z^{-d} u(z) + n(z) , \tag{10.3.1}$$

where the disturbance $n(k)$ is assumed to be a white noise with $\mathrm{E}\{n(k)\} = 0$ and variance σ_n^2. Then, the bias is according to (9.1.65) given as

$$\mathrm{E}\{\boldsymbol{b}(N+1)\} = -\mathrm{E}\{\boldsymbol{R}^{-1}(N+1)\} \underbrace{\left(\begin{array}{c|c} \boldsymbol{I} & \boldsymbol{0} \\ \hline \boldsymbol{0} & \boldsymbol{0} \end{array}\right)}_{\boldsymbol{S}} \boldsymbol{\theta}_0 \sigma_n^2 , \tag{10.3.2}$$

where $\boldsymbol{\theta}_0$ denotes the exact parameters. This bias is now used to correct the parameter estimates $\hat{\boldsymbol{\theta}}_{\mathrm{LS}}$, compare (9.1.39),

$$\begin{aligned}\hat{\boldsymbol{\theta}}_{\mathrm{CLS}}(N+1) &= \hat{\boldsymbol{\theta}}_{\mathrm{LS}}(N+1) - \boldsymbol{b}(N+1) \\ &= \hat{\boldsymbol{R}}^{-1}(N-1)\frac{1}{N+1}\boldsymbol{\Psi}^{\mathrm{T}}(N+1)\boldsymbol{y}(N+1) \\ &\quad + \hat{\boldsymbol{R}}^{-1}(N+1)\boldsymbol{S}\hat{\boldsymbol{\theta}}_{\mathrm{CLS}}(N+1)\sigma_{\mathrm{n}}^2\end{aligned} \tag{10.3.3}$$

and from there

$$\hat{\boldsymbol{\theta}}_{\mathrm{CLS}}(N+1) = \big(\boldsymbol{R}(N-1) - \boldsymbol{S}\sigma_{\mathrm{n}}^2\big)^{-1}\frac{1}{N+1}\boldsymbol{\Psi}^{\mathrm{T}}(N+1)\boldsymbol{y}(N+1)\,. \tag{10.3.4}$$

The variance σ_{n}^2 follows for a known system model from (10.3.1) and with

$$n(z) = y(z) - \frac{B(z^{-1})}{A(z^{-1})}z^{-d}u(z) \tag{10.3.5}$$

or, based on the the difference equation,

$$n(k) = y(k) - \boldsymbol{\psi}^{\mathrm{T}}(k)\hat{\boldsymbol{\theta}}(k) - \boldsymbol{\psi}_{\mathrm{n}}^{\mathrm{T}}(k)\boldsymbol{S}\hat{\boldsymbol{\theta}}(k) \tag{10.3.6}$$

as

$$\sigma_{\mathrm{n}}^2(N+1) = \mathrm{E}\big\{n^2(k)\big\} = \frac{1}{N+1-2m}\boldsymbol{n}^{\mathrm{T}}(N+1)\boldsymbol{n}(N+1)\,. \tag{10.3.7}$$

(10.3.4) and (10.3.7) can be used iteratively. A different way for the calculation of σ_{n}^2 is given by Stoica and Söderström (1982). There, it has also been shown that the estimates are not better than those for the method of instrumental variables. A method for the partial correction of a bias for colored errors $e(k)$ is presented by Kumar and Moore (1979).

10.4 Method of Total Least Squares (TLS)

The method of least squares uses a model of the form

$$\boldsymbol{y} - \boldsymbol{e} = \boldsymbol{\Psi}\hat{\boldsymbol{\theta}} \tag{10.4.1}$$

and determines the parameters from

$$\hat{\boldsymbol{\theta}} = \arg\min \|\boldsymbol{e}\|_2^2\,. \tag{10.4.2}$$

Here, it was assumed that only the output was disturbed by noise and hence only the distance between the model output $\boldsymbol{y}_M$ and the measurements $\boldsymbol{y}$ had to be minimized, see Fig. 10.3.

For the method of *total least squares* as proposed by Golub and Reinsch (1970) and Golub and van Loan (1980), the model is now given as

$$\boldsymbol{y} + \boldsymbol{e} = (\boldsymbol{\Psi} + \boldsymbol{F})\hat{\boldsymbol{\theta}}\,, \tag{10.4.3}$$

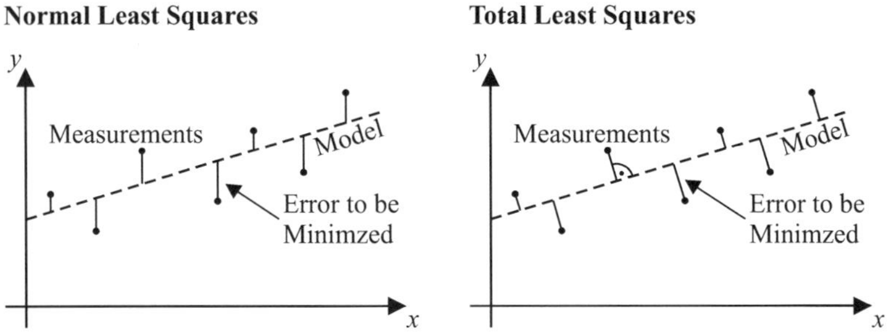

Fig. 10.3. Calculation of the error between measurement and model for the one dimensional case $y = \theta_1 x$ for the normal method of least squares and the method of total least squares

i.e. one now assumes an error not only in the measurements in the output vector $\boldsymbol{y}$, but also in the measurement contained in the data matrix $\boldsymbol{\Psi}$. The above model can be rewritten as

$$\Big(\underbrace{(\boldsymbol{\Psi}, \boldsymbol{y})}_{\boldsymbol{C}} + \underbrace{(\boldsymbol{F}, \boldsymbol{e})}_{\boldsymbol{\Delta}}\Big)\begin{pmatrix} \hat{\boldsymbol{\theta}} \\ -1 \end{pmatrix} = 0\,. \tag{10.4.4}$$

The method of total least squares will in the following be introduced in two ways: First, a more practical way will be presented and then, the correct mathematical derivation will be outlined.

The matrix $\boldsymbol{C}$ of the disturbed measurements in (10.4.4) will have the dimensions $N \times (m+1)$ with $N \gg m+1$ and for disturbed measurements will have full rank $m+1$. In order to determine the m parameters of the model, the rank must be reduced to m. For rank m, one column of the $m+1$ columns is linearly depend on the other columns and hence the system of equations can be solved unambiguously for the vector of parameter estimates $\hat{\boldsymbol{\theta}}$. This rank defect can be realized by utilizing the singular value decomposition. One carries out a singular value decomposition and then removes the smallest eigenvalue. This procedure will in the following be motivated from a mathematical point of view as well.

The underlying goal in the method of total least squares is to minimize the entries of the augmented error matrix $\boldsymbol{\Delta} = (\boldsymbol{F}, \boldsymbol{e})$. This minimization is done in the sense of the *Frobenius norm*, which is given as

$$\|\boldsymbol{\Delta}\|_{\mathrm{F}}^2 = \sum_{i=1}^{M}\sum_{j=1}^{N} \boldsymbol{\Delta}_{ij}^2\,, \tag{10.4.5}$$

where i and j are the row and column index of $\boldsymbol{\Delta}$ respectively. For real-valued measurements and in a non-mathematical way, this norm can be interpreted as the extension of the Euclidian vector norm to a matrix by taking the vector norm not only of the output error vector $\boldsymbol{e}$, but also of all columns of the data error matrix $\boldsymbol{F}$. With the Frobenius norm, the parameter estimates are then given as

$$\hat{\boldsymbol{\theta}} = \arg\min \|\boldsymbol{\Delta}\|_{\mathrm{F}}^2\,. \tag{10.4.6}$$

In order to solve the problem of total least squares, one will make use of the fact that the Frobenius norm of a matrix does not change under multiplication with a orthonormal matrix U (that satisfies $U^{-1} = U^T$),

$$\|\boldsymbol{U}\boldsymbol{A}\|_{\mathrm{F}}^2 = \|\boldsymbol{A}\|_{\mathrm{F}}^2 \,. \tag{10.4.7}$$

This fact will now be exploited in the development of the method of total least squares. One tries to minimize the augmented error matrix $\boldsymbol{\Delta}$, or equivalently, find a matrix $\tilde{\boldsymbol{C}}$ of rank m that best approximates the matrix $\boldsymbol{C}$,

$$\|\boldsymbol{\Delta}\|_{\mathrm{F}}^2 = \|\boldsymbol{C} - \tilde{\boldsymbol{C}}\|_{\mathrm{F}}^2 \,. \tag{10.4.8}$$

The matrix $\boldsymbol{C}$ can without a loss of generality be written as

$$\boldsymbol{C} = \boldsymbol{U}\boldsymbol{\Sigma}\boldsymbol{V}^{\mathrm{T}} \tag{10.4.9}$$

where

$$\boldsymbol{\Sigma} = \mathrm{diag}\big(\sigma_1\ \sigma_2\ \ldots\ \sigma_{n+1}\big) \text{ with } \sigma_1 \geq \sigma_2 \geq \ldots \geq \sigma_{n+1} \,, \tag{10.4.10}$$

which merely represents the Singular Value Decomposition (SVD) of $\boldsymbol{C}$. The matrices $\boldsymbol{U}$ and $\boldsymbol{V}$ are orthonormal. The matrix $\tilde{\boldsymbol{C}}$ can be written as

$$\tilde{\boldsymbol{C}} = \boldsymbol{U}\boldsymbol{\Sigma}\boldsymbol{V}^{\mathrm{T}} \Leftrightarrow \boldsymbol{U}^{-1}\tilde{\boldsymbol{C}}\big(\boldsymbol{V}^{\mathrm{T}}\big)^{-1} = \boldsymbol{U}^{\mathrm{T}}\tilde{\boldsymbol{C}}\boldsymbol{V} = \boldsymbol{S} \tag{10.4.11}$$

with an arbitrary matrix $\boldsymbol{S}$. Then, the cost function can be rewritten as

$$\|\boldsymbol{C} - \tilde{\boldsymbol{C}}\|_{\mathrm{F}}^2 = \|\boldsymbol{U}\boldsymbol{\Sigma}\boldsymbol{V}^{\mathrm{T}} - \boldsymbol{U}\boldsymbol{S}\boldsymbol{V}^{\mathrm{T}}\|_{\mathrm{F}}^2 = \|\boldsymbol{\Sigma} - \boldsymbol{S}\|_{\mathrm{F}}^2 \,, \tag{10.4.12}$$

since $\boldsymbol{U}$ and $\boldsymbol{V}$ are orthonormal. Because the matrix $\boldsymbol{\Sigma}$ is a diagonal matrix, it is obvious that the matrix $\boldsymbol{S}$ should also be diagonal. All non-diagonal elements of $\boldsymbol{S}$ must be zero to ensure that the cost function indeed reaches a minimum. The matrix $\boldsymbol{S}$ will now be written as

$$\boldsymbol{S} = \mathrm{diag}\big(s_1\ s_2\ \ldots\ s_{n+1}\big) \,, \tag{10.4.13}$$

and the cost function becomes

$$V = \|\boldsymbol{\Sigma} - \boldsymbol{S}\|_{\mathrm{F}}^2 = \sum_{i=1}^{m+1} \big(\sigma_i - s_i\big)^2 \,. \tag{10.4.14}$$

As the matrix $\tilde{\boldsymbol{C}}$ should be of rank m, the matrix $\boldsymbol{S}$ can have at least m non-zero entries s_i on the diagonal and must have one element with $s_i = 0$. In the interest of minimizing the cost function, one should choose $s_i = \sigma_i$ for $i = 1, \ldots, n$ and $s_{n+1} = 0$. The cost function then becomes

$$V = \|\boldsymbol{\Sigma} - \boldsymbol{S}\|_{\mathrm{F}}^2 = \sigma_{n+1}^2 \,. \tag{10.4.15}$$

The matrix $\tilde{\boldsymbol{C}}$ is hence given as

$$\hat{\boldsymbol{C}} = \boldsymbol{U} \operatorname{diag}\bigl(\sigma_1\ \sigma_2 \ldots \sigma_n\ 0\bigr)\boldsymbol{V}^{\mathrm{T}} . \tag{10.4.16}$$

Now that the rank n approximation $\hat{\boldsymbol{C}}$ has been determined, one can solve (10.4.4), which is repeated here as

$$\hat{\boldsymbol{C}}\begin{pmatrix} \hat{\boldsymbol{\theta}} \\ -1 \end{pmatrix} = \boldsymbol{0} . \tag{10.4.17}$$

This is equal to determining the null space of $\hat{\boldsymbol{C}}$, defined as

$$\hat{\boldsymbol{C}}\boldsymbol{x} = \boldsymbol{0} . \tag{10.4.18}$$

The SVD delivers an orthonormal basis for the null space, therefore the solution to the above equation can easily be given. The columns of the matrix $\boldsymbol{V}$ that are associated with singular values being zero form an *orthonormal basis for the null space* of the matrix $\tilde{\boldsymbol{C}}$. As the last singular value of the matrix $\hat{\boldsymbol{C}}$ has been set to zero, the last column of $\boldsymbol{V}$ forms the orthonormal basis of the null space of $\tilde{\boldsymbol{C}}$, hence

$$\begin{pmatrix} \hat{\boldsymbol{\theta}} \\ -1 \end{pmatrix} = \alpha\boldsymbol{x} = \alpha\begin{pmatrix} V_{12} \\ V_{22} \end{pmatrix} . \tag{10.4.19}$$

From the requirement $\alpha V_{22} = -1$, one can fix α as $\alpha = -V_{22}^{-1}$ and hence the parameter estimates can be determined as

$$\hat{\boldsymbol{\theta}} = -V_{22}^{-1} V_{12} . \tag{10.4.20}$$

For the derivation of this solution, see also (Goedecke, 1987; Zimmerschied, 2008). This approach is also termed *errors-in-variables* and *orthogonal regression*. A detailed survey of the method of total least squares can be found in the monograph by van Huffel and Vandewalle (1991) or the survey by Markovsky and van Huffel (2007).

If errors have different variance, then one can resort to the problem of *generalized total least squares (GTLS)*, which allows to incorporate the variance of the different columns of $\boldsymbol{\Psi}$. In this case, one introduces a scaling matrix $\boldsymbol{G}$ into the problem, so that the cost function

$$V = \|\boldsymbol{\Delta G}\|_{\mathrm{F}}^2 \tag{10.4.21}$$

is minimized, where $\boldsymbol{G} = \operatorname{diag}(1/\sigma_1, 1/\sigma_2, \ldots, 1/\sigma_{n+1})$ and σ_i is the error standard deviation of the corresponding regressor or output respectively.

By an appropriate scaling, one can also deal with correlation between the different columns of $\boldsymbol{\Delta}$. The covariance matrix $\boldsymbol{C}$ of the error of $\boldsymbol{\Delta}$ must be known up to a scaling factor. The scaling matrix can then be chosen as

$$\boldsymbol{G} = \boldsymbol{R}_{\mathrm{C}}^{-1} , \tag{10.4.22}$$

where $\boldsymbol{R}_{\mathrm{C}}$ is given such that $\boldsymbol{C} = \boldsymbol{R}_{\mathrm{C}}^{\mathrm{T}}\boldsymbol{R}_{\mathrm{C}}$.

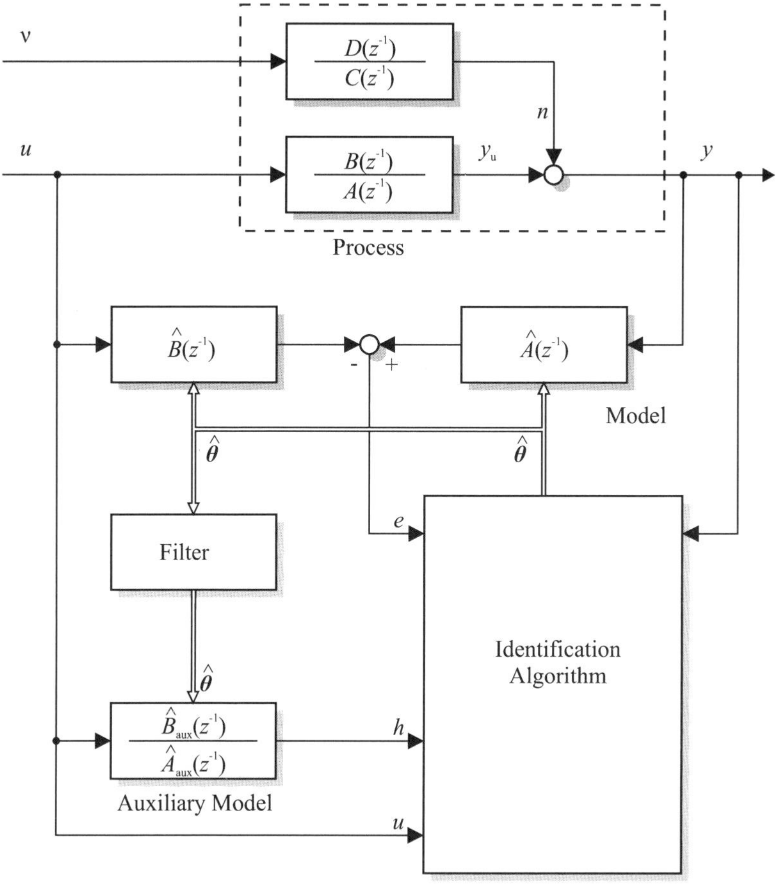

Fig. 10.4. Block diagram of the method of instrumental variables (IV)

It is pointed out by Markovsky and van Huffel (2007) and Söderström (2007) that the method of total least squares may in its native form not be well suited for the identification of dynamic systems since the elements of $\boldsymbol{\Psi}$ and $\boldsymbol{y}$ are often coupled and especially since $\boldsymbol{\Psi}$ is often a Hankel matrix. Here, the STLS (structured total least squares) method can be of better suitability (Markovsky et al, 2005).

The method of total least squares closely relates to the principal component analysis (PCA) which is used in statistics to find correlation in data sets and to reduce the dimensionality of data sets. The PCA determines and keeps those subspaces of measured data that have the largest variance, see e.g. the book by Jolliffe (2002).

10.5 Instrumental Variables Method (IV)

10.5.1 Non-Recursive Method of Instrumental Variables (IV)

A direct method to avoid the problem of biased estimates is the introduction of the so-called *Instrumental Variables*. This method goes back to Reiersøl (1941), Durbin (1954), and Kendall and Stuart (1961). It is also based on the model equation of the equation error as

$$\boldsymbol{e} = \boldsymbol{y} - \boldsymbol{\Psi}\boldsymbol{\theta} \,. \tag{10.5.1}$$

This equation is now multiplied on both sides with the transpose of an *instrumental variable matrix* $\boldsymbol{W}$ as

$$\boldsymbol{W}^{\mathrm{T}}\boldsymbol{e} = \boldsymbol{W}^{\mathrm{T}}\boldsymbol{y} - \boldsymbol{W}^{\mathrm{T}}\boldsymbol{\Psi}\boldsymbol{\theta} \,. \tag{10.5.2}$$

If the elements of $\boldsymbol{W}$, the so-called *instrumental variables* are chosen such that

$$\operatorname*{plim}_{N\to\infty} \boldsymbol{W}^{\mathrm{T}}\boldsymbol{e} = \boldsymbol{0} \tag{10.5.3}$$

$$\operatorname*{plim}_{N\to\infty} \boldsymbol{W}^{\mathrm{T}}\boldsymbol{\Psi} \text{ positive definite} \,, \tag{10.5.4}$$

then it follows from (10.5.2)

$$\operatorname*{plim}_{N\to\infty} \boldsymbol{W}^{\mathrm{T}}\boldsymbol{\Psi}\boldsymbol{\theta} = \operatorname*{plim}_{N\to\infty} \boldsymbol{W}^{\mathrm{T}}\boldsymbol{y} \tag{10.5.5}$$

and the estimation equation is given as

$$\hat{\boldsymbol{\theta}} = \left(\boldsymbol{W}^{\mathrm{T}}\boldsymbol{\Psi}\right)^{-1}\boldsymbol{W}^{\mathrm{T}}\boldsymbol{y} \,. \tag{10.5.6}$$

According to Theorem 9.2, this equation yields asymptotic bias-free (consistent) estimates if in addition

$$\operatorname*{plim}_{N\to\infty} \boldsymbol{e} = \boldsymbol{0} \,. \tag{10.5.7}$$

The main problem now is to find appropriate instrumental variables. (10.5.3) and (10.5.4) suggest to choose the instrumental variables $w_i(k)$ so that they are as much as possible

- uncorrelated with the disturbance $n(k)$
- correlated with the useful signals $u(k)$ and $y_{\mathrm{u}}(k)$

In the matched case with $\hat{\boldsymbol{\theta}} = \boldsymbol{\theta}_0$, $e(k)$ depends only on $n(k)$ so that (10.5.3) is satisfied and with the useful signals in $\boldsymbol{W}$, also $\boldsymbol{W}^{\mathrm{T}}\boldsymbol{\Psi}$ will be positive definite, see Sect. 9.1.4.

The input signal was chosen as instrumental variables by Joseph et al (1961), i.e.

$$\boldsymbol{w}^{\mathrm{T}} = \big(u(k-1-\delta) \ldots u(k-m-\delta)\big|u(k-d-1) \ldots u(k-d-m)\big) \,, \tag{10.5.8}$$

because these instrumental variables are easy to obtain and are correlated with $\boldsymbol{\Psi}$. One can choose δ such that the elements of the covariance matrix $R_{\boldsymbol{w}\boldsymbol{\psi}}(\tau)$ are maximal.

A stronger correlation between $\boldsymbol{W}$ and $\boldsymbol{\Psi}$ can be obtained if $\boldsymbol{W}$ contains the undisturbed signals of $\boldsymbol{\Psi}$. One therefore has to estimate the undisturbed output signals $h(k) = \hat{y}_{\mathrm{u}}(k)$. Then, one can set up the instrumental variable matrix as

$$\boldsymbol{w}^{\mathrm{T}} = \big(-h(k-1) \ldots -h(k-m)\big|u(k-d-1) \ldots u(k-d-m)\big) . \quad (10.5.9)$$

This was proposed by Wong and Polak (1967) and Young (1970). The estimates of the undisturbed output can be obtained by means of the known input signals and the estimated parameters according to (9.1.18) as

$$h(k) = \hat{y}_{\mathrm{u}}(k) = \boldsymbol{\psi}^{\mathrm{T}}(k)\hat{\boldsymbol{\theta}}(k) . \quad (10.5.10)$$

This can be considered as an auxiliary model, providing auxiliary parameters $\boldsymbol{\theta}_{\mathrm{aux}}$, with the goal to reconstruct the useful output $y_{\mathrm{u}}(k)$, see Fig. 10.4. The matrix of instrumental variables is then given as

$$\boldsymbol{W} = \left(\begin{array}{ccc|ccc} -h(m+d+1) & \ldots & -h(d) & u(m-1) & \ldots & u(0) \\ -h(m+d) & \ldots & -h(d+1) & u(m) & \ldots & u(1) \\ \vdots & & \vdots & \vdots & & \vdots \\ -h(m+d+N+1) & \ldots & -h(d+N) & u(m+N-1) & \ldots & u(N) \end{array}\right) . \quad (10.5.11)$$

For the non-recursive application of this method, one uses the following approach (Young, 1970):

1. In the first iteration, one uses the instrumental variables from (10.5.8) or one uses the normal method of least squares according to (9.1.18)
2. From the parameter estimates $\hat{\boldsymbol{\theta}}_1$, the improved instrumental variables are determined according to (10.5.10) and estimates of the new parameter vector $\hat{\boldsymbol{\theta}}_2$ are obtained
3. Step 2 is repeated until the estimated parameters do not change significantly from iteration to iteration

In general, a few iterations are already sufficient to obtain suitable estimates. Furthermore, experience has shown that the instrumental variables do not have to match the undisturbed signals very precisely. The start with the normal method of least squares has proven very useful (Baur, 1976).

The covariance of the parameter estimates is in analogy to (9.1.69) given as

$$\operatorname{cov} \Delta\boldsymbol{\theta} = \mathrm{E}\big\{(\hat{\boldsymbol{\theta}}-\boldsymbol{\theta}_0)(\hat{\boldsymbol{\theta}}-\boldsymbol{\theta}_0^{\mathrm{T}})\big\} = \mathrm{E}\big\{(\boldsymbol{W}^{\mathrm{T}}\boldsymbol{\Psi})^{-1}\boldsymbol{W}^{\mathrm{T}}\boldsymbol{e}\boldsymbol{e}^{\mathrm{T}}\boldsymbol{W}(\boldsymbol{W}^{\mathrm{T}}\boldsymbol{\Psi})^{-1}\big\} . \quad (10.5.12)$$

In this equation, $\boldsymbol{W}$ and $\boldsymbol{e}$ are statistically independent, but this is not true for $\boldsymbol{\Psi}$ and $\boldsymbol{e}$, since $\boldsymbol{e}$ is correlated. Therefore, this equation cannot be simplified immediately.

If the parameters of the auxiliary model converge (10.5.10) to the true process parameters,

$$\operatorname*{plim}_{N\to\infty} \hat{\boldsymbol{\theta}}_{\mathrm{aux}} = \operatorname*{plim}_{N\to\infty} \hat{\boldsymbol{\theta}} = \boldsymbol{\theta}_0 , \quad (10.5.13)$$

one can assume that for large N

$$\frac{1}{N+1}\mathrm{E}\{\boldsymbol{W}^{\mathrm{T}}\boldsymbol{\Psi}\} \approx \frac{1}{N+1}\mathrm{E}\{\boldsymbol{W}^{\mathrm{T}}\boldsymbol{W}\}\,. \qquad (10.5.14)$$

Then, it follows from (10.5.12),

$$\operatorname{cov}\Delta\boldsymbol{\theta} \approx \mathrm{E}\Big\{\big(\boldsymbol{W}^{\mathrm{T}}\boldsymbol{W}\big)^{-1}\boldsymbol{W}^{\mathrm{T}}\Big\}\mathrm{E}\{\boldsymbol{e}\boldsymbol{e}^{\mathrm{T}}\}\mathrm{E}\Big\{\boldsymbol{W}\big(\boldsymbol{W}^{\mathrm{T}}\boldsymbol{W}\big)^{-1}\Big\}\,. \qquad (10.5.15)$$

One can show by the appropriate manipulations of the above equation that the covariance diminishes with $1/\sqrt{N+1}$, if $e(k)$ is stationary.

So far, only the small signal behavior of the input and output, i.e.

$$u(k) = U(k) = U_{00},\quad y(k) = Y(k) - Y_{00}\,,$$

was considered. Here, Y_{00} is typically unknown. If however $\mathrm{E}\{U(k)\} = 0$, then Y_{00} has no influence, if the output of the auxiliary model is also governed by $\mathrm{E}\{h(k)\} = 0$, since in (10.5.6), the values of $\boldsymbol{y}(k)$ are not correlated with the $h(k)$.

Theorem 10.1 (Conditions for a Consistent Parameter Estimation by the Method of Instrumental Variables).

The parameters $\boldsymbol{\theta}$ can be estimated consistently in the mean square, if the following conditions are satisfied:

1. *m and d are known exactly*
2. *$u(k) = U(k) - U_{00}$ is known exactly*
3. *$e(k)$ is not correlated with the instrumental variables $\boldsymbol{w}^{\mathrm{T}}(k)$*
4. *$\mathrm{E}\{e(k)\} = 0$*

□

From this follows that

1. $\mathrm{E}\{u(k-\tau)n(k)\} = 0$ for $|\tau| \geq 0$
2. Y_{00} must not be known if $\mathrm{E}\{u(k)\} = 0$ and $\mathrm{E}\{h(k)\} = 0$ with $h(k) = 0$ according to (10.5.9)
3. If either $\mathrm{E}\{n(k)\} = 0$ and $\mathrm{E}\{u(k)\} = \text{const.}$ or $\mathrm{E}\{u(k)\} = 0$ and $\mathrm{E}\{n(k)\} = 0$

A big advantage of the method of instrumental variables is that no special assumptions must be made about the noise and its form filter. The noise $n(k)$ can be an arbitrary stationary colored noise, i.e. it can be described by

$$n(z) = \frac{D(z^{-1})}{C(z^{-1})}\nu(z)\,. \qquad (10.5.16)$$

Then the polynomials $D(z^{-1})$ and $C(z^{-1})$ can, if they have stable roots, be arbitrary and independent of $A(z^{-1})$ and $B(z^{-1})$. The IV method per se does not provide a model of the noise. This can however be derived as is described in the next section. A detailed analysis of IV methods is given by Söderström and Stoica (1983). Although proposed for frequency domain identification, the following idea can also be applied in the time domain: In the case of multiple measurements, Pintelon and Schoukens (2001) propose to use measurements from a different experiment as instrumental variables as these are strongly correlated with the measurements and practically uncorrelated with the noise. The instrumental variable approach can also be combined with weighting of the estimation equations (Stoica and Jansson, 2000).

10.5.2 Recursive Method of Instrumental Variables (RIV)

According to the recursive method of least squares, one can also provide recursive equations for the method of instrumental variables (Wong and Polak, 1967; Young, 1968):

$$\hat{\boldsymbol{\theta}}(k+1) = \hat{\boldsymbol{\theta}}(k) + \boldsymbol{\gamma}(k)\big(y(k+1) - \boldsymbol{\psi}^{\mathrm{T}}(k+1)\hat{\boldsymbol{\theta}}(k)\big) \tag{10.5.17}$$

$$\boldsymbol{\gamma}(k) = \frac{1}{\boldsymbol{\psi}^{\mathrm{T}}(k+1)\boldsymbol{P}(k)\boldsymbol{\psi}(k+1)+1}\boldsymbol{P}(k)\boldsymbol{w}(k+1) \tag{10.5.18}$$

$$\boldsymbol{P}(k+1) = \big(\boldsymbol{I} - \boldsymbol{\gamma}(k)\boldsymbol{\psi}^{\mathrm{T}}(k+1)\big)\boldsymbol{P}(k) \tag{10.5.19}$$

Here,

$$\boldsymbol{P}(k) = \big(\boldsymbol{W}^{\mathrm{T}}(k)\boldsymbol{\Psi}(k)\big)^{-1} \tag{10.5.20}$$

$$\boldsymbol{w}^{\mathrm{T}}(k) \text{ and } h(k) \text{ see (10.5.9) and (10.5.10)}\,.$$

A block diagram of this method is shown in Fig. 10.4.

To avoid a strong correlation between the instrumentals $h(k)$ and the current error signal $e(k)$, it is suggested by Wong and Polak (1967) to introduce a dead time q between the estimated parameters and the parameter set used for the auxiliary model, where q should be chosen such that $e(k+q)$ is independent of $e(k)$.

Young (1970) furthermore used a discrete-time low-pass filter so that

$$\hat{\boldsymbol{\theta}}_{\mathrm{aux}}(k) = (1-\beta)\hat{\boldsymbol{\theta}}_{\mathrm{aux}}(k-1) + \beta\hat{\boldsymbol{\theta}}(k-q)\,. \tag{10.5.21}$$

In this case, the choice of q is less critical and the parameter estimates are smoothed, such that fast parameter changes of the auxiliary model are avoided. β should be chosen as $0.01 \leq \beta \leq 0.1$ (Baur, 1976).

As initial values, one chooses in analogy to the normal method of least squares the matrix $\boldsymbol{P}(0) = \alpha\boldsymbol{I}$ as a diagonal matrix with large elements and the parameter vector $\hat{\boldsymbol{\theta}} = \boldsymbol{0}$. In the starting phase, one might also want to supervise the convergence of the auxiliary model. It has been proven useful to employ the recursive method of least squares in the starting phase of the algorithm (Baur, 1976).

Since one does not automatically obtain a model of the disturbance, one can proceed as follows (Young, 1970):

1. First, the noise $n(k)$ is determined as

   $$n(k) = y(k) - \hat{y}_{\mathrm{u}}(k) = y(k) - h(k)\,,$$

   where $y(k)$ is the measured process output and $h(k)$ the output of the auxiliary model.
2. Then one uses a suitable parameter estimation technique to determine the parameters of an ARMA signal process given by

   $$n(z) = \frac{D(z^{-1})}{C(z^{-1})}\nu(z)\,.$$

   One can employ e.g. the recursive method of least squares as described in Sect. 9.4.

10.6 Method of Stochastic Approximation (STA)

The methods of stochastic approximation are recursive parameter estimation techniques, which are computationally less demanding than the recursive method of least squares. The minimum of the cost function is determined by a gradient-based method, which can be applied to both deterministic and stochastic models.

The method of stochastic approximation goes back to work by Robbins and Monro (1951), Kiefer and Wolfowitz (1952), Blum (1954), and Dvoretzky (1956). A survey of the methods can be found in (Sakrison, 1966; Albert and Gardner, 1967; Sage and Melsa, 1971) and recently e.g. in (Kushner and Yin, 2003).

10.6.1 Robbins-Monro Algorithm

As an introductory example for the gradient based methods, the one-dimensional case, i.e. the estimation of a single parameter, is presented. This parameter θ satisfies the equation

$$g(\theta) = g_0 , \tag{10.6.1}$$

where $g(0)$ must be exactly measurable and g_0 must be a known constant. Then, the unknown parameter θ, i.e. the root of (10.6.1), can be determined iteratively by the gradient as

$$\theta(k+1) = \theta(k) - \varrho(k)\big(g(\theta(k)) - g_0\big) . \tag{10.6.2}$$

Here, the weighting factors $\varrho(k)$ must be chosen appropriately to ensure convergence of the algorithm. If $g(\theta(k)) - g_0 = 0$, then $\theta(k+1)$ is the exact solution.

Now, it is assumed that $g(\theta)$ cannot be measured and that one can only measure the disturbed quantity

$$f(\theta, n) = g(\theta) + n , \tag{10.6.3}$$

where n is a stochastic quantity with $\mathrm{E}\{n\} = 0$ and finite variance. Then, also $f(\theta, n)$ is a stochastic quantity and (10.6.2) cannot be used to determine θ as $g(\theta)$ is not known. Since

$$\mathrm{E}\big\{f(\theta, n)\big\} = g(\theta) , \tag{10.6.4}$$

one would expect that the algorithm in (10.6.2) after replacing $g(\theta)$ by $f(\theta, n)$ and becoming a stochastic algorithm

$$\hat{\theta}(k+1) = \hat{\theta}(k) - \varrho(k)\big(f(\hat{\theta}(k), n(k)) - g_0\big) \tag{10.6.5}$$

would still converge towards the true value θ_0 after a sufficient number of iterations. This algorithm is termed *Robbins-Monro algorithm*.

The new value of the parameter estimate is then obtained by subtracting of the error as determined by the disturbed equation (10.6.1) from the old parameter estimate weighted with a correction factor $\varrho(k)$

$$e(k) = f(\hat{\theta}(k), n(k)) - g_0 . \tag{10.6.6}$$

Theorem 10.2 (Robbins-Monro Algorithm). *The Robins-Monro algorithm converges in the mean squared error sense*

$$\lim_{k\to\infty} \mathrm{E}\Big\{\big(\hat{\theta}(k) - \theta_0\big)^2\Big\} = 0$$

under the following conditions:

1. *(10.6.1) has a unique solution*
2. *The stochastic quantities $f(k)$ must have an equal probability density function and must be statistically independent*
3. $\lim_{k\to\infty} \varrho(k) = 0, \sum_{k=1}^{\infty} \varrho(k) = \infty, \sum_{k=1}^{\infty} \varrho^2(k) < \infty$

□

The proof is presented e.g. in (Sakrison, 1966). Some weighting factors $\varrho(k)$ which satisfy the above stated conditions are

$$\varrho(k) = \frac{\alpha}{\beta + k} \text{ or } \varrho(k) = \frac{\alpha}{k} \,. \tag{10.6.7}$$

The choice of α and β is arbitrary. If α is sufficiently large, one can expect a good convergence for large k.

10.6.2 Kiefer-Wolfowitz Algorithm

A second stochastic approximation algorithm can be stated, if a parameter θ has to be determined such that a function $g(\theta)$ reaches an extremal point, i.e.

$$\frac{\mathrm{d}}{\mathrm{d}\theta} g(\theta) = 0 \tag{10.6.8}$$

is satisfied. The deterministic, gradient-based algorithm is in this case given as

$$\theta(k+1) = \theta(k) - \varrho(k)\frac{\mathrm{d}}{\mathrm{d}\theta} g(\theta) \,. \tag{10.6.9}$$

If $g(\theta)$ cannot be measured exactly and the only possible measurement is governed by (10.6.3), then one can derive in analogy to (10.6.5) the following stochastic algorithm

$$\hat{\theta}(k+1) = \hat{\theta}(k) - \varrho(k)\frac{\mathrm{d}}{\mathrm{d}\theta} f(\hat{\theta}(k), n(k)) \,, \tag{10.6.10}$$

which is termed the *Kiefer-Wolfowitz algorithm.*

If the function $f(\hat{\theta}(k), n(k))$ is not differentiable everywhere or if the determination of the derivative is too difficult, one can resort to replacing the first derivative by the difference quotient and one obtains

$$\hat{\theta}(k+1) = \hat{\theta}(k) - \frac{\varrho(k)}{2\Delta\theta(k)}\big(f(\hat{\theta}(k) + \Delta\theta(k), n(k)) - f(\hat{\theta}(k) - \Delta\theta(k), n(k))\big) \,. \tag{10.6.11}$$

This leads directly to

Theorem 10.3 (Kiefer-Wolfowitz Algorithm). *The Kiefer-Wolfowitz algorithm converges in the mean squared error under the following conditions:*

1. *$g(\theta)$ has a single extremal point*
2. *The stochastic quantities $f(k)$ must have an equal probability density function and must be statistically independent*
3. $\lim_{k\to\infty} \Delta\theta(k) = 0,\ \lim_{k\to\infty} \varrho(k) = 0,\ \sum_{k=1}^{\infty} \varrho(k)\Delta\theta(k) < \infty,$ $\sum_{k=1}^{\infty}\left(\frac{\varrho(k)}{\Delta\theta(k)}\right)^2 < \infty$

□

In order to simultaneously estimate more than one parameter of the scalar function $g(\theta)$, one can replace the scalar quantity θ in (10.6.5) and (10.6.10) by a parameter vector $\boldsymbol{\theta}$.

The method of the stochastic approximation shall now be applied to the estimation of the parameters of a difference equation according to (9.1.5) or (9.1.7) respectively. For this, one is interested in determining the minimum of the cost function

$$V(k) = e^2(k)\ . \tag{10.6.12}$$

For determining the minimum, which is not known a priori, one will now employ the Kiefer-Wolfowitz algorithm. The following relations are in line with Sect. 9.4. The error for a given sample can be determined as

$$e(k+1) = y(k+1) - \boldsymbol{\psi}^{\mathrm{T}}(k+1)\hat{\boldsymbol{\theta}}(k)\ , \tag{10.6.13}$$

compare (9.4.13). Then, the derivative of the cost function in (10.6.12) becomes

$$\frac{\partial V(k+1)}{\partial \boldsymbol{\theta}} = -2\boldsymbol{\psi}(k+1)\big(y(k+1) - \boldsymbol{\psi}^{\mathrm{T}}(k+1)\hat{\boldsymbol{\theta}}(k)\big)\ . \tag{10.6.14}$$

Then (10.6.10) is given as

$$\begin{aligned} \underbrace{\hat{\boldsymbol{\theta}}(k+1)}_{\substack{\text{New Parameter}\\ \text{Estimate}}} &= \underbrace{\hat{\boldsymbol{\theta}}(k)}_{\substack{\text{Old Parameter}\\ \text{Estimate}}} + \underbrace{2\varrho(k+1)\boldsymbol{\psi}(k+1)}_{\substack{\text{Correction}\\ \text{Vector}}} \\ &\quad \big(\underbrace{y(k+1)}_{\substack{\text{New}\\ \text{Measurement}}} - \underbrace{\boldsymbol{\psi}^{\mathrm{T}}(k+1)\hat{\boldsymbol{\theta}}(k)}_{\substack{\text{Predicted Measurement based}\\ \text{on Last Parameter Estimate}}}\big)\ . \end{aligned} \tag{10.6.15}$$

It is often suggested to choose the weighting factor as

$$2\varrho(k+1) = \frac{1}{k+1}\frac{1}{\kappa} \text{ with } \kappa > 0\ . \tag{10.6.16}$$

This stochastic algorithm matches with the algorithm of recursive least squares in (9.4.11) up to the correction vector, which is defined differently. The difference is

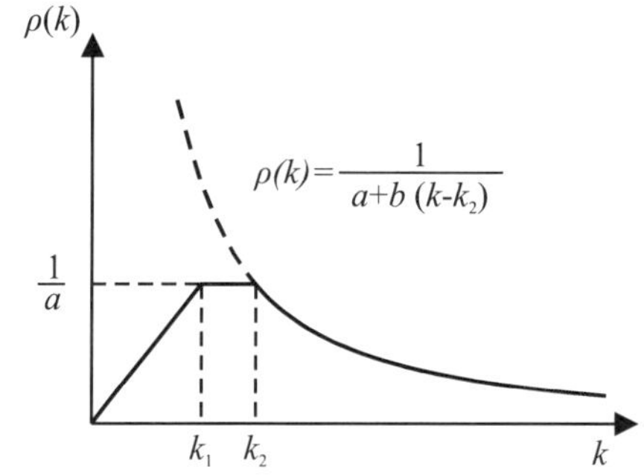

Fig. 10.5. Suggested time behavior of $\varrho(k)$ for the stochastic approximation

that the recursive method of least squares determines the update as a function of the variance of the data. In the vectorial case, one will use the scalar correction factor $2\varrho(k+1)$ for the stochastic approximation and the parameter error covariance matrix $\boldsymbol{P}(k+1)$ in the case of the recursive least squares, which weights the latest equation error based on the current accuracy of the parameter estimates. Hence, the method of stochastic approximation can be interpreted as a strongly simplified version of the recursive method of least squares.

According to Theorem 10.3 and (10.6.12), $e^2(k)$ must be statistically independent for a consistent estimation. Since this can in most applications not be guaranteed, one will typically not obtain bias-free estimates. Saridis and Stein (1968) show that the bias can be corrected if the statistic properties of the measured signals are known exactly. One can use the method of stochastic approximation also for the estimation of non-parametric models (Saridis and Stein, 1968; Isermann and Baur, 1973).

It should also be mentioned that the convergence can be improved by certain modifications of the calculation of the weighting factor $\varrho(k)$. If the factor is chosen according to the suggestion in (10.6.16), then it becomes very large at the start of the algorithm, meaning that the error $e(k)$ is emphasized too strongly. A choice of $\varrho(k)$ according to Fig. 10.5 leads to a damped change of the parameter estimates and leads to better convergence as was shown by Isermann and Baur (1973). The method of stochastic approximation is only used in very limited areas of application as the convergence is unreliable and also the higher computational effort for the recursive method of least squares can in most cases easily be handled nowadays.

Example 10.1 (Identification of a First Order Process with the Kiefer-Wolfowitz Algorithm and the Normalized Least Mean Squares).

Example 9.2 shall now be used for a comparison of the RLS method and the Kiefer-Wolfowitz (termed KW from now on) method. One can see a comparison of the convergence of the parameter estimates in Fig. 10.6 and can see that the RLS method converges much faster. Figure 10.7 shows how the methods converge towards the optimal parameter set, One can see that the choice of the (in this case constant) factor ϱ has tremendous influence on the convergence behavior and speed. The ellipsoids are contours of the cost function V, i.e. the sum of squared errors $\sum e^2(k)$.

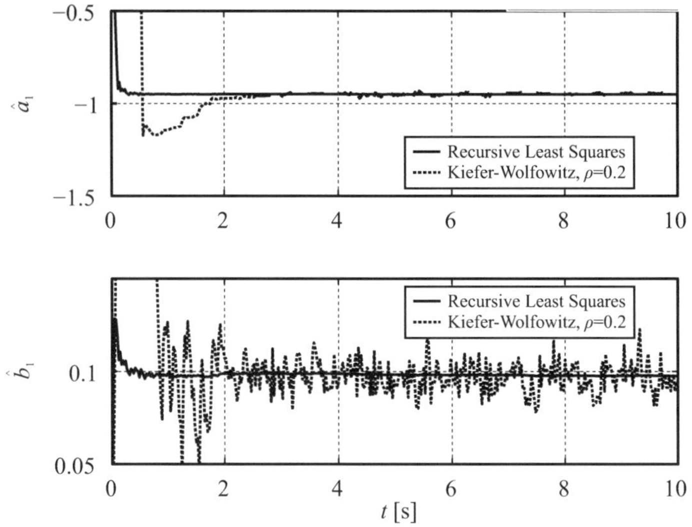

Fig. 10.6. Parameter estimation for a first order system comparing the method of recursive least squares with the Kiefer-Wolfowitz algorithm with $\varrho = 0.2$

In this example, also the results of the normalized least mean squares (NLMS) have been shown. The algorithm will be discussed in the next section.

Example 10.2 (Identification of the Three-Mass Oscillator by Means of the Kiefer-Wolfowitz Algorithm).

The Kiefer-Wolfowitz algorithm has now been applied to the three-mass oscillator, where the system was excited with a PRBS signal (see Sect. 6.3). The measurements were the same as in Fig. 9.6 to be able to compare the results of the different methods.

As can be seen from Fig. 10.8, the parameters need a long time to converge, compare this with the time of 15 seconds that the parameter estimates needed to settle in case of the RLS method, as could be witnessed in Fig. 9.7. An important issue for convergence is the factor $\varrho(k)$ which was chosen according to the approach presented in Fig. 10.5. The value of $\varrho(k)$ as a function of time has been graphed in Fig. 10.10. Even at the end of the experiment, the frequency response does not match totally with the theoretical model, see Fig. 10.9. During the experiments, it has been shown also that the factor $\varrho(k)$ which on the one hand is critical for convergence, is difficult to choose on the other hand. □

10.7 (Normalized) Least Mean Squares (NLMS)

Similar to the update equation of the Kiefer-Wolfowitz algorithm (10.6.15), one can write

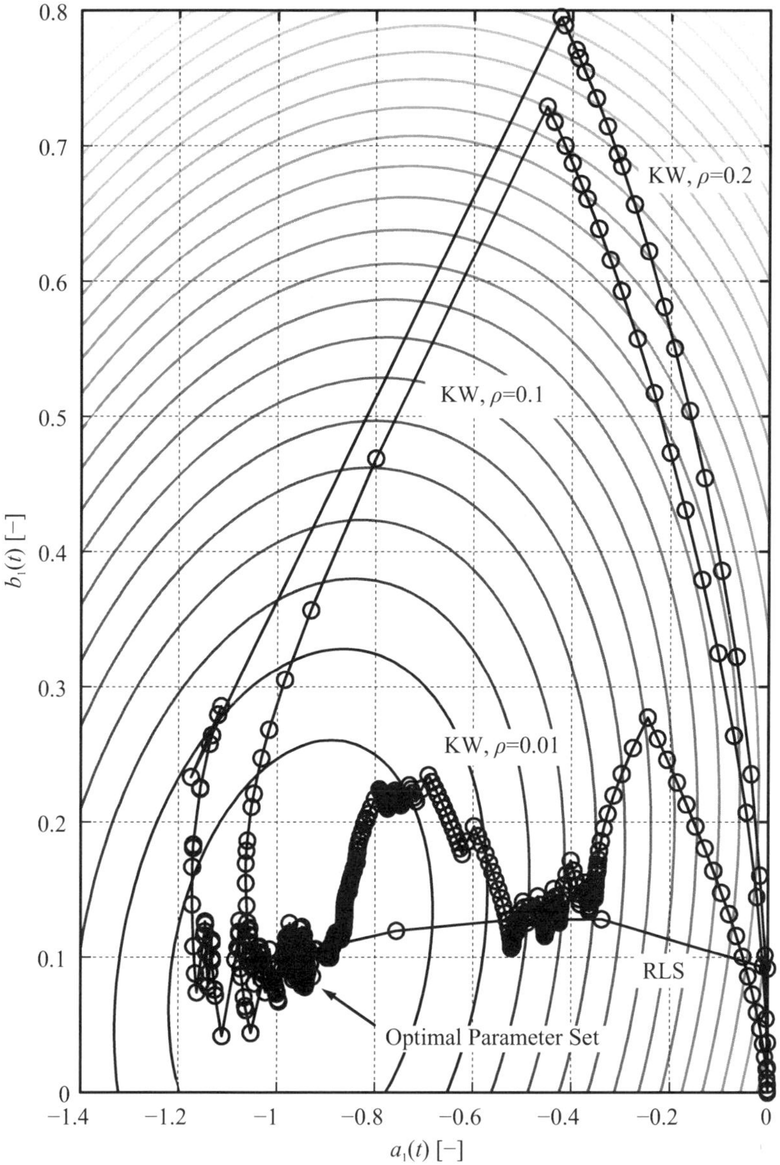

Fig. 10.7. Convergence by the RLS method, the KW method with different constant ϱ and the NLMS method and cost function contours, $\hat{\boldsymbol{\theta}}(0) = \mathbf{0}$. Note that superior performance of the method of recursive least squares as shown by the thick line.

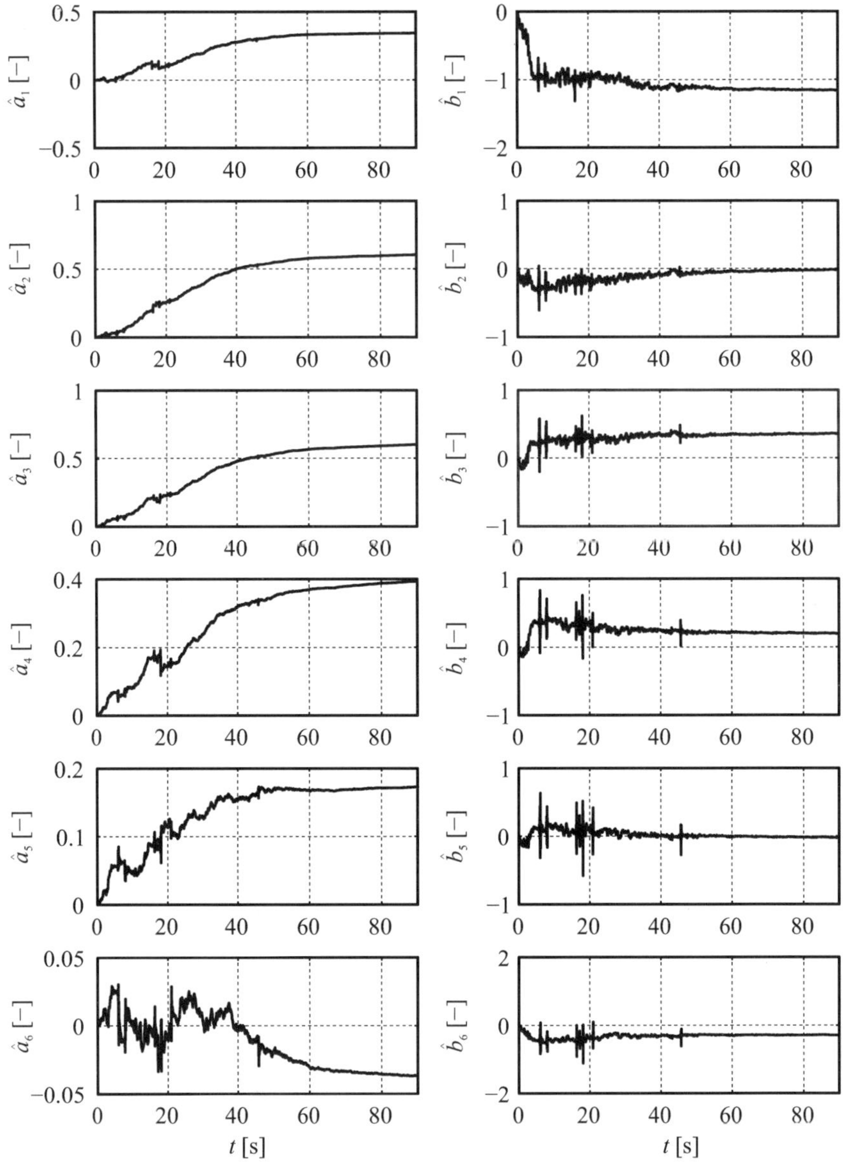

Fig. 10.8. Parameter estimation by means of the Kiefer-Wolfowitz gradient method for a discrete-time model of the Three-Mass Oscillator, parameter estimates

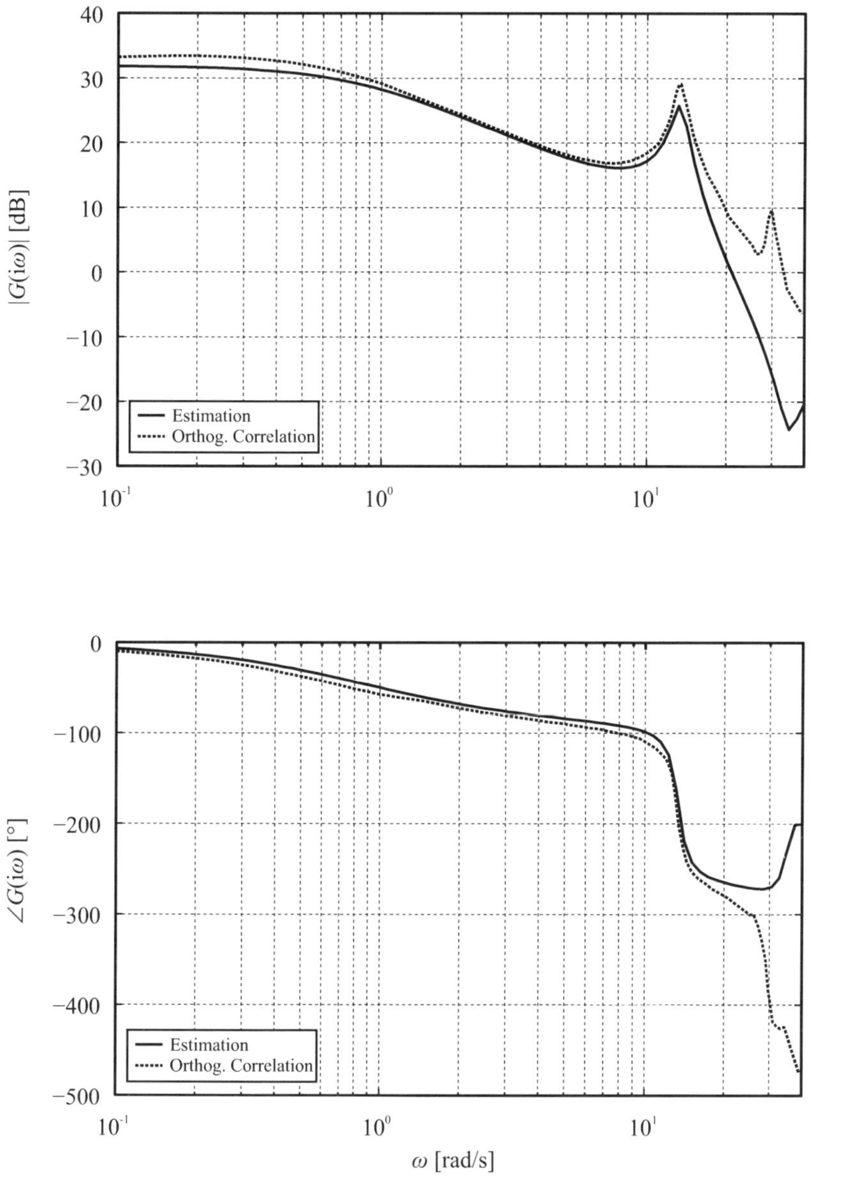

Fig. 10.9. Parameter estimation by means of the Kiefer-Wolfowitz gradient method for a discrete-time model of the Three-Mass Oscillator, comparison of frequency responses

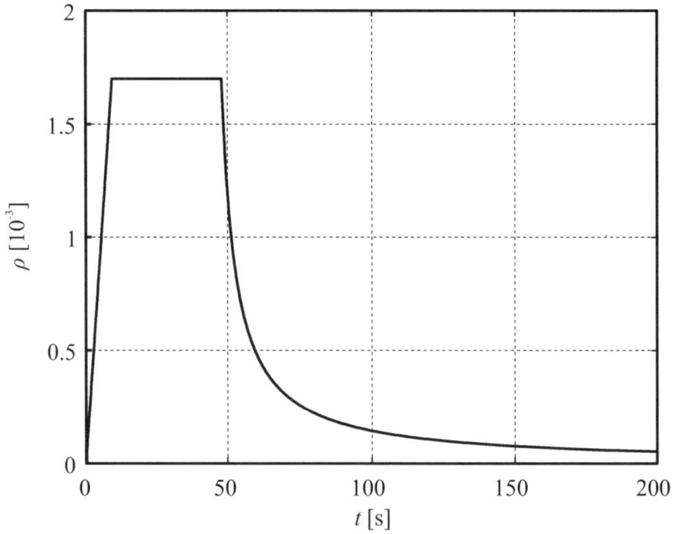

Fig. 10.10. Factor $\varrho(k)$ for the Kiefer-Wolfowitz gradient method for Fig. 10.8

$$\begin{aligned}\underbrace{\hat{\boldsymbol{\theta}}(k+1)}_{\substack{\text{New Parameter}\\ \text{Estimate}}} &= \underbrace{\hat{\boldsymbol{\theta}}(k)}_{\substack{\text{Old Parameter}\\ \text{Estimate}}} + \underbrace{\beta\boldsymbol{\psi}(k+1)}_{\substack{\text{Correction}\\ \text{Vector}}}\\ &\quad \Big(\underbrace{y(k+1)}_{\substack{\text{New}\\ \text{Measurement}}} - \underbrace{\boldsymbol{\psi}^{\mathrm{T}}(k+1)\hat{\boldsymbol{\theta}}(k)}_{\substack{\text{Predicted Measurement based}\\ \text{on Last Parameter Estimate}}}\Big)\ ,\end{aligned} \tag{10.7.1}$$

where the weighting factor for the correction vector has been written as $\beta(k+1)$. This yields the *least mean squares* algorithm (Haykin and Widrow, 2003). Here, β is interpreted as a *learning rate*. While for the Kiefer-Wolfowitz algorithm, the factor $\varrho(k+1)$ was steered by the number of time steps k only, the learning rate β will now be expressed as a function of the measured data.

In the noise free case, the algorithm should converge to the true values in one sample step. Therefore, with the parameter update $\hat{\boldsymbol{\theta}}(k+1)$, the model output should match with the measurement, i.e.

$$y(k+1) = \boldsymbol{\psi}^{\mathrm{T}}(k+1)\hat{\boldsymbol{\theta}}(k+1)\ . \tag{10.7.2}$$

Combining (10.7.1) and (10.7.2) yields the ideal step size as

$$\begin{aligned}& y(k+1) = \boldsymbol{\psi}^{\mathrm{T}}(k+1)\Big(\hat{\boldsymbol{\theta}}(k) + \beta\boldsymbol{\psi}(k+1)\big(y(k+1) - \boldsymbol{\psi}^{\mathrm{T}}(k+1)\hat{\boldsymbol{\theta}}(k)\big)\Big)\\ & \Leftrightarrow \beta = \frac{1}{\boldsymbol{\psi}^{\mathrm{T}}(k+1)\boldsymbol{\psi}(k+1)}\ .\end{aligned} \tag{10.7.3}$$

Hence, the learning rate should be in the interval

$$0 < \beta < \frac{1}{\psi^{\mathrm{T}}(k+1)\psi(k+1)} . \tag{10.7.4}$$

The big disadvantage of the least mean squares algorithm is the fact that the actual fault reduction varies from step to step. Hence the algorithm is normalized and one uses the update equation

$$\hat{\theta}(k+1) = \hat{\theta}(k) + \frac{\tilde{\beta}}{\psi^{\mathrm{T}}(k+1)\psi(k+1)}\psi(k+1)\big(y(k+1) - \psi^{\mathrm{T}}(k+1)\hat{\theta}(k)\big) , \tag{10.7.5}$$

see (Brown and Harris, 1994). The algorithm was also presented in Example 10.1 in Fig. 10.7. One can see that the convergence has not improved much compared to the KW algorithm. A further problem can be that one divides by $\psi^{\mathrm{T}}(k+1)\psi(k+1)$, which can become zero. Goodwin and Sin (1984) present two remedial actions: The first idea is to augment the vector by $\psi(k)$ by the constant "1" as the last element. This would at the same time allow to estimate an operation point dependent DC value. The other idea is to divide by the factor $\psi^{\mathrm{T}}(k+1)\psi(k+1) + c$, where c is a constant with $c > 0$.

10.8 Summary

Modifications to and alternative solutions of the method of least squares have been presented in this chapter. The modifications to the classical method of least squares had the goal to avoid the bias that exists if the method of least squares is applied to identify linear dynamic discrete-time systems with considerable noise at the output. Different assumptions about the noise signal model have been made, such as AR, MA, or ARMA. The quality of the estimates depends strongly on the assumptions about the noise. The method of RELS has found more widespread use due to the simplicity of its application. Other methods accept the existence of a bias and try to correct the bias after the estimation has taken place. The method of total least squares is different from the previous methods as it allowed to account for noise both on the input and the output.

The method of instrumental variables has also been presented in this chapter. It is another attractive parameter estimation method as it can provide bias-free estimates. The convergence of the method however depends strongly on the choice of the instrumental variables and can be problematic if the method is applied for closed loop parameter estimation, see Chap. 13. The choice of the instrumental variables has also been discussed in this chapter.

Finally, the method of stochastic approximation has been presented in this chapter. It is an easy to realize method that has its origin in the gradient based optimization methods, see Chap. 19, and can be interpreted as a severely simplified RLS parameter estimation method. The stochastic approximation is not frequently used in practical applications as the additional computational expense of the RLS method does not matter nowadays in most applications.

Problems

10.1. GLS, ELS, and TLS
What are the assumptions made on the noise model for the three methods generalized least squares, extended least squares, and total least squares?

10.2. Instrumental Variables Method I
What are the instrumental variables? What requirements do they have to satisfy to provide a bias-free estimate. What are typical approaches to choose the instrumental variables?

10.3. Instrumental Variables Method II
If several experiments with the same test signal have been recorded, how could one choose the instrumental variables in this case.

10.4. Robbins-Monro and Kiefer-Wolfowitz Algorithm
What equation does the estimated parameter θ in the one-dimensional case satisfy? Which of the two methods can be used to determine the extremal point of a function? Explain in your own words, why the method of RLS in general converges faster than the Kiefer-Wolfowitz algorithm.

References

Albert AE, Gardner LA Jr (1967) Stochastic approximation and non-linear regression. MIT Press, Cambridge, MA

Baur U (1976) On-Line Parameterschätzverfahren zur Identifikation linearer, dynamischer Prozesse mit Prozeßrechnern: Entwicklung, Vergleich, Erprobung: KfK-PDV-Bericht Nr. 65. Kernforschungszentrum Karlsruhe, Karlsruhe

Blum J (1954) Multidimensional stochastic approximation procedures. Ann Math Statist 25(4):737–744

Brown M, Harris C (1994) Neurofuzzy adaptive modelling and control. Prentice-Hall international series in systems and control engineering, Prentice Hall, New York

Clarke DW (1967) Generalized least squares estimation of the parameters of a dynamic model. In: Preprints of the IFAC Symposium on Identification, Prag

Durbin J (1954) Errors in variables. Revue de l'Institut International de Statistique / Review of the International Statistical Institute 22(1/3):23–32

Dvoretzky A (1956) On stochastic approximation. In: Proceedings of the 3rd Berkeley Symposium on Mathematical Statistics and Probability, Berkeley, CA, USA

Goedecke W (1987) Fehlererkennung an einem thermischen Prozeß mit Methoden der Parameterschätzung: Fortschr.-Ber. VDI Reihe 8 Nr. 130. VDI Verlag, Düsseldorf

Golub GH, van Loan C (1980) An analysis of the total least squares problems. SIAM J Numer Anal 17(6):883–893

Golub GH, Reinsch C (1970) Singular value decomposition and least squares solutions. Numer Math 14(5):403–420

Goodwin GC, Sin KS (1984) Adaptive filtering, prediction and control. Prentice-Hall information and system sciences series, Prentice-Hall, Englewood Cliffs, NJ

Hastings-James R, Sage MW (1969) Recursive generalized least-squares procedure for on-line identification of process parameters. Proc IEE 116(12):2057–2062

Haykin S, Widrow B (2003) Least-mean-square adaptive filters. Wiley series in adaptive and learning systems for signal processing, communication and control, Wiley-Interscience, Hoboken, NJ

van Huffel S, Vandewalle J (1991) The total least squares problem: Computational aspects and analysis, Frontiers in applied mathematics, vol 9. SIAM, Philadelphia, PA

Isermann R (1974) Prozessidentifikation: Identifikation und Parameterschätzung dynamischer Prozesse mit diskreten Signalen. Springer, Heidelberg

Isermann R, Baur U (1973) Results of testcase A. In: Proceedings of the 3rd IFAC Symposium on System Identification, North Holland Publ. Co., Amsterdam, Netherlands

Jolliffe IT (2002) Principal component analysis, 2nd edn. Springer series in statistics, Springer, New York

Joseph P, Lewis J, Tou J (1961) Plant identification in the presence of disturbances and application to digital control systems. Trans AIEE (Appl and Ind) 80:18–24

Kendall MG, Stuart A (1961) The advanced theory of statistics. Volume 2. Griffin, London, UK

Kiefer J, Wolfowitz J (1952) Statistical estimation of the maximum of a regression function. Ann Math Stat 23(3):462–466

Kumar PR, Moore JB (1979) Towards bias elimination in least squares system identification via detection techniques. In: Proceedings of the 5th IFAC Symposium on Identification and System Parameter Estimation Darmstadt, Pergamon Press, Darmstadt, Germany

Kushner HJ, Yin GG (2003) Stochastic approximation and recursive algorithms and applications, Applications of mathematics, vol 35, 2nd edn. Springer, New York, NY

Ljung L (1999) System identification: Theory for the user, 2nd edn. Prentice Hall Information and System Sciences Series, Prentice Hall PTR, Upper Saddle River, NJ

Markovsky I, van Huffel S (2007) Overview of total least-squares methods. Signal Proc 87(10):2283–2302

Markovsky I, Willems JC, van Huffel S, de Bart M, Pintelon R (2005) Application of structured total least squares for system identification and model reduction. IEEE Trans Autom Control 50(10):1490–1500

Panuska V (1969) An adaptive recursive least squares identification algorithm. In: Proceedings of the IEEE Symposium in Adaptive Processes, Decision and Control

Pintelon R, Schoukens J (2001) System identification: A frequency domain approach. IEEE Press, Piscataway, NJ

Reiersøl O (1941) Confluence analysis by means of lag moments and other methods of confluence analysis. Econometrica 9(1):1–24

Robbins H, Monro S (1951) A stochastic approximation method. Ann Math Statist 22(3):400–407

Sage AP, Melsa JL (1971) System identification. Academic Press, New York

Sakrison DJ (1966) Stochastic approximation. Adv Commun Syst 2:51–106

Saridis GN, Stein G (1968) Stochastic approximation algorithms for linear discrete-time system identification. IEEE Trans Autom Control 13(5):515–523

Söderström T (2007) Errors-in-variables methods in system identification. Automatica 43(6):939–958

Söderström T, Stoica PG (1983) Instrumental variable methods for system identification, Lecture notes in control and information sciences, vol 57. Springer, Berlin

Steiglitz K, McBride LE (1965) A technique for the identification of linear systems. IEEE Trans Autom Control 10:461–464

Stoica P, Jansson M (2000) On the estimation of optimal weights for instrumental variable system identification methods. In: Proccedings of the 12th IFAC Symposium on System Identification, Santa Barbara, CA, USA

Stoica PG, Söderström T (1977) A method for the identification of linear systems using the generalized least squares principle. IEEE Trans Autom Control 22(4):631–634

Stoica PG, Söderström T (1982) Bias correction in least squares identification. Int J Control 35(3):449–457

Wong K, Polak E (1967) Identification of linear discrete time systems using the instrumental variable method. IEEE Trans Autom Control 12(6):707–718

Young P (1968) The use of linear regression and related procedures for the identification of dynamic processes. In: Proceedings of the 7th IEEE Symposium in Adaptive Processes, Los Angeles, CA, USA

Young P (1970) An instrumental variable method for real-time identification of a noisy process. Automatica 6(2):271–287

Zimmerschied R (2008) Identifikation nichtlinearer Prozesse mit dynamischen lokalaffinen Modellen : Maßnahmen zur Reduktion von Bias und Varianz. Fortschr.-Ber. VDI Reihe 8 Nr. 1150. VDI Verlag, Düsseldorf

11

Bayes and Maximum Likelihood Methods

While the parameter estimation methods presented so far assumed that the parameters $\boldsymbol{\theta}$ and the observations of the output $\boldsymbol{y}$ are deterministic values, the parameters themselves and/or the output will now be seen in a stochastic view as a series of random variables. In Bayes estimation, the parameter vector has the probability density function $p(\boldsymbol{\theta})$ and the output can be described by the conditional probability density function $p(\boldsymbol{y}|\boldsymbol{\theta})$. One can then derive a solution to the parameter estimation problem based on this statistical information. As especially information about the probability density function of the parameters, $p(\boldsymbol{\theta})$, is seldom available in practical applications, the maximum likelihood estimator will be derived subsequently. It is based on the probability density function of the observed output $p(\boldsymbol{y}|\boldsymbol{\theta})$.

11.1 Bayes Method

For a given set of measurements $\boldsymbol{y}$, one can infer the parameters from the conditional probability density function $p(\boldsymbol{\theta}|\boldsymbol{y})$. This conditional probability density function can only be determined once the experiment has been conducted since it obviously depends on the measurements. Hence, it is an a posteriori probability density function. Based on this a posteriori probability density function, one is now interested in finding "best" parameter estimates $\hat{\boldsymbol{\theta}}$. To judge the optimality, once again an optimality criterion has to be introduced as $W(\hat{\boldsymbol{\theta}}, \boldsymbol{\theta})$, for which a cost function must then be minimized,

$$\min_{\hat{\boldsymbol{\theta}}} \int_m W(\hat{\boldsymbol{\theta}}, \boldsymbol{\theta}) p(\boldsymbol{\theta}|\boldsymbol{y}) \mathrm{d}^m \boldsymbol{\theta} , \tag{11.1.1}$$

and to find the minimum

$$\frac{\partial}{\partial \hat{\boldsymbol{\theta}}} \int_m W(\hat{\boldsymbol{\theta}}, \boldsymbol{\theta}) p(\boldsymbol{\theta}|\boldsymbol{y}) \mathrm{d}^m \boldsymbol{\theta} = \mathbf{0} , \tag{11.1.2}$$

where $\int_m$ is the m-dimensional integral over all components $\mathrm{d}\theta_1, \mathrm{d}\theta_2, \ldots, \mathrm{d}\theta_m$ of $\boldsymbol{\theta}$. The optimality criterion can for example be a quadratic function such as

R. Isermann, M. Münchhof, *Identification of Dynamic Systems*,
DOI 10.1007/978-3-540-78879-9_11, © Springer-Verlag Berlin Heidelberg 2011

$$W = (\hat{\boldsymbol{\theta}} - \boldsymbol{\theta})^{\mathrm{T}} (\hat{\boldsymbol{\theta}} - \boldsymbol{\theta}) \ . \tag{11.1.3}$$

In the one dimensional case, one can then write (11.1.1) as

$$\min_{\hat{\theta}} \int (\hat{\theta} - \theta)^2 p(\theta|\boldsymbol{y}) \mathrm{d}\theta \ . \tag{11.1.4}$$

Taking the first derivative to determine the optimal value $\hat{\theta}$ leads to

$$\hat{\theta} = \int \theta p(\theta|\boldsymbol{y}) \mathrm{d}\theta \ . \tag{11.1.5}$$

which is just the expected value of the parameter θ for the given probability density function $p(\theta|\boldsymbol{y})$.

A different approach is to choose the most likely value as indicated by the probability density function i.e. to choose the maximum of the probability density function as the estimate,

$$\hat{\boldsymbol{\theta}} = \arg\max_{\theta} p(\boldsymbol{\theta}|\boldsymbol{y}) \ . \tag{11.1.6}$$

In this setting, the PDF is termed *likelihood function.*

The key issue for both approaches is the determination of the conditional probability density function $p(\boldsymbol{\theta}|\boldsymbol{y})$, which can be determined by Bayes rule (Papoulis, 1962) as

$$p(\boldsymbol{\theta}, \boldsymbol{y}) = p(\boldsymbol{\theta}|\boldsymbol{y}) p(\boldsymbol{y}) \ . \tag{11.1.7}$$

Here, $p(\boldsymbol{\theta}, \boldsymbol{y})$ is the joint PDF and $p(\boldsymbol{y})$ is the a posteriori PDF which follows from the measurements conducted during the experiment. Furthermore

$$p(\boldsymbol{\theta}, \boldsymbol{y}) = p(\boldsymbol{y}|\boldsymbol{\theta}) p(\boldsymbol{\theta}) \ . \tag{11.1.8}$$

Hence, it follows from (11.1.7) that

$$p(\boldsymbol{\theta}|\boldsymbol{y}) = \frac{p(\boldsymbol{\theta}, \boldsymbol{y})}{p(\boldsymbol{y})} \ , \tag{11.1.9}$$

and with (11.1.8) that

$$p(\boldsymbol{\theta}|\boldsymbol{y}) = \frac{p(\boldsymbol{y}|\boldsymbol{\theta}) p(\boldsymbol{\theta})}{p(\boldsymbol{y})} \ , \tag{11.1.10}$$

where the PDF of $\boldsymbol{\theta}$ must be known a priori. If this is the case, then one could for example solve (11.1.2) directly.

Similarly, (11.1.6) can be written using the above results as

$$\hat{\boldsymbol{\theta}} = \arg\max_{\theta} p(\boldsymbol{y}|\boldsymbol{\theta}) p(\boldsymbol{\theta}) \ . \tag{11.1.11}$$

If no assumption can be made about $\boldsymbol{\theta}$ and hence it is assumed to be distributed uniformly over the parameter space, then

$$\hat{\boldsymbol{\theta}} = \arg\max_{\theta} p(\boldsymbol{y}|\boldsymbol{\theta}) \tag{11.1.12}$$

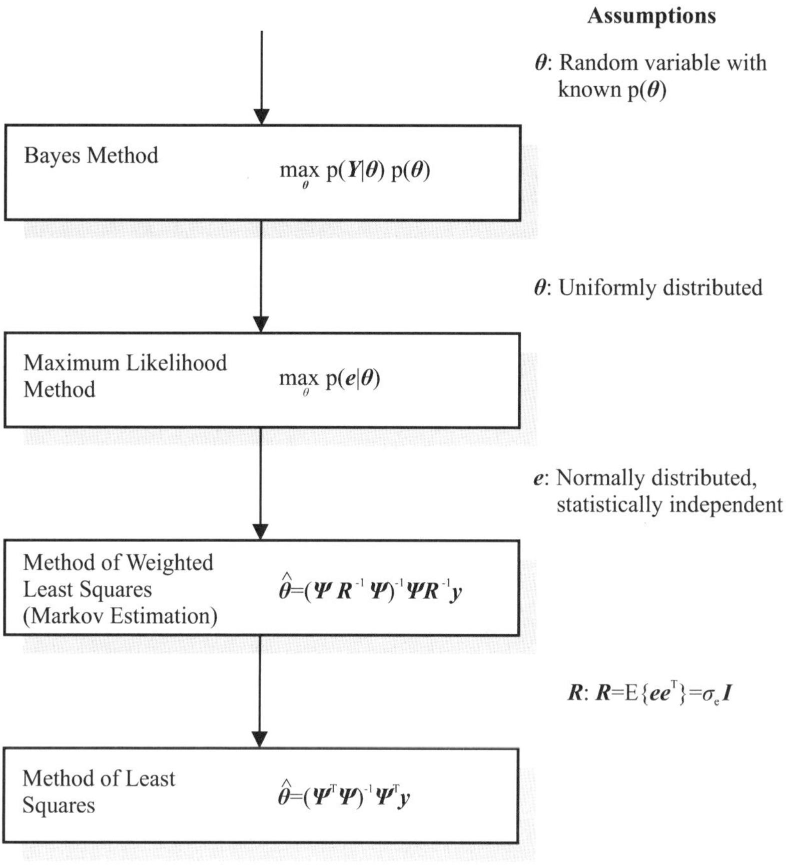

Fig. 11.1. Derivation of different parameter estimation methods from the Bayes method through specializing assumptions

results, which is the *maximum likelihood* estimate, that has been introduced in Sect. 8.5. In those cases, where the prior PDF has negligible influence on the estimation results, the maximum a posteriori estimation is also close to the maximum likelihood estimation (Ljung, 1999).

The main drawback is the fact that the Bayes estimation necessitates knowledge of the probability density function of the parameters $\boldsymbol{\theta}$ and that the conditional probability density function can only be established under a high mathematical burden. Hence, the Bayes estimation has little relevance for practical applications in the area of system identification. It can however be seen as the most comprehensive parameter estimation technique and it serves as a starting point for the development of many other algorithms, such as e.g. the maximum likelihood estimate, which is covered in the following section and can be seen as a specialization of the Bayes estimation.

The maximum likelihood method in turn can be brought into liaison with the least squares parameter estimation under certain assumptions about the noise (Isermann, 1992), see Fig. 11.1. For further reading, one can consult e.g. (Lee, 1964; Nahi, 1969; Eykhoff, 1974; Peterka, 1981; Ljung, 1999). The Bayes rule is also often applied in classification problems (e.g. Isermann, 2006) .

There number of publications about the application of Bayes method for parameter estimation is relatively sparse. This can mainly be attributed to the problems in the computational problems in determining the conditional probability density functions and the fact that the probability density functions of the parameters are typically unknown. Therefore, the Bayes estimation is mainly of theoretical value. It can be regarded as the most general and most comprehensive estimation method. Other fundamental estimation methods can be derived from this starting point by making certain assumptions or specializations.

This is depicted in Fig. 11.1. Upon the assumption of uniformly distributed parameters, i.e. $p(\boldsymbol{\theta}_0) = \text{const}$, then the Bayes estimation (11.1.11) becomes a maximum likelihood estimation (11.1.12). As will be shown in the following derivation of the maximum likelihood estimation for dynamic systems, one uses the equation error $\boldsymbol{e}$ instead of the measured signals $\boldsymbol{y}$ due to the easier treatment, i.e.

$$\hat{\boldsymbol{\theta}} = \arg\max_{\theta} p(\boldsymbol{e}|\boldsymbol{\theta}) \,. \tag{11.1.13}$$

If one furthermore assumes that the error $\boldsymbol{e}$ is statistically independent, Gaussian distributed with $\mathrm{E}\{\boldsymbol{e}\} = 0$, and has the error covariance matrix $\boldsymbol{R} = \mathrm{E}\{\boldsymbol{e}\,\boldsymbol{e}^{\mathrm{T}}\}$, then

$$p(\boldsymbol{e}|\boldsymbol{\theta}) = \frac{1}{(2\pi)^{N/2}(\det \boldsymbol{R})^{1/2}} \exp\left(-\frac{1}{2}\boldsymbol{e}^{\mathrm{T}}\boldsymbol{R}^{-1}\boldsymbol{e}\right) \tag{11.1.14}$$

results, which will be derived in Sect. 11.2 in a more detailed way. From there follows

$$\ln p(\boldsymbol{e}|\boldsymbol{\theta}) = -\frac{1}{2}\boldsymbol{e}^{\mathrm{T}}\boldsymbol{R}^{-1}\boldsymbol{e} + \text{const} \tag{11.1.15}$$

and

$$\frac{\partial}{\partial\boldsymbol{\theta}} = -\frac{\partial}{\partial\boldsymbol{\theta}}\boldsymbol{e}^{\mathrm{T}}\boldsymbol{R}^{-1}\boldsymbol{e} = \boldsymbol{0} \,. \tag{11.1.16}$$

Hence, one has to minimize the quadratic cost function (11.1.15), where the errors are weighted with the inverse of their covariance matrix. According to (9.5.6) and (9.5.4) this is the method of weighted least squares with minimal variance of the parameter estimates,

$$\hat{\boldsymbol{\theta}} = \left(\boldsymbol{\Psi}^{\mathrm{T}}\boldsymbol{R}^{-1}\boldsymbol{\Psi}\right)^{-1}\boldsymbol{\Psi}^{\mathrm{T}}\boldsymbol{R}^{-1}\boldsymbol{y} \,, \tag{11.1.17}$$

and therefore is a Markov estimator. For uncorrelated errors, one obtains

$$\boldsymbol{R} = \sigma_{\boldsymbol{e}}^2\boldsymbol{I} \,, \tag{11.1.18}$$

so that (11.1.17) results in the estimation equation for the method of least squares as

$$\hat{\boldsymbol{\theta}} = \left(\boldsymbol{\Psi}^{\mathrm{T}}\boldsymbol{\Psi}\right)^{-1}\boldsymbol{\Psi}^{\mathrm{T}}\boldsymbol{y} \,. \tag{11.1.19}$$

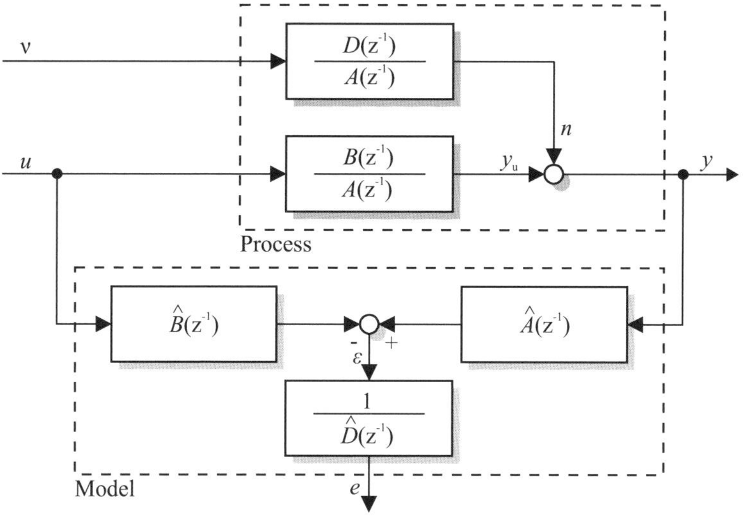

Fig. 11.2. Schematic diagram of the arrangement for the maximum likelihood method for dynamic systems

The maximum-likelihood method, which is described in Sect. 11.2, can be regarded as a method of least squares as it is assumed that $\boldsymbol{e}$ is uniformly distributed and statistically independent. The estimation equation must be solved iteratively due to the non-linear relation between $\boldsymbol{e}$ and the coefficients of the noise form filter polynomial $D(z^{-1})$.

11.2 Maximum Likelihood Method (ML)

In the following, the maximum likelihood estimator for linear dynamic discrete-time systems will first be formulated in a non-recursive formulation. It is then shown that under certain simplifying assumptions, it can also be formulated in a recursive fashion.

11.2.1 Non-Recursive Maximum Likelihood Method

While the maximum likelihood method has been introduced in Sect. 8.5 for static systems, it shall now be applied to linear dynamic systems in discrete-time. The process shall be governed by the model

$$A(z^{-1})y(z) - B(z^{-1})u(z) = D(z^{-1})e(z) \tag{11.2.1}$$

with

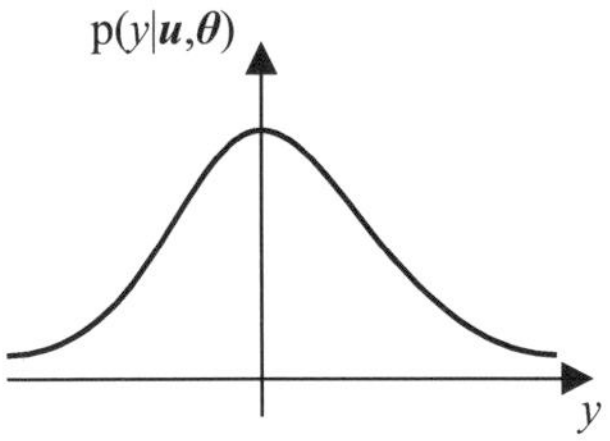

Fig. 11.3. Conditional probability density function of the observed signal $y(k)$

$$A(z^{-1}) = 1 + a_1 z^{-1} + \ldots + a_m z^{-m} \tag{11.2.2}$$
$$B(z^{-1}) = b_1 z^{-1} + \ldots + b_m z^{-m} \tag{11.2.3}$$
$$D(z^{-1}) = 1 + d_1 z^{-1} + \ldots + d_m z^{-m} \,, \tag{11.2.4}$$

where $e(k)$ shall be a Gaussian distributed, statistically independent signal with $(0, \sigma_e)$ and all roots of $D(z^{-1})$ shall lie within the unit circle. Compared to the method of least squares as introduced in Sect. 9.1, the model in (11.2.1) filters the equation error $\varepsilon(k)$ with the filter $1/\hat{D}(z^{-1})$, i.e.

$$\varepsilon(z) = \hat{D}(z^{-1}) e(z) \Leftrightarrow e(z) = \frac{1}{\hat{D}(z^{-1})} \varepsilon(z) \,. \tag{11.2.5}$$

The equation error $\varepsilon(k)$ is hence assumed to be a correlated signal, which by means of the filter is converted into an uncorrelated error $e(k)$, see Fig. 11.2.

In order to derive the maximum likelihood estimator (see also the development in Sect. 8.5), the probability density function of the measured, disturbed output has to be considered. In the following, it shall be assumed that the measured output $y(k)$ follows a Gaussian distribution, which allows to analytically treat the resulting equations.

The conditional probability density function of the observed signal samples $\{y(k)\}$ for a given input signal $\{u(k)\}$ and for given process parameters

$$\boldsymbol{\theta} = \begin{pmatrix} a_1 \ldots a_m \big| b_1 \ldots b_m \big| d_1 \ldots d_m \end{pmatrix} \tag{11.2.6}$$

shall be denoted as

$$p(\{y(k)\}|\{u(k)\}, \boldsymbol{\theta}) = p(\boldsymbol{y}|\boldsymbol{u}, \boldsymbol{\theta}) \,. \tag{11.2.7}$$

and shall be known, see Fig. 11.3. One can now insert the measured values $y_p(k)$ and $u_p(k)$ into the above equation. Then, one obtains the *likelihood function*

$$p(\boldsymbol{y}_{\mathrm{P}}|\boldsymbol{u}_{\mathrm{P}}, \boldsymbol{\theta}) \,, \tag{11.2.8}$$

which is analyzed in dependence of the unknown parameters θ_i, see Fig. 11.4.

As the parameters θ_i are constants and hence no stochastic variables, the likelihood function is not a probability density function of the parameters. The underlying principle of the maximum likelihood estimation is the idea that the best estimates for the unknown parameters θ_i are those values that attribute maximum possibility (or

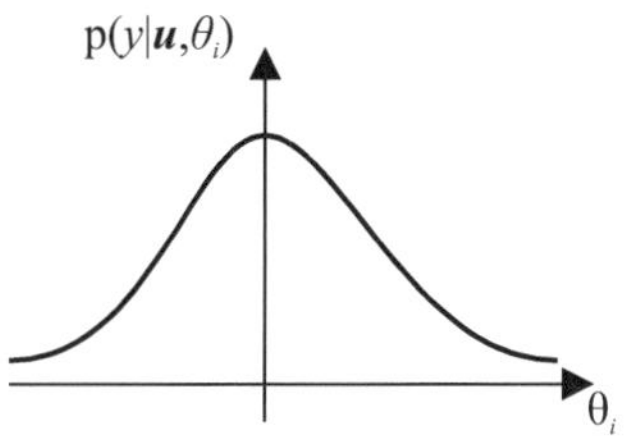

Fig. 11.4. Likelihood function for a single parameter θ_i

likelihood) to the observed results. Mathematically speaking, one is looking for those values of θ_i that maximize the likelihood function. Hence, the parameters $\boldsymbol{\theta}$ can be determined by locating the maximum of the likelihood function or correspondingly by taking the first derivative and equating it to zero

$$\frac{\partial}{\partial \boldsymbol{\theta}} p(\boldsymbol{y}|\boldsymbol{u},\boldsymbol{\theta})\bigg|_{\boldsymbol{\theta}=\hat{\boldsymbol{\theta}}} = \mathbf{0} \,. \tag{11.2.9}$$

Since the individual measurements $y(k)$ are not statistically independent, the probability density function is difficult to calculate. Hence, the following derivation will be based on the error $e(k)$ which is assumed to be Gaussian distributed and statistically independent. In this case, one can consider the likelihood function

$$p(\boldsymbol{e}|\boldsymbol{u},\boldsymbol{\theta}) \tag{11.2.10}$$

and can determine the estimates by

$$\frac{\partial}{\partial \boldsymbol{\theta}} p(\boldsymbol{e}|\boldsymbol{u},\boldsymbol{\theta})\bigg|_{\boldsymbol{\theta}=\hat{\boldsymbol{\theta}}} = \mathbf{0} \,. \tag{11.2.11}$$

Because $e(k)$ is assumed to be statistically independent, one can now write the probability density function $p(\boldsymbol{e}|\boldsymbol{u},\boldsymbol{\theta})$ as

$$p(\boldsymbol{e}|\boldsymbol{u},\boldsymbol{\theta}) = \prod_{k=1}^{N} p(e(k)|\boldsymbol{u},\boldsymbol{\theta}) \,. \tag{11.2.12}$$

As the individual errors $e(k)$ are assumed to be Gaussian distributed, it is beneficial to take the logarithm of the likelihood function as

$$\begin{aligned} L(\boldsymbol{\theta}) &= \ln\left(\left(\frac{1}{\sigma_e\sqrt{2\pi}}\right)^N \prod_{k=1}^{N} \mathrm{e}^{-\frac{1}{2}\frac{e^2(k)}{\sigma_e^2}}\right) \\ &= -\frac{1}{2\sigma_e^2}\sum_{k=1}^{N} e^2(k) - N\ln\sigma_e - \frac{N}{2}\ln 2\pi \,. \end{aligned} \tag{11.2.13}$$

One can see that maximizing the log-likelihood function (which should more precisely be called ln-likelihood function) with respect to the parameters $\boldsymbol{\theta}$ is the same as minimizing the sum of squared errors,

$$V = \sum_{k=1}^{N} e^2(k) \,. \tag{11.2.14}$$

Hence, for a Gaussian distributed error $e(k)$, the maximum likelihood and the least squares estimator yield identical results for the system structure as shown in Fig. 11.2.

The solution can only be determined iteratively, since the cost function is linear in the parameters of $A(z^{-1})$ and $B(z^{-1})$, but non-linear in the parameters $D(z^{-1})$. Åström and Bohlin (1965) employed a Newton-Raphson algorithm to solve this optimization problem. The first and second order derivatives will be denoted as

$$\boldsymbol{V}_{\boldsymbol{\theta}}^{\mathrm{T}}(\boldsymbol{\theta}) = \left(\frac{\partial V}{\partial \boldsymbol{\theta}}\right)^{\mathrm{T}} = \left(\frac{\partial V}{\partial \theta_1}\;\frac{\partial V}{\partial \theta_2}\;\cdots\;\frac{\partial V}{\partial \theta_p}\right) \tag{11.2.15}$$

with the Hesse matrix being

$$\boldsymbol{V}_{\boldsymbol{\theta\theta}}(\boldsymbol{\theta}) = \frac{\partial^2 V}{\partial \boldsymbol{\theta}^{\mathrm{T}} \partial \boldsymbol{\theta}} = \begin{pmatrix} \dfrac{\partial^2 V}{\partial \theta_1 \partial \theta_1} & \cdots & \dfrac{\partial^2 V}{\partial \theta_p \partial \theta_1} \\ \vdots & & \vdots \\ \dfrac{\partial^2 V}{\partial \theta_1 \partial \theta_p} & \cdots & \dfrac{\partial^2 V}{\partial \theta_p \partial \theta_p} \end{pmatrix} . \tag{11.2.16}$$

The corresponding partial derivatives can be summarized as

$$\frac{\partial V}{\partial \theta_i} = \sum_{k=1}^{N} e(k) \frac{\partial e(k)}{\partial \theta_i} \tag{11.2.17}$$

$$\frac{\partial^2 V}{\partial \theta_i \partial \theta_j} = \sum_{k=1}^{N} \frac{\partial e(k)}{\partial \theta_i} \frac{\partial e(k)}{\partial \theta_j} + \sum_{k=1}^{N} e(k) \frac{\partial^2 e(k)}{\partial \theta_i \partial \theta_j} \,. \tag{11.2.18}$$

One therefore needs the partial derivatives of the error $e(k)$ with respect to the individual parameters, which can be provided as follows

$$D(q^{-1}) \frac{\partial e(k)}{\partial a_i} = y(k) q^{-i} \tag{11.2.19}$$

$$D(q^{-1}) \frac{\partial e(k)}{\partial b_i} = -u(k) q^{-i} \tag{11.2.20}$$

$$D(q^{-1}) \frac{\partial e(k)}{\partial d_i} = -e(k) q^{-i} \tag{11.2.21}$$

$$D(q^{-1}) \frac{\partial^2 e(k)}{\partial a_i \partial d_j} = -q^{-j} \frac{\partial e(k)}{\partial a_i} = -q^{-i-j+1} \frac{\partial e(k)}{\partial a_1} \tag{11.2.22}$$

$$D(q^{-1}) \frac{\partial^2 e(k)}{\partial b_i \partial d_j} = -q^{-j} \frac{\partial e(k)}{\partial b_i} = -q^{-i-j+1} \frac{\partial e(k)}{\partial b_1} \tag{11.2.23}$$

$$D(q^{-1}) \frac{\partial^2 e(k)}{\partial d_i \partial d_j} = -2q^{-j} \frac{\partial e(k)}{\partial d_i} = -2q^{-i-j+1} \frac{\partial e(k)}{\partial d_1} \,, \tag{11.2.24}$$

where the time shift operator q has been introduced and defined as

$$y(k)q^{-l} = y(k-l) \tag{11.2.25}$$

Furthermore

$$D(q^{-1})\frac{\partial^2 e(k)}{\partial a_i \partial a_j} = 0 \tag{11.2.26}$$

$$D(q^{-1})\frac{\partial^2 e(k)}{\partial a_i \partial b_j} = 0 \tag{11.2.27}$$

$$D(q^{-1})\frac{\partial^2 e(k)}{\partial b_i \partial b_j} = 0\,. \tag{11.2.28}$$

Since the update equation for the optimization algorithm is given as

$$\begin{aligned}\boldsymbol{\theta}(k+1) &= \boldsymbol{\theta}(k) - \Big(\frac{\partial^2 V}{\partial \boldsymbol{\theta}^{\mathrm{T}} \partial \boldsymbol{\theta}}\Big)^{-1}\Big|_{\boldsymbol{\theta}=\boldsymbol{\theta}(k)} \Big(\frac{\partial V}{\partial \boldsymbol{\theta}}\Big)\Big|_{\boldsymbol{\theta}=\boldsymbol{\theta}(k)} \\ &= \boldsymbol{\theta}(k) - \boldsymbol{V}_{\boldsymbol{\theta}\boldsymbol{\theta}}\big(\boldsymbol{\theta}(k)\big)^{-1} \boldsymbol{V}_{\boldsymbol{\theta}}(\boldsymbol{\theta}(k))\,,\end{aligned} \tag{11.2.29}$$

the term $D(q^{-1})$ in this case cancels out.

A prerequisite for the convergence of the maximum likelihood estimate are appropriate initial values. It is suggested to set $D(z^{-1}) = 1$, i.e. $d_i = 0$, in the first iteration, which leads to the normal method of least squares and allows to obtain (biased) initial values by the direct solution of the least squares problem.

Theorem 11.1 (Convergence of the Maximum Likelihood Estimate).

The Maximum Likelihood estimator delivers for an ARMAX process as depicted in Fig. 11.2 consistent asymptotically efficient parameter estimates, that fulfill the Cramér-Rao bound (Åström and Bohlin, 1965; van der Waerden, 1969; Deutsch, 1965), if the following conditions are met:

- $u(k) = U(k) - U_{00}$ *is exactly known*
- Y_{00} *is exactly known and belongs to* U_{00}
- *The elements of* $e(k)$ *are statistically independent and Gaussian distributed*
- *The roots of* $D(z) = 0$ *lie within the unit circle*
- *Appropriate initial values* $\hat{\boldsymbol{\theta}}(0)$ *are known*

□

The described method should also converge for many other noise distributions, but will in most cases not be asymptotically efficient any longer.

The maximum likelihood estimation for dynamic systems has also been outlined in (Raol et al, 2004). Here, the maximum likelihood estimation is applied to the output error model, partial derivatives of the cost function with respect to the parameters have been determined by finite differencing and a corresponding perturbation of the parameters. The monograph by van den Bos (2007) also discusses the maximum likelihood estimation in combination with non-linear optimization algorithms. The maximum likelihood estimation can according to Ljung (1999) also be interpreted as a maximum entropy or minimum information distance estimate.

11.2.2 Recursive Maximum Likelihood Method (RML)

The recursive Maximum Likelihood method can be derived by an approximation of the partial derivatives of the non-recursive method (Söderström, 1973; Fuhrt and Carapic, 1975). For the derivation, the process model in (11.2.1) is first expressed as

$$y(k) = \boldsymbol{\psi}^{\mathrm{T}}(k)\boldsymbol{\theta} + \nu(k) \tag{11.2.30}$$

with

$$\begin{aligned}\boldsymbol{\psi}^{\mathrm{T}}(k) = \big(&-y(k-1)\ \dots\ -y(k-m)\big|u(k-d-1)\ \dots\ u(k-d-m)\big|\\ &\nu(k-1)\ \dots\ \nu(k-m)\big)\end{aligned} \tag{11.2.31}$$

and

$$\boldsymbol{\theta}^{\mathrm{T}} = \big(a_1\ \dots\ a_m\big|b_1\ \dots\ b_m\big|d_1\ \dots\ d_m\big)\,. \tag{11.2.32}$$

The cost function is given as

$$V(k+1,\hat{\boldsymbol{\theta}}) = V(k,\hat{\boldsymbol{\theta}}) + \frac{1}{2}e^2(k+1,\hat{\boldsymbol{\theta}})\,. \tag{11.2.33}$$

The first and second partial derivative are then given as

$$\boldsymbol{V}_{\boldsymbol{\theta}}(\hat{\boldsymbol{\theta}},k+1) = \underbrace{\boldsymbol{V}_{\boldsymbol{\theta}}(\hat{\boldsymbol{\theta}},k)}_{\approx\boldsymbol{0}} + e(\hat{\boldsymbol{\theta}},k+1)\frac{\partial e(\boldsymbol{\theta},k+1)}{\partial\boldsymbol{\theta}}\Big|_{\boldsymbol{\theta}=\hat{\boldsymbol{\theta}}} \tag{11.2.34}$$

and

$$\begin{aligned}\boldsymbol{V}_{\boldsymbol{\theta\theta}}(\hat{\boldsymbol{\theta}},k+1) = \boldsymbol{V}_{\boldsymbol{\theta\theta}}(\hat{\boldsymbol{\theta}},k) &+ \Big(\frac{\partial e(\boldsymbol{\theta},k+1)}{\partial\boldsymbol{\theta}}\Big)^{\mathrm{T}}\Big|_{\boldsymbol{\theta}=\hat{\boldsymbol{\theta}}}\Big(\frac{\partial e(\boldsymbol{\theta},k+1)}{\partial\boldsymbol{\theta}}\Big)\Big|_{\boldsymbol{\theta}=\hat{\boldsymbol{\theta}}}\\ &\underbrace{+\, e(\hat{\boldsymbol{\theta}},k+1)\Big(\frac{\partial^2 e(\boldsymbol{\theta},k+1)}{\partial\boldsymbol{\theta}^2}\Big)\Big|_{\boldsymbol{\theta}=\hat{\boldsymbol{\theta}}}}_{\approx\boldsymbol{0}}\,,\end{aligned} \tag{11.2.35}$$

where the indicated terms have been approximated by zero (Söderström, 1973). These equations now allow to formulate the estimation algorithm as

$$\hat{\boldsymbol{\theta}}(k+1) = \hat{\boldsymbol{\theta}}(k) + \boldsymbol{\gamma}(k)e(k+1) \tag{11.2.36}$$

with

$$\boldsymbol{\gamma}(k) = \boldsymbol{P}(k+1)\boldsymbol{\varphi}(k+1) = \frac{\boldsymbol{P}(k)\boldsymbol{\varphi}(k+1)}{1+\boldsymbol{\varphi}^{\mathrm{T}}(k+1)\boldsymbol{P}(k)\boldsymbol{\varphi}(k+1)} \tag{11.2.37}$$

$$\boldsymbol{P}(k) = \boldsymbol{V}_{\boldsymbol{\theta\theta}}^{-1}(\hat{\boldsymbol{\theta}}(k-1),k) \tag{11.2.38}$$

$$\boldsymbol{P}(k+1) = \big(\boldsymbol{I} - \boldsymbol{\gamma}(k)\boldsymbol{\varphi}^{\mathrm{T}}(k+1)\big)\boldsymbol{P}(k) \tag{11.2.39}$$

$$\boldsymbol{\varphi}(k+1) = -\frac{\partial e(\boldsymbol{\theta}(k),k+1)}{\partial\boldsymbol{\theta}}\Big|_{\boldsymbol{\theta}=\hat{\boldsymbol{\theta}}} \tag{11.2.40}$$

$$e(k+1) = y(k+1) - \hat{\boldsymbol{\psi}}^{\mathrm{T}}(k+1)\hat{\boldsymbol{\theta}}(k) \tag{11.2.41}$$

$$\hat{\nu}(k+1) = \hat{e}(k+1) \tag{11.2.42}$$

and hence one approximates $\boldsymbol{\psi}^{\mathrm{T}}$ from (11.2.31) by

$$\hat{\boldsymbol{\psi}}^{\mathrm{T}}(k+1) = \big(-y(k-1)\ \dots\ -y(k-m)\big|u(k-d-1)\ \dots\ u(k-d-m)\big| \\ e(k-1)\ \dots\ e(k-m)\big)\,. \tag{11.2.43}$$

The elements of the vector $\boldsymbol{\varphi}^{\mathrm{T}}(k+1)$ can now be determined as

$$\boldsymbol{\varphi}^{\mathrm{T}}(k+1) = -\bigg(\frac{\partial e(k+1)}{\partial a_1}\ \dots\ \frac{\partial e(k+1)}{\partial a_m}\ \frac{\partial e(k+1)}{\partial b_1}\ \dots\ \frac{\partial e(k+1)}{\partial b_m} \\ \frac{\partial e(k+1)}{\partial d_1}\ \dots\ \frac{\partial e(k+1)}{\partial d_m}\bigg) \tag{11.2.44}$$

with $e(k) = \hat{v}(k)$ and (11.2.1) are given as

$$z\frac{\partial e(z)}{\partial a_i} = \frac{1}{\hat{D}(z^{-1})}y(z)z^{-(i-1)} = y'(z)z^{-(i-1)} \tag{11.2.45}$$

$$z\frac{\partial e(z)}{\partial b_i} = -\frac{1}{\hat{D}(z^{-1})}u(z)z^{-(i-1)}z^{-d} = -u'(z)z^{-(i-1)}z^{-d} \tag{11.2.46}$$

$$z\frac{\partial e(z)}{\partial d_i} = -\frac{1}{\hat{D}(z^{-1})}e(z)z^{-(i-1)} = -e'(z)z^{-(i-1)} \tag{11.2.47}$$

for $i = 1, \dots, m$. These entries can be understood as filtered signals

$$\hat{\boldsymbol{\varphi}}^{\mathrm{T}}(k+1) = \big(-y'(k-1)\ \dots\ -y'(k-m)\big|u'(k-d-1)\ \dots \\ u'(k-d-m)\big|e'(k-1)\ \dots\ e'(k-m)\big) \tag{11.2.48}$$

which can be generated by the difference equation

$$y'(k) = y(k) - \hat{d}_1 y'(k-1) - \dots - \hat{d}_m y'(k-m) \tag{11.2.49}$$

$$u'(k-d) = u(k-d) - \hat{d}_1 u'(k-d-1) - \dots - \hat{d}_m u'(k-d-m) \tag{11.2.50}$$

$$e'(k) = e(k) - \hat{d}_1 e'(k-1) - \dots - \hat{d}_m e'(k-m)\,. \tag{11.2.51}$$

For the $\hat{d}_i$, one can use the current estimates $\hat{d}_i(k)$. Due to the simplifying approximations at the beginning of the derivation, one will only obtain an approximation of the solution of the non-recursive maximum likelihood method.

As initial values, one can use

$$\hat{\boldsymbol{\theta}}(0) = \mathbf{0},\ \ \boldsymbol{P}(0) = \alpha\boldsymbol{I},\ \ \boldsymbol{\varphi}(0) = \mathbf{0}\,. \tag{11.2.52}$$

The convergence criteria are identical to those of the non-recursive maximum likelihood estimation. In particular, the roots of $D(z) = 0$ must be within the unit circle, so that (11.2.49), (11.2.50), and (11.2.51) are stable.

11.2.3 Cramér-Rao Bound and Maximum Precision

The Cramér-Rao bound (Eykhoff, 1974), see (8.5.14), can also be evaluated for the maximum likelihood estimation of linear dynamic systems. In the case of multiple parameters, the Cramér-Rao bound is given as

$$\operatorname{cov} \Delta\hat{\boldsymbol{\theta}} = \mathrm{E}\big\{(\hat{\boldsymbol{\theta}} - \boldsymbol{\theta}_0)(\hat{\boldsymbol{\theta}} - \boldsymbol{\theta}_0)^{\mathrm{T}}\big\} \geq \boldsymbol{J}^{-1} \tag{11.2.53}$$

with the *information matrix*

$$\boldsymbol{J} = \mathrm{E}\left\{\left(\frac{\partial L}{\partial \boldsymbol{\theta}_0}\right)\left(\frac{\partial L}{\partial \boldsymbol{\theta}_0}\right)^{\mathrm{T}}\right\} = -\mathrm{E}\left\{\frac{\partial^2 L}{\partial \boldsymbol{\theta}_0 \partial \boldsymbol{\theta}_0^{\mathrm{T}}}\right\}. \tag{11.2.54}$$

Here, $\boldsymbol{\theta}_0$ denotes the true parameters. For a Gaussian distributed error $e(k)$, one can equate

$$\frac{\partial L}{\partial \boldsymbol{\theta}_0} = -\frac{1}{\sigma_{\mathrm{e}}^2}\frac{\partial V}{\partial \boldsymbol{\theta}_0} \tag{11.2.55}$$

and hence

$$\boldsymbol{J} = \frac{1}{\sigma_{\mathrm{e}}^4}\mathrm{E}\left\{\left(\frac{\partial V}{\partial \boldsymbol{\theta}_0}\right)\left(\frac{\partial V}{\partial \boldsymbol{\theta}_0}\right)^{\mathrm{T}}\right\} = \frac{1}{\sigma_{\mathrm{e}}^2}\mathrm{E}\left\{\frac{\partial^2 L}{\partial \boldsymbol{\theta}_0 \partial \boldsymbol{\theta}_0^{\mathrm{T}}}\right\}. \tag{11.2.56}$$

From this follows for the covariance of the parameter estimates

$$\operatorname{cov} \Delta\hat{\boldsymbol{\theta}} \geq \frac{2V}{N}\mathrm{E}\big\{V_{\boldsymbol{\theta}\boldsymbol{\theta}}^{-1}\big\}. \tag{11.2.57}$$

This result shows that under the given assumptions, there is no other unbiased estimator that delivers estimates with a smaller variance than the maximum likelihood estimator. The maximum likelihood estimate is hence *asymptotically efficient.*

If the Cramér-Rao bound is applied to the fundamental equation of the least squares parameter estimation, (9.1.12),

$$\boldsymbol{y} = \boldsymbol{\Psi}\hat{\boldsymbol{\theta}} + \boldsymbol{e}, \tag{11.2.58}$$

then the ln-likelihood function is given as

$$L(\boldsymbol{\theta}) = -\frac{1}{2\sigma_{\mathrm{e}}^2}\boldsymbol{e}^{\mathrm{T}}\boldsymbol{e} + \text{const} \tag{11.2.59}$$

and the information matrix is given as

$$J = \frac{1}{\sigma_{\mathrm{e}}^2}\mathrm{E}\big\{\boldsymbol{\Psi}^{\mathrm{T}}\boldsymbol{\Psi}\big\}, \tag{11.2.60}$$

and hence, compare (9.1.24),

$$\operatorname{cov} \Delta\hat{\boldsymbol{\theta}} \geq \sigma_{\mathrm{e}}^2\mathrm{E}\Big\{\big(\boldsymbol{\Psi}^{\mathrm{T}}\boldsymbol{\Psi}\big)^{-1}\Big\}. \tag{11.2.61}$$

The lower bound is thus identical with (9.1.69). A further comparison with (9.5.7) shows that for the case of a non-correlated error signal and a model according to (9.1.12) or (11.2.58) respectively, the estimation by means of the method of least squares, by the Markov estimation and by the maximum likelihood method all yield parameter estimates with the smallest possible variance. A comparison of the Cramér-Rao bound with simulation results by van den Boom (1982) shows a good match for the best parameter estimation methods. Ninness (2009) discussed the error quantification also for finite (and especially short length) data sequences.

11.3 Summary

This chapter has presented the Bayes estimator and the maximum likelihood estimator, which were now specifically tailored to the identification of linear dynamic systems in discrete-time. The Bayes estimator treats the parameters as random variables and incorporates information about their probability density functions into the solution of the parameter estimation problem. As this information however is seldom available in practical applications, the Bayes estimator is limited in its applicability for parameter estimation from experimental data. Still, it can be shown that the maximum likelihood estimator and the least squares estimator can both be derived from the Bayes estimator, see Fig. 11.1.

The maximum likelihood estimation is based on a stochastic treatment of the measured signals. The parameter estimates are determined based on the probability density function of the observed measurements. For an ARMAX model structure and a normally distributed, statistically independent error signal, a maximum likelihood estimation technique has been derived for linear dynamic discrete-time systems, which can be solved by a non-linear optimization algorithm. After certain simplifying approximations, also a recursive maximum likelihood estimator could be formulated. While the computational burden for the maximum likelihood estimator is high, it can on the other hand be shown that the estimator is asymptotically efficient, i.e. that it reaches the Cramér-Rao bound and yields estimates with the smallest possible variance for specified conditions. Maximum likelihood estimators can also be formulated for many other settings, e.g. for frequency domain identification (McKelvey, 2000).

Problems

11.1. Bayes Estimation
How can the Bayes rule be used to determine the conditional probability density function of the parameters for a given set of observed measurements $p(\boldsymbol{\theta}\,|\,\boldsymbol{y})$?

11.2. Bayes Estimation, Maximum Likelihood, and Least Squares
How do these parameter estimation methods relate to each other. Which assumptions lead from one estimator to the other?

11.3. Cramér-Rao Lower Bound
Derive the Cramér-Rao inequality for one parameter θ_0. (Solution can be found in (, p. 14 Isermann, 1992)

References

Åström KJ, Bohlin T (1965) Numerical identification of linear dynamic systems from normal operating records. In: Proceedings of the IFAC Symposium Theory of Self-Adaptive Control Systems, Teddington

van den Boom AJW (1982) System identification - on the variety and coherence in parameter- and order erstimation methods. Ph. D. thesis. TH Eindhoven, Eindhoven

van den Bos A (2007) Parameter estimation for scientists and engineers. Wiley-Interscience, Hoboken, NJ

Deutsch R (1965) Estimation theory. Prentice-Hall, Englewood Cliffs, NJ

Eykhoff P (1974) System identification: Parameter and state estimation. Wiley-Interscience, London

Fuhrt BP, Carapic M (1975) On-line maximum likelihood algorithm for the identification of dynamic systems. In: 4th IFAC-Symposium on Identification, Tbilisi, USSR

Isermann R (1992) Identifikation dynamischer Systeme: Besondere Methoden, Anwendungen (Vol 2). Springer, Berlin

Isermann R (2006) Fault-diagnosis systems: An introduction from fault detection to fault tolerance. Springer, Berlin

Lee KI (1964) Optimal estimation, identification, and control, Massachusetts Institute of Technology research monographs, vol 28. MIT Press, Cambridge, MA

Ljung L (1999) System identification: Theory for the user, 2nd edn. Prentice Hall Information and System Sciences Series, Prentice Hall PTR, Upper Saddle River, NJ

McKelvey T (2000) Frequency domain identification. In: Proccedings of the 12th IFAC Symposium on System Identification, Santa Barbara, CA, USA

Nahi NE (1969) Estimation theory and applications. J. Wiley, New York, NY

Ninness B (2009) Some system identification challenges and approaches. In: Proceedings of the 15th IFAC Symposium on System Identification, Saint-Malo, Francc

Papoulis A (1962) The Fourier integral and its applications. McGraw Hill, New York

Peterka V (1981) Bayesian approach to system identification. In: Trends and progress in system identification, Pergamon Press, Oxford

Raol JR, Girija G, Singh J (2004) Modelling and parameter estimation of dynamic systems, IEE control engineering series, vol 65. Institution of Electrical Engineers, London

Söderström T (1973) An on-line algorithm for approximate maximum likelihood identification of linear dynamic systems. Report 7308. Dept. of Automatic Control, Lund Inst of Technology, Lund

van der Waerden BL (1969) Mathematical statistics. Springer, Berlin

12

Parameter Estimation for Time-Variant Processes

For many real processes, the parameters of the governing linear difference equations are not constant. They rather vary over time due to internal or external influences. Also, quite often non-linear processes can only be linearized in a small interval around the current operating point. If the operating point changes, also the linearized dynamics will change in this case. For slow changes of the operating point, one can obtain good results with linear difference equations that contain time-varying parameters. The method of recursive least squares (see Chap. 9) can also be used to identify time-varying parameters. Different methods are introduced in the following that allow to track the changes of time varying parameters with the method of least squares.

12.1 Exponential Forgetting with Constant Forgetting Factor

In connection with the method of weighted least squares, a technique was suggested in Sect. 9.6 which allowed the identification of slowly time-varying processes by choosing the weights $w(k)$ as

$$w(k) = \lambda^{N'-k} . \tag{12.1.1}$$

This particular way of choosing $w(k)$ to rate the error is termed *exponential forgetting*.

The recursive estimation equations (9.6.11), (9.6.12), and (9.6.13) for the method of weighted least squares with exponential forgetting had been given as

$$\hat{\boldsymbol{\theta}}(k+1) = \hat{\boldsymbol{\theta}}(k) + \boldsymbol{\gamma}(k)\big(y(k+1) - \boldsymbol{\psi}^{\mathrm{T}}(k+1)\hat{\boldsymbol{\theta}}(k)\big) \tag{12.1.2}$$

$$\boldsymbol{\gamma}(k) = \frac{1}{\boldsymbol{\psi}^{\mathrm{T}}(k+1)\boldsymbol{P}(k)\boldsymbol{\psi}(k+1) + \lambda}\boldsymbol{P}(k)\boldsymbol{\psi}(k+1) \tag{12.1.3}$$

$$\boldsymbol{P}(k+1) = \big(\boldsymbol{I} - \boldsymbol{\gamma}(k)\boldsymbol{\psi}^{\mathrm{T}}(k+1)\big)\boldsymbol{P}(k)\frac{1}{\lambda} . \tag{12.1.4}$$

R. Isermann, M. Münchhof, *Identification of Dynamic Systems*,
DOI 10.1007/978-3-540-78879-9_12, © Springer-Verlag Berlin Heidelberg 2011

The influence of the forgetting factor λ can be recognized directly from the inverse of the covariance matrix (9.6.6)

$$\boldsymbol{P}^{-1}(k+1) = \lambda \boldsymbol{P}^{-1}(k) + \boldsymbol{\psi}(k+1)\boldsymbol{\psi}^{\mathrm{T}}(k+1)\,. \tag{12.1.5}$$

$\boldsymbol{P}^{-1}$ is proportional to the information matrix $\boldsymbol{J}$ (11.2.60) given by

$$\boldsymbol{J} = \frac{1}{\sigma_e^2}\mathrm{E}\{\boldsymbol{\Psi}^{\mathrm{T}}\boldsymbol{\Psi}\} = \frac{1}{\sigma_e^2}\mathrm{E}\{\boldsymbol{P}^{-1}\}\,, \tag{12.1.6}$$

see (Eykhoff, 1974; Isermann, 1992).

By taking $\lambda < 1$, the information of the last step is reduced or the covariances are increased respectively. This means a worse quality of the estimates is pretended, such that the new measurements get more weight.

For $\lambda = 1$, one obtains

$$\lim_{k\to\infty} \mathrm{E}\{\boldsymbol{P}(k)\} = \boldsymbol{0} \tag{12.1.7}$$

$$\lim_{k\to\infty} \mathrm{E}\{\boldsymbol{\gamma}(k)\} = \lim_{k\to\infty} \mathrm{E}\{\boldsymbol{P}(k+1)\boldsymbol{\psi}(k+1)\} = \boldsymbol{0}\,. \tag{12.1.8}$$

For large times k, the measurements have practically no influence on $\hat{\boldsymbol{\theta}}(k+1)$. Then the elements of $\boldsymbol{P}^{-1}(k+1)$ tend to infinity, (12.1.5).

If, however, one uses a forgetting factor $\lambda < 1$, then, from (12.1.5), follows

$$\boldsymbol{P}^{-1}(k) = \lambda^k \boldsymbol{P}^{-1}(0) + \sum_{i=0}^{k} \lambda^{k-i}\boldsymbol{\psi}(i)\boldsymbol{\psi}^{\mathrm{T}}(i)\,. \tag{12.1.9}$$

For large values α of the initial matrix $\boldsymbol{P}(0) = \alpha\boldsymbol{I}$, the first term in (12.1.9) vanishes. As for $\lambda < 1$

$$\lim_{k\to\infty} \sum_{i=1}^{k} \lambda^{k-i} = \lim_{k\to\infty} \sum_{i=0}^{k-1} \lambda^{i} < \infty \tag{12.1.10}$$

(convergent series with positive elements) and hence $\boldsymbol{P}^{-1}(k)$ converges to fixed values

$$\lim_{k\to\infty} \mathrm{E}\{\boldsymbol{P}^{-1}(k)\} = \boldsymbol{P}^{-1}(\infty) \tag{12.1.11}$$

and does not approach infinity. Hence,

$$\lim_{k\to\infty} \mathrm{E}\{\boldsymbol{P}(k)\} = \boldsymbol{P}(\infty) \tag{12.1.12}$$

as well as

$$\lim_{k\to\infty} \mathrm{E}\{\boldsymbol{\gamma}(k)\} = \boldsymbol{\gamma}(\infty) \tag{12.1.13}$$

are finite and nonzero. Therefore, the new measurements get a constant weight for large k and the estimator remains sensible to parameter changes and can follow slow changes of the process. This is in contrast to the case $\lambda = 1$, where the weight or influence of new measurements gets smaller and smaller as k increases. Because of the smaller effective averaging time for the case of exponential forgetting, the noise influence increases and also the variances.

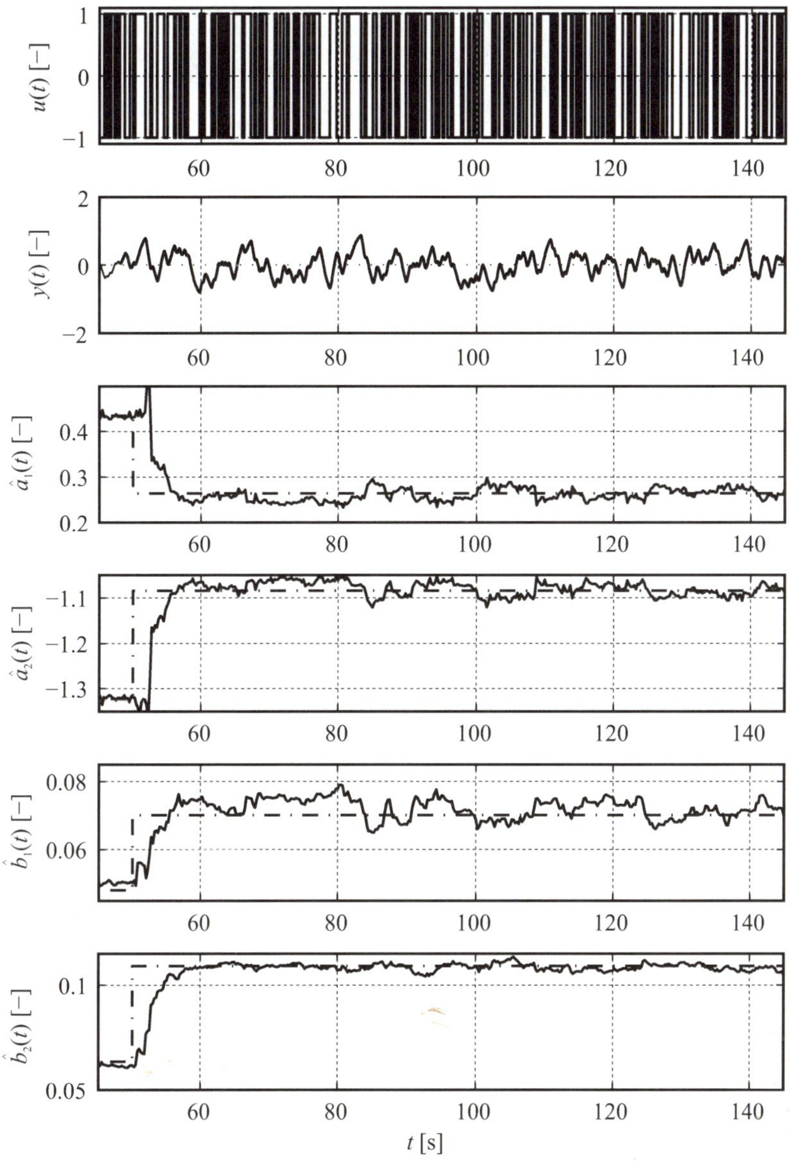

Fig. 12.1. Parameter estimation for a second order system with a change in the system parameters, $\lambda = 0.9$. True parameter values (*dash-dotted line*)

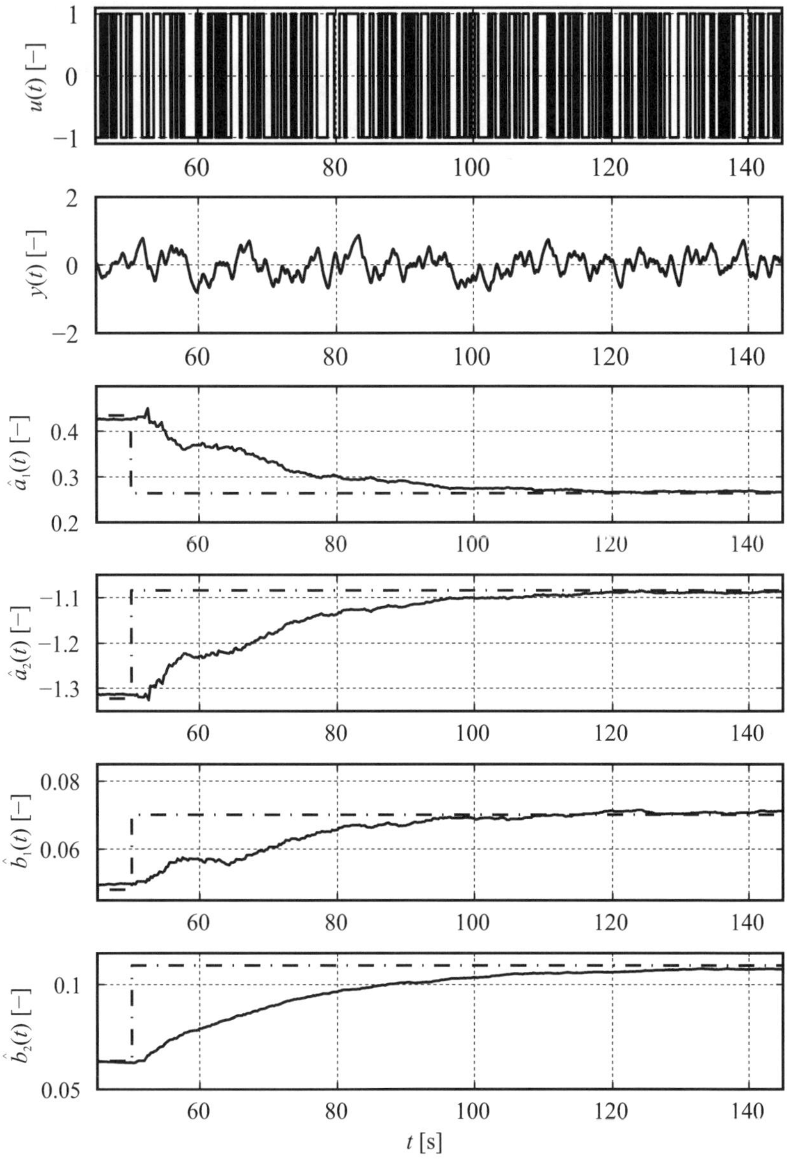

Fig. 12.2. Parameter estimation for a second order system with a change in the system parameters, $\lambda = 0.99$. True parameter values (*dash-dotted line*)

Example 12.1 (Parameter Estimation with a Constant Forgetting Factor).

The system is given as a second order system governed by

$$G(s) = \frac{K}{(1 + T_1 s)(1 + T_2 s)} \tag{12.1.14}$$

with $K = 1$, $T_1 = 0.75\,\text{s}$ and $T_{2,0} = 0.5\,\text{s}$. At time $t = 50\,\text{s}$, the system parameters are changed and the second time constant becomes $T_{2,1} = 0.25\,\text{s}$. The system is excited by a PRBS signal with amplitude $a = 1$ and cycle time $T = 0.25\,\text{s}$. The system has been discretized with a sample time of $T_0 = 0.25\,\text{s}$ and the discrete-time transfer functions have been determined as

$$G_0(z) = \frac{0.06347z + 0.04807}{z^2 - 1.323z + 0.4346} \tag{12.1.15}$$

and

$$G_1(z) = \frac{0.1091z + 0.07004}{z^2 - 1.084z + 0.2636} \,. \tag{12.1.16}$$

A Gaussian white noise has been superpositioned with $(0, 0.0045)$.

Now, the system has been identified first with $\lambda = 0.9$, see Fig. 12.1, and then with $\lambda = 0.99$, see Fig. 12.2. One can see that for smaller values of λ, the changes of the system parameters are tracked faster, but at the same time, the system parameters have a larger variance. □

The forgetting factor λ has to be selected as follows:

- λ small, if the speed of parameter changes is large (say $\lambda = 0.90$). Then only small noise is allowed
- λ large, if the speed of parameter changes is small (say $\lambda = 0.98$). Then the noise can be larger

Goodwin and Sin (1984) suggested to introduce a dead zone and hence only update the parameter estimation if the correction vector exceeds a certain threshold. This effectively cancels the small parameter variations that otherwise occur if the forgetting factor is chosen small in order to be able to track time-variant processes sufficiently fast.

As the RML and RELS methods converge more slowly during the starting phase, because of the unknown $e(k) = \hat{\nu}(k)$, the convergence can be accelerated by smaller weights at the beginning.

Söderström et al (1974) suggest the following way of choosing a time-variant forgetting factor

$$\lambda(k) = \lambda_0 \lambda(k-1) + 1 - \lambda_0 \tag{12.1.17}$$

with $\lambda_0 < 1$ and $\lambda(0) < 1$. Mikleš and Fikar (2007) proposed to choose the initial conditions as

$$\lambda(0) = \lambda_0 = 0.95 \ldots 0.99 \,. \tag{12.1.18}$$

The forgetting factor asymptotically goes to 1 and hence only initial data are forgotten over time.

One can also combine this start-up technique with the exponential forgetting by varying λ as

$$\lambda(k+1) = \lambda_0\lambda(k) + \lambda(1-\lambda_0) . \tag{12.1.19}$$

According to the parameters λ_0 and λ, one can achieve a smaller weighting of the error during the starting phase, and then, for large k, obtain the classical exponential forgetting as

$$\lim_{k\to\infty} \lambda(k+1) = \lambda . \tag{12.1.20}$$

Parameter estimation algorithms with *constant forgetting factor* are suited for processes with small parameter changes and persistent input excitation. Also, if the process parameters are constant, good results are obtainable if the noise with regard to the memory length $M = 1/(1-\lambda)$ is not too large. However, problems may arise if in the case of a constant forgetting factor $\lambda < 1$, the input is not sufficiently exciting. Then the values $\boldsymbol{P}^{-1}(k+1)$ decrease because $\boldsymbol{\psi}(k+1) \approx \boldsymbol{0}$, see (12.1.5), or the elements of $\boldsymbol{P}(k+1)$ increase continuously (covariance matrix blows up). As the correcting vector is

$$\boldsymbol{\gamma}(k) = \boldsymbol{P}(k+1)\boldsymbol{\psi}(k+1) , \tag{12.1.21}$$

the estimator becomes more and more sensitive. Then a small disturbance or a numerical error may suffice to generate sudden large changes of the parameter estimates. The estimator then becomes unstable. This situation can be observed with adaptive control systems. Therefore, the input excitation has to be monitored or the forgetting factor has to be time-variant.

12.2 Exponential Forgetting with Variable Forgetting Factor

In order to match the forgetting factor λ to the current situation, one can control the forgetting factor λ as a function of the estimation quality, e.g. by watching the a posteriori error. If $e_0(k)$ is small, then the estimation either is in good accordance with the process or the the process is not excited. In both cases, one should choose $\lambda(k) \approx 1$. On the other hand, if the error is large, then one should reduce $\lambda(k)$ to allow for fast changes of the model coefficients to track the process behavior.

One could also use the weighted sum of the a posteriori error (Fortescue et al, 1981), which is given as

$$\Sigma(k) = \lambda(k)\Sigma(k-1) + \big(1 - \boldsymbol{\psi}^{\mathrm{T}}(k)\boldsymbol{\gamma}(k-1)\big)e^2(k) , \tag{12.2.1}$$

where $\lambda(k)$ is now chosen such that the weighted sum of the a posteriori error remains constant $\Sigma(k) = \Sigma(k-1) = \Sigma_0$ and hence $\lambda(k) =$ is given as

$$\lambda(k) = 1 - \frac{1}{\Sigma_0}\big(1 - \boldsymbol{\psi}^{\mathrm{T}}(k)\boldsymbol{\gamma}(k-1)\big)e^2(k) . \tag{12.2.2}$$

Isermann et al (1992) discussed the choice of Σ_0. There, it is proposed to choose

$$\Sigma_0 = \sigma_n^2 N_0 \tag{12.2.3}$$

with σ_n^2 being the variance of the noise and

$$N_0 = \frac{1}{1 - \lambda_0} \,. \tag{12.2.4}$$

A small value of N_0 results in a sensitive estimator (λ_0 small) and hence allows a fast adaptation to parameter changes and vice versa. In addition, one must define a lower bound λ_{min}.

Example 12.2 (Parameter Estimation with a Time-Varying Forgetting Factor).

The same process as in Example 12.1 is used again to illustrate the effects of the time-varying forgetting factor. Figure 12.3 shows the implementation and illustrates, how the forgetting factor is adjusted during the change of the system parameters, where the forgetting factor is reduced to $0.4 \leq \lambda(k) \leq 1$. □

A practical problem is the choice of Σ_0. If it is chosen too small, then also $\lambda(k)$ will vary too much, even if there are no parameter changes. Another disadvantage is the fact that $\lambda(k)$ changes also according to the variance of the noise σ_n^2 even if the process parameters do not change. Furthermore, $\lambda(k)$ gets smaller as σ_n^2 increases which is contradictionary to an useful adaption of $\lambda(k)$ to the noise.

The performance of the algorithm can be increased, if the variance σ_n^2 is estimated (Siegel, 1985). This can for example be done by the recursive equation

$$\hat{\sigma}_n^2(k) = \kappa\hat{\sigma}_n^2(k-1) + (1-\kappa)e^2(k) \text{ with } \kappa < 1 \,. \tag{12.2.5}$$

If the following thresholds

$$\hat{\sigma}_n^2(k) \geq \sigma_{n0}^2 \tag{12.2.6}$$

$$|\hat{\sigma}_n^2(k) - \hat{\sigma}_n^2(k-1)| \geq \Delta\sigma_n^2 \tag{12.2.7}$$

are exceeded, then it is assumed that the parameters have changed and hence one chooses N_{02} small such that

$$\Sigma_0 = \hat{\sigma}_n^2 N_{02} \,. \tag{12.2.8}$$

If the value of $\hat{\sigma}_n^2(k)$ did not exceed the thresholds, one sets

$$\Sigma_0 = \hat{\sigma}_n^2 N_{01} \tag{12.2.9}$$

on the contrary with $N_{01} > N_{02}$ (e.g. $N_{01} = 10N_{02}$).

12.3 Manipulation of Covariance Matrix

The methods based on adjusting the forgetting factor λ are well suited for slow parameter changes only, since the correction vector $\boldsymbol{\gamma}(k)$ depends on the covariance matrix $\boldsymbol{P}(k)$ that only changes (exponentially) slow, see (12.1.2) though (12.1.5).

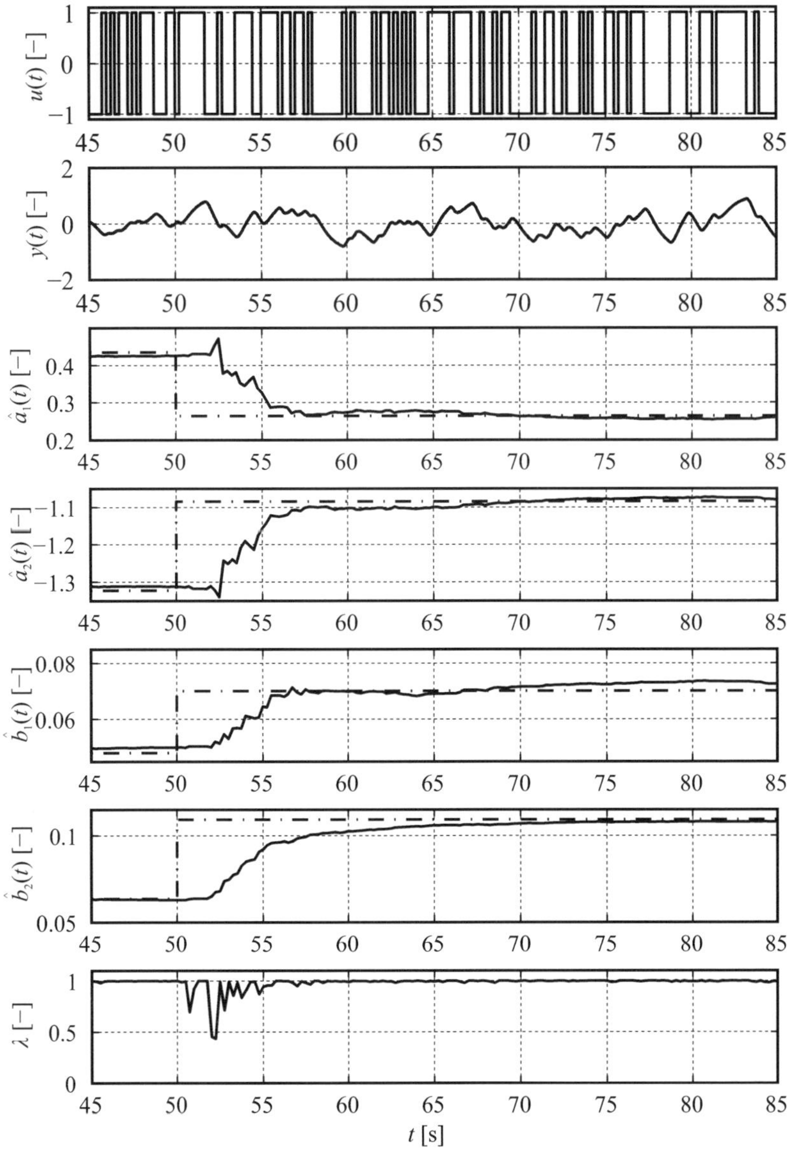

Fig. 12.3. Parameter estimation for a second order system with a change in the system parameters, variable forgetting factor $\lambda(k)$. True parameter values (*dash-dotted line*)

However, for fast parameter changes, $\boldsymbol{\gamma}(k)$ and also $\boldsymbol{P}(k)$ must change fast, which can achieved by adding a matrix $\boldsymbol{R}(k)$ to $\boldsymbol{P}(k)$ as

$$\boldsymbol{P}(k+1) = \left(\boldsymbol{I} - \boldsymbol{\gamma}(k)\boldsymbol{\psi}^{\mathrm{T}}(k+1)\boldsymbol{P}(k)\frac{1}{\lambda}\right) + \boldsymbol{R}(k) \,. \tag{12.3.1}$$

Increasing the elements of the covariance matrix allows the parameters to change much more rapidly than just changing the forgetting factor. If one only considers diagonal matrices as e.g. for the choice of the initial values of $\boldsymbol{P}(0)$, then one can choose e.g.

$$\boldsymbol{R}(k) = \beta \frac{e^2(k)}{\hat{\sigma}_{\mathrm{n}}^2(k)} \boldsymbol{I} \,, \tag{12.3.2}$$

if the thresholds in (12.2.6) and (12.2.7) are exceeded. Below the thresholds, one sets $\boldsymbol{R}(k) = \boldsymbol{0}$.

A disadvantage is that the diagonal elements of $\boldsymbol{P}(k)$ are all increased by the same value. One can therefore think about introducing the relation

$$\boldsymbol{R}(k) = \alpha_{\mathrm{R}} \boldsymbol{P}(k) \,, \tag{12.3.3}$$

which relates the change to the current values of $\boldsymbol{P}(k)$. One should chose $\alpha_{\mathrm{R}} \gg 1$, e.g. $\alpha_{\mathrm{R}} = 100, \ldots, 1\,000$. This can be seen as a *restart*.

Example 12.3 (Parameter Estimation of a Time-Variant Process with Manipulation of the Covariance Matrix).

The process from Example 12.1 is considered again. Figure 12.4 shows the control of the adaptation speed by means of a manipulation of the covariance matrix. Because the covariance is increased after passing the threshold $\Delta\sigma_n^2(k) > 0.0002$ (gradient of the estimated variance of the disturbance or error signal according to (12.2.7)), the covariance is increased by $\alpha_{\mathrm{R}} = 100$. Therefore, the adaptation speed becomes faster, however, with passing higher variances. □

12.4 Convergence of Recursive Parameter Estimation Methods

Convergence of the parameter estimates has been exhaustively analyzed, where a classical results has been presented by Lai and Wei (1982) and revisited recently by Hu and Ljung (2008). This classical analysis of convergence has also been outlined by Isermann (1992) and Isermann et al (1992) and shall not be repeated here. If one however interprets the parameter estimates $\hat{\boldsymbol{\theta}}$ as state variables, then characteristics of discrete-time state feedback and state observers can be transferred to the recursive parameter estimation as was shown by Kofahl (1988) and Isermann (1992) and will be outlined in the following.

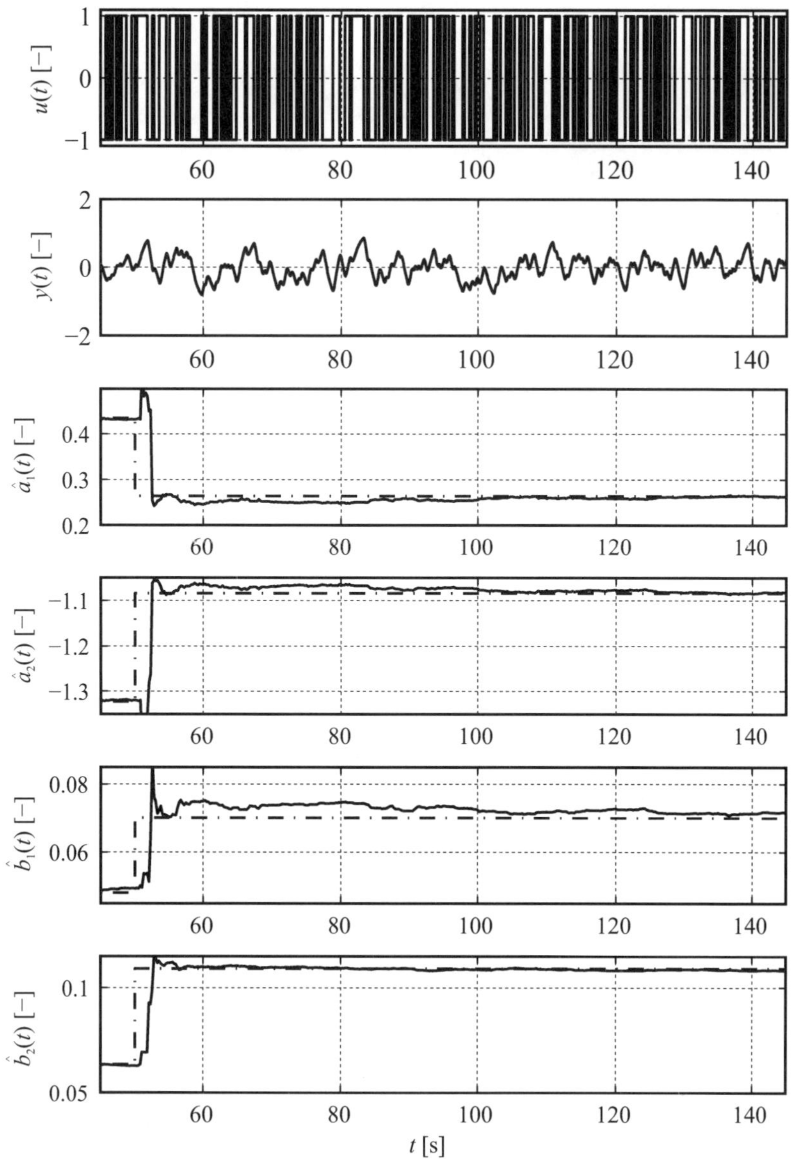

Fig. 12.4. Parameter estimation for a second order system with a change in the system parameters, covariance matrix manipulation. After passing the threshold $\Delta\sigma_n^2(k) > 0.0002$, the covariance is increased by $\alpha_R = 100$. True parameter values (*dash-dotted line*)

12.4.1 Parameter Estimation in Observer Form

For a process that is represented by a state space model (2.2.24) and (2.2.25), the equation for a state observer of this SISO system is given as

$$\hat{\boldsymbol{x}}(k+1) = \boldsymbol{A}\hat{\boldsymbol{x}}(k) + \boldsymbol{b}u(k) + \boldsymbol{h}\big(y(k) - \boldsymbol{c}^{\mathrm{T}}\hat{\boldsymbol{x}}(k)\big)\,. \tag{12.4.1}$$

The error of the state estimate,

$$\tilde{\boldsymbol{x}}(k) = \boldsymbol{x}(k) - \hat{\boldsymbol{x}}(k) \tag{12.4.2}$$

is then governed by the difference equation

$$\tilde{\boldsymbol{x}}(k+1) = \big(\boldsymbol{A} - \boldsymbol{h}\boldsymbol{c}^{\mathrm{T}}\big)\tilde{\boldsymbol{x}}(k)\,. \tag{12.4.3}$$

In order for the error $\tilde{\boldsymbol{x}}(k)$ to vanish for $k \to \infty$, i.e.

$$\lim_{k\to\infty} \tilde{\boldsymbol{x}}(k) = \boldsymbol{0}\,, \tag{12.4.4}$$

(12.4.3) must be asymptotically stable. The characteristic equation of the observer,

$$\det\big(z\boldsymbol{I} - \boldsymbol{A} - \boldsymbol{h}\boldsymbol{c}^{\mathrm{T}}\big) = (z - z_1)(z - z_2)\ldots(z - z_m) \tag{12.4.5}$$

may hence only have poles with $|z_i| < 1,\ i = 1, 2, \ldots, m$. If one now interprets the time-variant parameters $\boldsymbol{\theta}$ of the process as state variables, then one can derive a *parameter state model* as

$$\boldsymbol{\theta}(k+1) = \boldsymbol{I}\boldsymbol{\theta}(k) + \boldsymbol{\eta}(k) \tag{12.4.6}$$

$$y(k+1) = \boldsymbol{\psi}^{\mathrm{T}}(k+1)\boldsymbol{\theta}(k) + n(k+1)\,. \tag{12.4.7}$$

Here $\boldsymbol{\eta}(k)$ denotes a (deterministic) parameter change. The resulting block diagram is shown in Fig. 12.5.

The recursive least squares parameter estimation algorithm is then according to (9.4.17) given as

$$\hat{\boldsymbol{\theta}}(k+1) = \hat{\boldsymbol{\theta}}(k) + \boldsymbol{\gamma}(k)e(k+1) \tag{12.4.8}$$

$$e(k+1) = y(k+1) - \boldsymbol{\psi}^{\mathrm{T}}(k+1)\hat{\boldsymbol{\theta}}(k) \tag{12.4.9}$$

with

$$\boldsymbol{\psi}^{\mathrm{T}}(k+1) = \big(-y(k)\ \ldots\ -y(k-m+1)\big|u(k-d)\ \ldots\ u(k-d-m+1)\big)\,. \tag{12.4.10}$$

This estimation is shown in Fig. 12.5 in the lower part. The block diagram corresponds to a state observer (12.4.1) without input ($\boldsymbol{b} = \boldsymbol{0}$) and the following equivalent terms

$$\boldsymbol{A} \to \boldsymbol{I},\ \ \boldsymbol{h} \to \boldsymbol{\gamma}(k),\ \ \boldsymbol{c} \to \boldsymbol{\psi}^{\mathrm{T}}(k+1)\,. \tag{12.4.11}$$

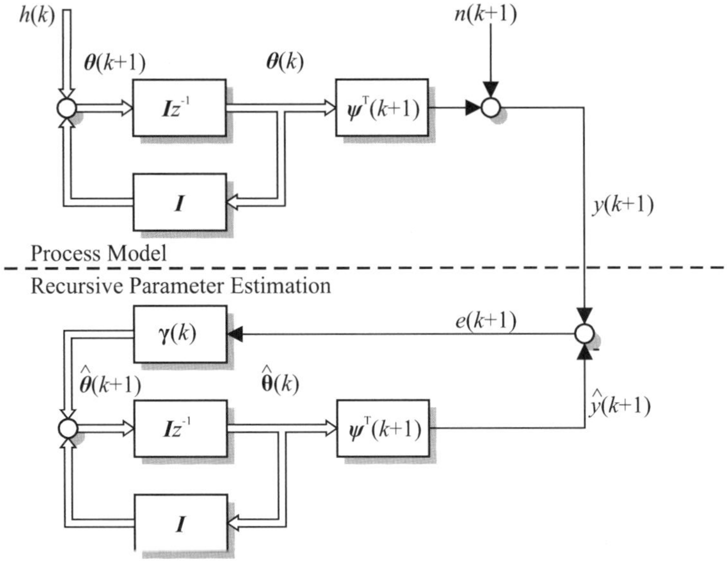

Fig. 12.5. Block diagram of the recursive parameter estimation by the method of least squares in a state space framework

This *parameter-state observer* has a time-variant feedback gain $\boldsymbol{\gamma}(k)$ and a time-variant output vector $\boldsymbol{\psi}^{\mathrm{T}}(k+1)$. For the error of the parameter estimates, one obtains with (12.4.6), (12.4.8), (12.4.9),

$$\begin{aligned} \boldsymbol{e}_{\boldsymbol{\theta}}(k+1) &= \boldsymbol{\theta}(k+1) - \hat{\boldsymbol{\theta}}(k+1) \\ &= \big(\boldsymbol{I} - \boldsymbol{\gamma}(k)\boldsymbol{\psi}^{\mathrm{T}}(k+1)\big)\boldsymbol{e}_{\boldsymbol{\theta}}(k+1) + \boldsymbol{\eta}(k) - \boldsymbol{\gamma}(k)n(k+1)\,. \end{aligned} \tag{12.4.12}$$

This corresponds to the homogeneous vector difference equation in (12.4.3) of the state observer, with the difference that $\boldsymbol{\gamma}$ and $\boldsymbol{\psi}^{\mathrm{T}}$ are time varying quantities that depend on the measured signals. Furthermore, the disturbances $\boldsymbol{\eta}(k)$ as well as $\boldsymbol{\gamma}(k)n(k+1)$ are acting on the system. These however vanish for a time invariant process and $n(k) = 0$.

In order for the parameter error in (12.4.12) not to diverge, the homogeneous part of the difference equation must be asymptotically stable. This part however contains time variant terms, which is in contrast to the classical state observer. Assuming that these time variant parameters can be frozen, one can determine the characteristic equation in a similar way to (12.4.5) as

$$\det\big(z\boldsymbol{I} - \boldsymbol{I} + \boldsymbol{\gamma}(k)\boldsymbol{\psi}^{\mathrm{T}}(k+1)\big) = 0\,. \tag{12.4.13}$$

In analogy to the observer, one can now determine the eigenvalues from this equation. According to Kofahl (1988), one obtains with $\boldsymbol{\gamma}(k) = \boldsymbol{\gamma}$ and $\boldsymbol{\psi}^{\mathrm{T}}(k+1) = \boldsymbol{\psi}^{\mathrm{T}}$ being treated as constant quantities,

$$\begin{aligned}\det\big(z\boldsymbol{I}-\boldsymbol{I}+\boldsymbol{\gamma}\boldsymbol{\psi}^{\mathrm{T}}\big)&=\det\big((z\boldsymbol{I}-\boldsymbol{I})(\boldsymbol{I}+(z\boldsymbol{I}-\boldsymbol{I})^{-1}\boldsymbol{\gamma}\boldsymbol{\psi}^{\mathrm{T}})\big)\\&=\det\big(z\boldsymbol{I}-\boldsymbol{I}\big)\det\big(\boldsymbol{I}+(z\boldsymbol{I}-\boldsymbol{I})^{-1}\boldsymbol{\gamma}\boldsymbol{\psi}^{\mathrm{T}}\big)\,.\end{aligned}\tag{12.4.14}$$

By employing the relation $\det(\boldsymbol{A}+\boldsymbol{u}\boldsymbol{v}^{\mathrm{T}})=\det\boldsymbol{A}(1+\boldsymbol{v}^{\mathrm{T}}\boldsymbol{A}^{-1}\boldsymbol{u})$ (Gröbner, 1966), one can now write

$$\begin{aligned}&\det\big(z\boldsymbol{I}-\boldsymbol{I}\big)\det(\boldsymbol{I})\big(1+\boldsymbol{\psi}^{\mathrm{T}}\boldsymbol{I}^{-1}(z\boldsymbol{I}-\boldsymbol{I})^{-1}\boldsymbol{\gamma}\big)\\&\quad=(z-1)^n\big(1+\boldsymbol{\psi}^{\mathrm{T}}(z-1)^{-1}\boldsymbol{\gamma}\big)\\&\quad=(z-1)^{n-1}\big(z-1+\boldsymbol{\psi}^{\mathrm{T}}\boldsymbol{\gamma}\big)\end{aligned}\tag{12.4.15}$$

and then determine the eigenvalues as

$$z_i=1\text{ for }i=1,\dots,n-1\tag{12.4.16}$$

$$z_n=1-\boldsymbol{\psi}^{\mathrm{T}}(k+1)\boldsymbol{\gamma}(k+1)\,.\tag{12.4.17}$$

The eigenvalue z_n that was assumed to be constant depends on the time varying quantities $\boldsymbol{\gamma}(k)$ and $\boldsymbol{\psi}^{\mathrm{T}}(k+1)$ and is hence time variant. It shall be called the "time variant eigenvalue" of the parameter estimation equation.

From this follows, see also (Isermann et al, 1992),

Theorem 12.1 (Dynamics of the Recursive Parameter Estimation).

The recursive method of least squares has for n parameters

- $(n-1)$ *constant eigenvalues at*

$$z_i=1,\ i=0,1,\dots,n-1\tag{12.4.18}$$

- *One "time variant eigenvalue" at*

$$\begin{aligned}z_n&=1-\boldsymbol{\psi}^{\mathrm{T}}(k+1)\boldsymbol{\gamma}(k)\\&=\big(\lambda+\boldsymbol{\psi}^{\mathrm{T}}(k+1)\boldsymbol{P}(k)\boldsymbol{\psi}(k+1)\big)^{-1}\end{aligned}\tag{12.4.19}$$

with

$$0<z_n(k)\le\lambda\,,\tag{12.4.20}$$

where the right side in (12.4.19) follows from (9.4.18). □

As discussed by Kofahl (1988), the "time-variant eigenvalue" tends to

$$z_n(k)\to 0\tag{12.4.21}$$

for a sudden excitation. In case of no excitation, the eigenvalue approaches

$$\lim_{k\to\infty}z_n(k)\to\lambda\,.\tag{12.4.22}$$

This behavior can be witnessed in the following example.

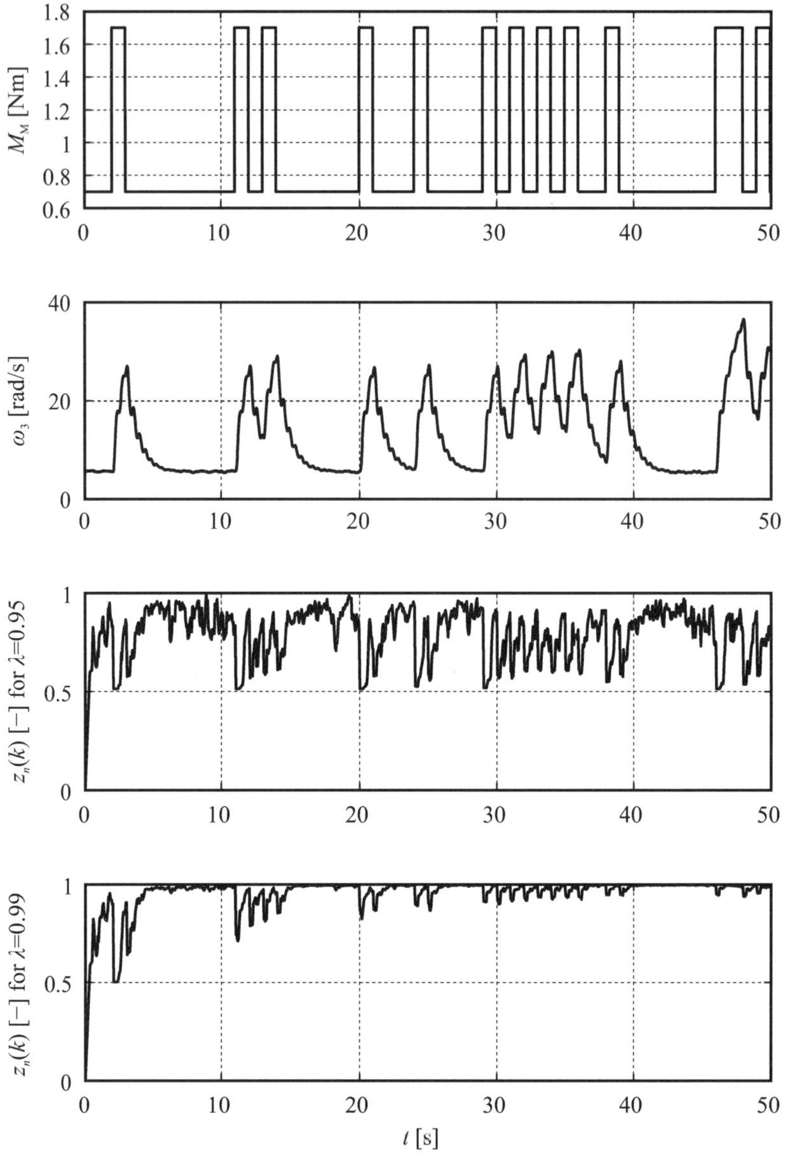

Fig. 12.6. "Time-varying eigenvalue" for the parameter estimation for the Three-Mass Oscillator excited by a PRBS signal

Table 12.1. Different cases for the estimation of time-varying processes

Disturbance / Noise	Speed of Parameter Change		
	Slow	Fast	Variable
Small	A $(\rightarrow \lambda_1)$	C	B $(\rightarrow \lambda = f(\sigma_{\mathrm{e}}^2))$
Large	A $(\rightarrow \lambda_2 > \lambda_1)$	-	-
Variable	B $(\rightarrow \lambda = f(\sigma_{\mathrm{n}}^2))$	-	B $(\rightarrow \lambda = f(\sigma_{\mathrm{n}}^2, k))$

Method A: Constant Forgetting Factor λ
Method B: Variable Forgetting Factor $\lambda(k)$
Method C: Correction of the Covariance Matrix $\boldsymbol{P}(k)$

Example 12.4 (Eigenvalues of the Recursive Parameter Estimation).
Figure 12.6 shows the behavior of the "time-varying eigenvalue" $z_n(k)$ of the parameter estimation for $\lambda = 0.95$ and $\lambda = 1$. In the case of a sudden excitation, the eigenvalue gets smaller and in the intervals of little or no excitation, it tends to the chosen value of $\lambda = 0.95$ or $\lambda = 1$ respectively. □

The "eigenvalue" $z_n(k)$ hence is a measure for the amount of excitation and can be used to control or supervise the time variant parameter estimation, e.g. for adaptive control applications (Kofahl, 1988).

12.5 Summary

For the parameter estimation of time-variant processes, one can discern different partially contrary cases as illustrated in Table 12.1. The speed of the parameter changes can be fast or slow or both (variable). Further, the *signal-to-noise ratio* can be small or large or both (variable). In all of these cases, one must find a compromise with respect to the parameter estimation algorithm and the capabilities of

- fast tracking of parameter changes
- good elimination of disturbances

The simplest case is the combination of slow parameter changes and small disturbances, the most difficult to handle case is the combination of fast parameter changes and large disturbances.

For slow parameter changes, one can use recursive parameter estimation with a constant forgetting factor λ in the presence of either small or large disturbances. For faster parameter changes and small disturbance levels, one should correct the covariance matrix. If the speed of the parameter changes varies and the noise level is small and/or variable, then the forgetting factor should be variable as well.

As an alternative, one can model the time behavior of the parameters (Isermann, 1992; Young, 2009)

Problems

12.1. Exponential Forgetting with Constant Forgetting Factor I
What is the trade-off that has to be observed in the choice of λ? How is λ chosen if the noise is high?

12.2. Exponential Forgetting with Constant Forgetting Factor II
Why can it be beneficial to reduce the forgetting factor λ during the starting phase?

12.3. Exponential Forgetting with Variable Forgetting Factor
Why might it be necessary to steer the forgetting factor?

12.4. Manipulation of Covariance Matrix
Why might it be necessary to manipulate the covariance matrix? Describe in own words, why this allows to track parameter changes.

12.5. No Excitation
What happens with the parameter estimates of a time-variant process if the input excitation tends to zero?

References

Eykhoff P (1974) System identification: Parameter and state estimation. Wiley-Interscience, London

Fortescue TR, Kershenbaum LS, Ydstie BE (1981) Implementation of self-tuning regulators with variable forgetting factor. Automatica 17(6):831–835

Goodwin GC, Sin KS (1984) Adaptive filtering, prediction and control. Prentice-Hall information and system sciences series, Prentice-Hall, Englewood Cliffs, NJ

Gröbner W (1966) Matrizenrechnung. BI-Hochschultaschenbücher Verlag, Mannheim

Hu XL, Ljung L (2008) New convergence results for the least squares identification algorithm. In: The International Federation of Automatic Control (ed) Proceedings of the 17th IFAC World Congress, Seoul, Korea, pp 5030–5035

Isermann R (1992) Identifikation dynamischer Systeme: Besondere Methoden, Anwendungen (Vol 2). Springer, Berlin

Isermann R, Lachmann KH, Matko D (1992) Adaptive control systems. Prentice Hall international series in systems and control engineering, Prentice Hall, New York, NY

Kofahl R (1988) Robuste parameteradaptive Regelungen: Fachberichte Messen, Steuern, Regeln Nr. 19. Springer, Berlin

Lai TC, Wei CZ (1982) Least squares estimates in stochastic regression models with applications to identification and control of dynamic systems. Ann Stat 10(1):154–166

Mikleš J, Fikar M (2007) Process modelling, identification, and control. Springer, Berlin

Siegel M (1985) Parameteradaptive Regelung zeitvarianter Prozesse. Studienarbeit. Institut für Regelungstechnik, TH Darmstadt, Darmstadt

Söderström T, Ljung L, Gustavsson I (1974) A comparative study of recursive identification methods. Report 7427. Dept. of Automatic Control, Lund Inst of Technology, Lund

Young PC (2009) Time variable parameter estimation. In: Proceedings of the 15th IFAC Symposium on System Identification, Saint-Malo, France

13

Parameter Estimation in Closed-Loop

In certain applications, the process can only be identified in closed loop. For example, in biological and economical systems, the controller is an integrated and by no means detachable part of the system. For technical systems, e.g. in the area of adaptive control systems, the process model must also be updated while the system is under closed-loop control. Furthermore, integral acting processes can typically only be operated reliably in closed-loop control to account for the influence of disturbances acting on the system, thereby avoiding a drift of the system. Also, for many safety critical systems it may be too dangerous to disconnect the controller. In production systems on the other hand, it may not be possible to maintain the necessary quality of the product without closed-loop control.

For the application of the identification methods presented so far, one must first check, whether the convergence criteria allow operation in closed loop. For the correlation analysis for example, it was required that the input $u(k)$ and the disturbance $n(k)$ are not correlated. A feedback loop however will cause such a correlation. For the method of least squares on the other hand, it was required that the error $e(k)$ is not correlated with the data vector $\boldsymbol{\psi}^{\mathrm{T}}(k)$. Here, it must also be checked if the feedback loop will cause such a correlation.

For the identification, one can in general discern two cases, see also Figs. 13.1 and 13.2:

- *Case a: Indirect Process Identification*: A model of the closed-loop process is identified. The controller must be known and the process model is derived from the closed-loop model.
- *Case b: Direct Process Identification*: The process is identified directly, that is without the intermediate step of determining the closed-loop model. The controller must therefore not be known.

Further issues that can be identified are

- *Case c*: Only the output $y(k)$ is measured.
- *Case d*: Input $u(k)$ and output $y(k)$ are both measured.
- *Case e*: No injection of an additional test signal.

R. Isermann, M. Münchhof, *Identification of Dynamic Systems*,
DOI 10.1007/978-3-540-78879-9_13, © Springer-Verlag Berlin Heidelberg 2011

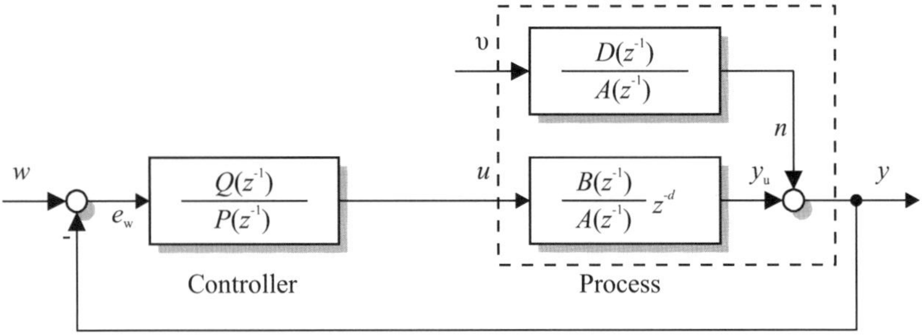

Fig. 13.1. Block diagram of the process to be identified in closed loop without an additional test signal

- *Case f*: Additional test signal $u_S(k)$ (either not measurable or measurable)
- *Case g*: Additional measurable test signal $u_S(k)$ is used for identification.

As will be shown in the following, several combinations are possible. Section 13.1 will cover the cases a+c+e and b+d+e. Sections 13.2 and 13.3 will then deal with the cases a+g and b+d+f.

13.1 Process Identification Without Additional Test Signals

According to Fig. 13.1, a linear time invariant process with the transfer function

$$G_P = \frac{y_u(z)}{u(z)} = \frac{B(z^{-1})}{A(z^{-1})} z^{-d} = \frac{b_1 z^{-1} + \ldots + b_{m_b} z^{-m_b}}{1 + a_1 z^{-1} + \ldots + a_{m_a} z^{-m_a}} z^{-d} \tag{13.1.1}$$

and the noise form filter

$$G_\nu(z) = \frac{n(z)}{\nu(z)} = \frac{D(z^{-1})}{C(z^{-1})} \tag{13.1.2}$$

shall be identified in closed-loop operation. By assuming $C(z^{-1}) = A(z^{-1})$ in the form filter describing the noise, the identification without an additional test signal is severely simplified. The form filter is then given as

$$G_\nu(z) = \frac{n(z)}{\nu(z)} = \frac{D(z^{-1})}{A(z^{-1})} = \frac{1 + d_1 z^{-1} + \ldots + d_{m_d} z^{-m_d}}{1 + a_1 z^{-1} + \ldots + a_{m_a} z^{-m_a}} . \tag{13.1.3}$$

The controller is given by the transfer function

$$G_C = \frac{u(z)}{e_w(z)} = \frac{Q(z^{-1})}{P(z^{-1})} = \frac{q_0 + q_1 z^{-1} + \ldots + q_\nu z^{-\nu}}{1 + p_1 z^{-1} + \ldots + p_\mu z^{-\mu}} . \tag{13.1.4}$$

Furthermore, the output and the control deviation are given as

$$y(z) = y_u(z) + n(z)$$
$$e_w(z) = w(z) - y(z) \,.$$

In general, it shall be assumed that $w(z) = 0$, i.e. $e_w(z) = -y(z)$, $\nu(z)$ shall be a non-measurable, statistically independent noise with $\mathrm{E}\{\nu(k)\} = 0$ and the variance σ_ν^2.

13.1.1 Indirect Process Identification (Case a+c+e)

The disturbance transfer function of the closed-loop process is given as

$$\begin{aligned}\frac{y(u)}{\nu(z)} &= \frac{G_\nu(z)}{1+G_C(z)G_P(z)}\\ &= \frac{D(z^{-1})P(z^{-1})}{A(z^{-1})P(z^{-1})+B(z^{-1})z^{-d}Q(z^{-1})}\\ &= \frac{1+\beta_1 z^{-1}+\ldots+\beta_r z^{-r}}{1+\alpha_1 z^{-1}+\ldots+\alpha_l z^{-l}} = \frac{\mathcal{B}(z^{-1})}{\mathcal{A}(z^{-1})}\,.\end{aligned} \tag{13.1.5}$$

The output $y(k)$ hence is an ARMA process. By the control loop assuming the role of a form filter, the output $y(k)$ is generated from the statistically independent noise $\nu(k)$. The orders of the polynomials are

$$l = \max(m_a + \mu, m_b + \nu + d) \tag{13.1.6}$$
$$r = m_d + \mu \,. \tag{13.1.7}$$

If only the output $y(k)$ of the control loop is analyzed, then the estimates of the ARMA process

$$\boldsymbol{\theta}_{\alpha,\beta}^{\mathrm{T}} = \left(\hat{\alpha}_1 \,\ldots\, \hat{\alpha}_l \middle| \hat{\beta}_1 \,\ldots\, \hat{\beta}_r\right) \tag{13.1.8}$$

can be determined, e.g. by the recursive method of ELS that was covered in Sect. 10.2, if all poles of $\mathcal{A}(z^{-1})$ are inside the unit circle and the polynomials $D(z^{-1})$ and $\mathcal{A}(z^{-1})$ have no common roots.

The further task of the indirect identification method is to determine the unknown process parameters

$$\boldsymbol{\theta}^{\mathrm{T}} = \left(\hat{a}_1 \,\ldots\, \hat{a}_{m_a} \middle| \hat{b}_1 \,\ldots\, \hat{b}_{m_b} \middle| \hat{d}_1 \,\ldots\, \hat{d}_{m_d}\right) \tag{13.1.9}$$

from the estimated parameters $\hat{\alpha}_i$ and $\hat{\beta}_i$. To ensure that the parameters can be determined unambiguously, certain conditions on the identifiability must be satisfied.

Conditions for the Identifiability in Closed Loop

A process shall be called *parameter identifiable*, if its parameter estimates are consistent upon the application of a suitable parameter estimation method. In the following, conditions are formulated, which must be satisfied if only the output $y(k)$ is measurable.

Identifiability Condition 1

In a condensed notation, the input/output behavior of the control loop from (13.1.5) is given as

$$\left(A + B\frac{Q}{P}\right)y = D\nu\,. \tag{13.1.10}$$

This equation can now be manipulated by adding and subtracting the polynomial $S(z^{-1})$.

$$\left(A + S + B\frac{Q}{P} - S\right)y = D\nu \tag{13.1.11}$$

$$\left(A + S + \left(B - \frac{P}{Q}S\right)\frac{Q}{P}\right)y = D\nu \tag{13.1.12}$$

$$\left(Q(A + S) + (QB - PS)\frac{Q}{P}\right)y = QD\nu \tag{13.1.13}$$

$$\left(A^* + B^*\frac{Q}{P}\right)y = D^*\nu\,. \tag{13.1.14}$$

Comparison with (13.1.10) shows that a control loop with

$$\frac{B^*}{A^*} = \frac{BQ - PS}{AQ + SQ} \text{ and } \frac{D^*}{A^*} = \frac{DQ}{AQ + SQ} \tag{13.1.15}$$

and the controller Q/P has the same input/output behavior as the originally considered control loop from (13.1.5). Since S can be arbitrary, the process cannot be identified unambiguously from the input/output behavior y/ν even if the controller Q/P is known exactly unless the order of the polynomials $B(z^{-1})z^{-d}$ and $A(z^{-1})$ is also known exactly (Bohlin, 1971). Hence, the *identifiability condition 1* is: *The model order must be known a priori.*

Identifiability Condition 2

From (13.1.5) follows that the $m_a + m_b$ unknown parameters $\hat{a}_i$ and $\hat{b}_i$ must be determined from the l parameters $\hat{\alpha}_i$. If the polynomials D and $\mathcal{A}$ have no common roots, then, for the unambiguous determination of the process parameters, it is required that $l = m_a + m_b$ or

$$\max(m_a + \mu, m_b + \nu + d) \geq m_a + m_b \tag{13.1.16}$$

$$\max(\mu - m_b, \nu + d - m_a) \geq 0\,. \tag{13.1.17}$$

The *identifiability condition 2* therefore is: *The controller order has to be large enough and has to satisfy*

$$\text{if } \nu > \mu - d + m_a - m_b \Rightarrow \nu \geq m_a - d \tag{13.1.18}$$

or

$$\text{if } \nu < \mu - d + m_a - m_b \Rightarrow \mu \geq m_b \,. \tag{13.1.19}$$

In the case $d = 0$, the controller order must either be $\nu \geq m_a$ or $\mu \geq m_b$. If $d > 0$, then either $\nu \geq m_a - d$ or $\mu \geq m_b$. It is indifferent whether the dead time belongs to the process or to the controller. Hence, the identifiability condition can also be satisfied by a controller with dead time $d = m_a$ and the order $\nu = 0, \mu = 0$.

The parameters $\hat{d}_i$ (13.1.3) can be determined unambiguously from $\hat{\beta}_i$ (13.1.5) if $r \geq m_d$ or

$$\mu \geq 0 \,. \tag{13.1.20}$$

The estimation of the parameters d_i is hence possible with any arbitrary controller, as long as $D(z^{-1})$ and $\mathcal{A}(z^{-1})$ have no common roots.

If $D(z^{-1})$ and $\mathcal{A}(z^{-1})$ have p common roots, then those cannot be identified and hence only $l - p$ parameters $\hat{\alpha}_i$ and $r - p$ parameters $\hat{\beta}_i$ are available. Identifiability condition 2 for the process parameters $\hat{a}_i$ and $\hat{b}_i$ is then given as

$$\max(\mu - m_b, \nu + d - m_a) \geq p \,. \tag{13.1.21}$$

A note should be made that only the common roots of $D(z^{-1})$ and $\mathcal{A}(z^{-1})$ matter and not those of $\mathcal{B}(z^{-1})$ and $\mathcal{A}(z^{-1})$ as $\mathcal{B} = DP$ and P is known.

If the order of the controller is not large enough, one can conduct the identification in closed loop with two different sets of controller parameters (Gustavsson et al, 1974; Kurz and Isermann, 1975; Gustavsson et al, 1977).

Example 13.1 (Identification in Closed Loop with External Disturbance).

The parameters of a process of first order with $m_\mathrm{a} = m_\mathrm{b} = m = 1$

$$y(k) + ay(k-1) = bu(k-1) + \nu(k) + d\nu(k-1) \tag{13.1.22}$$

shall be identified in closed loop. For this endeavor, different controllers will be considered.

1. One P controller: $u(k) = -q_0 y(k) (\nu = 0, \mu = 0)$.
   With this controller, one obtains the ARMA process equation

   $$y(k) + (a + bq_0)y(k-1) = \nu(k) + d\nu(k-1) \tag{13.1.23}$$

   or

   $$y(k) + \alpha y(k-1) = \nu(k) + \beta \nu(k-1) \,. \tag{13.1.24}$$

   By equating coefficients, one obtains

   $$\hat{\alpha} = \hat{a} + \hat{b} q_0 \tag{13.1.25}$$

   $$\hat{\beta} = \hat{d} \tag{13.1.26}$$

   and can immediately see that no unique solution for $\hat{a}$ and $\hat{b}$ can be obtained. This is in line with the conditions in (13.1.19), which are clearly violated as $\nu \geq 1$ or $\mu \geq 1$.

2. One PD-controller: $u(k) = -q_0 y(k) - q_1 y(k-1) (\nu = 1, \mu = 0)$
   The ARMA process equation is now of second order

$$y(k) + (a + bq_0)y(k-1) + bq_1 y(k-2) = \nu(k) + d\nu(k-1) \quad (13.1.27)$$

$$y(k) + \alpha_1 y(k-1) + \alpha_2 y(k-2) = \nu(k) + \beta\nu(k-1)\,. \quad (13.1.28)$$

   By equations coefficients, one obtains

$$\hat{a} = \hat{\alpha}_1 + \hat{b}q_0 \quad (13.1.29)$$

$$\hat{b} = \hat{\alpha}_2 / q_1 \quad (13.1.30)$$

$$\hat{d} = \hat{\beta}\,. \quad (13.1.31)$$

   The process parameters can hence be identified.
3. Two P controllers: $u(k) = -q_{01} y(k), u(k) = -q_{02} y(k)$.
   With these controllers, one obtains two ARMA process equations, which by equating coefficients, yield

$$\hat{\alpha}_{11} = \hat{a} + \hat{b}q_{01} \quad (13.1.32)$$

$$\hat{\alpha}_{12} = \hat{a} + \hat{b}q_{02}\,. \quad (13.1.33)$$

   From there follows

$$\hat{a} = \frac{\hat{\alpha}_{11} - \frac{q_{01}}{q_{02}}\hat{\alpha}_{12}}{1 - \frac{q_{01}}{q_{02}}} \quad (13.1.34)$$

$$\hat{b} = \frac{1}{q_{02}}(\hat{\alpha}_{12} - \hat{a})\,. \quad (13.1.35)$$

   The process parameters are hence identifiable, provided that $q_{01} \neq q_{02}$. □

In the general case, one obtains the process parameters $\hat{\boldsymbol{\theta}}$ from the parameters $\hat{\alpha}_1, \ldots, \hat{\alpha}_l$ of the ARMA model by equating coefficients in (13.1.3) observing the above stated conditions for identifiability. If $d = 0$, $m_a = m_b$, and the controller order is governed by $\nu = m$ and $\mu \leq m$, hence $l = 2m$, and thus satisfying the condition in (13.1.19), then it follows with $p_0 = 1$ that

$$\begin{array}{llllllll}
a_1 & & & +b_1 q_0 & & & = \alpha_1 - p_1 \\
a_1 p_1 & +a_2 & & +b_1 q_1 & +b_2 q_0 & & = \alpha_2 - p_2 \\
\vdots & \vdots & & \vdots & \vdots & & \vdots \\
a_1 p_{j-1} & +a_2 p_{j-2} \ldots & +a_m p_{j-m} & +b_1 q_{j-1} & +b_2 q_{j-2} \ldots & b_m q_{j-m} & = \alpha_j - p_j
\end{array} \quad (13.1.36)$$

In matrix form, this set of equations can be written as

$$\underbrace{\begin{pmatrix} 1 & 0 & \ldots & 0 & q_0 & 0 & \ldots & 0 \\ p_1 & 1 & \ldots & 0 & q_1 & q_0 & \ldots & 0 \\ \vdots & p_1 & \ldots & 0 & q_2 & q_1 & \ldots & \vdots \\ p_\mu & \vdots & & 1 & & \vdots & & q_0 \\ 0 & p_\mu & & p_1 & q_m & & & q_1 \\ 0 & 0 & & \vdots & 0 & q_m & & \\ \vdots & \vdots & & p_\mu & \vdots & \vdots & & \vdots \\ 0 & 0 & \ldots & 0 & 0 & 0 & \ldots & q_m \end{pmatrix}}_{\boldsymbol{S}} \underbrace{\begin{pmatrix} a_1 \\ a_2 \\ a_3 \\ \vdots \\ a_m \\ b_1 \\ b_2 \\ \vdots \\ b_m \end{pmatrix}}_{\boldsymbol{\theta}} = \underbrace{\begin{pmatrix} \alpha_1 - p_1 \\ \alpha_2 - p_2 \\ \alpha_3 - p_3 \\ \vdots \\ \alpha_\mu - p_\mu \\ \alpha_{\mu+1} \\ \alpha_{\mu+2} \\ \vdots \\ \alpha_{2m} \end{pmatrix}}_{\boldsymbol{\alpha}^*} . \tag{13.1.37}$$

Since the matrix $\boldsymbol{S}$ is a square matrix, one can determine the process parameters by

$$\hat{\boldsymbol{\theta}} = \boldsymbol{S}^{-1}\boldsymbol{\alpha}^* . \tag{13.1.38}$$

This shows again that for an unambiguous solution of (13.1.37) the matrix $\boldsymbol{S}$ must have the rank $r = 2m$, hence $\nu = m$ or $\mu = m$. If $\nu > m$ or $\mu > m$, one can solve the overdetermined system of equations by the pseudo-inverse. As will be discussed in Sect. 13.3, the process parameters for the indirect process identification converge very slowly for constant controller parameters. A further disadvantage is also that in some practical applications, the controller is not linear. Even a standard industrial PID controller can have delimiters, anti-windup and other non-linear effects like dead zones, making the application of the indirect methods infeasible if one cannot stay within the linear range of operation (Forssell and Ljung, 1999).

13.1.2 Direct Process Identification (Case b+d+e)

In the last section, it was assumed that the output $y(k)$ was measurable and the controller was known. Then, one could calculate the plant input $u(k)$ by the controller equation and hence the measurement of $u(k)$ would in theory not provide any new information about the process. However, if one measures the input $u(k)$, then the process can be identified directly, that is without the intermediate step of identifying the closed loop dynamics. Furthermore, knowledge of the controller is no longer necessary.

If one would use methods for the identification of non-parametric models, such as e.g. the correlation analysis on the control loop according to Fig. 13.1 and hence the measured signals $u(k)$ and $y(k)$, then, because of

$$\frac{u(z)}{\nu(z)} = -\frac{-G_{\mathrm{C}}(z)G_\nu(z)}{1 + G_{\mathrm{C}}(z)G_{\mathrm{P}}(z)} \tag{13.1.39}$$

and

$$\frac{y(z)}{\nu(z)} = -\frac{-G_\nu(z)}{1 + G_{\mathrm{C}}(z)G_{\mathrm{P}}(z)} , \tag{13.1.40}$$

one would identify the process with the transfer function

$$\frac{y(z)}{u(z)} = \frac{y(z)/\nu(z)}{u(z)/\nu(z)} = -\frac{1}{G_{\mathrm{C}}(z)} , \tag{13.1.41}$$

which is the negative reciprocal transfer function of the controller. For the identification, one should employ the useful signal $y_{\mathrm{u}}(k) = y(k) - n(k)$, because then

$$\frac{y_{\mathrm{u}}(z)}{u(z)} = \frac{y(z) - n(z)}{u(z)} = \frac{y(z)/\nu(z) - n(z)/\nu(z)}{u(z)/\nu(z)} = G_{\mathrm{P}}(z) \tag{13.1.42}$$

i.e. the process transfer function can be identified. This points out that the form filter $n(z)/\nu(z)$ must be known. Therefore, the process model from (13.1.1) and (13.1.3) will be used, resulting in

$$\hat{A}(z^{-1})y(z) = \hat{B}(z^{-1})z^{-d}u(z) + \hat{D}(z^{-1})\nu(z) . \tag{13.1.43}$$

This process model also contains the form filter of the disturbance.

As shown in Fig. 13.1, the process is operated in closed loop with

$$\frac{u(z)}{e_{\mathrm{w}}(z)} = \frac{Q(z^{-1})}{P(z^{-1})} \Leftrightarrow Q(z^{-1})y(z) = -P(z^{-1})u(z) . \tag{13.1.44}$$

Inserting the control law into (13.1.43) yields

$$\hat{A}(z^{-1})P(z^{-1})y(z) - \hat{B}(z^{-1})z^{-d}P(z^{-1})u(z) = \hat{D}(z^{-1})P(z^{-1})\nu(z) . \tag{13.1.45}$$

After canceling $P(z^{-1})$, one can see that this is identical to the open loop model of the process (13.1.43). The only difference to open loop operation of the process is that the input $u(k)$ cannot be chosen freely, but depends on the output $y(k)$ according to the control law in (13.1.44).

The identifiability conditions can be derived from the requirement that the cost function V has a unique minimum. It can be shown that with $e(k) = \nu(k)$ in case of convergence, the *same identifiability conditions hold* that have already been introduced in the last section on indirect process identification (Isermann, 1992).

Finally, it shall be investigated if the same methods for direct parameter estimation can be used that have already successfully been applied to open loop identification.

The method of least squares as well as the method of extended least squares is based on the error

$$e(k) = y(k) - \hat{y}(k|k-1) = y(k) - \boldsymbol{\psi}^{\mathrm{T}}(k)\hat{\theta}(k-1) . \tag{13.1.46}$$

A condition for convergence is that $e(k)$ is statistically independent from the elements of $\boldsymbol{\psi}^{\mathrm{T}}(k)$. For the method of least squares, one must ensure that

$$\boldsymbol{\psi}^{\mathrm{T}}(k) = \big(-y(k-1) \ldots \big| u(k-d-1) \ldots\big) \tag{13.1.47}$$

and for the method of extended least squares

$$\boldsymbol{\psi}^{\mathrm{T}}(k) = \big(-y(k-1) \ldots \big| u(k-d-1) \ldots \big| \hat{\nu}(k-1) \ldots\big) \tag{13.1.48}$$

are statistically independent from $e(k)$. Upon convergence, one may assume that $e(k) = \nu(k)$. Since however $\nu(k)$ does only influence $y(k), y(k+1), \ldots$ and these values do not turn up in $\psi^{\mathrm{T}}(k)$, one can conclude that $e(k)$ is independent from the elements of $\psi^{\mathrm{T}}(k)$.

If a feedback loop is introduced, basically nothing changes. Also in closed loop, the error $e(k)$ does not depend on the elements of $\psi^{\mathrm{T}}(k)$. Therefore, all methods that are based on the prediction error $e(k)$ as formulated in (13.1.46), will also deliver consistent estimates in closed loop as long as the previously formulated identifiability conditions are met. These methods can hence be applied to the signals $u(k)$ and $y(k)$ regardless of the presence of the feedback loop. The suitability of other parameter identification methods is treated in Sect. 13.3.

The most important results for the *identification in closed loop without an injected test signal* and a linear, time-invariant, noise-free controller can be summarized as follows:

1. For the indirect process identification (measurement of $y(k)$ only) as well as the direct process identification (measurement of $y(k)$ and $u(k)$) with parameter estimation methods, the *identifiability conditions 1 and 2* in Sect. 13.1.1 must be satisfied.
2. Since for the indirect identification, a signal process with $l \geq m_a + m_b$ parameters in the denominator and $r = m_d + \mu$ parameters in the numerator and in the case of direct identification only a process with m_a parameters in the denominator and m_b parameters in the numerator must be identified, one may expect better results if the direct identification methods are applied, especially, if the process has a high order.
3. For the direct identification methods in closed loop, one can use the open loop methods that are based on the prediction error, if the identifiability conditions are met. The controller must not be known.
4. If the controller does not satisfy identifiability condition 2, because the order of the controller is too low, one can still obtain identifiability by the following measures:
   a) Switching between two controllers with different parameters (Gustavsson et al, 1977; Kurz, 1977)
   b) Including a dead time in the feedback loop with $d \geq m_a - \nu + p$
   c) Use of a non-linear or time-variant controller
5. The underlying system does not need to be stable (Forssell and Ljung, 1999)

13.2 Process Identification With Additional Test Signals

Now, an external test signal shall be injected into the control loop that was introduced at the beginning of this chapter, see Fig. 13.2. In this case, the input $u(k)$ to the process is given as

$$u(k) = u_{\mathrm{C}}(k) + u_{\mathrm{S}}(k) \,, \tag{13.2.1}$$

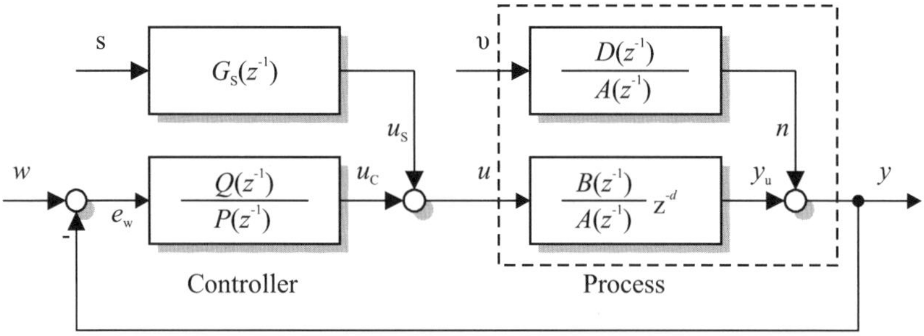

Fig. 13.2. Block diagram of the process to be identified in closed loop with an additional test signal

where

$$u_C(z) = -\frac{Q(z^{-1})}{R(z^{-1})} y(z) . \qquad (13.2.2)$$

The additional signal $u_S(z)$ can be generated by a special filter from the signal $s(z)$,

$$u_S(z) = G_S(z) s(z) . \qquad (13.2.3)$$

This setup allows to treat different experimental setups simultaneously: If $G_S(z) = G_R(z)$, then $s(k) = w(k)$ is the setpoint. $s(k)$ can also be a disturbance that is caused by the controller as e.g. in the case of non-technical controllers. Finally, if a test signal is acting directly on the process input, then $G_S(z) = 1$ and $s(k) = u_S(k)$.

$u_S(k)$ can be generated in different ways. For the following derivations, it must however only be ensured that $u_S(k)$, which is an external signal acting on the control loop, is uncorrelated with the disturbance $\nu(k)$. The additional signal $s(k)$ does in general not need to be measurable for now.

The process can again be identified indirectly based on measurements of $y(k)$ or directly with measurements of $u(k)$ as well as $y(k)$. The following derivations will be confined to the direct identification as the indirect identification does not provide any benefits.

The closed-loop transfer function is given as

$$y(z) = \frac{DP}{AP + Bz^{-d}Q} \nu(z) + \frac{Bz^{-d}P}{AP + Bz^{-d}Q} u_S(z) . \qquad (13.2.4)$$

From this follows

$$\left(AP + Bz^{-d}Q\right) y(z) = DP\nu(z) + Bz^{-d}Pu_S(z) . \qquad (13.2.5)$$

Considering (13.2.1), one obtains

$$A(z^{-1})P(z^{-1})y(z) - B(z^{-1})z^{-d}P(z^{-1})u(z) = D(z^{-1})P(z^{-1})\nu(z) . \qquad (13.2.6)$$

After canceling the polynomial $P(z^{-1})$, the same relation as for the open loop, i.e.

$$A(z^{-1})y(z) - B(z^{-1})z^{-d}u(z) = D(z^{-1})\nu(z) \qquad (13.2.7)$$

result. In contrast to (13.1.43), $u(z)$ is not only generated by the controller, but rather according to (13.2.1) also by an external test signal $u_S(z)$. Therefore, the difference equation is given as

$$\begin{aligned} u(k-d-1) = &- p_1 u(k-d-2) - \dots - p_\mu u(k-\mu-d-1) \\ &- q_0 y(k-d-1) - \dots - q_\nu y(k-\nu-d-1) \\ &+ u_S(k-d-1) - \dots - p_1 u_S(k-d-2) \\ &\dots + p_\mu u_S(k-\mu-d-1) \end{aligned} \qquad (13.2.8)$$

considering (13.2.1) and (13.2.2). If $u_S(k) \neq 0$, then $u(k-1)$ is for any arbitrary order μ and ν not linearly dependent on $\boldsymbol{\psi}^T(k)$. Hence *the process* according to (13.2.7) is *directly identifiable*, as long as the external signal $u_S(k)$ is sufficiently exciting the interesting process dynamics. Also note that it was not assumed that the additional test signal $u_S(z)$ is measurable. For an external test signal $u_S(k)$, the identifiability condition 2 that was formulated in the last section is no longer relevant. *Identifiability condition 1 however still has to be fulfilled.*

As in the last section, one can use the same open loop identification methods that are based on the prediction error $e(k)$ also in closed-loop identification, if an external test signal $u_S(k)$ is injected into the control loop. The controller must not be known and also the additional signal u_S does not need to be measurable. These results are also valid for an arbitrary noise form filter $D(z^{-1})/C(z^{-1})$.

For the identification of non-parametric transfer function models, a similar approach is suggested by Schoukens et al (2009). It is proposed to identify the transfer function $\hat{G}(\mathrm{i}\omega_k)$ by

$$\hat{G}(\mathrm{i}\omega_k) = \frac{\hat{G}_{\mathrm{wy}}(\mathrm{i}\omega_k)}{\hat{G}_{\mathrm{wu}}(\mathrm{i}\omega_k)}\,, \qquad (13.2.9)$$

where $w(\mathrm{i}\omega)$ denotes the setpoint and $\hat{G}_{\mathrm{wy}}(\mathrm{i}\omega_k)$ the estimate of the transfer function from $w(\mathrm{i}\omega)$ to $y(\mathrm{i}\omega)$. Both intermediate transfer functions show a measurement error only in the "output" and can hence be identified asymptotically bias-free with appropriate methods for the identification of frequency responses.

13.3 Methods for Identification in Closed Loop

In this section, some concluding remarks shall be made on the applicability of online identification methods to closed-loop process identification. For the application, one has to observe the identifiability conditions that have been stated in Sects. 13.1 and 13.2.

13.3.1 Indirect Process Identification Without Additional Test Signals

If the process is identified indirectly, i.e. only by measurement of the output $y(k)$ and if no additional signal is injected into the control loop, then one can determine the parameters α_i and β_i of the ARMA process in (13.1.5) by the RLS method for stochastic signals, see Sect. 9.4.2. In a second step, the process parameters a_i and b_i have to be determined from (13.1.37) by e.g. (13.1.38), if the identifiability conditions are met. A different method is based on the correlation in combination with least squares (RCOR-LS) (Kurz and Isermann, 1975).

The parameter estimates converge very slowly, which can be attributed to both the large number of parameters $l + r$ to be estimated and the fact that the input signal $v(k)$ is also unknown and must be estimated as well. If the process input $u(k)$ is measurable, one should therefore always prefer the direct method, as discussed in the next section.

13.3.2 Indirect Process Identification With Additional Test Signals

If the controller is completely known, it is theoretically also possible to identify a model from setpoint $w(k)$ to output $y(k)$ and then determine the transfer function of the plant from knowledge of the identified transfer function of the entire closed-loop process and the transfer function of the controller. As $w(k)$ and $y(k)$ are uncorrelated, one can in this case use all methods that are capable of identifying open-loop systems (Forssell and Ljung, 1999).

13.3.3 Direct Process Identification Without Additional Test Signals

As shown in Sect. 13.1, parameter estimation methods that are based on the prediction error $e(k)$ are on principle also suited for identification in closed loop. Therefore, especially the methods RLS, RELS, and RML are applicable. If the identifiability conditions 1 and 2 are satisfied, then these methods can be applied to the signals $u(k)$ and $y(k)$ just as in the open-loop case. They yield unbiased and consistent estimates, if the form filter of the noise has the form $1/A$ in the case of RLS and the form D/A in the case of RELS and RML.

For the unbiased estimation with the RIV method, the vector of instrumental variables $\boldsymbol{w}^{\mathrm{T}}(k)$ (10.5.9) may not be correlated with the error $e(k)$ and hence the disturbance $n(k)$. The process input $u(k-\tau)$ is due to the feedback loop, however, for $\tau \geq 0$ correlated with $n(k)$. The method RIV will henceforth yield biased estimates in closed loop. The correlation between $u(k-\tau)$ and $e(k)$ would vanish for $\tau \geq 1$ only, if the form filter of the noise has the form $1/A$ and $e(k)$ is an uncorrelated signal. Special adaptations of the instrumental variable method to closed-loop identification have been proposed (e.g. Gilson and van den Hof, 2005; Gilson et al, 2009).

13.3.4 Direct Process Identification With Additional Test Signals

If, as described in Sect. 13.2, an external signal is injected into the control loop, then one must only observe identifiability condition 1 and can abolish identifiability condition 2. If $u(k)$ and $y(k)$, but not the additional signal $u_{\mathrm{S}}(k)$, are utilized for parameter estimation, then one can use the methods RLS, RELS, and RML. If measurable, the additional signal $u_S(k)$ can be used to form instrumental variables for RIV. Then, this method can be applied to the same form filters of the noise as in the open loop case.

The application of the RCOR-LS method in closed loop to the three cases considered in this chapter has been described by Kurz and Isermann (1975). It is suitable if the parameter estimates are not required after each sample step, but only in larger time intervals.

13.4 Summary

The identification of processes in closed loop can be accomplished even *without an additional external test signal*. In case of the indirect method, one measures the output signal $y(k)$ only, estimates an intermediate ARMA model of the closed-loop process and then determines the parameters of the open-loop process from this intermediate model. The direct parameter estimation method makes use of the measurements of both the input $u(k)$ and output $y(k)$ as in the open loop case. In both cases, the process is identifiable under two identifiability conditions that have been formulated in Sect. 13.1.1. The order of controller and process must be known exactly and the feedback must be of sufficiently high order. If the identifiability condition 2 is not fulfilled by a constant controller, one can identify in closed loop by switching between *two different controllers* or between *two different parameter sets for the same controller*. Kurz (1977) has shown that the variance of the parameter estimates can be reduced if the switching time is reduced to $(5 \dots 10)T_0$. A quantification of the variance-error for identified poles and zeros is shown in (Mårtensson and Hjalmarsson, 2009) and is also discussed for the closed loop case. Often, closed loop identification shall be used to improve the control quality of the closed loop system by subsequently designing an improved (adaptive) controller. This gives raise to different issues that are discussed in (Hjalmarsson, 2005).

For the identification *with an additional external test signal*, the second identifiability condition does not need to be considered any longer. Since the indirect methods converge very slowly, one will in general prefer the direct methods. Here, one can apply the typical open-loop methods that have already been introduced in the previous chapters. As a lookahead to subspace methods (Chap. 16), it shall be noted here that subspace methods will typically yield biased estimates when applied to closed-loop identification (Ljung, 1999).

Problems

13.1. Indirect Process Identification

Given a process of order $n = 2$ and $d = 1$. What is the required order of a linear controller for indirect process identification in closed loop?

13.2. Direct Process Identification
Why should you employ information about both the input $u(k)$ and the output $y(k)$ whenever possible?

13.3. Direct Process Identification
A colleague suggests to identify a system in closed loop without an external test signal by measuring both the input $u(k)$ and the output $y(k)$ and determining the process transfer function by $G(\mathrm{i}\omega) = y(\mathrm{i}\omega)/u(\mathrm{i}\omega)$. Why will this not provide the expected results? If one can inject an external test signal, will the approach then work?

References

Bohlin T (1971) On the problem of ambiguities in maximum likelihood identification. Automatica 7(2):199–210

Forssell U, Ljung L (1999) Closed-loop identification revisited. Automatica 35(7):1215–1241

Gilson M, van den Hof PMJ (2005) Instrumental variable methods for closed-loop system identification. Automatica 41:241–249

Gilson M, Garnier H, Young PC, van den Hof PMJ (2009) Refined instrumental variable methods for closed-loop system identification. In: Proceedings of the 15th IFAC Symposium on System Identification, Saint-Malo, France

Gustavsson I, Ljung L, Söderström T (1974) Identification of linear multivariable process dynamics using closed-loop experiments. Report 7401. Dept. of Automatic Control, Lund Inst of Technology, Lund

Gustavsson I, Ljung L, Söderström T (1977) Identification in closed loop – identifiability and accuracy aspects. Automatica 13(1):59–75

Hjalmarsson H (2005) From experiment design to closed-loop control. Automatica 41(3):393–438

Isermann R (1992) Identifikation dynamischer Systeme: Besondere Methoden, Anwendungen (Vol 2). Springer, Berlin

Kurz H (1977) Recursive process identification in closed loop with switching regulators. In: Proceedings of the 4th IFAC Symposium Digital Computer Applications to Process Control, Amsterdam, Netherlands

Kurz H, Isermann R (1975) Methods for on-line process identification in closed loop. In: Proceedings of the 6th IFAC Congress, Boston, MA, USA

Ljung L (1999) System identification: Theory for the user, 2nd edn. Prentice Hall Information and System Sciences Series, Prentice Hall PTR, Upper Saddle River, NJ

Mårtensson J, Hjalmarsson H (2009) Variance-error quantification for identified poles and zeros. Automatica 45(11):2512–2525

Schoukens J, Vandersteen K, Barbé, Pintelon R (2009) Nonparametric preprocessing in system identification: A powerful tool. In: Proceedings of the European Control Conference 2009 - ECC 09, Budapest, Hungary, pp 1–14

Part IV

Identification with Parametric Models — Continuous Time Signals

14

Parameter Estimation for Frequency Responses

This chapter presents parameter estimation methods which use the non-parametric frequency response function as an intermediate model. Using this intermediate model can provide many advantages: One can use methods such as the orthogonal correlation to record the frequency response function even under very adverse (noise) conditions. Furthermore, the experimental data are in most cases condensed by smoothing the frequency response function before the parameter estimation method is applied. Also, the non-parametric frequency response function can give hints on the model order to choose, the presence of a dead-time, resonances, and so forth. Juang (1994) also points out that e.g. in modal analysis, the frequency response is much better established representation than the plot of measurement data over time and thus more familiar to the test engineers.

14.1 Introduction

In the following, it is assumed that $N + 1$ points of the frequency response function have been determined as

$$G(\mathrm{i}\omega_\nu) = |G(\mathrm{i}\omega_\nu)|\mathrm{e}^{-\mathrm{i}\varphi(\omega_\nu)} = \mathrm{Re}\{G(\mathrm{i}\omega_\nu)\} + \mathrm{Im}\{G(\mathrm{i}\omega_\nu)\} \tag{14.1.1}$$

by using a direct frequency response measurement technique, see Chap. 4 and 5.

This frequency response, which is given in non-parametric form and represents an intermediate model, shall now be approximated by a parametric transfer function

$$G(\mathrm{i}\omega) = \frac{b_0 + b_1\mathrm{i}\omega + \ldots + b_m(\mathrm{i}\omega)^m}{1 + a_1\mathrm{i}\omega + \ldots + a_n(\mathrm{i}\omega)^n}\,. \tag{14.1.2}$$

While in the past, often graphical methods have been used, see e.g. the overview by Strobel (1968), nowadays mainly analytical methods are used (e.g. Isermann, 1992; Pintelon and Schoukens, 2001).

In the following, methods shall be presented that allow to extract a parametric model from the frequency response. Some of the methods covered in the following

R. Isermann, M. Münchhof, *Identification of Dynamic Systems*,
DOI 10.1007/978-3-540-78879-9_14, © Springer-Verlag Berlin Heidelberg 2011

use only one coordinate of the frequency response, i.e. magnitude or phase or real or imaginary part respectively, as under certain prerequisites, the two components are coupled. This mutual dependency is given as follows:

Real and imaginary part of a stable, realizable ($m \leq n$) system are coupled by means of the Hilbert transform $g(y) = \mathfrak{H}(f(x))$, see e.g. (Unbehauen, 2008; Kammeyer and Kroschel, 2009),

$$g(y) = \mathfrak{H}(f(x)) = \frac{1}{\pi} \int_{-\infty}^{\infty} \frac{f(x)}{y-x} \mathrm{d}x \tag{14.1.3}$$

as

$$R(\omega) = +\frac{1}{\pi} \int_{-\infty}^{\infty} \frac{I(u)}{\omega - u} \mathrm{d}u + R(\infty) \tag{14.1.4}$$

$$I(\omega) = -\frac{1}{\pi} \int_{-\infty}^{\infty} \frac{R(u)}{\omega - u} \mathrm{d}u\;. \tag{14.1.5}$$

Hence, if the course of the imaginary part of the frequency response is given, one can determine the real part under the above stated prerequisites and vice versa. As a side-note, the satisfaction of the Hilbert transform is a necessary and also a sufficient condition for causality of a system.

A similar relation is given for the magnitude and phase,

$$\ln|G(\mathrm{i}\omega)| - \ln|G(\mathrm{i}\infty)| = -\frac{1}{\pi} \int_{-\infty}^{\infty} \frac{\varphi(u) - \varphi(\omega)}{u - \omega} \mathrm{d}u \tag{14.1.6}$$

$$\varphi(\omega) = \frac{1}{\pi} \int_{-\infty}^{\infty} \frac{\ln|G(\mathrm{i}u)| - \ln|G(\mathrm{i}\omega)|}{u - \omega} \mathrm{d}u\;. \tag{14.1.7}$$

For a given shape of the magnitude $|G(\mathrm{i}\omega)|$, one can thus easily determine the phase and vice versa. For those methods that only approximate the shape of the magnitude of the frequency response, one must ensure that the system does not contain a dead-time or an all-pass. If a dead-time is present, one should first determine the rational part of the transfer function by approximating the magnitude of the frequency response and then in the next step determine the dead time from the phase difference.

14.2 Method of Least Squares for Frequency Response Approximation (FR-LS)

Typically, there are much more points of the frequency response function measured than there are parameters of the (parametric) transfer function to be identified. In this case, one can use parameter estimation methods to minimize the error between the measured frequency response function and the model. The process shall be given by

$$G(\mathrm{i}\omega) = R(\omega) + \mathrm{i}I(\omega) \tag{14.2.1}$$

and the model by

$$\hat{G}(\mathrm{i}\omega) = \frac{\hat{B}(\mathrm{i}\omega)}{\hat{A}(\mathrm{i}\omega)} = \frac{\hat{B}_\mathrm{R}(\omega) + \mathrm{i}\hat{B}_\mathrm{I}(\omega)}{\hat{A}_\mathrm{R}(\omega) + \mathrm{i}\hat{A}_\mathrm{I}(\omega)} \,. \tag{14.2.2}$$

The task is now to determine the parameters $\hat{a}_i$ and $\hat{b}_i$ from $N+1$ measurements

$$G(\mathrm{i}\omega_n) = R(\omega_n) + \mathrm{i}I(\omega_n),\ n = 0, 1, \ldots, N \,. \tag{14.2.3}$$

For this task, one could use the output error as shown in Fig. 1.8 and defined as

$$e(\mathrm{i}\omega_n) = G(\mathrm{i}\omega_n) - \frac{\hat{B}(\mathrm{i}\omega_n)}{\hat{A}(\mathrm{i}\omega_n)} \,. \tag{14.2.4}$$

A cost function would then be given by

$$V = \frac{1}{N+1} \sum_{k=0}^{N} \frac{|G(\mathrm{i}\omega_k) - \hat{G}(\mathrm{i}\omega_k, \boldsymbol{\theta})|^2}{\sigma_\mathrm{G}^2(k)} \,, \tag{14.2.5}$$

where in this case, already the measurement error has been introduced as a weight for the individual errors (Pintelon and Schoukens, 2001).

The output error once again is non-linear in the parameters and hence necessitates the use of an iterative optimization algorithm, see Chap. 19. Here, the identification of $G(\mathrm{i}\omega_k)$ might have been carried out by dividing the DFT of the output $y(\mathrm{i}\omega_k)$ by the the DFT of the input $u(\mathrm{i}\omega_k)$. In most applications, the sampling is fast enough and the transfer function shows a sufficient decay of the amplitude at high frequencies so that $G(\mathrm{i}\omega_k)$ approximates the continuous-time transfer function $G(s)$ sufficiently well. In other cases, one must use approximating techniques as described e.g. by Gillberg and Ljung (2010).

Multiplying (14.2.5) with $U(s)$, one can derive the cost function

$$V = \sum_{k=1}^{N} |Y(\mathrm{i}\omega_k) - \hat{Y}(\mathrm{i}\omega_k, \boldsymbol{\theta})|^2 \,, \tag{14.2.6}$$

which still leads to a non-linear optimization problem (Pintelon and Schoukens, 2001)

One can also use the generalized equation error, which in this case is given as

$$\varepsilon(\mathrm{i}\omega_n) = \hat{A}(\mathrm{i}\omega_n)e(\mathrm{i}\omega_n) = \hat{A}(\mathrm{i}\omega_n)G(\mathrm{i}\omega_n) - \hat{B}(\mathrm{i}\omega_n) \,, \tag{14.2.7}$$

see (Levy, 1959; Sawaragi et al, 1981; Pintelon and Schoukens, 2001). As a cost function, one can now use the sum of the weighted squared error as

$$V = \sum_{n=0}^{N} w_n |\varepsilon(\mathrm{i}\omega_n)|^2 \,, \tag{14.2.8}$$

where the w_n are weighting factors. Inserting (14.2.3) and (14.2.7), one obtains

$$\varepsilon(\mathrm{i}\omega) = \hat{A}(\mathrm{i}\omega)e(\mathrm{i}\omega) = \big(R(\omega)\hat{A}_{\mathrm{R}}(\omega) - I(\omega)\hat{A}_{\mathrm{I}}(\omega) - \hat{B}_{\mathrm{R}}(\mathrm{i}\omega)\big) + \mathrm{i}\big(R(\omega)\hat{A}_{\mathrm{I}}(\omega) + I(\omega)\hat{A}_{\mathrm{R}}(\omega) - \hat{B}_{\mathrm{I}}(\mathrm{i}\omega)\big)\,, \tag{14.2.9}$$

which then yields

$$\begin{aligned} V &= \sum_{n=0}^{N}\Big(w_n\big(R(\omega_n)\hat{A}_{\mathrm{R}}(\omega_n) - I(\omega_n)\hat{A}_{\mathrm{I}}(\omega_n) - \hat{B}_{\mathrm{R}}(\omega_n)\big)^2 \\ &\quad + w_n\big(R(\omega_n)\hat{A}_{\mathrm{I}}(\omega_n) + I(\omega_n)\hat{A}_{\mathrm{R}}(\omega_n) - \hat{B}_{\mathrm{I}}(\omega_n)\big)^2\Big) \\ &= \sum_{n=0}^{N} w_n\big(L_n^2 + M_n^2\big)\,. \end{aligned} \tag{14.2.10}$$

With the polynomials

$$A_{\mathrm{R}}(\omega) = 1 - a_2\omega^2 + a_4\omega^4 - a_6\omega^6 + \ldots \tag{14.2.11}$$

$$A_{\mathrm{I}}(\omega) = a_1\omega - a_3\omega^3 + a_5\omega^5 - \ldots \tag{14.2.12}$$

$$B_{\mathrm{R}}(\omega) = b_0 - b_2\omega^2 + b_4\omega^4 - b_6\omega^6 + \ldots \tag{14.2.13}$$

$$B_{\mathrm{I}}(\omega) = b_1\omega - b_3\omega^3 + b_5\omega^5 - \ldots\,, \tag{14.2.14}$$

one can now formulate the data matrix, parameter vector, and output vector for the method of least squares. As the cost function (14.2.10) contains a sum of two squared terms for each frequency ω_n, the matrices and vectors will contain two rows for each frequency ω_n. For a certain frequency, the addend $\Delta V(\omega_n)$ for the cost function V is given as

$$\Delta V(\omega_n) = w_n\varepsilon_{\mathrm{R}}(\omega_n)^2 + w_n\varepsilon_{\mathrm{I}}(\omega_n)^2\,. \tag{14.2.15}$$

Therefore,

$$\boldsymbol{\varepsilon}_n = \begin{pmatrix} \varepsilon_{\mathrm{R}}(\omega_n) \\ \varepsilon_{\mathrm{I}}(\omega_n) \end{pmatrix}. \tag{14.2.16}$$

The errors for the real part can be written as

$$\begin{aligned} \varepsilon_{\mathrm{R}}(\omega_n) = R(\omega_n) &- a_2\omega_n^2 R(\omega_n) + a_4\omega_n^4 R(\omega_n) - \ldots \\ &- a_1\omega_n I(\omega_n) + a_3\omega_n^3 I(\omega_n) - \ldots - b_0 + b_2\omega_n^2 - b_4\omega_n^4 - \ldots \end{aligned} \tag{14.2.17}$$

and the imaginary part as

$$\begin{aligned} \varepsilon_{\mathrm{I}}(\omega_n) = I(\omega_n) &- a_2\omega_n^2 I(\omega_n) + a_4\omega_n^4 I(\omega_n) - \ldots \\ &+ a_1\omega_n R(\omega_n) - a_3\omega_n^3 R(\omega_n) + \ldots - b_1\omega_n + b_3\omega_n^3 - \ldots\,. \end{aligned} \tag{14.2.18}$$

These equations can now be split up into the data matrix

$$\boldsymbol{\Psi}_n^{\mathrm{T}} = \begin{pmatrix} -\omega_n I(\omega_n) & -\omega_n^2 R(\omega_n) & \omega_n^3 I(\omega_n) & \omega_n^4 R(\omega_n) & \ldots & -1 & 0 & \omega_n^2 & 0 & \ldots \\ +\omega_n R(\omega_n) & -\omega_n^2 I(\omega_n) & -\omega_n^3 R(\omega_n) & \omega_n^4 I(\omega_n) & \ldots & 0 & -\omega_n & 0 & \omega_n^3 & \ldots \end{pmatrix} \tag{14.2.19}$$

and the output vector

$$y_n = \begin{pmatrix} R(\omega_n) \\ I(\omega_n) \end{pmatrix} . \tag{14.2.20}$$

For this decomposition, the parameter vector is given as

$$\boldsymbol{\theta}^{\mathrm{T}} = \begin{pmatrix} a_1 \; a_2 \; a_3 \; a_4 \ldots b_0 \; b_1 \; b_2 \; b_3 \; b_4 \ldots \end{pmatrix} \tag{14.2.21}$$

and the weighting matrix is given as

$$\boldsymbol{W}_n = \begin{pmatrix} w(\omega_n) & 0 \\ 0 & w(\omega_n) \end{pmatrix} . \tag{14.2.22}$$

From these parts, one can now construct

$$\boldsymbol{\Psi} = \begin{pmatrix} \boldsymbol{\psi}_0^{\mathrm{T}} \\ \boldsymbol{\psi}_1^{\mathrm{T}} \\ \vdots \\ \boldsymbol{\psi}_N^{\mathrm{T}} \end{pmatrix} \tag{14.2.23}$$

and

$$\boldsymbol{y} = \begin{pmatrix} y_0 \\ y_1 \\ \vdots \\ y_N \end{pmatrix} , \tag{14.2.24}$$

as well as

$$\boldsymbol{W} = \begin{pmatrix} \boldsymbol{W}_0 & \boldsymbol{0} & & \boldsymbol{0} \\ \boldsymbol{0} & \boldsymbol{W}_1 & & \boldsymbol{0} \\ & & \ddots & \\ \boldsymbol{0} & \boldsymbol{0} & & \boldsymbol{W}_N \end{pmatrix} , \tag{14.2.25}$$

and can solve the problem of weighted least squares by e.g.

$$\hat{\theta} = \left(\boldsymbol{\Psi}^{\mathrm{T}} \boldsymbol{W} \boldsymbol{\Psi}\right)^{-1} \boldsymbol{\Psi}^{\mathrm{T}} \boldsymbol{W} \boldsymbol{y} , \tag{14.2.26}$$

see (9.5.4).

For the choice of the weighting factors w_n, one can consider the accuracy of the frequency response measurements by weighting with the relative error

$$w_n = c \frac{|G(\mathrm{i}\omega_n)|^2}{|\Delta G(\mathrm{i}\omega_n)|^2} , \tag{14.2.27}$$

where $|\Delta G(\mathrm{i}\omega_n)|$ is a given absolute value. This is often an appealing choice as the relative error of the measurements increases at higher frequencies. Measurements at high frequencies are often of smaller accuracy and hence by choosing the reciprocal, a high relative error will lead to a small weight. Furthermore, as pointed out by

Pintelon et al (1994), the cost function in (14.2.7) is a polynomial in ω_n and measurement errors are weighted with ω_n^{2m}. Therefore, the individual terms of the cost function will have a smaller weight as ω increases.

By formulating the equation error (14.2.7), one does from a mathematical point of view weight the output error $e(\omega_n)$ with $A(\mathrm{i}\omega_n)$. This weighting can be compensated by choosing w_n as

$$w_n = \frac{1}{|A(\mathrm{i}\omega_n)|^2} \,. \tag{14.2.28}$$

As $A(\mathrm{i}\omega_n)$ is seldom known a priori, one must use an iterative approach to solve the parameter estimation problem. One can also combine both of the above weights.

A modified method of least squares was proposed by Strobel (1968, 1975). Here, the cost function is formulated by the weighted relative frequency response error

$$V = \sum_{n=0}^{N} w_n \left| \frac{\Delta G(\mathrm{i}\omega_n)}{G(\mathrm{i}\omega_n)} \right|^2 = \sum_{n=0}^{N} w_n \left| \frac{G(\mathrm{i}\omega_n) - \hat{G}(\mathrm{i}\omega_n)}{G(\mathrm{i}\omega_n)} \right|^2 \tag{14.2.29}$$

with

$$\Delta G(\mathrm{i}\omega_n) = G(\mathrm{i}\omega_n) - \hat{G}(\mathrm{i}\omega_n) \tag{14.2.30}$$

being the deviation between measured and estimated frequency response.

Example 14.1 (Identification of a Discrete-Time Dynamic Model from the Frequency Response of the Three-Mass Oscillator).

This technique has now been applied to the Three-Mass Oscillator and a transfer function has been estimated from the frequency response that was recorded by means of the orthogonal correlation, see Sect. 5.5.2. The results have been shown in Fig. 14.1 and illustrate a good match between the parametric transfer function model and the non-parametric frequency response. □

If the input $u(k)$ and output $y(k)$ have been brought independently to the frequency domain by the Fourier transform, one can formulate the cost function

$$V = \sum_{k=1}^{N} |\hat{A}(\mathrm{i}\omega_k, \boldsymbol{\theta})Y(\mathrm{i}\omega_k) - \hat{B}(\mathrm{i}\omega_k, \boldsymbol{\theta})U(\mathrm{i}\omega_k)|^2 \tag{14.2.31}$$

in analogy to the above equation error. This formulation is also linear in parameters and can be solved by the direct and recursive method of least squares (Pintelon and Schoukens, 2001; Ljung, 1999).

14.3 Summary

It has been shown how one can estimate the parameters of the transfer function from the directly measured frequency response as an intermediate model. Determining the frequency response as a non-parametric model before estimating the parameters of

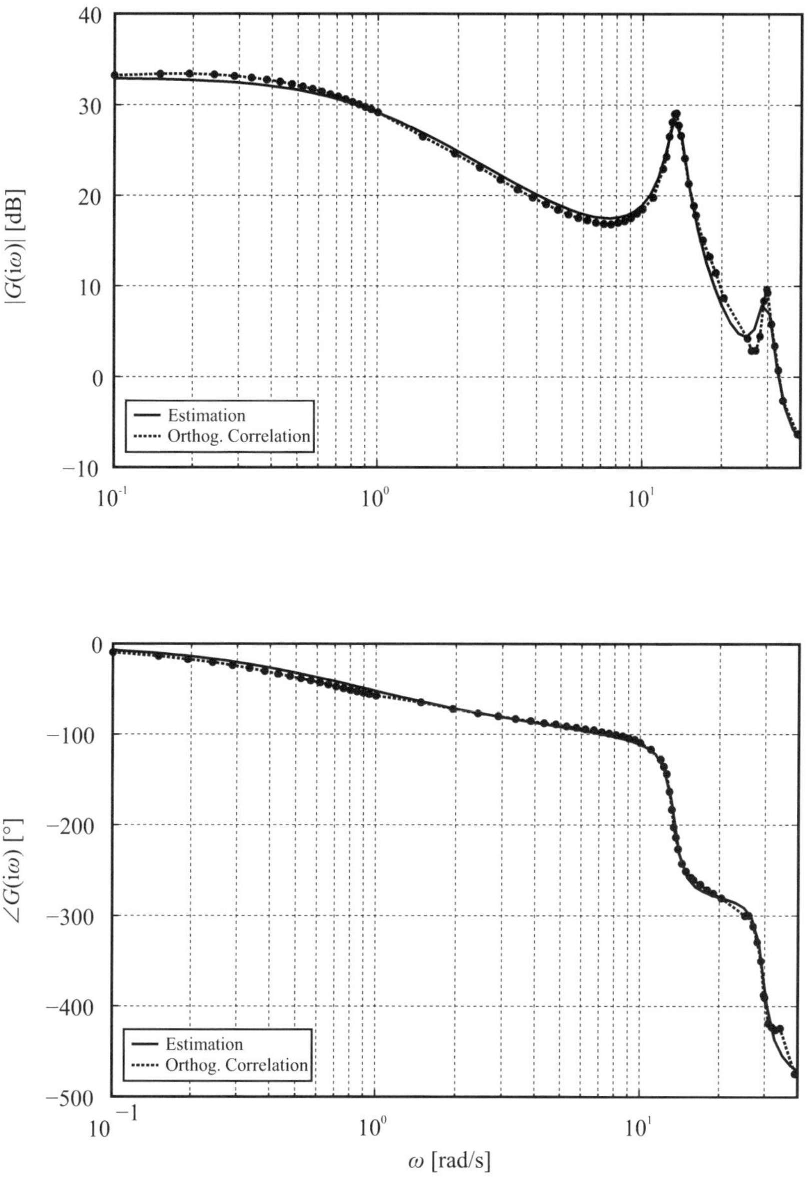

Fig. 14.1. Estimation of the transfer function of the Three-Mass Oscillator from the frequency response measurement by means of the orthogonal correlation with $N = 68$ data points in the frequency range $0.1\,\text{rad/s} \leq \omega \leq 40\,\text{rad/s}$

the transfer function has several advantages. First, the frequency response can e.g. be determined by means of the orthogonal correlation, which is capable of processing very noisy signals. Secondly, as the frequency response function in this case is a non-parametric model, it can be used to verify the assumptions about the model order and dead time before deriving the parametric model. Also, the transfer function can easily be converted into a set of ordinary differential equations governing the dynamics of the process in continuous-time. As pointed out in Chap. 15, the coefficients of ordinary differential equations can often easily be converted to process parameters. Pintelon et al (1994) also stress the easy noise reduction capabilities by simply leaving out non or slightly excited and hence especially noisy frequencies and the possibility the combine data from different experiments. Hence, the parameters can more easily be interpreted than those of discrete-time models. Other estimators, such as e.g. maximum likelihood estimators based on different assumptions about the noise have been formulated for identification in the frequency domain, (Pintelon and Schoukens, 1997; McKelvey, 2000). However, these can often only be solved with iterative optimization algorithms and hence can become computationally expensive.

Problems

14.1. Identification from Frequency Response
Name three advantages of using the non-parametric frequency response function as an intermediate model. What methods do you know to obtain the frequency response function?

14.2. Minimum Phase System
Why is it for a minimum phase system sufficient to match the real part of the recorded non-parametric frequency response and the parametric transfer function?

14.3. Minimum Phase System with Additional Dead Time
Given is a minimum phase system with an unknown dead time due to the measurement setup. Why is it possible to identify this system based on the magnitude of the frequency response function neglecting the dead time? Derive a cost function that would identify the parametric transfer function based on the magnitude. How can one identify the dead time of the measurement setup?

References

Gillberg J, Ljung L (2010) Frequency domain identification of continuous-time output error models, Part I: Uniformly sampled data and frequency function approximation. Automatica 46(1):1–10

Isermann R (1992) Identifikation dynamischer Systeme: Besondere Methoden, Anwendungen (Vol 2). Springer, Berlin

Juang JN (1994) Applied system identification. Prentice Hall, Englewood Cliffs, NJ

Kammeyer KD, Kroschel K (2009) Digitale Signalverarbeitung: Filterung und Spektralanalyse mit MATLAB-Übungen, 7th edn. Teubner, Wiesbaden

Levy EC (1959) Complex curve fitting. IRE Trans Autom Control 4:37–43

Ljung L (1999) System identification: Theory for the user, 2nd edn. Prentice Hall Information and System Sciences Series, Prentice Hall PTR, Upper Saddle River, NJ

McKelvey T (2000) Frequency domain identification. In: Proccedings of the 12th IFAC Symposium on System Identification, Santa Barbara, CA, USA

Pintelon R, Schoukens J (1997) Identification of continuous-time systems using arbitrary signals. Automatica 33(5):991–994

Pintelon R, Schoukens J (2001) System identification: A frequency domain approach. IEEE Press, Piscataway, NJ

Pintelon R, Guillaume P, Rolain Y, Schoukens J, van Hamme H (1994) Parametric identification of transfer functions in the frequency domain: A survey. IEEE Trans Autom Control 39(11):2245–2260

Sawaragi V, Soeda T, Nakamizo T (1981) Classical methods and time series estimation. In: Trends and progress in system identification, Pergamon Press, Oxford

Strobel H (1968) Systemanalyse mit determinierten Testsignalen. Verlag Technik, Berlin

Strobel H (1975) Experimentelle Systemanalyse. Elektronisches Rechnen und Regeln, Akademie Verlag, Berlin

Unbehauen H (2008) Regelungstechnik, 15th edn. Vieweg + Teubner, Wiesbaden

15

Parameter Estimation for Differential Equations and Continuous Time Processes

Parameter estimation methods for dynamic processes were first developed for process models in discrete-time in combination with digital control systems. For some applications, e.g. the validation of theoretical models or for fault diagnosis, however, parameter estimation methods for models with continuous-time signals are needed.

Furthermore, continuous-time models are often better to interpret as the model parameters can in many cases be transformed to physical parameters. A comparison of discrete-time and continuous-time methods to identify a continuous-time model has been carried out by Rao and Garnier (2002). Garnier and Wang (2008) and Rao and Unbehauen (2006) suggest to discern two fundamentally different approaches: For *direct* approaches, a discrete-time model is set up with the same parameterization as the continuous time model, for *indirect* approaches, the model is estimated in discrete time with a different parameterization and then transformed back to continuous time. Both methods will be treated in this chapter. First, it will be discussed, how models with the continuous time parameterization can be estimated directly, where various techniques are presented to determine the derivatives. Then, the approach to convert a discrete time model to continuous time parameters will also be discussed. The first survey paper on this topic was published in 1981 by Young (1981), a recent survey was given by Rao and Unbehauen (2006). Also, the book by Garnier and Wang (2008) gives an overview over different developments in this field.

15.1 Method of Least Squares

15.1.1 Fundamental Equations

A stable process with lumped parameters is considered, which can be described by the linear, time invariant differential equation

$$\begin{aligned} &a_n y_u^{(n)}(t) + a_{n-1} y_u^{(n-1)}(t) + \ldots + a_1 y_u^{(1)}(t) + y_u(t) \\ &\quad = b_m u^{(m)}(t) + b_{m-1} u^{(m-1)}(t) + \ldots + b_1 u^{(1)}(t) + b_0 u(t) \,, \end{aligned} \qquad (15.1.1)$$

R. Isermann, M. Münchhof, *Identification of Dynamic Systems*,
DOI 10.1007/978-3-540-78879-9_15, © Springer-Verlag Berlin Heidelberg 2011

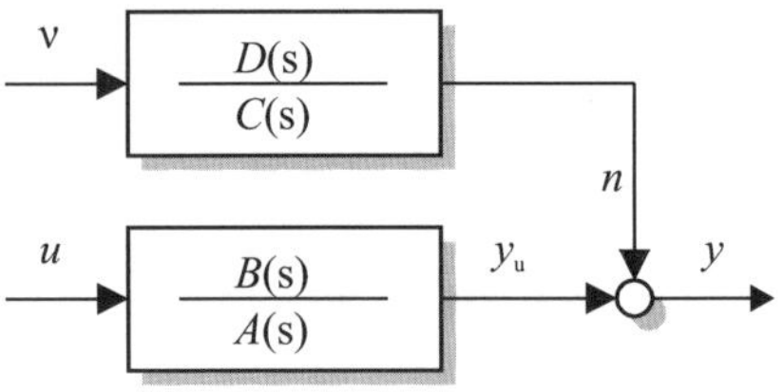

Fig. 15.1. Linear process with continuous-time signals

where $m < n$. It is assumed that the derivatives of the output signal

$$y^{(j)}(t) = \frac{\mathrm{d}^j y(t)}{\mathrm{d}t^j} \text{ with } j = 1, 2, \ldots, n \tag{15.1.2}$$

and of the input signal $u(t)$ for $j = 1, 2, \ldots, m$ exist. $u(t)$ and $y(t)$ are the deviations

$$\begin{aligned} u(t) &= U(t) - U_{00} \\ y(t) &= Y(t) - Y_{00} \end{aligned} \tag{15.1.3}$$

of the absolute signals $U(t)$ and $Y(t)$ from the operating point described by U_{00} and Y_{00}. The transfer function corresponding to the ODE given in (15.1.1) is

$$G_{\mathrm{P}}(s) = \frac{y_{\mathrm{u}}(s)}{u(s)} = \frac{B(s)}{A(s)} = \frac{b_0 + b_1 s + \ldots + b_{m-1} s^{m-1} + b_m s^m}{1 + a_1 s + \ldots + a_{n-1} s^{n-1} + a_n s^n} , \tag{15.1.4}$$

see Fig. 15.1.

The measurable signal $y(t)$ contains an additional disturbance signal $n(t)$

$$y(t) = y_{\mathrm{u}}(t) + n(t) . \tag{15.1.5}$$

Substituting (15.1.5) into (15.1.1) and introducing an *equation error* $e(t)$ yields

$$y(t) = \boldsymbol{\psi}^{\mathrm{T}}(t)\boldsymbol{\theta} + e(t) \tag{15.1.6}$$

with

$$\boldsymbol{\psi}^{\mathrm{T}}(t) = \big(-y^{(1)}(t) \ldots -y^{(n)}(t) \big| u(t) \ldots u^{(m)}(t)\big) \tag{15.1.7}$$

$$\boldsymbol{\theta} = \big(a_1 \ldots a_n \big| b_0 \ldots b_m\big)^{\mathrm{T}} \tag{15.1.8}$$

in similarity to Sect. 9.1. All methods presented in this chapter work with the equation error. If one wants to use the output, i.e. simulation error, one has to resort to the methods described in Chap. 19.

One can also set up the model

$$y^{(n)}(t) = \boldsymbol{\psi}^{\mathrm{T}}(t)\boldsymbol{\theta} + e(t) \tag{15.1.9}$$

with the appropriate entries of $\boldsymbol{\psi}$ and $\boldsymbol{\theta}$ (Young, 1981). As one should try to select the most noisy variable as the output y of the model for identification methods that assume noise only on the model output, this can be an appealing alternative. One can however also use other methods, such as the method of total least squares (TLS), see Sect. 10.4, that assume noise not only on the model output, but also on the regressors.

The input and output signals are now measured at discrete-time samples $t = kT_0$ with $k = 0, 1, 2, \ldots, N$ with sampling time T_0 and the derivatives are determined. Based on these data, then $N + 1$ equations can be written down

$$y(k) = \boldsymbol{\psi}^{\mathrm{T}}(k)\boldsymbol{\theta} + e(k) \text{ for } k =, 1, 2, \ldots, N \; . \tag{15.1.10}$$

This system of equations can be written in matrix notation as

$$\boldsymbol{y} = \boldsymbol{\Psi}\boldsymbol{\theta} + \boldsymbol{e} \tag{15.1.11}$$

with

$$\boldsymbol{y}^{\mathrm{T}} = \big(y(0)\; y(1)\; \ldots\; y(N)\big) \tag{15.1.12}$$

$$\boldsymbol{\Psi} = \left(\begin{array}{ccc|ccc} -y^{(1)}(0) & \ldots & -y^{(n)}(0) & u(0) & \ldots & u^{(m)}(0) \\ -y^{(1)}(1) & \ldots & -y^{(n)}(1) & u(1) & \ldots & u^{(m)}(1) \\ \vdots & & \vdots & \vdots & & \vdots \\ -y^{(1)}(N) & \ldots & -y^{(n)}(N) & u(1) & \ldots & u^{(m)}(N) \end{array}\right) . \tag{15.1.13}$$

Both, Ljung and Wills (2008) as well as Larsson and Söderström (2002) point out that an irregular sample time can much easier be accommodated by continuous-time models. If the derivatives can be measured directly, one could obviously evaluate the differential equations for any time t. Otherwise, algorithms for determining the derivatives of the measured signals numerically have to be adapted to the varying sample time (e.g. Larsson and Söderström, 2002).

Minimizing the cost function

$$V = \boldsymbol{e}^{\mathrm{T}}\boldsymbol{e} = \sum_{k=0}^{N} e^2(k) \tag{15.1.14}$$

yields with $\mathrm{d}V/\mathrm{d}\hat{\boldsymbol{\theta}} = \boldsymbol{0}$ and $\boldsymbol{\theta} = \hat{\boldsymbol{\theta}}$, as previously shown in Sect. 9.1, the vector of parameter estimates for the least squares method as

$$\hat{\boldsymbol{\theta}} = \left(\boldsymbol{\Psi}^{\mathrm{T}}\boldsymbol{\Psi}\right)^{-1}\boldsymbol{\Psi}^{\mathrm{T}}\boldsymbol{y} \; . \tag{15.1.15}$$

The existence of a unique solution requires that the matrix $\boldsymbol{\Psi}^{\mathrm{T}}(N)\boldsymbol{\Psi}(N)$ is non-singular. It can be seen that this approach is very similar to the least squares method for models with discrete-time signals. Hence, a lot of the derivations can be transferred directly, such as the recursive formulation and the numerically improved versions in Chap. 22. However, particular problems arise concerning the convergence and the evaluation of the required derivatives of the signals.

15.1.2 Convergence

It is now assumed that the output is disturbed by a stationary stochastic signal $n(t)$. Similarly to the derivations in Sect. 9.1.2, the expected value of the parameter estimates can now be determined.

After inserting

$$y(k) = \boldsymbol{\psi}^{\mathrm{T}}(k)\boldsymbol{\theta}_0 + e(k) \tag{15.1.16}$$

into (15.1.15), i.e. assuming that the model parameters $\hat{\boldsymbol{\theta}}$ match with the true parameters $\boldsymbol{\theta}_0$, one obtains the expected values

$$\mathrm{E}\{\hat{\boldsymbol{\theta}}\} = \boldsymbol{\theta}_0 + \mathrm{E}\{(\boldsymbol{\Psi}^{\mathrm{T}}\boldsymbol{\Psi})^{-1}\boldsymbol{\Psi}^{\mathrm{T}}\boldsymbol{e}\}\,, \tag{15.1.17}$$

where

$$\boldsymbol{b} = \mathrm{E}\{(\boldsymbol{\Psi}^{\mathrm{T}}\boldsymbol{\Psi})^{-1}\boldsymbol{\Psi}^{\mathrm{T}}\boldsymbol{e}\} \tag{15.1.18}$$

is a bias. For the bias to vanish, it is required that

$$\mathrm{E}\{\boldsymbol{\Psi}^{\mathrm{T}}\boldsymbol{e}\} = \mathbf{0} \tag{15.1.19}$$

With (15.1.5), an unbiased estimate will only result if

$$\hat{\boldsymbol{R}}_{\mathrm{y}^{(j)}\mathrm{e}} = \hat{\boldsymbol{R}}_{\mathrm{n}^{(j)}\mathrm{e}} \tag{15.1.20}$$

and since $e(t)$ and $u(t)$ are uncorrelated, an unbiased estimate can only be obtained if

$$\mathrm{E}\begin{Bmatrix} -\hat{\boldsymbol{R}}_{\mathrm{n}^{(1)}\mathrm{e}}(0) & -\hat{\boldsymbol{R}}_{\mathrm{n}^{(2)}\mathrm{e}}(0) & \ldots & -\hat{\boldsymbol{R}}_{\mathrm{n}^{(n)}\mathrm{e}}(0) \\ 0 & & & \\ \vdots & & & \\ 0 & & & \end{Bmatrix} = \mathbf{0}^{\mathrm{T}}\,. \tag{15.1.21}$$

This leads to the differential equation

$$e(t) = a_n n^{(n)}(t) + \ldots + a_1 n^{(1)}(t) + n(t)\,, \tag{15.1.22}$$

so that the disturbance $n(t)$ is generated from the equation error $e(t)$ by the form filter

$$G_{\mathrm{F}}(s) = \frac{n(s)}{e(s)} = \frac{1}{1 + a_1 s + \ldots + a_n s^n}\,. \tag{15.1.23}$$

After multiplication of (15.1.22) with $n^{(j)}(t-\tau)$ and taking the expected value, one obtains

$$\hat{\boldsymbol{R}}_{\mathrm{n}^{(j)}\mathrm{e}}(\tau) = a_n \hat{\boldsymbol{R}}_{\mathrm{n}^{(1)}\mathrm{n}^{(n)}}(\tau) + \ldots + a_1 \hat{\boldsymbol{R}}_{\mathrm{n}^{(1)}\mathrm{n}^{(1)}}(\tau) + a_n \hat{\boldsymbol{R}}_{\mathrm{n}^{(1)}\mathrm{n}}(\tau)\,. \tag{15.1.24}$$

Now, one can write

$$\hat{\boldsymbol{R}}_{\mathrm{n}^{(j)}\mathrm{e}}(\tau) = \frac{\mathrm{d}^j}{\mathrm{d}\tau}\hat{\boldsymbol{R}}_{\mathrm{ne}}(\tau) = \frac{\mathrm{d}^j}{\mathrm{d}\tau}\mathrm{E}\{n(t)e(t+\tau)\}\,. \tag{15.1.25}$$

It is now assumed that $e(t)$ is a Gaussian white noise with

$$R_{\mathrm{ee}}(\tau) = \lambda\delta(t)\,. \tag{15.1.26}$$

The cross-correlation function of the form filter is then given as

$$R_{\mathrm{ne}}(\tau) = g_{\mathrm{F}}(\tau) = \mathfrak{L}^{-1}\{G_{\mathrm{F}}(s)\} \tag{15.1.27}$$

and the elements of (15.1.21) are hence given as

$$\begin{aligned}\hat{R}_{\mathrm{n}^{(j)}\mathrm{e}}(0) &= \lim_{\tau\to 0}\left(s^{j+1}G_{\mathrm{F}}(s) - s^{j}g_{\mathrm{F}}(0+) - s^{j-1}g_{\mathrm{F}}^{(1)}(0+) - \ldots - sg_{\mathrm{F}}^{(j-1)}(0+)\right)\\ &\begin{cases} = 0 \text{ for } j = 1, 2, \ldots, n-2 \\ \neq 0 \text{ for } j = n-1, n \end{cases}\,.\end{aligned} \tag{15.1.28}$$

Hence (15.1.21) is not satisfied, and therefore, biased estimates will result. In contrast to the method of least squares for discrete-time linear dynamic systems, one will not obtain bias-free estimates if the error $e(t)$ is a Gaussian distributed white noise. Henceforth, the method of least squares for continuous-time systems should only be used if the signal-to-noise ratio is very favorable.

15.2 Determination of Derivatives [1]

If the required derivatives of the signals are directly measurable (e.g. as for vehicle applications), these values can be entered directly into the data matrix $\boldsymbol{\Psi}$ and therefore the correlation functions in the matrix $(\boldsymbol{\Psi}^{\mathrm{T}}\boldsymbol{\Psi})/(N+1)$ can easily be calculated directly. On the contrary, if the derivatives are not measurable, the derivatives have to be determined from the sampled signals $u(t)$ and $y(t)$. For this, one basically has a choice between numerical differentiation and state variable filtering. A short look shall be thrown on the determination of the derivatives by means of FIR filters, but especially if a large number of derivatives are required, the computational effort can exceed that of the state variable filter extensively.

15.2.1 Numerical Differentiation

The *numerical differentiation* in combination with interpolation approaches (e.g. splines) is usually not able to suppress noise due to disturbance signals for higher derivatives which limits this technique to the application of second or third order derivatives as a maximum. Often however, only first order derivatives can be determined reliably. Approaches to use the derivative of an interpolation function have been discussed e.g. by (Söderström and Mossberg, 2000).

One can in general the following methods of determining the derivatives:

[1] Based on the diploma thesis by Michael Vogt (1998)

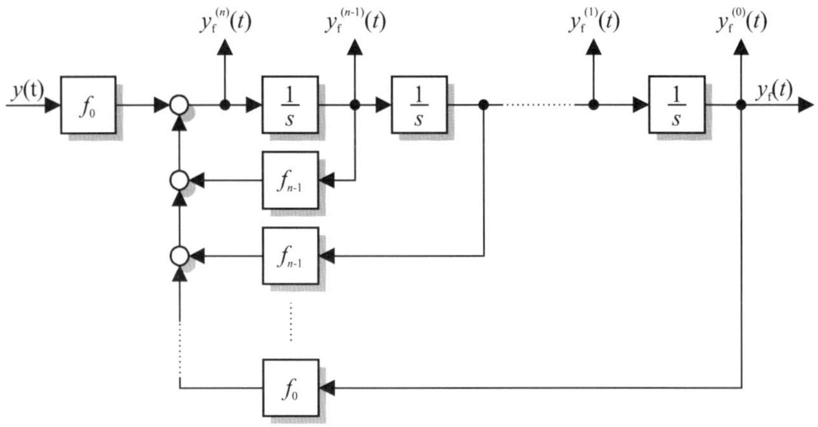

Fig. 15.2. State variable filter for filtering and determining the derivatives of a signal simultaneously

$$\text{Forward Differential Quotient: } \hat{\dot{x}}(k) = \frac{x(k+1) - x(k)}{T_0} \tag{15.2.1}$$

$$\text{Backward Differential Quotient: } \hat{\dot{x}}(k) = \frac{x(k) - x(k-1)}{T_0} \tag{15.2.2}$$

$$\text{Central Differential Quotient: } \hat{\dot{x}}(k) = \frac{x(k+1) - x(k-1)}{2T_0}. \tag{15.2.3}$$

All of them have distinct advantages and disadvantages: While the forward and the central differential quotient depend on future values, the backward differential quotient introduces a delay of half a sample step.

15.2.2 State Variable Filters

The principle of state variable filters is to use a low-pass filter that dampens out the higher frequent noise and transform it into a state-space representation, such that the states are the derivatives of the filter output, i.e. the filtered signal. They are designed to suppress any noise above the cut-off frequency, i.e. for $f > f_c$.

State variable filters (SVF), see Fig. 15.2,

$$G_F(s) = \frac{y_F(s)}{y(s)} = \frac{f_0}{f_0 + f_1 s + \ldots + f_{n-1}s^{n-1} + s^n} \tag{15.2.4}$$

have proven to yield good results in identification of continuous-time systems. A state variable filter is an analog filter with a certain topology that is subsequently discretized for the realization on a digital computer. The input signal $u(t)$ and the output signal $y(t)$ must both be filtered with the same state variable filter, see Fig. 15.5. The choice of the filter parameters f_i is relatively free. For the discretization, different

Table 15.1. Polynomials $B(s)$ of the normalized Butterworth filter

Order n	Polynomial $B(s)$ of the Butterworth Filter
$n=1$	$s+1$
$n=2$	$s^2+1.4142s+1$
$n=3$	s^3+2s^2+2s+1
$n=4$	$s^4+2.6131s^3+3.4142s^2+2.6131s+1$
$n=5$	$s^5+3.2361s^4+5.2361s^3+5.2361s^2+3.2361s+1$
$n=6$	$s^6+3.8637s^5+7.4641s^4+9.1416s^3+7.4641s^2+3.8637s+1$

approximation techniques can be employed, see also the detailed analysis by Vogt (1998).

According to Young (1981), one can choose the filter coefficients as $f_i = \hat{a}_i$, which yields an adaptive low-pass filter. However, also a fixed parameterization is possible. For linear systems, one can e.g. choose a Butterworth low-pass filter or any other type of low-pass filter, e.g. Bessel, Chebyshev, etc. (Kammeyer and Kroschel, 2009; Hamming, 2007; Tietze et al, 2010). For non-linear processes one should resort to other filters, such as Bessel, that do not exhibit oscillations in the time domain.

The *Butterworth filter* is designed such that its transfer function has a constant amplitude in the passband as long as possible. The transfer function of the normalized Butterworth filter ($\omega_c = 1$) is given as

$$G_F(s) = \frac{K}{\prod_i \left(1+\alpha_i s+\beta_i s^2\right)} = \frac{K}{B(s)} \tag{15.2.5}$$

with the coefficients being

- For order n even:

$$\left.\begin{aligned} \alpha_i &= 2\cos\tfrac{(2i-1)\pi}{2n} \\ \beta_i &= 1 \end{aligned}\right\} i = 1,\ldots,\frac{n}{2} \tag{15.2.6}$$

- For order n odd:

$$\left.\begin{aligned} \alpha_1 &= 1;\ \alpha_i = 2\cos\tfrac{(i-1)\pi}{n} \\ \beta_1 &= 0;\ \beta_i = 1 \end{aligned}\right\} i = 2,\ldots,\frac{n+1}{2} \tag{15.2.7}$$

The coefficients of the Butterworth filter are listed in Table 15.1. for low orders of n. The frequency response is shown in Fig. 15.3. If the filter shall be designed for an arbitrary cut-off frequency ω_C, then the transfer function is given as

$$G_F(s) = \frac{1}{\prod_i \left(1+\alpha_i \frac{s}{\omega_C}+\beta_i \frac{s^2}{\omega_C^2}\right)}, \tag{15.2.8}$$

where typically, also $K=1$ is chosen. This transfer function can then be multiplied out, yielding

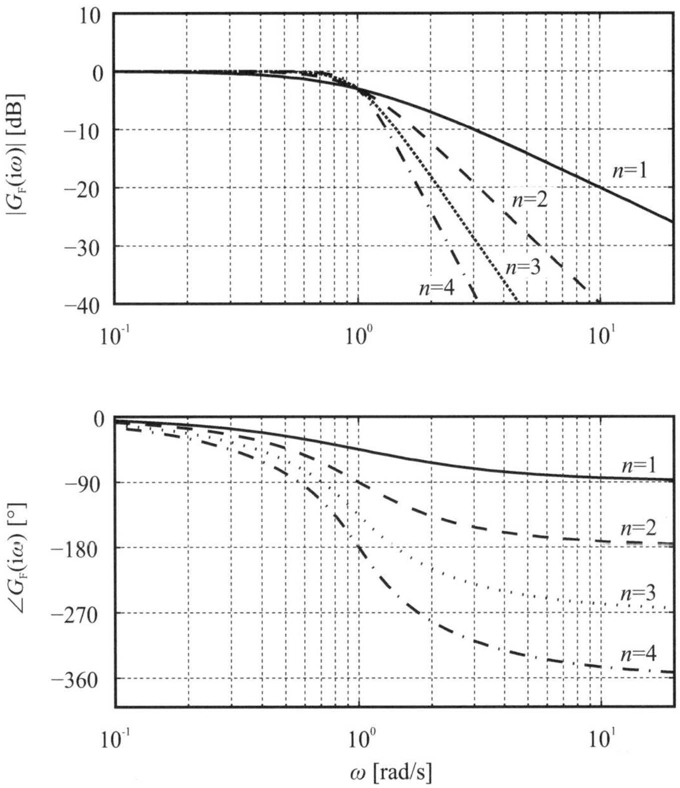

Fig. 15.3. Frequency response of the normalized Butterworth filter with $\omega_C = 1$

$$G_F(s) = \frac{1}{1 + a_1 s + \ldots + a_n s^n} \; . \tag{15.2.9}$$

This filter can now be brought to the controllable canonical form, which is given by the state space system

$$\frac{d}{dt}\boldsymbol{x}_F(t) = \underbrace{\begin{pmatrix} 0 & 1 & 0 & \ldots & 0 \\ 0 & 0 & 1 & \ldots & 0 \\ \vdots & \vdots & \vdots & \ddots & \vdots \\ 0 & 0 & 0 & \ldots & 1 \\ -\frac{1}{a_n} & -\frac{a_1}{a_n} & -\frac{a_2}{a_n} & \ldots & -\frac{a_{n-1}}{a_n} \end{pmatrix}}_{\boldsymbol{A}} \boldsymbol{x}_F(t) + \underbrace{\begin{pmatrix} 0 \\ 0 \\ \vdots \\ 0 \\ \frac{1}{a_n} \end{pmatrix}}_{\boldsymbol{b}} x(t) \; . \tag{15.2.10}$$

The output equation is negligible since the states of the state variable filter already provide all necessary quantities. A comparison with the structure in Fig. 15.2 reveals that

$$\left.\begin{aligned} f_0 &= -\tfrac{1}{a_n} \\ f_1 &= -\tfrac{a_1}{a_n} \\ &\vdots \\ f_{n-1} &= -\tfrac{a_{n-1}}{a_n}\ . \end{aligned}\right\} \tag{15.2.11}$$

This filter will be implemented on a digital computer and hence must be discretized. Therefore, one must derive the discrete-time state space representation which can be obtained by

$$\boldsymbol{x}(k+1) = \boldsymbol{A}_\mathrm{d}\boldsymbol{x}(k) + \boldsymbol{b}_\mathrm{d}u(k) \tag{15.2.12}$$

$$y(k) = \boldsymbol{c}_\mathrm{d}^\mathrm{T}\boldsymbol{x}(k) + d_\mathrm{d}u(k)\ , \tag{15.2.13}$$

see (2.2.24) and (2.2.25), where the index d shall denote the discrete-time variables. The continuous-time system can be transformed into the discrete-time representation by means of the the relations

$$\begin{aligned} \boldsymbol{x}(k) &= \mathrm{e}^{\boldsymbol{A}T_0}\boldsymbol{x}(k-1) + \int_{(k-1)T_0}^{(k)T_0} \mathrm{e}^{\boldsymbol{A}(kT_0-\tau)}\boldsymbol{b}u(\tau)\mathrm{d}\tau \\ &= \mathrm{e}^{\boldsymbol{A}T_0}\boldsymbol{x}(k-1) + \int_{(k-1)T_0}^{(k)T_0} \mathrm{e}^{\boldsymbol{A}(\tau)}\boldsymbol{b}u(kT_0-\tau)\mathrm{d}\tau\ , \end{aligned} \tag{15.2.14}$$

compare (2.1.27). One can see already that the resulting discrete-time implementation will have the state matrix

$$\boldsymbol{A}_\mathrm{d} = \mathrm{e}^{\boldsymbol{A}T_0} = \sum_{k=0}^{\infty}\frac{1}{k}(\boldsymbol{A}T_0)^k\ , \tag{15.2.15}$$

where the infinite sum can be approximated by a sum of low order. As an alternative, one can employ special algorithms for the calculation of the matrix exponential (e.g. Moler and van Loan, 2003).

The shape of $u(\tau)$ inside the integration interval $(k-1)T_0 < \tau < kT_0$ is unknown as the signal is sampled only at integer multiples $t = kT_0$. The input signal $u(t)$ shall in the following be approximated by a polynomial as

$$u(kT_0-\tau) \approx p(\tau) = \sum_{i=0}^{r}\kappa_i\left(\frac{\tau}{T_0}\right)^i\ , \tag{15.2.16}$$

see (Wolfram and Vogt, 2002). Inserting this polynomial approximation into (15.2.14), one obtains

$$\begin{aligned} \boldsymbol{x}(k) &\approx \mathrm{e}^{\boldsymbol{A}T_0}\boldsymbol{x}(k-1) + \sum_{i=0}^{r}\kappa_i \underbrace{\int_0^{T_0} \mathrm{e}^{\boldsymbol{A}(\tau)}\boldsymbol{b}\left(\frac{\tau}{T_0}\right)^i \mathrm{d}\tau}_{\gamma_i} \\ &= \mathrm{e}^{\boldsymbol{A}T_0}\boldsymbol{x}(k-1) + \boldsymbol{\Gamma}\begin{pmatrix}\kappa_0\\ \vdots \\ \kappa_r\end{pmatrix}. \end{aligned} \tag{15.2.17}$$

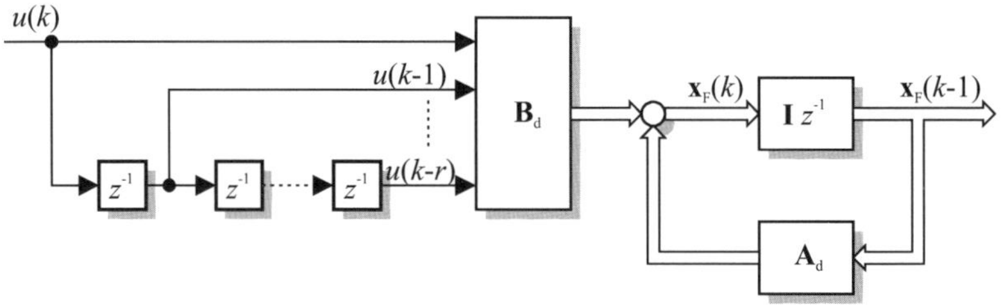

Fig. 15.4. Discrete time state variable filter with interpolator

The columns of $\boldsymbol{\Gamma} = (\boldsymbol{\gamma}_0\ \boldsymbol{\gamma}_1\ \ldots\ \boldsymbol{\gamma}_r)$, can be determined as

$$\boldsymbol{\gamma}_0 = \boldsymbol{A}^{-1}\big(\mathrm{e}^{\boldsymbol{A}T_0} - \boldsymbol{I}\big)\boldsymbol{b} \tag{15.2.18}$$

$$\boldsymbol{\gamma}_i = \boldsymbol{A}^{-1}\left(\mathrm{e}^{\boldsymbol{A}T_0}\boldsymbol{b} - \frac{i}{T_0}\boldsymbol{\gamma}_{i-1}\right) \text{ for } i = 1, \ldots, r\,. \tag{15.2.19}$$

A way to determine the polynomial coefficients κ_i is to mandate that the polynomial and the signal match for the last $r+1$ samples (Peter, 1982), i.e.

$$p(lT_0) \stackrel{!}{=} u\big((k-l)T_0\big) \text{ for } l = 0, \ldots, r\,, \tag{15.2.20}$$

which leads to the system of equations

$$\underbrace{\begin{pmatrix} 1 & 0 & \ldots & 0 \\ \vdots & \vdots & & \vdots \\ 1 & r-1 & \ldots & (r-1)^r \\ 1 & r & \ldots & r^r \end{pmatrix}}_{\boldsymbol{V}} \begin{pmatrix} \kappa_0 \\ \kappa_1 \\ \vdots \\ \kappa_r \end{pmatrix} = \begin{pmatrix} u(kT_0) \\ u((k-1)T_0) \\ \vdots \\ u((k-r)T_0) \end{pmatrix}. \tag{15.2.21}$$

This system of equations can be solved by the inversion of Vandermonde matrix $\boldsymbol{V}$ (e.g. Eisinberg and Fedele, 2006). With

$$\boldsymbol{B} = \boldsymbol{\Gamma}\boldsymbol{V}^{-1}\,, \tag{15.2.22}$$

one can derive the structure shown in Fig. 15.4. $\boldsymbol{V}^{-1}$ for $r = 1, 2, 3$ is given in Table 15.2.

For higher orders of r, the polynomial can show ripples. An alternative for the determination of the coefficients of the approximation polynomial is the use of a Taylor series expansion. For this approximation, one equates the first r derivatives of the polynomial and the input sequence for k as

$$p^{(l)}(0) = u^{(l)}(kT_0) \text{ for } l = 0, \ldots, r\,. \tag{15.2.23}$$

The derivatives are given as

Table 15.2. Matrices $\boldsymbol{V}^{-1}$ and $\boldsymbol{T}$ for low orders of r

Order n	Matrix $\boldsymbol{V}^{-1}$	Matrix T
$r=1$	$\begin{pmatrix} 1 & 0 \\ -1 & 1 \end{pmatrix}$	$\begin{pmatrix} 1 & 0 \\ 1 & -1 \end{pmatrix}$
$r=2$	$\begin{pmatrix} 1 & 0 & 0 \\ -\frac{3}{2} & 2 & -\frac{1}{2} \\ \frac{1}{2} & -1 & \frac{1}{2} \end{pmatrix}$	$\begin{pmatrix} 1 & 0 & 0 \\ 1 & -1 & 0 \\ \frac{1}{2} & -1 & \frac{1}{2} \end{pmatrix}$
$r=3$	$\begin{pmatrix} 1 & 0 & 0 & 0 \\ -\frac{11}{6} & 3 & -\frac{3}{2} & \frac{1}{3} \\ 1 & -\frac{5}{2} & 2 & -\frac{1}{2} \\ -\frac{1}{6} & \frac{1}{2} & -\frac{1}{2} & \frac{1}{6} \end{pmatrix}$	$\begin{pmatrix} 1 & 0 & 0 & 0 \\ 1 & -1 & 0 & 0 \\ \frac{1}{2} & -1 & \frac{1}{2} & 0 \\ \frac{1}{6} & -\frac{1}{2} & \frac{1}{2} & -\frac{1}{6} \end{pmatrix}$

$$p^{(0)} = \left.\frac{\mathrm{d}^l p(\tau)}{\mathrm{d}\tau^l}\right|_{\tau=0} = \frac{l!}{T_0^l}\kappa_l \tag{15.2.24}$$

and

$$u^{(l)} = \left.\frac{\mathrm{d}^l u(kT_0-\tau)}{\mathrm{d}\tau^l}\right|_{\tau=0} \approx \frac{1}{T_0^l}\Delta^l u(kT_0) = \frac{1}{T_0^l}\sum_{i=0}^{l}(-1)^i u\big((k-i)T_0\big)\,, \tag{15.2.25}$$

which leads to the polynomial coefficients as

$$\kappa_i = \sum_{k=0}^{l}\frac{(-1)^i}{l!}\binom{l}{i}u((k-i)T_0) = \sum_{k=0}^{l}\frac{(-1)^i}{i!\,(l-i)!} \tag{15.2.26}$$

or, in matrix form,

$$\boldsymbol{\kappa} = \boldsymbol{T}\boldsymbol{u}(kT_0)\,. \tag{15.2.27}$$

For the structure in Fig. 15.4, the matrix $\boldsymbol{B}_\mathrm{d}$ is now given as

$$\boldsymbol{B}_\mathrm{d} = \boldsymbol{\Gamma}\boldsymbol{T}\,. \tag{15.2.28}$$

The entries of $\boldsymbol{T}$ for $r = 1, 2, 3$ are given in Table 15.2.

Another alternative to obtain a discrete-time realization is by means of the bilinear transform (Wolfram and Vogt, 2002). Here, one replaces s by

$$s = \frac{2}{T_0}\frac{z-1}{z+1} \text{ or } \frac{1}{s} = \frac{T_0}{2}\frac{z+1}{z-1}\,. \tag{15.2.29}$$

The bilinear transform can now be used to transform the state equation

$$s\boldsymbol{x}(s) = \boldsymbol{A}\boldsymbol{x}(s) + \boldsymbol{b}u(s)\,, \tag{15.2.30}$$

which is the Laplace transform of the continuous-time state equation into the z-domain, which is better suited for later discrete-time processing. The bilinear transform is given as

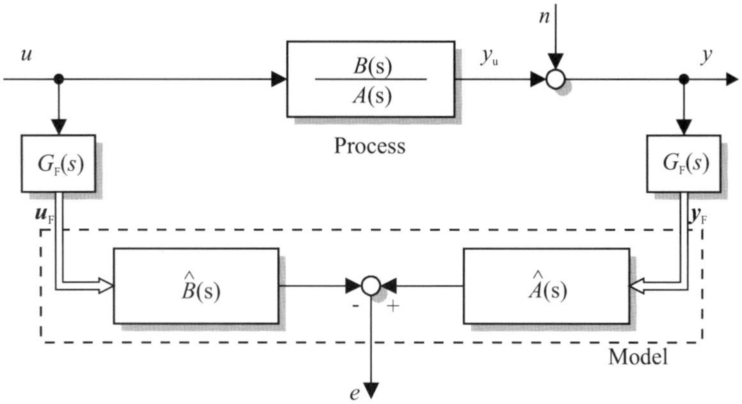

Fig. 15.5. State variable filter employed for parameter estimation

$$\frac{2}{T_0}\frac{z-1}{z+1}\boldsymbol{x}(z) = \boldsymbol{A}\boldsymbol{x}(z) + \boldsymbol{b}u(z)\,. \tag{15.2.31}$$

To bring the above equation into the form of a normal state equation, new states have to be introduced. These new states can be chosen in two different ways. The first choice is given as

$$\tilde{\boldsymbol{x}} = \frac{1}{T_0}\left(\boldsymbol{I} - \frac{T_0}{2}\boldsymbol{A}\right)\boldsymbol{x} - \frac{1}{2}\boldsymbol{b}u\,, \tag{15.2.32}$$

which leads to

$$\boldsymbol{A}_{\mathrm{d}} = \left(\boldsymbol{I} + \frac{T_0}{2}\boldsymbol{A}\right)\left(\boldsymbol{I} - \frac{T_0}{2}\boldsymbol{A}\right)^{-1} \tag{15.2.33}$$

$$\boldsymbol{b}_{\mathrm{d}} = \left(\boldsymbol{I} - \frac{T_0}{2}\right)^{-1}\boldsymbol{b} \tag{15.2.34}$$

$$\boldsymbol{C}_{\mathrm{d}} = T_0\left(\boldsymbol{I} - \frac{T_0}{2}\boldsymbol{A}\right)^{-1} \tag{15.2.35}$$

$$\boldsymbol{d}_{\mathrm{d}} = \frac{T_0}{2}\left(\boldsymbol{I} - \frac{T_0}{2}\boldsymbol{A}\right)^{-1}\boldsymbol{b}\,. \tag{15.2.36}$$

Here, $\boldsymbol{C}_{\mathrm{d}}$ and $\boldsymbol{d}_{\mathrm{d}}$ are dense matrices.

A different choice of the state variables $\tilde{\boldsymbol{x}}$ leads to $\boldsymbol{C}_{\mathrm{d}}$ being an identity matrix. The state variables are chosen as

$$\tilde{\boldsymbol{x}} = \boldsymbol{x} - \frac{T_0}{2}\left(\boldsymbol{I} - \frac{T_0}{2}\boldsymbol{A}\right)\boldsymbol{b}u \tag{15.2.37}$$

leading to

$$A_{\mathrm{d}} = \left(\boldsymbol{I} - \frac{T_0}{2}\boldsymbol{A}\right)^{-1}\left(\boldsymbol{I} + \frac{T_0}{2}\boldsymbol{A}\right) \tag{15.2.38}$$

$$\boldsymbol{b}_{\mathrm{d}} = T_0\left(\boldsymbol{I} - \frac{T_0}{2}\boldsymbol{A}\right)^{-2}\boldsymbol{b} \tag{15.2.39}$$

$$\boldsymbol{C}_{\mathrm{d}} = \boldsymbol{I} \tag{15.2.40}$$

$$\boldsymbol{d}_{\mathrm{d}} = \frac{T_0}{2}\left(\boldsymbol{I} - \frac{T_0}{2}\boldsymbol{A}\right)^{-1}\boldsymbol{b}\,. \tag{15.2.41}$$

The squared term in $\boldsymbol{b}_{\mathrm{d}}$ is critical from a numerical point of view. If one does not need all derivatives, one could leave out the corresponding rows in the calculation. An investigation of the frequency response has shown that the bilinear transform is also beneficial in terms of the damping in the stop-band as the bilinear transform introduces an additional pole at $T_0/2$.

Due to the good damping characteristics, the discretization by means of the bilinear transform is in general suggested for the derivation of state variable filters. If one however wants to use Taylor series or polynomial approximation, one should, especially for higher derivatives, choose the Taylor series approximation as it avoids the problems associated with interpolation with polynomials (Wolfram and Vogt, 2002).

To ensure that the results of the parameter estimation match with the original signal, the input and output must be filtered with the same filter, see Fig. 15.5. Then, the use of the state variable filter should in general not change the transfer function of a linear process to be estimated, however it still affects the estimation results as the noise characteristics are altered (Ljung, 1999).

15.2.3 FIR Filters

Due to the infinitely long impulse response of the state variable filter (which is an IIR filter), it is difficult to predict the influence of disturbances on the filtered signal. Furthermore, one has to take stability issues into account as an IIR filter can theoretically become unstable. Other aspects of the comparison of FIR and IIR filters can be seen in Table 15.3, see also (Wolfram and Vogt, 2002).

These problems can be avoided by using FIR filters, i.e. filters with an impulse response of finite length. In order to determine the derivatives of the filtered signal, the appropriate derivative of the impulse response of a low-pass filter is convolved with the signal itself. This is possible, because the time-derivative of the convolution of an arbitrary signal $u(t)$ with an FIR filter with the final time T_{F} of the impulse response can be determined as

$$\begin{aligned} \frac{\mathrm{d}y(t)}{\mathrm{d}t} &= \frac{\mathrm{d}}{\mathrm{d}t}\int_{t-T_{\mathrm{F}}}^{t} g(t-\tau)u(\tau)\mathrm{d}\tau \\ &= \int_{t-T_{\mathrm{F}}}^{t} \frac{\partial}{\partial t} g(t-\tau)u(\tau)\mathrm{d}\tau + g(0)u(t) - g(T_{\mathrm{F}})u(t-T_{\mathrm{F}})\,. \end{aligned} \tag{15.2.42}$$

If the impulse response vanishes outside the interval $0 < t < T_{\mathrm{f}}$, then the time derivative is given as

Table 15.3. Properties of FIR and IIR filters

IIR (Infinite Impulse Response)	FIR (Finite Impulse Response)
Impulse response is of infinite duration	Impulse response is of finite duration
A steady state is theoretically only reached for $k \to \infty$	For filter order m, the steady state is reached after $k = m + 1$ sample steps
If implemented on computers with finite word length, the filter can become unstable or limit cycles can occur	Filters cannot become unstable and limit cycles cannot occur
Can typically only be implemented on floating point units	Implementation is also possible on fixed point units
No constant group delays and hence non-linear phase	Group delay is constant for symmetric impulse responses and hence linear phase
One filter provides all derivatives	Separate filter for each derivative
Typically small filter order	Large filter order necessary

$$\dot{y}(t) = \int_0^{T_F} \dot{g}(\tau)u(t-\tau)\mathrm{d}\tau \; , \tag{15.2.43}$$

see (Oppenheim and Schafer, 2009; Wolfram and Moseler, 2000; Wolfram and Vogt, 2002). The main problem however is the design of the FIR filter, where it might be necessary to use high filter orders and/or fast sample rates. High filters orders are required to have good damping characteristics. Long filter length unfortunately go along with a large phase shift. Furthermore, for each derivative, a separate filter must be designed, a disadvantage that can become important if higher order models shall be identified. This usually leads to a higher computational effort compared with the state variable filter. However, FIR filters are non-recursive and hence cannot become unstable. Furthermore, they have a linear phase if designed appropriately. A comparison of the magnitude of the filter transfer functions is shown in Fig. 15.6. Here, one can see that the FIR filter shows a certain ripple in the stop-band and also has a less distinct fall-off at the corner frequency.

As the parameter estimation of continuous-time processes using the method of least squares in the end goes back to the discrete-time algorithms because the estimation is finally implemented on a digital computer, one can analogously apply all results obtained in part III and part VII of this book, especially the modifications and extensions that have been provided. These include

- recursive implementations of the estimation algorithms
- time variant processes
- numerically improved methods
- determination of the model order
- choice of the input signal

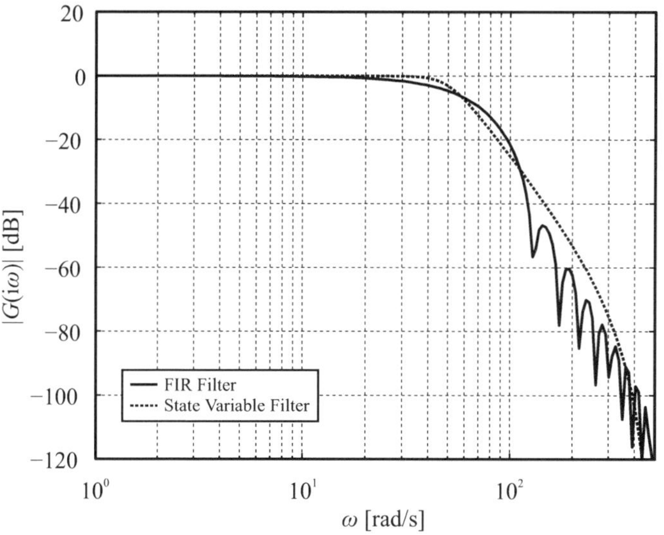

Fig. 15.6. Comparison of a state variable filter (order $n = 4$, corner frequency $f_C = 50\,\text{Hz}$, sample time $T_0 = 1\,\text{ms}$) and an FIR filter with (order $m = 24$, sample time $T_0 = 1\,\text{ms}$) (Wolfram and Vogt, 2002)

Example 15.1 (Estimation of a Continuous-Time Model of the Three-Mass Oscillator).

This technique has been applied to the Three-Mass Oscillator as well and a continuous-time transfer function model could be obtained, see Fig. 15.7. A state variable filter with Butterworth characteristics has been used to generate the derivatives of the input and output signal. The parameters of the filter were a cut-off frequency of $\omega_C = 37.9\,\text{rad/s}$ and a filter order of $n = 9$. Figure 15.8 shows the frequency response of the state variable filter that was used to filter the input and output. □

15.3 Consistent Parameter Estimation Methods

15.3.1 Method of Instrumental Variables

For favorable signal-to-noise ratios, the above presented least squares method has been shown to yield good results. For larger noises, consistent parameter estimation methods should be employed such as *the instrumental variables method*, see Sect. 10.5.

As was already described, instrumental variables have to be found that on the one hand are strongly correlated with the useful signals, on the other hand are little or not at all correlated with the noise. One can use an adaptive model analogously to the setup shown in Fig. 10.4 to reconstruct the useful signal as

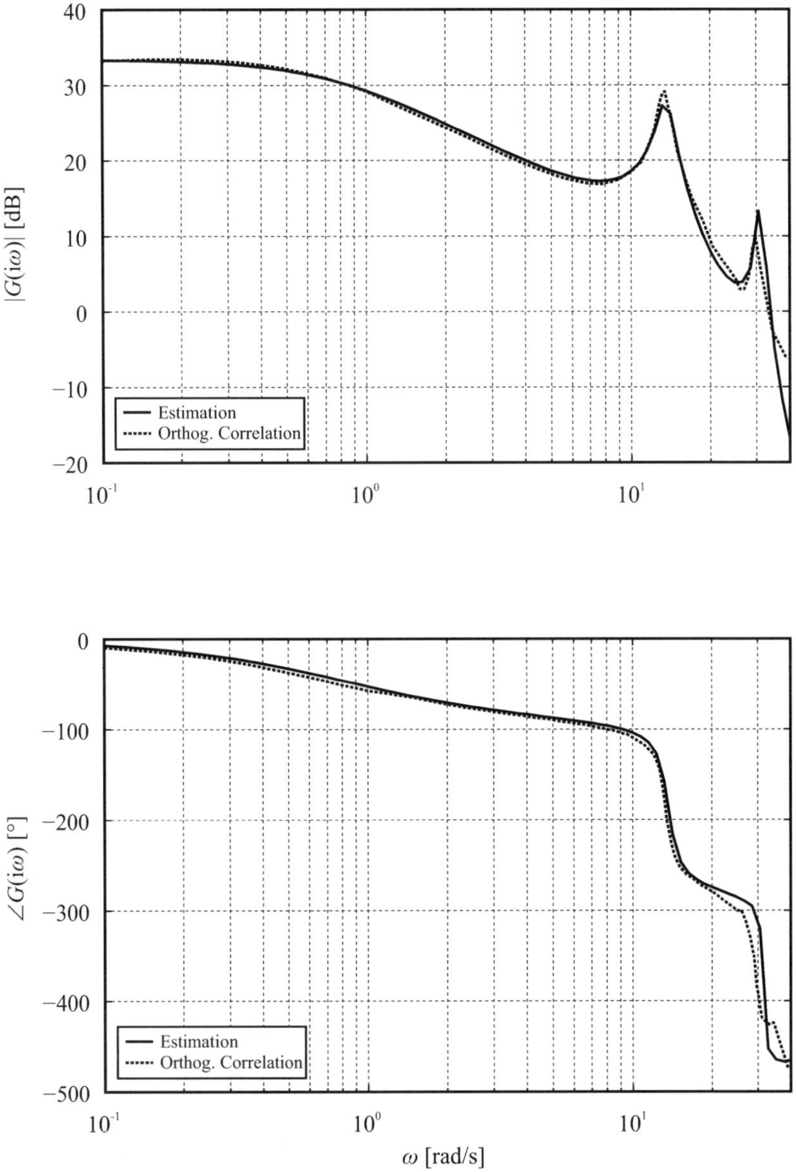

Fig. 15.7. Estimated frequency response of the Three-Mass Oscillator based on a continuous-time model

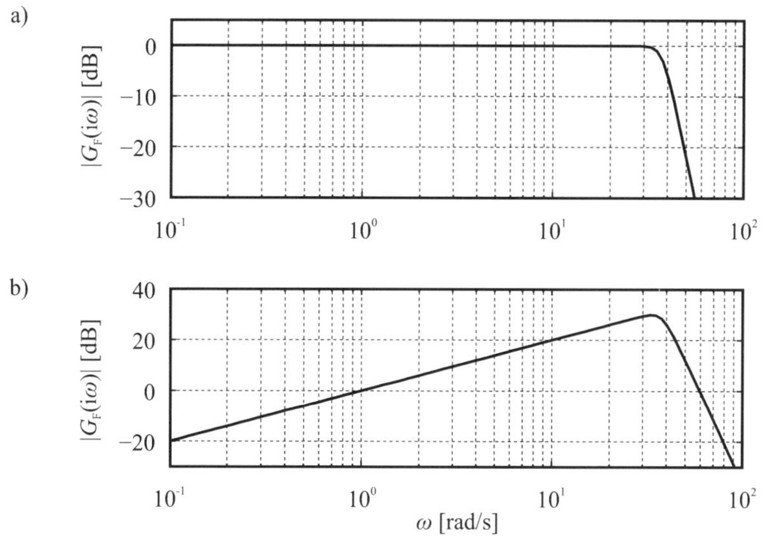

Fig. 15.8. Frequency response of the state variable filter for the generation of the (**a**) filtered input ($y_F(t)$) and (**b**) first time derivative of the filtered input ($\dot{y}_F(t)$)

$$\hat{y}_u(s) = \frac{\hat{B}(s)}{\hat{A}(s)} u(s) . \tag{15.3.1}$$

The estimated system output in the time domain $\hat{y}_u(s)$ together with the input signal $u(t)$ and the corresponding derivatives can then be used to form the vector of instrumental variables as

$$\boldsymbol{w}^{\mathrm{T}}(t) = \left(-y_u^{(n)}(t) \ldots -y_u^{(1)}(t) \; -y_u(t) \middle| u^{(m)}(t) \ldots u^{(1)}(t) \; u(t) \right) , \tag{15.3.2}$$

which can then be used together with the estimation equations. A big advantage is the fact that no assumptions must be made about the noise. In conjunction with the state variable filters used for determining the derivatives, the results can be improved (Young, 1981, 2002).

15.3.2 Extended Kalman Filter, Maximum Likelihood Method

The Extended Kalman Filter as introduced in Chap. 21 can be used to estimate parameters of a system and can also be applied to continuous-time systems. One can also use a combination of the Kalman filter for state generation and a parameter estimation method for parameter estimation (e.g. Raol et al, 2004; Ljung and Wills, 2008).

15.3.3 Correlation and Least Squares

The combination of correlation methods for the generation of a non-parametric intermediate model and the subsequent application of the method of least squares can also be adapted to work with continuous-time models. If one multiplies the differential equation in (15.1.1) with $u(t-\tau)$ and determines the expected values of all products, then one obtains

$$\begin{aligned} &a_n R_{\mathrm{uy}^{(n)}}(\tau) + a_{n-1} R_{\mathrm{uy}^{(n-1)}}(\tau) + \ldots + a_1 R_{\mathrm{uy}^{(1)}}(\tau) + R_{\mathrm{uy}}(\tau) \\ &\quad = b_m R_{\mathrm{uu}^{(m)}}(\tau) + b_{m-1} R_{\mathrm{uu}^{(m-1)}}(\tau) + \ldots + b_1 R_{\mathrm{uu}^{(1)}}(\tau) + b_0 R_{\mathrm{uu}^{(0)}}(\tau) \,. \end{aligned} \tag{15.3.3}$$

As

$$\begin{aligned} \frac{\mathrm{d}^j R_{\mathrm{uy}}(\tau)}{\mathrm{d}\tau^j} &= \frac{\mathrm{d}^j}{\mathrm{d}\tau^j} \mathrm{E}\{y(t+\tau)u(t)\} = \mathrm{E}\left\{\frac{\mathrm{d}^j}{\mathrm{d}\tau^j} y(t+\tau)u(t)\right\} \\ &= \mathrm{E}\left\{\left(\frac{\mathrm{d}^j}{\mathrm{d}t^j} y(t+\tau)\right)u(t)\right\} = R_{\mathrm{uy}^{(n)}}(\tau) \,, \end{aligned} \tag{15.3.4}$$

one obtains

$$R_{\mathrm{uy}^{(n)}}(\tau) = \frac{\mathrm{d}^j R_{\mathrm{uy}}(\tau)}{\mathrm{d}\tau^j} \,. \tag{15.3.5}$$

With this result, (15.3.3) can now be written as a differential equation of the correlation functions

$$\begin{aligned} &a_n \frac{\mathrm{d}^n}{\mathrm{d}\tau^n} R_{\mathrm{uy}}(\tau) + a_{n-1} \frac{\mathrm{d}^{n-1}}{\mathrm{d}\tau^{n-1}} R_{\mathrm{uy}}(\tau) + \ldots + a_1 \frac{\mathrm{d}}{\mathrm{d}\tau} R_{\mathrm{uy}}(\tau) + R_{\mathrm{uy}}(\tau) \\ &\quad = b_m \frac{\mathrm{d}^m}{\mathrm{d}\tau^m} R_{\mathrm{uu}}(\tau) + b_{m-1} \frac{\mathrm{d}^{m-1}}{\mathrm{d}\tau^{m-1}} R_{\mathrm{uu}}(\tau) + \ldots + b_1 \frac{\mathrm{d}}{\mathrm{d}\tau} R_{\mathrm{uu}}(\tau) + b_0 R_{\mathrm{uu}}(\tau) \,. \end{aligned} \tag{15.3.6}$$

One can now proceed as for the COR-LS method in discrete-time. One determines the correlation functions and its derivative for different times $\tau = \nu\tau_0$ where τ_0 is the sample time of the correlation function values. After introducing an equation error setup, one can obtain the estimation equation as

$$
\underbrace{\begin{pmatrix} R_{\mathrm{uy}}(-P\tau_0) \\ \vdots \\ R_{\mathrm{uy}}(-\tau_0) \\ R_{\mathrm{uy}}(0) \\ R_{\mathrm{uy}}(\tau_0) \\ \vdots \\ R_{\mathrm{uy}}(M\tau_0) \end{pmatrix}}_{\boldsymbol{\Phi}} =
$$

$$
\underbrace{\begin{pmatrix}
R_{\mathrm{uy}}^{(1)}(-P\tau_0) & \dots & R_{\mathrm{uy}}^{(n)}(-P\tau_0) & R_{\mathrm{uu}}(-P\tau_0) & \dots & R_{\mathrm{uu}}^{(m)}(-P\tau_0) \\
\vdots & & \vdots & \vdots & & \vdots \\
-R_{\mathrm{uy}}^{(1)}(-\tau_0) & \dots & -R_{\mathrm{uy}}^{(n)}(-\tau_0) & R_{\mathrm{uu}}(-\tau_0) & \dots & R_{\mathrm{uu}}^{(m)}(-\tau_0) \\
-R_{\mathrm{uy}}^{(1)}(0) & \dots & -R_{\mathrm{uy}}^{(n)}(0) & R_{\mathrm{uu}}(0) & \dots & R_{\mathrm{uu}}^{(m)}(0) \\
-R_{\mathrm{uy}}^{(1)}(\tau_0) & \dots & -R_{\mathrm{uy}}^{(n)}(\tau_0) & R_{\mathrm{uu}}(\tau_0) & \dots & R_{\mathrm{uu}}^{(m)}(\tau_0) \\
\vdots & & \vdots & \vdots & & \vdots \\
-R_{\mathrm{uy}}^{(1)}(M\tau_0) & \dots & -R_{\mathrm{uy}}^{(n)}(M\tau_0) & R_{\mathrm{uu}}(M\tau_0) & \dots & R_{\mathrm{uu}}^{(m)}(M\tau_0)
\end{pmatrix}}_{\boldsymbol{S}}
\underbrace{\begin{pmatrix} a_1 \\ \vdots \\ a_n \\ b_0 \\ \vdots \\ b_m \end{pmatrix}}_{\boldsymbol{\theta}}
$$

$$
+ \underbrace{\begin{pmatrix} e(-P\tau_0) \\ \vdots \\ e(-\tau_0) \\ e(0) \\ e(\tau_0) \\ \vdots \\ e(M\tau_0) \end{pmatrix}}_{\boldsymbol{e}} , \tag{15.3.7}
$$

where as in the discrete-time case, the correlation functions are known in the time interval $PT_0 \leq \tau \leq MT_0$. Minimizing the cost function $V = \boldsymbol{e}^{\mathrm{T}}\boldsymbol{e}$ yields the estimates as

$$
\hat{\boldsymbol{\theta}} = \left(\boldsymbol{S}^{\mathrm{T}}\boldsymbol{S}\right)^{-1}\boldsymbol{S}^{\mathrm{T}}\boldsymbol{\Phi} \, . \tag{15.3.8}
$$

The COR-LS method can hence be applied with the following steps:

1. The correlation functions $R_{\mathrm{uy}}(\tau)$ and $R_{\mathrm{uu}}(\tau)$ are determined for $-P\tau_0 \leq \tau \leq M\tau_0$ either recursively or non-recursively
2. The derivatives of the correlation functions are determined by numerical methods, e.g. spline approximation
3. The parameter estimates can be determined according to (15.3.8) by any of the methods presented for the solution of the least squares problem

The method yields consistent estimates provided that the correlation functions converge as was shown in Sect 9.3.2. A major advantage is the fact that one does not have to determine the derivatives of the original signals but rather of the correlation functions, where the influence of the noise has already been reduced drastically. Also, one does not only employ correlation function values for $\tau = 0$ (as does the method of least squares), but rather for various τ. Further advantages are the easy determination of model order and dead time because of the smaller matrices to be operated on (data reduction).

15.3.4 Conversion of Discrete-Time Models

Many parameter estimation methods for discrete-time process models have been developed and several software packages are available for this task. Hence, an obvious approach is to estimate the (unbiased) parameters of a model in the discrete-time first and then determine the parameters of a continuous-time model by an appropriate transformation. One method shall be presented that allows to determine the parameters of the continuous-time model

$$\dot{\boldsymbol{x}}(t) = \boldsymbol{A}\boldsymbol{x}(t) + \boldsymbol{B}\boldsymbol{u}(t) \tag{15.3.9}$$

$$y(t) = \boldsymbol{C}\boldsymbol{x}(t) \tag{15.3.10}$$

from the parameters of the discrete-time model

$$\boldsymbol{x}(k+1) = \boldsymbol{F}\boldsymbol{x}(k) + \boldsymbol{G}\boldsymbol{u}(k) \tag{15.3.11}$$

$$y(k) = \boldsymbol{C}\boldsymbol{x}(k)\,, \tag{15.3.12}$$

see (Sinha and Lastman, 1982). The methods vary in the assumptions that are made on the time lapse of the input signals between the sample points.

One method is based on the integration of (15.3.9) (Hung et al, 1980) as

$$\boldsymbol{x}(k+1) = \boldsymbol{x}(k) + \boldsymbol{A}\int_{kT_0}^{(k+1)T_0} \boldsymbol{x}(t)\mathrm{d}t + \boldsymbol{B}\int_{kT_0}^{(k+1)T_0} \boldsymbol{u}(t)\mathrm{d}t \tag{15.3.13}$$

and its approximation according to the trapezoidal rule

$$\boldsymbol{x}(k+1) = \boldsymbol{x}(k) + \frac{1}{2}\boldsymbol{A}T_0\big(\boldsymbol{x}(k+1) + \boldsymbol{x}(k)\big) + \frac{1}{2}\boldsymbol{B}T_0\big(\boldsymbol{u}(k+1) + \boldsymbol{u}(k)\big)\,. \tag{15.3.14}$$

The parameters $\boldsymbol{A}$ and $\boldsymbol{B}$ can now be determined e.g. by the method of least squares, if the input signals as well as the state variables are known (Sinha and Lastman, 1982).

The conversion of a model from a discrete-time representation to a continuous-time representation can require a high computational effort and can become cumbersome, and frequently one can only provide a solution for special cases. Hence, other methods should be preferred if one is interested in the identification based on continuous-time models.

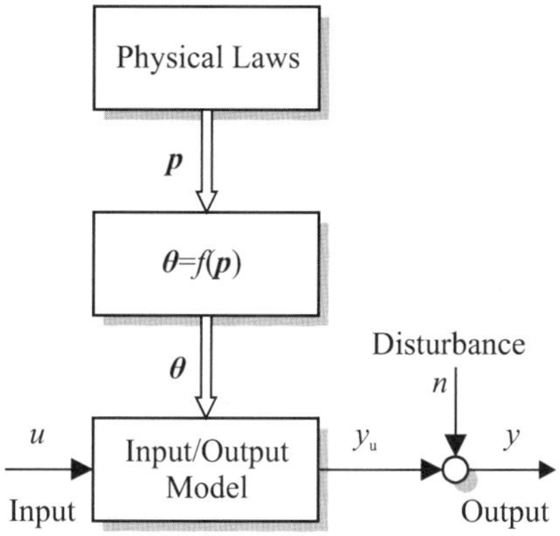

Fig. 15.9. Dynamic process model with model parameters $\boldsymbol{\theta}$ and physical process coefficients $\boldsymbol{p}$

15.4 Estimation of Physical Parameters

For certain tasks, one is not only interested in determining the parameters $\boldsymbol{\theta}$ of an input/output model of the process (e.g. a differential equation), but is rather interested in determining the physically defined parameters $\boldsymbol{p}$ of the governing physical laws, see Fig. 15.9. These physical parameters shall be termed *process coefficients* to distinguish them from the *model parameters* $\boldsymbol{\theta}$ of the input/output model. For many typical tasks in control engineering, such as the design of fixed parameter controllers or adaptive controllers, knowledge of the model parameters $\boldsymbol{\theta}$ is sufficient. However, for the following tasks, one must also know the process coefficients $\boldsymbol{p}$:

- Determination of not directly measurable coefficients in all fields of natural sciences
- Verification of specifications of technical processes
- Fault detection and diagnosis during operation
- Quality control of production plants

The model parameters can in general be determined from the process coefficients by means of algebraic relations as

$$\boldsymbol{\theta} = f(\boldsymbol{p}) \,. \tag{15.4.1}$$

These often non-linear relations are known from theoretical modeling, see Sect. 1.1. The question which shall now be discussed is how one can determine the process coefficients $\boldsymbol{p}$ from measurements of the input signal $u(t)$ and the (disturbed) output signal $y(t)$. This typically leads to models where the error is not linear in the parameters, hence iterative methods must be applied, see Chap. 19. This direct way of estimating process coefficients leads to offline solutions.

A different approach is to first estimate model parameters $\boldsymbol{\theta}$ and then by means of the inverse relation

$$\boldsymbol{p} = f^{-1}(\boldsymbol{\theta}) \,. \tag{15.4.2}$$

determine the process coefficients $\boldsymbol{p}$. However, this gives rise to the following questions:

- Can the process coefficients $\boldsymbol{p}$ be determined uniquely?
- Which test signals must be applied and which signals must be measured so that the process coefficients $\boldsymbol{p}$ are identifiable?
- Can one employ a priori knowledge as e.g. a priori known process coefficients p_i or model parameters θ_i to improve the estimation results of $\boldsymbol{\theta}$ or $\boldsymbol{p}$?

For the solution of these questions, one must combine theoretical and experimental modeling techniques. The first step is to set up balance equations, constitutive equations (physical or chemical equations of state), phenomenological equations and the corresponding interconnection equations. These yield a physically motived process model in form of a set of equations or a block diagram.

It is assumed that the process is linear and possesses M measurable signals $\eta_j(t)$ and N non-measurable signals $\xi_j(t)$. For L elements the Laplace transformed *process element equations* become

$$\sum_{j=1}^{M} g_{ij}\eta_j(s) = \sum_{j=1}^{N} h_{ij}\xi_j(s),\ i = 1, 2, \ldots, L \tag{15.4.3}$$

with

$$g_{ij} \in \{0, \pm 1, \alpha_{ij}, s^\kappa\} \tag{15.4.4}$$
$$h_{ij} \in \{0, \pm 1, \beta_{ij}, s^\kappa\} \tag{15.4.5}$$
$$\kappa \in \{-1, 0, 1\}\ . \tag{15.4.6}$$

The system of L equations can now be written in vector/matrix-notation as

$$\boldsymbol{G}\boldsymbol{q} = \boldsymbol{H}\boldsymbol{\xi} \tag{15.4.7}$$

and

$$\boldsymbol{\eta}^{\mathrm{T}} = \begin{pmatrix} \eta_1 & \eta_2 & \ldots & \eta_M \end{pmatrix} \tag{15.4.8}$$
$$\boldsymbol{\xi}^{\mathrm{T}} = \begin{pmatrix} \xi_1 & \xi_2 & \ldots & \xi_N \end{pmatrix}, \tag{15.4.9}$$

where $\boldsymbol{G}$ has the dimensions $L \times M$ and the matrix $\boldsymbol{H}$ has the dimensions $L \times N$.

It is assumed that the individual process element equations are linearly independent.

The parameter estimation requires a model structure, where only measurable signals turn up. In order to eliminate the non-measured signals, the system of equations in (15.4.7) must be transformed such that the matrix $\boldsymbol{H}$ has an upper triangular form. After applying the inverse Laplace transform, one obtains an ordinary differential equation that governs the input/output behavior as

$$a_n^* y^{(n)}(t) + \ldots + a_1^* y^{(1)}(t) + a_0^* y(t) = b_0^* u(t) + b_1^* u^{(1)}(t) + \ldots + b_m^* u^{(m)}(t)\ . \tag{15.4.10}$$

In this equation, all variables are connected with a parameter. For the parameter estimation, one however needs the form

$$a_n y^{(n)}(t) + \ldots + a_1 y^{(1)}(t) + y(t) = b_0 u(t) + b_1 u^{(1)}(t) + \ldots + b_m u^{(m)}(t) \,. \tag{15.4.11}$$

All starred quantities a_i^* and b_i^* must therefore be multiplied with the factor $1/a_0^*$.

In the fundamental equation (15.4.7), the process coefficients p_i show up as individual values in their original form. Therefore, the elements of $\boldsymbol{G}$ and $\boldsymbol{H}$ are given as

$$g_{ij} = p_{gij} s^\kappa ,\ h_{ij} = p_{hij} s^\kappa ,\ \kappa \in (-1, 0, 1) \,. \tag{15.4.12}$$

After transforming into the upper triangular form, then only the last row has to be considered to set up the input/output model. In this row, the parameters show up as

$$\begin{aligned} \theta_i &= \sum_{\mu=1}^{q} c_{i\mu} \prod_{\nu=1}^{l} p_\nu^{\varepsilon_{\mu\nu}} \\ &= c_{i1} p_1^{\varepsilon_{11}} p_2^{\varepsilon_{12}} \ldots p_t^{\varepsilon_{1t}} + c_{i2} p_1^{\varepsilon_{21}} p_2^{\varepsilon_{22}} \ldots p_t^{\varepsilon_{2t}} \\ &\quad + \ldots + c_{iq} p_1^{\varepsilon_{q1}} p_2^{\varepsilon_{q2}} \ldots p_t^{\varepsilon_{qt}} \,. \end{aligned} \tag{15.4.13}$$

For this equation, it was assumed that a_0^* contains the process coefficients in a simple product form, such that e.g.

$$a_0^* = \prod_{\nu=1}^{l} p_\nu^{\varepsilon_{\mu\nu}} \,. \tag{15.4.14}$$

With

$$z_\mu = \prod_{\nu}^{l} p_\nu^{\varepsilon_{\mu\nu}} \,, \tag{15.4.15}$$

one then obtains

$$\theta_i = \sum_{\mu=1}^{q} c_{i\mu} z_\mu \,, \tag{15.4.16}$$

which provides the parameters as algebraic functions of the process coefficients. The parameters z_μ are here abbreviations for all appearing products and single occurrences of the process coefficients $p_1, \ldots, p_i$. For the exponents, one often has $\varepsilon_{\mu\nu} = 1$ or $\varepsilon_{\mu\nu} = -1$.

The relation between the model parameters and the process coefficients are then given in vector/matrix notation as

$$\underbrace{\begin{pmatrix} \theta_1 \\ \theta_2 \\ \vdots \\ \theta_r \end{pmatrix}}_{\boldsymbol{\theta}} = \underbrace{\begin{pmatrix} c_{11} & \ldots & c_{1q} \\ c_{21} & \ldots & c_{2q} \\ \vdots & & \vdots \\ c_{r1} & \ldots & c_{rq} \end{pmatrix}}_{\boldsymbol{C}} \underbrace{\begin{pmatrix} z_1 \\ z_2 \\ \vdots \\ z_q \end{pmatrix}}_{\boldsymbol{z}} \,. \tag{15.4.17}$$

Th vector $\boldsymbol{z}$ contains the products z_μ of the process coefficients as well as the process coefficients individually and hence the dimension of $\boldsymbol{z}$ is $q \geq l$.

In the following, it is assumed that the parameters θ_i have already been determined by the application of parameter estimation methods and one is now interested in determining the process coefficients p_ν.

As the relation $\boldsymbol{\theta} = f(\boldsymbol{p})$ is in most cases non-linear, the above questions cannot be answered in general. Once the r model parameters θ_i have been determined by any of the parameter estimation methods presented in this book, one is now interested in determining the l process coefficients p_j. These can be determined by solving the system of non-linear equations $\boldsymbol{\theta} = f(\boldsymbol{p})$ for the unknown process coefficients p_j, i.e. determine

$$\boldsymbol{p} = f^{-1}(\boldsymbol{\theta}) \tag{15.4.18}$$

For models of low order, i.e. first or second order, the above system of equations can typically be solved with moderate effort. However, for process models with a higher number of parameters, only some general considerations can be given.

Here, it is often helpful to rewrite the system of equations as an implicit equation

$$\underbrace{\begin{pmatrix} q_1 \\ q_2 \\ \vdots \\ q_r \end{pmatrix}}_{\boldsymbol{q}} = \underbrace{\begin{pmatrix} \theta_1 \\ \theta_2 \\ \vdots \\ \theta_r \end{pmatrix}}_{\boldsymbol{\theta}} - \underbrace{\begin{pmatrix} c_{11} & \dots & c_{1q} \\ c_{21} & \dots & c_{2q} \\ \vdots & & \vdots \\ c_{r1} & \dots & c_{rq} \end{pmatrix}}_{\boldsymbol{C}} \underbrace{\begin{pmatrix} z_1 \\ z_2 \\ \vdots \\ z_q \end{pmatrix}}_{\boldsymbol{z}} = \underbrace{\begin{pmatrix} 0 \\ 0 \\ \vdots \\ 0 \end{pmatrix}}_{\boldsymbol{0}} , \tag{15.4.19}$$

which leads to

$$\boldsymbol{q} = \boldsymbol{\theta} - \boldsymbol{C}\boldsymbol{z} = \boldsymbol{0} . \tag{15.4.20}$$

Here,

$$\boldsymbol{z} = g(\boldsymbol{p}) \tag{15.4.21}$$

with the required process coefficients,

$$\boldsymbol{p} = \begin{pmatrix} p_1 & p_2 & \dots & p_t \end{pmatrix}^{\mathrm{T}} . \tag{15.4.22}$$

According to the *implicit function theorem*, the non-linear system of equations can be solved for the l unknowns $\boldsymbol{p}$ in a neighborhood around the solution $\boldsymbol{p}_0$ if for the Jacobian holds

$$\det \boldsymbol{Q}_{\boldsymbol{p}} \neq 0 , \tag{15.4.23}$$

where $\boldsymbol{Q}_{\boldsymbol{p}}$ is given as

$$\boldsymbol{Q}_{\boldsymbol{p}} = \frac{\partial \boldsymbol{q}^{\mathrm{T}}}{\partial \boldsymbol{p}} = \begin{pmatrix} \frac{\partial q_1}{\partial p_1} & \frac{\partial q_2}{\partial p_1} & \dots & \frac{\partial q_r}{\partial p_1} \\ \frac{\partial q_1}{\partial p_2} & \frac{\partial q_2}{\partial p_2} & \dots & \frac{\partial q_r}{\partial p_2} \\ \vdots & \vdots & & \vdots \\ \frac{\partial q_l}{\partial p_1} & \frac{\partial q_l}{\partial p_2} & \dots & \frac{\partial q_l}{\partial p_r} \end{pmatrix} \tag{15.4.24}$$

Here, $\boldsymbol{q}$ must be continuously differentiable and furthermore $r = l$. From here, one can derive conditions for the *identifiability of process coefficients* $\boldsymbol{p}$.

Necessary conditions for the unambiguous determination of l process coefficients p_ν from r model parameters θ_i are that $r = l$ and that the Jacobian ($\det \boldsymbol{Q}_p$) does not vanish in a neighborhood of the solution $\boldsymbol{p}_0$ (Isermann, 1992).

The identifiability condition does only state, whether the problem can be solved in general. A certain solution cannot be derived from this condition. One still has to determine the unknowns by a successive solution of the system of equations for the unknown process parameters. For this, it can be helpful to use computer algebra programs (Schumann, 1991). For worked out examples, see also (Isermann, 1992).

A different approach is based on similarity transforms, see (17.2.9) through (17.2.11). Here, one first estimates a black box model in state space representation and in the next step, one determines a similarity transform to transform to system into a structure that matches a physically derived state space model and determines the parameters of that model (Parillo and Ljung, 2003; Xie and Ljung, 2002).

Example 15.2 (Determination of Physical Coefficients from the Model Parameters of a Continuous-Time Model of the Three-Mass Oscillator).

From a transfer function estimate, it is impossible to deduce the parameters of the Three-Mass Oscillator directly as the symbolic transfer function, which could be used for a comparison of coefficients is a highly complex and non-linear function of the physical system parameters. With full state information, it is however possible to parameterize the full state matrix and from there determine all physical parameters.

As is shown in detail in Chap. 17, the equations of motion of the Three-Mass Oscillator can be written as

$$\ddot{\varphi}_1(t) = \underbrace{\frac{c_1}{J_1}}_{\theta_{11}} \big(\varphi_2(t) - \varphi_1(t)\big) + \underbrace{\frac{-d_1}{J_1}}_{\theta_{12}} \dot{\varphi}_1(t) + \underbrace{\frac{1}{J_1}}_{\theta_{13}} M_{\mathrm{M}}(t) \tag{15.4.25}$$

$$\ddot{\varphi}_2(t) = \underbrace{\frac{c_1}{J_2}}_{\theta_{21}} \big(\varphi_1(t) - \varphi_2(t)\big) + \underbrace{\frac{c_2}{J_2}}_{\theta_{22}} \big(\varphi_3(t) - \varphi_2(t)\big) + \underbrace{\frac{-d_2}{J_2}}_{\theta_{23}} \dot{\varphi}_2(t) \tag{15.4.26}$$

$$\ddot{\varphi}_3(t) = \underbrace{\frac{c_2}{J_3}}_{\theta_{31}} \big(\varphi_2(t) - \varphi_3(t)\big) + \underbrace{\frac{-d_3}{J_3}}_{\theta_{32}} \dot{\varphi}_3(t)\,. \tag{15.4.27}$$

From the model coefficient estimates $\hat{\theta}_{11}, \ldots, \hat{\theta}_{32}$, one can determine the physical parameters as follows

$$\hat{J}_1 = \frac{1}{\hat{\theta}_{13}} = 12.2 \times 10^{-3}\,\mathrm{kg\,m^2} \tag{15.4.28}$$

$$\hat{d}_1 = -\hat{\theta}_{12}\hat{J}_1 = 0.0137\,\frac{\mathrm{s\,Nm}}{\mathrm{rad}} \tag{15.4.29}$$

$$\hat{c}_1 = \hat{\theta}_{11}\hat{J}_1 = 2.4955\,\frac{\mathrm{Nm}}{\mathrm{rad}} \tag{15.4.30}$$

$$\hat{J}_2 = \frac{\hat{c}_1}{\hat{\theta}_{21}} = 7.8 \times 10^{-3}\,\mathrm{kg\,m^2} \tag{15.4.31}$$

$$\hat{d}_2 = \hat{\theta}_{23}\hat{J}_2 = 0.0017\frac{\mathrm{s\,Nm}}{\mathrm{rad}} \tag{15.4.32}$$

$$\hat{c}_2 = \hat{\theta}_{22}\hat{J}_2 = 1.9882\frac{\mathrm{Nm}}{\mathrm{rad}} \tag{15.4.33}$$

$$\hat{J}_3 = \frac{\hat{c}_2}{\hat{\theta}_{31}} = 13.7 \times 10^{-3}\,\mathrm{kg\,m^2} \tag{15.4.34}$$

$$\hat{d}_3 = \hat{\theta}_{32}\hat{J}_3 = 0.0024\frac{\mathrm{s\,Nm}}{\mathrm{rad}}\ . \tag{15.4.35}$$

It was luck that in this case there were as many estimated model coefficients as physical parameters. This however does not always have to be so. □

15.5 Parameter Estimation for Partially Known Parameters

It is now assumed that some parameters a_i and b_i of the parameter vector

$$\boldsymbol{\theta} = \left(a_1\ a_2\ \ldots\ a_n \middle| b_0\ b_1\ \ldots b_m\right)^{\mathrm{T}} \tag{15.5.1}$$

of the differential equations (15.1.1) are known a priori. These known parameters will be denoted with a_i'' and b_i''.

Then, the parameter vector $\boldsymbol{\theta}$ can be split up into a vector of known parameters $\boldsymbol{\theta}''$and a vector of remaining unknown parameters $\boldsymbol{\theta}'$ as

$$\boldsymbol{\theta} = \begin{pmatrix} \boldsymbol{\theta}' \\ \boldsymbol{\theta}'' \end{pmatrix}. \tag{15.5.2}$$

In the corresponding system of equations, one can bring the known parameters along with the corresponding elements of the data vector to the left side,

$$y(t) - \boldsymbol{\psi}''^{T}(t)\boldsymbol{\theta}'' = \boldsymbol{\psi}'^{T}(t)\boldsymbol{\theta}'(t) + e(t)\ . \tag{15.5.3}$$

With the short hand notation

$$\tilde{y}(t) = y(t) - \boldsymbol{\psi}''^{T}(t)\boldsymbol{\theta}''\ , \tag{15.5.4}$$

one can then write

$$\tilde{y}(t) = \boldsymbol{\psi}'^{T}(t)\boldsymbol{\theta}'(t) + e(t) \tag{15.5.5}$$

and in vector form

$$\tilde{\boldsymbol{y}} = \boldsymbol{\Psi}'\boldsymbol{\theta}' + \boldsymbol{e}\ , \tag{15.5.6}$$

leading to

$$\hat{\boldsymbol{\theta}}' = \left(\boldsymbol{\Psi}'^{\mathrm{T}}\boldsymbol{\Psi}'\right)^{-1}\boldsymbol{\Psi}'^{\mathrm{T}}\tilde{\boldsymbol{y}}''\ . \tag{15.5.7}$$

Hence, the parameter estimation methods can be applied to the reduced parameter vector $\boldsymbol{\theta}'$, the reduced data matrix $\boldsymbol{\Psi}'$, and the augmented output vector $\tilde{\boldsymbol{y}}$.

The influence of individual a priori known parameters on the quality of the parameter estimates was investigated by Rentzsch (1988) with several processes of second order, which were governed by

$$a_2 y^{(2)}(t) + a_1 y^{(1)}(t) + a_0 y(t) = b_1 u^{(1)}(t) + b_0 u(t) \,. \tag{15.5.8}$$

The results can be summarized as follows:

- *One parameter known*: The parameter estimation can be improved asymptotically by fixing those parameters that would have a relatively large estimation variance otherwise, hence especially a priori knowledge about a_2 and b_1 is helpful.
- *Multiple parameters known*:
  1. In comparison to the case of only one known parameter as outlined above, the asymptotic estimation quality does only improve if a_2 or b_1 are among the known parameters
  2. The convergence speed increases the more parameters are known

In comparison to the case of a totally unknown parameter vector, big improvements can be obtained even if only a priori knowledge about one single parameter is available. Especially, a priori knowledge of a_2 can cause a big enhancement. The a priori knowledge of further parameters is only beneficial if these parameters are relatively well known (for the example considered above, up to 5% error).

15.6 Summary

For the parameter estimation of processes governed by continuous-time models, one can in general use the same methods that have already been introduced for discrete-time models. However, a problem is the calculation of the derivatives. Numerical methods such as finite differencing have only proven helpful for low orders, in many cases even only first order. Therefore, the state variable filter has been introduced in this chapter as a tool to determine the derivatives of the (filtered) signal.

Its transfer function must be a low-pass filter such as e.g. the Butterworth filter. The cut-off frequency should be placed close to the highest process frequency and the order should be two higher than the order of the process, which in turn determines the highest derivative that is required. The sample frequency should be approximately 20 times the corner frequency. It must be kept in mind that the filter determines the derivatives of the filtered signal $u_\mathrm{F}(t)$, $y_\mathrm{F}(t)$ and not of the original signal $u(t)$, $y(t)$. Especially, the presence of larger disturbances can still lead to systematic errors in the estimation. In this cases, one should use consistent parameter estimation methods, such as the method of instrumental variables, correlation methods and iterative optimization methods. A case study comparing different approaches is shown in (Ljung, 2009).

An appealing alternative that also avoids the calculation of derivatives is to use a parallel model and evaluate the output error $y(t) - \hat{y}(t)$. This however necessitates the use of iterative optimization techniques, see Chap. 19. Subspace methods can also be applied to continuous-time systems, see Chap. 16.

One can also use frequency domain methods to determine a transfer function which then yields the coefficients of continuous-time differential equations, see also Chap. 4 and Chap. 14 in this book and (Pintelon and Schoukens, 2001). Here, one can also estimate a noise model together with the parametric transfer function (e.g. Pintelon et al, 2000).

Finally, one has to derive the process parameters from the model coefficients. Here, one typically has to solve a system of non-linear equations, for which some guidelines and necessary conditions for the solution have been stated in this chapter. As general results, one can conclude that much more process parameters can be identified from the dynamic behavior than from the static behavior. Furthermore, it also depends on the choice and availability of certain inputs and outputs of the process, whether process parameters can be identified from dynamic measurements.

Problems

15.1. Continuous-Time Models I
What can be reasons for the interest in working with a continuous-time model?

15.2. Continuous-Time Models II
Which methods allow to identify continuous-time models directly?

15.3. Derivative Information
Describe ways to provide derivative information to the parameter estimation methods described in this chapter and show this for a second order differential equation.

15.4. First Order System
Determine the gain and time constant for the thermometer governed in Example 2.1 by first estimating the parameters of a discrete-time first order model and then converting this model to a continuous-time model.

15.5. First Order System
Determine the gain and time constant for the thermometer governed in Example 2.1 by using a continuous-time first order model, matching the time response directly.

References

Eisinberg A, Fedele G (2006) On the inversion of the Vandermonde matrix. Applied Mathematics and Computation 174(2):1384–1397

Garnier H, Wang L (2008) Identification of continuous-time models from sampled data. Advances in Industrial Control, Springer, London

Hamming RW (2007) Digital filters, 3rd edn. Dover books on engineering, Dover Publications, Mineola, NY

Hung JC, Liu CC, Chou PY (1980) In: Proceedings of the 14th Asilomar Conf. Circuits Systems Computers, Pacific Grove, CA, USA

Isermann R (1992) Estimation of physical parameters for dynamic processes with application to an industrial robot. Int J Control 55(6):1287–1298

Kammeyer KD, Kroschel K (2009) Digitale Signalverarbeitung: Filterung und Spektralanalyse mit MATLAB-Übungen, 7th edn. Teubner, Wiesbaden

Larsson EK, Söderström T (2002) Identification of continuous-time AR processes from unevenly sampled data. Automatica 38(4):709–718

Ljung L (1999) System identification: Theory for the user, 2nd edn. Prentice Hall Information and System Sciences Series, Prentice Hall PTR, Upper Saddle River, NJ

Ljung L (2009) Experiments with identification of continuous time models. In: Proceedings of the 15th IFAC Symposium on System Identification, Saint-Malo, France

Ljung L, Wills A (2008) Issues in sampling and estimating continuous-time models with stochastic disturbances. In: Proceedings of the 17th IFAC World Congress, Seoul, Korea

Moler C, van Loan C (2003) Nineteen dubios ways to compute the exponential of a matrix, twenty-five years later. SIAM Rev 45(1):3–49

Oppenheim AV, Schafer RW (2009) Discrete-time signal processing, 3rd edn. Prentice Hall, Upper Saddle River, NJ

Parillo PA, Ljung L (2003) Initialization of physical parameter estimates. In: Proceedings of the 13th IFAC Symposium on System Identification, Rotterdam, The Netherlands, pp 1524–1529

Peter K (1982) Parameteradaptive Regelalgorithmen auf der Basis zeitkonitnuierlicher Prozessmodelle. Dissertation. TH Darmstadt, Darmstadt

Pintelon R, Schoukens J (2001) System identification: A frequency domain approach. IEEE Press, Piscataway, NJ

Pintelon R, Schoukens J, Rolain Y (2000) Box-Jenkins continuous-time modeling. Automatica 36(7):983–991

Rao GP, Garnier H (2002) Numerical illustrations of the relevance of direct continuous-time model identification. In: Proceedings of the 15th IFAC World Congress, Barcelona, Spain

Rao GP, Unbehauen H (2006) Identification of continuous-time systems. IEE Proceedings Control Theory and Applications 153(2):185–220

Raol JR, Girija G, Singh J (2004) Modelling and parameter estimation of dynamic systems, IEE control engineering series, vol 65. Institution of Electrical Engineers, London

Rentzsch M (1988) Analyse des Verhaltens rekursiver Parametershätzverfahren beim Einbringen von A-Priori Information. Diplomarbeit. Institut für Regelungstechnik, TH Darmstadt, Darmstadt, Germany

Schumann A (1991) INID - A computer software for experimental modeling. In: Proceedings of the 9tf IFAC Symposium on Identification, Budapest, Hungary

Sinha NK, Lastman GJ (1982) Identification of continuous time multivariable systems from sampled data. Int J Control 35(1):117–126

Söderström T, Mossberg M (2000) Performance evaluation of methods for identifying continuous-time autoregressive processes. Automatica 36(1):53–59

Tietze U, Schenk C, Gamm E (2010) Halbleiter-Schaltungstechnik, 13th edn. Springer, Berlin

Vogt M (1998) Weiterentwicklung von Verfahren zur Online-Parameterschätzung und Untersuchung von Methoden zur Erzeugung zeitlicher Ableitungen. Diplomarbeit. Institut für Regelungstechnik, TU Darmstadt, Darmstadt

Wolfram A, Moseler O (2000) Design and application of digital FIR differentiators using modulating functions. In: Proccedings of the 12th IFAC Symposium on System Identification, Santa Barbara, CA, USA

Wolfram A, Vogt M (2002) Zeitdiskrete Filteralgorithmen zur Erzeugung zeitlicher Ableitungen. at 50(7):346–353

Xie LL, Ljung L (2002) Estimate physical parameters by black-box modeling. In: Proceedings of the 21st Chinese Control Conference, Hangzhou, China, pp 673–677

Young P (1981) Parameter estimation for continuous time models – a survey. Automatica 17(1):23–39

Young PC (2002) Optimal IV identification and estimation of continuous time TF models. In: Proceedings of the 15th IFAC World Congress, Barcelona, Spain

16

Subspace Methods

Subspace methods can identify state space models if only input and output measurements are available. Since no measurements of the states are required and employed, it is not possible to come up with a unique structure, hence the model is only known up to a similarity transform $\boldsymbol{T}$. Interesting features of the state space identification are the fact that subspace identification allows to determine the suitable model oder as part of the identification process and furthermore, the method is from the very beginning formulated to cover MIMO systems as well. A short history of subspace methods can e.g. be found in the editorial note by Viberg and Stoica (1996).

After some preliminaries, the subspace identification will be introduced. Two different approaches will be presented. First, the case of a purely deterministic system will be discussed, where there is no noise acting on the system. This allows to easily convey the notions and ideas behind subspace identification. Then, the more practical case of measurements which are more or less affected by noise will be discussed.

16.1 Preliminaries

In order to derive the subspace identification approach, first the discrete-time state space equations shall be recalled from Chap. 2. The recursive solutions (2.2.32) and (2.2.33) are restated here for convenience,

$$\boldsymbol{x}(k) = \boldsymbol{A}^k \boldsymbol{x}(0) + \sum_{i=0}^{k-1} \boldsymbol{A}^{k-i-1} \boldsymbol{B} \boldsymbol{u}(i) \tag{16.1.1}$$

$$\boldsymbol{y}(k) = \boldsymbol{C}\boldsymbol{x}(k) + \boldsymbol{D}\boldsymbol{u}(k)\,. \tag{16.1.2}$$

Based on these equations, it is possible to derive equations for the input/output behavior of the process. These equations can then be used to derive an identification technique that provides state space models based on measurements of the input and output only.

For a stream of k samples of the input sequence $\boldsymbol{u}$, the corresponding output $\boldsymbol{y}$ can be written as

R. Isermann, M. Münchhof, *Identification of Dynamic Systems*,
DOI 10.1007/978-3-540-78879-9_16, © Springer-Verlag Berlin Heidelberg 2011

$$\begin{pmatrix} \boldsymbol{y}(0) \\ \boldsymbol{y}(1) \\ \boldsymbol{y}(2) \\ \vdots \\ \boldsymbol{y}(k-1) \end{pmatrix} = \underbrace{\begin{pmatrix} \boldsymbol{C} \\ \boldsymbol{C}\boldsymbol{A} \\ \boldsymbol{C}\boldsymbol{A}^2 \\ \vdots \\ \boldsymbol{C}\boldsymbol{A}^{k-1} \end{pmatrix}}_{\boldsymbol{Q}_{\mathrm{Bk}}} \boldsymbol{x}(0)$$

$$+ \underbrace{\begin{pmatrix} \boldsymbol{D} & \boldsymbol{0} & \boldsymbol{0} & \dots & 0 \\ \boldsymbol{C}\boldsymbol{B} & \boldsymbol{D} & \boldsymbol{0} & \dots & \boldsymbol{0} \\ \boldsymbol{C}\boldsymbol{A}\boldsymbol{B} & \boldsymbol{C}\boldsymbol{B} & \boldsymbol{D} & \dots & \boldsymbol{0} \\ \vdots & \vdots & \vdots & & \boldsymbol{0} \\ \boldsymbol{C}\boldsymbol{A}^{k-2}\boldsymbol{B} & \boldsymbol{C}\boldsymbol{A}^{k-3}\boldsymbol{B} & \boldsymbol{C}\boldsymbol{A}^{k-4}\boldsymbol{B} & \dots & \boldsymbol{D} \end{pmatrix}}_{\mathcal{T}_k} \begin{pmatrix} \boldsymbol{u}(0) \\ \boldsymbol{u}(1) \\ \boldsymbol{u}(2) \\ \vdots \\ \boldsymbol{u}(k-1) \end{pmatrix} . \tag{16.1.3}$$

The matrices $\boldsymbol{D}$, $\boldsymbol{C}\boldsymbol{B}$, $\boldsymbol{C}\boldsymbol{A}\boldsymbol{B}$, up to $\boldsymbol{C}\boldsymbol{A}^{k-2}\boldsymbol{B}$ are called the *Markov parameters* (Juang, 1994; Juang and Phan, 2006).

The above equations are also valid for a shift in time of d samples, hence

$$\begin{pmatrix} \boldsymbol{y}(d) \\ \boldsymbol{y}(d+1) \\ \vdots \\ \boldsymbol{y}(d+k-1) \end{pmatrix} = \boldsymbol{Q}_{\mathrm{Bk}}\boldsymbol{x}(d) + \mathcal{T}_k \begin{pmatrix} \boldsymbol{u}(d) \\ \boldsymbol{u}(d+1) \\ \vdots \\ \boldsymbol{u}(d+k-1) \end{pmatrix} . \tag{16.1.4}$$

In the above equations, the matrix $\boldsymbol{Q}_{\mathrm{Bk}}$ denotes the *extended observability matrix*

$$\boldsymbol{Q}_{\mathrm{Bk}} = \begin{pmatrix} \boldsymbol{C} \\ \boldsymbol{C}\boldsymbol{A} \\ \boldsymbol{C}\boldsymbol{A}^2 \\ \vdots \\ \boldsymbol{C}\boldsymbol{A}^{k-1} \end{pmatrix} . \tag{16.1.5}$$

$\boldsymbol{Q}_{\mathrm{Bk}}$ is assumed to have full rank, which is tantamount to the assumption that the system is observable. $\boldsymbol{Q}_{\mathrm{Bk}}$ is called extended observability matrix, because $k > n$. Also, the *reversed extended controllability matrix* $\boldsymbol{Q}_{\mathrm{Sk}}$ is defined as

$$\boldsymbol{Q}_{\mathrm{Sk}} = \left(\boldsymbol{A}^{k-1}\boldsymbol{B} \ \boldsymbol{A}^{k-2}\boldsymbol{B} \ \dots \ \boldsymbol{A}\boldsymbol{B} \ \boldsymbol{B} \right) . \tag{16.1.6}$$

This matrix is also assumed to be of full rank, which means that the system is assumed to be controllable. Finally, the matrix $\mathcal{T}_k$ is given as

$$\mathcal{T}_k = \begin{pmatrix} \boldsymbol{D} & \mathbf{0} & \mathbf{0} & \dots & 0 \\ \boldsymbol{CB} & \boldsymbol{D} & \mathbf{0} & \dots & \mathbf{0} \\ \boldsymbol{CAB} & \boldsymbol{CB} & \boldsymbol{D} & \dots & \mathbf{0} \\ \vdots & \vdots & \vdots & & \vdots \\ \boldsymbol{CA}^{k-2}\boldsymbol{B} & \boldsymbol{CA}^{k-3}\boldsymbol{B} & \boldsymbol{CA}^{k-4}\boldsymbol{B} & \dots & \boldsymbol{D} \end{pmatrix} \tag{16.1.7}$$

and contains the so-called *Markov parameters* of the system, see Chap. 17.

In the interest of a compact notation, the input $\boldsymbol{u}(k)$ and output $\boldsymbol{y}(k)$ are now both grouped as

$$\boldsymbol{U} = \left(\frac{\boldsymbol{U}_-}{\boldsymbol{U}_+}\right) = \left(\frac{\boldsymbol{U}_{0|k-1}}{\boldsymbol{U}_{k|2k-1}}\right) = \left(\begin{array}{ccccc} u_0 & u_1 & u_2 & \dots & u_{N-1} \\ u_1 & u_2 & u_3 & \dots & u_N \\ \vdots & \vdots & \vdots & & \vdots \\ u_{k-1} & u_k & u_{k+1} & \dots & u_{k+N-2} \\ \hline u_k & u_{k+1} & u_{k+2} & \dots & u_{k+N-1} \\ u_{k+1} & u_{k+2} & u_{k+3} & \dots & u_{k+N} \\ \vdots & \vdots & \vdots & & \vdots \\ u_{2k-1} & u_{2k} & u_{2k+1} & \dots & u_{2k+N-2} \end{array}\right), \tag{16.1.8}$$

for the input and

$$\boldsymbol{Y} = \left(\frac{\boldsymbol{Y}_-}{\boldsymbol{Y}_+}\right) = \left(\frac{\boldsymbol{Y}_{0|k-1}}{\boldsymbol{Y}_{k|2k-1}}\right) \tag{16.1.9}$$

for the output, where the subscript "−" denotes the *past* values and the subscript "+" the *future* values. Matrices with such a special structure are termed *Hankel* matrices, they have constant values along the counter (block) diagonals (Golub and van Loan, 1996).

The above relations and definitions can now be used to relate the past and future values of the input $\boldsymbol{u}$ and output $\boldsymbol{y}$ as

$$\boldsymbol{Y}_- = \boldsymbol{Q}_{\mathrm{Bk}}\boldsymbol{X}_- + \mathcal{T}_k\boldsymbol{U}_- \tag{16.1.10}$$

$$\boldsymbol{Y}_+ = \boldsymbol{Q}_{\mathrm{Bk}}\boldsymbol{X}_+ + \mathcal{T}_k\boldsymbol{U}_+ , \tag{16.1.11}$$

where $\boldsymbol{X}_-$ denotes the *past state matrix* as

$$\boldsymbol{X}_- = \big(\boldsymbol{x}(0)\ \boldsymbol{x}(1)\ \dots\ \boldsymbol{x}(k-1)\big) . \tag{16.1.12}$$

and $\boldsymbol{X}_+$ is analogously defined as the *future state matrix*,

$$\boldsymbol{X}_+ = \big(\boldsymbol{x}(k)\ \boldsymbol{x}(k+1)\ \dots\ \boldsymbol{x}(2k)\big) . \tag{16.1.13}$$

The state matrices $\boldsymbol{X}_+$ and $\boldsymbol{X}_-$ are related as

$$\boldsymbol{X}_+ = \boldsymbol{A}^k\boldsymbol{X}_- + \boldsymbol{Q}_{\mathrm{Sk}}\boldsymbol{U}_- . \tag{16.1.14}$$

The following assumptions have to be made

- $\operatorname{rank} X_k = n \Rightarrow$ The system is excited sufficiently and more samples k than the state matrix dimension n have been recorded
- $\operatorname{rank} U = 2kn$, with $k > n \Rightarrow$ The system is persistently excited of order $2k$
- $\operatorname{span} X_k \cap \operatorname{span} U_+ = \emptyset \Rightarrow$ The row vectors of X_k and U_+ are linearly independent, hence the system must be operated in open loop

The major drawback is that the method by itself is not suitable for closed loop system identification, where the feedback causes a correlation between the input and the output. Similarly to what is presented in Chap. 13, there are different approaches. The first approach is to neglect the effect of the feedback loop and hope that the resulting error due to a possible correlation between the input and the output is acceptable. The second approach is to identify the closed-loop dynamics and then determine the open loop dynamics by considering the controller transfer function. Finally, one can once again employ joint input-output techniques (Katayama and Tanaka, 2007).

There are also modifications to the subspace identification that allow operation in closed loop. One idea is presented by Ljung and McKelvey (1996) and is based on using a high-dimensional intermediate ARX model. Other work is presented in (Verhaegen, 1993; Chou and Verhaegen, 1999; van Overshee and de Moor, 1997; Lin et al, 2005; Jansson, 2005). The consistency of subspace identification algorithms for closed-loop system identification is treated by Chiuso and Picci (2005).

From (16.1.10), one can determine the states X_- as

$$X_- = Q_{\text{Bk}}{}^{\dagger} Y_- - Q_{\text{Bk}}{}^{\dagger} \mathcal{T}_k U_- \,. \tag{16.1.15}$$

With (16.1.14), one obtains

$$X_+ = \underbrace{\left(Q_{\text{Sk}} - A^k Q_{\text{Bk}}{}^{\dagger} \mathcal{T}_k \quad A^k Q_{\text{Bk}}{}^{\dagger} \right)}_{L_-} \underbrace{\begin{pmatrix} U_- \\ Y_- \end{pmatrix}}_{W_-} , \tag{16.1.16}$$

where the past (and future) inputs and outputs are grouped as

$$W_- = \begin{pmatrix} U_- \\ Y_- \end{pmatrix} . \tag{16.1.17}$$

From (16.1.11)

$$Y_+ = Q_{\text{Bk}} X_+ + \mathcal{T}_k U_+ = Q_{\text{Bk}} L_- W_- + \mathcal{T}_k U_+ \,, \tag{16.1.18}$$

one can already determine the number of states of the system. The state matrix X has rank n, since it has the dimension $n \times k$ with $k \geq n$ and the rank of any $m \times n$ matrix A is at most the minimum of m and n, i.e. $\operatorname{rank} A \leq \min(m, n)$.

Unfortunately, X_+ is not known. Instead, one only knows U and Y. If the term $\mathcal{T}_k U_+$ could be eliminated from Y_+, then an analysis of the rank of Y_+ would directly provide the rank of X, since the product of that above $m \times n$ matrix A and an $l \times m$ matrix C satisfies $\operatorname{rank} CA = \operatorname{rank} A$, provided that C has rank m. This can be applied to the problem at hand, The matrix Q_{Bk} has full rank as the system was assumed to be observable. Hence, the remainder of $\boldsymbol{Y}_+$ without the effect of the input matrix $\boldsymbol{U}_+$ will also have rank n.

The question now is how one can eliminate the influence of $\boldsymbol{U}_+$ on $\boldsymbol{Y}_+$. This can be done by a so-called *subspace based approach*, which will be presented after a short introduction to subspaces in general.

16.2 Subspace

Before the identification method is further developed, the notion of the subspace of a matrix along with the projection into the subspace of a matrix shall be introduced. Any real valued matrix $\boldsymbol{A}$ of dimension $m \times n$ can be written as

$$\boldsymbol{A} = \begin{pmatrix} \boldsymbol{r}_1 \\ \boldsymbol{r}_2 \\ \vdots \\ \boldsymbol{r}_m \end{pmatrix} = \begin{pmatrix} \boldsymbol{c}_1 \ \boldsymbol{c}_2 \ \dots \ \boldsymbol{c}_n \end{pmatrix} \tag{16.2.1}$$

with the row vectors $\boldsymbol{r}_k \in \mathbb{R}^{1 \times n}$ or column vectors $\boldsymbol{c}_k \in \mathbb{R}^{m \times 1}$. Then, the *row space* of the matrix is spanned by the row vectors $\boldsymbol{r}_k$, i.e. the space contains all points that can be represented as a linear combination of the row vectors. The *column space* is correspondingly spanned by the column vectors $\boldsymbol{c}_k$. Both of these spaces are *subspaces* of the matrix $\boldsymbol{A}$.

Now, the *orthogonal projection* into the subspace of a matrix is introduced. For the projection of the vector $\boldsymbol{f}$ into the row space of the matrix $\boldsymbol{A}$, one thus wants to express the vector $\boldsymbol{f}$ by a linear combination of the row vectors $\boldsymbol{r}_k$ of the matrix $\boldsymbol{A}$, i.e.

$$\tilde{\boldsymbol{f}} = \sum_{i=1}^{m} g_i \boldsymbol{r}_i = \boldsymbol{g}\boldsymbol{A} \ , \tag{16.2.2}$$

where the projection was denoted as $\tilde{\boldsymbol{f}}$. This projection is given as

$$\boldsymbol{f}/\boldsymbol{A} = \boldsymbol{g}\boldsymbol{A} = \boldsymbol{f}\boldsymbol{A}^{\mathrm{T}}\big(\boldsymbol{A}\boldsymbol{A}^{\mathrm{T}}\big)^{-1}\boldsymbol{A} \tag{16.2.3}$$

The extension from a vector $\boldsymbol{f}$ to a matrix $\boldsymbol{F}$ is straightforward.

The *oblique projection* is a projection, where the vector $\boldsymbol{f}$ is first projected onto the joint space of $\boldsymbol{A}$ and $\boldsymbol{B}$ and only the part laying in the subspace of $\boldsymbol{A}$ is retained,which formally means that the vector $\boldsymbol{f}$ is projected along the row space of $\boldsymbol{B}$ into the row space of $\boldsymbol{A}$. This operation is written as

$$\begin{aligned} \boldsymbol{f}/_{\boldsymbol{B}}\boldsymbol{A} = \boldsymbol{g}\boldsymbol{A} &= \boldsymbol{f}\begin{pmatrix}\boldsymbol{A}^{\mathrm{T}} \ \boldsymbol{B}^{\mathrm{T}}\end{pmatrix}\begin{pmatrix}\boldsymbol{A}\boldsymbol{A}^{\mathrm{T}} & \boldsymbol{A}\boldsymbol{B}^{\mathrm{T}} \\ \boldsymbol{B}\boldsymbol{A}^{\mathrm{T}} & \boldsymbol{B}\boldsymbol{B}^{\mathrm{T}}\end{pmatrix}^{\dagger}\begin{pmatrix}\boldsymbol{A} \\ \not{\boldsymbol{B}}\end{pmatrix} \\ &= \boldsymbol{f}\begin{pmatrix}\boldsymbol{A}^{\mathrm{T}} \ \boldsymbol{B}^{\mathrm{T}}\end{pmatrix}\begin{pmatrix}\boldsymbol{A}\boldsymbol{A}^{\mathrm{T}} & \boldsymbol{A}\boldsymbol{B}^{\mathrm{T}} \\ \boldsymbol{B}\boldsymbol{A}^{\mathrm{T}} & \boldsymbol{B}\boldsymbol{B}^{\mathrm{T}}\end{pmatrix}^{\dagger}\begin{pmatrix}\boldsymbol{A} \\ \boldsymbol{0}\end{pmatrix} , \end{aligned} \tag{16.2.4}$$

where the zero matrix $\mathbf{0}$ has the dimensions of $\boldsymbol{B}$ and $\not{\boldsymbol{B}}$ means that this term is taken out. This is tantamount to

$$f/_{\boldsymbol{B}}\boldsymbol{A} = \boldsymbol{g}\boldsymbol{A} = \boldsymbol{f}\begin{pmatrix}\boldsymbol{A}^{\mathrm{T}} & \boldsymbol{B}^{\mathrm{T}}\end{pmatrix}\left(\begin{pmatrix}\boldsymbol{A}\boldsymbol{A}^{\mathrm{T}} & \boldsymbol{A}\boldsymbol{B}^{\mathrm{T}} \\ \boldsymbol{B}\boldsymbol{A}^{\mathrm{T}} & \boldsymbol{B}\boldsymbol{B}^{\mathrm{T}}\end{pmatrix}^{\dagger}\right)_{\text{Retain first } m \text{ columns}} \boldsymbol{A}\,, \tag{16.2.5}$$

where $\boldsymbol{A}$ is a $m \times n$ matrix. Once again, the extension from the vector to a matrix is straightforward.

16.3 Subspace Identification

Provided that inputs and outputs have been grouped as in (16.1.8) and (16.1.9), one can proceed in the (almost) noise free case as follows and calculates

$$\boldsymbol{Y}_{+}/_{\boldsymbol{U}_{+}}\begin{pmatrix}\boldsymbol{U}_{-} \\ \boldsymbol{Y}_{-}\end{pmatrix} = \boldsymbol{Q}_{\mathrm{Bk}}\boldsymbol{L}_{-}\begin{pmatrix}\boldsymbol{U}_{-} \\ \boldsymbol{Y}_{-}\end{pmatrix} = \boldsymbol{Q}_{\mathrm{Bk}}\boldsymbol{X}_{+} = \boldsymbol{P}\,. \tag{16.3.1}$$

This is the basic approach to the N4SID (*Numerical algorithms for Subspace State Space IDentification*) (van Overshee and de Moor, 1994).

This oblique projection can also be calculated using a QR decomposition. With the notation

$$\begin{pmatrix}\boldsymbol{U}_{+} \\ \boldsymbol{W}_{-} \\ \boldsymbol{Y}_{+}\end{pmatrix} = \begin{pmatrix}\boldsymbol{R}_{11} & \mathbf{0} & \mathbf{0} \\ \boldsymbol{R}_{21} & \boldsymbol{R}_{22} & \mathbf{0} \\ \boldsymbol{R}_{31} & \boldsymbol{R}_{32} & \boldsymbol{R}_{33}\end{pmatrix}\begin{pmatrix}\boldsymbol{Q}_1^{\mathrm{T}} \\ \boldsymbol{Q}_2^{\mathrm{T}} \\ \boldsymbol{Q}_3^{\mathrm{T}}\end{pmatrix}, \tag{16.3.2}$$

the solution is given as

$$\boldsymbol{P} = \boldsymbol{Y}_{+}/_{\boldsymbol{U}_{+}}\begin{pmatrix}\boldsymbol{U}_{-} \\ \boldsymbol{Y}_{-}\end{pmatrix} = \boldsymbol{R}_{32}\boldsymbol{R}_{22}^{\dagger}\boldsymbol{W}_{-}\,. \tag{16.3.3}$$

This is the basic idea of the MOESP (*Multi-variable Output Error State sPace*) algorithm (Verhaegen, 1994).

The matrix $\boldsymbol{P}$ contains all information about the extended observability matrix $\boldsymbol{Q}_{\mathrm{Bk}}$ as well as the state matrix $\boldsymbol{X}_{+}$. Furthermore, one can determine the number of states n from $\boldsymbol{P}$, because $\boldsymbol{P}$ has the same rank as the state matrix $\boldsymbol{X}_{+}$. To unveil all of this information, a singular value decomposition of $\boldsymbol{P}$ is carried out, yielding

$$\boldsymbol{P} = \boldsymbol{U}\boldsymbol{\Sigma}\boldsymbol{V}^{\mathrm{T}} = \begin{pmatrix}\boldsymbol{U}_1 & \boldsymbol{U}_2\end{pmatrix}\begin{pmatrix}\boldsymbol{\Sigma}_1 & \mathbf{0} \\ \mathbf{0} & \boldsymbol{\Sigma}_2\end{pmatrix}\begin{pmatrix}\boldsymbol{V}_1^{\mathrm{T}} \\ \boldsymbol{V}_2^{\mathrm{T}}\end{pmatrix} \text{ with } \Sigma_{1ii} \gg \Sigma_{2jj} \text{ for all } i, j\,. \tag{16.3.4}$$

$\boldsymbol{\Sigma}_1$ and $\boldsymbol{\Sigma}_2$ are diagonal matrices and the entries on the diagonal of $\boldsymbol{\Sigma}_1$ are much larger than those of $\boldsymbol{\Sigma}_2$, which, in the noise free case, should be zero, i.e. $\boldsymbol{\Sigma}_2 = \mathbf{0}$ and in the noisy case will deviate more or less from zero. The dimension of the $n \times n$ matrix $\boldsymbol{\Sigma}_1$ determines the number of states x_i of the system.

To divide the eigenvalues σ_i into "larger" and "smaller" ones, the values of σ_i can be sorted in descending value and then a threshold can be selected by the user. Once the threshold is determined, the number of states of the system is fixed.

Next, the extended observability matrix $\boldsymbol{Q}_{\mathrm{Bk}}$ and the state matrix $\boldsymbol{X}_+$ can be extracted from the measured data. Up to a similarity transform $\boldsymbol{T}$, the extended observability matrix and the state matrix are given as

$$\boldsymbol{Q}_{\mathrm{Bn}} = \boldsymbol{U}_1 \boldsymbol{\Sigma}_1^{1/2} \boldsymbol{T} \tag{16.3.5}$$

$$\boldsymbol{X}_+ = \boldsymbol{T}^{-1} \boldsymbol{\Sigma}_1^{1/2} \boldsymbol{V}_1^{\mathrm{T}} \,. \tag{16.3.6}$$

The existence of the similarity transform $\boldsymbol{T}$ can easily be explained because from sole measurement of the input $\boldsymbol{u}(k)$ and output $\boldsymbol{y}(k)$, one does only obtain information about the number of states n, but not about how the states relate to the inputs and outputs.

There are now several ways on how $\boldsymbol{Q}_{\mathrm{Bn}}$ and $\boldsymbol{X}_+$ can be related to the matrices $\boldsymbol{A}$, $\boldsymbol{B}$, $\boldsymbol{C}$, and $\boldsymbol{D}$ of the state space model. Two ways shall be presented in the following, making however no claim on completeness.

Looking at the structure of $\boldsymbol{Q}_{\mathrm{Bk}}$ with

$$\boldsymbol{Q}_{\mathrm{Bk}} = \begin{pmatrix} \boldsymbol{C} \\ \boldsymbol{C}\boldsymbol{A} \\ \boldsymbol{C}\boldsymbol{A}^2 \\ \vdots \\ \boldsymbol{C}\boldsymbol{A}^{k-1} \end{pmatrix} , \tag{16.3.7}$$

one notes the following: The first p rows $\boldsymbol{Q}_{\mathrm{Bk}}$ provide directly the matrix $\boldsymbol{C}$. Furthermore,

$$\underline{\boldsymbol{Q}_{\mathrm{B}_k}} = \overline{\boldsymbol{Q}_{\mathrm{B}k}} \boldsymbol{A} \text{ with } \underline{\boldsymbol{Q}_{\mathrm{B}_k}} = \begin{pmatrix} \cancel{\boldsymbol{C}} \\ \boldsymbol{C}\boldsymbol{A} \\ \boldsymbol{C}\boldsymbol{A}^2 \\ \vdots \\ \boldsymbol{C}\boldsymbol{A}^{k-1} \end{pmatrix} \text{ and } \overline{\boldsymbol{Q}_{\mathrm{B}k}} = \begin{pmatrix} \boldsymbol{C} \\ \boldsymbol{C}\boldsymbol{A} \\ \boldsymbol{C}\boldsymbol{A}^2 \\ \vdots \\ \cancel{\boldsymbol{C}\boldsymbol{A}^{k-1}} \end{pmatrix} , \tag{16.3.8}$$

where the first r rows have been eliminated by the notation $\cancel{\boldsymbol{C}}$ respectively last r rows have been eliminated by $\cancel{\boldsymbol{C}\boldsymbol{A}^{k-1}}$. This allows to determine the matrix $\boldsymbol{A}$ as

$$\boldsymbol{A} = \overline{\boldsymbol{Q}_{\mathrm{B}k}}^{\dagger} \underline{\boldsymbol{Q}_{\mathrm{B}_k}} \,. \tag{16.3.9}$$

The matrices $\boldsymbol{B}$ and $\boldsymbol{D}$ can then be determined by the following approach: In addition to the projection onto the column space, the projection onto the orthogonal column space is defined as $\boldsymbol{Q}_{\mathrm{Bk}}{}^{\perp} = \boldsymbol{I} - \boldsymbol{Q}_{\mathrm{Bk}}(\boldsymbol{Q}_{\mathrm{Bk}}{}^{\mathrm{T}}\boldsymbol{Q}_{\mathrm{Bk}})^{-1}\boldsymbol{Q}_{\mathrm{Bk}}{}^{\mathrm{T}}$. This allows to derive

$$\boldsymbol{Q}_{\mathrm{Bk}}{}^{\perp} \boldsymbol{Y}_+^{\mathrm{T}} = \boldsymbol{Q}_{\mathrm{Bk}}{}^{\perp} \mathcal{T}_k = \boldsymbol{Q}_{\mathrm{Bk}}{}^{\perp} \begin{pmatrix} \boldsymbol{I} & \boldsymbol{0} \\ \boldsymbol{0} & \underline{\boldsymbol{Q}_{\mathrm{Bk}}} \end{pmatrix} , \tag{16.3.10}$$

which allows to determine the matrices $\boldsymbol{B}$ and $\boldsymbol{D}$ by the method of least squares.

An alternative solution is based on the state equations (16.1.1) and (16.1.2). For one sample step, the state equation and output equation are given as

$$\boldsymbol{X}^{+} = \boldsymbol{A}\boldsymbol{X} + \boldsymbol{B}\boldsymbol{U} \tag{16.3.11}$$

$$\boldsymbol{Y} = \boldsymbol{C}\boldsymbol{X} + \boldsymbol{D}\boldsymbol{U} \,, \tag{16.3.12}$$

with

$$\boldsymbol{U} = \begin{pmatrix} \boldsymbol{u}_k & \boldsymbol{u}_{k+1} & \boldsymbol{u}_{k+2} & \ldots & \boldsymbol{u}_{k+N-1} \end{pmatrix} \tag{16.3.13}$$

$$\boldsymbol{Y} = \begin{pmatrix} \boldsymbol{y}_k & \boldsymbol{y}_{k+1} & \boldsymbol{y}_{k+2} & \ldots & \boldsymbol{y}_{k+N-1} \end{pmatrix} \tag{16.3.14}$$

$$\boldsymbol{X} = \begin{pmatrix} \boldsymbol{x}_k & \boldsymbol{x}_{k+1} & \boldsymbol{x}_{k+2} & \ldots & \boldsymbol{x}_{k+N-1} \end{pmatrix} \tag{16.3.15}$$

$$\boldsymbol{X}^{+} = \begin{pmatrix} \boldsymbol{x}_{k+1} & \boldsymbol{x}_{k+2} & \boldsymbol{x}_{k+3} & \ldots & \boldsymbol{x}_{k+N} \end{pmatrix} . \tag{16.3.16}$$

The only unknown are the states $\boldsymbol{X}^{+}$ at time step $k+1$. Based on

$$\boldsymbol{Y}_{k+1|k+N} = \boldsymbol{Q}_{\mathrm{B,N-1}}\boldsymbol{X}^{+} - \mathcal{T}_{N-1}\boldsymbol{U}_{k+1|k+N} \,, \tag{16.3.17}$$

these can be determined as

$$\boldsymbol{X}^{+} = \boldsymbol{Q}_{\mathrm{B,N-1}}{}^{\dagger}\boldsymbol{Y}_{k+1|k+N} - \boldsymbol{Q}_{\mathrm{B,N-1}}{}^{\dagger}\mathcal{T}_{N-1}\boldsymbol{U}_{k+1|k+N} \,. \tag{16.3.18}$$

Now that all entries are known, the system of equations

$$\begin{pmatrix} \boldsymbol{X}^{+} \\ \boldsymbol{Y} \end{pmatrix} = \begin{pmatrix} \boldsymbol{A} & \boldsymbol{B} \\ \boldsymbol{C} & \boldsymbol{D} \end{pmatrix} \begin{pmatrix} \boldsymbol{X} \\ \boldsymbol{U} \end{pmatrix} \tag{16.3.19}$$

can be solved by the method of least squares and all parameter matrices of the state space model be determined.

In the case of excessive noise, an augmented state space model with noise should be chosen as the underlying model structure, i.e.

$$\boldsymbol{x}(k+1) = \boldsymbol{A}\boldsymbol{x}(k) + \boldsymbol{B}\boldsymbol{u}(k) + \boldsymbol{w}(k) \tag{16.3.20}$$

$$\boldsymbol{y}(k) = \boldsymbol{C}\boldsymbol{x}(k) + \boldsymbol{D}\boldsymbol{u}(k) + \boldsymbol{v}(k) \tag{16.3.21}$$

with the noise having the statistical properties

$$\mathrm{E}\left\{\begin{pmatrix} \boldsymbol{w}(k) \\ \boldsymbol{v}(k) \end{pmatrix} \begin{pmatrix} \boldsymbol{w}^{\mathrm{T}}(k) & \boldsymbol{v}^{\mathrm{T}}(k) \end{pmatrix}\right\} = \begin{pmatrix} \boldsymbol{Q} & \boldsymbol{S} \\ \boldsymbol{S}^{\mathrm{T}} & \boldsymbol{R} \end{pmatrix} \delta(k) \,. \tag{16.3.22}$$

The first algorithm is easy to understand but biased. First, one uses the oblique projection to determine

$$\boldsymbol{P}_k = \boldsymbol{Y}_{+}/_{\boldsymbol{U}_{+}}\boldsymbol{W}_{-} \tag{16.3.23}$$

$$\boldsymbol{P}_{k+1} = \boldsymbol{Y}_{+}^{-}/_{\boldsymbol{U}_{+}^{-}}\boldsymbol{W}_{-}^{+} \,, \tag{16.3.24}$$

where the superscript "–" and "+" denote the matrices where the boundary has been moved by one row in the respective direction to lengthen respectively shorten the matrix. Then, the singular value decomposition

$$\boldsymbol{P}_k = \boldsymbol{U}\boldsymbol{\Sigma}\boldsymbol{V}^{\mathrm{T}} \tag{16.3.25}$$

is used to determine

$$\boldsymbol{Q}_{\mathrm{Bn}} = \boldsymbol{U}_1\boldsymbol{\Sigma}_1^{1/2}\boldsymbol{T} \tag{16.3.26}$$

$$\boldsymbol{Q}_{\mathrm{B,N-1}} = \underline{\boldsymbol{Q}_{\mathrm{Bn}}} \tag{16.3.27}$$

and from there determine the state sequences as

$$\boldsymbol{X}_n = \boldsymbol{Q}_{\mathrm{Bn}}{}^{\dagger}\boldsymbol{P}_n \tag{16.3.28}$$

$$\boldsymbol{X}_{n+1} = \boldsymbol{Q}_{\mathrm{B,N-1}}{}^{\dagger}\boldsymbol{P}_{n+1} \,. \tag{16.3.29}$$

Next, one can determine the parameters of the state space model by first solving the equation

$$\begin{pmatrix} \boldsymbol{X}(n+1) \\ \boldsymbol{Y} \end{pmatrix} = \begin{pmatrix} \boldsymbol{A} & \boldsymbol{B} \\ \boldsymbol{C} & \boldsymbol{D} \end{pmatrix} \begin{pmatrix} \boldsymbol{X}(n) \\ \boldsymbol{U} \end{pmatrix} + \begin{pmatrix} \boldsymbol{\varrho}_{\mathrm{w}} \\ \boldsymbol{\varrho}_{\mathrm{v}} \end{pmatrix} , \tag{16.3.30}$$

where the parameters of the noise model can be determined from the residuals $\boldsymbol{\varrho}$ as

$$\begin{pmatrix} \boldsymbol{Q} & \boldsymbol{S} \\ \boldsymbol{S}^{\mathrm{T}} & \boldsymbol{R} \end{pmatrix} = \mathrm{E}\left\{ \begin{pmatrix} \boldsymbol{\varrho}_{\mathrm{w}} \\ \boldsymbol{\varrho}_{\mathrm{v}} \end{pmatrix} \begin{pmatrix} \boldsymbol{\varrho}_{\mathrm{w}}^{\mathrm{T}} & \boldsymbol{\varrho}_{\mathrm{v}}^{\mathrm{T}} \end{pmatrix} \right\} \tag{16.3.31}$$

with the expected value being determined as the time average for the number of samples going to infinity. This algorithm as presented by (van Overshee and de Moor, 1994) is unfortunately biased in most applications. It is only unbiased if the measurement time goes to infinity, the system is purely deterministic (i.e. noise-free) or the input $\boldsymbol{u}$ is white noise. This can in most cases not be ensured.

Other algorithms which are less sensitive to noise are also presented by van Overshee and de Moor (1996b). One of them shall be outlined in the following: If the future outputs are projected onto the past inputs and outputs as well as the future inputs, one obtains

$$\boldsymbol{\mathcal{Z}}_k = \boldsymbol{Y}_+ / \begin{pmatrix} \boldsymbol{W}_- \\ \boldsymbol{U}_+ \end{pmatrix} = \boldsymbol{Q}_{\mathrm{Bk}}\hat{\boldsymbol{X}}_k + \boldsymbol{\mathcal{T}}_k\boldsymbol{U}_+ \,. \tag{16.3.32}$$

The same is true for the next sample step, hence

$$\boldsymbol{\mathcal{Z}}_{k+1} = \boldsymbol{Y}_+^- / \begin{pmatrix} \boldsymbol{W}_-^+ \\ \boldsymbol{U}_+^- \end{pmatrix} = \boldsymbol{Q}_{\mathrm{B(k-1)}}\hat{\boldsymbol{X}}_{k+1} + \boldsymbol{\mathcal{T}}_{k-1}\boldsymbol{U}_+^- \,. \tag{16.3.33}$$

The estimated states could e.g. be provided by an observer and should only resemble the dynamics of the deterministic part, i.e. should obey the differential equation

$$\hat{\boldsymbol{X}}_{k+1} = \boldsymbol{A}\hat{\boldsymbol{X}}_k + \boldsymbol{B}\boldsymbol{U}_k + \boldsymbol{\varrho}_{\mathrm{w}} \tag{16.3.34}$$

$$\hat{\boldsymbol{Y}}_k = \boldsymbol{C}\hat{\boldsymbol{X}}_k + \boldsymbol{D}\boldsymbol{U}_k + \boldsymbol{\varrho}_{\mathrm{v}} \,. \tag{16.3.35}$$

From (16.3.32) and (16.3.33), one obtains

$$\hat{X}_k = \mathcal{Q}_{\mathrm{Bk}}{}^{\dagger}(\mathcal{Z}_k - \mathcal{T}_k U_+) \tag{16.3.36}$$

$$\hat{X}_{k+1} = \mathcal{Q}_{\mathrm{B(k-1)}}{}^{\dagger}(\mathcal{Z}_{k+1} - \mathcal{T}_{k-1} U_+^-) \, . \tag{16.3.37}$$

With (16.3.34) and (16.3.35), one obtains

$$\begin{pmatrix} \mathcal{Q}_{\mathrm{B(k-1)}}{}^{\dagger}\mathcal{Z}_{k+1} \\ Y \end{pmatrix} = \begin{pmatrix} A \\ C \end{pmatrix} \mathcal{Q}_{\mathrm{Bk}}{}^{\dagger}\mathcal{Z}_k + \underbrace{\begin{pmatrix} \left(B \middle| \mathcal{Q}_{\mathrm{B(k-1)}}{}^{\dagger}\mathcal{T}_{k-1}\right) - A\,\mathcal{Q}_{\mathrm{Bk}}{}^{\dagger}\mathcal{T}_k \\ \left(D \middle| 0\right) - C\,\mathcal{Q}_{\mathrm{Bk}}{}^{\dagger}\mathcal{T}_k \end{pmatrix}}_{\mathcal{K}} U + \begin{pmatrix} \varrho_{\mathrm{w}} \\ \varrho_{\mathrm{v}} \end{pmatrix} . \tag{16.3.38}$$

This equation can be solved by the method of least squares for A, C, and $\mathcal{K}$, where the noise ϱ_{w} and ϱ_{v} is first seen as a residual and not identified at the present time. From knowledge of $\mathcal{K}$, one can then determine B and D. In the last step, one can determine the residuals ϱ_{w} and ϱ_{v} and from there, the respective noise covariances.

16.4 Identification from Impulse Response

Especially in modal analysis of mechanical systems, one can often measure the impulse response of systems. Here, for example, a system is excited with a hammer and then the resulting accelerations at different parts of the structure are measured. The Ho-Kalman method (Ho and Kalman, 1966) allows to directly work with such measurements of the impulse response, see also (Juang, 1994).

The impulse response of a MIMO system in state space representation is given as

$$G_k = \begin{cases} D \text{ for } k = 0 \\ CA^{k-1}B \text{ for } k = 1, 2, \ldots \end{cases} . \tag{16.4.1}$$

The recorded impulse response can now be used to determine the model parameters of a state space model. Obviously, the matrix D can directly be determined as

$$D = G_0 \, . \tag{16.4.2}$$

The parameters A, B, and C, can now be determined based on a Hankel matrix representation. The *Hankel matrix* $\mathcal{H}$ is in this case defined as

$$\mathcal{H}_{k,l} = \begin{pmatrix} G_1 & G_2 & G_3 & \ldots & G_l \\ G_2 & G_3 & G_4 & \ldots & G_{l+1} \\ G_3 & G_4 & G_5 & \ldots & G_{l+2} \\ \vdots & \vdots & \vdots & & \vdots \\ G_k & G_{k+1} & G_{k+2} & \ldots & G_{k+l+1} \end{pmatrix} \tag{16.4.3}$$

It can be shown that the Hankel matrix can be written as the product of controllability and observability matrix, i.e.

$$\mathcal{H}_{k,l} = \boldsymbol{Q}_{\mathrm{B}k} \boldsymbol{Q}_{\mathrm{S}l} \,. \tag{16.4.4}$$

The Hankel matrix can now be decomposed by means of a singular value decomposition as

$$\mathcal{H}_{k,l} = \boldsymbol{U}\boldsymbol{\Sigma}\boldsymbol{V}^{\mathrm{T}} = \begin{pmatrix} \boldsymbol{U}_1 & \boldsymbol{U}_2 \end{pmatrix} \begin{pmatrix} \boldsymbol{\Sigma}_1 & \boldsymbol{0} \\ \boldsymbol{0} & \boldsymbol{\Sigma}_2 \end{pmatrix} \begin{pmatrix} \boldsymbol{V}_1^{\mathrm{T}} \\ \boldsymbol{V}_2^{\mathrm{T}} \end{pmatrix} \text{ with } \boldsymbol{\Sigma}_2 = \boldsymbol{0} \,. \tag{16.4.5}$$

The singular value decomposition yields the observability and controllability matrix up to a similarity transform $\boldsymbol{T}$ as

$$\boldsymbol{Q}_{\mathrm{B}} = \boldsymbol{U}_1 \boldsymbol{\Sigma}_1^{1/2} \boldsymbol{T} \tag{16.4.6}$$

$$\boldsymbol{Q}_{\mathrm{S}} = \boldsymbol{T}^{-1} \boldsymbol{\Sigma}_1^{1/2} \boldsymbol{V}_1^{\mathrm{T}} \,. \tag{16.4.7}$$

From there, one can determine the wanted matrices as

$$\boldsymbol{A} = \underline{\boldsymbol{Q}_{\mathrm{B}k}}^{\dagger} \overline{\boldsymbol{Q}_{\mathrm{B}k}} \tag{16.4.8}$$

$$\boldsymbol{B} = \boldsymbol{Q}_{\mathrm{S}}(1:n, 1:m) \tag{16.4.9}$$

$$\boldsymbol{C} = \boldsymbol{Q}_{\mathrm{B}}(1:p, 1:n) \,, \tag{16.4.10}$$

where the matrices $\boldsymbol{B}$ and $\boldsymbol{C}$ are cut out of the observability and controllability matrix respectively.

16.5 Some Modifications to the Original Formulations

Many modifications have been introduced to original formulation, allowing e.g. to solve the subspace identification problem recursively (Houtzager et al, 2009). Here, the large number of parameters can still object a real-time application of the algorithms. Massiono and Verhaegen (2008) show a method for large-scale systems.

The subspace identification can also be used for frequency domain identification (van Overshee and de Moor, 1996a). Here, the idea is that in the frequency domain the state space model is given as

$$s\boldsymbol{x}(s) = \boldsymbol{A}\boldsymbol{x}(s) + \boldsymbol{B}\boldsymbol{I}_m \tag{16.5.1}$$

$$\boldsymbol{g}(s) = \boldsymbol{C}\boldsymbol{x}(s) + \boldsymbol{D}\boldsymbol{I}_m \,. \tag{16.5.2}$$

Since the input $\boldsymbol{u}(s)$ in the frequency domain is chosen as the unit matrix $\boldsymbol{I}_m$ of appropriate dimensions, the output $\boldsymbol{y}(s)$ represents the impulse response of the system. Then, the data matrices are set up as

$$\mathcal{G} = \begin{pmatrix} \boldsymbol{g}(\mathrm{i}\omega_1) & \boldsymbol{g}(\mathrm{i}\omega_2) & \dots & \boldsymbol{g}(\mathrm{i}\omega_N) \\ (\mathrm{i}\omega_1)\boldsymbol{g}(\mathrm{i}\omega_1) & (\mathrm{i}\omega_2)\boldsymbol{g}(\mathrm{i}\omega_2) & \dots & (\mathrm{i}\omega_N)\boldsymbol{g}(\mathrm{i}\omega_N) \\ \vdots & \vdots & & \vdots \\ (\mathrm{i}\omega_1)^{2k-1}\boldsymbol{g}(\mathrm{i}\omega_1) & (\mathrm{i}\omega_2)^{2k-1}\boldsymbol{g}(\mathrm{i}\omega_2) & \dots & (\mathrm{i}\omega_N)^{2k-1}\boldsymbol{g}(\mathrm{i}\omega_N) \end{pmatrix} , \tag{16.5.3}$$

and

$$\mathcal{I} = \begin{pmatrix} \boldsymbol{I}_m & \boldsymbol{I}_m & \dots & \boldsymbol{I}_m \\ (\mathrm{i}\omega_1)\boldsymbol{I}_m & (\mathrm{i}\omega_2)\boldsymbol{I}_m & \dots & (\mathrm{i}\omega_N)\boldsymbol{I}_m \\ \vdots & \vdots & & \vdots \\ (\mathrm{i}\omega_1)^{2k-1}\boldsymbol{I}_m & (\mathrm{i}\omega_2)^{2k-1}\boldsymbol{I}_m & \dots & (\mathrm{i}\omega_N)^{2k-1}\boldsymbol{I}_m \end{pmatrix}, \tag{16.5.4}$$

as well as

$$\mathcal{X} = \left(\boldsymbol{x}(\mathrm{i}\omega_1)\ \boldsymbol{x}(\mathrm{i}\omega_2)\ \dots\ \boldsymbol{x}(\mathrm{i}\omega_N)\right). \tag{16.5.5}$$

With these matrices, one can then use the classical approach as presented above, i.e. use the SVD of

$$\mathcal{H}/\mathcal{I}^{\perp} = \boldsymbol{U}\boldsymbol{\Sigma}\boldsymbol{V}^{\mathrm{T}} = \left(\boldsymbol{U}_1\ \boldsymbol{U}_2\right)\begin{pmatrix} \boldsymbol{\Sigma}_1 & \boldsymbol{0} \\ \boldsymbol{0} & \boldsymbol{\Sigma}_2 \end{pmatrix}\begin{pmatrix} \boldsymbol{V}_1^{\mathrm{T}} \\ \boldsymbol{V}_2^{\mathrm{T}} \end{pmatrix} \text{ with } \Sigma_{1ii} \gg \Sigma_{2jj} \text{ for all } i, j \tag{16.5.6}$$

and proceed as above. As the matrices might be badly conditioned, an alternative solution is also shown in (van Overshee and de Moor, 1996a). Asymptotic properties of the subspace estimation method have been discussed e.g. in (Bauer, 2005).

Example 16.1 (Subspace Identification of the Three-Mass Oscillator).

In this example, the N4SID algorithm is applied to identify a model of the Three-Mass Oscillator.

The process has been excited with a PRBS signal (see Sect. 6.3). The process input is the torque of the motor M_{M} acting on the first mass. The output is the rotational speed of the third mass, $\omega_3 = \dot{\varphi}_3$, as shown in Fig. 16.1. An important issue in the estimation of discrete-time models is the sample rate. The data for the Three-Mass Oscillator have been sampled with a sample time of $T_0 = 0.003\,\mathrm{s}$. This sample rate was too high to obtain reasonable results, thus the data have been downsampled by a factor of $N = 6$, leading to $T_0 = 18\,\mathrm{ms}$.

In order to judge the quality of the estimated model, the frequency response of the discrete-time model has been graphed against the frequency response determined by direct measurement with the orthogonal correlation (see Sect. 5.5.2). This comparison in Fig. 16.2 demonstrates the good fidelity of the estimated model. □

16.6 Application to Continuous Time Systems

The above-described approach can in general also be applied to continuous time systems, see (Rao and Unbehauen, 2006). One can use the equation

$$\boldsymbol{Y}_i = \boldsymbol{T}_{1,i}\boldsymbol{X} + \boldsymbol{T}_{2,i}\boldsymbol{U}_i \tag{16.6.1}$$

as a basis with

$$\boldsymbol{X} = \left(\boldsymbol{x}(t_1)\ \boldsymbol{x}(t_2)\ \dots\ \boldsymbol{x}(t_N)\right) \tag{16.6.2}$$

and

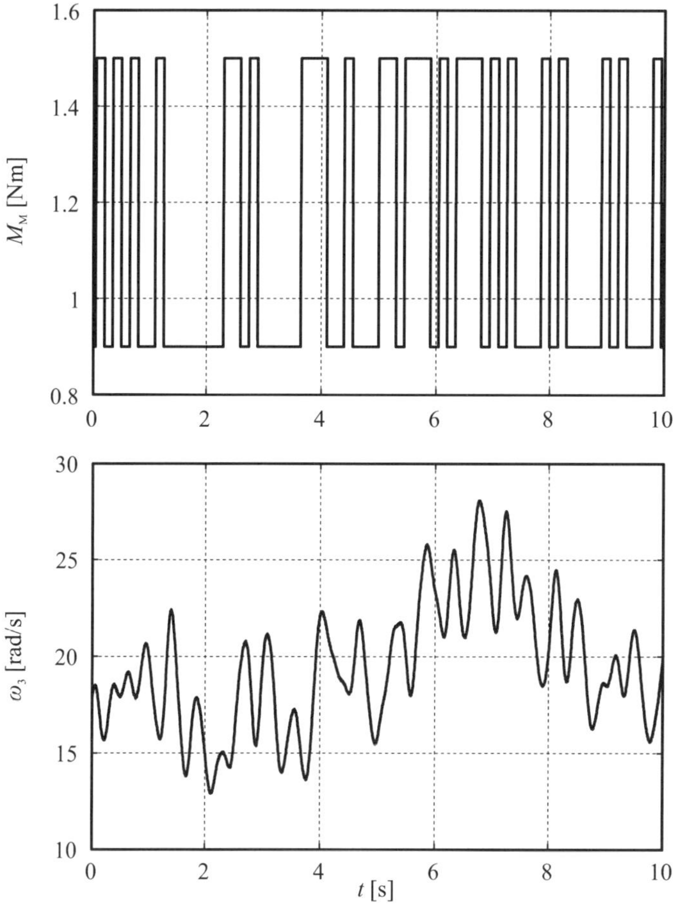

Fig. 16.1. Input and output signals for the parameter estimation of a discrete-time model of the Three-Mass Oscillator based on N4SID

$$
U_i = \begin{pmatrix} u(t_1) & \dots & u(t_N) \\ \dot{u}(t_1) & \dots & \dot{u}(t_N) \\ \vdots & & \vdots \\ u^{(r-1)}(t_1) & \dots & u^{(r-1)}(t_N) \end{pmatrix} \tag{16.6.3}
$$

as well as

$$
Y_i = \begin{pmatrix} y(t_1) & \dots & y(t_N) \\ \dot{y}(t_1) & \dots & \dot{y}(t_N) \\ \vdots & & \vdots \\ y^{(i-1)}(t_1) & \dots & y^{(i-1)}(t_N) \end{pmatrix} . \tag{16.6.4}
$$

Here, one can already see the problematic point of this approach, which is to obtain the derivative of the appropriate order. The extended observability matrix is then

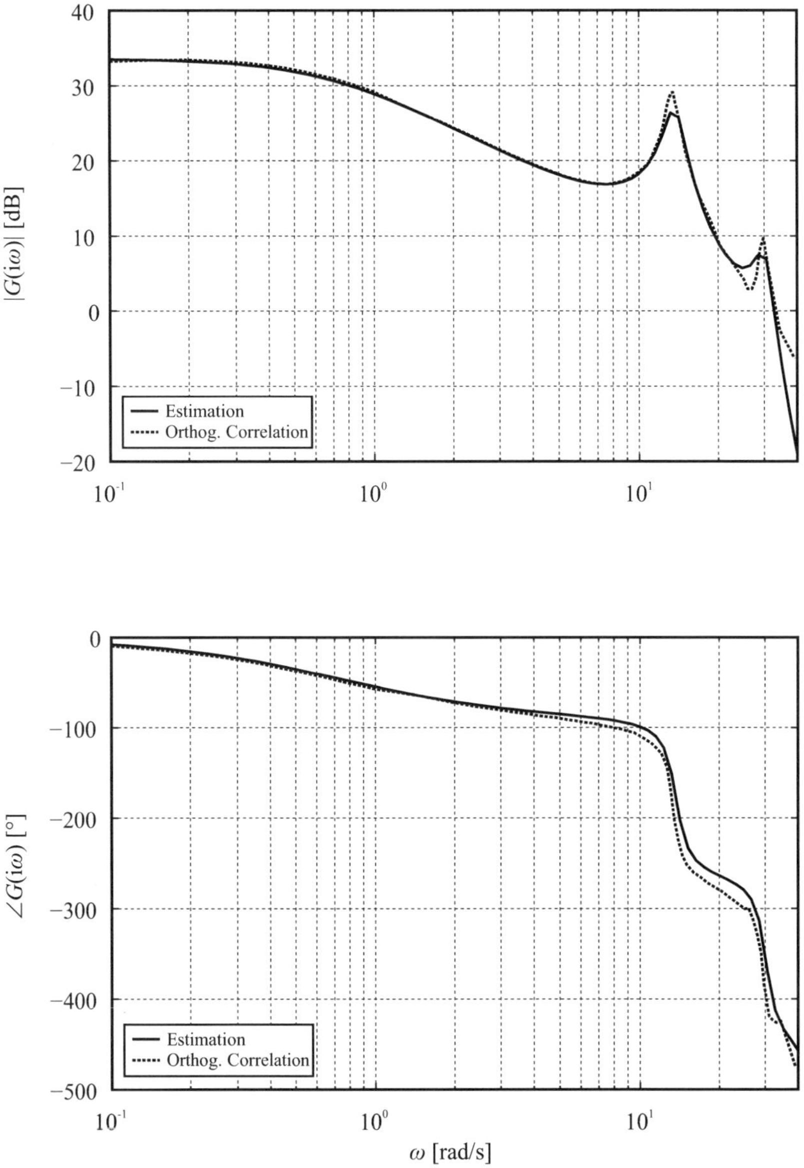

Fig. 16.2. Calculation of the frequency response based on the subspace identification using N4SID for a discrete-time model of the Three-Mass Oscillator, comparison with direct measurement of the frequency responses

formed as

$$Q_\mathrm{B} = \begin{pmatrix} C \\ CA \\ \vdots \\ CA^{i-1} \end{pmatrix} \tag{16.6.5}$$

and the extended controllability matrix as

$$Q_\mathrm{S} = \begin{pmatrix} D & 0 & \dots & 0 \\ CB & D & \dots & 0 \\ \vdots & \vdots & & \vdots \\ CA^{i-2}B & CA^{i-3}B & \dots & D \end{pmatrix} . \tag{16.6.6}$$

With these matrices, one could now employ the subspace based approach as outlined above.

16.7 Summary

In this chapter, subspace based methods, so-called subspace methods, have been presented. They allow to identify not only single, but also multiple input/output (MIMO) systems, see Chap. 17. Their biggest advantage is that no a priori assumptions must be made and the model order is determined as part of the identification process. Furthermore, a state space model can be identified without the need to know the initial conditions of the states and without the need to know the relation between the states and the input/output behavior of the system. This comes at the expense that the states and the matrices of the state space model are only known up to a similarity transform $\boldsymbol{T}$. A priori knowledge is hard to introduce into the model apart from an appropriate, but difficult to realize choice of $\boldsymbol{T}$. The method is not directly suitable for non-linear models and furthermore, in the fundamental formulation, the system must be operated in open loop. What speaks in favor of the subspace methods is that a solution is provided without iterations, similar to the direct method of least squares. It is suggested to choose k two or three times larger than the model order n. Some first investigation of user choices in subspace identification is presented in (Ljung, 2009).

Problems

16.1. Subspace Identification 1
Why is the state space model that is identified by subspace methods only known up to a similarity transform $\boldsymbol{T}$?

16.2. Subspace Identification 2
Does the original subspace identification algorithm as introduced in this chapter work in closed loop? Why or why not?

References

Bauer D (2005) Asymptotic properties of subspace estimators. Automatica 41(3):359–376

Chiuso A, Picci G (2005) Consistency analysis of some closed-loop subspace identification methods. Automatica 41(3):377–391

Chou CT, Verhaegen M (1999) Closed-loop identification using canonical correlation analysis. In: Proceedings of the European Control Conference 1999, Karlsruhe, Germany

Golub GH, van Loan CF (1996) Matrix computations, 3rd edn. Johns Hopkins studies in the mathematical sciences, Johns Hopkins Univ. Press, Baltimore

Ho BL, Kalman RE (1966) Effective construction of linear state variable models from input/output functions. Regelungstechnik 14:545–548

Houtzager I, van Wingerden JW, Verhaegen M (2009) Fast-array recursive closed-loop subspace model identification. In: Proceedings of the 15th IFAC Symposium on System Identification, Saint-Malo, France

Jansson M (2005) A new subspace identification method for closed loop data. In: Proceedings of the 16th IFAC World Congress

Juang JN (1994) Applied system identification. Prentice Hall, Englewood Cliffs, NJ

Juang JN, Phan MQ (2006) Identification and control of mechanical systems. Cambridge University Press, Cambridge

Katayama T, Tanaka H (2007) An approach to closed-loop subspace identification by orthogonal decomposition. Automatica 43(9):1623–1630

Lin W, Qin SJ, Ljung L (2005) Comparison of subspace identification methods for systems operating in closed-loop. In: Proceedings of the 16th IFAC World Congress

Ljung L (2009) Aspects and experiences of user choices in subspace identification methods. In: Proceedings of the 15th IFAC Symposium on System Identification, Saint-Malo, France, pp 1802–1807

Ljung L, McKelvey T (1996) Subspace identification from closed loop data. Signal Process 52(2):209–215

Massiono P, Verhaegen M (2008) Subspace identification of a class of large-scale systems. In: Proceedings of the 17th IFAC World Congress, Seoul, Korea, pp 8840–8845

van Overshee P, de Moor B (1994) N4SID: Subspace algorithms for the identification of combined deterministic-stochastic systems. Automatica 30(1):75–93

van Overshee P, de Moor B (1996a) Continuous-time frequency domain subspace system identification. Signal Proc 52(2):179–194

van Overshee P, de Moor B (1996b) Subspace identification for linear systems: Theory – implementation – applications. Kluwer Academic Publishers, Boston

van Overshee P, de Moor B (1997) Closed loop subspace systems identification. In: Proceedings of the 36th IEEE Conference on Decision and Control, San Diego, CA, USA

Rao GP, Unbehauen H (2006) Identification of continuous-time systems. IEE Proceedings Control Theory and Applications 153(2):185–220

Verhaegen M (1993) Application of a subspace model identification technique to identify LTI systems operating in closed-loop. Automatica 29(4):1027–1040

Verhaegen M (1994) Identification of the deterministic part of MIMO state space models given in innovations form from input-output data. Automatica 30(1):61–74

Viberg M, Stoica P (1996) Editorial note. Signal Proc 52(2):99–101

Part V

Identification of Multi-Variable Systems

17

Parameter Estimation for MIMO Systems

The choice of an appropriate model structure plays an important role in the identification of MIMO systems as it determines the number of parameters, the convergence and the computational effort. Hence, in this chapter, different model structures for MIMO systems will be presented.

Due to the large number of model structures and identification methods, there exists a huge variety of methods, which cannot be treated in full in this chapter. Rather, this chapter will concentrate on a few structures that have proven well in the identification of MIMO systems. The P-canonical and a simplified P-canonical structure will be introduced as structures which are well suited if input/output models shall be identified. Then, the state space representation of MIMO systems is introduced in various forms. In particular, the observable and the controllable canonical form are introduced. Furthermore, it is shown how these state space representations can be transformed into input-output models. Also, the representation by impulse responses and the associated Markov parameters are presented.

Many methods that have already been introduced for the identification of SISO systems can as well be used for the identification of MIMO systems. A first (yet very slow) approach is to excite one input after the other and identify each input-output dynamics separately as a SISO model. However, much time can be saved, if all inputs are excited in parallel. Furthermore, this will yield coherent models. The input signals must exhibit certain properties to be suitable for the identification methods to be presented. To account for this, a method is presented that allows to generate orthogonal test signals from a PRBS signal. Then, identification methods that can be applied to MIMO system identification will be presented, such as correlation analysis and parameter estimation based on the method of least squares.

17.1 Transfer Function Models

In the following, a linear process with p inputs u_j and r outputs y_i will be considered. If one describes each output y_i as a sum of the contributions of the partial

R. Isermann, M. Münchhof, *Identification of Dynamic Systems*,
DOI 10.1007/978-3-540-78879-9_17, © Springer-Verlag Berlin Heidelberg 2011

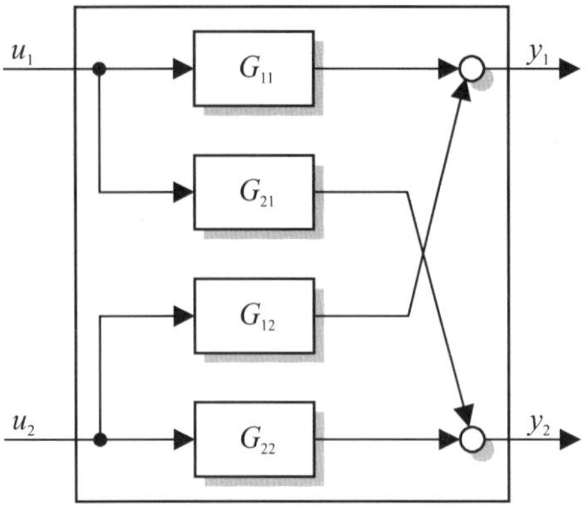

Fig. 17.1. MIMO process with $p = 2$ inputs and $r = 2$ outputs in transfer function representation in the P-canonical form

transfer functions G_{ij} driven by the inputs u_j, then one obtains the *generalized transfer function matrix* model

$$\underbrace{\begin{pmatrix} y_1 \\ y_2 \\ \vdots \\ y_r \end{pmatrix}}_{\boldsymbol{y}(z)} = \underbrace{\begin{pmatrix} G_{11} & G_{12} & \dots & G_{1p} \\ G_{21} & G_{22} & \dots & G_{2p} \\ \vdots & \vdots & & \vdots \\ G_{r1} & G_{r2} & \dots & G_{rp} \end{pmatrix}}_{\boldsymbol{G}(z)} \underbrace{\begin{pmatrix} u_1 \\ u_2 \\ \vdots \\ u_p \end{pmatrix}}_{\boldsymbol{u}(z)} . \tag{17.1.1}$$

For the same number of inputs and outputs $r = p$, $\boldsymbol{G}$ will be a square matrix. Figure 17.1 shows the corresponding P-canonical structure for $r = p = 2$. Other canonical forms, such as e.g. the V-canonical structure can be transformed to a P-canonical structure (Schwarz, 1967, 1971; Isermann, 1991).

If one assumes individual transfer functions for the entries G_{ij} of the transfer function matrix, then a P-canonical structure will result directly, regardless of the true system structure. In many cases, the individual transfer functions G_{ij} contain common parts, so that the transfer function matrix $\boldsymbol{G}$ contains too many parameters. Since (17.1.1) furthermore does not allow to formulate an equation error that is linear in the parameters, this very general transfer function matrix representation is not well suited for parameter estimation. For the determination of non-parametric models, it can however be used as a model structure.

A simplified structure, see Fig. 17.2, can be obtained if the transfer functions contributing to one output

$$y_i = \sum_{j=1}^{p} G_{ij} u_j = \sum_{j=1}^{p} \frac{B_{ij}}{A_{ij}} u_j \tag{17.1.2}$$

have a common denominator polynomial $A_{ij} = A_i$, such that

$$y_i = \frac{1}{A_i} \sum_{j=1}^{p} B_{ij} u_j . \tag{17.1.3}$$

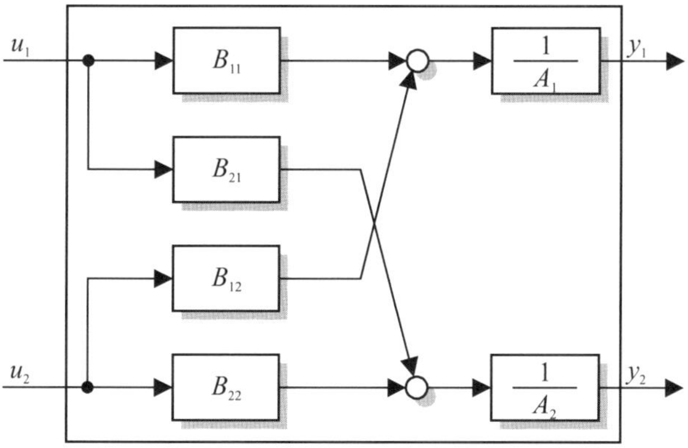

Fig. 17.2. MIMO process with $p = 2$ inputs and $r = 2$ outputs in transfer function representation in the simplified P-canonical form or matrix polynomial form

In this case, the simplified P-canonical transfer function model is given as

$$\underbrace{\begin{pmatrix} A_1 & 0 & \dots & 0 \\ 0 & A_2 & & 0 \\ \vdots & & \ddots & \vdots \\ 0 & 0 & & A_r \end{pmatrix}}_{\boldsymbol{A}(z^{-1})} \underbrace{\begin{pmatrix} y_1 \\ y_2 \\ \vdots \\ y_r \end{pmatrix}}_{\boldsymbol{y}(z)} = \underbrace{\begin{pmatrix} B_{11} & B_{12} & \dots & B_{1p} \\ B_{21} & B_{22} & \dots & B_{2p} \\ \vdots & \vdots & & \vdots \\ B_{r1} & B_{r2} & \dots & B_{rp} \end{pmatrix}}_{\boldsymbol{B}(z^{-1})} \underbrace{\begin{pmatrix} u_1 \\ u_2 \\ \vdots \\ u_p \end{pmatrix}}_{\boldsymbol{u}(z)} \tag{17.1.4}$$

or

$$\boldsymbol{y}(z) = \boldsymbol{A}^{-1}(z^{-1})\boldsymbol{B}(z^{-1})\boldsymbol{u}(z)\,, \tag{17.1.5}$$

such that

$$\boldsymbol{G}(z) = \boldsymbol{A}^{-1}(z^{-1})\boldsymbol{B}(z^{-1})\,. \tag{17.1.6}$$

If the output $\boldsymbol{y}(k)$ is affected by statistically independent noises,

$$\boldsymbol{n}(z) = \boldsymbol{G}_\nu(z)\boldsymbol{\nu}(z) \tag{17.1.7}$$

with

$$\boldsymbol{\nu}(z)^{\mathrm{T}} = \big(\nu_1(z)\ \nu_2(z)\ \dots\ \nu_r(z)\big)\,, \tag{17.1.8}$$

then a transfer function matrix model follows

$$\boldsymbol{y}(z) = \boldsymbol{G}(z)\boldsymbol{u}(z) + \boldsymbol{G}_\nu(z)\boldsymbol{\nu}(z)\,. \tag{17.1.9}$$

17.1.1 Matrix Polynomial Representation

An alternative to the transfer function representation is the *matrix polynomial model*

$$\boldsymbol{A}(z^{-1})\boldsymbol{y}(z) = \boldsymbol{B}(z^{-1})\boldsymbol{u}(z) \tag{17.1.10}$$

with the matrix polynomials

$$\boldsymbol{A}(z^{-1}) = \boldsymbol{A}_0 + \boldsymbol{A}_1 z^{-1} + \ldots + \boldsymbol{A}_m z^{-m} \tag{17.1.11}$$

$$\boldsymbol{B}(z^{-1}) = \boldsymbol{B}_0 + \boldsymbol{B}_1 z^{-1} + \ldots + \boldsymbol{B}_m z^{-m} \,. \tag{17.1.12}$$

If $\boldsymbol{A}(z^{-1})$ is a diagonal matrix polynomial, then (17.1.4) results and hence the system is in simplified P-canonical structure. One can also include the influence of disturbances as

$$\boldsymbol{A}(z^{-1})\boldsymbol{y}(z) = \boldsymbol{B}(z^{-1})\boldsymbol{u}(z) + \boldsymbol{D}(z^{-1})\boldsymbol{v}(z) \,. \tag{17.1.13}$$

17.2 State Space Models

In the following, different models in state space form will be presented. While the general state space form was already introduced and Sects. 2.1.2 as well as 2.2.1, and will only shortly be touched upon, the main focus will be on different structures for the parameter matrices and vectors of the state space models.

17.2.1 State Space Form

It is assumed that a linear time-invariant MIMO system can be determined by the discrete-time generalized state space representation

$$\boldsymbol{x}(k+1) = \boldsymbol{A}\boldsymbol{x}(k) + \boldsymbol{B}\boldsymbol{u}(k) \tag{17.2.1}$$

$$\boldsymbol{y}(k) = \boldsymbol{C}\boldsymbol{x}(k) \tag{17.2.2}$$

with

$\boldsymbol{x}(k)$	State Vector	$\dim \boldsymbol{x} = m \times 1$
$\boldsymbol{u}(k)$	Input Vector	$\dim \boldsymbol{u} = p \times 1$
$\boldsymbol{y}(k)$	Output Vector	$\dim \boldsymbol{y} = r \times 1$
$\boldsymbol{A}$	State Matrix	$\dim \boldsymbol{A} = m \times m$
$\boldsymbol{B}$	Input Matrix	$\dim \boldsymbol{B} = m \times p$
$\boldsymbol{C}$	Output Matrix	$\dim \boldsymbol{C} = r \times m$.

Here, the vectorial quantities are given as deviations from the DC values, i.e.

$$\boldsymbol{u}(k) = \boldsymbol{U}(k) - \boldsymbol{U}_{00} \text{ and } \boldsymbol{y}(k) = \boldsymbol{Y}(k) - \boldsymbol{Y}_{00} \,. \tag{17.2.3}$$

Such a model can for example result from a theoretical continuous-time model that was then discretized as e.g. described in Sect. 2.2.1. The structure of the state space representation depends on the system and the laws that describe its dynamics.

As was stated in Sect. 2.2.1, a system has the minimal realizable form if it is both controllable and observable. A system is controllable if the controllability matrix $\boldsymbol{Q}_\mathrm{S}$ with

$$\boldsymbol{Q}_\mathrm{S} = \begin{pmatrix} \boldsymbol{B} & \boldsymbol{A}\boldsymbol{B} & \ldots & \boldsymbol{A}^{m-1}\boldsymbol{B} \end{pmatrix} \tag{17.2.4}$$

has full rank (i.e. rank m). Similarly, a system is observable if the observability matrix $\boldsymbol{Q}_{\mathrm{B}}$ defined as

$$\boldsymbol{Q}_{\mathrm{B}} = \begin{pmatrix} \boldsymbol{C} \\ \boldsymbol{C}\boldsymbol{A} \\ \vdots \\ \boldsymbol{C}^{m-1}\boldsymbol{A} \end{pmatrix} \tag{17.2.5}$$

has full rank (i.e. rank m).

In state space representation, the model contains a maximum of $m^2 + mp + mr$ parameters. If the system furthermore has a direct feedthrough, the maximum number of parameters increases to $m^2 + mp + mr + pr$. For the description of the input/output behavior, one however typically needs much less parameters, which will be shown in the following.

By a linear transformation $\boldsymbol{T}$,

$$\boldsymbol{x}_{\mathrm{t}} = \boldsymbol{T}\boldsymbol{x} \;, \tag{17.2.6}$$

where $\boldsymbol{T}$ is a non-singular transform matrix, it follows that

$$\boldsymbol{x}_{\mathrm{t}}(k+1) = \boldsymbol{A}_{\mathrm{t}}\boldsymbol{x}_{\mathrm{t}}(k) + \boldsymbol{B}_{\mathrm{t}}\boldsymbol{u}(k) \tag{17.2.7}$$

$$\boldsymbol{y}(k) = \boldsymbol{C}_{\mathrm{t}}\boldsymbol{x}_{\mathrm{t}}(k) \tag{17.2.8}$$

with

$$\boldsymbol{A}_{\mathrm{t}} = \boldsymbol{T}\boldsymbol{A}\boldsymbol{T}^{-1} \tag{17.2.9}$$

$$\boldsymbol{B}_{\mathrm{T}} = \boldsymbol{T}\boldsymbol{B} \tag{17.2.10}$$

$$\boldsymbol{C}_{\mathrm{t}} = \boldsymbol{C}\boldsymbol{T}^{-1} \;. \tag{17.2.11}$$

The transfer function matrix of the state space model is

$$\boldsymbol{G}(z) = \boldsymbol{C}\big(z\boldsymbol{I} - \boldsymbol{A}\big)^{-1}\boldsymbol{B} \tag{17.2.12}$$

and for the transformed system

$$\boldsymbol{G}_{\mathrm{t}}(z) = \boldsymbol{C}_{\mathrm{t}}\big(s\boldsymbol{I} - \boldsymbol{A}_{\mathrm{t}}\big)^{-1}\boldsymbol{B}_{\mathrm{t}} = \boldsymbol{C}\big(z\boldsymbol{I} - \boldsymbol{A}\big)^{-1}\boldsymbol{B} = \boldsymbol{G}(z) \;, \tag{17.2.13}$$

as one can proof by inserting the transform relations that have just been introduced.

The choice of the transformation matrix hence has no influence on the input-output behavior. Therefore, there is no unambiguous realization $\boldsymbol{A}, \boldsymbol{B}, \boldsymbol{C}$ for a certain input-output behavior. One can now choose $\boldsymbol{T}$ such that as many elements of $\boldsymbol{A}$ are fixed to 0 or 1 as possible. This leads to special canonical forms. Two of them shall shortly be presented in the following.

Observable Canonical Form

The observable canonical form is characterized by a special form of the matrices $\boldsymbol{A}$ and $\boldsymbol{C}$. For the SISO case, one can obtain this canonical form by the transformation

$\boldsymbol{T} = \boldsymbol{Q}_{\mathrm{B}}$. In the MIMO case, one can formulate a similar transformation. First, the output matrix is divided into row vectors as

$$\boldsymbol{C} = \begin{pmatrix} \boldsymbol{c}_1^{\mathrm{T}} \\ \boldsymbol{c}_2^{\mathrm{T}} \\ \vdots \\ \boldsymbol{c}_r^{\mathrm{T}} \end{pmatrix} . \tag{17.2.14}$$

The transformation matrix is then constructed as

$$\boldsymbol{T}' = \left(\begin{array}{c} \left.\begin{array}{l} \boldsymbol{c}_1^{\mathrm{T}} \\ \boldsymbol{c}_1^{\mathrm{T}}\boldsymbol{A} \\ \vdots \\ \boldsymbol{c}_1^{\mathrm{T}}\boldsymbol{A}^{m_1-1} \end{array}\right\} m_1 \\ \hline \vdots \\ \hline \left.\begin{array}{l} \boldsymbol{c}_r^{\mathrm{T}} \\ \boldsymbol{c}_r^{\mathrm{T}}\boldsymbol{A} \\ \vdots \\ \boldsymbol{c}_r^{\mathrm{T}}\boldsymbol{A}^{m_r-1} \end{array}\right\} m_r \end{array} \right) . \tag{17.2.15}$$

This matrix must contain m linearly independent row vectors, since $\boldsymbol{T}'$ must be non-singular with the rank m. Hence, one should start with the first output and places $\boldsymbol{c}_1^{\mathrm{T}}, \boldsymbol{c}_1^{\mathrm{T}}\boldsymbol{A}, \ldots$ in the matrix until the first linearly dependent vector $\boldsymbol{c}_1^{\mathrm{T}}\boldsymbol{A}^{m_1}$ occurs. Then one does the same with the second output, etc. One can see that

$$\sum_{i=1}^{r} m_i = m . \tag{17.2.16}$$

The observable canonical form is then given as

$$\boldsymbol{x}'(k+1) = \boldsymbol{A}'\boldsymbol{x}'(k) + \boldsymbol{B}'\boldsymbol{u}(k) \tag{17.2.17}$$

$$\boldsymbol{y}'(k) = \boldsymbol{C}'\boldsymbol{x}'(k) \tag{17.2.18}$$

with

$$\boldsymbol{x}' = \boldsymbol{T}'\boldsymbol{x} \tag{17.2.19}$$

and

$$\boldsymbol{A}' = \begin{pmatrix} \boldsymbol{A}_{11}' & \boldsymbol{0} & \ldots & \boldsymbol{0} \\ \boldsymbol{A}_{21}' & \boldsymbol{A}_{22}' & \ldots & \boldsymbol{0} \\ \vdots & \vdots & & \vdots \\ \boldsymbol{A}_{r1}' & \boldsymbol{A}_{r2}' & \ldots & \boldsymbol{A}_{rr}' \end{pmatrix} , \tag{17.2.20}$$

where

$$A'_{ii} = \begin{pmatrix} 0 & 1 & 0 & \dots \\ 0 & 0 & 1 & \dots \\ \vdots & \vdots & \ddots & \\ 0 & 0 & 0 \dots & 1 \\ & \boldsymbol{a}'^{\mathrm{T}}_{ii} & & \end{pmatrix}, \; A'_{ij} = \begin{pmatrix} 0 & 0 & 0 & \dots \\ 0 & 0 & 0 & \dots \\ \vdots & \vdots & \ddots & \\ 0 & 0 & 0 \dots & 0 \\ & \boldsymbol{a}'^{\mathrm{T}}_{ij} & & \end{pmatrix} \tag{17.2.21}$$

$$i = 1, 2, \dots, r, \; j = 1, 2, \dots, r$$

$$\boldsymbol{a}'^{\mathrm{T}}_{ij} = \left(a'_{ij,m_j}, \dots, a'_{ij,1}\right), \; j = 1, \dots, i$$

and the output matrix as

$$\boldsymbol{c}' = \left.\begin{pmatrix} \overbrace{1\,0\dots0}^{m_1>0} & \overbrace{1\,0\dots0}^{m_2>0} & & \overbrace{1\,0\dots0}^{m_p>0} \\ 0\,0\dots0 & 0\,0\dots0 & \cdots & 0\,0\dots0 \\ \vdots\;\vdots\quad\vdots & \vdots\;\vdots\quad\vdots & & \vdots\;\vdots\quad\vdots \\ 0\,0\quad0 & 0\,0\quad0 & & 0\,0\quad0 \\ & \left.\begin{matrix} \boldsymbol{c}'^{\mathrm{T}}_{m+1} \\ \vdots \\ \boldsymbol{c}'^{\mathrm{T}}_{r} \end{matrix}\right\} m_i = 0 & & \end{pmatrix}\right\} r \tag{17.2.22}$$

with

$$\begin{aligned} \boldsymbol{c}'^{\mathrm{T}}_i &= \left(\boldsymbol{a}'^{\mathrm{T}}_{i1}, \boldsymbol{a}'^{\mathrm{T}}_{i2}, \dots, \boldsymbol{a}'^{\mathrm{T}}_{i,i-1}, 0, \dots, 0\right) \\ i &= m+1, \dots, r \,, \end{aligned} \tag{17.2.23}$$

see also (Goodwin and Sin, 1984).

The controllability matrix follows with $\boldsymbol{B}' = \boldsymbol{T}'\boldsymbol{B}$ as

$$\boldsymbol{B}' = \begin{pmatrix} \left.\begin{matrix} \boldsymbol{b}'^{\mathrm{T}}_{11} \\ \vdots \\ \boldsymbol{b}'^{\mathrm{T}}_{1m_1} \end{matrix}\right\} m_1 \\ \hline \vdots \\ \hline \left.\begin{matrix} \boldsymbol{b}'^{\mathrm{T}}_{r1} \\ \vdots \\ \boldsymbol{b}'^{\mathrm{T}}_{rm_r} \end{matrix}\right\} m_r \end{pmatrix} = \begin{pmatrix} \boldsymbol{c}_1^{\mathrm{T}}\boldsymbol{B} \\ \vdots \\ \boldsymbol{c}_1^{\mathrm{T}}\boldsymbol{A}^{m_1-1}\boldsymbol{B} \\ \hline \vdots \\ \hline \boldsymbol{c}_r^{\mathrm{T}}\boldsymbol{B} \\ \vdots \\ \boldsymbol{c}_r^{\mathrm{T}}\boldsymbol{A}^{m_r-1}\boldsymbol{B} \end{pmatrix} \tag{17.2.24}$$

with

$$\boldsymbol{b}'^{\mathrm{T}}_{ij} = \left(\boldsymbol{b}'_{ij1}, \dots, \boldsymbol{b}'_{ijp}\right), \text{ with } i = 1, \dots, r, \text{ and } j = 1, \dots, m_i \,. \tag{17.2.25}$$

Using the Markov parameters

$$\boldsymbol{M}(q) = \begin{pmatrix} \boldsymbol{m}_1^{\mathrm{T}}(q) \\ \vdots \\ \boldsymbol{m}_r^{\mathrm{T}}(q) \end{pmatrix} = \boldsymbol{C}\boldsymbol{A}^{q-1}\boldsymbol{B} = \boldsymbol{C}'\boldsymbol{A}'^{(q-1)}\boldsymbol{B}', \; q = 1, 2, \dots \,, \tag{17.2.26}$$

one can also write

$$b'_{ij} = m_i^{\mathrm{T}}(j) = c_i^{\mathrm{T}} A^{j-1} B \;. \tag{17.2.27}$$

The observable canonical form has the following properties:

- The blocks of the system matrix A have a triangular form, therefore the subsystems are only coupled in one direction, The ith subsystem is only coupled with the subsystems $1, 2, \ldots, i-1$
- On the main diagonal, one can find the subsystems with the order $m_1, m_2, \ldots, m_r$ in the observable canonical form for SISO systems
- The output matrix has an extremely simple form: Each output y_i is matched with one subsystem of order $m_1, m_2, \ldots$. The outputs are hence identical with the corresponding number of state variables.
- If outputs occur with $m_i = 0$, then the vectors a_{ij} show up in the matrix C
- The number of parameters gets the smallest, if one chooses as the first output y_1 the smallest system m_1, then the second smallest additional system m_2, etc.

The observable canonical form is treated extensively in (Popov, 1972; Guidorzi, 1975; Ackermann, 1988; Blessing, 1980; Goodwin and Sin, 1984).

Controllable Canonical Form

The controllable canonical form is the dual form the observable canonical form and characterized by a special form of the matrices A and B. For the SISO case, this canonical form can be determined by the transform $T = {Q_{\mathrm{S}}}^{-1}$. Once again, a similar transform will now be formulated for the MIMO case. The input matrix B is divided as

$$B = \begin{pmatrix} b_1 \; b_2 \; \ldots \; b_p \end{pmatrix} . \tag{17.2.28}$$

The transformation matrix is then constructed as the controllability matrix Q_{S}

$$(T'')^{-1} = \big(\underbrace{b_1, A b_1, \ldots, A_{m_1}^{m_1-1} b_1}\big| \ldots \big| \underbrace{b_p, A b_p, \ldots, A_{m_p}^{m_p-1} b_r}\big) = R \;. \tag{17.2.29}$$

This matrix must also contain m linearly independent column vectors as T'' must be nonsingular and of rank m. As for the observable canonical form, one now starts with the first input until the first linearly dependent vector $A^{m_1} b_1$ occurs. Then, one continues with the second input. The structural parameters, that can be interpreted as controllability indices, fulfill the relation

$$\sum_{i=1}^{p} m_i = m \;. \tag{17.2.30}$$

The controllable canonical form is then given as

$$x''(k+1) = A'' x''(k) + B'' u(k) \tag{17.2.31}$$

$$y''(k) = C'' x''(k) \tag{17.2.32}$$

with

$$x'' = T''x = R^{-1}x \tag{17.2.33}$$

and

$$A' = \begin{pmatrix} A''_{11} & A''_{12} & \dots & A''_{1p} \\ \mathbf{0} & A''_{22} & \dots & A''_{2p} \\ \vdots & \vdots & & \vdots \\ \mathbf{0} & \mathbf{0} & \dots & A''_{pp} \end{pmatrix}, \tag{17.2.34}$$

where

$$A''_{ii} = \begin{pmatrix} 0 \dots 0 & \\ 1 \dots 0 & \\ \vdots \ \ddots \ \ddots & a''_{ii} \\ 0 \dots 1 & \end{pmatrix}, \ A'_{ij} = \begin{pmatrix} 0 \dots 0 & \\ 1 \dots 0 & \\ \vdots \ \ddots \ \ddots & a''_{ij} \\ 0 \dots 1 & \end{pmatrix} \tag{17.2.35}$$

$$i = 1, 2, \dots, p, \ j = 1, 2, \dots, p$$
$$a''^{\mathrm{T}}_{ij} = \left(a''_{ij,m_j}, \dots, a''_{ij,1}\right), \ j = 1, \dots, i$$

and

$$B' = \begin{array}{c} \left.\begin{array}{c} 1\,0 \dots 0 \\ 0\,0 \dots 0 \\ \vdots\ \vdots \quad \vdots \\ 0\,0 \quad 0 \end{array}\right\} m_1 \\ \hline \left.\begin{array}{c} 1\,0 \dots 0 \\ 0\,0 \dots 0 \\ \vdots\ \vdots \quad \vdots \\ 0\,0 \quad 0 \end{array}\right\} m_2 \\ \hline \vdots\ \vdots \quad \vdots \\ \hline \left.\begin{array}{c} 1\,0 \dots 0 \\ 0\,0 \dots 0 \\ \vdots\ \vdots \quad \vdots \\ 0\,0 \quad 0 \end{array}\right\} m_p \end{array} . \tag{17.2.36}$$

Here, $m_i \neq 0$ was assumed. The output matrix follows from $C'' = CR$ as

$$\begin{aligned} C'' &= (c''_{11} \dots c''_{1m_1} \big| \dots \big| c''_{p1} \dots c''_{pm_p}) \\ &= (Cb_1 \dots CA^{m_1-1}b_1 \big| \dots \big| Cb_p \dots CA^{m_p-1}b_p) \end{aligned} \tag{17.2.37}$$

with

$$c''_{ij} = \left(c''_{ij1}, \dots, c''_{ijp}\right), i = 1, \dots, p, \ j = 1, \dots, m_i \ , \tag{17.2.38}$$

see also (Goodwin and Sin, 1984). In this equation, one can find the Markov parameters as

$$M(q) = \big(m_1(q), \dots, m_p(q)\big) = CA^{q-1}B = C''A''^{(q-1)}B'',\ q = 1, 2, \dots\ , \tag{17.2.39}$$

so that

$$c''_{ij} = m_i(j) = CA^{j-1}b_i\ . \tag{17.2.40}$$

The controllable canonical form has the following properties:

- The blocks of the system matrix A have an upper triangular form. Hence, subsystems are only coupled in one direction, the ith subsystem is only coupled with the subsystems $i+1, i+2, \dots, p$
- On the main diagonal, one can find the subsystems with the order $m_1, m_2, \dots, m_p$ in the controllable canonical form for SISO systems
- The input matrix has an extremely simple form. Each input u_j is assigned to one subsystem of order $m_1, m_2, \dots$

Besides these two canonical forms, one can also use the *Jordan canonical* or *block diagonal form* for identification. Stochastic disturbances are included into the state space structure as

$$x(k+1) = Ax(k) + Bu(k) + Dv(k) \tag{17.2.41}$$

$$y(k) = Cx(k) + v(k)\ . \tag{17.2.42}$$

17.2.2 Input/Output Models

The state space models that were just presented required knowledge of the states $x(k)$ to be able to determine the parameters, i.e. either A and C or A and B. The states are not always measured and in this case, need to be estimated as well. This leads to a non-linear estimation problem. One can apply an extended Kalman filter to this estimation problem, see Chap. 21, but the convergence often is slow. The estimation can even diverge. It is in many cases practical to bring the state space system to an input/output form, estimate the parameters and transform the model back to state space (Blessing, 1980; Schumann, 1982). Here, the observable canonical form is especially well suited. From (17.2.41) follows for $y(k+\nu)$ with $\nu = 0, 1, m-1$

$$\begin{pmatrix} y(k) \\ y(k+1) \\ \vdots \\ y(k+m-1) \end{pmatrix} = \begin{pmatrix} C \\ CA \\ \vdots \\ CA^{m-1} \end{pmatrix} x'(k) + \begin{pmatrix} 0 & \dots & 0 \\ 0 & \dots & CB \\ \vdots & & \vdots \\ CB & \dots & CA^{m-2}B \end{pmatrix} \begin{pmatrix} u(k) \\ u(k+1) \\ \vdots \\ u(k+m-2) \end{pmatrix}$$
$$+ \begin{pmatrix} v(k) \\ v(k+1) \\ \vdots \\ v(k+m-1) \end{pmatrix} + \begin{pmatrix} 0 & \dots & 0 \\ 0 & \dots & CD \\ \vdots & & \vdots \\ CD & \dots & CA^{m-2}D \end{pmatrix} \begin{pmatrix} v(k) \\ v(k+1) \\ \vdots \\ v(k+m-2) \end{pmatrix}\ . \tag{17.2.43}$$

Since $\boldsymbol{x}'$ is multiplied with the observability matrix and $\boldsymbol{T} = \boldsymbol{Q}_{\mathrm{B}}^{-1}$, an identity matrix will result. Hence, one can solve the system of equations as

$$\boldsymbol{x}'(k) = \boldsymbol{y}_{\mathrm{m}} - \boldsymbol{v}_{\mathrm{m}} - \boldsymbol{S}_{\mathrm{u}}u_{\mathrm{m}} - \boldsymbol{S}_{\nu}\boldsymbol{v}_{m} \qquad (17.2.44)$$

and by application of the z-transform, one obtains

$$\begin{aligned} A'_{ii}\big(y_i(z) - \nu_i(z)\big) &= \sum_{j=1}^{i-1} A'_{ij}(z^{-1})(y_i(z) - \nu_i(z)) \\ &+ \sum_{j=1}^{p} B'_{ij}(z^{-1})u_j(z) + \sum_{j=1}^{r} D'_{ij}(z^{-1})\nu_j(z) \end{aligned} \qquad (17.2.45)$$

with $i = 1, 2, \ldots, r$ (Schumann, 1982). The model shall be denoted as the *minimal input/output model*. It contains coupling terms of the outputs $y_i(k)$ which depend on $y_j(k)$ for $j < i$. If one successively removes these coupling terms, then a simplified P-canonical input-output model results as

$$A_{ii}y_i(z) = \sum_{j=1}^{p} B_{ij}(z^{-1})u_j(z) + D_{ii}(z^{-1})\nu_i(z) \,. \qquad (17.2.46)$$

These P-canonical models are however no longer minimal. A direct determination of the state space model in (17.2.41) is hence no longer possible.

17.3 Impulse Response Models, Markov Parameters

For the calculation of the outputs for given inputs, one can recursively evaluate (17.2.1) and together with (17.2.2) obtain the solution

$$\boldsymbol{y}(k) = \boldsymbol{C}\boldsymbol{A}^k\boldsymbol{x}(0) + \sum_{\nu=0}^{k-1} \boldsymbol{C}\boldsymbol{A}^{\nu}\boldsymbol{B}\boldsymbol{u}(k-\nu-1), k = 2, 3, \ldots \,, \qquad (17.3.1)$$

where $\boldsymbol{x}(0)$ are the initial values of the states and

$$\boldsymbol{G}(\nu) = \boldsymbol{C}\boldsymbol{A}^{\nu}\boldsymbol{B} \qquad (17.3.2)$$

is the impulse response matrix (Schwarz, 1967, 1971). The transfer function matrix is then given as

$$\boldsymbol{G}(z) = \sum_{\nu=0}^{\infty} \boldsymbol{G}(\nu)z^{-\nu} \,. \qquad (17.3.3)$$

One denotes

$$\boldsymbol{M}_{\nu} = \boldsymbol{G}(\nu) = \boldsymbol{C}\boldsymbol{A}^{\nu}\boldsymbol{B}, \; \nu = 0, 1, 2, \ldots \qquad (17.3.4)$$

as the Markov parameters of the MIMO system (Ho and Kalman, 1966).

From (17.3.1) and (17.3.3) follows for the transfer function matrix

$$G(z) = \sum_{\nu=0}^{\infty} M_\nu z^{-(\nu+1)} \tag{17.3.5}$$

and for the output

$$y(k) = M_k \beta_0 + \sum_{\nu=0}^{k-1} M_\nu u(k-\nu-1) \tag{17.3.6}$$

with

$$B\beta_0 = x(0) \text{ or } \beta_0 = \left(B^{\mathrm{T}} B\right)^{-1} B x(0) \ . \tag{17.3.7}$$

In vectorial notation, one gets

$$\begin{pmatrix} y(0) \\ y(1) \\ y(2) \\ \vdots \end{pmatrix} = \begin{pmatrix} M_0 & 0 & & \dots 0 \\ M_1 & M_0 & 0 & \dots 0 \\ M_2 & M_1 & M_0 & \dots 0 \\ \vdots & \vdots & \vdots & \vdots \end{pmatrix} \begin{pmatrix} u(-1) \\ u(0) \\ u(1) \\ \vdots \end{pmatrix} + \begin{pmatrix} M_0 & M_1 & \dots & M_{m-1} \\ M_1 & M_2 & \dots & M_m \\ M_2 & M_3 & \dots & M_{m+1} \\ \vdots & \vdots & & \vdots \end{pmatrix} \begin{pmatrix} \beta_0 \\ 0 \\ 0 \\ \vdots \end{pmatrix} \tag{17.3.8}$$

or

$$y = \mathcal{T}(0,\infty) u + \mathcal{H}(m,\infty) \beta_0 \ , \tag{17.3.9}$$

which is termed a *Hankel model*. Here, $\mathcal{T}$ is a Toeplitz matrix and $\mathcal{H}$ a Hankel matrix. The Hankel matrix is given as the product from controllability and observability matrix as

$$\mathcal{H} = Q_{\mathrm{B}} Q_{\mathrm{S}} = \begin{pmatrix} M_0 & M_1 & \dots & M_{m-1} \\ M_1 & M_2 & \dots & M_m \\ M_2 & M_3 & \dots & M_{m+1} \\ \vdots & \vdots & & \vdots \end{pmatrix} \tag{17.3.10}$$

with the Markov parameters M_ν as elements and rank $\mathcal{H} = m$. For the often encountered case of zero initial conditions, i.e. $\beta_0 = 0$, one obtains

$$\left(y(0), y(1), y(2), \dots\right) = \left(M_0, M_1, M_2, \dots\right) \begin{pmatrix} u(-1) & u(0) & u(1) & \dots \\ 0 & u(-1) & u(0) & \dots \\ 0 & 0 & u(-1) & \dots \\ \vdots & \vdots & & \vdots \end{pmatrix} , \tag{17.3.11}$$

which can be written as

$$Y = MU \tag{17.3.12}$$

with additional disturbances

$$Y = MU + N \ . \tag{17.3.13}$$

17.4 Subsequent Identification

If one excites the inputs $i = 1, 2, \ldots, p$ one after the other and measures all outputs $j = 1, 2, \ldots, r$, then one can identify a SIMO system for each input separately. If using a P-canonical model structure, one can use the classical SISO identification methods presented so far. However, upon the simultaneous excitation of multiple inputs (MIMO), one can save a lot of measurement time and in addition, one obtains a coherent model. Hence, MIMO identification methods will be discussed in the following. The positive effects of a parallel excitation of all inputs have also been proven in (Gevers et al, 2006).

17.5 Correlation Methods

The correlation analysis methods from Chaps. 6 and 7 will now be used to identify MIMO systems. Based on the correlation functions, one can once again use the de-convolution for MIMO systems.

17.5.1 De-Convolution

If one multiples the convolution sum

$$y(k) = \sum_{\nu=0}^{k-1} \boldsymbol{M}_\nu \boldsymbol{u}(k-\nu-1) \tag{17.5.1}$$

from the right with $\boldsymbol{u}^{\mathrm{T}}(k-\tau)$ and considers the expected value, then one obtains

$$\boldsymbol{R}_{\mathrm{uy}} = \sum_{\nu=0}^{k-1} \boldsymbol{M}_\upsilon \boldsymbol{R}_{\mathrm{uu}}(\tau-\nu-1) \, . \tag{17.5.2}$$

If one estimates the correlation functions from the measured signals as presented in Chap. 7, then one can determine the Markov parameters in analogy to (7.2.1) through (7.2.3) by means of the de-convolution. One sets up the system of equations as

$$\boldsymbol{R}_{\mathrm{uy}} = \boldsymbol{M}\boldsymbol{R}_{\mathrm{uu}} \tag{17.5.3}$$

and obtains

$$\boldsymbol{M} = \boldsymbol{R}_{\mathrm{uy}} \boldsymbol{R}_{\mathrm{uu}}^{-1} \, , \tag{17.5.4}$$

if $\boldsymbol{R}_{\mathrm{uu}}$ is a square matrix. If $\boldsymbol{R}_{\mathrm{uu}}$ is a non-square matrix, then one has to use the pseudo-inverse as

$$\boldsymbol{M} = \boldsymbol{R}_{\mathrm{uy}} \boldsymbol{R}_{\mathrm{uu}}^{\mathrm{T}} \left(\boldsymbol{R}_{\mathrm{uu}}^{\mathrm{T}} \boldsymbol{R}_{\mathrm{uu}}\right)^{-1} \, . \tag{17.5.5}$$

The computational effort is very high since a matrix of high dimensions has to be inverted. If the input signals $\boldsymbol{u}(k)$ consists of components that are white noise, hence

$$\boldsymbol{R}_{\mathrm{uu}}(\tau) = \boldsymbol{R}_{\mathrm{uu}}(0)\delta(\tau) \tag{17.5.6}$$

and these input signals are mutually uncorrelated, then one can determine the Markov parameters directly as

$$m_{k,i,j} = \frac{R_{\mathrm{u}_i \mathrm{y}_j}(k)}{R_{\mathrm{u}_i \mathrm{u}_i}(0)} \; . \tag{17.5.7}$$

This is also presented in (Juang, 1994).

17.5.2 Test Signals

For a simultaneous excitation of all inputs, the identification will become more easy with respect to both evaluation and convergence, if the input signals are uncorrelated. For the application of correlation methods and stable systems with $\boldsymbol{M}_\nu \approx \boldsymbol{0}$ for $\nu > \nu_{\max}$, the cross-correlation functions should vanish, i.e.

$$R_{\mathrm{u}_i \mathrm{u}_j}(\tau) = 0 \text{ for } |\tau| = 0, \ldots, \nu_{\max},\; i = 1, \ldots, p;\; j = 1, \ldots, p, i \neq j \; . \tag{17.5.8}$$

This means that the test signals must be orthogonal to each other. The PRBS signals that have been used in the SISO case however do not satisfy this requirement. They can however be modified to do so (Briggs et al, 1967; Tsafestas, 1977; Blessing, 1980).

For the generation of orthogonal test signals, one has to write the input signal as a product of two binary periodic signals as

$$u_i(k) = h_i(k)p(k) \tag{17.5.9}$$

$$h_i(k) = h_i(k + \nu N_{\mathrm{H}}),\; \nu = 1, 2, \ldots \tag{17.5.10}$$

$$p(k) = p(k + \nu N_{\mathrm{p}}) \; . \tag{17.5.11}$$

The period length of $u_i(k)$ is then given as $N = N_{\mathrm{p}} N_{\mathrm{H}}$. Here, $p(k)$ is a basis PRBS as described in Sect. 6.3. If $h_i(k)$ and $p(k)$ are statistically independent, then the cross-correlation is given as

$$R_{\mathrm{u}_i \mathrm{u}_j} = R_{\mathrm{h}_i \mathrm{h}_j} R_{\mathrm{pp}} \; i, j = 1, \ldots, p \; . \tag{17.5.12}$$

In order for the underlying PRBS signal to satisfy

$$R_{\mathrm{pp}}(\tau) = 0 \text{ for } |\tau| = 1, \ldots, N_{\mathrm{p}} - 1 \; , \tag{17.5.13}$$

one must use the amplitudes $+a$ and $-aP$, where

$$P = \frac{\sqrt{N_{\mathrm{p}} + 1} - 2}{\sqrt{N_{\mathrm{p}} + 1}} \; . \tag{17.5.14}$$

According to (17.5.8), the input signals must be uncorrelated, i.e.

$$R_{\mathrm{u}_i \mathrm{u}_j} = R_{\mathrm{h}_i \mathrm{h}_j} R_{\mathrm{pp}} = 0 \text{ for } i \neq j \; , \tag{17.5.15}$$

and therefore

$$R_{\mathrm{h}_i\mathrm{h}_j} = 0 \text{ for } i \neq j \ . \tag{17.5.16}$$

This can be fulfilled if the signals $h_i(k)$ with the period N_H are chosen from the elements of a *Hadamard matrix* $\boldsymbol{H}$ of the order $N_\mathrm{H} = 2^n$ (Brauer, 1953). A simple generation of a Hadamard matrix of order $N_\mathrm{H} = 2^n$ can be realized if its elements are Walsh functions (Briggs et al, 1967; Blessing, 1980). Then follows the recursive relation

$$\boldsymbol{H}_{(2^n)} = \begin{pmatrix} \boldsymbol{H}_{(2^{n-1})} & \boldsymbol{H}_{(2^{n-1})} \\ \boldsymbol{H}_{(2^{n-1})} & -\boldsymbol{H}_{(2^{n-1})} \end{pmatrix} \tag{17.5.17}$$

with $\boldsymbol{H}_1 = 1$ and $n = p - 1$. For the generation of p inputs, one has to set up a matrix $\boldsymbol{H}_{(N_H)}$ with $N_\mathrm{H} = 2^n = 2^{p-1}$. The periodic binary signals $h_i(k)$ can then be generated from the components of the ith row of $\boldsymbol{H}$ multiplied with the amplitude H_i. Since the first row of $\boldsymbol{H}$ only contains the value 1, $u_1(k)$ will be a pseudo-random binary signal with the amplitudes H_1 and $-H_1 P$. The other signals $u_2(k), u_3(k), \ldots, u_p(k)$ will assume the four different values $\pm H_i$ and $\pm H_i P$, so that one has four-valued signals (Blessing, 1980; Hensel, 1987). This means amplitude modulated PRBS signals (APRBS) have been generated. In (Pintelon and Schoukens, 2001), the same method is suggested for frequency domain identification.

The period length of the individual test signals is $N = N_\mathrm{H} N_\mathrm{P}$. For $2, 3, 4, \ldots$ input signals, one has $n = 1, 2, 3, \ldots$ and $N = 2N_\mathrm{P}, 4N_\mathrm{P}, 8N_\mathrm{P}$. This at the same time is the smallest possible sample length so that all p input signals are mutually uncorrelated. Figures 17.3 and 17.4 show an example of an orthogonal test signal.

17.6 Parameter Estimation Methods

Due to the large number of model structures and parameter estimation methods, there are many possibilities to estimate the parameters of MIMO systems. If in addition to the input and output signals, one can also measure the states of the system, one should use the state space representation (17.2.1) and (17.2.2), which can for example be obtained from physically modeling the system (typically continuous-time signals). This model must be both controllable and observable. One can then use the SISO parameter estimation methods, e.g. method of least squares, if one uses for each output signal the submodel

$$y_i(k) = \boldsymbol{\psi}_i^\mathrm{T} \boldsymbol{\theta}_i + e_i(k),\ i = 1, 2, \ldots, p \ . \tag{17.6.1}$$

In $\boldsymbol{\psi}_i^\mathrm{T}(k)$, one has to include all measured signals (i.e. also the states) which act on the corresponding $y_i(k)$. If the states are not measurable and one wants to avoid the concurrent estimation of states and parameters due to the non-linearity of the resulting estimation problem, then one has to resort to input/output models, see the overview in Fig. 17.5.

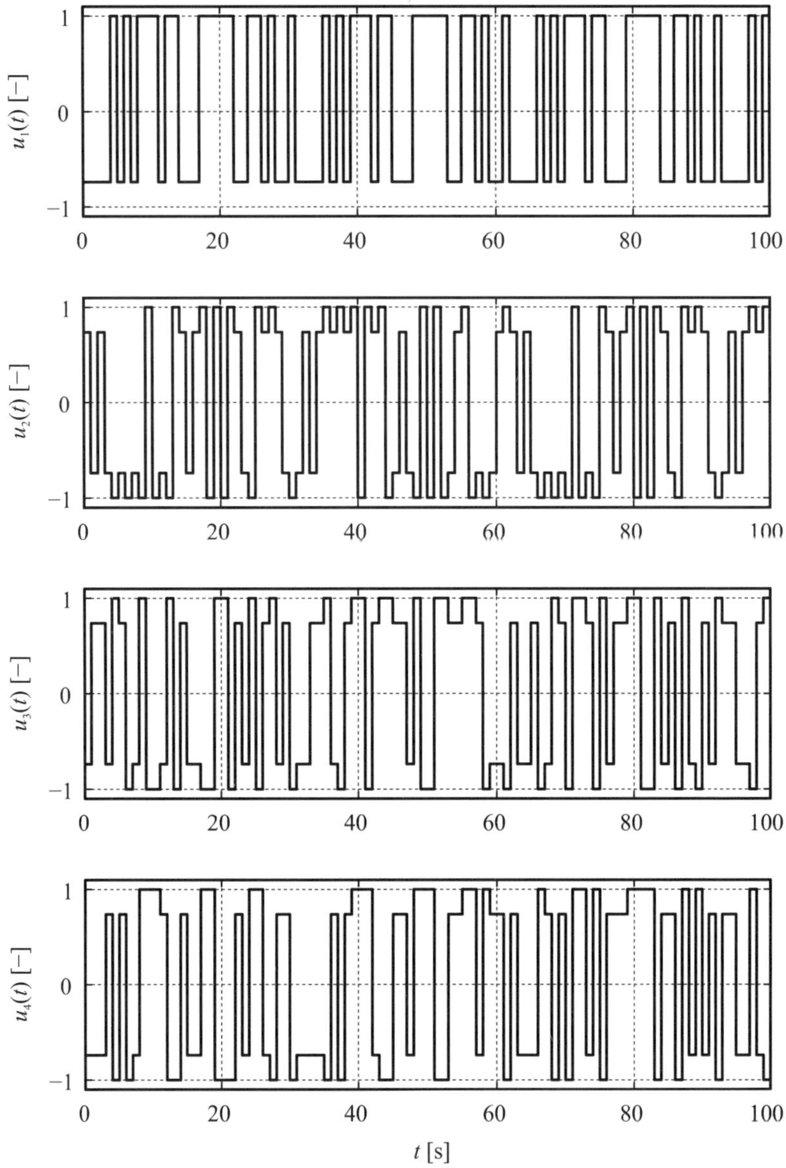

Fig. 17.3. Pseudo random binary signal generated by a shift register with $n = 5$ stages and $N_{\mathrm{P}} = 31$ and a Hadamard matrix of order $N_{\mathrm{H}} = 8$

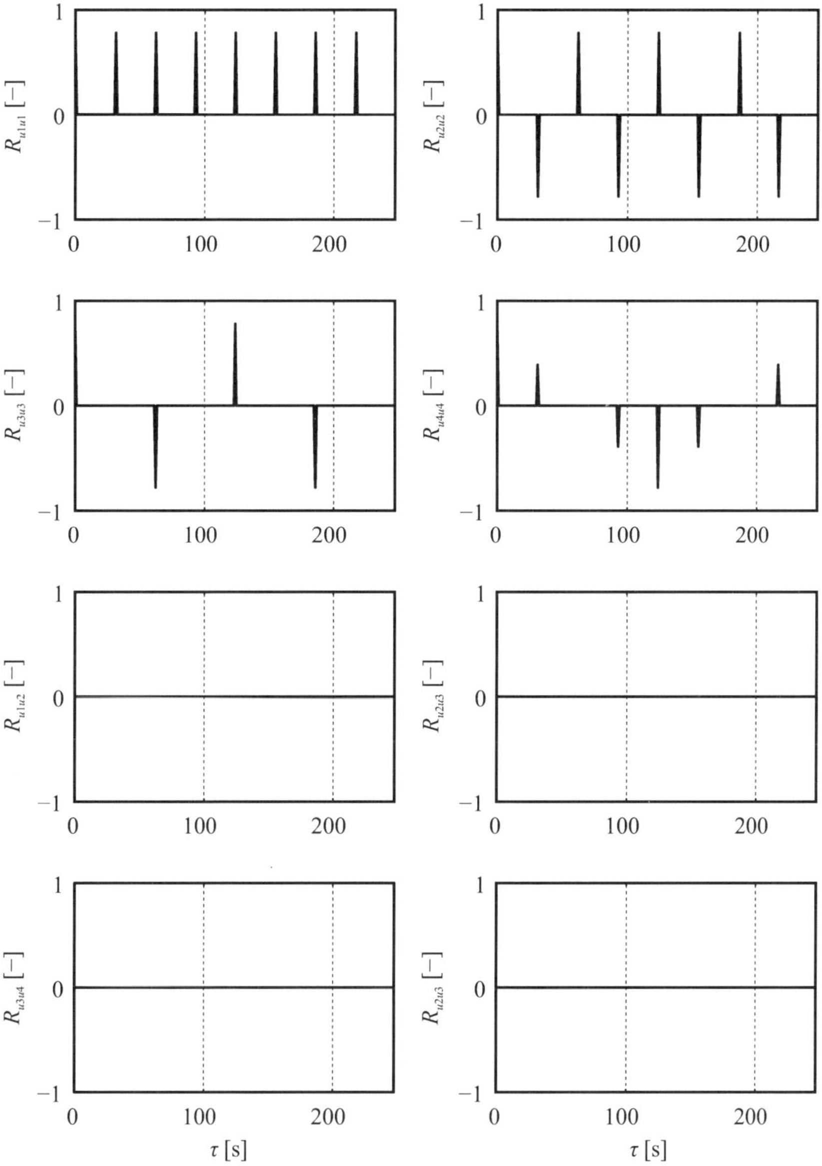

Fig. 17.4. Correlation function of the pseudo-random binary signal generated by a shift register with $n = 5$ stages and $N_P = 31$ and a Hadamard matrix of order $N_H = 8$

17.6.1 Method of Least Squares

The method of least squares can directly by applied to the following input/output models: Minimal I/O model, P-canonical I/O model, simplified P-canonical I/O model, and matrix polynomial I/O model, if for each output $y_i(k)$, a submodel according to (17.6.1) is set up. This means that the MIMO system is split up into MISO systems for the purpose of estimating the parameters. Then, the data vector and parameter vector are e.g. for the simplified P-canonical model given as

$$\begin{aligned}\boldsymbol{\psi}_i^{\mathrm{T}} &= \big(\big(-y_i(k-1)\,\ldots\,-y_i(k-m)\big|u_1(k-1)\,\ldots\,u_1(k-m_i)\big|\\ &\quad u_p(k-1)\,\ldots\,u_p(k-m_i)\big)\big) \qquad (17.6.2)\\ \boldsymbol{\theta}_i^{\mathrm{T}} &= \big(a_{i11}\,\ldots\,a_{i1m_i}\big|b_{i11}\,\ldots\,b_{i1m_i}\big|b_{ip1}\,\ldots\,b_{ipm_i}\big)\,, \qquad (17.6.3)\end{aligned}$$

see also (Schumann, 1982; Hensel, 1987). For the estimation of the parameters of such a model, one can use the parameter estimation methods for SISO systems either in non-recursive or recursive form.

17.6.2 Correlation Analysis and Least Squares

The advantages of the parameter estimation with a non-parametric intermediate model as introduced in Sect. 9.3 are even more beneficial, if the methods are applied to MIMO systems. This is especially true, if also the structure, i.e. model order and dead time must be determined from experimental data.

The method COR-LS, that was described in Sect. 9.3 can also be applied to the estimation of multiple I/O models of MIMO systems. For example, for a P-canonical I/O model, one can, based on

$$A_{ii}(q^{-1})y_i(q) = \sum_{j=1}^{p} B_{ij}(q^{-1})u_j(q) \qquad (17.6.4)$$

and multiplication with $u_j(q-\tau)$ and subsequent calculation of the expected values, obtain

$$A_{ii}(q^{-1})R_{\mathrm{u}_i\mathrm{y}_j}(\tau) = \sum_{j=1}^{p} B_{ij}(q^{-1})R_{\mathrm{u}_i\mathrm{u}_j}(q) \qquad (17.6.5)$$

and set up a system of equations as in (9.3.20), which can then be solved as in (9.3.22).

The computational effort can be reduced, if one determines the sum of the input signals (Hensel, 1987)

$$u_\Sigma(k) = \sum_{j=1}^{p} u_j(k)\,. \qquad (17.6.6)$$

If the input signals are uncorrelated, it holds that

$$R_{\mathrm{u}_\Sigma\mathrm{u}_j}(\tau) = R_{\mathrm{u}_j\mathrm{u}_j}(\tau)\,. \qquad (17.6.7)$$

The model is then given as

$$A_{ii}(q^{-1})R_{\mathrm{u}_\Sigma \mathrm{y}_j}(\tau) = \sum_{j=1}^{p} B_{ij}(q^{-1})R_{\mathrm{u}_\Sigma \mathrm{u}_j}(q) \, . \tag{17.6.8}$$

The correlation functions can be determined recursively to obtain a reduction of the storage space, (9.3.16).

Blessing (1979, 1980) has applied the method COR-LS to a minimal I/O model which is based on the observable-canonical form $(\boldsymbol{A}, \boldsymbol{B}, \boldsymbol{C})$ and hence can be converted to this structure. In a second step, also the parameters of a noise model have been estimated.

For the determination of model order and dead time of I/O models and the structural indices for state space models, one can use in principle the same methods as for SISO systems, see Sect. 23.3. It is shown in (Blessing, 1980) how one can determine the structural index $\hat{m}_i$ which belongs to the output y_i by means of an analysis of the cost function and of the eigenvalues of the information matrix. The determinant ratio test from (Woodside, 1971) was applied to the information matrix of the simplified P-canonical model by Hensel (1987).

17.7 Summary

Figure 17.5 gives an overview over the different model structures for MIMO systems as well as the suitable identification methods for the case that only the inputs and outputs of the system, but not the states, can be measured and that multiple inputs are excited at once.

The *state space model* follows directly from a theoretical model of the system. For identification, the system as well as the model must be controllable and observable, hence it must be a minimal realization. By transformation into a *canonical state space model*, one obtains a model with the minimum number of parameters and can furthermore work with a model structure that can be employed for e.g. state feedback controller design or observer design. Furthermore, such a model can be transformed into other models.

If one wants to avoid non-linear estimation problems, then one has to eliminate the states, so that the states and the parameters must no longer be estimated simultaneously. Hence, one must eliminate state variables and use suitable *input/output models*. From the canonical state space models, one can obtain directly a *minimal input/output model* and by elimination of the couplings between the outputs, one obtains the *simplified P-canonical input/output model*. This can also be written as a *matrix polynomial input/output model*.

The *parameter estimation* can be carried out with any of the input/output models. One obtains the parameters of the model that was taken as a basis for the formulation of the parameter estimation problem. If one wants to obtain a canonical state space model as result, one can obtain this from the minimal input/output model by simple

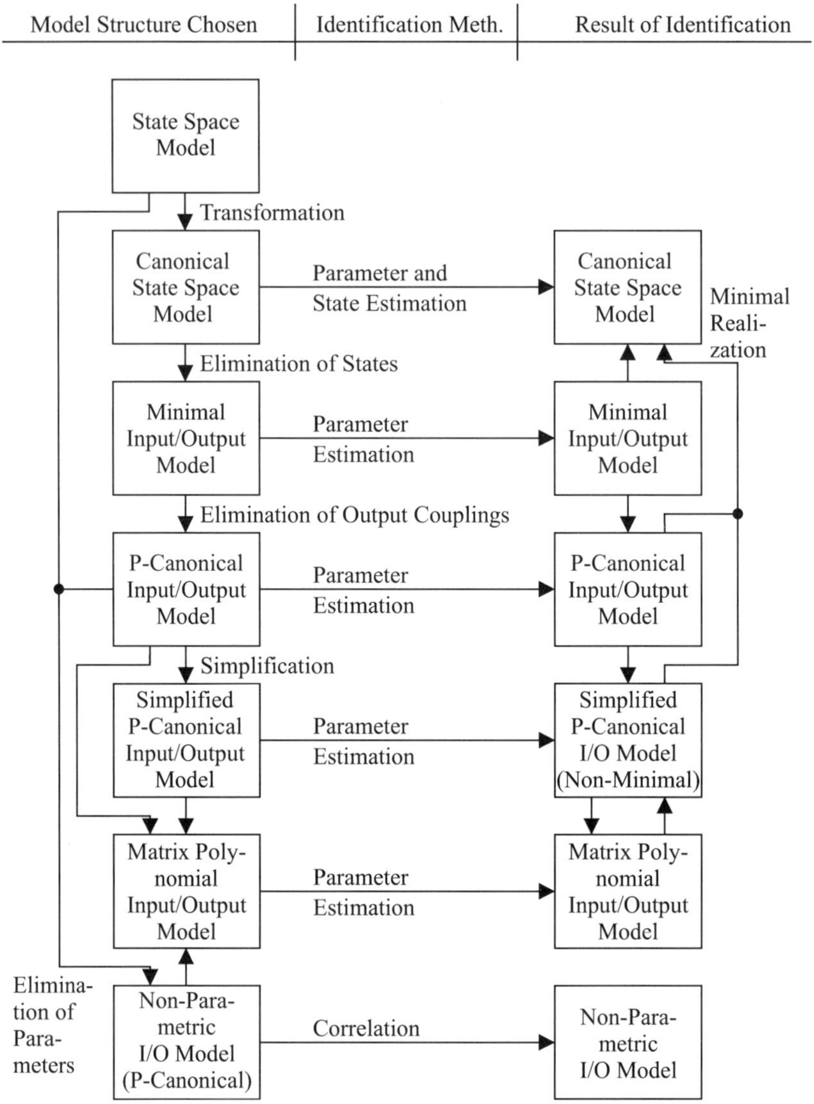

Fig. 17.5. Structures of linear MIMO systems and suitable identification methods

conversion and from the P-canonical input/output models by means of a minimal realization.

For the use of correlation functions, one can use non-parametric input/output models with Markov parameters.

The identification of non-parametric models with correlation methods and the identification of parametric models by means of parameter estimation methods itself can upon the concurrent excitation of all inputs be done with the same methods as for the SISO case. In order to simplify the evaluation of the resulting equations and to speed up the convergence, one should use always orthogonal test signals, which can be generated from a single PRBS signal by the approach shown in this chapter.

The Markov parameters can be determined by the de-convolution of the correlation functions. For an overview of model structure suitable for parameter estimation, one should consider Fig. 17.5. The method of least squares and the modifications to it that have been presented in preceding chapters can be applied to multiple I/O models. The method COR-LS is especially well suited for the identification of MIMO systems. The determination of model order and dead time or other structural indices can also be incorporated into the identification of MIMO systems, here one can apply many techniques that have been presented for SISO systems in a straightforward way. An alternative to the identification of MIMO systems are subspace methods, see Chap. 16, which allow to determine the model order during the identification and hence need no a priori assumptions.

One attractive application for the simultaneous excitation of several input signals is the identification of MISO models for internal combustion engines on test benches for the calibration of electronic control systems (Schreiber and Isermann, 2009).

Problems

17.1. MIMO System Identification Based on SISO Models
Which of the above model structures can be used in conjunction with SISO identification methods?

17.2. Two Input/Two Output Model 1
Assuming that all 4 transfer functions of the P-canonical structure are first order processes, determine the various possibilities for identification.

17.3. Two Input/Two Output Model 2
What changes follow for $G_{12}(s) = 0$ in Fig. 17.1 for the identification?

References

Ackermann J (1988) Abtastregelung, 3rd edn. Springer, Berlin

Blessing P (1979) Identification of the input/output and noise-dyanmics of linear multi-variable systems. In: Proceedings of the 5th IFAC Symposium on Identification and System Parameter Estimation Darmstadt, Pergamon Press, Darmstadt, Germany

Blessing P (1980) Ein Verfahren zur Identifikation von linearen, stochastisch gestörten Mehrgrößensystemen: KfK-PDV-Bericht. Kernforschungszentrum Karlsruhe, Karlsruhe

Brauer A (1953) On a new class of Hadamard determinants. Math Z 58(1):219–225

Briggs PAN, Godfrey KR, Hammond PH (1967) Estimation of process dynamic characteristics by correlation methods using pseudo random signals. In: Proceedings of the IFAC Symposium Identification, Prag, Czech Republic

Gevers M, Miskovic L, Bonvin D, Karimi A (2006) Identification of multi-input systems: Variance analysis and input design issues. Automatica 42(4):559–572

Goodwin GC, Sin KS (1984) Adaptive filtering, prediction and control. Prentice-Hall information and system sciences series, Prentice-Hall, Englewood Cliffs, NJ

Guidorzi R (1975) Canonical structures in the identification of multivariable systems. Automatica 11(4):361–374

Hensel H (1987) Methoden des rechnergestützten Entwurfs und Echtzeiteinsatzes zeitdiskreter Mehrgrößenregelungen und ihre Realisierung in einem CAD-System. Fortschr.-Ber. VDI Reihe 20 Nr. 4. VDI Verlag, Düsseldorf

Ho BL, Kalman RE (1966) Effective construction of linear state variable models from input/output functions. Regelungstechnik 14:545–548

Isermann R (1991) Digital control systems, 2nd edn. Springer, Berlin

Juang JN (1994) Applied system identification. Prentice Hall, Englewood Cliffs, NJ

Pintelon R, Schoukens J (2001) System identification: A frequency domain approach. IEEE Press, Piscataway, NJ

Popov VM (1972) Invariant description of linear time-variant controllable systems. SIAM J Control 10:252–264

Schreiber A, Isermann R (2009) Methods for stationary and dynamic measurement and modeling of combustion engines. In: Proceedings of the 3rd International Symposium on Development Methodology, Wiesbaden, Germany

Schumann R (1982) Digitale parameteradaptive Mehrgrößenregelung - KfK-PDV-Bericht Nr. 217. Kernforschungszentrum Karlsruhe, Karlsruhe

Schwarz H (1967, 1971) Mehrfach-Regelungen, vol 1. Springer, Berlin

Tsafestas SG (1977) Multivariable control system identification using pseudo random test input. Int J Control Theory and Applic 5:58–66

Woodside CM (1971) Estimation of the order of linear systems. Automatica 7(6):727–733

Part VI

Identification of Non-Linear Systems

18

Parameter Estimation for Non-Linear Systems

Due to the many structural possibilities of non-linear relations between the input and output of dynamic systems, one cannot expect to be able to identify many types of non-linear system with only a few model classes. However, for certain types of non-linear systems, models can be formulated that match well with the requirements on the model structure of known identification methods. In this sense, some model structures and suitable parameter identification methods will be covered in the following. First, dynamic systems with continuously differentiable non-linearities will be discussed, then dynamic systems with non-continuously differentiable non-linearities, such as friction and dead zone will be treated.

18.1 Dynamic Systems with Continuously Differentiable Non-Linearities

Classical methods for the identification of dynamic systems are mostly based on polynomial approximators. One distinguishes between general approaches, e.g. Volterra series or Kolmogorov-Gabor polynomials, and approaches that involve special structural assumptions such as Hammerstein, Wiener or non-linear difference equation (NDE) models (Eykhoff, 1974; Haber and Unbehauen, 1990; Isermann et al, 1992).

Certain static polynomial approximators have the advantage of being linear in the parameters. This advantage can be maintained for certain dynamic polynomial models. This way, computationally expensive iterative optimization methods can be avoided.

Hence, in the following, a special focus will be placed on examples of classical non-linear dynamic models that are based on a representation of the non-linearity by polynomials together with a dynamic part modeled as a linear discrete-time system. Note that the linear difference equation is written with the shift operator q^{-1}, where $y(k)q^{-i} = y(k-i)$

$$A(q^{-1})\, y(k) = B\,(q^{-1})\, q^{-d}\, u(k) + D\,(q^{-1})\, v(k) \tag{18.1.1}$$

R. Isermann, M. Münchhof, *Identification of Dynamic Systems*,
DOI 10.1007/978-3-540-78879-9_18, © Springer-Verlag Berlin Heidelberg 2011

according to (10.2.2).

18.1.1 Volterra Series

In allusion to the convolution integral

$$y(t) = \int_0^t g(\tau)u(t-\tau)\mathrm{d}\tau \;, \tag{18.1.2}$$

one can describe the input-output relation of systems with continuously differentiable non-linearities by means of a *Volterra series*

$$\begin{aligned} y(t) =& g_0' + \int_0^t g_1'(\tau_1)u(t-\tau_1)\mathrm{d}\tau_1 + \int_0^t\int_0^t g_2'(\tau_1,\tau_2)u(t-\tau_1)u(t-\tau_2)\mathrm{d}\tau_1\mathrm{d}\tau_2 \\ &+ \int_0^t\int_0^t\int_0^t g_3'(\tau_1,\tau_2,\tau_3)u(t-\tau_1)u(t-\tau_2)u(t-\tau_3)\mathrm{d}\tau_1\mathrm{d}\tau_2\mathrm{d}\tau_3 + \ldots \;, \end{aligned} \tag{18.1.3}$$

see (Volterra, 1959; Gibson, 1963; Eykhoff, 1974; Schetzen, 1980). This infinite functional power series contains symmetric *Volterra kernels* $g_n'(\tau_1,\ldots,\tau_n)$ of the order n, which are also termed impulse response of order n. The condition of causality implies that

$$g_n'(\tau_1,\ldots,\tau_n) = 0 \text{ for } \tau_i < 0,\; i = 1,2,\ldots,n \;. \tag{18.1.4}$$

With $n = 1$, one obtains the convolution integral for linear system. This model is suitable for continuous-time processes. However, typically the discrete-time form is used.

For discrete-time systems, the Volterra series is given as

$$\begin{aligned} y(k) = g_0 &+ \sum_{\tau_1=0}^{k} g_1(\tau_1)u(k-\tau_1) + \sum_{\tau_1=0}^{k}\sum_{\tau_2=0}^{k} g_2(\tau_1,\tau_2)u(k-\tau_1)u(k-\tau_2) \\ &+ \sum_{\tau_1=0}^{k}\sum_{\tau_2=0}^{k}\sum_{\tau_3=0}^{k} g_3(\tau_1,\tau_2,\tau_3)u(k-\tau_1)u(k-\tau_2)u(k-\tau_3) + \ldots \;, \end{aligned} \tag{18.1.5}$$

These Volterra series are non-parametric models, whose identification however necessitates the determination of the function values of the kernels. As non-linear models can also describe the system behavior for large deviations from the operating point, the large signal values $U(k)$ and $Y(k)$ will be used in the following. If one limits the significant elements of the impulse responses to the time $k \leq M$, then one can write the above Volterra series up to order p by

$$y(k) = c_{00} + \sum_{n=1}^{p} v_M^n(k) \tag{18.1.6}$$

$$v_M^n = \sum_{i_1=0}^{M} \cdots \sum_{i_n=1}^{M} \alpha_n(i_1, \ldots, i_n) u(k-i_1) \ldots u(k-i_n) \tag{18.1.7}$$

and determine all coefficients by the method of least squares as they appear linearly in the above function (Doyle et al, 2002). However, the number of parameters can grow quite fast, as typically long sequences of the impulse response have to be covered. Such a model is called non-linear finite impulse response (NFIR) model.

An alternative is to approximate the discrete Volterra series, which is limited to the order p, by a parametric model as

$$\begin{aligned} A(q^{-1})y(k) &= c_{00} + B_1(q^{-1})u(k-d) \\ &+ \sum_{\beta_1=0}^{h} B_{2\beta_1}(q^{-1})u(k-d)u(k-d-\beta_1) + \ldots \\ &+ \sum_{\beta_1=0}^{h} \sum_{\beta_2=\beta_1}^{h} \cdots \sum_{\beta_{p-1}=\beta_{p-2}}^{h} B_{\mathrm{p}\beta_1\beta_2\ldots\beta_{p-1}}(q^{-1})u(k-d) \prod_{\xi=1}^{p-1} u(k-d-\beta_\xi) + \ldots , \end{aligned} \tag{18.1.8}$$

see (Bamberger, 1978; Lachmann, 1983), where the dead time $d = T_\mathrm{D}/T_0$ has been introduced into the formulation. This non-linear difference equation now allows to approximate the Volterra series by a finite number of parameters and is called an AR-Volterra series.

As a limiting case, one can derive special non-linear parametric models, the so-called *Hammerstein models*. The following derivation is based on Lachmann (1983, 1985), see also (Isermann, 1992).

18.1.2 Hammerstein Model

If no time shift of the input signals $u(k)$ is allowed, i.e. $h = 0$, then one obtains the *generalized Hammerstein model as*

$$\begin{aligned} A(q^{-1})y(k) =& c_{00} + B_1^H(q^{-1})u(k-d) + B_2^H(q^{-1})u^2(k-d) \\ &+ \ldots + B_p^H(q^{-1})u^p(k-d) . \end{aligned} \tag{18.1.9}$$

The most well known Hammerstein model is the *simple Hammerstein Model*, see Fig. 18.1, which consists of a static non-linearity governed by a polynomial of order p,

$$x^*(k) = r_0 + r_1 u(k) + r_2 u^2(k) + \ldots + r_p u^p(k) \tag{18.1.10}$$

and a linear dynamic system given by

$$A(q^{-1})y(k) = B^*(q^{-1})q^{-d}x^*(k) \tag{18.1.11}$$

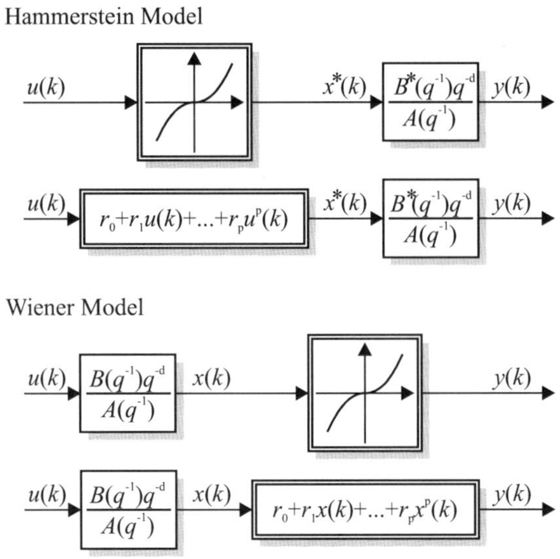

Fig. 18.1. Non-Linear Models

with

$$A(q^{-1}) = 1 + a_1 q^{-1} + \ldots + a_m q^{-m} \tag{18.1.12}$$

$$B^*(q^{-1}) = b_1^* q^{-1} + \ldots + b_m^* q^{-m} \; , \tag{18.1.13}$$

see (Hammerstein, 1930). Hence

$$\begin{aligned} A(q^{-1})y(k) =& r_{00} + B_1^*(q^{-1})u(k-d) \\ &+ \ldots + B_p^*(q^{-1})u^p(k-d) + D(q^{-1})\nu(k) \; . \end{aligned} \tag{18.1.14}$$

with

$$r_{00} = r_0 \sum_{j=1}^{m} b_j^* \tag{18.1.15}$$

and

$$B_i^*(q^{-1}) = r_i B^*(q^{-1}) \text{ with } i = 1, 2, \ldots, p \; . \tag{18.1.16}$$

Here, the subsequent linear system can be interpreted as a MISO system, where each power of $u(k)$ feeds an input of the MISO system (Chang and Luus, 1971). If the linear part is assumed to be an MA model and furthermore identical for all powers of $u(k)$, then one obtains a finite Hammerstein model.

If an arbitrary static non-linearity drives a linear SISO system, then such a model can only be identified by a non-linear optimization algorithm. However, the number of parameters to be optimized in the non-linear search can be reduced drastically by the method of separable least squares. Here, the parameters of the non-linear subsystem are determined by non-linear optimization and the parameters of the linear subsystem are then determined by the method of least squares directly as knowledge of the non-linear model parameters allows to provide estimates for $\hat{x}^*(k)$.

If the static non-linearity is linear in parameters, one can also employ the direct method of least squares to estimate the parameters assuming that the linear model shall be known and hence it is possible to provide estimates for $\hat{x}^*(k)$ based on the measured output and knowledge of the linear system. Then, with the estimated model of the linear part, one can provide an estimate for $\hat{x}^*(k)$ based on the measured input and the model of the non-linearity. This can be employed to refine the model of the linear part and so on (Liu and Bai, 2007). An example is given in Sect. 18.1.5.

Another special model structure is obtained if several finite Hammerstein models are connected in parallel, leading to a so-called *Uryson model*.

18.1.3 Wiener Model

A Hammerstein model describes a sequential concatenation of a static non-linearity at the input followed by a linear dynamic system, whereas a Wiener model consists of a linear dynamic model followed by a static non-linearity. The generalized Wiener model is given as

$$A_1(q^{-1})y(k) + A_2(q^{-1})y^2(k) + \ldots + A_l(q^{-1})y^l(k) = c_{00} + B(q^{-1})u(k-d)\,. \tag{18.1.17}$$

If the linear transfer function

$$A(q^{-1})x(k) = B(q^{-1})q^{-d}u(k) \tag{18.1.18}$$

and the static non-linearity, given as a polynomial of order p,

$$y(k) = r_0 + r_1 x(k) + r_2 x^2(k) + \ldots + r_l x^l(k) \tag{18.1.19}$$

are connected in series, then one obtains the *simple Wiener Model* as

$$y(k) = r_0 + r_1 \frac{B(q^{-1})q^{-d}}{A(q^{-1})}u(k) + r_2\left(\frac{B(q^{-1})q^{-d}}{A(q^{-1})}\right)^2 u^2(k) + \ldots\,, \tag{18.1.20}$$

see Fig. 18.1. A finite Wiener model consequently results, if the linear model again is an MA model. A parallel connection of several finite Wiener models is called *projection-pursuit model*.

If the non-linearity is positioned between two linear transfer functions, then one obtains a *Wiener-Hammerstein model*. A *Hammerstein-Wiener model* describes the opposite case, where a linear dynamic system is enframed by two non-linearities.

For parameter estimation, those models are especially well suited that are linear in parameters. This condition is met for the parametric Volterra model and the Hammerstein model, but not for the Wiener model. Hence, Wiener models can e.g. be identified based on a non-linear optimization problem, where the squared output error $(y(k)-\hat{y}(k))^2$ is minimized. Therefore, Lachmann (1983) has proposed a model, that has a non-linearity on the output side, but still is linear in its parameters and is governed in the next section.

Crama et al (2003) presented a method to obtain initial estimates for iterative optimization algorithms. The basic idea is to first identify the linear transfer function by using the input/output data. Once an initial estimate for the linear transfer function has been obtained, one can calculate the input and output signal of the static non-linearity and hence obtain an initial estimate for the model parameters of the static non-linearity. The approach can be repeated multiple times. Hagenblad et al (2008) developed a maximum likelihood cost function that can account for both noise acting on the output as well as noise disturbing the intermediate quantity $x(k)$.

18.1.4 Model According to Lachmann

If the generalized Hammerstein model is augmented by time-delayed products of the output, then one obtains (Lachmann, 1983)

$$\begin{aligned} &A(q^{-1})y(k) + \sum_{\beta_1=0}^{h} A_{2\beta_1}(q^{-1})y(k)y(k-\beta_1) + \ldots \\ &+ \sum_{\beta_1=0}^{h}\sum_{\beta_2=\beta_1}^{h}\cdots\sum_{\beta_{p-1}=\beta_{p-2}}^{h} A_{\mathrm{p}\beta_1\beta_2\ldots\beta_{p-1}}(q^{-1})y(k)\prod_{\xi=1}^{p-1} y(k-\beta_\xi) \\ &= B(q^{-1})u(k-d) + c_{00}\,, \end{aligned} \tag{18.1.21}$$

in similarity to the parametric Volterra model with

$$A_{\mathrm{p}\beta_1\beta_2\ldots\beta_{p-1}}(q^{-1}) = a_{\mathrm{p}\beta_1\beta_2\ldots\beta_{p-1}1}q^{-1} + \ldots + a_{\mathrm{p}\beta_1\beta_2\ldots\beta_{p-1}m}q^{-m}\,. \tag{18.1.22}$$

One can see that the model is the mirrored Volterra model.

18.1.5 Parameter Estimation

If, for a non-linear process, the underlying model structure is linear in parameters, then a linear estimation problem results that can be solved by any direct estimation method, such as e.g. the method of least squares and its modifications, see Chap. 9 and Chap. 10. The following model structures are linear in parameters:

- Parametric Volterra model
- General and simple Hammerstein model
- Model according to Lachmann

All of these models have the form

$$A(q^{-1})y(k) = NL(u, y, q^{-1})\,, \tag{18.1.23}$$

For other model structures, such as e.g. the Wiener model, the estimation equation is upon introduction of the equation error, non-linear in parameters. Then, one must employ iterative parameter estimation methods, see Chap. 19.

For the direct estimation methods, one can rewrite the model in the form

$$y(k) = \boldsymbol{\psi}^{\mathrm{T}}(k)\hat{\boldsymbol{\theta}}(k-1) + e(k) , \tag{18.1.24}$$

compare (9.1.11), and apply methods such as the LS or RLS. The data vector contains the following signal values ($d = 0$):

- *Parametric Volterra model*

$$\begin{aligned}\boldsymbol{\psi}^{\mathrm{T}}(k) = \big(&-y(k-1), \ldots, u(k-1), \ldots u^2(k-1), \ldots \\ &u(k-1)u(k-2), \ldots, u^3(k-1), \ldots, u(k-1)u^2(k-2), \ldots\big)\end{aligned} \tag{18.1.25}$$

- *Generalized Hammerstein model*

$$\boldsymbol{\psi}^{\mathrm{T}}(k) = \big(-y(k-1), \ldots, u(k-1), \ldots u^2(k-1), \ldots, u^3(k-1), \ldots\big) \tag{18.1.26}$$

- *Model according to Lachmann*

$$\begin{aligned}\boldsymbol{\psi}^{\mathrm{T}}(k) = \big(&-y(k-1), \ldots, -y^2(k-1), \ldots, -y(k-1)y(k-2), \ldots \\ &-y^3(k-1), \ldots, -y(k), y^2(k-1), \ldots, u(k-1), \ldots\big)\end{aligned} \tag{18.1.27}$$

The models can be augmented by a noise form filter so that from (18.1.23) follows

$$A(q^{-1})y(k) = NL(u, y, q^{-1}) + D(q^{-1})\nu(k) . \tag{18.1.28}$$

Then for example, also the ELS method can be applied.

The conditions for an unbiased estimate by the method LS and ELS are for the parametric Volterra model and the generalized Hammerstein model the same as for linear models, i.e. there is no bias if the noise is generated by a form filter with the transfer function $1/A(q^{-1})$ or $D(q^{-1})/A(q^{-1})$ respectively. For the model according to Lachmann, in general biased estimates will occur if $n(k) \neq 0$ (Lachmann, 1983).

The conditions for parameter identifiability once again result from the condition that $\boldsymbol{\Psi}^{\mathrm{T}}\boldsymbol{\Psi}$ must be positive definite. For the identification of the generalized Hammerstein model with $p = 2$ for example, the matrix

$$\boldsymbol{H}_{22} = \begin{pmatrix} R_{\mathrm{uu}}(0) \ldots R_{\mathrm{uu}}(m-1) & R_{\mathrm{uu}^2}(0) & \ldots & R_{\mathrm{uu}^2}(m-1) \\ \ddots & & & \vdots \\ R_{\mathrm{uu}}(0) & & & \\ & R_{\mathrm{u}^2\mathrm{u}^2}(0) & & \\ & & \ddots & \\ & & & R_{\mathrm{u}^2\mathrm{u}^2}(0) \end{pmatrix} \tag{18.1.29}$$

must be positive definite. Hence the auto-correlation functions

$$R_{\mathrm{u}^i\mathrm{u}^j}(\tau) = \mathrm{E}\big\{u^i(k)u^j(k-\tau)\big\}, \text{ for } i = 1, \ldots, p,\ j = 1, \ldots, p \tag{18.1.30}$$

must be such that $\det \boldsymbol{H}_{22} > 0$. This condition is satisfied by some multi-level pseudo-random binary signals (Godfrey, 1986; Dotsenko et al, 1971; Bamberger, 1978; Lachmann, 1983). Examples of the parameter estimation for non-linear processes with continuously differentiable non-linearities can be found e.g. in (Bamberger, 1978; Haber, 1979; Lachmann, 1983). While often, these non-linear models are used for discrete-time processes, it is possible to employ the same methods to continuous-time processes as well (Rao and Unbehauen, 2006).

Example 18.1 (Parameter Estimation for the Hammerstein Model).

The non-linearity is assumed to be of second order

$$x^*(k) = r_0 + r_1 u(k) + r_2 u^2(k) \tag{18.1.31}$$

and the dynamic model of first order

$$y(k) = -a_1 y(k-1) + b_1 u(k-1) \,. \tag{18.1.32}$$

The Hammerstein model then follows as

$$y(k) = -a_1 y(k-1) + b_1 r_0 + b_1 r_1 u(k-1) + b_1 r_2 u^2(k-1) \tag{18.1.33}$$

and

$$y(k) = \boldsymbol{\psi}^{\mathrm{T}}(k)\hat{\boldsymbol{\theta}} \text{ with } \boldsymbol{\psi}^{\mathrm{T}}(k) = \big(-y(k-1)\ 1\ u(k-1)\ u^2(k-1)\big) \tag{18.1.34}$$

The parameters $\hat{\boldsymbol{\theta}}$ can directly be determined by

$$\hat{\boldsymbol{\theta}} = \big(a_1\ b_1 r_0\ b_1 r_1\ b_1 r_2\big) = \big(a_1\ b_0^*\ b_1^*\ b_2^*\big) \tag{18.1.35}$$

As can be seen, the parameter b_1 cannot be determined unambiguously and hence will be fixed to $b_1 = 1$. The parameters r_0 through r_2 can then be determined as $r_0 = b_0^*/b_1$, $r_1 = b_1^*/b_1$, and $r_2 = b_2^*/b_1$. □

18.2 Dynamic Systems with Non-Continuously Differentiable Non-Linearities

Non-continuously differentiable non-linear processes appear in mechanical systems, especially in the form of friction and backlash and in electrical systems with magnetization hysteresis. They typically have to be modelled in the time domain as the resulting differential equation is non-linear and cannot be handled by the Laplace or z-transformation.

18.2.1 Systems with Friction

In many mechanical processes, dry and viscous friction appears. If a mechanical oscillator according to Fig. 18.2 is considered, the equation of motion is

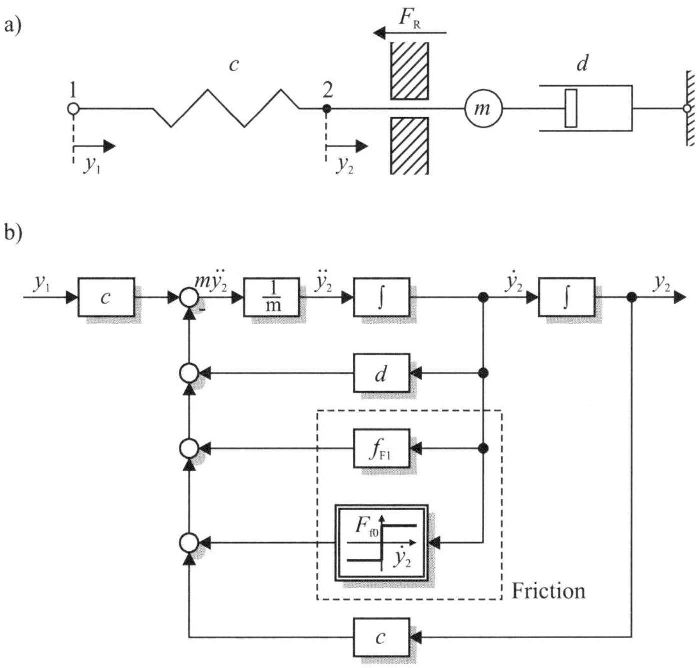

Fig. 18.2. Mechanical oscillator with friction. (**a**) schematic set-up. (**b**) block diagram

$$m\ddot{y}_2(t) + d\,\dot{y}_2(t) + cy_2(t) + F_\mathrm{F}(t) = cy_1(t)\,. \tag{18.2.1}$$

The friction force follows a Stribeck curve, see Fig. 18.3. With $f_\mathrm{F} = F_\mathrm{F}/F_\mathrm{N}$, where F_N is the normal force, it holds

$$f_\mathrm{F} = -\mu_\mathrm{C}\,\mathrm{sign}\,v + f_\mathrm{v}v + f_\mathrm{m}\mathrm{e}^{-c|v|}\,\mathrm{sign}\,v\,, \tag{18.2.2}$$

where the Friction force is then given by multiplication with the normal force F_N.

During standstill, i.e. $v = 0$, the static (adhesive) friction

$$|F_\mathrm{Fs}| \leq F_\mathrm{Fmax} = \mu_\mathrm{Smax}F_\mathrm{N} \tag{18.2.3}$$

is acting on the process. The static friction force is always as large as the attacking force and opposing in direction (and hence sign) to the attacking force. Once, the maximum force F_Fmax is exceeded, the object will start to move suddenly.

The friction forces can often be approximated by

$$F_\mathrm{F}(t) = F_\mathrm{F0}\,\mathrm{sign}\,\dot{y}(t) + f_\mathrm{F1}\dot{y}(t)\,, \tag{18.2.4}$$

where F_F0 is the velocity independent term of Coulomb or dry friction and $f_\mathrm{F1}\dot{y}(t)$ denotes the amount of velocity proportional, viscous friction (Isermann, 2005). One can incorporate this model directly into the dynamic equation of the system. However, one can also use the following identification technique:

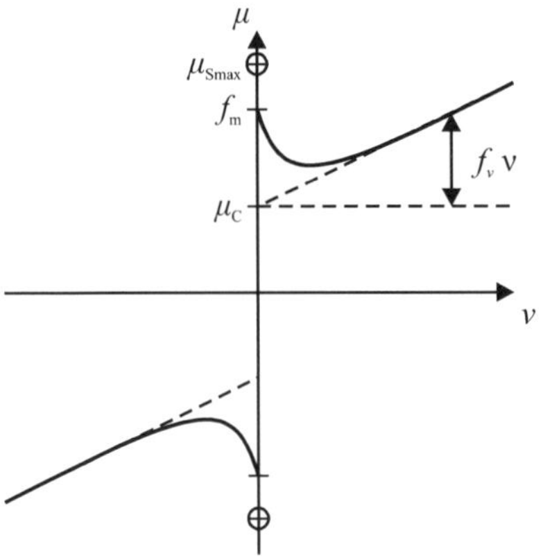

Fig. 18.3. Friction curve according to Stribeck for the dynamic friction μ_C dry or Coulomb friction, $f_v \dot{y}(t)$ viscous friction, f_m maximum amount of friction forces for $\dot{y} \rightarrow 0+$, μ_{smax} maximum holding force

For the identification of processes with friction, the hysteresis curve can directly be found pointwise by slow continuous or stepwise changes of the input signal $u(t) - y_1(t)$ and the measurement of $y(t) - y_2(t)$.

If the hysteresis curves are described by

$$\begin{aligned} y_+(u) &= K_{0+} + K_{1+}u \\ y_-(u) &= K_{0-} + K_{1-}u \; , \end{aligned} \tag{18.2.5}$$

then the parameters can be estimated from $\nu = 1, 2, \ldots, N-1$ measured points with the least squares method

$$\hat{K}_{1\pm} = \frac{N \sum u(\upsilon) y_\pm(\upsilon) - \sum u(\upsilon) \sum y_\pm(\upsilon)}{N \sum u^2(\upsilon) - \sum u(\upsilon) \sum u(\upsilon)} \tag{18.2.6}$$

$$\hat{K}_{0\pm} = \frac{1}{N}\left(\sum y_\pm(\upsilon) - \hat{K}_{1\pm} \sum u(\upsilon) \right) \tag{18.2.7}$$

As the differential equations are linear in the parameters, direct methods of parameter estimation can be applied for processes with dry and viscous friction in motion. For this, both differential and difference equations are well-suited process models. In some cases, it is expedient not only to use velocity-dependent dry friction but also velocity direction-dependent dynamic parameters, e.g. in the form of difference equations

$$y(k) = -\sum_{i=1}^{m} a_{1+} y(k-1) + \sum_{i=1}^{m} b_{i+} u(k-i) + K_{0+} \tag{18.2.8}$$

$$y(k) = -\sum_{i=1}^{m} a_{i-} y(k-i) + \sum_{i=1}^{m} b_{1-} u(k-1) + K_{0-} \tag{18.2.9}$$

K_{0+} and K_{0-} can be understood as direction-dependent offsets or DC values. Then, the following methods can be applied for the estimation of these offsets:

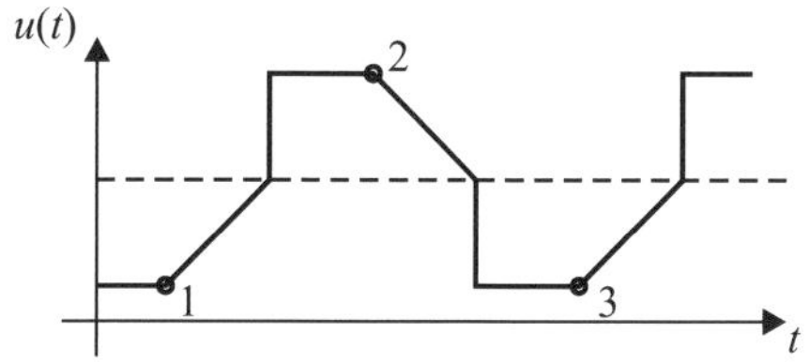

Fig. 18.4. Test signal for parameter estimation of processes with dry friction

- Implicit estimation of the offset parameters K_{0+} and K_{0-};
- Explicit estimation of the offset parameters K_{0+} and K_{0-} with generation of differences $\Delta y(k)$ and $\Delta u(k)$ and parameter estimation for

$$\Delta y(k) = -\sum_{i=1}^{m} \hat{a}_i \Delta y(k-i) + \sum_{i=1}^{m} \hat{b}_i \Delta u(k-i) \tag{18.2.10}$$

with the assumption of velocity-independent dynamic parameters $\hat{a}_i$ and $\hat{b}_i$. Then, for each direction, the parameters $\hat{K}_{0+}$ and $\hat{K}_{0-}$ have to be estimated separately.

For this parameter estimation method with a direction-dependent model, an additional identification requirement has to be considered, which is that the motion takes place in only one direction without reversal. This means that the motion has to satisfy

$$\dot{y}(t) > 0 \text{ or } \dot{y}(t) < 0 \,, \tag{18.2.11}$$

which can be tested by

$$\Delta y(k) > \varepsilon \text{ or } \Delta y(k) < -\varepsilon \tag{18.2.12}$$

for all k.

A test signal for proportional acting processes fulfilling this condition was proposed by Maron (1991), Fig. 18.4. The motion in one direction with a certain velocity is generated by a linear ascent. Then, this is followed by a step for the excitation of higher frequencies and a transition to a steady state condition. In the case of a reversal of motion, the parameter estimation has to be stopped (in Fig. 18.4 the points 1, 2, 3, ...) and has either to be restarted or continued with values according to the same direction.

The hysteresis curve can be computed from the static behavior of the model (18.2.8), (18.2.9) as

$$y_+(u) = \frac{\hat{K}_{0+}}{1+\sum \hat{a}_{i+}} + \frac{\sum \hat{b}_{i+}}{1+\sum \hat{a}_{i+}} u \tag{18.2.13}$$

$$y_-(u) = \frac{\hat{K}_{0-}}{1+\sum \hat{a}_{i-}} + \frac{\sum \hat{b}_{i-}}{1+\sum \hat{a}_{i-}} u \,. \tag{18.2.14}$$

For the verification of the parameter estimation based on the dynamic behavior, the computed characteristic curve can be compared with the measured curve resulting directly from the measured static behavior.

For rotary drives, Held (1989, 1991) has developed a special parameter estimation method that correlates the measured torque with the rotational acceleration and estimates the moment of inertia. Following from that, the characteristic curve of the friction torque can be estimated in a non-parametric form.

The methods described above for the identification of processes with friction have been successfully tested in practical applications and applied to digital control with friction compensation by Maron (1991) and Raab (1993). Further treatment is given by Armstrong-Hélouvry (1991) and Canudas de Wit (1988). The estimation of friction of ball bearings in robot drives was shown by Freyermuth (1991, 1993) and the friction of automotive suspension shock absorbers by Bußhardt (1995) and Weispfenning (1997).

18.2.2 Systems with Dead Zone

As an example again, an oscillator with backlash or dead zone of width $2y_t$ is considered, Fig. 18.5. For the oscillator without backlash, it is

$$m\ddot{y}_2(t) + d\dot{y}_2(t) + cy_2(t) = cy_3(t)\ . \tag{18.2.15}$$

The backlash can be described as follows

$$y_3(t) = \begin{cases} y_1(t) - y_t \text{ for } y_1(t) > y_t \\ 0 \text{ for } -y_t \leq y_1(t) \leq y_t \\ y_1(t) + y_t \text{ for } y_1(t) < -y_t \end{cases}\ . \tag{18.2.16}$$

This equation leads to the non-linear characteristic curve shown in Fig. 18.5b. In the case where the backlash is at one restriction with $y_1(t) > y_t$, it is

$$m\ddot{y}_2(t) + d\dot{y}_2(t) + cy_2(t) + cy_\mathrm{t} = cy_1(t) \tag{18.2.17}$$

and for the other restriction with $y_1(t) < y_\mathrm{t}$

$$m\ddot{y}_2(t) + d\dot{y}_2(t) + cy_2(t) - cy_\mathrm{t} = cy_1(t)\ . \tag{18.2.18}$$

The backlash appears as a constant with a sign depending on the sign of $y_1(t)$. For the range inside the backlash, it is $y_3(t) = 0$ and it holds that the system eigenbehavior is

$$m\ddot{y}_2(t) + d\dot{y}_2(t) + cy_2(t) = 0\ , \tag{18.2.19}$$

if point 3 (for instance, because of friction not modeled) is fixed. If point 3 is not fixed and can move arbitrarily inside the backlash, the spring forces do not apply. Then, one has to set $y_2 = y_3$ and in (18.2.19) $c = 0$.

The parameter estimation of the dead zone parameter y_t is only possible outside the dead zone, i.e. $y_1(t) < -y_\mathrm{t}$ and $y_1(t) > y_\mathrm{t}(t)$, similar as for the dry friction according to (18.2.8) and (18.2.9). As test signal, a slow motion in both directions can be applied in order to obtain the hysteresis curve (Maron, 1991). The estimation of dead zones in robot drives was motivated by Specht (1986), see also (Isermann, 2005, 2006).

Summarizing, one obtains a simplified block diagram for the regions outside the backlash shown in Fig. 18.6. The effect of the backlash in these regions can be interpreted as an offset shift of the input signal with changing sign.

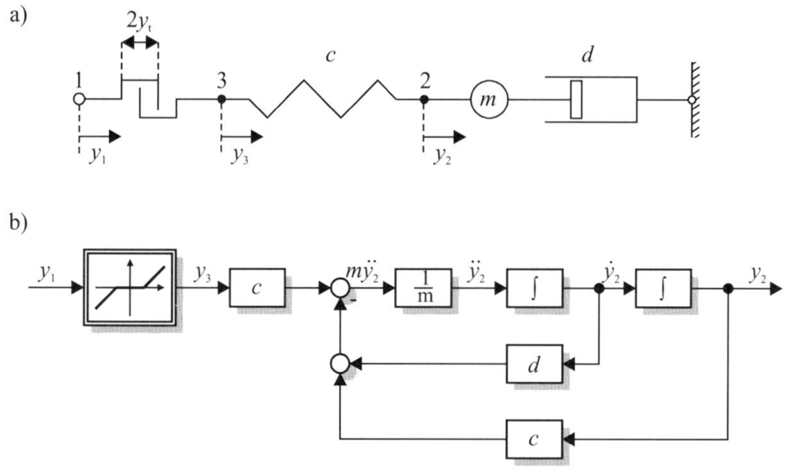

Fig. 18.5. Mechanical oscillator with backlash (dead zone). (**a**) schematic set-up. (**b**) block diagram for the cases $y_1(t) > y_t$ and $y_1(t) < -y_t$

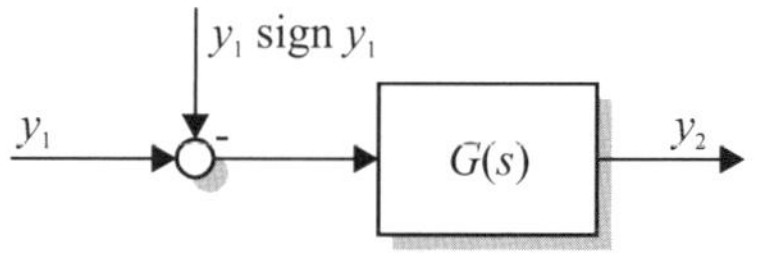

Fig. 18.6. Simplified block diagram for a linear system with backlash for $|y_1(t)| > |y_t|$

18.3 Summary

The parameter estimation methods originally developed for linear systems can as well be applied to non-linear systems, if model structures can be used that are linear in the parameters. For continuously differentiable systems, one can e.g. use Volterra series or Hammerstein models, etc. For non-linear systems which have models that are non-linear in parameters, iterative parameter estimation methods must be employed, which numerically minimize the cost function, such as e.g. the maximum likelihood cost. Here, e.g. Ninness (2009) suggested to use particle filtering to be able to determine the probability density function and hence be able to e.g. apply prediction error or maximum likelihood methods for parameter estimation. The optimality of the two-stage algorithm for Hammerstein system identification is analyzed by Wang et al (2009).

Problems

18.1. Hammerstein and Wiener Model
What are differences between the two model setups? Which model setup(s) can

be identified with direct parameter estimation methods such as the method of least squares.

18.2. Friction 1
How can the Stribeck curve be simplified in such a way so that its parameters can be identified directly by the method of least squares?

18.3. Friction 2
How can the estimation of dry and viscous friction be included in the equation for a mechanical of 2nd order? Which experiments and parameter identification method are applicable for estimation.

18.4. Backlash
Which conditions must be satisfied to be able to estimate backlash parameters?

References

Armstrong-Hélouvry B (1991) Control of machines with friction, The Kluwer international series in engineering and computer science. Robotics, vol 128. Kluwer Academic Publishers, Boston

Bamberger W (1978) Verfahren zur On-line-Optimierung des statischen Verhaltens nichtlinear, dynamisch träger Prozesse: KfK-PDV-Bericht 159. Kernforschungszentrum Karlsruhe, Karlsruhe

Bußhardt J (1995) Selbsteinstellende Feder-Dämpfer-Systeme für Kraftfahrzeuge. Fortschr.-Ber. VDI Reihe 12 Nr. 240. VDI Verlag, Düsseldorf

Chang F, Luus R (1971) A noniterative method for identification using Hammerstein model. IEEE Trans Autom Control 16(5):464–468

Crama P, Schoukens J, Pintelon R (2003) Generation of enhanced initial estimates for Wiener systems and Hammerstein systems. In: Proceedings of the 13th IFAC Symposium on System Identification, Rotterdam, The Netherlands

Dotsenko VI, Faradzhev RG, Charkartisvhili GS (1971) Properties of maximal length sequences with p-levels. Automatika i Telemechanika H 9:189–194

Doyle FJ, Pearson RK, Ogunnaike BA (2002) Identification and control using Volterra models. Communications and Control Engineering, Springer, London

Eykhoff P (1974) System identification: Parameter and state estimation. Wiley-Interscience, London

Freyermuth B (1991) An approach to model based fault diagnosis of industrial robots. In: Proceedings of the IEEE International Conference on Robotics and Automation 1991, pp 1350–1356

Freyermuth B (1993) Wissensbasierte Fehlerdiagnose am Beispiel eines Industrieroboters: Fortschr.-Ber. VDI Reihe 8 Nr. 315. VDI Verlag, Düsseldorf

Gibson JE (1963) Nonlinear automatic control. McGraw-Hill, New York, NY

Godfrey KR (1986) Three-level m-sequeunces. Electron Let 2 pp 241–243

Haber R (1979) Eine Identifikationsmethode zur Parameterschätzung bei nichtlinearen dynamischen Modellen für Prozessrechner: KfK-PDV-Bericht Nr. 175. Kernforschungszentrum Karlsruhe, Karlsruhe

Haber R, Unbehauen H (1990) Structure identification of nonlinear dynamic systems – a survey on input/output approaches. Automatica 26(4):651–677

Hagenblad A, Ljung L, Wills A (2008) Maximum likelihood identification of Wiener models. Automatica 44(11):2697–2705

Hammerstein A (1930) Nichtlineare Integralgleichungen nebst Anwendungen. Acta Math 54(1):117–176

Held V (1989) Identifikation der Trägheitsparameter von Industrierobotern. Robotersysteme 5:11–119

Held V (1991) Parameterschätzung und Reglersynthese für Industrieroboter. Fortschr.-Ber. VDI Reihe 8 Nr. 275. VDI Verlag, Düsseldorf

Isermann R (1992) Identifikation dynamischer Systeme: Besondere Methoden, Anwendungen (Vol 2). Springer, Berlin

Isermann R (2005) Mechatronic Systems: Fundamentals. Springer, London

Isermann R (2006) Fault-diagnosis systems: An introduction from fault detection to fault tolerance. Springer, Berlin

Isermann R, Lachmann KH, Matko D (1992) Adaptive control systems. Prentice Hall international series in systems and control engineering, Prentice Hall, New York, NY

Lachmann KH (1983) Parameteradaptive Regelalgorithmen für bestimmte Klassen nichtlinearer Prozesse mit eindeutigen Nichtlinearitäten. Fortschr.-Ber. VDI Reihe 8 Nr. 66. VDI Verlag, Düsseldorf

Lachmann KH (1985) Selbsteinstellende nichtlineare Regelalgorithmen für eine bestimmte Klasse nichtlinearer Prozesse. at 33(7):210–218

Liu Y, Bai EW (2007) Iterative identification of Hammerstein systems. Automatica 43(2):346–354

Maron C (1991) Methoden zur Identifikation und Lageregelung mechanischer Prozesse mit Reibung. Fortschr.-Ber. VDI Reihe 8 Nr. 246. VDI Verlag, Düsseldorf

Ninness B (2009) Some system identification challenges and approaches. In: Proceedings of the 15th IFAC Symposium on System Identification, Saint-Malo, France

Raab U (1993) Modellgestützte digitale Regelung und Überwachung von Kraftfahrzeugen. Fortschr.-Ber. VDI Reihe 8 Nr. 313. VDI Verlag, Düsseldorf

Rao GP, Unbehauen H (2006) Identification of continuous-time systems. IEE Proceedings Control Theory and Applications 153(2):185–220

Schetzen M (1980) The Volterra and Wiener theories of nonlinear systems. John Wiley and Sons Ltd, New York

Specht R (1986) Ermittlung von Getriebelose und Getriebereibung bei Robotergelenken mit Gleichstromantrieben. VDI-Bericht Nr. 589. VDI Verlag, Düsseldorf

Volterra V (1959) Theory of functionals and of integral and integro-differential equations. Dover Publications, London, UK

Wang J, Zhang Q, Ljung L (2009) Optimality analysis of the two-stage algorithm for Hammerstein system identification. In: Proceedings of the 15th IFAC Symposium on System Identification, Saint-Malo, France

Weispfenning T (1997) Fault detection and diagnosis of components of the vehicle vertical dynamics. Meccanica 32(5):459–472

Canudas de Wit CA (1988) Adaptive control for partially known systems : theory and applications, Studies in automation and control, vol 7. Elsevier, Amsterdam

19

Iterative Optimization

In this chapter, numerical optimization algorithms are presented, which allow to minimize cost functions even though they are not linear in parameters.

19.1 Introduction

For the development of identification methods for parametric continuous-time models, the use of analog computers played an important role in the past. Their models were tunable and the *model adjustment techniques* or *model reference adaptive identification methods* were developed. Nowadays, these models are no longer realized on analog computers, but can rather be realized as part of computer programs or in special software tools. In the following, numerical optimization algorithms are presented, which allow to adjust parameters of models such that the model matches best with recorded measurements. So far, the parameter estimation methods have been limited mainly to models whose cost function had been linear in the parameters. In the following, methods will be presented that can also deal with cost functions that are non-linear in the parameters. This gives a great latitude in the design of the cost function and e.g. allows to directly determine physical parameters in non-linear process models instead of e.g. transfer function coefficients, where physical parameters are frequently lumped together. Also, constraints can be included such as the stability of the resulting system or the requirement that certain physical parameters are positive.

Depending on the arrangement of the model, one has different ways of determining the error, as was shown in Fig. 1.8. The *output error*

$$e(s) = y(s) - y_{\mathrm{M}}(s) = y(s) - \frac{B_{\mathrm{M}}(s)}{A_{\mathrm{M}}(s)} u(s) \tag{19.1.1}$$

leads to a *parallel model*, the *equation error*

$$e(s) = A_{\mathrm{M}}(s) y(s) - B_{\mathrm{M}}(s) u(s) \tag{19.1.2}$$

R. Isermann, M. Münchhof, *Identification of Dynamic Systems*,
DOI 10.1007/978-3-540-78879-9_19, © Springer-Verlag Berlin Heidelberg 2011

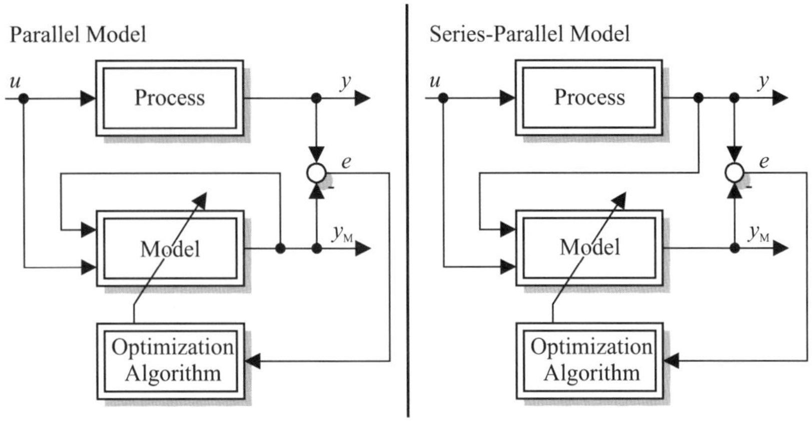

Fig. 19.1. Model setups for iterative optimization

to a *series-parallel model* and the input error

$$e(s) = \frac{A_{\mathrm{M}}(s)}{B_{\mathrm{M}}(s)} y(s) - u(s) \tag{19.1.3}$$

would lead to a *series model* or *reciprocal model*. Similar setups can be formulated in the time domain and for non-linear systems, see Fig. 19.1. A big advantage of the series-parallel model is the fact that it cannot become unstable as it does not contain a feedback loop. On the other hand, it cannot be guaranteed that the model obtained by the series-parallel setup can run as a stand-alone simulation model. Especially, for very small sample times compared to the system dynamics, the series-parallel model can pretend a model fidelity that is well above the true model fidelity. For very small sample times compared to the process dynamics, the model often collapses to simply $y_{\mathrm{M}}(k) \approx y(k-1)$.

As a cost function, one can choose any even function of the error, e.g. $f(\boldsymbol{\theta}, e) = e^2$ or $f(\boldsymbol{\theta}, e) = |e|$. One can also think about combined cost functions that rate small errors differently than large to moderate the influence of outliers. As was already discussed in Chap. 8, the use of the quadratic cost function over-emphasizes the effect of outliers. To mitigate this effect, several other cost functions have been proposed. They are tabulated in Table 19.1, see also Fig. 19.2 and (Rousseeuw and Leroy, 1987; Kötter et al, 2007).

The cost function $V(\boldsymbol{\theta}, e)$ is typically a (weighted) sum of the individual errors and the parameters are determined such that the cost function is minimized. Additional constraints, such as boundaries of the parameter space or stability of the resulting system can be formulated. In the following, it will be assumed that the cost function has a unique minimum. At the minimum, the error e does not have to vanish. The following sections will describe algorithms that allow to determine the minimum of a (non-linear) function numerically.

Table 19.1. Different cost functions, see also Fig. 19.2

Name	Cost Function
Least Squares	e^2
Huber	$\begin{cases} e^2/2 \text{ for } \lvert r\rvert \leq c \\ c(\lvert e\rvert - c/2) \text{ for } \lvert r\rvert > c \end{cases}$
Bisquare	$\begin{cases} c^2/6\Big(1-\big(1-(r/c)^2\big)^3\Big) \text{ for } \lvert r\rvert \leq c \\ c^2/6 \text{ for } \lvert r\rvert > c \end{cases}$
L1-L2	$2\left(\sqrt{1+e^2/2}-1\right)$
Absolute Value	$\lvert e\rvert$

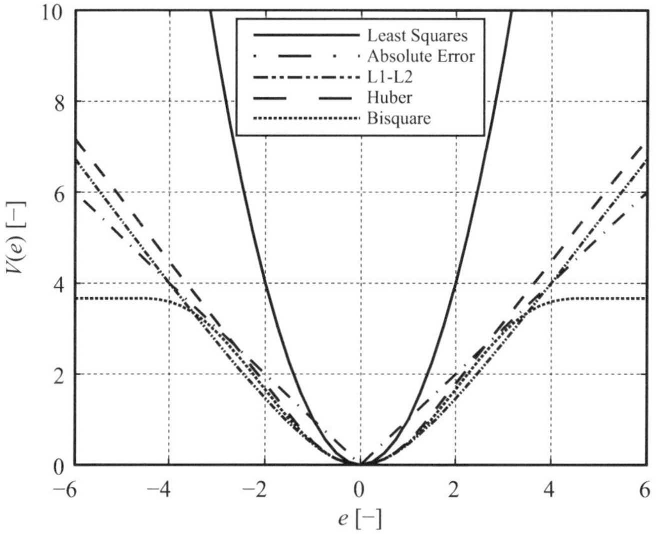

Fig. 19.2. Figure of different cost functions with $c = 1.345$ for Huber and $c = 4.6851$ for Bisquare, see also Table 19.1

19.2 Non-Linear Optimization Algorithms

There exists a huge number of algorithms for the optimization of non-linear functions. A small selection of those algorithms shall be presented in the following, focusing on those algorithms that are easy to implement and have proven to be practical for system identification. A thorough treatment of the subject of optimization algorithms can e.g. be found in the books by Vanderplaats (2005), Nocedal and Wright (2006), Snyman (2005), Ravindran et al (2006), and Boyd and Vandenberghe (2004) or with a focus on the computer implementation in the book by Press et al (2007). In these books, the reader can find detailed derivations of the algorithms. The aim of this chapter merely is to give some background information on how the different algorithms work so that one can choose the appropriate algorithm to solve a given parameter estimation problem. A survey of these as well as stochastic optimization techniques can also be found in (Nelles, 2001).

The non-linear optimization problem can in general be formulated as

$$\begin{aligned} &\min_{\boldsymbol{x}} f(\boldsymbol{x}) \\ &\text{s.t. } \boldsymbol{g}(\boldsymbol{x}) \leq \boldsymbol{0} \ , \\ &\quad\ \ \boldsymbol{h}(\boldsymbol{x}) = \boldsymbol{0} \end{aligned} \tag{19.2.1}$$

where "s.t." stands for "subject to" and the notation has been used as it is typical for optimization problems. One wants to minimize the value of $f(\boldsymbol{x})$ by adjusting the individual variables in the vector $\boldsymbol{x}$ accordingly. The optimization is subject to (s.t.) the constraints, i.e. the requirements that $\boldsymbol{g}(\boldsymbol{x}) \leq \boldsymbol{0}$ and $\boldsymbol{h}(\boldsymbol{x}) = \boldsymbol{0}$. To adapt it to the system identification framework, one must set $\boldsymbol{x} = \boldsymbol{\theta}$ and $f(\boldsymbol{x}) = V(\boldsymbol{\theta})$. In numerical optimization the parameter vector is called vector of *design variables* $\boldsymbol{x}$ and the cost or merit function is termed *objective function* f.

The constraints can be derived from conditions that have been stated for the design variables. Note that constraints are always formulated such that the resulting term satisfies the condition $g_i(\boldsymbol{x}) < 0$ or $h_j(\boldsymbol{x}) = 0$ respectively. If one for example has to ensure that $x_1 < 4$, then one will set up the the inequality constraint $g(\boldsymbol{x}) = x_1 + 4 < 0$. Similarly, if one wants to guarantee that $x_1 x_2 = x_3^2$, then one will formulate the equality constraint $h(\boldsymbol{x}) = x_1 x_2 - x_3^2 = 0$.

Many optimization algorithms are iterative in the form

$$\boldsymbol{x}(k+1) = \boldsymbol{x}(k) + \alpha \boldsymbol{s}(k) \ , \tag{19.2.2}$$

where $\boldsymbol{x}(k)$ with $k = 0, 1, \ldots$ is the *minimizing sequence*. The vector $\boldsymbol{s}(k)$ is termed the *search vector* and the value α is a measure for how far the algorithm proceeds in direction of the search vector to obtain $\boldsymbol{x}(k+1)$. The *optimal point* that should finally be approached is denoted as $\boldsymbol{x}^*$.

Obviously, one needs an *initial guess* $\boldsymbol{x}(0)$ as a starting point. The initial guess should be close to the optimum if possible. One can for example use process parameters from the specifications/data sheets or results from previously applied identification methods as starting points. If no information is available, one can also choose a random starting point or even restart the algorithm from several randomly chosen or equidistantly positioned starting points.

One must also define conditions, when the iterative algorithm should stop. A local optimum of the unconstrained function would have been reached if

$$\nabla f(\boldsymbol{x}) = \boldsymbol{0} \ , \tag{19.2.3}$$

where $\nabla f(\boldsymbol{x})$ is the gradient defined as

$$\nabla f(\boldsymbol{x}) = \left(\frac{\partial f(\boldsymbol{x})}{\partial \boldsymbol{x}}\right) = \left(\frac{\partial f(\boldsymbol{x})}{\partial x_1}\ \frac{\partial f(\boldsymbol{x})}{\partial x_2} \cdots \frac{\partial f(\boldsymbol{x})}{\partial x_p}\right)^{\mathrm{T}} . \tag{19.2.4}$$

Furthermore, the Hessian matrix $\boldsymbol{H}(\boldsymbol{x}) = \nabla^2 f(\boldsymbol{x})^2$ defined as

$$\nabla^2 f(\boldsymbol{x}) = \frac{\partial^2 f(\boldsymbol{x})}{\partial \boldsymbol{x}^{\mathrm{T}} \partial \boldsymbol{x}} = \begin{pmatrix} \frac{\partial^2 f(\boldsymbol{x})}{\partial x_1 \partial x_1} & \cdots & \frac{\partial^2 f(\boldsymbol{x})}{\partial x_p \partial x_1} \\ \vdots & & \vdots \\ \frac{\partial^2 f(\boldsymbol{x})}{\partial x_1 \partial x_p} & \cdots & \frac{\partial^2 f(\boldsymbol{x})}{\partial x_p \partial x_p} \end{pmatrix} \tag{19.2.5}$$

must be positive definite to guarantee that (at least a local) minimum has been reached. However, due to numerical inaccuracies or due to the presence of constraints, one may not reach the minimum exactly. Hence, one often uses *termination criteria* based on the size of the update step and the improvement of the cost function as

$$\|\boldsymbol{x}(k+1) - \boldsymbol{x}(k)\| \leq \varepsilon_{\mathrm{x}} \quad f(\boldsymbol{x}(k)) - f(\boldsymbol{x}(k+1)) \leq \varepsilon_{\mathrm{f}} . \tag{19.2.6}$$

One can also formulate relative convergence criteria.

The *feasible area* is the area of the design space, where all constraints are satisfied. The *usable* area is the area that leads to a reduction in the objective function. The *feasible, usable* area is where one should continue to search for the optimum in the next iteration.

Depending on what information about the cost function is used, one can differentiate

- *Zeroth order methods*: These methods only evaluate the cost function $f(\boldsymbol{x})$
- *First order methods*: Here the cost function $f(\boldsymbol{x})$ and its gradient $\partial f / \partial \boldsymbol{x}$ are employed
- *Second order methods*: They utilize the cost function $f(\boldsymbol{x})$, its gradient $\partial f / \partial \boldsymbol{x}$ and the Hessian $\partial^2 f / \partial \boldsymbol{x}^{\mathrm{T}} \partial \boldsymbol{x}$ (or an approximation to it)

The partial derivatives can be provided either analytically or by finite differencing, see Sect. 19.7.

19.3 One-Dimensional Methods

First, one-dimensional optimization techniques will be covered, which can be used to solve optimization problems with only one variable x to be chosen such that f becomes optimal. Although they are very elementary methods, yet they represent an extremely important type of optimization algorithm as even many multi dimensional optimization methods will employ a subsequent one dimensional search to determine the value α in (19.2.2) once the search vector has been established. The determination of α is called *line search*.

Point Estimation Algorithm (Zeroth Order Method)

If a function is unimodal and continuous, it can be approximated by a polynomial, which can then be used to determine the minimum. For the approximation polynomial, one can use the standard methods from calculus to determine the optimal point.

Given are the three points (x_1, f_1), (x_2, f_2), and (x_3, f_3). They can now be matched to a quadratic function

$$f(x) = a_0 + a_1(x - x_1) + a_2(x - x_1)(x - x_2) . \tag{19.3.1}$$

With the parameters being given as

$$a_0 = f_1 \tag{19.3.2}$$

$$a_1 = \frac{f_2 - f_1}{x_2 - x_1} \tag{19.3.3}$$

$$a_2 = \frac{\frac{f_3 - f_1}{x_3 - x_1} - \frac{f_2 - f_1}{x_2 - x_1}}{x_3 - x_2} , \tag{19.3.4}$$

the optimal solution follows with $\partial f(x)/\partial x = 0$ as

$$x^* = \frac{x_2 - x_1}{2} - \frac{a_1}{2a_2} . \tag{19.3.5}$$

This algorithm has the big advantage that only a few function evaluations are required. On the other hand, one can give no guarantee concerning the quality of the estimate, which can become a problem especially for highly non-linear functions.

One can also use higher order approximations, such as e.g. a cubic approximation, but with an increase in the order of the approximating polynomial, the computational effort for finding the minimum also grows. Furthermore, the number of local extrema increases as the polynomial order increases. Although point estimation methods should theoretically be superior to region elimination methods, this has not always proven true in practical applications. Region elimination methods benefit from their high robustness in practical applications and should be used for an initial interval refinement. The point estimation can then be used to precisely find the exact solution, once the solution has been bounded sufficiently well.

Region Elimination Algorithm (Zeroth Order Method)

Region elimination methods eliminate in each iteration certain subintervals of the region of interest and hence reduce the interval that has to be searched through to find the optimum.

The optimum is assumed to be bounded by x_L and x_R and the objective function values f_L and f_R are assumed to be known. Then

1. Evaluate f_M at the midpoint x_M given by

$$x_M = \frac{x_L + x_R}{2}$$

2. Determine the points x_1 and x_2 as

$$x_1 = x_L + \frac{x_R - x_L}{4} \text{ and } x_2 = x_R - \frac{x_R - x_L}{4}$$

3. Determine f_1 and f_2
4. Compare f_1 and f_M.
   If $f_1 < f_M$ then $x_R = x_M$, $x_M = x_1$
   Else compare f_2 and f_M
   If $f_2 < f_M$ then $x_L = x_M$, $x_M = x_2$
   If $f_2 > f_M$ then $x_L = x_1$, $x_R = x_2$
5. Check for convergence, else repeat from step 2

As the algorithm can eliminate half of the bounding interval in each iteration, the necessary number of steps can be determined a priori, if a termination tolerance on the interval length is given as ε_x.

Golden Section Search (Zeroth Order Method)

The golden section search is the most popular region elimination algorithm and is similar to the bisection algorithm presented thereafter in this section. It consists of the following steps:

1. Evaluate

$$x_1 = (1 - \tau)x_L + \tau x_R$$
$$x_2 = \tau x_L + (1 - \tau)x_R$$

   and determine the corresponding values of the objective function f_1 and f_2
2. If $f_1 > f_2$, then $x_L = x_1$, $x_1 = x_2$, $f_1 = f_2$, and

$$x_2 = \tau x_L + (1 - \tau)x_R \ .$$

   Then evaluate f_2 at x_2.
   If $f_1 < f_2$, then $x_R = x_2$, $x_2 = x_1$, $f_2 = f_1$, and

$$x_1 = (1 - \tau)x_L + \tau x_R \ .$$

   Then evaluate f_1 at x_1
3. Repeat from step 2 until converged

For the golden section search algorithm, $\tau = 0.38197$ and $1 - \tau = 0.61803$ have been chosen. These numbers guarantee that the ratio of the distance of the points remains constant. Although a choice of $\tau = 0.5$ would bound the maximum tighter in the same number of iterations, one often prefers the golden section ratio as it is felt to be beneficial because of its more conservative bounding of the optimal point.

Bisection Algorithm (First Order Method)

The bisection algorithm is a simple method that is based on evaluating the objective function and the first derivative of it. The first derivative shall be denoted as $f' = \partial f/\partial x$ in the interest of a compact notation of the algorithm. The method is very simple, it cuts the starting interval (x_L, x_R) in half in each iteration. The objective function must be unimodal and it must be differentiable so that f' exists. One only needs the first derivative in contrast to the Newton-Raphson algorithm, which will be presented later.

1. Find two points x_L and x_R that bound the minimum such that $f'(x_L) < 0$ and $f'(x_R) > 0$
2. Find the mid-point
$$x_M = \frac{x_L + x_R}{2}$$
3. Evaluate $f'(x_M)$
4. If $f'(x_M) > 0$, then eliminate the right half of the interval $(x_{R,new} = x_{M,old})$, otherwise eliminate the left half of the interval $(x_{L,new} = x_{M,old})$.
5. Check for convergence and if necessary, continue with step 2

Newton-Raphson Algorithm (Second Order Method)

The *Newton-Raphson* algorithm requires knowledge of the objective function as well as the first and second derivative of it with respect to the unknown parameters. The method is very efficient, if started close to the optimum. Further away from the optimum, it has the danger of diverging away from the optimum under adverse conditions.

1. Start from a point x
2. Move on to the new point
$$x^* = x - \frac{f'(x)}{f''(x)}$$
3. Evaluate $f^*(x^*)$
4. Check for convergence and if necessary, continue with step 2

19.4 Multi-Dimensional Optimization

Most often, one is confronted with multi-dimensional optimization problems, where the objective function f depends on a vector of design variables $\boldsymbol{x}$. In this case, the following methods can be applied. All of the following methods will in their original formulation solve unconstrained optimization problems. The inclusion of constraints will be treated later, see Sect. 19.5.

19.4.1 Zeroth Order Optimizers

Zeroth order methods do only employ information about the function variable. They are well suited in cases, where information about the gradient is not available. Furthermore, they have the advantage of being robust and simple to implement. For a survey on derivative free optimization see e.g. (Conn et al, 2009). Other zeroth order methods include genetic algorithms (Mitchell, 1996), simulated annealing (Schneider and Kirkpatrick, 2006), and biologically inspired algorithms such as swarm optimization (Kennedy et al, 2001) to name some techniques.

Downhill Simplex Algorithm (Nelder-Mead)

In an n-dimensional space, a simplex is a polyhedron of $N + 1$ equidistant points from its vertices. For example, in a two-dimensional space, it is a triangle. In each iteration of the downhill simplex algorithm, the worst point is projected through its respective opposing vertex, which results in a new simplex. Also, the size of the simplex can be adjusted to better bound the optimum. The downhill simplex algorithm, although very simple, has also proven to be very robust and to perform well in the presence of noise. Note that there is also a simplex algorithm used in linear programming, which should not be confused with this downhill simplex algorithm.

1. Order the points according to their ascending values of their objective function values as $f(\boldsymbol{x}_1) \leq f(\boldsymbol{x}_2) \leq \ldots \leq f(\boldsymbol{x}_{n+1})$
2. *Reflexion*: Mirror the worst performing point through the center of gravity of the remaining points $\boldsymbol{x}_0$ as $\boldsymbol{x}_\mathrm{R} = \boldsymbol{x}_0 + (\boldsymbol{x}_0 - \boldsymbol{x}_{n+1})$. If the objective function for the reflected point $f(\boldsymbol{x}_\mathrm{R})$ is better than $f(\boldsymbol{x}_n)$, but not better than $f(\boldsymbol{x}_1)$ then replace $\boldsymbol{x}_{n+1}$ by $\boldsymbol{x}_\mathrm{R}$ and go to step 6
3. *Expansion*: If the reflected point is the new optimal point, i.e. better than $f(\boldsymbol{x}_1)$, then it might be beneficial to extend the search in this direction by expanding the simplex as $\boldsymbol{x}_\mathrm{E} = \boldsymbol{x}_0 + \gamma(\boldsymbol{x}_0 - \boldsymbol{x}_{n+1})$ with $\gamma > 1$. If the expanded point $\boldsymbol{x}_\mathrm{E}$ yields an even lower objective function value, then expand the simplex and keep this point by replacing $\boldsymbol{x}_{n+1}$ by $\boldsymbol{x}_\mathrm{E}$ and go to step 6, else keep the size of the simplex and replace $\boldsymbol{x}_{n+1}$ by $\boldsymbol{x}_\mathrm{R}$ and go to step 6
4. *Contraction*: The objective function value could not yet be reduced as even the reflected point is still the worst point $f(\boldsymbol{x}_n)$. Therefore, one should try not to probe too far and contract the simplex by determining the point $\boldsymbol{x}_\mathrm{C} = \boldsymbol{x}_0 + \varrho(\boldsymbol{x}_0 - \boldsymbol{x}_{n+1})$ with $\varrho < 1$. If the contracted point $\boldsymbol{x}_\mathrm{C}$ is better than x_{n+1}, then contract the simplex and keep this point by replacing $\boldsymbol{x}_{n+1}$ by $\boldsymbol{x}_\mathrm{C}$ and go to step 6.
5. *Reduction*: At this point is seems likely that the optimum already is inside the simplex. Hence its size must be reduced to bound the optimum. All but the best point are replaced by $\boldsymbol{x}_i = \boldsymbol{x}_1 + \sigma(\boldsymbol{x}_i - \boldsymbol{x}_1)$ for $i = 2, \ldots, n+1$
6. Terminate if the convergence tolerances are met, else go to step 1.

Example 19.1 (Downhill Simplex Algorithm for Determining the Parameters of a First Order System).

In order to illustrate the general functioning principle of the downhill simplex algorithm, a two dimensional parameter estimation problem has been chosen, where the gain K and time constant T of the transfer function

$$G(s) = \frac{K}{Ts + 1} \tag{19.4.1}$$

are determined based on the output error $e(t) = y(t) - \hat{y}(t)$. The expansion of the simplex has been disabled for illustrative purposes.

The cost function has been set up as

$$V = \sum_{k=0}^{N-1} (y(k) - \hat{y}(k))^2 \,. \tag{19.4.2}$$

For the determination of the model response $\hat{y}(k)$, the process is simulated on a computer, hence the index (k) is used, even though the model is formulated in continuous-time. As a side-note, it should be mentioned that this cost function is the same as for the maximum likelihood estimation provided that a white noise is disturbing the output $y(k)$ and that the parallel model is used. This also provides the theoretically optimal solution (Ljung, 2009).

For the process model, the parameters $K = 2$ and $T = 0.5\,\mathrm{s}$ have been used. The input and output are shown in Fig. 19.3. Figure 19.4 shows how the algorithm approaches the minimum of the objective function from a given starting point, which had been chosen arbitrarily as $K = 0.5$, $T = 1.8\,\mathrm{s}$. As the minimum is approached, the simplex is reduced several times. This example will also be solved with the gradient descent algorithm to allow a comparison between zeroth order and first order methods. □

19.4.2 First Order Optimizers

First order methods also employ information about the gradient of the objective function, denoted as $\nabla f(\boldsymbol{x}) = \partial f(\boldsymbol{x})/\partial \boldsymbol{x}$. The gradient can be provided analytically or by means of finite differencing.

Gradient Descent (First Order Method)

The gradient descent proceeds in each search step in the negative direction of the gradient $\nabla f(\boldsymbol{x})$ as

$$\boldsymbol{x}(k+1) = \boldsymbol{x}(k) + \alpha \boldsymbol{s}(k) \tag{19.4.3}$$

$$\boldsymbol{s}(k) = -\frac{\nabla f(\boldsymbol{x}(k))}{\|\nabla f(\boldsymbol{x}(k))\|_2} \,, \tag{19.4.4}$$

where $\boldsymbol{s}(k)$ is the search vector and the step size α can be determined by a subsequent one dimensional search in the negative direction of the gradient, see Sect. 19.3. The

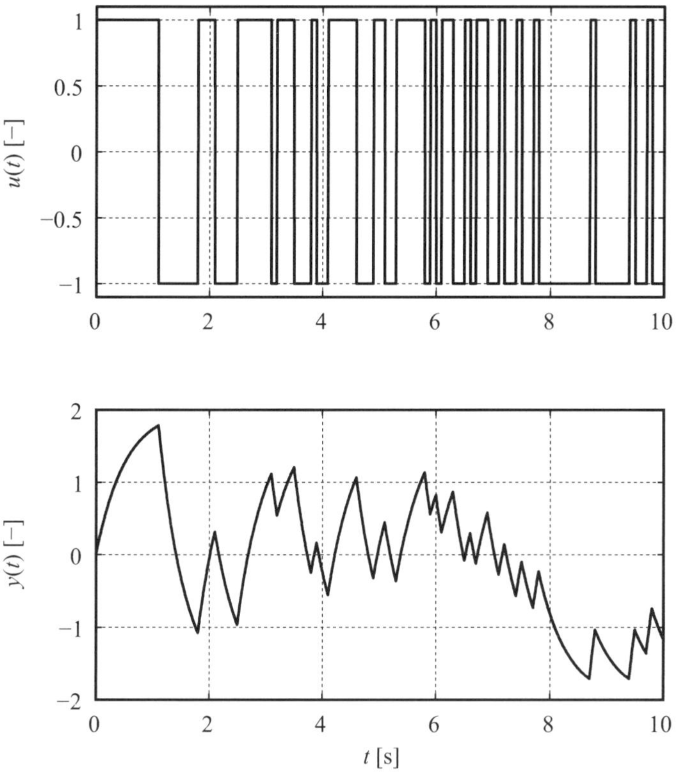

Fig. 19.3. Input $u(k)$, process output $y(k)$, and model output $\hat{y}(k)$ for the example first order system

gradient is typically normalized to unit length, i.e. divided by the Euclidian norm of the vector, i.e. $\|\nabla f(\boldsymbol{x}(k))\|_2$. The method is often called *steepest descent* as well. Its main disadvantage is the fact that it can become very slow e.g. for badly scaled design variables (Press et al, 2007).

An extension is the Fletcher-Reeves algorithm that has quadratic convergence for quadratic objective functions. After a first search in the direction of the steepest descent, the subsequent search vectors are given by

$$\boldsymbol{s}(k) = -\nabla f(\boldsymbol{x}(k)) + \beta(k)\boldsymbol{s}(k-1) \tag{19.4.5}$$

with

$$\beta(k) = \frac{\big(\nabla f(\boldsymbol{x}(k))\big)^{\mathrm{T}}\big(\nabla f(\boldsymbol{x}(k))\big)^{\mathrm{T}}}{\big(\nabla f(\boldsymbol{x}(k-1))\big)^{\mathrm{T}}\big(\nabla f(\boldsymbol{x}(k-1))\big)^{\mathrm{T}}}\,. \tag{19.4.6}$$

Besides the quadratic convergence for quadratic functions, it is advantageous that the method in general reuses information from past steps. It is also reliable if employed far from the optimum and accelerates as the optimum is approached. Due to

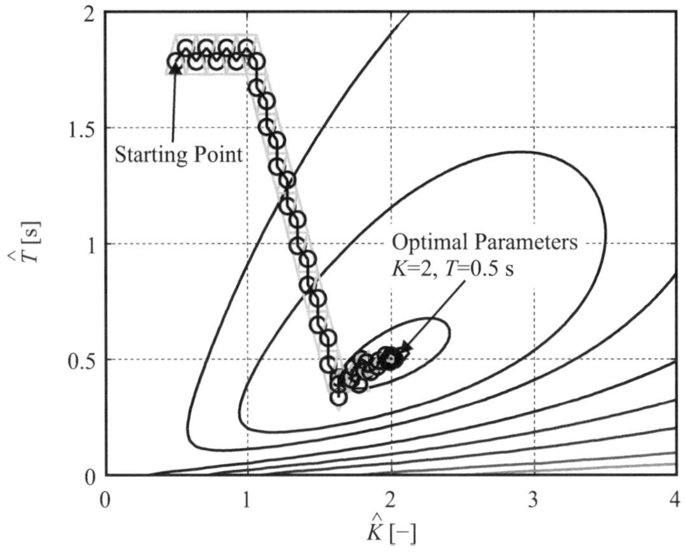

Fig. 19.4. Convergence of the downhill simplex algorithm with contours of the cost function

numerical imprecision, it might be necessary to restart the Fletcher-Reeves algorithm during the optimization run from time to time. In this case, one starts with a step in the direction of the gradient. A modification of the algorithm is given by

$$\beta(k) = \frac{\left(\nabla f(\boldsymbol{x}(k)) - \nabla f(\boldsymbol{x}(k-1))\right)^{\mathrm{T}}\left(\nabla f(\boldsymbol{x}(k))\right)}{\left(\nabla f(\boldsymbol{x}(k-1))\right)^{\mathrm{T}}\left(\nabla f(\boldsymbol{x}(k-1))\right)^{\mathrm{T}}}\,, \tag{19.4.7}$$

which is the *Polak-Ribiere algorithm.*

Example 19.2 (Gradient Descent for Determining the Parameters of a First Order System).

Figure 19.5 shows the convergence of the gradient descent algorithm applied to the example first order system that was already identified using a downhill simplex algorithm. One can see that the gradient descent algorithm converges in much less iterations, however the calculation of the gradient can become cumbersome for larger and more complex problems. Once again, the system parameters are identified correctly but in much less iterations. □

19.4.3 Second Order Optimizers

Second order methods are very fast and efficient, but suffer from the disadvantage that the second order derivatives must be known to determine the Hessian matrix. For many problems they cannot be supplied analytically because of the complexity of the calculation of the respective second partial derivatives of the cost function with respect to the individual parameters. To overcome this disadvantage, several

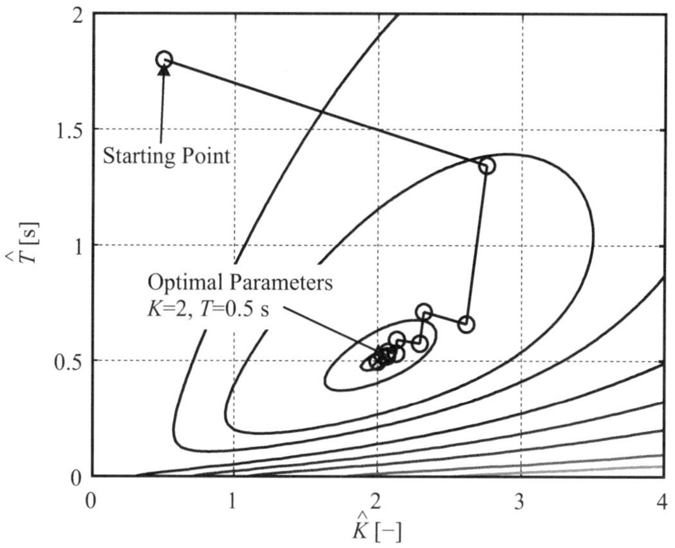

Fig. 19.5. Convergence of the steepest descent with contours of the cost function

approximation techniques have been proposed as will be described directly after the derivation of the Newton algorithm. These approximation techniques try to approximate the Hessian based on gradient information alone.

Newton algorithm

The Newton algorithm, sometimes also referred to as Newton-Raphson algorithm for finding zeros of a function, can be derived by considering the second order approximation of $f(\boldsymbol{x})$ at $\boldsymbol{x}_0$ with

$$f(\boldsymbol{x}_0+\Delta\boldsymbol{x}) \approx f(\boldsymbol{x}_0)+\nabla f(\boldsymbol{x}_0)\Delta\boldsymbol{x}+\frac{1}{2}\Delta\boldsymbol{x}^{\mathrm{T}}\nabla^2 f(\boldsymbol{x}_0)\Delta\boldsymbol{x}\;. \tag{19.4.8}$$

For the optimal point, the gradient with respect to the parameter vector $\boldsymbol{x}$ must vanish, hence

$$\begin{aligned}\frac{\partial f(\boldsymbol{x}_0+\Delta\boldsymbol{x})}{\partial\Delta\boldsymbol{x}} &= \nabla f(\boldsymbol{x}_0)+\nabla^2 f(\boldsymbol{x}_0)\Delta\boldsymbol{x}=0\\ \Leftrightarrow \Delta\boldsymbol{x} &= -\left(\nabla^2 f(\boldsymbol{x}_0)\right)^{-1}\nabla f(\boldsymbol{x}_0)\;,\end{aligned} \tag{19.4.9}$$

which yields the Newton step. If the function $f(\boldsymbol{x})$ is indeed quadratic, then the Newton step

$$\boldsymbol{x}(k+1)=\boldsymbol{x}(k)+\boldsymbol{s}(k) \tag{19.4.10}$$

$$\boldsymbol{s}(k)=-\left(\nabla^2 f(\boldsymbol{x}(k))\right)^{-1}\nabla f(\boldsymbol{x}(k)) \tag{19.4.11}$$

will lead directly to the optimum $\boldsymbol{x}^*$ with $\alpha = 1$. As pointed out in Sect. 22.3, the direct inversion of a matrix ($\nabla^2 f(\boldsymbol{x}(k))$ in the case at hand) can be numerically problematic. Furthermore, the Hessian matrix is seldom available, see also the Quasi-Newton algorithms in the next section.

If the function can be approximated sufficiently well by a quadratic function, then this approach will yield a point close to the optimum. The algorithm has to be carried out iteratively. A big advantage of the second order methods is their speed and efficiency. However, the second oder derivatives must be known and it must be ensured that the Hessian matrix is positive definite, which may not be the case far from the optimal point. In contrast to the first order (i.e. gradient based) algorithms, the size of the update step is automatically determined as $\alpha = 1$ for the truly quadratic problem and hence the unit step length must in theory not be determined by a subsequent one dimensional line search. If it is however determined by a subsequent line search, then $\alpha = 1$ is a good initial value for the line search.

Quasi-Newton algorithms

The main disadvantage of the Newton algorithm is that information about the Hessian matrix is seldom available. Hence, many Quasi-Newton algorithms have been derived that try to approximate the Hessian from first order information alone.

Here, the update step is given as

$$\boldsymbol{x}(k+1) = \boldsymbol{x}(k) + \boldsymbol{s}(k) \tag{19.4.12}$$

$$\boldsymbol{s} = -\boldsymbol{H}(k)\,\nabla f(\boldsymbol{x}(k))\,, \tag{19.4.13}$$

where $\boldsymbol{H}(k)$ is *not* the Hessian, but an approximation that approaches the *inverse* of the Hessian matrix. The matrix $\boldsymbol{H}(k)$ is initialized as $\boldsymbol{H}(0) = \boldsymbol{I}$. The update step is then given as

$$\boldsymbol{H}(k+1) = \boldsymbol{H}(k) + \boldsymbol{D}(k) \tag{19.4.14}$$

with

$$\begin{aligned}\boldsymbol{D}(k) = {} & \frac{\sigma + \theta\tau}{\sigma^2}\boldsymbol{p}(k)\boldsymbol{p}(k)^{\mathrm{T}} + \frac{\theta - 1}{\tau}\boldsymbol{H}(k)\boldsymbol{y}(\boldsymbol{H}(k)\boldsymbol{y})^{\mathrm{T}} \\ & - \frac{\theta}{\sigma}\left(\boldsymbol{H}(k)\boldsymbol{y}\boldsymbol{p}^{\mathrm{T}} + \boldsymbol{p}(\boldsymbol{H}(k)\boldsymbol{y})^{\mathrm{T}}\right)\end{aligned} \tag{19.4.15}$$

and the parameters

$$\sigma = \boldsymbol{p}^{\mathrm{T}}\boldsymbol{y} \tag{19.4.16}$$

$$\tau = \boldsymbol{y}^{\mathrm{T}}\boldsymbol{H}(k)\boldsymbol{y} \tag{19.4.17}$$

$$\boldsymbol{p} = \boldsymbol{x}(k) - \boldsymbol{x}(k-1) \tag{19.4.18}$$

$$\boldsymbol{y} = \nabla f(\boldsymbol{x}(k)) - \nabla f(\boldsymbol{x}(k-1))\,. \tag{19.4.19}$$

The parameter θ allows to switch between different modifications. For $\theta = 0$, one obtains the DFP (Davidon, Fletcher, Powell) and for $\theta = 1$ the BFGS (Broyden, Fletcher, Goldfarb, Shanno) algorithm, which vary on how they approximate the inverse Hessian matrix.

Gauss-Newton algorithm

For an objective function that is based on the sum of squares of an error function $e_k(\boldsymbol{x})$, one can find a special approximation of the Hessian matrix which leads to the Gauss-Newton algorithm. The objective function shall be given as

$$f(\boldsymbol{x}) = \sum_{k=1}^{N} e_k^2(\boldsymbol{x}) \,. \tag{19.4.20}$$

Then, the elements of the gradient $\partial f(\boldsymbol{x})/\partial \boldsymbol{x}$ are given as

$$\frac{\partial f(\boldsymbol{x})}{\partial x_i} = \frac{\partial}{\partial x_i}\left(\sum_{k=1}^{N} e_k^2(\boldsymbol{x})\right) = \sum_{k=1}^{N} \frac{\partial}{\partial x_i} e_k^2(\boldsymbol{x}) = \sum_{k=1}^{N} 2e_k(\boldsymbol{x})\frac{\partial}{\partial x_i} e_k(\boldsymbol{x}) \tag{19.4.21}$$

and the elements of the Hessian matrix $\boldsymbol{H}(\boldsymbol{x})$ are given as

$$\begin{aligned} H_{i,j}(\boldsymbol{x}) &= \frac{\partial^2 f(\boldsymbol{x})}{\partial x_i \partial x_j} = \frac{\partial}{\partial x_j}\left(\sum_{k=1}^{N} 2e_k^2(\boldsymbol{x})\frac{\partial}{\partial x_i} e_k(\boldsymbol{x})\right) \\ &= 2\sum_{k=1}^{N}\left(\frac{\partial}{\partial x_j} e_k(\boldsymbol{x})\right)\left(\frac{\partial}{\partial x_i} e_k(\boldsymbol{x})\right) + \underbrace{2\sum_{k=1}^{N} \frac{\partial^2}{\partial x_i \partial x_j} e_k(\boldsymbol{x})}_{\approx 0} \\ &\approx 2\sum_{k=1}^{N}\left(\frac{\partial}{\partial x_j} e_k(\boldsymbol{x})\right)\left(\frac{\partial}{\partial x_i} e_k(\boldsymbol{x})\right) . \end{aligned} \tag{19.4.22}$$

Here, the Hessian can be approximated by first order information alone.

Using the Jacobian, defined as

$$\boldsymbol{J} = \frac{\partial \boldsymbol{e}}{\partial \boldsymbol{x}} \,, \tag{19.4.23}$$

one can write the update step as

$$\boldsymbol{x}(k+1) = \boldsymbol{x}(k) + \boldsymbol{s}(k) \tag{19.4.24}$$

$$\boldsymbol{s}(k) = -\left(\boldsymbol{J}^{\mathrm{T}}\boldsymbol{J}\right)^{-1}\left(\boldsymbol{J}^{\mathrm{T}}\boldsymbol{e}\right) . \tag{19.4.25}$$

Then, the approximate of the Hessian matrix is given as $\boldsymbol{H} \approx 2\boldsymbol{J}^{\mathrm{T}}\boldsymbol{J}$ and the gradient as $\nabla^2 f(\boldsymbol{x}(k)) = 2\boldsymbol{J}^{\mathrm{T}}\boldsymbol{e}$. Once again, the direct inversion is numerically critical and should be avoided. Hence, one can also determine the search vector from the equation

$$\left(\boldsymbol{J}^{\mathrm{T}}\boldsymbol{J}\right)^{-1}\boldsymbol{s}(k) = -\boldsymbol{J}^{\mathrm{T}}\boldsymbol{e} \tag{19.4.26}$$

with the approaches that have been used in Sect. 22.3 to solve the problem of least squares.

Levenberg-Marquart algorithm

A modification to the Gauss-Newton is employed by the *Levenberg-Marquart algorithm*, which is given as

$$\boldsymbol{x}(k+1) = \boldsymbol{x}(k) + \boldsymbol{s}(k) \tag{19.4.27}$$

$$\boldsymbol{s}(k) = -\left(\boldsymbol{J}^{\mathrm{T}}\boldsymbol{J} + \lambda \boldsymbol{I}\right)^{-1}\left(\boldsymbol{J}^{\mathrm{T}}\boldsymbol{e}\right), \tag{19.4.28}$$

where $\boldsymbol{I}$ is the unit matrix. This technique is also called *trust-region algorithm*. The additional term in the denominator allows to rotate the update search vector $\boldsymbol{s}(k)$ in the direction of the steepest descent. For optimal results, the parameter λ has to be updated for each search step. However, typically, a heuristic approach is used and λ is increased if the algorithm is divergent and is decreased again if the algorithm is convergent. By the introduction of the term $\lambda \boldsymbol{I}$ and appropriate choice, one can enforce descending function values in the optimization sequence. Furthermore, the term $\lambda \boldsymbol{I}$ can be used to increase the numerical stability of the algorithm.

19.5 Constraints

Often, constraints have to be imposed as part of the optimization problem. One typical constraint is stability of the model as an unstable model may not be simulated correctly and may not be useful for the later application. Also, in many cases the model is made up of physical parameters such as e.g. spring stiffnesses or masses, which can for physical reasons not be negative and hence such physically meaningless models should be excluded from the design space. Also, the inclusion of constraints limits the flexibility of the model and hence can reduce the variance of the model at the price of an increased bias, see also the bias-variance dilemma in Sect. 20.2.

19.5.1 Sequential Unconstrained Minimization Technique

One approach to incorporate constraints into the optimization problem is the SUMT technique, i.e. *sequential unconstrained minimization technique*. Here, a pseudo-objective function is formulated as

$$\Phi(\boldsymbol{x}, r_{\mathrm{P}}) = f(\boldsymbol{x}) + r_{\mathrm{P}}\, p(\boldsymbol{x}), \tag{19.5.1}$$

where $p(\boldsymbol{x})$ is termed the *penalty function* and introduces the constraints into the cost function. r_{P} denotes the scalar *penalty multiplier* and is typically increased as the optimization goes on to put more and more emphasis on avoiding constraint violations. Several approaches are possible as described in the following.

Exterior Penalty Function Method

The exterior penalty function method is given by

$$p(\boldsymbol{x}) = \sum_{i=1}^{m} \left(\max(0, g_j(\boldsymbol{x}))\right)^2 + \sum_{k=1}^{l} \left(h_k(\boldsymbol{x})\right)^2 . \tag{19.5.2}$$

As was stated in the introduction, the entries $g_j(\boldsymbol{x})$ represent inequality constraints, i.e. must satisfy the condition $g_j(\boldsymbol{x}) \leq \mathbf{0}$ and the entries $h_i(\boldsymbol{x})$ represent equality constraints, i.e. $h_i(\boldsymbol{x}) = \mathbf{0}$.

Here, a penalty is only imposed if the corresponding constraint is violated, hence the term *exterior*. The optimal design is therefore always approached from the infeasible side, premature stopping will cause an infeasible and hence often unusable solution. One typically starts with small values of the penalty multiplier to allow infeasible designs at the beginning and hence to also permit the initial exploration of infeasible regions. r_{P} can start with e.g. 0.1 and approach values of 10 000 ... 100 000.

Interior Penalty Function Method

The interior penalty function method is given by

$$p(\boldsymbol{x}) = r'_{\mathrm{P}} \sum_{i=1}^{m} \frac{-1}{g_j(\boldsymbol{x})} + r_{\mathrm{P}} \sum_{k=1}^{l} (h_k(\boldsymbol{x}))^2 . \tag{19.5.3}$$

The penalty term is positive as long as the solution is feasible and approaches infinity as a constraint is violated. Here, both terms are weighted differently by r_{P} and r'_{P}. The interior penalty function should be used if one can start in the feasible area at a point not too far from the optimum. Then, the algorithm approaches the optimal solution from the feasible region. The exterior penalty function should be used if the initial guess is infeasible and/or far from the optimum. In contrast, one should never starts the interior penalty function algorithm from an infeasible initial guess as the algorithm *likes* infeasible solutions.

Extended Interior Penalty Function Method

The extended interior penalty function method is governed by

$$p(\boldsymbol{x}) = r'_{\mathrm{P}} \sum_{i=1}^{m} \tilde{g}_j(\boldsymbol{x})$$
$$\text{with } \tilde{g}_j = \begin{cases} -\dfrac{1}{g_j(\boldsymbol{x})} & \text{if } g_j(\boldsymbol{x}) \leq -\varepsilon \\ -\dfrac{2\varepsilon - g_j(\boldsymbol{x})}{\varepsilon^2} & \text{if } g_j(\boldsymbol{x}) > -\varepsilon \end{cases} , \tag{19.5.4}$$

where ε is a small positive number and the penalty function is no longer discontinuous at the constraint boundary.

Quadratic Extended Interior Penalty Function Method

The quadratic extended interior penalty function method calculates the penalty function as

$$p(\boldsymbol{x}) = r'_{\mathrm{P}} \sum_{i=1}^{m} \tilde{g}_j(\boldsymbol{x})$$
$$\text{with } \tilde{g}_j = \begin{cases} -\dfrac{1}{g_j(\boldsymbol{x})} \text{ if } g_j(\boldsymbol{x}) \leq \varepsilon \\ -\dfrac{1}{\varepsilon}\left(\left(\dfrac{g_j(\boldsymbol{x})}{\varepsilon}\right)2 - 3\dfrac{g_j(\boldsymbol{x})}{\varepsilon} + 3\right) \text{ if } g_j(\boldsymbol{x}) > \varepsilon \,, \end{cases} \tag{19.5.5}$$

where now also second order minimization techniques can be applied as the second derivatives are now also continuous. However, this comes at the expense that the degree of non-linearity of the penalty function increases.

Other Penalty Functions

Table 19.2 lists some penalty functions that can also be used.

Example 19.3 (Downhill Simplex Algorithm for Constrained Identification of the Frequency Response of the Three-Mass Oscillator based on a Physical Model).

In this example, the physical parameters of the Three-Mass Oscillator shall be determined from the frequency response that has been identified by means of the orthogonal correlation.

The identification has been carried out in three steps: First, the frequency response has been determined by the orthogonal correlation as described in Sect. 5.5.2. Then, the amplitude of the experimentally determined frequency response has been used to determine the physical parameters of the state space model, i.e. the vector of estimated parameters was given as

$$\boldsymbol{\theta}^{\mathrm{T}} = \begin{pmatrix} J_1 & J_2 & J_3 & d_1 & d_2 & d_3 & c_1 & c_2 \end{pmatrix} . \tag{19.5.6}$$

As a cost function, the squared error between the amplitude $|G(\mathrm{i}\omega_n)|$ of the recorded frequency response and the amplitude of the model $|\hat{G}(\mathrm{i}\omega_n)|$ was chosen. Hence

$$V = \sum_{n=1}^{N} \left(|G(\mathrm{i}\omega_n)| - |\hat{G}(\mathrm{i}\omega_n)|\right)^2 . \tag{19.5.7}$$

This allows to determine the parameters of the minimum-phase system. Constraints have been introduced that guarantee that all parameters are non-negative to ensure that one obtains physically meaningful parameter estimates.

The parameter estimates result as

$$\begin{array}{lll} \hat{J}_1 = 0.0184\,\mathrm{kg\,m^2} & \hat{J}_2 = 0.0082\,\mathrm{kg\,m^2} & \hat{J}_3 = 0.0033\,\mathrm{kg\,m^2} \\ \hat{c}_1 = 1.3545\,\frac{\mathrm{Nm}}{\mathrm{rad}} & \hat{c}_2 = 1.9307\,\frac{\mathrm{Nm}}{\mathrm{rad}} & \\ \hat{d}_1 = 9.8145 \times 10^{-5}\,\frac{\mathrm{Nm\,s}}{\mathrm{rad}} & \hat{d}_2 = 1.1047 \times 10^{-7}\,\frac{\mathrm{Nm\,s}}{\mathrm{rad}} & \hat{d}_3 = 0.0198\,\frac{\mathrm{Nm\,s}}{\mathrm{rad}} \end{array}$$

Table 19.2. Some penalty functions

Name, Shape and Equation	Name, Shape and Equation
Parabolic Penalty 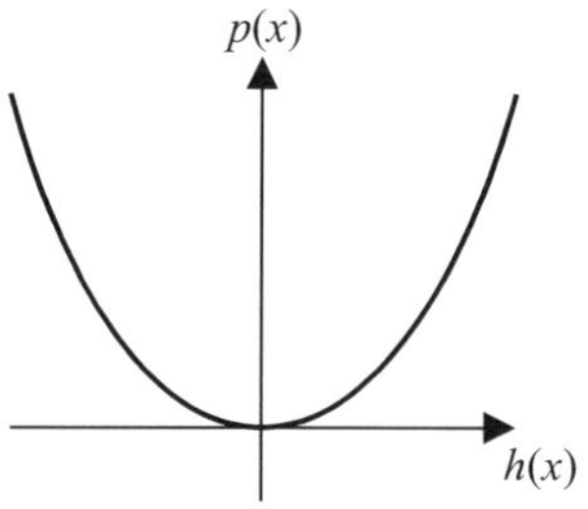 $p(\boldsymbol{x}) = r_P\big(h(\boldsymbol{x})\big)^2$ see Exterior Penalty Function Method	**Semiparabolic Penalty** 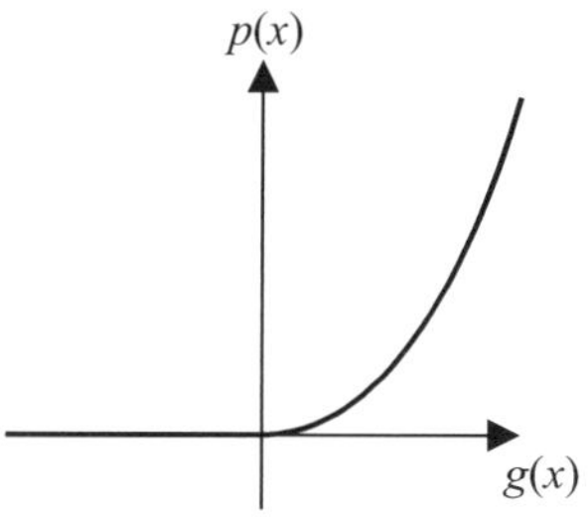 $p(\boldsymbol{x}) = \begin{cases} r_P\big(g(\boldsymbol{x})\big)^2 \text{ if } g(\boldsymbol{x}) > 0 \\ 0 \text{ else} \end{cases}$ see Exterior Penalty Function Method
Infinite Barrier Penalty 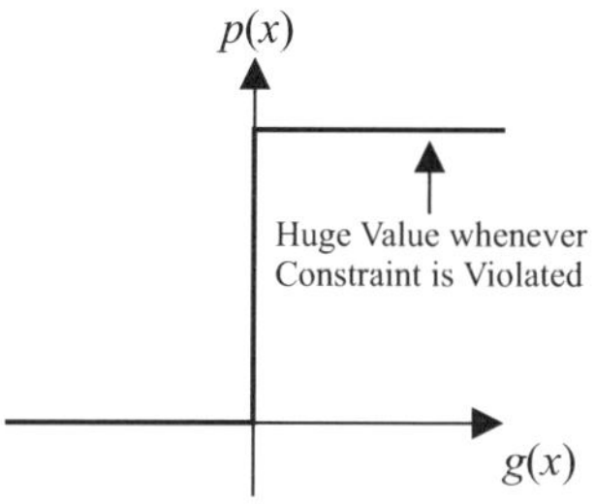 $p(\boldsymbol{x}) = \begin{cases} 10^{20} \text{ if } g(\boldsymbol{x}) > 0 \\ 0 \text{ else} \end{cases}$	**Log Penalty** 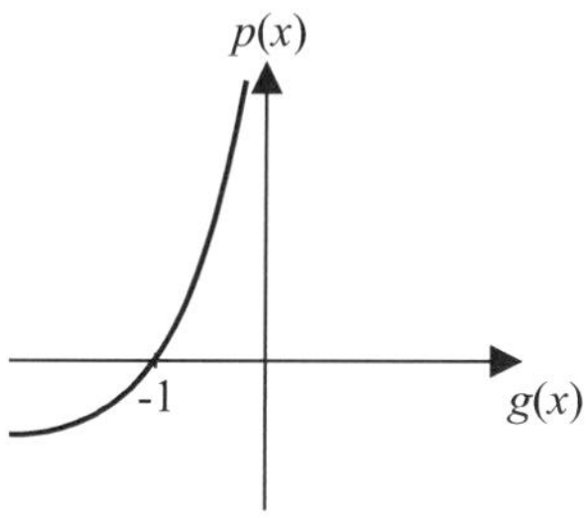 $p(\boldsymbol{x}) = -r_P \log\big(-g(\boldsymbol{x})\big)$
Inverse Penalty 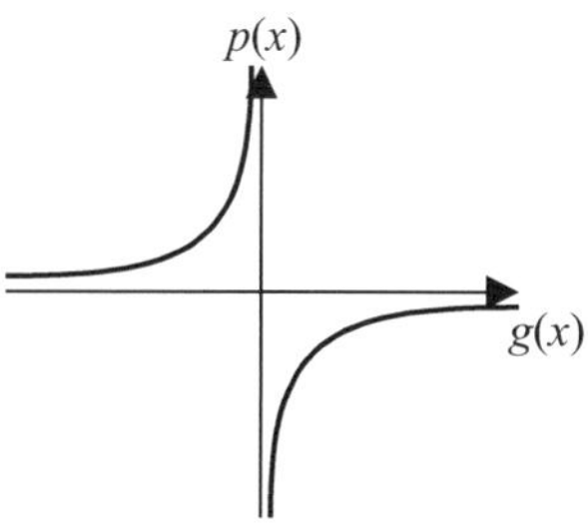 $p(\boldsymbol{x}) = -r_P \frac{1}{g(\boldsymbol{x})}$	

In the last step, also the phases have been matched, $\angle G(\mathrm{i}\omega_n)$ and $\angle \hat{G}(\mathrm{i}\omega_n)$ for all n. This allowed to detect an additional dead time that comes from the sampling and subsequent signal conditioning.

The additional dead time has been determined as

$$T_{\mathrm{D}} = 0.0047\,\mathrm{s}\;.$$

Finally, the frequency response of the model with this dead time is shown in Fig. 19.6. □

Example 19.4 (Downhill Simplex Algorithm for Constrained Identification based on the Output Error of the Three-Mass Oscillator with a Physical Model).

In this example, the physical parameters of the Three-Mass Oscillator shall be determined again, this time from a measurement in the time domain. The Three-Mass Oscillator model was augmented with dry friction (Bähr et al, 2009) as the iterative optimization techniques can deal also with non-linear models.

The cost function will be based on the output error

$$V = \sum_{k=1}^{N} \big(y(k) - \hat{y}(k)\big)^2 \tag{19.5.8}$$

and will yield the parameter vector

$$\boldsymbol{\theta}^{\mathrm{T}} = \big(J_1 \; J_2 \; J_3 \; d_1 \; d_2 \; d_3 \; d_{0,1} \; d_{0,2} \; d_{0,3} \; c_1 \; c_2 \big)\;. \tag{19.5.9}$$

Constraints have once again been introduced that guarantee that all parameters are positive to ensure that one obtains meaningful parameter estimates. The results of the time domain model with the parameter estimates are presented in Fig. 19.7.

$$\begin{array}{lll}
\hat{J}_1 = 0.0188\,\mathrm{kg\,m^2} & \hat{J}_2 = 0.0076\,\mathrm{kg\,m^2} & \hat{J}_3 = 0.0031\,\mathrm{kg\,m^2} \\
\hat{c}_1 = 1.3958\,\frac{\mathrm{Nm}}{\mathrm{rad}} & \hat{c}_2 = 1.9319\,\frac{\mathrm{Nm}}{\mathrm{rad}} & \\
\hat{d}_1 = 2.6107 \times 10^{-4}\,\frac{\mathrm{Nm\,s}}{\mathrm{rad}} & \hat{d}_2 = 0.001\,\frac{\mathrm{Nm\,s}}{\mathrm{rad}} & \hat{d}_3 = 0.0295\,\frac{\mathrm{Nm\,s}}{\mathrm{rad}} \\
\hat{d}_{10} = 0.0245\;\mathrm{Nm} & \hat{d}_{20} \approx 0.0\;\mathrm{Nm} & \hat{d}_{30} = 0.6709\;\mathrm{Nm} \\
\hat{T}_{\mathrm{D}} = 0.0018\;\mathrm{s} & &
\end{array}$$

Comparing these values with the results from Example. 19.3, where a purely linear model was used, one can see that the results match very well. As the friction model was refined, the identification results for the friction model of course have changed. □

One can also derive constrained direct search techniques. However, for these techniques the reader is referred to the before mentioned books that deal exclusively with numerical optimization techniques.

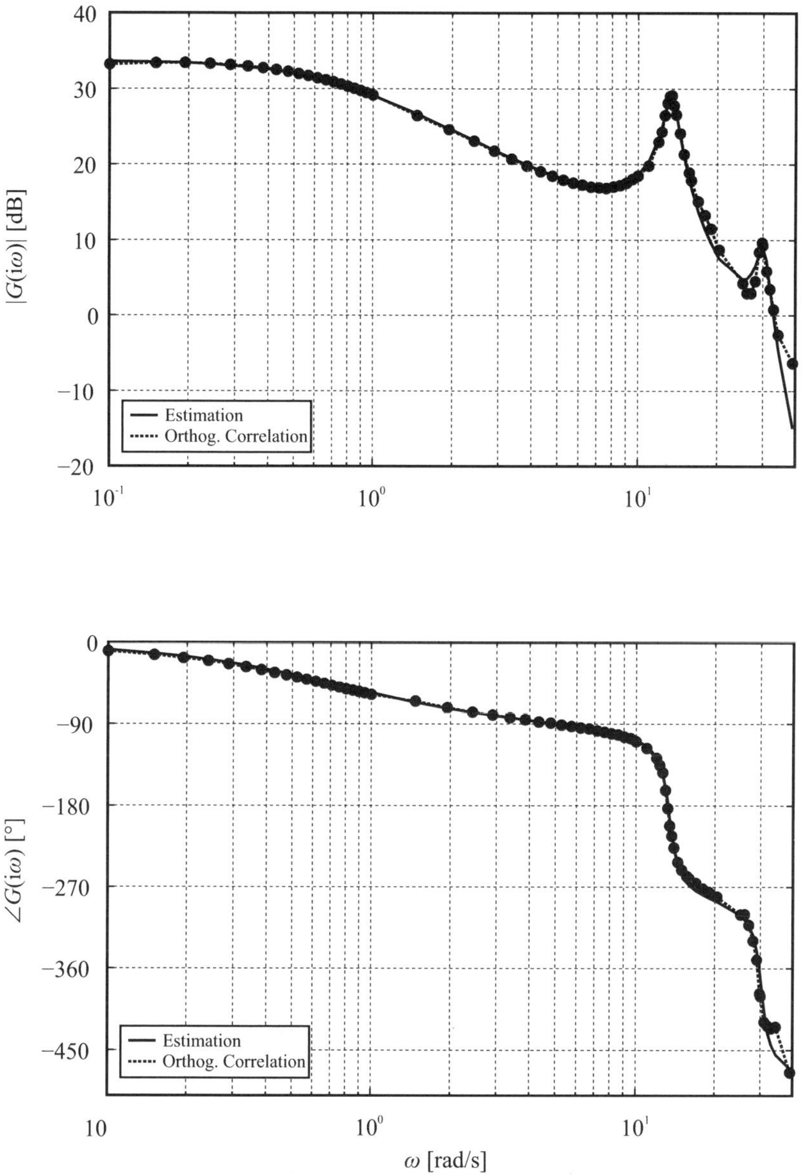

Fig. 19.6. Identification of the Three-Mass Oscillator based on frequency response measurements and the constrained downhill simplex algorithm

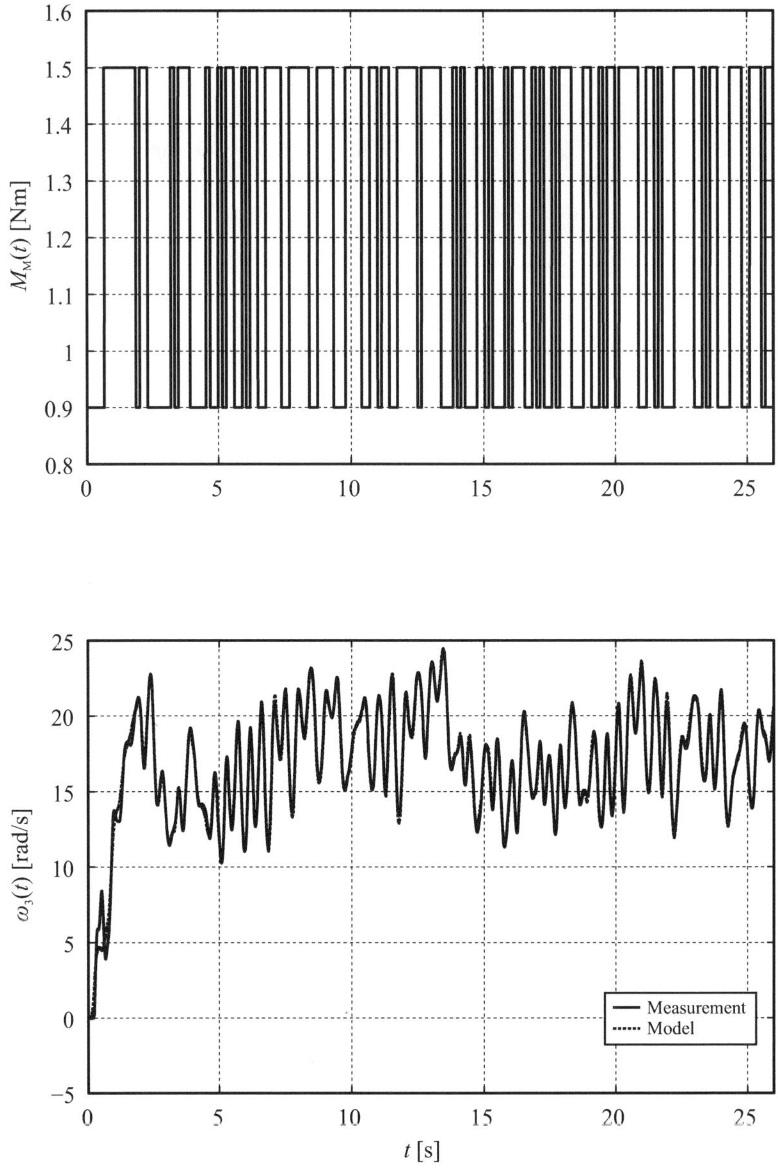

Fig. 19.7. Identification of the Three-Mass Oscillator based on time domain measurements

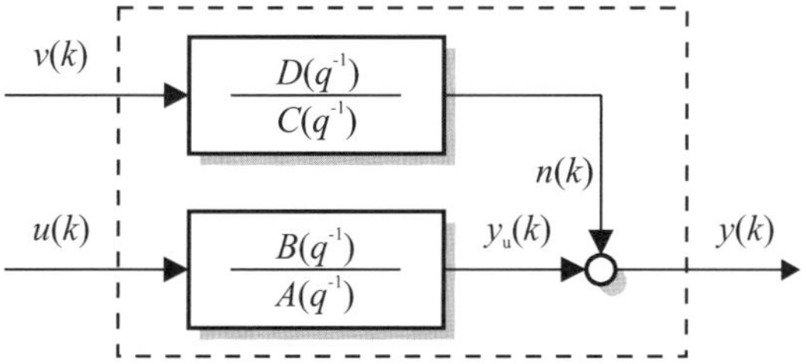

Fig. 19.8. Basis block diagram for development of predicition error method (PEM)

19.6 Prediction Error Methods using Iterative Optimization

The identification of linear systems as discussed in Chap. 9 was limited to special model classes as the problem had to be linear in parameters. Hence, only an ARX model could be estimated using the direct method of least squares. Several modifications to this basic approach for linear dynamic processes were presented in Chap. 10 and allowed to also identify certain other model classes. In this section, the *prediction error* method will now be introduced, which uses the Gauss-Newton or Levenberg-Marquart method to solve a non-linear least squares problem. As the problem now does not have to be linear in the parameters, a much larger variety of different model classes can be identified (Ljung, 1999).

The class of *prediction error methods (PEM)* shall now be developed. Their common approach is to predict the current process output $\hat{y}(k)$ based on measurement of the input up to time step $k-1$. The algorithm will be developed for the most generic case depicted in Fig. 19.8 and can easily be adapted to more specific cases by equating the appropriate terms.

The output $y(k)$ is given by the sum of the response due to the model input $u(k)$ and the disturbing noise $n(k)$ as

$$y(k) = y_{\mathrm{u}}(k) + n(k)\,. \tag{19.6.1}$$

If the white noise $\nu(k)$ that is driving the form filter used to generate the disturbance $n(k)$ would be known, one could easily calculate the true values of the disturbance $n(k)$ as a function of time by

$$\sum_{i=0}^{m_{\mathrm{c}}} c_i \nu(k) q^{-i} = \sum_{i=0}^{m_{\mathrm{d}}} d_i n(k) q^{-i}\,, \tag{19.6.2}$$

which is just the transfer function of the form filter written in the time domain, see Fig. 19.9a. Here and in the following $c_0 = 1$, $d_0 = 1$.

However, since the true value of the white noise at time step k is unknown, it must be estimated as $\hat{\nu}(k|k-1)$ based on measurements up to time step $k-1$. (19.6.2) can be written as

$$\sum_{i=0}^{m_{\mathrm{c}}} c_i \hat{\nu}(k) q^{-i} = \sum_{i=0}^{m_{\mathrm{d}}} d_i n(k) q^{-i}\,. \tag{19.6.3}$$

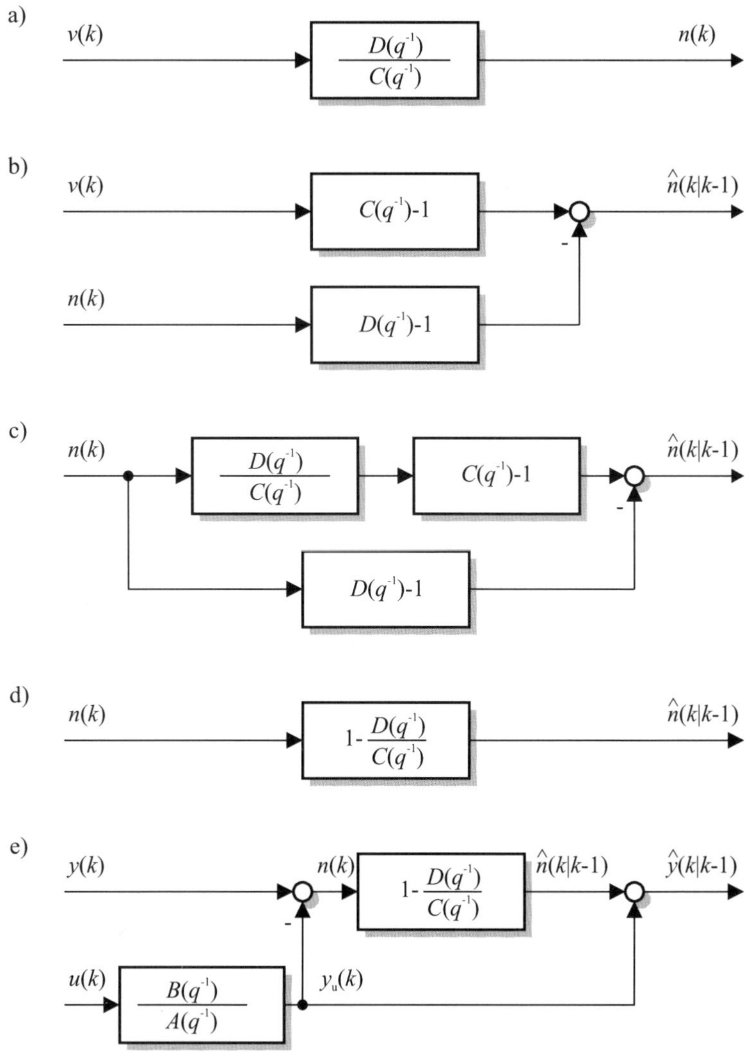

Fig. 19.9. Prediction of the noise $\hat{n}(k)$ based on measurements of $n(i)$ for $i \leq k-1$

The past values of $\nu(i)$ with $i \leq k-1$ can be determined from the past values of $n(i)$ with $i \leq k-1$ and are hence assumed to be known. The current value $\nu(k)$ is unknown and must be assumed properly. Since $\nu(k)$ is a zero-mean white noise sequence, each value of $\nu(k)$ is uncorrelated with all other values and the expected value is $\mathrm{E}\{\nu(k)\} = 0$. Therefore, the estimate $\hat{\nu}(k)$ should be chosen as

$$\hat{\nu}(k|k-1) = \mathrm{E}\big\{\nu(k)\big\} = 0\,. \tag{19.6.4}$$

Then, the estimate $\hat{n}(k|k-1)$ is given as

$$\sum_{i=1}^{m_c} c_i \hat{v}(k)q^{-1} - \sum_{i=1}^{m_d} d_i n(k)q^{-1} = \hat{n}(k|k-1)$$
$$\Leftrightarrow \left(\sum_{i=0}^{m_c} c_i \hat{v}(k)q^{-i} - \hat{v}(k)\right) - \left(\sum_{i=0}^{m_d} d_i n(k)q^{-i} - n(k)\right) = \hat{n}(k|k-1) \,. \tag{19.6.5}$$

The resulting block diagram is shown in Fig. 19.9c. This block diagram can be transformed to the form in Fig. 19.9d, which then provides the prediction of $\hat{n}(k|k-1)$ as

$$\hat{n}(k|k-1) = \left(1 - \frac{D(q^{-1})}{C(q^{-1})}\right) n(k) \,. \tag{19.6.6}$$

This model already allows to estimate noise models, i.e. estimate stochastic models.

In the next step, the deterministic part of the process model in Fig. 19.8 will be attached. Since

$$y(k) = y_u(k) + n(k) \tag{19.6.7}$$

and

$$y_u(k) = \frac{B(q^{-1})}{A(q^{-1})} u(k) \,, \tag{19.6.8}$$

one can first subtract $y_u(k)$ to obtain $n(k)$, then provide the estimate $\hat{n}(k|k-1)$ by (19.6.6) and then add $y_u(k)$ to obtain $\hat{y}(k|k-1)$. This leads to the prediction equation

$$\hat{y}(k|k-1) = \left(1 - \frac{D(q^{-1})}{C(q^{-1})}\right) y(k) + \frac{D(q^{-1})\,B(q^{-1})}{C(q^{-1})\,A(q^{-1})} u(k) \,, \tag{19.6.9}$$

see also Fig. 19.9e.

The predicted output $\hat{y}(k|k-1)$ can now be used to determine the prediction error $e(k)$ as

$$e(k) = y(k) - \hat{y}(k|k-1) \,. \tag{19.6.10}$$

Based on this error, one can now use a quadratic cost function as

$$f(\boldsymbol{x}) = \sum_{k=1}^{N} e_k^2(\boldsymbol{x}) \,, \tag{19.6.11}$$

which can be solved by the Gauss-Newton algorithm presented in Sec. 19.4.3.

For the model in (19.6.9), analytical gradients can be determined as

$$\frac{\partial}{\partial a_i} \hat{y}(k|k-1) = -\frac{D(q^{-1})\,B(q^{-1})}{C(q^{-1})\,A(q^{-1})\,A(q^{-1})} u(k-i) \tag{19.6.12}$$

$$\frac{\partial}{\partial b_i} \hat{y}(k|k-1) = \frac{D(q^{-1})}{C(q^{-1})\,A(q^{-1})} u(k-i) \tag{19.6.13}$$

$$\frac{\partial}{\partial c_i}\hat{y}(k|k-1) = -\frac{D(q^{-1})\,B(q^{-1})}{C(q^{-1})\,C(q^{-1})\,A(q^{-1})}u(k-i) + \frac{D(q^{-1})}{C(q^{-1})\,C(q^{-1})}y(k-i) \tag{19.6.14}$$

$$\frac{\partial}{\partial d_i}\hat{y}(k|k-1) = \frac{B(q^{-1})}{C(q^{-1})\,A(q^{-1})}u(k-i) - \frac{1}{C(q^{-1})}y(k-i)\,. \tag{19.6.15}$$

19.7 Determination of Gradients

In the previous section, first and second order methods have been presented, which require knowledge of the corresponding derivatives. One can use the standard finite differencing techniques. In the one dimensional case, the first derivative of f with respect to x can be approximated by

$$\frac{\mathrm{d}f(x)}{\mathrm{d}x} = \frac{f(x+\Delta x) - f(x)}{\Delta x}\,, \tag{19.7.1}$$

where Δx must be chosen appropriately. The second derivative can be determined as

$$\frac{\mathrm{d}^2 f}{\mathrm{d}x^2} = \frac{f(x+\Delta x) - 2f(x) + f(x-\Delta x)}{\Delta x^2}\,. \tag{19.7.2}$$

The approach can be extended to the multi-dimensional case in a straightforward way.

For state space systems, also an algorithm for determining the derivatives of the output with respect to a parameter of the input or state matrix shall be derived. As this approach can also be used for the identification of input/output models for linear systems based on the differential equation in the output error setting, this derivation is of high relevance.

For the state space model given in (2.1.24) and (2.1.25)

$$\dot{\boldsymbol{x}}(t) = \boldsymbol{A}\boldsymbol{x}(t) + \boldsymbol{b}u(t) \tag{19.7.3}$$

$$y(t) = \boldsymbol{c}^{\mathrm{T}}\boldsymbol{x}(t) + du(t)\,, \tag{19.7.4}$$

one now wants to determine the partial derivative of the states $\boldsymbol{x}(t)$ and the output $y(t)$ with respect to the entry $a_{i,j}$ of the matrix $\boldsymbol{A}$. There one can determine $\partial\boldsymbol{x}(t)/\partial a_{i,j}$ by the differential equation

$$\frac{\partial}{\partial a_{i,j}}\dot{\boldsymbol{x}}(t) = \frac{\partial \boldsymbol{A}}{\partial a_{i,j}}\boldsymbol{x}(t) + \boldsymbol{A}\frac{\partial \boldsymbol{x}(t)}{\partial a_{i,j}}\,. \tag{19.7.5}$$

Hence, one can solve the augmented state space system

$$\begin{pmatrix} \dot{\boldsymbol{x}}(t) \\ \dfrac{\partial}{\partial a_{i,j}}\dot{\boldsymbol{x}}(t) \end{pmatrix} = \begin{pmatrix} \boldsymbol{A} & \boldsymbol{0} \\ \dfrac{\partial \boldsymbol{A}}{\partial a_{i,j}} & \boldsymbol{A} \end{pmatrix} \begin{pmatrix} \boldsymbol{x}(t) \\ \dfrac{\partial}{\partial a_{i,j}}\boldsymbol{x}(t) \end{pmatrix} + \begin{pmatrix} \boldsymbol{b} \\ \boldsymbol{0} \end{pmatrix} u(t)\,. \tag{19.7.6}$$

to obtain the states and the partial derivatives of the state with respect to the entry $a_{i,j}$ at the same time. From there, one can determine the partial derivative of the output as

$$\frac{\partial}{\partial a_{i,j}} y(t) = \boldsymbol{c}^{\mathrm{T}} \frac{\partial \boldsymbol{x}(t)}{\partial a_{i,j}} . \tag{19.7.7}$$

With a similar approach, one can determine the partial derivative of the states $\boldsymbol{x}(t)$ and the output $y(t)$ with respect to the parameter b_i of the input vector. Here, the augmented state space system is given as

$$\begin{pmatrix} \dot{\boldsymbol{x}}(t) \\ \dfrac{\partial}{\partial b_i} \dot{\boldsymbol{x}}(t) \end{pmatrix} = \begin{pmatrix} \boldsymbol{A} & \boldsymbol{0} \\ \boldsymbol{0} & \boldsymbol{A} \end{pmatrix} \begin{pmatrix} \boldsymbol{x}(t) \\ \dfrac{\partial}{\partial b_i} \boldsymbol{x}(t) \end{pmatrix} + \begin{pmatrix} \boldsymbol{b} \\ \dfrac{\partial \boldsymbol{b}}{\partial b_i} \end{pmatrix} u(t) \tag{19.7.8}$$

and

$$\frac{\partial}{\partial b_i} y(t) = \boldsymbol{c}^{\mathrm{T}} \frac{\partial \boldsymbol{x}(t)}{\partial b_i} . \tag{19.7.9}$$

Partial derivatives with respect to the entries c_i of the output distribution vector manifest themselves only in the output equation (2.1.25) as the states are not a function of the c_i. Hence

$$\frac{\partial}{\partial c_i} y(t) = \frac{\partial \boldsymbol{c}^{\mathrm{T}}}{\partial c_i} \boldsymbol{x}(t) . \tag{19.7.10}$$

From these partial derivatives, one can easily determine the partial derivatives of the cost function with respect to the parameters to be estimated in the linear case. A derivation of the partial derivatives can also be found in (Verhaegen and Verdult, 2007) and (van Doren et al, 2009). The major disadvantage is that for a large number of parameters θ_i, a large number of filters must be employed (Ninness, 2009), where also alternatives for high dimensional parameter estimation problems are discussed. Note that one can speed up the calculation by first solving the differential equation $\dot{\boldsymbol{x}}(t) = \boldsymbol{A}\boldsymbol{x}(t) + \boldsymbol{b}u(t)$ and then solving the differential equation for the partial derivatives of the states in a second step.

The book by van den Bos (2007) also contains an extensive treatment of nonlinear optimization algorithms for parameter estimation and shows the derivation of the Newton step for some other cost functions, such as the maximum likelihood, as well.

19.8 Model Uncertainty

With an approach similar to Sect. 9.1.3, one can now also determine the parameter covariance for the solutions found by the iterative optimization. One approach, based on a Taylor series expansion is shown in (Ljung, 1999). Here, a different approach will be pursued based on the rules of error propagation. The model output shall be denoted as

$$\hat{y} = f(\hat{\boldsymbol{\theta}}) . \tag{19.8.1}$$

The covariance of the model output can be determined from

$$\operatorname{cov}\hat{y} = \mathrm{E}\{(\hat{y} - \mathrm{E}\{\hat{y}\})^2\}\,, \tag{19.8.2}$$

which for finite sample times can be determined as

$$\operatorname{cov}\hat{y} \approx \frac{1}{N}\sum_{k=0}^{N-1}\big(\hat{y}(k) - y(k)\big)^2\,, \tag{19.8.3}$$

where it was assumed that the model error, i.e. $\hat{y}(k) - y(k)$ is zero-mean. If it was not zero mean, then one would not have obtained the minimum possible error metric. Furthermore, from error propagation, one knows that

$$\operatorname{cov} y = \left(\left.\frac{\partial y(\boldsymbol{\theta})}{\partial \boldsymbol{\theta}}\right|_{\boldsymbol{\theta}=\hat{\boldsymbol{\theta}}}\right)^{\mathrm{T}} \operatorname{cov}\Delta\boldsymbol{\theta}\left(\left.\frac{\partial y(\boldsymbol{\theta})}{\partial \boldsymbol{\theta}}\right|_{\boldsymbol{\theta}=\hat{\boldsymbol{\theta}}}\right), \tag{19.8.4}$$

which can now be solved for the covariance of the parameters as

$$\operatorname{cov}\Delta\boldsymbol{\theta} = \left(\left.\frac{\partial^2 y(\boldsymbol{\theta})}{\partial\boldsymbol{\theta}\,\partial\boldsymbol{\theta}^{\mathrm{T}}}\right|_{\boldsymbol{\theta}=\hat{\boldsymbol{\theta}}}\right)^{-1}\left(\left.\frac{\partial y(\boldsymbol{\theta})}{\partial \boldsymbol{\theta}}\right|_{\boldsymbol{\theta}=\hat{\boldsymbol{\theta}}}\right)^{\mathrm{T}} \operatorname{cov} y \left(\left.\frac{\partial y(\boldsymbol{\theta})}{\partial \boldsymbol{\theta}}\right|_{\boldsymbol{\theta}=\hat{\boldsymbol{\theta}}}\right)\left(\left.\frac{\partial^2 y(\boldsymbol{\theta})}{\partial\boldsymbol{\theta}\,\partial\boldsymbol{\theta}^{\mathrm{T}}}\right|_{\boldsymbol{\theta}=\hat{\boldsymbol{\theta}}}\right)^{-1}. \tag{19.8.5}$$

This equation allows to determine the parameter error covariance for arbitrary non-linear models.

19.9 Summary

The numerical optimization algorithms presented in this chapter allow to solve parameter estimation problems even and especially if they are not linear in the parameters. Due to the iterative approach of these algorithms, they hardly qualify for real-time implementations. However, their big advantage is the fact that these methods can solve a much larger variety of estimation problems than the direct methods and that in addition constraints on the parameters can easily be introduced into the problem formulation.

The parallel and the series-parallel model are introduced as two means to determine the error between model and measurement. The optimization problem statement has been introduced incorporating both inequality and equality constraints. Although it may seem impeding at first to formulate constraints, it is often advisable to do so. Constraints can help to steer the optimizer towards the optimum as they exclude certain areas of the design space from the search. Furthermore, equality constraints can be used to reduce the number of parameters that have to be determined by the optimizer. Also, constraints can be necessary to ensure the subsequent applicability of the model. For example, physical parameters can often not be negative and hence such useless solutions can be excluded from the very beginning.

A big advantage is the fact that the model must not be brought to a form where it is linear in parameters, i.e. any model can be used for system identification regardless of being linear or non-linear, continuous-time or discrete-time, linear in parameters or not.

The major disadvantage is the fact that it is not guaranteed that the global optimum is found. As a matter of fact, convergence is not guaranteed at all. Also, the number of iterations that is necessary to reach an optimal point is not known a priori for most algorithms, thus impeding real-time application. And finally, the computational effort is much larger than for the direct methods presented so far.

Although the solution of many of these parameter estimation problems are in theory unbiased and efficient, no guarantee is given that the numerical algorithms, employed for the solution of the problem, in fact approach the global optimum. However, properties such as efficiency and unbiasedness are only satisfied for the globally optimal solution. Therefore, one might obtain results that are of less quality than those results obtained by the direct methods such as e.g. the method of least squares.

One should try to impose constraints on the parameters whenever easily possible. Often, an interval with lower and upper bound can be specified for the parameters. Stability is another typical constraint, especially since unstable systems can hardly be simulated in parallel to the process. A very easy approach is to incorporate the constraints into the cost function by means of a penalty function. Many details have been spared in this overview, such as design variable scaling, constraint scaling, etc.

The starting point of the optimization run, which is called *initial guess*, plays an important role for the convergence of the algorithms. One can try to start the algorithm from several randomly chosen starting points to ensure that the optimizer approaches the global optimum and does not get stuck in a local optimum. Also, some penalty functions require that the initial guess is feasible, i.e. satisfies all constraints. For the determination of starting values, the simple methods presented in Sect. 2.5 that allowed to determine characteristic values of the system can be helpful. A further approach is to first identify a model using a non-iterative method such as e.g. the method of least squares (see Chap. 9) or subspace methods to derive the parameters of a black-box model. This model is either given in a state space representation or can be brought to such a form. Then, a similarity transform (see (17.2.9) through (17.2.11)) can be determined that transforms the identified state space model to the structure of a physically derived state space model and at the same time allows to determine the physical parameters, (Parillo and Ljung, 2003; Xie and Ljung, 2002), see Chap. 15.

For such an initial model determination, a zeroth order method together with the incorporation of constraints by means of a penalty function seems a very favorable way as zeroth order methods are very robust and the number and type of constraints changes often. Also, for zeroth order methods, one does not have to provide information about the gradient and possibly Hessian matrix. This is very beneficial at the beginning of the process identification, where the structure of the model might be changed or refined often due to first identification results. One might even start without any constraints and add constraints as they become necessary to steer the optimizer away from meaningless solutions. Here, one can always use the same unconstrained optimization algorithms regardless of the presence or absence of constraints.

If the model as parameterized by the above non-linear parameter estimation method, has been proven well in representing the process dynamics (see Sect. 23.8, for methods to validate the results), then one should try to formulate a model that is linear in parameters and can be parametrized by direct methods as this will guarantee that the estimator converges and that the global optimum will be obtained. Also, the computational effort is much smaller, which is beneficial if e.g. models of multiple similar processes must be parameterized or the parameter estimation shall be implemented in a real-time application, e.g. for adaptive control (Isermann et al, 1992) or for fault detection and diagnosis (Isermann, 2006).

Problems

19.1. Parallel and Serial-Parallel Model
What are advantages and disadvantages of the two model setups? Why does the series-parallel model become extremely problematic for small sample times compared to the system dynamics?

19.2. Rosenbrock Function
Using different n-dimensional optimization algorithms that have been presented in this chapter, try to determine the optimum of the Rosenbrock function

$$f(\boldsymbol{x}) = (1 - x_1)^2 + 100(x_2 - x_1^2)^2$$

also called *banana function*.

19.3. Zeroth, First, and Second Order Methods
What are the advantages and disadvantages of using an algorithm that requires gradient or Hessian information?

19.4. Objective Functions
Set up suitable objective functions for identification in the time domain and the frequency domain. Are these objective functions linear in parameters or not? How would you determine the optimal parameters, i.e. find the minimum?

19.5. Constraints
What are typical constraints in the area of system identification that could be introduced into the optimization problem?

19.6. System Identification
Determine the optimization algorithm for a first order process

$$y(k) + a_1 y(k-1) = b_1 u(k-1)$$

by forming a quadratic cost function and apply

- a gradient search algorithm
- a Newton-Raphson algorithm

to identify the parameters a_1 and b_1.

References

Bähr J, Isermann R, Muenchhof M (2009) Fault management for a three mass torsion oscillator. In: Proceedings of the European Control Conference 2009 - ECC 09, Budapest, Hungary

van den Bos A (2007) Parameter estimation for scientists and engineers. Wiley-Interscience, Hoboken, NJ

Boyd S, Vandenberghe L (2004) Convex optimization. Cambridge University Press, Cambridge, UK

Conn AR, Scheinberg K, Vicente LN (2009) Introduction to derivative-free optimization, MPS-SIAM series on optimization, vol 8. SIAM, Philadelphia

van Doren JFM, Douma SG, van den Hof PMJ, Jansen JD, Bosgra OH (2009) Identifiability: From qualitative analysis to model structure approximation. In: Proceedings of the 15th IFAC Symposium on System Identification, Saint-Malo, France

Isermann R (2006) Fault-diagnosis systems: An introduction from fault detection to fault tolerance. Springer, Berlin

Isermann R, Lachmann KH, Matko D (1992) Adaptive control systems. Prentice Hall international series in systems and control engineering, Prentice Hall, New York, NY

Kennedy J, Eberhart RC, Shi Y (2001) Swarm intelligence. The Morgan Kaufmann series in evolutionary computation, Morgan Kaufmann, San Francisco, CA

Kötter H, Schneider F, Fang F, Gußner T, Isermann R (2007) Robust regressor and outlier-detection for combustion engine measurements. In: Röpke K (ed) Design of experiments (DoE) in engine development - DoE and other modern development methods, Expert Verlag, Renningen, pp 377–396

Ljung L (1999) System identification: Theory for the user, 2nd edn. Prentice Hall Information and System Sciences Series, Prentice Hall PTR, Upper Saddle River, NJ

Ljung L (2009) Experiments with identification of continuous time models. In: Proceedings of the 15th IFAC Symposium on System Identification, Saint-Malo, France

Mitchell M (1996) An introduction to genetic algorithms. MIT Press, Cambridge, MA

Nelles O (2001) Nonlinear system identification: From classical approaches to neural networks and fuzzy models. Springer, Berlin

Ninness B (2009) Some system identification challenges and approaches. In: Proceedings of the 15th IFAC Symposium on System Identification, Saint-Malo, France

Nocedal J, Wright SJ (2006) Numerical optimization, 2nd edn. Springer series in operations research, Springer, New York

Parillo PA, Ljung L (2003) Initialization of physical parameter estimates. In: Proceedings of the 13th IFAC Symposium on System Identification, Rotterdam, The Netherlands, pp 1524–1529

Press WH, Teukolsky SA, Vetterling WT, Flannery BP (2007) Numerical recipes: The art of scientific computing, 3rd edn. Cambridge University Press, Cambridge, UK

Ravindran A, Ragsdell KM, Reklaitis GV (2006) Engineering optimization: Methods and applications, 2nd edn. John Wiley & Sons, Hoboken, NJ

Rousseeuw PJ, Leroy AM (1987) Robust regression and outlier detection. Wiley Series in Probability and Statistics, Wiley, New York

Schneider JJ, Kirkpatrick S (2006) Stochastic optimization. Springer, Berlin

Snyman JA (2005) Practical mathematical optimization: An introduction to basic optimization theory and classical and new gradient-based algorithms. Springer, New York

Vanderplaats GN (2005) Numerical optimization techniques for engineering design, 4th edn. Vanderplaats Research & Development, Colorado Springs, CO

Verhaegen M, Verdult V (2007) Filtering and system identification: A least squares approach. Cambridge University Press, Cambridge

Xie LL, Ljung L (2002) Estimate physical parameters by black-box modeling. In: Proceedings of the 21st Chinese Control Conference, Hangzhou, China, pp 673–677

20

Neural Networks and Lookup Tables for Identification

Many processes show a non-linear static and dynamic behavior, especially if wide areas of operation are considered. Therefore, the identification of non-linear processes is of increasing interest. Examples are vehicles, aircraft, combustion engines, and thermal plants. In the following, models of such non-linear systems will be derived based on artificial neural networks, that had first been introduced as universal approximators of non-linear static functions.

20.1 Artificial Neural Networks for Identification

For a general identification approach, methods of interest are those that do not require specific knowledge of the process structure and hence are widely applicable. Artificial neural networks fulfill these requirements. They are composed of mathematically formulated neurons. At first, these neurons were used to describe the behavior of biological neurons (McCulloch and Pitts, 1943). The interconnection of neurons in networks allowed the description of relationships between input and output signals (Rosenblatt, 1958; Widrow and Hoff, 1960). In the sequel of this chapter, artificial neural networks (ANNs) are considered that map input signals u to output signals y, Fig. 20.1. Usually, the adaptable parameters of neural networks are unknown. As a result, they have to be adapted by processing measured signals u and y (Hecht-Nielsen, 1990; Haykin, 2009). This is termed "training" or "learning".

One can discern two steps in the design of a neural net. The first is the *training*, where the weights or other parameters of the neural net are optimized. The second step then is the *generalization*, where the net is used to simulate new data, that have not been part of the training data and allow to judge the performance of the net for unknown data. The goal is to obtain the smallest possible error for both training and generalization. The model error can be split in two parts as

$$\underbrace{\mathrm{E}\{(y_0-\hat{y})^2\}}_{\text{(Model Error)}^2} = \underbrace{\mathrm{E}\{(y_0-\mathrm{E}\{\hat{y}\})^2\}}_{\text{(Bias Error)}^2} + \underbrace{\mathrm{E}\{(\hat{y}-\mathrm{E}\{\hat{y}\})^2\}}_{\text{(Variance Error)}} . \quad (20.1.1)$$

R. Isermann, M. Münchhof, *Identification of Dynamic Systems*,
DOI 10.1007/978-3-540-78879-9_20, © Springer-Verlag Berlin Heidelberg 2011

Fig. 20.1. System with P inputs and M outputs, which has to be approximated by an artificial neural network (ANN)

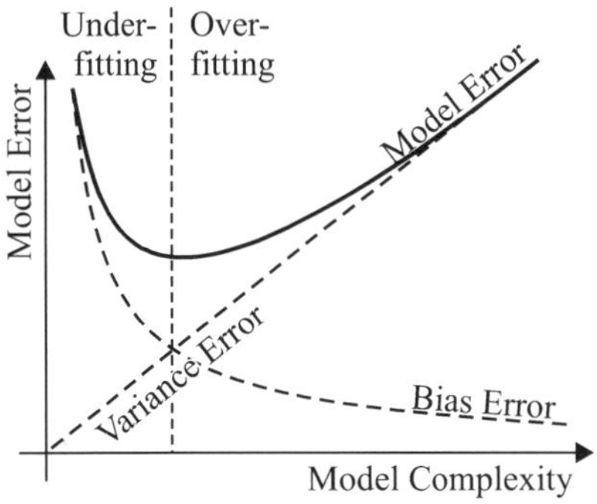

Fig. 20.2. Trade-off between bias error and variance error

The *bias error* is a systematic deviation between the true system output and the expected model output, that appears when the model does not have enough flexibility to fit the real process (*Underfitting*). Consequently, the bias error decreases as the model complexity increases. The *variance error* is the deviation between the model output and the expected model output. The variance error increases as the number of degrees of freedom of the model increases. The model is more and more adapted to the specific peculiarities of the training data set such as noise and outliers. Hence, in choosing the model structure, there is always a trade-off between the bias error and the variance error, which is termed *bias-variance dilemma*, see Fig. 20.2 (German et al, 1992; Harris et al, 2002).

In identification, one is interested in approximating the static or dynamic behavior of processes by means of (non)-linear functions. On the contrary, if inputs and outputs are gathered into groups or clusters, a classification task in connection with e.g. pattern recognition is given (Bishop, 1995). In the following, the problem of non-linear system identification is considered (supervised learning). Thereby, the capability of ANNs to approximate non-linear relationships to any desired degree of accuracy is utilized. Firstly, ANNs for describing static behavior (Hafner et al, 1992; Preuß and Tresp, 1994), will be investigated, which will then be extended to dynamic behavior (Ayoubi, 1996; Nelles et al, 1997; Isermann et al, 1997).

20.1.1 Artificial Neural Networks for Static Systems

Neural networks are universal approximators for static non-linearities and are consequently an alternative to polynomial approaches. Their advantages are the requirement of only little a priori knowledge about the process structure and the uniform treatment of single-input and multi-input processes. In the following, it is assumed

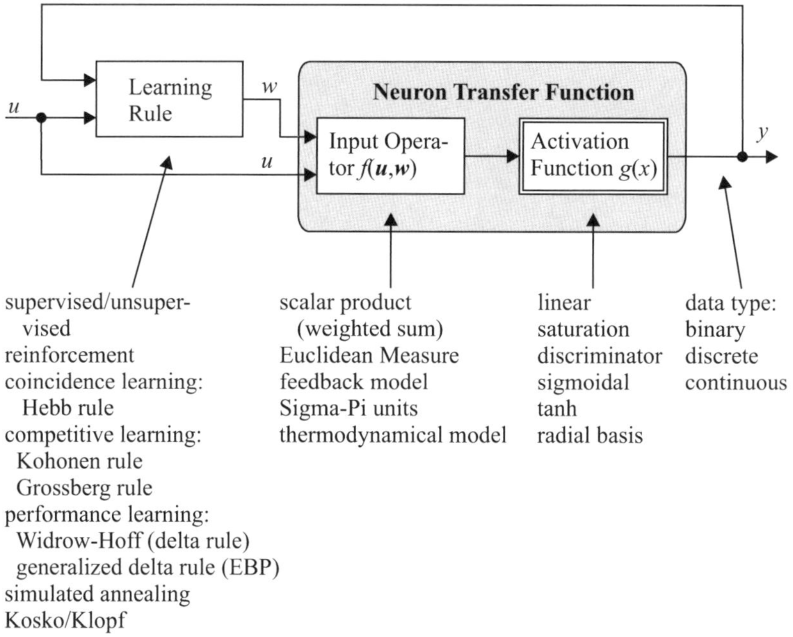

Fig. 20.3. General neuron model for process modeling with measured inputs and outputs

that a non-linear system with P inputs and M outputs has to be approximated, see Fig. 20.1.

Neuron Model

Figure 20.3 shows the block diagram of a neuron. In the input operator (synaptic function), a similarity measure between the input vector $\boldsymbol{u}$ and the (stored) weight vector $\boldsymbol{w}$ is formed, e.g. by the scalar product

$$x = \boldsymbol{w}^{\mathrm{T}}\boldsymbol{u} = \sum_{i=1}^{P} w_i u_i = |\boldsymbol{w}^{\mathrm{T}}||\boldsymbol{u}| \cos\varphi \tag{20.1.2}$$

or the Euclidean distance

$$x = \|\boldsymbol{u} - \boldsymbol{w}\|^2 = \sum_{i=1}^{P} (u_i - w_i)^2 \,. \tag{20.1.3}$$

If $\boldsymbol{w}$ and $\boldsymbol{u}$ are similar, the resulting scalar quantity x will be large in the first case and small in the second case. The quantity x, also called the *activation* of the neuron, affects the activation function and consequently the output value

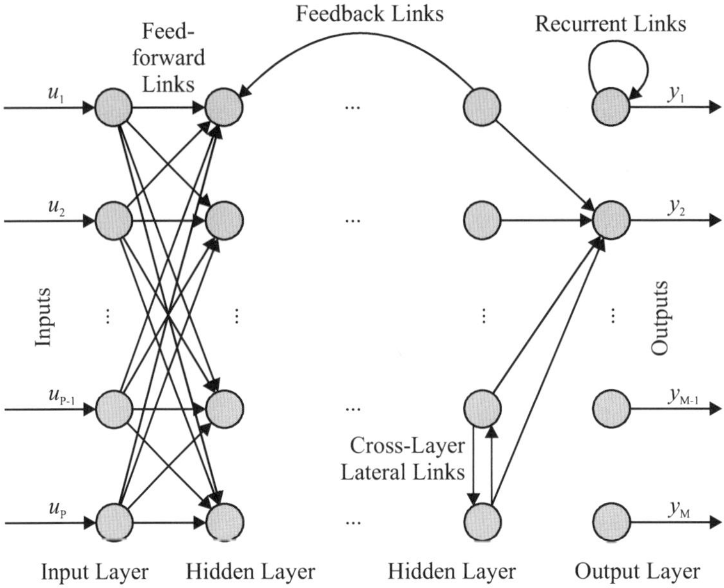

Fig. 20.4. Network structure: Layers and links in a neural network

$$y = \gamma(x - c) \ . \tag{20.1.4}$$

Table 20.1 shows several examples of those in general non-linear functions. The threshold c is a constant causing a parallel shift in the x-direction.

Network Structure

The single neurons are interconnected to a network structure, Fig. 20.4. Hence, one has to distinguish between different *layers* with neurons arranged in parallel: the input layer, the first, second, ... hidden layer and the output layer. Generally, the input layer is used to scale the input signals and is often not counted as a separate layer. Then, the real network structure begins with the first hidden layer. Figure 20.4 shows the most important types of internal links between neurons: feedforward, feedback, lateral and recurrent. With respect to their range of values, the input signals can be either binary, discrete, or continuous. Binary and discrete signals are used especially for classification, while continuous signals are used for identification tasks.

Multi Layer Perceptron (MLP) Network

The neurons of an MLP network are called *perceptrons*, Fig. 20.5, and follow directly from the general neuron model, shown in Fig. 20.3. Typically, the input operator is realized as a scalar product, while the activation functions are realized by sigmoidal

Table 20.1. Examples of activation functions

Name, Shape and Equation	Name, Shape and Equation
Hyberbolic Tangens (Tangens Hyperbolics) 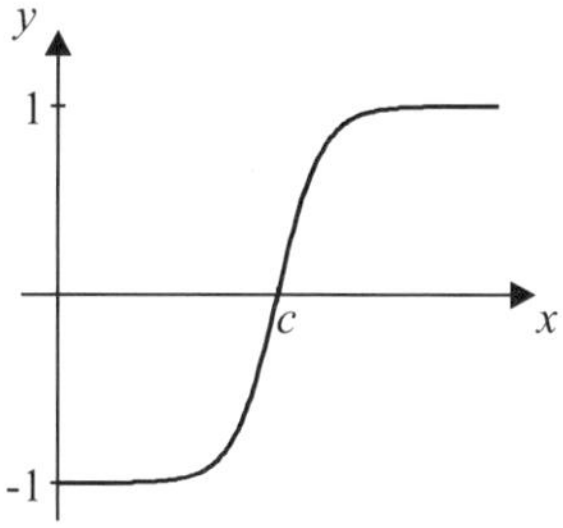 $y = \frac{e^{(x-c)} - e^{-(x-c)}}{e^{(x-c)} + e^{-(x-c)}}$	Neutral Zone 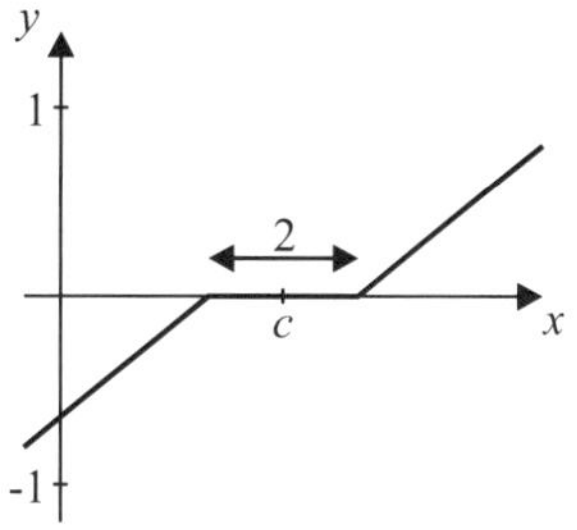 $y = \begin{cases} x-c-1 \text{ for } x-c \geq 1 \\ 0 \text{ for } \lvert x-c \rvert < 1 \\ x-c+1 \text{ for } x-c \leq -1 \end{cases}$
Sigmoid Function 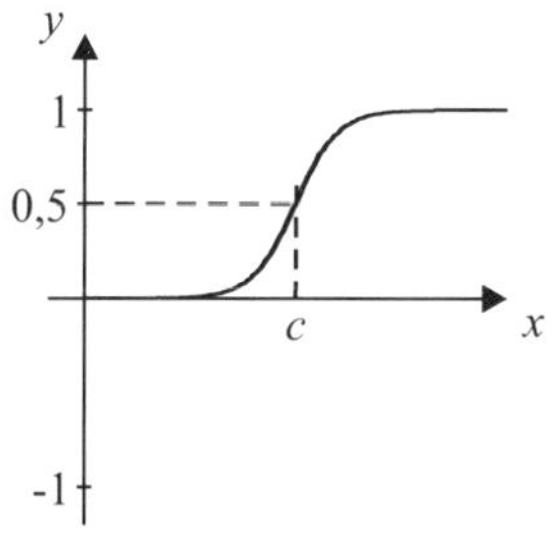 $y = \frac{1}{1+e^{-(x-c)}}$	Gauss Function 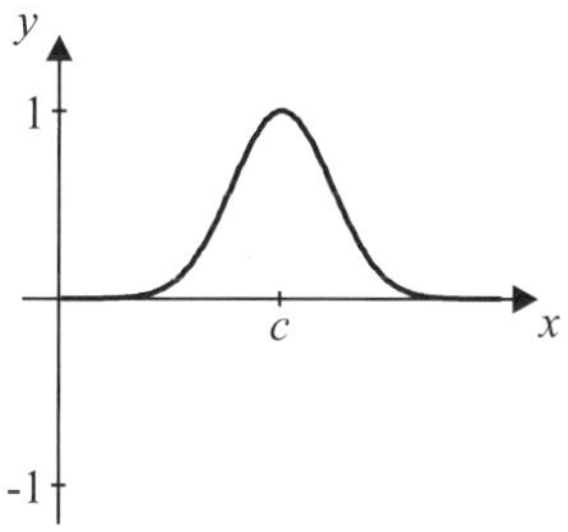 $y = e^{-(x-c)^2}$
Limiter 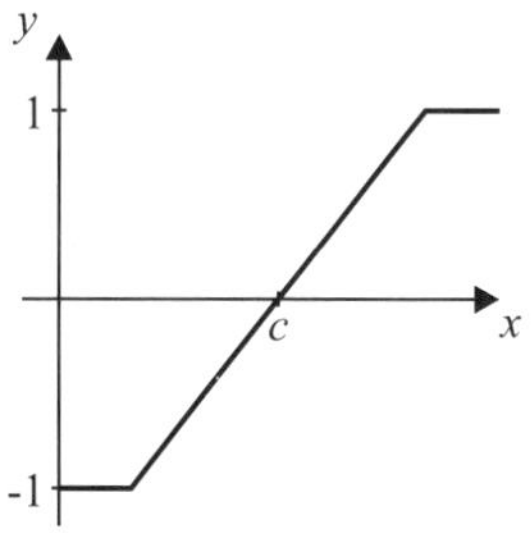 $y = \begin{cases} 1 \text{ for } (x-c) \geq 1 \\ x-c \text{ for } \lvert x-c \rvert < 1 \\ -1 \text{ for } x-c \leq -1 \end{cases}$	Binary Function 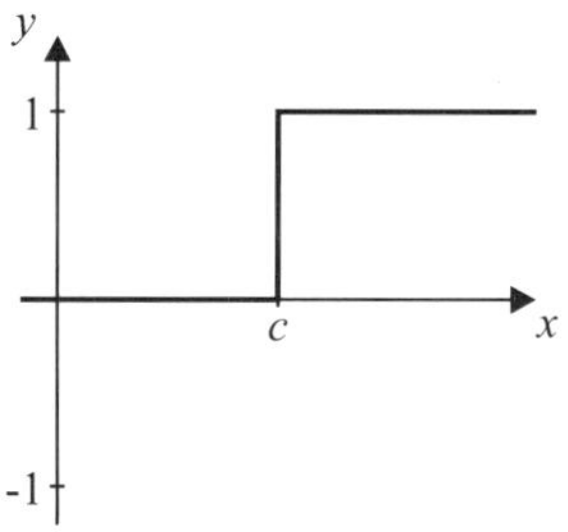 $y = \begin{cases} 0 \text{ for } x-c < 0 \\ 1 \text{ for } x-c \geq 0 \end{cases}$

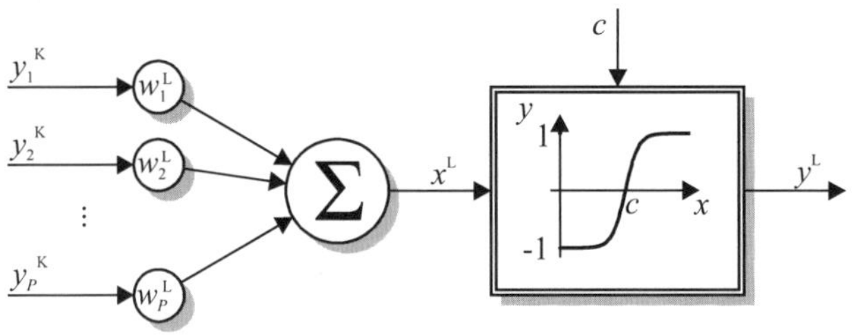

Fig. 20.5. Perceptron neuron with weights w_i, summation of input signals (scalar product) and non-linear activation function for layer L

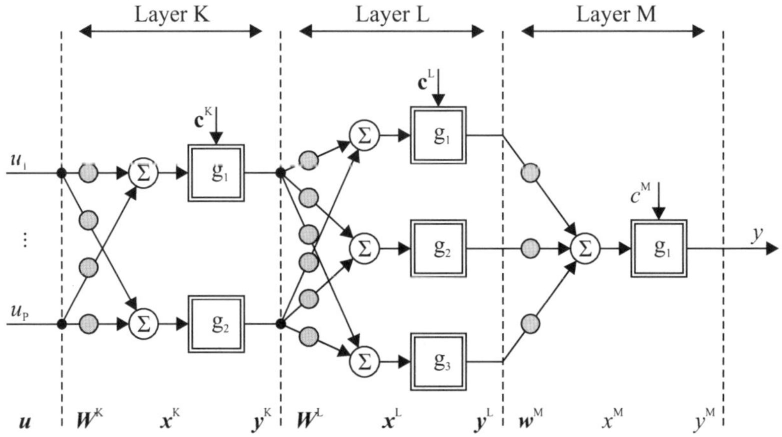

Fig. 20.6. Feedforward Multi Layer Perceptron network (MLP network). Three layers with $(2 \cdot 3 \cdot 1)$ perceptrons. $< K >$ is the first hidden layer

or hyperbolic tangent functions. The latter ones are a multiple of differentiable functions yielding a neuron output with $y = 0$ in a wide range. As however, the output is also non-zero in a wide range as well, they have a global effect with extrapolation capability. The weights w_i are assigned to the input operator and lie in the signal flow before the activation function.

The perceptrons are connected in parallel and are arranged in consecutive layers to a feedforward MLP network, Fig. 20.6. Each of the P inputs affects each perceptron in such a way that in a layer with P inputs and K perceptrons there exist $(K \cdot P)$ weights w_{kp}. The output neuron is most often a perceptron with a linear activation function, Fig. 20.7.

The adaptation of the weights w_i based on measured input and output signals is usually realized by the minimization of the quadratic cost function.

$$V(\boldsymbol{w}) = \frac{1}{2} \sum_{k=0}^{N-1} e^2(k) \tag{20.1.5}$$

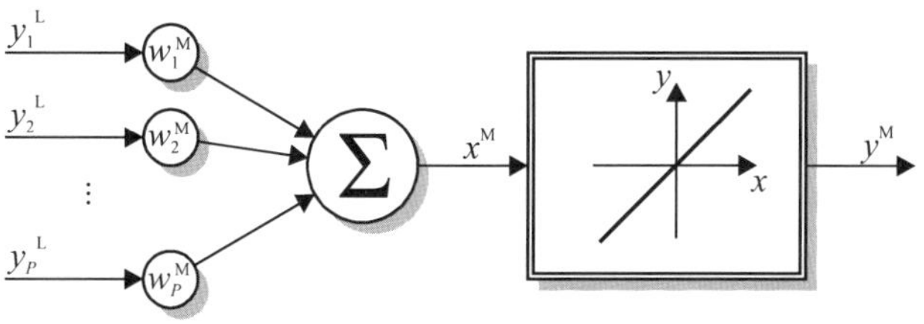

Fig. 20.7. Output neuron as Perceptron with linear activation function for layer M

with

$$e(k) = y(k) - \hat{y}(k) , \tag{20.1.6}$$

where $e(k)$ is the model error, $y(k)$ is the measured output signal, and $\hat{y}(k)$ is the network output after the output layer M.

As in the case of parameter estimation with the least squares method, one equates the first derivative of the cost function with respect to the parameters of the net to zero, i.e.

$$\frac{\mathrm{d}V(\boldsymbol{w})}{\mathrm{d}\boldsymbol{w}} = \boldsymbol{0} . \tag{20.1.7}$$

Due to the non-linear dependency, a direct solution is not possible. Therefore, e.g. gradient-based methods for numerical optimization are applied, see also Chap. 19. Because of the necessary back-propagation of the errors through all hidden layers, the method is called *error back-propagation* or also *delta-rule*. The so-called *learning rate* η has to be chosen (tested) suitably. In principle, gradient-based methods allow only slow convergence in the case of a large number of unknown parameters.

Radial Basis Function (RBF) Network

The neurons of RBF networks, Fig. 20.8, compute the Euclidean distance in the input operator

$$x = \|\boldsymbol{u} - \boldsymbol{c}\|^2 \tag{20.1.8}$$

and feed it to the activation function

$$y_m = \gamma_m\big(\|\boldsymbol{u} - \boldsymbol{c}_m\|^2\big) . \tag{20.1.9}$$

The activation function is given by radial basis functions usually in the form of Gaussian functions with

$$\gamma_m = \exp\left(-\frac{1}{2}\left(\frac{(u_1 - c_{m1})^2}{\sigma_{m1}^2} + \frac{(u_2 - c_{m2})^2}{\sigma_{m2}^2} + \ldots + \frac{(u_P - u_{mP})^2}{\sigma_{mP}^2}\right)\right) . \tag{20.1.10}$$

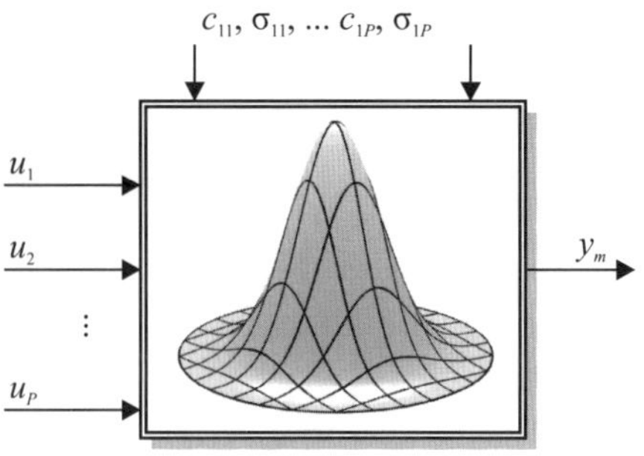

Fig. 20.8. Neuron with Radial Basis Function (RBF)

The centers c_m and the standard deviations σ_m are determined a priori so that the Gaussian functions are spread, e.g. uniformly in the input space. The activation function determines the distances of each input signal to the center of the corresponding basis function. However, radial basis functions contribute to the model output only locally, namely in the vicinity of their centers. They possess less extrapolation capability, since their output values tend to go to zero with a growing distance to their centers.

Usually, radial basis function networks consist of two layers, Fig. 20.9. The outputs y_i are weighted and added up in a neuron of the perceptron type, Fig. 20.7, so that

$$\hat{y} = \sum_{m=1}^{M} w_m \gamma_m \big(\|\boldsymbol{u} - \boldsymbol{c}_m\|^2\big) \, . \tag{20.1.11}$$

Since the output layer weights are located behind the non-linear activation functions in the signal flow, the error signal is linear in these parameters and consequently, the least squares method in its explicit form can be applied. In comparison to MLP networks with gradient-based methods, a significantly faster convergence can be obtained. However, if the centers and standard deviations have to be optimized too, non-linear numerical optimization methods are required again.

Local Linear Model Networks

The local linear model tree network (LOLIMOT) is an extended radial basis function network (Nelles et al, 1997; Nelles, 2001). It is extended by replacing the output layer weights with a linear function of the network inputs (20.1.12). Furthermore, the RBF network is normalized, such that the sum of all basis functions is always one. Thus, each neuron represents a local linear model with its corresponding validity function, see Fig. 20.10. The validity functions determine the regions of the input space where each neuron is active. The general architecture of local model networks is extensively discussed in (Murray-Smith and Johansen, 1997a).

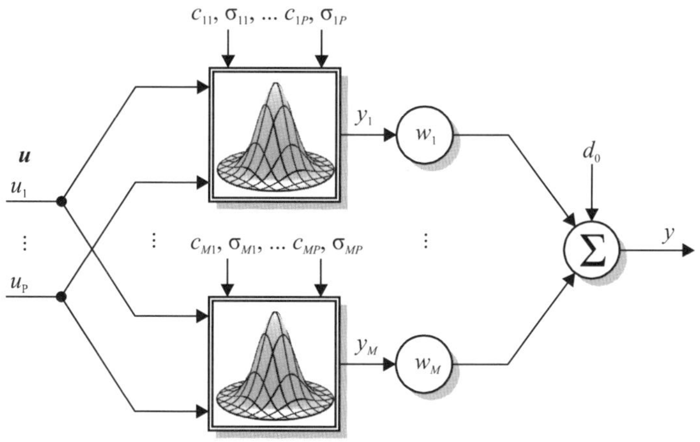

Fig. 20.9. Feedforward Radial Basis Function (RBF) network

The kind of local model network discussed here utilizes normalized Gaussian validity functions (20.1.10) and an axis-orthogonal partitioning of the input space. Therefore, the validity functions can be composed of one-dimensional membership functions and the network can be interpreted as a Takagi-Sugeno fuzzy model.

The output of the local linear model is calculated by

$$\hat{y} = \sum_{m=1}^{M} \Phi_m(\boldsymbol{u})\big(w_{m,0} + w_{m,1}u_m + \ldots + w_{m,p}u_p\big) \tag{20.1.12}$$

with the normalized Gaussian validity functions

$$\Phi_m(\boldsymbol{u}) = \frac{\mu_m(\boldsymbol{u})}{\sum_{i=1}^{M} \mu_i(\boldsymbol{u})} \tag{20.1.13}$$

with

$$\mu_i(\boldsymbol{u}) = \prod_{j=1}^{p} \exp\left(-\frac{1}{2}\left(\frac{\left(u_j - c_{i,j}\right)^2}{\sigma_{i,j}^2}\right)\right). \tag{20.1.14}$$

The centers $c_{i,j}$ and standard deviations $\sigma_{i,j}$ are non-linear parameters, while the local model parameters w_m are linear parameters. The local linear model tree (LOLIMOT) algorithm is applied for the training. It consists of an outer loop, in which the input space is decomposed by determining the parameters of the validity functions, and a nested inner loop in which the parameters of the local linear models are optimized by locally weighted least squares estimation.

The input space is decomposed in an axis-orthogonal manner, yielding hyper-rectangles. In their centers, Gaussian validity functions $\mu_i(\boldsymbol{u})$ are placed. The standard deviations of these Gaussians are chosen proportionally to the extension of

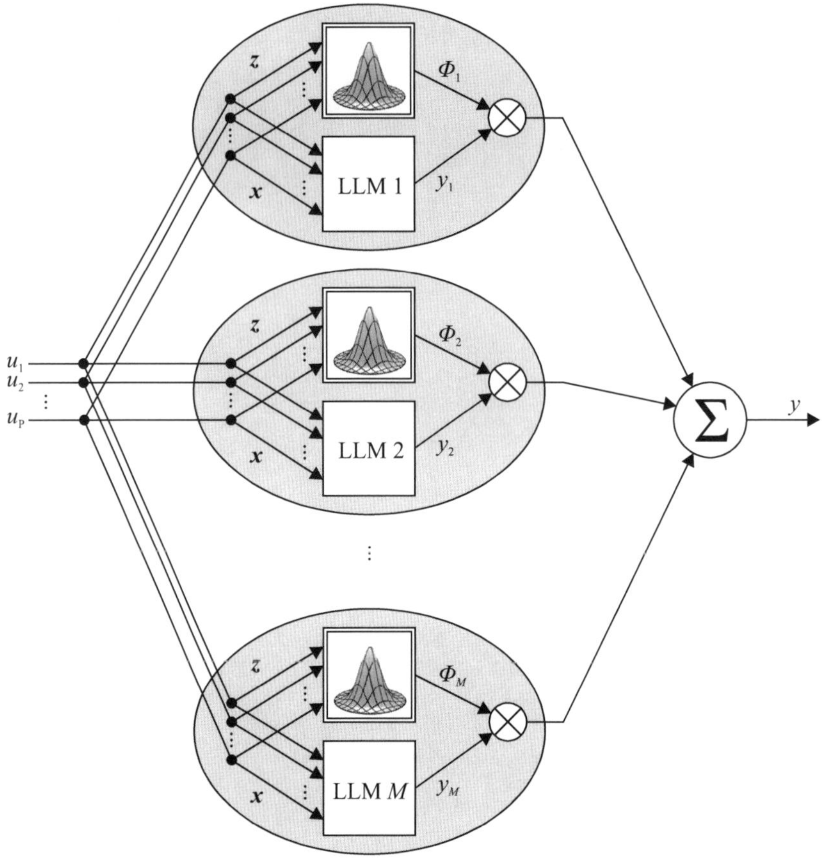

Fig. 20.10. Local Linear Model Network (LOLIMOT)

hyper-rectangles to account for the varying granularity. Thus, the non-linear parameters $c_{i,j}$ and $\sigma_{i,j}$ are determined by a heuristic-avoiding explicit non-linear optimization. LOLIMOT starts with a single linear model that is valid for the whole input space. In each iteration, it splits one local linear model into two new sub-models. Only the (locally) worst performing local model is considered for further refinement. Splits along all input axes are compared and the best performing split is carried out, see Fig. 20.11.

The main advantages of this local model approach are the inherent structure identification and the very fast and robust training algorithm. The model structure is adapted to the complexity of the process. However, explicit application of time-consuming non-linear optimization algorithms can be avoided.

Another local linear model architecture, the so-called *hinging hyperplane trees*, is presented by Töpfer (1998, 2002a,b). These models can be interpreted as an extension of the LOLIMOT networks with respect to the partitioning scheme. While the

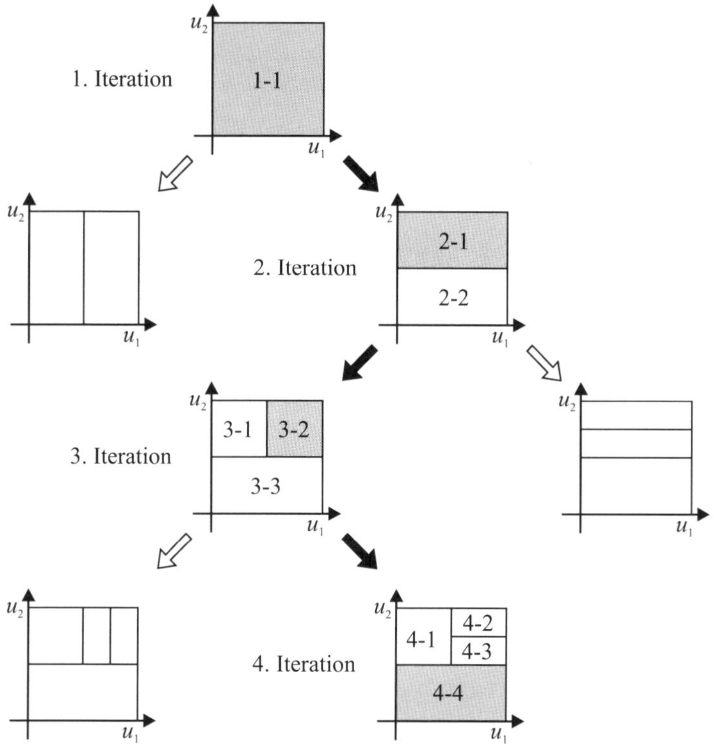

Fig. 20.11. Tree construction (axis orthogonal) of the LOLIMOT algorithm

LOLIMOT algorithm is restricted to axis-orthogonal splits, the hinging hyperplane trees allow an axis-oblique decomposition of the input space. These more complex partitioning strategies lead to an increased effort in model construction. However, this feature has advantages in the case of strongly non-linear model behavior and higher-dimensional input spaces.

The fundamental structures of three artificial neural networks have been described. These models are very well suited to the approximation of measured input/output data of static processes, compare also (Hafner et al, 1992; Preuß and Tresp, 1994). For this, the training data has to be chosen in such a way that the considered input space is as evenly as possible covered with data. After the training procedure, a parametric mathematical model of the static process behavior is available. Consequently, direct computation of the output values $\hat{y}$ for arbitrary input combinations $\boldsymbol{u}$ is possible.

An advantage of the automatic training procedure is the possibility of using arbitrarily distributed data in the training data set. There is no necessity to know data at exactly defined positions, as in the case of grid-based look-up table models, see Sect. 20.2. This clearly decreases the effort required for measurements in practical applications.

Example 20.1 (Artificial Neural Network for the Static Behavior of an Internal Combustion Engine).

As an example, the engine characteristics of a six-cylinder SI (spark-ignition) engine is used. Here, the engine torque has to be identified and is dependent on the throttle angle and the engine speed. Figure 20.12 shows the 433 available data points that were measured on an engine test stand.

For the approximation, an MLP network is applied. After the training, an approximation for the measurement data shown in Fig. 20.13 is given. For that purpose, 31 parameters are required. Obviously, the neural network possesses good interpolation and extrapolation capabilities. This also means that in areas with only few training data, the process behavior can be approximated quite well (Holzmann et al, 1997).

□

20.1.2 Artificial Neural Networks for Dynamic Systems

The memoryless static networks can be extended with dynamic elements to dynamic neural networks. One can distinguish between neural networks with external and internal dynamics (Nelles et al, 1997; Isermann et al, 1997). ANNs with external dynamics are based on static networks, e.g. MLP or RBF networks. The discrete-time input signals $u(k)$ are passed to the network through additional filters $F_i(q^{-1})$. In the same way, either the measured output signals $y(k)$ or the NN outputs $\hat{y}(k)$ are passed to the network through filters $G_i(q^{-1})$. The operator q^{-1} denotes a time shift

$$y(k)q^{-1} = y(k-1)\ . \tag{20.1.15}$$

In the simplest case, the filters are pure time delays, Fig. 20.14a, such that the time-shifted sampled values are the network input signals, i.e.

$$\hat{y}(k) = f_{\mathrm{NN}}\big(u(k), u(k-1), \dots, \hat{y}(k-1), \hat{y}(k-2), \dots\big)\ . \tag{20.1.16}$$

The structure in Fig. 20.14a shows a parallel model (equivalent to the output error model for parameter estimation of linear models). In Fig. 20.14b, the measured output signal is passed to the network input. This represents, the series-parallel model (equivalent to the equation error model for parameter estimation of linear models). One advantage of the external dynamic approach is the possibility of using the same adaptation methods as in the case of static networks. However, the drawbacks are the increased dimensionality of the input space, possible stability problems, and an iterative way of computing the static model behavior, namely through simulation of the model.

ANNs with internal dynamics realize dynamic elements inside the model structure. According to the kind of included dynamic elements, one can distinguish between recurrent networks, partially recurrent networks and locally recurrent globally feedforward networks (LRGF) (Nelles et al, 1997). The LRGF networks maintain the structure of static networks except that dynamic neurons are utilized, see Fig. 20.15.

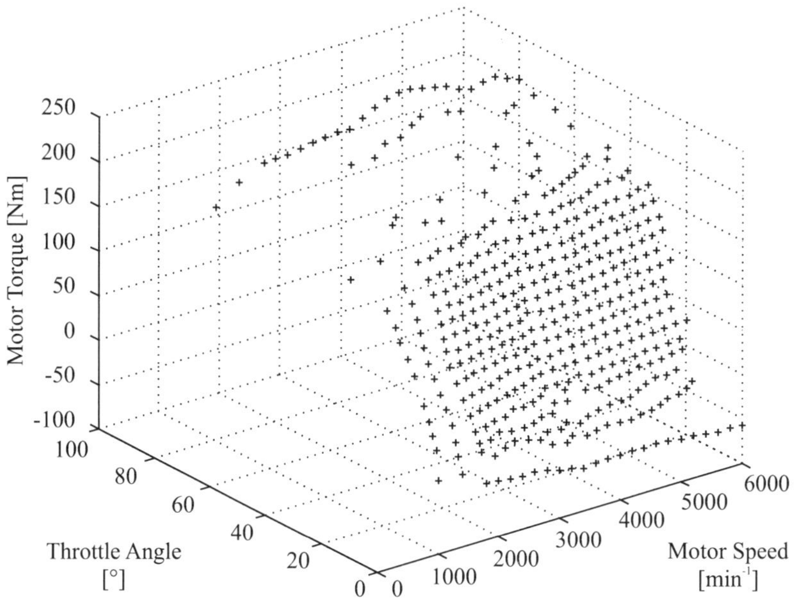

Fig. 20.12. Measured SI engine data (2.5 l, V6 cyl.): Unevenly distributed, 433 measurement data points

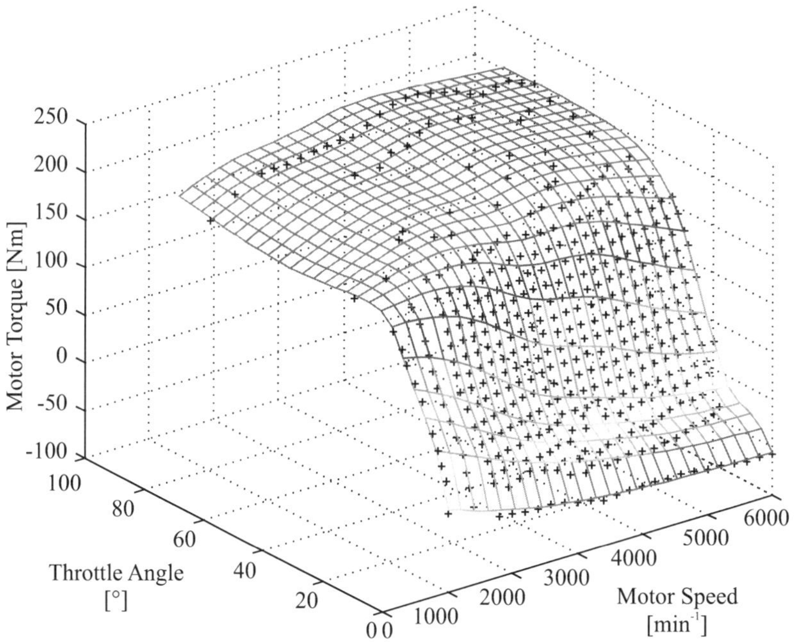

Fig. 20.13. Approximation of measured engine data (+) with an MLP network $(2 \cdot 6 \cdot 1)$: 31 parameters

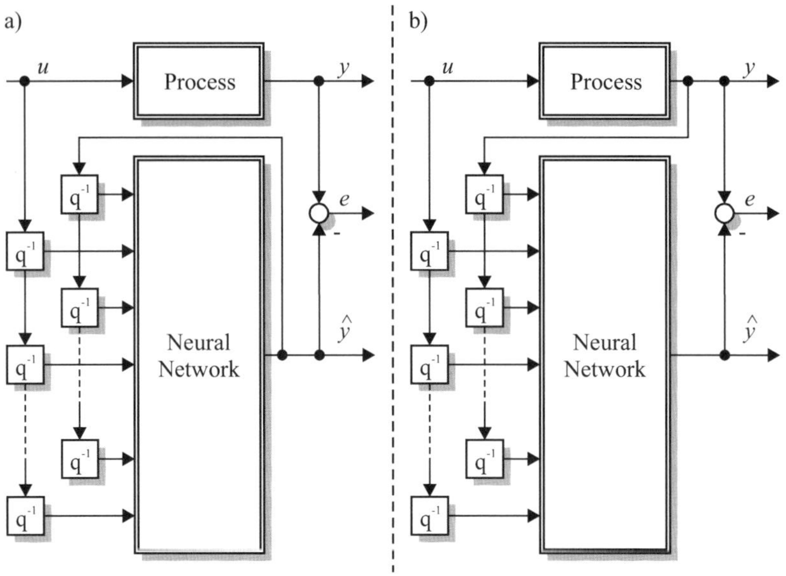

Fig. 20.14. Artificial neural network with external dynamics. (**a**) parallel model. (**b**) series-parallel model

The following can be distinguished: Local synapse feedback, local activation feedback, and local output feedback. The simplest case is the local activation feedback (Ayoubi, 1996). Here, each neuron is extended by a linear transfer function, most often of first or second order, see Fig. 20.16. The dynamic parameters a_i and b_i are adapted. Static and dynamic behavior can be easily distinguished and stability can be guaranteed.

Usually, MLP networks are used in LRGF structures. However, RBF networks with dynamic elements in the output layer can be applied as well, if a Hammerstein-structure of the process can be assumed (Ayoubi, 1996). Usually, the adaptation of these dynamic NNs is based on extended gradient methods (Nelles et al, 1997).

Based on the basic structure of ANNs, special structures with particular properties can be built. If, for example, the local linear model network (LOLIMOT) is combined with the external dynamic approach, a model structure with locally valid linear input/output models result. The excitation of non-linear processes requires the application of multi-valued input signals, e.g. amplitude modulated random binary signals (e.g. APRBS or AGRBS), see Sect. 6.3.

20.1.3 Semi-Physical Local Linear Models

Frequently the static or dynamic behavior of processes depends on the operating point, described by the variables $\boldsymbol{z}$. Then all the inputs have to be separated into manipulated variables $\boldsymbol{u}$ and operating point variables $\boldsymbol{z}$. By this separation, local

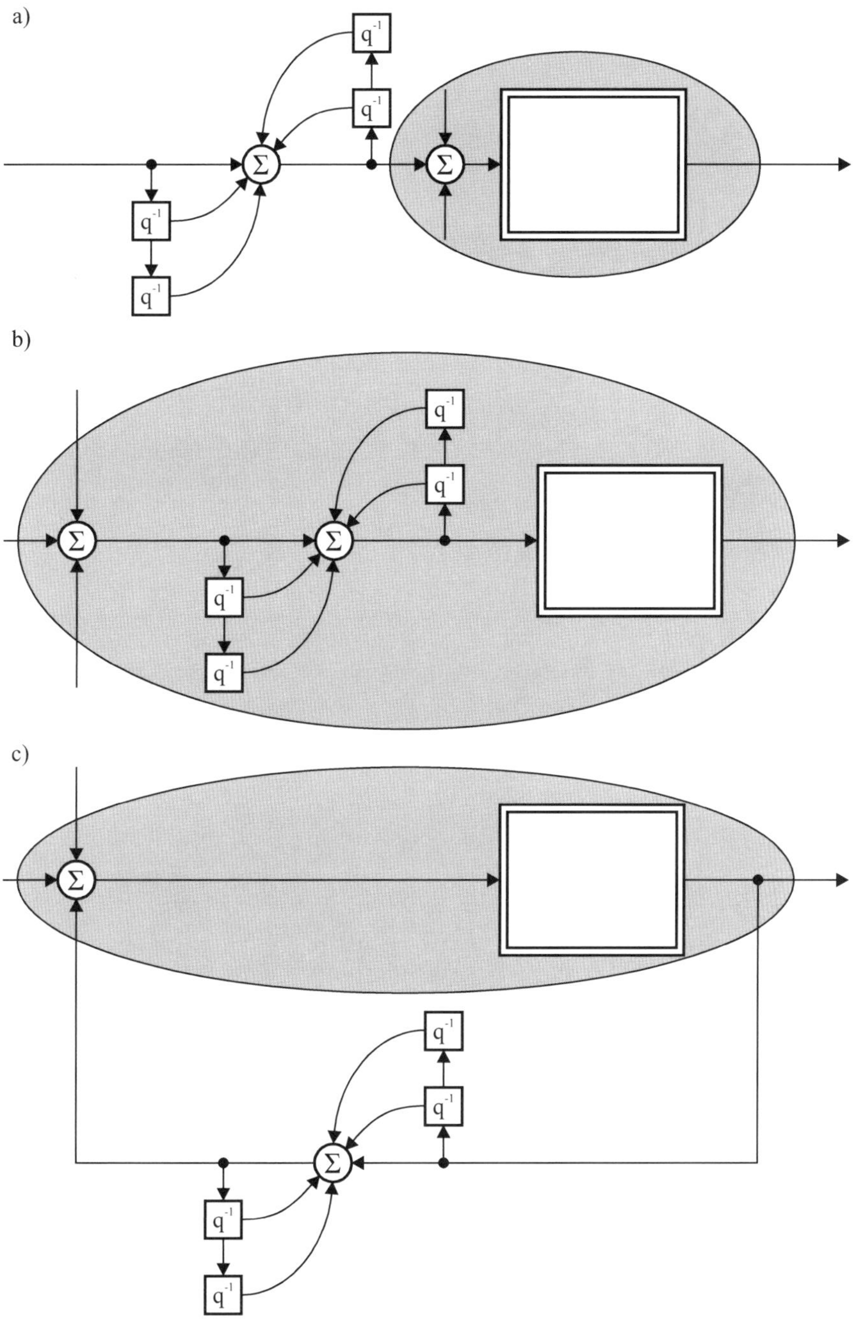

Fig. 20.15. Dynamic neurons for neural networks with internal dynamics. (**a**) local synapse feedback. (**b**) local activation feedback. (**c**) local output feedback

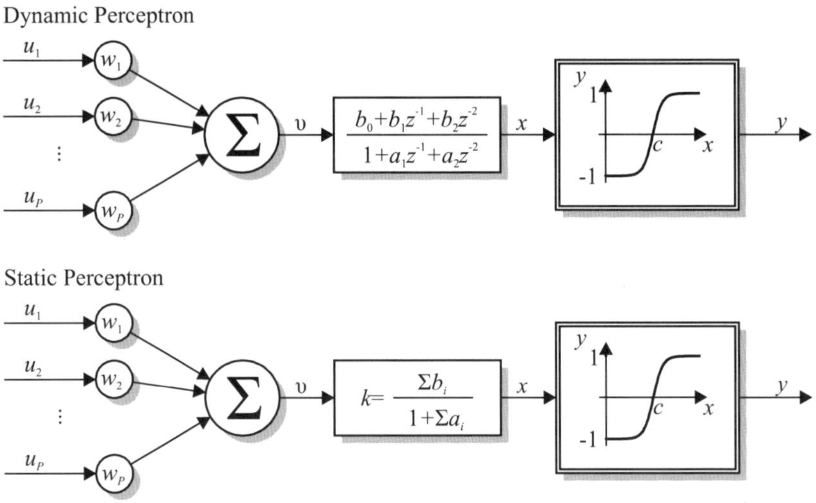

Fig. 20.16. Dynamic and static perceptron (Ayoubi, 1996)

linear models can be identified with varying parameters depending on the operating point, also called linear parameter variable models (LPVM) (Ballé, 1998).

A non-linear discrete-time dynamic model with p inputs u_i and one output y can be described by

$$y(k) = f(\boldsymbol{x}(k)) \tag{20.1.17}$$

with

$$\begin{aligned}\boldsymbol{x}^{\mathrm{T}}(k) = \big(&u_1(k-1) \ldots u_1(k-n_{u1}) \ldots u_p(k-1) \ldots u_p(k-n_{up}) \\ &y(k-1) \ldots y(k-n_y)\big) .\end{aligned} \tag{20.1.18}$$

For many types of non-linearities this non-linear (global) overall model can be represented as a combination of locally active submodels

$$\hat{y} = \sum_{m=1}^{M} \Phi_m(\boldsymbol{u}) g_m(\boldsymbol{u}) . \tag{20.1.19}$$

The validity of each submodel g_m is given by its corresponding weighting function Φ_m (also called activation or membership function). These weighting functions describe the partitioning of the input space and determine the transition between neighboring submodels (Nelles and Isermann, 1995; Babuška and Verbruggen, 1996; Murray-Smith and Johansen, 1997a; Nelles, 2001). Different local models result from the way of partitioning the input space $\boldsymbol{u}$, e.g. grid structure, axis-orthogonal cuts, axis-oblique cuts, etc., as well as the local model structure and the transition between submodels (Töpfer, 2002b).

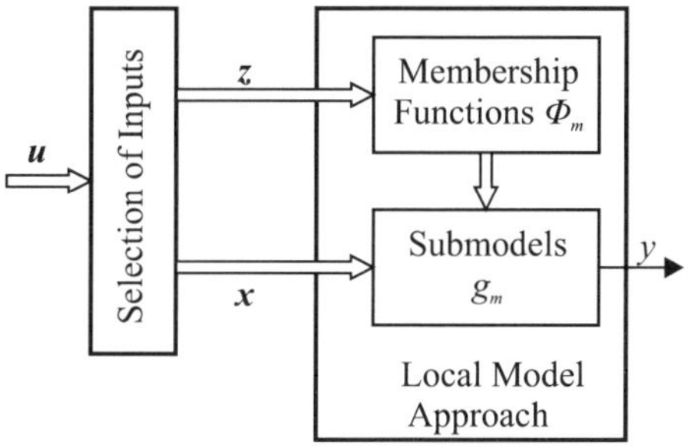

Fig. 20.17. Structure of local model approaches with distinct inputs spaces for local submodels and membership functions

Due to their transparent structure, local models offer the possibility of adjusting the model structure to the process structure in terms of physical law based relationships. Such an incorporation of physical insight improves the training and the generalization behavior considerably and reduces the required model complexity in many cases.

According to (20.1.19), identical input spaces for the local submodels $g_m(\boldsymbol{u})$ and the membership functions $\Phi(\boldsymbol{u})$ have been assumed. However, local models allow the realization of *distinct input spaces*, Fig. 20.17, with

$$\hat{y} = \sum_{m=1}^{M} \Phi_m(\boldsymbol{z}) g_m(\boldsymbol{x}) \,. \tag{20.1.20}$$

The input vector $\boldsymbol{z}$ of the weighting functions contains only those inputs of the vector $\boldsymbol{u}$ having significant non-linear effects that cannot be captured sufficiently well by the local submodels. Only those directions require a subdivision into different parts. The decisive advantage of this procedure is the considerable reduction of the number of inputs in $\boldsymbol{z}$. Thus, the difficult task of structure identification can be simplified.

The use of separate input spaces for the local models (vector $\boldsymbol{x}$) and the membership functions (vector $\boldsymbol{z}$) becomes more precise by considering another representation of the structure in (20.1.20). As normally, local model approaches are assumed to be linear with respect to their parameters according to

$$g_i(\boldsymbol{x}) = w_{i0} + w_{i1}x_1 + \ldots + w_{in_x}x_{n_x} \,. \tag{20.1.21}$$

One can arrange (20.1.20) as

$$\hat{y} = w_0(\boldsymbol{z}) + w_1(\boldsymbol{z})x_1 + \ldots + w_{n_x}(\boldsymbol{z})x_{n_x} \text{ with } w_j(\boldsymbol{z}) = \sum_{m=1}^{M} w_{mj}\Phi_m(\boldsymbol{z}) \,, \tag{20.1.22}$$

where typically a constant term w_{m0} is added to each local model making it an affine local model rather than a purely linear local model. The term w_{m0} is used to model the operation point dependent DC value of the large signal values.

Thus, the specified local model approaches can be interpreted as linear in the parameters with operating point dependent parameters $w_j(\boldsymbol{z})$, whereupon these parameters depend on the input values in the vector $\boldsymbol{z}$. Consequently, the process co-

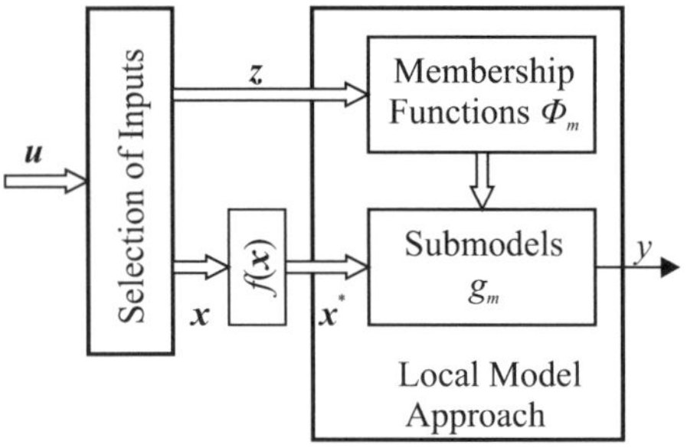

Fig. 20.18. Pre-processing of input variables $\boldsymbol{x}$ for incorporation of prior knowledge into the submodel structure

efficients $w_j(\boldsymbol{z})$ still have a physical meaning. To account for this, these models are called *semi-physical models* (Töpfer et al, 2002).

The choice of *approximate submodel structures* always requires a compromise between submodel complexity and the number of submodels. The most often applied linear submodels have the advantage of being a direct extension of the well known linear models. However, under certain conditions, more complex submodels may be reasonable. If the main non-linear influence of input variables can be described qualitatively by a non-linear transformation of the input variables (e.g. $f_1(x) = (x_1^2,\ x_1 x_2, \ldots)$), then the incorporation of that knowledge into the submodels leads to a considerable reduction of the required number of submodels. Generally, this approach can be realized by a pre-processing of the input variables $\boldsymbol{x}$ to the non-linearly transformed variables, Fig. 20.18,

$$\boldsymbol{x}^* = F(\boldsymbol{x}) = \big(f_1(\boldsymbol{x})\ f_2(\boldsymbol{x}) \ldots f_{n_{x*}}(\boldsymbol{x})\big)^{\mathrm{T}} . \tag{20.1.23}$$

Besides those heuristically determined model structures, local model approaches also enable the incorporation of fully physically determined models. Furthermore, local models allow the employment of inhomogeneous models. Consequently, different local submodel structures are valid within the different operating regimes.

20.1.4 Local and Global Parameter Estimation[1]

For parameter estimation of local dynamic models, some additional care has to be taken in comparison to the stationary case. In general, there are two possibilities to estimate the model parameters: They can either be estimated by means of a *local loss function* or by a *global loss function*. The *local loss function* is given as

$$V_{m,\text{local}} = \frac{1}{2} \sum_{k=1}^{N} (y - \hat{y}_m)^{\mathrm{T}} \boldsymbol{W}_m (y - \hat{y}_m) , \tag{20.1.24}$$

where $\boldsymbol{W}_m$ is the diagonal weighting matrix of the m-th local model resulting from the activation function $\Phi_m(\boldsymbol{z})$. This loss functions is set up and minimized individually for each local model $m = 1, 2, \ldots M$. According to the least squares solution (9.5.4), the model parameters calculate to

[1] compiled by Heiko Sequenz

$$w_m = (X^{\mathrm{T}} W_m X)^{-1} X^{\mathrm{T}} W_m y \ . \tag{20.1.25}$$

It can be seen, that the local parameters are independent with respect to each other. Thus they can be estimated consecutively. The local models can be regarded as local linearizations of the process, independent of the neighboring linearizations. Therefore, an implicit regularization of the parameters is given by the local parameter estimation.

In contrast to that, the *global loss function* can be written as

$$V_{\text{global}} = \frac{1}{2} \sum_{k=1}^{N} (y - \hat{y})^{\mathrm{T}} (y - \hat{y}) \ , \tag{20.1.26}$$

where the global model output is calculated as in (20.1.12) by a superposition of local models

$$\hat{y} = \sum_{m=1}^{M} W_m \hat{y}_m \ . \tag{20.1.27}$$

Inserting (20.1.27) in (20.1.26), it can be seen that the weighting matrix is introduced quadratically in contrast to the local case (20.1.24), where the error is weighted with a single weighting matrix. Again, the model parameters can be calculated by the least squares solution

$$w = \left(X_{\mathrm{g}}^{\mathrm{T}} X_{\mathrm{g}}\right)^{-1} X_{\mathrm{g}}^{\mathrm{T}} y \ , \tag{20.1.28}$$

with the global regression matrix

$$X_{\mathrm{g}} = \begin{pmatrix} W_1 X & W_2 X & \dots & W_M X \end{pmatrix} . \tag{20.1.29}$$

Here w is the global parameter vector consisting of all local model parameters. The local model parameters w_m are therefore coupled with each other by the matrix X_{g}. This makes the global model more flexible as the transition regions of the local models are also used to adapt the process. On the other hand, the model loses its regularization effect. Due to the higher flexibility, the variance-error increases. Furthermore, the model can no longer be interpreted locally and the parameter estimation becomes more complex.

For these reasons, it is recommended to use the local loss function for parameter estimation if possible. This is in general possible for stationary models, but is not possible for some dynamic models as is shown in the following section.

20.1.5 Local Linear Dynamic Models[2]

This section is organized in four parts. First, the choice of z-regressors for dynamic models is discussed. The second part presents a variation of the model structure based on parameters varying at different points of time. The following part presents the analogy of the presented transfer function structure to the state space structure. Finally, the parameter estimation is discussed for different error configurations. It is shown for which model structures, a minimization of the global loss function becomes necessary.

[2] compiled by Heiko Sequenz based on the dissertation by Ralf Zimmerschied (2008)

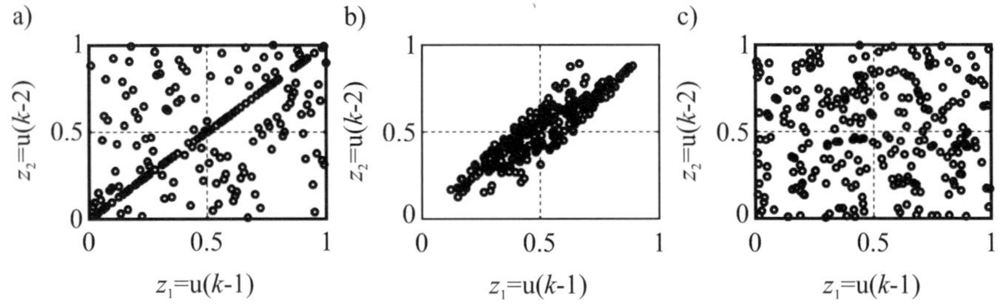

Fig. 20.19. Distribution in the z-regressor space for z-regressor from the same inputs showing (**a**) an APRBS input signal (**b**) the output of a PT_1 first order process (with the APRBS as input) and (**c**) a white noise input signal.

Choice of z-regressors

For dynamic models, the number of parameters is a multiple of the number of parameters in the stationary case. To lower the computational effort, the input spaces of the local models and the activation functions can be distinct as was already shown in Sect. 20.1.3. There, the separation of the input spaces was motivated by a priori knowledge, and only those inputs with significant non-linear influence were utilized for the activation function. The inputs of the activation function are denoted as z-regressors and the inputs of the local models as x-regressors. The z- and the x-regressors consist of process inputs u and process outputs y, since both u and y are regarded as model inputs for dynamic models.

Instead of a priori knowledge, the separation of the input spaces will here be motivated from the point of view of identification. Therefore, the distribution in the z-regressor space will be regarded for commonly used input signals. Figure 20.19a shows the distribution of an APRBS (see Sect. 6.3) with the z-regressors chosen as $z_1 = u(k-1)$ and $z_2 = u(k-2)$. It can be seen that most of the data is placed on the diagonal, which is due to the holding time of an APRBS. The data is only distributed around the diagonal if a step in the input signal occurs. The example shows the utilization of a single input $u(k)$ with two different time delays, but this can be generalized to any z-regressor vector arising from the same input with different time delays. The distribution is even worse if the process output is regarded as a z-regressor, see Fig. 20.19b. Here, a simple PT_1 first order process is considered. Due to its filter characteristics, most of the data is placed around the diagonal. It can be seen that broad areas of the input space are not covered.

The poor distribution in Fig. 20.19a can be encountered with different input signals. An equally distributed white noise as is shown in Fig. 20.19c covers the input space quite well. However, even this input signal does not improve the poor distribution shown in Fig. 20.19b. Furthermore, a white noise input signal might not be an appropriate system excitation as it excites the high frequencies and has a poor signal to noise ratio.

Because of these reasons, it is recommended to use at most one regressor of each model input/output as z-regressor. Meaning for the considered example that either

$z_1 = u(k-1)$ can be used as input for the activation function or $z_1 = u(k-2)$ but not both. This limits the model flexibility and thus increases the bias-error but at the same time decreases the variance-error. Furthermore, a bad coverage of the input space arising from z-regressors of the same model input/output as was shown in this section can easily be avoided.

Delayed Parameter Variation

The choice of z-regressors can be further improved with a delayed parameter variation. So far, the model parameters depend on the z-regressor for a definite time, such as $\boldsymbol{z}(k) = u(k-1)$. Meaning that for a second order SISO system with $\boldsymbol{z}(k) = u(k-1)$ as the only z-regressor, the output calculates to

$$\begin{aligned}\hat{y}(k) &= \sum_{m=1}^{M} \Phi_m(\boldsymbol{z}(k))\big(b_1 u(k-1) + b_2 u(k-2) + a_1 y(k-1) + a_2 y(k-2)\big)\\ &= b_1(\boldsymbol{z}(k))u(k-1) + b_2(\boldsymbol{z}(k))u(k-2) + a_1(\boldsymbol{z}(k))y(k-1) + a_2(\boldsymbol{z}(k))y(k-2)\end{aligned} \tag{20.1.30}$$

with the parameters in (20.1.30) expressed with respect to the z-regressor

$$b_i(\boldsymbol{z}(k)) = \sum_{m=1}^{M} \Phi_m(\boldsymbol{z}(k)) b_{i,m} \tag{20.1.31}$$

and the parameters a_i analog. Notice that all parameters vary with a fixed delay time of the z-regressor, which is $\boldsymbol{z}(k) = u(k-1)$ in this example. Therefore, model inputs can be found, which are further in the past than their parameters' activation function, such as $u(k-2)$ with the parameter b_2 depending on the activation functions $\Phi_m(\boldsymbol{z}(k)) = \Phi_m(u(k-1))$. This means in general, that inputs $u(k-i)$ can be coupled with parameters $b(k-j)$ depending on different times $k-i$ and $k-j$ respectively.

The transfer function can also be written in the general form

$$\hat{y}(k) = \sum_{m=1}^{M} \Phi_m(\boldsymbol{z}(k)) \left(\sum_{i=1}^{n_u} b_i u(k-i) + \sum_{i=1}^{n_y} a_i y(k-i) \right) \tag{20.1.32}$$

with the dynamic orders n_u and n_y for the input and output respectively.

It is therefore recommended to change the parameters individually with their inputs. Then, for the inputs $u(k-1)$ and $y(k-1)$, the parameters b_1 and a_1 depend on the z-regressor $\boldsymbol{z}(k-1) = u(k-1)$ and for the input $u(k-2)$ and $y(k-2)$ the parameters b_2 and a_2 depend on $\boldsymbol{z}(k-2) = u(k-2)$. This varied model structure does not change the model flexibility but has some desirable theoretical properties as it can be transformed to a state space structure, which is shown in the following section. The model given in (20.1.30) can then be written as

$$\begin{aligned}\hat{y}(k) &= b_1(z(k-1))u(k-1) + b_2(z(k-2))u(k-2) \\ &\quad + a_1(z(k-1))y(k-1) + a_2(z(k-2))y(k-2)\,.\end{aligned} \tag{20.1.33}$$

Notice the difference of the individually delayed parameters in (20.1.33) to the not individually delayed form in (20.1.30).

The model can also be written in the general form with the dynamic orders n_u and n_y

$$\hat{y}(k) = \sum_{i=1}^{n_u} b_i(z(k-i))u(k-i) + \sum_{i=1}^{n_y} a_i(z(k-i))y(k-i) \tag{20.1.34}$$

and the parameters b_i as sum of the $m = 1, 2, \ldots, M$ local parameters $b_{i,m}$

$$b_i(z(k-i)) = \sum_{m=1}^{M} \Phi_m(z(k-i))b_{i,m}\,. \tag{20.1.35}$$

Notice again the difference of the non-delayed form in (20.1.32) and the delayed form in (20.1.34).

Analogy to the State Space Structure

Given the model as a transfer function with the previously presented delayed parameter variation, it can be transformed to a state space structure. It shall be mentioned that the transformation from local linear transfer functions to local state space models is in general not obvious and a meaningful analog local state space model might not even exist. However, given the model in the presented structure (20.1.34), the analog state space model can be stated in the following observable canonical form:

$$\boldsymbol{x}(k+1) = \begin{pmatrix} 0 & \ldots & 0 & -a_n(z(k)) \\ 1 & & 0 & -a_{n-1}(z(k)) \\ \vdots & \ddots & \vdots & \vdots \\ 0 & \ldots & 1 & -a_1(z(k)) \end{pmatrix} \boldsymbol{x}(k) + \begin{pmatrix} b_n(z(k)) \\ b_{n-1}(z(k)) \\ \vdots \\ b_1(z(k)) \end{pmatrix} u(k) \tag{20.1.36}$$

$$y(k) = \begin{pmatrix} 0 & \ldots & 0 & 1 \end{pmatrix} \boldsymbol{x}(k)\,. \tag{20.1.37}$$

To prove the equivalence of (20.1.37) and (20.1.34), the transformation from the state space structure to the transfer function will be sketched in the following. Writing the matrix equation (20.1.37) row by row, the states could be written as

$$\begin{aligned} x_1(k+1) &= - a_n(z(k))x_n(k) + b_n(z(k))u(k) \\ x_2(k+1) &= x_1(k) - a_{n-1}(z(k))x_n(k) + b_{n-1}(z(k))u(k) \\ &\vdots \\ x_n(k+1) &= x_{n-1}(k) - a_1(z(k))x_n(k) + b_{n-1}(z(k))u(k)\,. \end{aligned} \tag{20.1.38}$$

Substituting the first row delayed by one time step ($x_1(k)$) in the second row gives

$$\begin{aligned} x_2(k+1) = & - a_n(z(k-1))x_n(k-1) + b_n(z(k-1))u(k-1) \ldots \\ & - \quad a_{n-1}(z(k))x_n(k) \quad + \quad b_{n-1}(z(k))u(k) \,. \end{aligned} \tag{20.1.39}$$

This equation can again be delayed and inserted for $x_2(k)$ in the third line

$$\begin{aligned} x_3(k+1) = & \; x_2(k) - a_{n-2}(z(k))x_n(k) + \quad b_{n-2}(z(k))u(k) \\ = & \; -a_n(z(k-2))x_n(k-2) + b_n(z(k-2))u(k-2) \ldots \\ & -a_{n-1}(z(k-1))x_n(k-1) + b_{n-1}(z(k-1))u(k-1) \ldots \\ & -a_{n-2}(z(k))x_n(k) \quad + \quad b_{n-2}(z(k))u(k) \end{aligned} \tag{20.1.40}$$

and so on until the last but one line ($x_{n-1}(k)$) is inserted into the last line ($x_n(k+1)$). Finally, with $y(k) = x_n(k)$, the transfer function (20.1.34) is obtained, what proves the equivalence. Therefore, every function given in the delayed parameter variation form can be transformed to a state space realization in observable canonical form.

Parameter Estimation

As mentioned before, the estimation of parameters depends for dynamic models on the error configuration. The two common error configurations are given by the parallel and by the series-parallel model (see Fig. 20.14). The latter corresponds in the linear case to the equation error, which is presented in Sect. 9.1 with the corresponding model denoted as ARX models. Accordingly, a local linear dynamical model in series-parallel configuration as presented here is called *NARX(Non-Linear ARX)* model. The model is denoted as non-linear, since the local linear ARX models compose to a global non-linear ARX model. This model depends on measured inputs and measured outputs and is given by the equation

$$\hat{y}_{\mathrm{NARX}}(k) = \sum_{m=1}^{M}\left(\sum_{m=1}^{n_u} b_i(z(k-i))u(k-i) + \sum_{m=1}^{n_y} a_i(z(k-i))y(k-i)\right) . \tag{20.1.41}$$

In an analog way, the parallel model configuration corresponds to the output error which again corresponds to the OE model in the linear case. The global non-linear model in parallel configuration which is composed by local linear OE models is therefore denoted as *NOE* (*Non-Linear OE*) model. This model depends on measured inputs and modeled outputs

$$\hat{y}_{\mathrm{NOE}}(k) = \sum_{m=1}^{M}\left(\sum_{m=1}^{n_u} b_i(z(k-i))u(k-i) + \sum_{m=1}^{n_y} a_i(z(k-i))\hat{y}(k-i)\right) . \tag{20.1.42}$$

Notice the difference of the models in the delayed process outputs on the right hand side of (20.1.41) and (20.1.42), which are either measured or modeled in the NARX or the NOE model respectively.

The local model output of the NARX model can therefore be written in dependence of measured values only

$$\hat{y}_{\text{NARX},m}(k) = \sum_{m=1}^{n_u} z(k-i)u(k-i) + \sum_{m=1}^{n_y} z(k-i)y(k-i) \,. \qquad (20.1.43)$$

In contrast to that, the NOE model uses the modeled output also as a model input

$$\hat{y}_{\text{NOE},m}(k) = \sum_{m=1}^{n_u} z(k-i)u(k-i) + \sum_{m=1}^{n_y} z(k-i)\hat{y}(k-i) \,. \qquad (20.1.44)$$

For the NOE model, the outputs must therefore be simulated up to $\hat{y}(k-1)$ to determine $\hat{y}(k)$. This means in particular, that all local models use the global model output as input and are therefore dependent on one another. Hence for the NOE model, the parameters need to be estimated all at once. Thus a global parameter estimation, as presented in Sect. 20.1.4, is necessary. Since the error is no longer linear in the parameters, non-linear optimization algorithms, such as the Levenberg-Marquardt algorithm (see Sect. 19.4.3) are applied. This makes the training of the NOE model computationally costly. Furthermore, the convergence to the global minimum can no longer be guaranteed as convergence depends on the initial values.

In contrast to that, the local models of the NARX model can be regarded independently as the measured outputs are available. Therefore, either the local or the global parameter estimation can be applied. As the local estimation has some desirable properties, such as an implicit regularization, and is computationally much faster, this is usually preferred.

Despite the higher computational effort, the NOE model is often preferred if the model shall be used for simulation. The NOE model minimizes the simulation error and is therefore better suited for simulation than the NARX model. This model on the other hand minimizes the prediction error and would therefore be preferred if the process output shall be predicted on the base of previous measured outputs. Surely, the NARX model can also be used for simulation but as it was not trained in that configuration, it is expected to have a bias-error (Ljung, 1999).

A major drawback of the NOE model is its lost local interpretability. That drawback can be met with an explicit regularization, which is described in more detail in (Zimmerschied and Isermann, 2008, 2009).

20.1.6 Local Polynomial Models with Subset Selection[3]

In this section, a further development of the LOLIMOT algorithm (Sect. 20.1.1) is presented. The extensions are based on the algorithm presented in (Sequenz et al, 2009), where some parts are described in more detail. The algorithm is primarily suited for stationary processes, but can also be used for dynamic processes. As was already shown in Sect. 20.1.1, the LOLIMOT algorithm consists of an outer loop to adapt to the non-linearities of the process and an inner loop for local model estimation. So far the local models are restricted to be linear or affine.

[3] compiled by Heiko Sequenz

To overcome the associated limitations, here the assumption of a general local function is made. The local function is approximated by a power series up to a predefined order o. This allows a reduction of the number of local models, since now, the local models are more complex and have thus a wider validity region. As the number of x-regressors grows fast with the order o and the number of inputs p, a selection algorithm of the significant regressors is introduced. This algorithm eliminates regressors without process information and hence decreases the variance error.

The structure of the model is illustrated in Fig. 20.20. The local models are denoted as LPM (*Local Polynomial Model*) and are weighted by Gaussian activation functions. The decomposition of the input space is attained by a tree construction algorithm with axis orthogonal partitions, see Fig. 20.11. Because this outer loop is the same as for the LOLIMOT algorithm, in the following only the estimation of the local models will be illustrated. Further the term regressors will be used for the x-regressors as the z-regressors are not further regarded.

The idea for the local model approach is as follows: Basically all steady functions appearing in real applications can be approximated arbitrarily well by a power series in a neighborhood of its center. This center point is denoted by $\boldsymbol{u}_0$, and hence a multi-dimensional power series for a general non-linear function can be written as (e.g. Bronstein et al, 2008)

$$f(u_0 + \Delta u) = f(u_0) + \sum_{i=1}^{o} \frac{1}{i!}\left(\frac{\partial}{\partial u_1}\Delta u_1 + \ldots + \frac{\partial}{\partial u_p}\Delta u_p\right)^i f(u_0) + R_o \,, \tag{20.1.45}$$

where o is the order of the power series expansion and R_o the remainder. The remainder decreases with the order, but the number of regressors increases. Therefore, an order $o = 3$ is recommended as a good trade-off between accuracy and computational effort (Hastie et al, 2009). In order to eliminate negligible regressors, a selection algorithm needs to be performed.

Motivated by the power series, an admissible set of regressors is given as

$$\mathcal{A} = \left\{u_1, u_2, u_1^2, u_2^2, u_1 u_2, \ldots\right\} \tag{20.1.46}$$

from which the best subset of regressors needs to be selected. A guarantee to find the best subset can only be given by checking all possible subsets of regressors. As this is not feasible even for relatively small admissible sets, a heuristic selection algorithm is introduced.

This selection algorithm combines *forward selection*, *backward elimination*, and *replacement of regressors*. It is sketched in Fig. 20.21. The algorithm starts with an empty set and adds successively regressors to that selection set from the admissible set by a forward selection. After adding a new regressor to the selection set, some previously selected regressors might become negligible. Therefore, a backward elimination is performed to delete regressors from the selection set. The replacement of regressors tries to swap regressors between the selection set and the admissible set to exchange significant regressors by even more significant regressors. A detailed overview of selection algorithms can be found in (Miller, 2002). It shall be men-

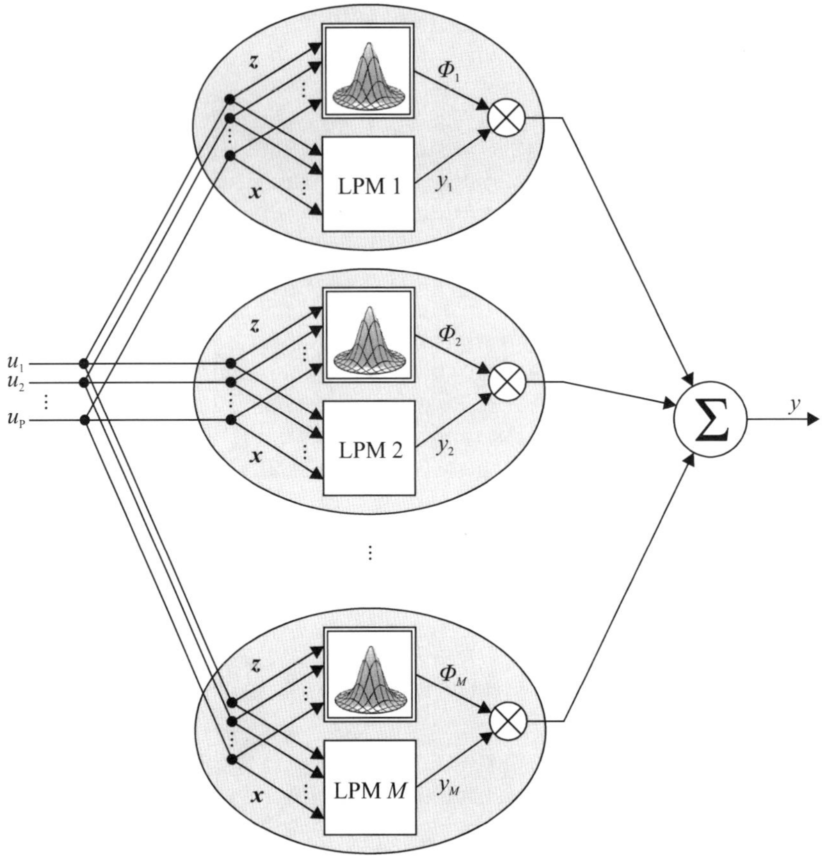

Fig. 20.20. Local Polynomial Model (LPM) Network — a further development of the LOLIMOT structure

tioned, that the combination of the selection steps should depend on the size of the admissible set, as it can become computationally costly as well.

For comparing models with regressor sets of different sizes, a criterion of fit is required to rate the models. Possible criteria are among others Mallows' C_p-statistic, Akaike's AIC, and Bayesian Information Criterion (BIC). Mallows' C_p-statistic can be calculated as

$$C_p = \frac{\sum_{i=1}^{N}(y_i - \hat{y}_i)^2}{\hat{\sigma}^2} - N + 2n \, , \tag{20.1.47}$$

where $\hat{\sigma}^2$ is an estimation of the residual variance, $\hat{y}$ the output of the considered model with n parameters fitted and N the size of the dataset. Similar, the AIC corrected for finite sets is given as

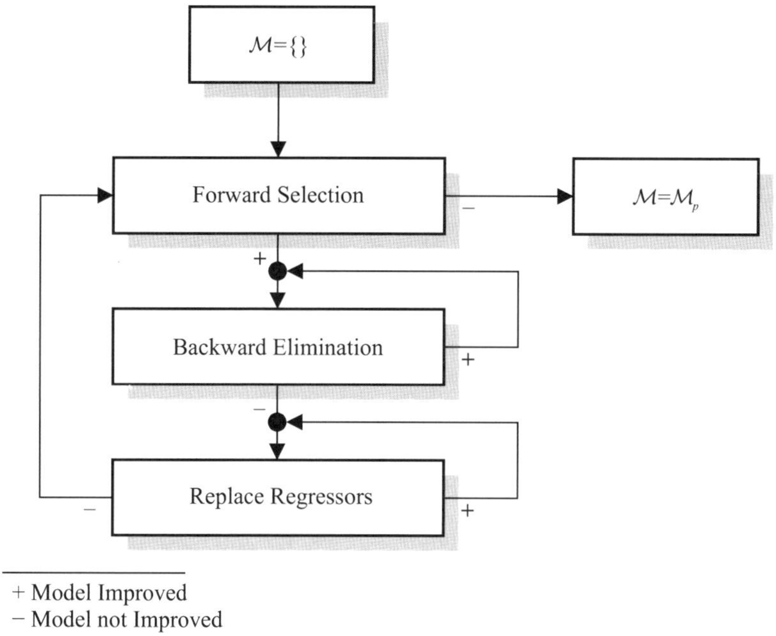

Fig. 20.21. Selection algorithm to find the set of most significant regressors starting from the empty set

$$\mathrm{AIC}_c = N \ln\left(\frac{\sum_{i=1}^{N}\left(y_i - \hat{y}_i\right)^2}{N}\right) + \frac{2N}{N-n-1} n \; . \tag{20.1.48}$$

Both give a trade-off between the number of regressors n, which increases with the model complexity, and the error $\sum_{i=1}^{N}(y_i - \hat{y}_i)^2$, that is decreasing with the model complexity. This trade-off is an estimator of the minimum of the model error decomposed into bias and variance error, see Fig. 20.2. Other criteria such as the BIC work in a similar way and are detailed described in (Mallows, 1973; Burnham and Anderson, 2002; Stoica and Selen, 2004; Loader, 1999).

As these criteria are designed for global models, and since here local models with weighting functions are regarded, these criteria must be adapted. A local version of Mallows' C_p is given by Loader (1999) as

$$C_{p,\mathrm{local},j} = \frac{(\boldsymbol{y} - \hat{\boldsymbol{y}}_j)' \boldsymbol{W}_j (\boldsymbol{y} - \hat{\boldsymbol{y}}_j)}{\hat{\sigma}^2} + \mathrm{tr}(\boldsymbol{W}_j) + 2 n_{\mathrm{eff},j} \; , \tag{20.1.49}$$

where the numerator is the local model error instead of the global model error as in (20.1.47). The term $n_{\mathrm{eff},j}$ describes the effective number of parameters in the jth local model. The effective number of parameters is smaller than the total number of parameters n. This is caused by the overlapping of the local models, which introduces a regularization on the local models and thus reduces the total number of degrees of freedom. These degrees of freedom are denoted for the local models as the effective number of parameters. Further literature about the effective number of parameters can be found in (Moody, 1992) or (Murray-Smith and Johansen, 1997b). It can be calculated by the trace of the HAT matrix (see Sect. 23.7.2) and is given as

$$n_{\mathrm{eff},j} = \mathrm{tr}\Big(\boldsymbol{W}_j \boldsymbol{X}\big(\boldsymbol{X}'\boldsymbol{W}_j\boldsymbol{X}\big)^{-1}\boldsymbol{X}'\boldsymbol{W}_j\Big) = \mathrm{tr}\,\boldsymbol{H}_j\;. \tag{20.1.50}$$

The residual variance, indicated as $\hat{\sigma}^2$ in (20.1.49), can be estimated as the weighted squared error sum divided by the degrees of freedom

$$\hat{\sigma}^2 = \frac{\displaystyle\sum_{j=1}^{M}(y-\hat{y}_j)'\boldsymbol{W}_j(y-y_j)}{N-\displaystyle\sum_{j=1}^{M}n_{\mathrm{eff},j}}\;. \tag{20.1.51}$$

To evaluate the local model output $\hat{y}$, the parameters have to be determined first. This can for the stationary local model simply be done by the least squares solution. Notice that regressors, like u_1u_2 are themselves non-linear, but their parameters c_i are linear in the model equation

$$\hat{y}_j = \sum_{m=1}^{M} W_j\big(c_1u_1 + c_2u_1^2 + c_3u_1u_2 + \ldots\big)\;. \tag{20.1.52}$$

Given this criterion, local models with differently sized regressor sets can be rated. Thus for each local model, the best regressor set can be selected, which allows an individual local adaption to the process non-linearities.

Besides for local regressor selection, the introduced criteria (20.1.47) or (20.1.48) can be used to select the best global model in the partitioning algorithm (outer loop). As the input space is divided in a tree construction algorithm, several partitions are selectable. Regarding the training error, the model with the most partitions will be best. However, as the variance error increases with the number of parameters and therefore with the number of local models, again a trade-off has to be found. This can be attained by minimizing the global version of the presented criteria.

Regarding the local model structure, a global version of the C_p-statistic is given as

$$C_{p,\mathrm{global}} = \frac{\displaystyle\sum_{j=1}^{M}(y-\hat{y}_j)'\boldsymbol{W}_j(y-y_j)}{\hat{\sigma}^2} - N + 2\left(\sum_{j=1}^{M}n_{\mathrm{eff},j} + M\right)\;. \tag{20.1.53}$$

The M additional parameters in the last term are added to account for the M local variance estimations. With this criterion given, the partitioning algorithm can be terminated if no further improvement is achieved by additional partitions of the input

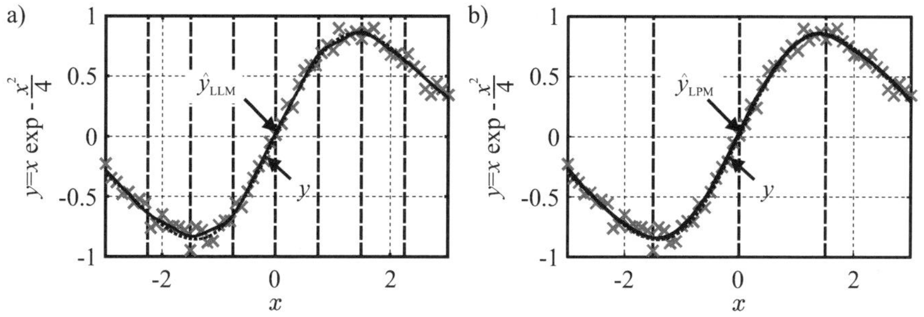

Fig. 20.22. Example of identification of a non-linear static process with (**a**) Local Linear Models (LLM) and (**b**) Local Polynomial Models (LPM)

Table 20.2. NRMSE training and validation error using local linear models (LLM) and local polynomial models (LPM) averaged over 100 simulations

	Training	Validation	# Regressors
LLM	0.073	0.034	16
LPM	0.076	0.032	13

space. It is also possible to divide the input space up to a predefined number of local polynomial models or until a predefined accuracy is reached and then select the best model by (20.1.53).

To illustrate the advantages of the presented algorithm in comparison to the LOLIMOT algorithm, a known non-linear process shall be identified. Figure 20.22 shows a realization of the function

$$y = x\mathrm{e}^{\frac{-x^2}{4}} + \nu \ , \tag{20.1.54}$$

where the true process output is disturbed by a white measurement noise ν. The simulated data is marked by crosses in Fig. 20.22. This data is used to identify the process with the LOLIMOT algorithm and with the modified algorithm showing in the left and the right plots respectively. The training error and the error with respect to the true process are given in Table 20.2.

It can well be seen that both algorithms are able to adapt to the non-linear structure. The LOLIMOT algorithm needs 8 partitions, whereas the modified algorithm needs only half of the partitions as can be seen in Fig. 20.22. Furthermore, the modified algorithm was able to achieve a slightly better quality on the true process with an even smaller total number of regressors, as is shown in Table 20.2. A further advantage of the modified algorithm is, that less transition regions of local models exist. A drawback however is, that the local models become more complex and can no longer be interpreted as linearizations but rather as local power series realizations.

Further examples of the presented algorithm can be found in (Sequenz et al, 2010) and (Mrosek et al, 2010), where the emissions of a Common-Rail Diesel engine are modeled.

It can be summarized, that the modified algorithm is especially useful if only few measurements are available, as only the most significant parameters and less local models have to be estimated. Given a large dataset, both algorithms are able to adapt to the process structure. Furthermore, with the new algorithm, a selection criterion for the best global model is provided.

It shall be mentioned that the same algorithm can be used for identification of dynamic processes. However, depending on the error configuration, the estimation of the parameters might become computationally costly (see *global parameter estimation* in Sect. 20.1.5). Hence the selection algorithm must then be reduced to a very simple one, such as a simple forward selection. Some recent work on selection procedures for dynamic processes can be found in (Piroddi and Spinelli, 2003; Farina and Piroddi, 2009).

Example 20.2 (Dynamic Model of the Three-Mass Oscillator).

A dynamic model of the Three-Mass Oscillator has been identified using the LOLIMOT neural network, see Fig. 20.23. As can be seen from the diagrams, the dynamic behavior has been modeled quite well. Bähr et al (2009) used a LOLIMOT model to model the friction effects. □

20.2 Look-Up Tables for Static Processes

In this section, a further non-linear model architecture besides the polynomial-based models, neural networks, and fuzzy systems is presented. Grid-based look-up tables (data maps) are the most common type of non-linear static models used in practice. Especially in the field of non-linear control, look-up tables are widely accepted as they provide a transparent and flexible representation of non-linear relationships. Electronic control units of modern automobiles, for example, contain up to several hundred such grid-based look-up tables, in particular for combustion engine and emission control (Robert Bosch GmbH, 2007).

In automotive applications, due to cost reasons, computational power and storage capacity are strongly restricted. Furthermore, constraints of real-time operation have to be met. Under these conditions, grid-based look-up tables represent a suitable means of storing non-linear static mappings. The models consist of a set of data points or nodes positioned on a multi-dimensional grid. Each node comprises two components. The scalar data point heights are estimates of the approximated non-linear function at their corresponding data point position. All nodes located on grid lines, as shown in Fig. 20.25, are stored, e.g. in the ROM of the control unit. For model generation, usually all data point positions are fixed a priori. The most widely applied method of obtaining the data point heights is to position measurement data points directly on the grid.

In the following, the most common two-dimensional case will be considered. The calculation of the desired output Z for given input values X and Y consists of

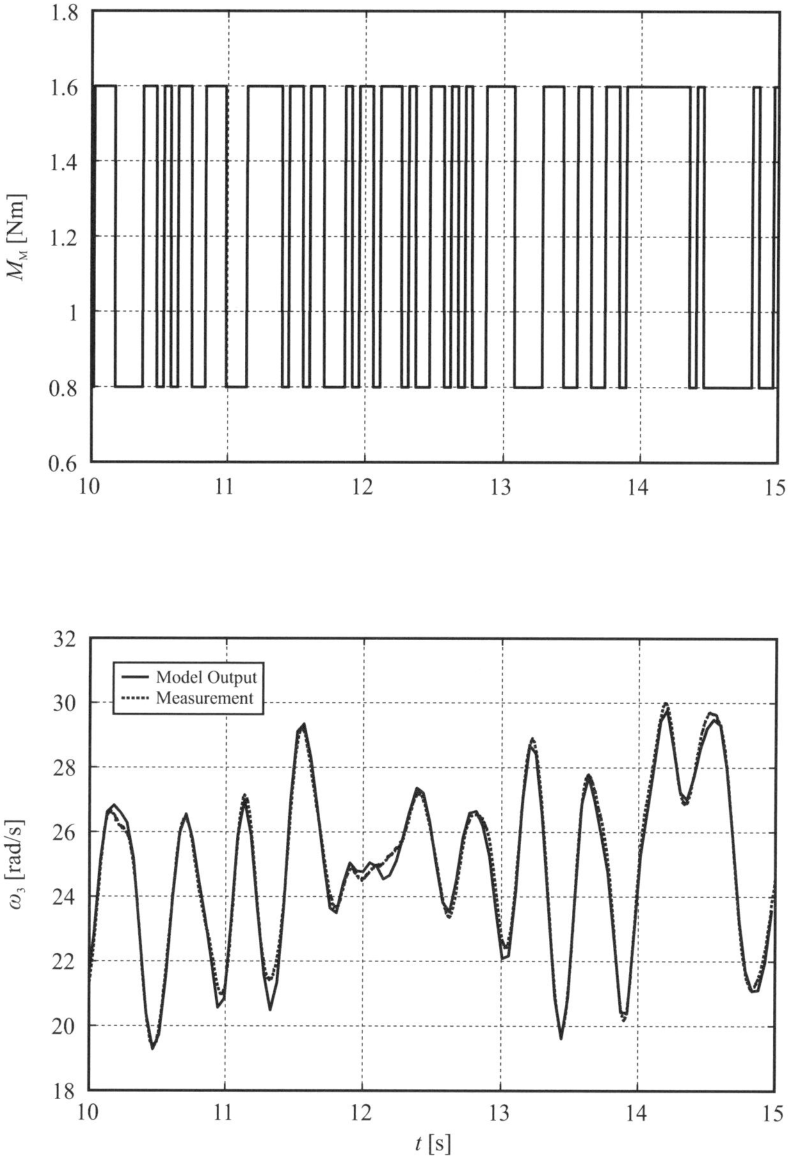

Fig. 20.23. Application of the local linear model estimation algorithm LOLIMOT for the dynamics of the Three-Mass Oscillator. Input: torque $M_{\mathrm{M}}(t)$, output: rotational velocity ω_3, number of local linear models $N = 36$, sample time $T_0 = 0.048$ s, model order $m = 6$

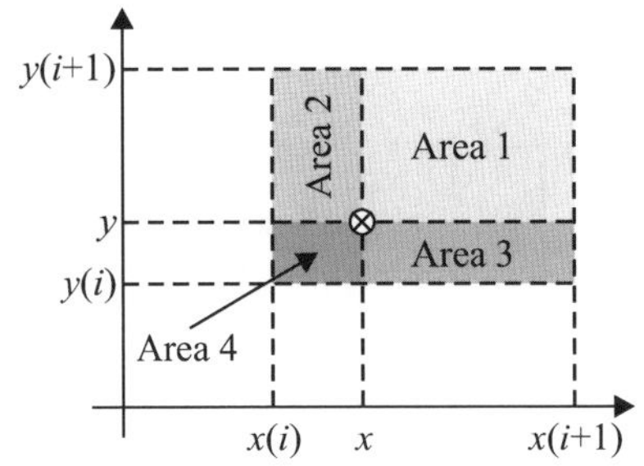

Fig. 20.24. Areas for interpolation within a look-up table

two steps. In the first step, the indices of the enclosing four data points have to be selected. Then, a bilinear area interpolation is performed (Schmitt, 1995). For this, four areas have to be calculated, as shown in Fig. 20.24 (Schmitt, 1995; Töpfer, 2002b).

For the calculation of the desired output Z, the four selected data point heights are weighted with the opposite areas and added up. Finally, this result has to be divided by the total area,

$$\begin{aligned} Z(X,Y) = \Big(&\big(Z(i,j)\ \underbrace{(X(i+j)-X)(Y(j+1)-Y)}_{\text{area 1}}\big) \\ &+\big(Z(i+1,j)\ \underbrace{(X-X(i))(Y(j+1)-Y)}_{\text{area 2}}\big) \\ &+\big(Z(i,j+1)\ \underbrace{(X(i+1)-X)(Y-Y(j))}_{\text{area 3}}\big) \\ &+\big(Z(i+1,j+1)\ \underbrace{(X-X(i))(Y-Y(j))}_{\text{area 4}}\big)\Big) \\ &/\Big(\underbrace{(X(i+1)-X(i))(Y(j+1)-Y(j))}_{\text{overall area}}\Big)\,. \end{aligned} \tag{20.2.1}$$

Because of the relatively simple computational algorithm, area interpolation rules are widely applied, especially in real-time applications. The accuracy of the method depends on the number of grid positions. For the approximation of "smooth" mappings, a small number of data points is sufficient, while for stronger non-linear behavior a finer grid has to be chosen.

The area interpolation is based on the assumption that all data point heights are available in the whole range covered by the grid. However, this condition is often not fulfilled.

Grid-based look-up tables belong to the class of non-parametric models. The described model structure has the advantage that a subsequent adaptation of single data point heights due to changing environmental conditions is easy to realize. However, the main disadvantage of this look-up table is the exponential growth of the number

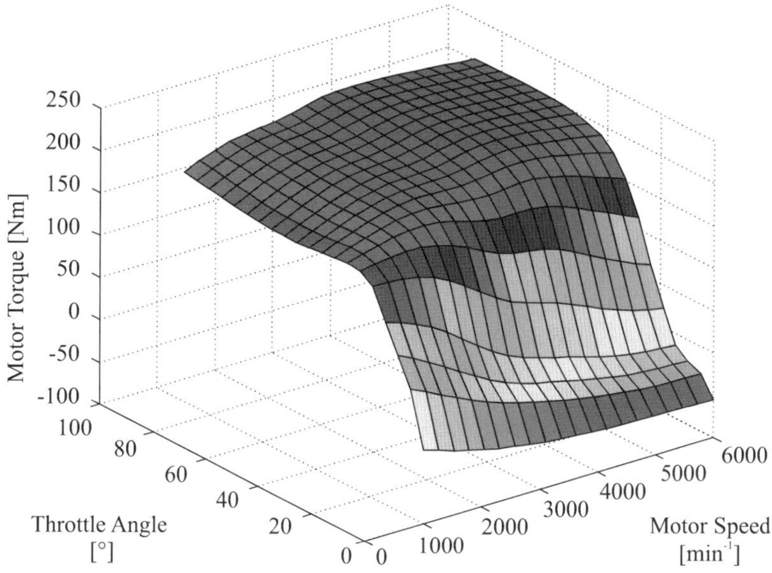

Fig. 20.25. Grid-based look-up table (data map) of a six-cylinder SI engine

of data points with an increasing number of inputs. Therefore, grid-based look-up tables are restricted to one- and two-dimensional input spaces in practical applications. If more inputs have to be considered, this can be handled with nested look-up tables. Determination of the heights of the look-up table based on measurements at arbitrary coordinates with parameter estimation methods is treated by Müller (2003).

Example 20.3 (Grid-Based Look-up Table for the Static Behavior of an Internal Combustion Engine).

The engine characteristics of a six-cylinder SI (spark-ignition) engine already presented in Example 20.1 is used again. This time, the 433 available data points that were measured on an engine test stand and had been shown in Fig. 20.12 were used to generate a grid-based look-up table. The resulting two-dimensional look-up table is shown in Fig. 20.25. □

Another alternative are parametric model representations, like polynomial models, neural networks or fuzzy models, which clearly require less model parameters to approximate a given input/output relationship. Therefore, the storage demand of these models is much lower. However, in contrast to area interpolation, the complexity of the computation of the output is much higher, since non-linear functions for each neuron have to be computed. On the other hand, grid-based look-up tables are not suitable for the identification and modeling of dynamic process behavior.

A detailed overview of model structures for non-linear system identification is given by Nelles (2001).

20.3 Summary

This chapter has discussed the application of neural networks to the identification of non-linear processes. Neural networks are universal static approximators. The Multi Layer Perceptron and the Radial Basis Function network are prominent network structures. The MLP can only be trained by a non-linear iterative optimization algorithm. While the weights of the RBF nets can be determined directly by the method of least squares, one has the problem that the placement of Gaussian basis functions cannot be optimized during the training with the least squares method. Also, the placement of the Gaussian basis functions can be quite difficult for higher dimension, as the basis functions can no longer be evenly spread in the input space.

By means of external dynamics, these universal static approximators can also be used to identify dynamic systems. However, the main disadvantage is the fact that the resulting models cannot be interpreted well as the structure of the neural nets in general does not allow a physical interpretation. However, better interpretability of the models is possible with local linear models, which are weighted by radial basis functions for different operating points. The LOLIMOT net, which represents such a local model approach, has been presented.

Finally, also look-up tables have been presented. They avoid the complex evaluation of the neurons and can also approximate functions universally. The approximation quality is however determined by the spacing of the data grid. Higher dimensional models can be realized by nested look-up tables. However, then the storage space typically increases exponentially.

Problems

20.1. Neural Network Structure
Name the different types of links in a neural network.
How are the different layers named?

20.2. Multi Layer Perceptron
Draw the structure of a neuron.
Give examples of activation functions.
How can the parameters of the net be determined?
How does the complexity of the net increase with additional inputs?

20.3. Radial Basis Function Networks
Draw the structure of a neuron.
Give examples of activation functions.
How can the parameters of the net be determined?
How does the complexity of the net increase with additional inputs?
What is in this context meant by the *curse of dimensionality*?

20.4. Dynamic Neural Networks
What are two possibilities to use a neural net with external dynamics?
What are the advantages and disadvantages of the two approaches?

20.5. Comparison Neural Networks and Look-Up Tables
How is the interpolation performance for neural networks and look-up tables.

20.6. Look-Up Tables
Draw a look-up table for the fuel consumption of an automobile with gasoline engine in dependence on (constant) speed and mass. Discuss the selection of a grid.

References

Ayoubi M (1996) Nonlinear system identification based on neural networks with locally distributed dynamics and application to technical processes . Fortschr.-Ber. VDI Reihe 8 Nr. 591. VDI Verlag, Düsseldorf

Babuška R, Verbruggen H (1996) An overview of fuzzy modeling for control. Control Eng Pract 4(11):1593–1606

Bähr J, Isermann R, Muenchhof M (2009) Fault management for a three mass torsion oscillator. In: Proceedings of the European Control Conference 2009 - ECC 09, Budapest, Hungary

Ballé P (1998) Fuzzy-model-based parity equations for fault isolation. Control Eng Pract 7(2):261–270

Bishop CM (1995) Neural networks for pattern recognition. Oxford University Press, Oxford

Bronstein IN, Semendjajew KA, Musiol G, Mühlig H (2008) Taschenbuch der Mathematik. Harri Deutsch, Frankfurt a. M.

Burnham KP, Anderson DR (2002) Model selection and multimodel inference: A practical information-theoretic approach, 2nd edn. Springer, New York

Farina M, Piroddi L (2009) Simulation error minimization–based identification of polynomial input–output recursive models. In: Proceedings of the 15th IFAC Symposium on System Identification, Saint-Malo, France

German S, Bienenstock E, Doursat R (1992) Neural networks and the bias/variance dilemma. Neural Comput 4(1):1–58

Hafner S, Geiger H, Kreßel U (1992) Anwendung künstlicher neuronaler Netze in der Automatisierungstechnik. Teil 1: Eine Einführung. atp 34(10):592–645

Harris C, Hong X, Gan Q (2002) Adaptive modelling, estimation and fusion from data: A neurofuzzy approach. Advanced information processing, Springer, Berlin

Hastie T, Tibshirani R, Friedman J (2009) The elements of statistical learning : data mining, inference, and prediction, 2nd edn. Springer, New York

Haykin S (2009) Neural networks and learning machines, 3rd edn. Prentice-Hall, New York, NY

Hecht-Nielsen R (1990) Neurocomputing. Addison-Wesley, Reading, MA

Holzmann H, Halfmann C, Germann S, Würtemberger M, Isermann R (1997) Longitudinal and lateral control and supervision of autonomous intelligent vehicles. Control Eng Pract 5(11):1599–1605

Isermann R, Ernst (Töpfer) S, Nelles O (1997) Identificaion with dynamic neural networks : architecture, comparisons, applications. In: Proceedings of the 11th IFAC Symposium on System Identification, Fukuoka, Japan

Ljung L (1999) System identification: Theory for the user, 2nd edn. Prentice Hall Information and System Sciences Series, Prentice Hall PTR, Upper Saddle River, NJ

Loader C (1999) Local regression and likelihood. Springer, New York

Mallows C (1973) Some comments on C P. Technometrics 15(4):661–675

McCulloch W, Pitts W (1943) A logical calculus of the ideas immanent in nervous activity. Bull Math Biophys 5(4):115–133

Miller AJ (2002) Subset selection in regression, 2nd edn. CRC Press, Boca Raton, FL

Moody JE (1992) The effective number of parameters: An analysis of generalization and regularization in nonlinear learning systems. Adv Neural Inf Process Syst 4:847–854

Mrosek M, Sequenz H, Isermann R (2010) Control oriented NOx and soot models for Diesel engines. In: Proceedings of the 6th IFAC Symposium Advances in Automotive Control, Munich, Germany

Müller N (2003) Adaptive Motorregelung beim Ottomotor unter Verwendung von Brennraumdruck-Sensoren: Fortschr.-Ber. VDI Reihe 12 Nr. 545. VDI Verlag, Düsseldorf

Murray-Smith R, Johansen T (1997a) Multiple model approaches to modelling and control. The Taylor and Francis systems and control book series, Taylor & Francis, London

Murray-Smith R, Johansen TA (1997b) Multiple model approaches to modelling and control. Taylor & Francis, London

Nelles O (2001) Nonlinear system identification: From classical approaches to neural networks and fuzzy models. Springer, Berlin

Nelles O, Isermann R (1995) Identification of nonlinear dynamic systems – classical methods versus radial basis function networks. In: Proceedings of the 1995 American Control Conference (ACC), Seattle, WA, USA

Nelles O, Hecker O, Isermann R (1997) Automatic model selection in local linear model trees (LOLIMOT) for nonlinear system identifcation of a transport delay process. In: Proceedings of the 11th IFAC Symposium on System Identification, Fukuoka, Japan

Piroddi L, Spinelli W (2003) An identification algorithm for polynomial NARX models based on simulation error minimization. Int J Control 76(17):1767–1781

Preuß HP, Tresp V (1994) Neuro-Fuzzy. atp 36(5):10–24

Robert Bosch GmbH (2007) Automotive handbook, 7th edn. Bosch, Plochingen

Rosenblatt E (1958) The perceptron: A probabilistic model for information storage & organisation in the brain. Psychol Rev 65:386–408

Schmitt M (1995) Untersuchungen zur Realisierung mehrdimensionaler lernfähiger Kennfelder in Großserien-Steuergeräten. Fortschr.-Ber. VDI Reihe 12 Nr. 246. VDI Verlag, Düsseldorf

Sequenz H, Schreiber A, Isermann R (2009) Identification of nonlinear static processes with local polynomial regression and subset selection. In: Proceedings of the 15th IFAC Symposium on System Identification, Saint-Malo, France

Sequenz H, Mrosek M, Isermann R (2010) Stationary global-local emission models of a CR-Diesel engine with adaptive regressor selection for measurements of air-path and combustion. In: Proceedings of the 6th IFAC Symposium Advances in Automotive Control, Munich, Germany

Stoica P, Selen Y (2004) Model-order selection: A review of information criterion rules. IEEE Signal Process Mag 21(4):36–47

Töpfer S (2002a) Approximation nichtlinearer Prozesse mit Hinging Hyperplane Baummodellen. at 50(4):147–154

Töpfer S (2002b) Hierarchische neuronale Modelle für die Identifikation nichtlinearer Systeme. Fortschr.-Ber. VDI Reihe 10 Nr. 705. VDI Verlag, Düsseldorf

Töpfer S, Wolfram A, Isermann R (2002) Semi-physical modeling of nonlinear processes by means of local model approaches. In: Proceedings of the 15th IFAC World Congress, Barcelona, Spain

Töpfer S geb Ernst (1998) Hinging hyperplane trees for approximation and identification. In: Proceedings of the 37th IEEE Conference on Descision and Control, Tampa, FL, USA

Widrow B, Hoff M (1960) Adaptive switching circuits. IRE Wescon Convention Records pp 96–104

Zimmerschied R (2008) Identifikation nichtlinearer Prozesse mit dynamischen lokalaffinen Modellen : Maßnahmen zur Reduktion von Bias und Varianz. Fortschr.-Ber. VDI Reihe 8 Nr. 1150. VDI Verlag, Düsseldorf

Zimmerschied R, Isermann R (2008) Regularisierungsverfahren für die Identifikation mittels lokal-affiner Modelle. at 56(7):339–349

Zimmerschied R, Isermann R (2009) Nonlinear system identification of block-oriented systems using local affine models. In: Proceedings of the 15th IFAC Symposium on System Identification, Saint-Malo, France

21

State and Parameter Estimation by Kalman Filtering

Often, one is interested in providing an estimate $\hat{\boldsymbol{x}}(k)$ of the states of a discrete-time system at the time step k based on measurements of the input $\boldsymbol{u}(l)$ and the output $\boldsymbol{y}(l)$ up to the point in time j, see Sects. 2.1.2 and 2.2.1 and Fig. 21.1. Different cases can be discriminated based on the choice of k and j. The state estimation can then be given different names (Tomizuka, 1998)

$$\begin{aligned} k > j &\quad n\text{-step (ahead) prediction problem with } n = k - j \\ k = j &\quad \text{filtering problem} \\ k < j &\quad \text{smoothing problem} . \end{aligned}$$

The one-step (ahead) prediction problem will be considered in the following as this is the typical setting for state and parameter estimation problems.

While the classical approach for filtering, smoothing and prediction, which was developed by Wiener and Kolmogorov (e.g. Hänsler, 2001; Papoulis and Pillai, 2002), was based on a design in the frequency domain, the Kalman filter can completely be designed in the time domain. In Sect. 21.1, first the original Kalman filter for linear, time-invariant discrete-time systems will be developed. Following the lines of Kalmans original derivation, (Kalman, 1960), it is assumed that the state variables $\boldsymbol{x}(k)$ and the input $\boldsymbol{u}(k)$ are Gaussian distributed variables with zero mean.

For the time-invariant case, the filter gains will tend to constant values, which can be computed a priori, saving a lot of computations and making the filter easier to implement online. This is shown in Sect. 21.2. Furthermore, the Kalman filter can easily be formulated for linear time-varying discrete-time systems, see Sect. 21.3. In Sect. 21.4, the Kalman filter will be extended to cover non-linear, time-variant discrete-time systems. Although the extended Kalman filter is not an optimal estimator, it is nevertheless employed for many tasks. The extended Kalman filter can not only be used to estimate the system states, but can also be used to obtain parameter estimates at the same time. This is discussed in Sect. 21.5.

R. Isermann, M. Münchhof, *Identification of Dynamic Systems*,
DOI 10.1007/978-3-540-78879-9_21, © Springer-Verlag Berlin Heidelberg 2011

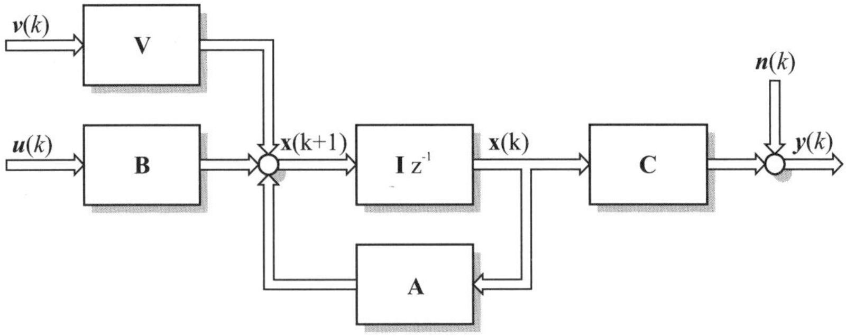

Fig. 21.1. State space representation of a discrete-time MIMO system

21.1 The Discrete Kalman Filter

The classical (discrete-time) Kalman filter shall be derived in the following, starting with the state space representation of a linear dynamic system as

$$\boldsymbol{x}(k+1) = \boldsymbol{A}\boldsymbol{x}(k) + \boldsymbol{B}\boldsymbol{u}(k) + \boldsymbol{V}\boldsymbol{\nu}(k) \tag{21.1.1}$$

$$\boldsymbol{y}(k) = \boldsymbol{C}\boldsymbol{x}(k) + \boldsymbol{n}(k)\,, \tag{21.1.2}$$

where $\boldsymbol{\nu}(k)$ and $\boldsymbol{n}(k)$ are uncorrelated white noise processes with zero mean and covariances

$$\mathrm{E}\big\{\boldsymbol{\nu}(k)\boldsymbol{\nu}^{\mathrm{T}}(k)\big\} = \boldsymbol{M} \tag{21.1.3}$$

$$\mathrm{E}\big\{\boldsymbol{n}(k)\boldsymbol{n}^{\mathrm{T}}(k)\big\} = \boldsymbol{N}\,. \tag{21.1.4}$$

These noises are acting on the system states and also on the system outputs. The assumed model does not have a direct feedthrough. Consequently, there is no direct feedthrough matrix $\boldsymbol{D}$ in the output equation (21.1.2).

One is now interested in finding the optimal linear filter such that the states are predicted with the smallest error possible. The optimality measure is therefore the expected value of the squared prediction error. The prediction error is measured by the vector 2-norm as

$$\begin{aligned} V &= \mathrm{E}\Big\{\big\|\hat{\boldsymbol{x}}(k+1) - \boldsymbol{x}(k+1)\big\|_2^2\Big\} \\ &= \mathrm{E}\Big\{\big(\hat{\boldsymbol{x}}(k+1) - \boldsymbol{x}(k+1)\big)^{\mathrm{T}}\big(\hat{\boldsymbol{x}}(k+1) - \boldsymbol{x}(k+1)\big)\Big\}\,. \end{aligned} \tag{21.1.5}$$

This cost function shall be minimized and is the basis for the development of the Kalman filter.

In the following derivation, the Kalman filter will be developed in a *predictor/corrector setting*, i.e. the states will first be predicted one time-step ahead into the future at $k+1$ and then will be corrected based on the available measurements of the output $\boldsymbol{y}(k+1)$. In the following, $\boldsymbol{x}(k)$ will be the true states at time k, $\hat{\boldsymbol{x}}(k+1|k)$

will be the predicted states based on measurements until k and $\hat{\boldsymbol{x}}(k+1|k+1)$ will the predicted and corrected estimates of the states at time step $k+1$. The notation $(k+1|k)$ means that the value at time-step $k+1$ is determined based on measurements up to the time-step k only. The matrix $\boldsymbol{P}(k)$ will be the covariance of the states, i.e.

$$\boldsymbol{P}(k) = \mathrm{E}\Big\{\big(\hat{\boldsymbol{x}}(k) - \boldsymbol{x}(k)\big)\big(\hat{\boldsymbol{x}}(k) - \boldsymbol{x}(k)\big)^{\mathrm{T}}\Big\}\,. \tag{21.1.6}$$

First, the *prediction step* will be derived. The true system dynamics will be governed by

$$\boldsymbol{x}(k+1) = \boldsymbol{A}\boldsymbol{x}(k) + \boldsymbol{B}\boldsymbol{u}(k) + \boldsymbol{V}\boldsymbol{\nu}(k)\,. \tag{21.1.7}$$

The actual noise $\boldsymbol{\nu}(k)$ is unknown. Since the noise $\boldsymbol{\nu}(k)$ is assumed to be zero mean, the estimates of the states are updated as

$$\hat{\boldsymbol{x}}(k+1|k) = \boldsymbol{A}\hat{\boldsymbol{x}}(k) + \boldsymbol{B}\boldsymbol{u}(k) \tag{21.1.8}$$

based on the measurements up to time-step k.

The new covariance matrix $\boldsymbol{P}^{-}(k+1)$ can then be determined as

$$\begin{aligned}
\boldsymbol{P}^{-}(k+1) &= \mathrm{E}\Big\{\big(\hat{\boldsymbol{x}}(k+1|k) - \boldsymbol{x}(k+1)\big)\big(\hat{\boldsymbol{x}}(k+1|k) - \boldsymbol{x}(k+1)\big)^{\mathrm{T}}\Big\}\\
&= \mathrm{E}\Big\{\big(\boldsymbol{A}\hat{\boldsymbol{x}}(k) - \boldsymbol{A}\boldsymbol{x}(k) + \boldsymbol{V}\boldsymbol{\nu}(k)\big)\big(\boldsymbol{A}\hat{\boldsymbol{x}}(k) - \boldsymbol{A}\boldsymbol{x}(k) + \boldsymbol{V}\boldsymbol{\nu}(k)\big)\Big\}\\
&= \boldsymbol{A}\mathrm{E}\Big\{\big(\hat{\boldsymbol{x}}(k) - \boldsymbol{x}(k)\big)\big(\hat{\boldsymbol{x}}(k) - \boldsymbol{x}(k)\big)^{\mathrm{T}}\Big\}\boldsymbol{A}^{\mathrm{T}}\\
&\quad + \boldsymbol{A}\mathrm{E}\Big\{\big(\hat{\boldsymbol{x}}(k) - \boldsymbol{x}(k)\big)\boldsymbol{\nu}^{\mathrm{T}}\Big\}\boldsymbol{V}^{\mathrm{T}}\\
&\quad + \boldsymbol{V}\mathrm{E}\Big\{\boldsymbol{\nu}(k)\big(\hat{\boldsymbol{x}}(k) - \boldsymbol{x}(k)\big)^{\mathrm{T}}\Big\}\boldsymbol{A}^{\mathrm{T}}\\
&\quad + \boldsymbol{V}\mathrm{E}\Big\{\boldsymbol{\nu}(k)\boldsymbol{\nu}^{\mathrm{T}}(k)\Big\}\boldsymbol{V}^{\mathrm{T}}
\end{aligned} \tag{21.1.9}$$

So, finally

$$\boldsymbol{P}^{-}(k+1) = \boldsymbol{A}\boldsymbol{P}(k)\boldsymbol{A}^{\mathrm{T}} + \boldsymbol{V}\boldsymbol{M}\boldsymbol{V}^{\mathrm{T}}\,. \tag{21.1.10}$$

Here the superscript $^{-}$ denotes the covariance matrix for the prediction step before the correction has taken place. For the derivation, it has been exploited that $\hat{\boldsymbol{x}}(k)$ as well as $\boldsymbol{x}(k)$ are uncorrelated with $\boldsymbol{\nu}(k)$ and furthermore $\boldsymbol{\nu}(k)$ is zero mean, leading to

$$\mathrm{E}\big\{\boldsymbol{x}(k)\boldsymbol{\nu}(k)^{\mathrm{T}}\big\} = \boldsymbol{0} \tag{21.1.11}$$

$$\mathrm{E}\big\{\hat{\boldsymbol{x}}(k)\boldsymbol{\nu}(k)^{\mathrm{T}}\big\} = \boldsymbol{0}\,. \tag{21.1.12}$$

Now, the *correction step* follows. A new measurement $\boldsymbol{y}(k+1)$ is available and will be used to correct the estimates by

$$\hat{\boldsymbol{x}}(k+1|k+1) = \hat{\boldsymbol{x}}(k+1|k) + \boldsymbol{K}(k+1)\big(\boldsymbol{y}(k+1) - \boldsymbol{C}\hat{\boldsymbol{x}}(k+1|k)\big)\,, \tag{21.1.13}$$

where now, the estimates are based on measurements up to time step $k+1$. The choice of the feedback gain $\boldsymbol{K}(k+1)$ determines, whether the prediction of the states $\hat{\boldsymbol{x}}(k+1|k)$ based on the internal model or the actual measurements $\boldsymbol{y}(k+1)$ get more weight in updating the estimated states $\hat{\boldsymbol{x}}(k+1|k+1)$. The observation error $\boldsymbol{y}(k+1)-\boldsymbol{C}\hat{\boldsymbol{x}}(k+1|k)$ is also termed *innovation* in the context of Kalman filtering. Now, the question arises, how the optimal feedback gain $\boldsymbol{K}(k+1)$ must be chosen for any given sample $k+1$. Therefore, the covariance matrix $\boldsymbol{P}(k+1)$ shall be derived. (21.1.13) is rewritten as

$$\hat{\boldsymbol{x}}(k+1|k+1)=\hat{\boldsymbol{x}}(k+1|k)+\boldsymbol{K}(k+1)\big(\boldsymbol{C}\boldsymbol{x}(k+1)+\boldsymbol{n}(k+1)-\boldsymbol{C}\hat{\boldsymbol{x}}(k+1|k)\big) \tag{21.1.14}$$

leading to

$$\boldsymbol{P}(k+1)=\mathrm{E}\Big\{\big(\hat{\boldsymbol{x}}(k+1|k+1)-\boldsymbol{x}(k+1)\big)\big(\hat{\boldsymbol{x}}(k+1|k+1)-\boldsymbol{x}(k+1)\big)^{\mathrm{T}}\Big\}\,. \tag{21.1.15}$$

The cost function (21.1.5) can be rewritten using the trace operator as

$$\begin{aligned}V&=\mathrm{E}\Big\{\big\|\hat{\boldsymbol{x}}(k+1)-\boldsymbol{x}(k+1)\big\|_2^2\Big\}\\&=\mathrm{E}\Big\{\mathrm{tr}\Big(\big(\hat{\boldsymbol{x}}(k+1|k+1)-\boldsymbol{x}(k+1)\big)\big(\hat{\boldsymbol{x}}(k+1|k+1)-\boldsymbol{x}(k+1)\big)^{\mathrm{T}}\Big)\Big\}\\&=\mathrm{tr}\,\mathrm{E}\Big\{\big(\hat{\boldsymbol{x}}(k+1|k+1)-\boldsymbol{x}(k+1|k)\big)\big(\boldsymbol{x}(k+1)-\hat{\boldsymbol{x}}(k+1|t)\big)^{\mathrm{T}}\Big\}\,,\end{aligned} \tag{21.1.16}$$

since the expected value and the trace are both linear operators, their order can be exchanged arbitrarily. The derivation of the ideal gain matrix $\boldsymbol{K}$ based on the trace operator and subsequent application of vector calculus is also presented by Heij et al (2007).

The argument of the trace operator is the expected value of the covariance between the true states $\boldsymbol{x}(k)$ and the estimated states $\hat{\boldsymbol{x}}(k)$, which is the matrix $\boldsymbol{P}(k+1)$. Hence

$$V=\mathrm{tr}\,\boldsymbol{P}(k+1)\,. \tag{21.1.17}$$

In the following, the indices will be dropped to obtain a more compact notation. Now, one can insert

$$\begin{aligned}\boldsymbol{P}&=\mathrm{E}\Big\{\Big(\hat{\boldsymbol{x}}-\boldsymbol{x}-\boldsymbol{K}\boldsymbol{C}\big(\hat{\boldsymbol{x}}-\boldsymbol{x}\big)+\boldsymbol{K}\boldsymbol{n}\Big)\Big(\hat{\boldsymbol{x}}-\boldsymbol{x}-\boldsymbol{K}\boldsymbol{C}\big(\hat{\boldsymbol{x}}-\boldsymbol{x}\big)+\boldsymbol{K}\boldsymbol{n}\Big)^{\mathrm{T}}\Big\}\\&=\mathrm{E}\Big\{\Big(\big(\boldsymbol{I}-\boldsymbol{K}\boldsymbol{C}\big)\big(\hat{\boldsymbol{x}}-\boldsymbol{x}\big)+\boldsymbol{K}\boldsymbol{n}\Big)\Big(\big(\boldsymbol{I}-\boldsymbol{K}\boldsymbol{C}\big)\big(\hat{\boldsymbol{x}}-\boldsymbol{x}\big)+\boldsymbol{K}\boldsymbol{n}\Big)^{\mathrm{T}}\Big\}\end{aligned} \tag{21.1.18}$$

and finally obtains

$$\boldsymbol{P}=\big(\boldsymbol{I}-\boldsymbol{K}\boldsymbol{C}\big)\boldsymbol{P}^{-}\big(\boldsymbol{I}-\boldsymbol{K}\boldsymbol{C}\big)^{\mathrm{T}}+\boldsymbol{K}\boldsymbol{N}\boldsymbol{K}^{\mathrm{T}}\,. \tag{21.1.19}$$

To determine the optimal choice of $\boldsymbol{K}(k+1)$, one can determine the first derivative of (21.1.17) with respect to $\boldsymbol{K}(k)$ and equate it to zero. This first derivative is given as

$$\frac{\partial}{\partial \boldsymbol{K}} \operatorname{tr} \boldsymbol{P} = \frac{\partial}{\partial \boldsymbol{K}} \operatorname{tr}\Big(\big(\boldsymbol{I} - \boldsymbol{K}\boldsymbol{C}\big)\boldsymbol{P}^{-}\big(\boldsymbol{I} - \boldsymbol{K}\boldsymbol{C}\big)^{\mathrm{T}} + \boldsymbol{K}\boldsymbol{N}\boldsymbol{K}^{\mathrm{T}}\Big)\,. \tag{21.1.20}$$

In order to determine the derivative of the trace with respect to the gain matrix $\boldsymbol{K}$, some aspects from matrix calculus shall be presented first. For arbitrary matrices $\boldsymbol{A}$, $\boldsymbol{B}$, and $\boldsymbol{X}$, the following rules for the derivative of the trace can be stated

$$\frac{\partial}{\partial \boldsymbol{X}} \operatorname{tr}\big(\boldsymbol{A}\boldsymbol{X}\boldsymbol{B}\big) = \boldsymbol{A}^{\mathrm{T}}\boldsymbol{B}^{\mathrm{T}} \tag{21.1.21}$$

$$\frac{\partial}{\partial \boldsymbol{X}} \operatorname{tr}\big(\boldsymbol{A}\boldsymbol{X}^{\mathrm{T}}\boldsymbol{B}\big) = \boldsymbol{B}\boldsymbol{A} \tag{21.1.22}$$

$$\frac{\partial}{\partial \boldsymbol{X}} \operatorname{tr}\big(\boldsymbol{A}\boldsymbol{X}\boldsymbol{B}\boldsymbol{X}^{\mathrm{T}}\boldsymbol{C}\big) = \boldsymbol{A}^{\mathrm{T}}\boldsymbol{C}^{\mathrm{T}}\boldsymbol{X}\boldsymbol{B}^{\mathrm{T}} + \boldsymbol{C}\boldsymbol{A}\boldsymbol{X}\boldsymbol{B} \tag{21.1.23}$$

$$\frac{\partial}{\partial \boldsymbol{X}} \operatorname{tr}\big(\boldsymbol{X}\boldsymbol{A}\boldsymbol{X}^{\mathrm{T}}\big) = \boldsymbol{X}\boldsymbol{A}^{\mathrm{T}} + \boldsymbol{X}\boldsymbol{A}\,, \tag{21.1.24}$$

see e.g. (Brookes, 2005). One can apply these rules to (21.1.19) as

$$\begin{aligned}\frac{\partial V}{\partial \boldsymbol{K}} &= \frac{\partial}{\partial \boldsymbol{K}} \operatorname{tr}\big(\boldsymbol{P}^{-} - \boldsymbol{K}\boldsymbol{C}\boldsymbol{P}^{-} - \boldsymbol{P}^{-}\boldsymbol{C}^{\mathrm{T}}\boldsymbol{K}^{\mathrm{T}} + \boldsymbol{K}\boldsymbol{C}\boldsymbol{P}^{-}\boldsymbol{C}^{\mathrm{T}}\boldsymbol{K}^{\mathrm{T}} + \boldsymbol{K}\boldsymbol{N}\boldsymbol{K}^{\mathrm{T}}\big)\\ &= \underbrace{-\frac{\partial}{\partial \boldsymbol{K}} \operatorname{tr} \boldsymbol{K}\boldsymbol{C}\boldsymbol{P}^{-}}_{(21.1.21),\ \boldsymbol{A}=\boldsymbol{I}\text{ and }\boldsymbol{B}=\boldsymbol{C}\boldsymbol{P}^{-}} \underbrace{-\frac{\partial}{\partial \boldsymbol{K}} \operatorname{tr} \boldsymbol{P}^{-}\boldsymbol{C}^{\mathrm{T}}\boldsymbol{K}^{\mathrm{T}}}_{(21.1.22),\ \boldsymbol{A}=\boldsymbol{P}^{-}\boldsymbol{C}^{\mathrm{T}}\text{ and }\boldsymbol{B}=\boldsymbol{I}} \underbrace{+\frac{\partial}{\partial \boldsymbol{K}} \operatorname{tr} \boldsymbol{K}\boldsymbol{C}\boldsymbol{P}^{-}\boldsymbol{C}^{\mathrm{T}}\boldsymbol{K}^{\mathrm{T}}}_{(21.1.24),\ \boldsymbol{A}=\boldsymbol{C}\boldsymbol{P}^{-}\boldsymbol{C}^{\mathrm{T}}}\\ &\quad \underbrace{+\frac{\partial}{\partial \boldsymbol{K}} \operatorname{tr} \boldsymbol{K}\boldsymbol{N}\boldsymbol{K}^{\mathrm{T}}}_{(21.1.24),\ \boldsymbol{A}=\boldsymbol{N}}\end{aligned} \tag{21.1.25}$$

to determine the derivative as

$$\frac{\partial V}{\partial \boldsymbol{K}} = -\boldsymbol{P}^{-}\boldsymbol{C}^{\mathrm{T}} - \boldsymbol{P}^{-}\boldsymbol{C}^{\mathrm{T}} + \boldsymbol{K}\boldsymbol{C}\boldsymbol{P}^{-}\boldsymbol{C}^{\mathrm{T}} + \boldsymbol{K}\boldsymbol{C}\boldsymbol{P}^{-}\boldsymbol{C}^{\mathrm{T}} + \boldsymbol{K}\boldsymbol{N}^{\mathrm{T}} + \boldsymbol{K}\boldsymbol{N} \stackrel{!}{=} \boldsymbol{0}\,. \tag{21.1.26}$$

The solution for this equation is given as

$$2\boldsymbol{K}\big(\boldsymbol{C}\boldsymbol{P}^{-1}\boldsymbol{C}^{\mathrm{T}} + \boldsymbol{N}\big) = 2\boldsymbol{P}^{-}\boldsymbol{C}^{\mathrm{T}} \tag{21.1.27}$$

$$\Leftrightarrow \boldsymbol{K} = \boldsymbol{P}^{-}\boldsymbol{C}^{\mathrm{T}}\big(\boldsymbol{C}\boldsymbol{P}^{-}\boldsymbol{C}^{\mathrm{T}} + \boldsymbol{N}\big)^{-1}\,. \tag{21.1.28}$$

Including the time index k again, yields

$$\boldsymbol{K}(k+1) = \boldsymbol{P}^{-}(k+1)\boldsymbol{C}^{\mathrm{T}}\big(\boldsymbol{C}\boldsymbol{P}^{-}(k+1)\boldsymbol{C}^{\mathrm{T}} + \boldsymbol{N}\big)^{-1}\,. \tag{21.1.29}$$

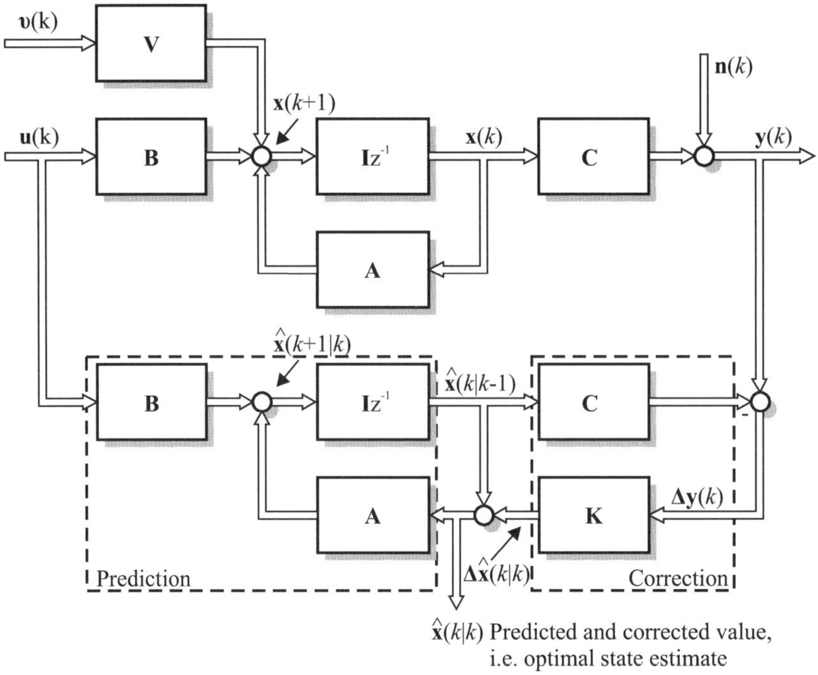

Fig. 21.2. Block diagram of the Kalman filter

In summary, the Kalman filter is given by the set of equations (21.1.8), (21.1.10), (21.1.13), (21.1.19), and (21.1.29) and leads to the following algorithm:

Prediction:

$$\hat{\boldsymbol{x}}(k+1|k) = \boldsymbol{A}\hat{\boldsymbol{x}}(k) + \boldsymbol{B}\boldsymbol{u}(k) \tag{21.1.30}$$

$$\boldsymbol{P}^{-}(k+1) = \boldsymbol{A}\boldsymbol{P}(k)\boldsymbol{A}^{\mathrm{T}} + \boldsymbol{V}\boldsymbol{M}\boldsymbol{V}^{\mathrm{T}} \tag{21.1.31}$$

Correction:

$$\boldsymbol{K}(k+1) = \boldsymbol{P}^{-}(k+1)\boldsymbol{C}^{\mathrm{T}}\big(\boldsymbol{C}\boldsymbol{P}^{-}(k+1)\boldsymbol{C}^{\mathrm{T}} + \boldsymbol{N}\big)^{-1} \tag{21.1.32}$$

$$\hat{\boldsymbol{x}}(k+1|k+1) = \hat{\boldsymbol{x}}(k+1|k) + \boldsymbol{K}(k+1)\big(\boldsymbol{y}(k+1) - \boldsymbol{C}\hat{\boldsymbol{x}}(k+1|k)\big) \tag{21.1.33}$$

$$\boldsymbol{P}(k+1) = \big(\boldsymbol{I} - \boldsymbol{K}(k+1)\boldsymbol{C}\big)\boldsymbol{P}^{-}(k+1)\,, \tag{21.1.34}$$

where the update equation for $\boldsymbol{P}(k+1)$ (21.1.34) is in this form *only* valid *if* the optimal Kalman gain $\boldsymbol{K}(k+1)$ (21.1.32) has been used for the feedback.

For the initial condition of the states, one typically chooses $\hat{\boldsymbol{x}}(0) = \boldsymbol{0}$. The matrix $\boldsymbol{P}(0)$ must be supplied by the user as the covariance of the states $\boldsymbol{x}(0)$. The corresponding block diagram is shown in Fig. 21.2.

In this setting, the Kalman filter is used as a one-step ahead predictor. For the derivation of the m-step ahead predictor, the reader is e.g. referred to (Heij et al,

2007). In order to speed up the calculation, one can interpret the vector of n measurements that are mutually uncorrelated as n subsequent scalar measurements. If data samples are missing, the filter can be run with $\boldsymbol{K} = \boldsymbol{0}$ and hence, only the prediction step, but not the correction using actual measurements, is carried out (Grewal and Andrews, 2008).

21.2 Steady-State Kalman Filter

The major drawback of the Kalman filter as described by (21.1.30) through (21.1.34) is the high computational expense, which is mainly caused by the update of the covariance matrix $\boldsymbol{P}(k)$ in (21.1.31) and (21.1.34) and the calculation of the filter gain $\boldsymbol{K}(k)$ in (21.1.32). For linear, time-invariant systems, it will now be shown that both $\boldsymbol{P}(k)$ and $\boldsymbol{K}(k)$ tend to constant values as $k \to \infty$. These constant values can be determined a priori during the design of the Kalman filter and can be left unchanged during operation of the filter. The a priori determination has the interesting side-effect that the filter will operate right from the start with the "ideal" filter gain $\boldsymbol{K}(k)$ as one does not have to wait for $\boldsymbol{P}(k)$ and $\boldsymbol{K}(k)$ to settle to their optimal steady-state values.

In order to determine $\boldsymbol{P}(k)$ for $k \to \infty$, one sets up the update equation for $\boldsymbol{P}^-(k+1)$ from (21.1.31) and (21.1.31) as

$$\begin{aligned}&\boldsymbol{P}^-(k+1)\\&= \boldsymbol{A}\boldsymbol{P}^-(k)\boldsymbol{A}^{\mathrm{T}} - \boldsymbol{A}\boldsymbol{P}^-(k)\boldsymbol{C}^{\mathrm{T}}\big(\boldsymbol{C}\boldsymbol{P}^-(k)\boldsymbol{C}^{\mathrm{T}} + \boldsymbol{N}\big)^{-1}\boldsymbol{C}\boldsymbol{P}^-(k)\boldsymbol{A}^{\mathrm{T}} + \boldsymbol{V}\boldsymbol{M}\boldsymbol{V}^{\mathrm{T}}\end{aligned} \tag{21.2.1}$$

For $k \to \infty$, the entries of the matrix $\boldsymbol{P}^-(k)$ settle to constant values, thus

$$\boldsymbol{P}^-(k) = \boldsymbol{P}^-(k+1) = \boldsymbol{P}^- \,. \tag{21.2.2}$$

(21.2.1) then becomes

$$\boldsymbol{P}^- = \boldsymbol{A}\boldsymbol{P}^-\boldsymbol{A}^{\mathrm{T}} - \boldsymbol{A}\boldsymbol{P}^-\boldsymbol{C}^{\mathrm{T}}\big(\boldsymbol{C}\boldsymbol{P}^-\boldsymbol{C}^{\mathrm{T}} + \boldsymbol{N}\big)^{-1}\boldsymbol{C}\boldsymbol{P}^-\boldsymbol{A}^{\mathrm{T}} + \boldsymbol{V}\boldsymbol{M}\boldsymbol{V}^{\mathrm{T}} \,. \tag{21.2.3}$$

This is nothing else than the more general discrete-time algebraic Riccati equation (DARE) (e.g. Arnold and Laub, 1984), which is given as

$$\boldsymbol{A}^{\mathrm{T}}\boldsymbol{X}\boldsymbol{A} - \boldsymbol{E}^{\mathrm{T}}\boldsymbol{X}\boldsymbol{E} - \big(\boldsymbol{A}^{\mathrm{T}}\boldsymbol{X}\boldsymbol{B} + \boldsymbol{S}\big)\big(\boldsymbol{B}^{\mathrm{T}}\boldsymbol{X}\boldsymbol{B} + \boldsymbol{R}\big)^{-1}\big(\boldsymbol{B}^{\mathrm{T}}\boldsymbol{X}\boldsymbol{A} + \boldsymbol{S}^{\mathrm{T}}\big) + \boldsymbol{Q} = \boldsymbol{0} \,. \tag{21.2.4}$$

Software packages are available, which allow to solve (21.2.1), see also (Arnold and Laub, 1984; Söderström, 2002). A comparison of (21.2.3) and (21.2.4) shows, how the coefficients must be passed to the DARE solver,

$$\underbrace{\boldsymbol{P}^-}_{\boldsymbol{E}=\boldsymbol{I}} = \underbrace{\boldsymbol{A}\boldsymbol{P}^-\boldsymbol{A}^{\mathrm{T}}}_{\boldsymbol{A}=\boldsymbol{A}^{\mathrm{T}}} - \underbrace{\boldsymbol{A}\boldsymbol{P}^-\boldsymbol{C}^{\mathrm{T}}\big(\boldsymbol{C}\boldsymbol{P}^-\boldsymbol{C}^{\mathrm{T}} + \boldsymbol{N}\big)^{-1}\boldsymbol{C}\boldsymbol{P}^-\boldsymbol{A}}_{\boldsymbol{A}=\boldsymbol{A}^{\mathrm{T}},\,\boldsymbol{B}=\boldsymbol{C}^{\mathrm{T}},\,\boldsymbol{S}=\boldsymbol{0},\,\boldsymbol{R}=\boldsymbol{N}} + \underbrace{\boldsymbol{V}\boldsymbol{M}\boldsymbol{V}^{\mathrm{T}}}_{\boldsymbol{Q}=\boldsymbol{V}\boldsymbol{M}\boldsymbol{V}^{\mathrm{T}}} \,. \tag{21.2.5}$$

The matrix $\boldsymbol{P}^-$ becomes the unknown matrix $\boldsymbol{X}$. This is very similar to the design of the linear quadratic regulator (LQR). The result of this calculation then is the steady-state value $\overline{\boldsymbol{P}}^-$ of $\boldsymbol{P}^-$.

Once the solution $\overline{\boldsymbol{P}}^-$ has been obtained, the steady-state gain of the Kalman filter can be determined as

$$\overline{\boldsymbol{K}} = \overline{\boldsymbol{P}}^- \boldsymbol{C}\left(\boldsymbol{C}\overline{\boldsymbol{P}}^- \boldsymbol{C}^{\mathrm{T}} + \boldsymbol{N}\right)^{-1} , \tag{21.2.6}$$

see also (Verhaegen and Verdult, 2007). Software packages in the control engineering domain often have dedicated functions for the design of Kalman filters and already return the gain matrix $\overline{\boldsymbol{K}}$. Once, $\overline{\boldsymbol{P}}^-$ and $\overline{\boldsymbol{K}}$ have been determined, the Kalman filter equations that have to be updated in real-time, reduce to

Prediction:

$$\hat{\boldsymbol{x}}(k+1|k) = \boldsymbol{A}\hat{\boldsymbol{x}}(k) + \boldsymbol{B}\boldsymbol{u}(k) \tag{21.2.7}$$

Correction:

$$\hat{\boldsymbol{x}}(k+1|k+1) = \hat{\boldsymbol{x}}(k+1|k) + \overline{\boldsymbol{K}}\big(\boldsymbol{y}(k+1) - \boldsymbol{C}\hat{\boldsymbol{x}}(k+1|k)\big) . \tag{21.2.8}$$

For the initial condition of the state estimates, set $\hat{\boldsymbol{x}}(0) = \boldsymbol{0}$.

To obtain a comparison to a state observer, the previous correction (21.2.8) is inserted in the prediction (21.2.7), leading to

$$\hat{\boldsymbol{x}}(k+1|k) = \boldsymbol{A}\hat{\boldsymbol{x}}(k|k-1) + \boldsymbol{B}\boldsymbol{u}(k) + \boldsymbol{A}\overline{\boldsymbol{K}}\big(\boldsymbol{y}(k) - \boldsymbol{C}\hat{\boldsymbol{x}}(k|k-1)\big) \tag{21.2.9}$$

A comparison with the observer, which is governed by

$$\hat{\boldsymbol{x}}(k+1) = \boldsymbol{A}\hat{\boldsymbol{x}}(k) + \boldsymbol{B}\boldsymbol{u}(k) + \boldsymbol{H}\big(\boldsymbol{y}(k) - \boldsymbol{C}\hat{\boldsymbol{x}}(k)\big) , \tag{21.2.10}$$

shows that, if the observer gain is chosen as

$$\boldsymbol{H} = \boldsymbol{A}\overline{\boldsymbol{K}} , \tag{21.2.11}$$

then the observer equation corresponds to the Kalman filter.

21.3 Kalman Filter for Time-Varying Discrete Time Systems

Similar to the above considerations, the Kalman filter can be extended to time-varying systems as the filter equations are independent of past values of the matrices $\boldsymbol{A}(k)$, $\boldsymbol{B}(k)$, and $\boldsymbol{C}(k)$. The time-varying process shall be governed by the state space model

$$\boldsymbol{x}(k+1|k) = \boldsymbol{A}(k)\boldsymbol{x}(k) + \boldsymbol{B}(k)\boldsymbol{u}(k) + \boldsymbol{V}(k)\boldsymbol{\nu}(k) \tag{21.3.1}$$

$$\boldsymbol{y}(k) = \boldsymbol{C}(k)\boldsymbol{x}(k) + \boldsymbol{n}(k) . \tag{21.3.2}$$

Then, the Kalman filter, given by the set of equations consisting of (21.1.8), (21.1.13), (21.1.10), (21.1.19), and (21.1.29), can easily be written in a form that is suitable for time-varying systems. The filter equations are now given by

Prediction:

$$\hat{\boldsymbol{x}}(k+1) = \boldsymbol{A}(k)\hat{\boldsymbol{x}}(k) + \boldsymbol{B}(k)\boldsymbol{u}(k) \tag{21.3.3}$$

$$\boldsymbol{P}^{-}(k+1) = \boldsymbol{A}(k)\boldsymbol{P}(k)\boldsymbol{A}^{\mathrm{T}}(k) + \boldsymbol{V}(k)\boldsymbol{M}(k)\boldsymbol{V}^{\mathrm{T}}(k) \tag{21.3.4}$$

Correction:

$$\boldsymbol{K}(k+1) = \boldsymbol{P}^{-}(k+1)\boldsymbol{C}^{\mathrm{T}}(k+1)\big(\boldsymbol{C}(k+1)\boldsymbol{P}^{-}(k+1)\boldsymbol{C}^{\mathrm{T}}(k+1) + \boldsymbol{N}(k)\big)^{-1} \tag{21.3.5}$$

$$\hat{\boldsymbol{x}}(k+1|k+1) = \hat{\boldsymbol{x}}(k+1|k) + \boldsymbol{K}(k+1)\big(\boldsymbol{y}(k+1) - \boldsymbol{C}(k+1)\hat{\boldsymbol{x}}(k+1|k)\big) \tag{21.3.6}$$

$$\boldsymbol{P}(k+1) = \big(\boldsymbol{I} - \boldsymbol{K}(k+1)\boldsymbol{C}(k+1)\big)\boldsymbol{P}^{-}(k+1)\,, \tag{21.3.7}$$

where the update equation for $\boldsymbol{P}(k+1)$ is again in this form only valid if the optimal Kalman gain $\boldsymbol{K}(k+1)$ has been used for the feedback.

21.4 Extended Kalman Filter

In many applications, one is confronted with non-linear system models of the form

$$\boldsymbol{x}(k+1) = \boldsymbol{f}_k(\boldsymbol{x}(k), \boldsymbol{u}(k)) + \boldsymbol{V}(k)\boldsymbol{v}(k) \tag{21.4.1}$$

$$\boldsymbol{y}(k) = \boldsymbol{g}_k(\boldsymbol{x}(k)) + \boldsymbol{n}(k)\,, \tag{21.4.2}$$

where the index k in $\boldsymbol{f}_k$ and $\boldsymbol{g}_k$ indicate that also the functions themselves can be time-varying.

For processes of this form, the *Extended Kalman Filter* (EKF) has been used in many applications. In a few first publications, this filter was called the Kalman-Schmidt-Filter (Grewal and Andrews, 2008). In the EKF, the update equation for the states is based on the "true" non-linear model, whereas the update for the error covariance matrix $\boldsymbol{P}(k)$ is based on a first order Taylor series expansion of (21.4.1) and (21.4.1). The prediction step for the states is hence given as

$$\hat{\boldsymbol{x}}(k+1|k) = \boldsymbol{f}_k\big(\hat{\boldsymbol{x}}(k), \boldsymbol{u}(k)\big) \tag{21.4.3}$$

The update of the covariance matrix requires the calculation of the Jacobian matrices in each update step. The Jacobian matrices are given as

$$\boldsymbol{F}(k) = \left.\frac{\partial \boldsymbol{f}_k(\boldsymbol{x}, \boldsymbol{u})}{\partial \boldsymbol{x}}\right|_{\boldsymbol{x}=\hat{\boldsymbol{x}}(k),\, \boldsymbol{u}=\boldsymbol{u}(k)} \tag{21.4.4}$$

$$\boldsymbol{G}(k+1) = \left.\frac{\partial \boldsymbol{g}_{k+1}(\boldsymbol{x})}{\partial \boldsymbol{x}}\right|_{\boldsymbol{x}=\hat{\boldsymbol{x}}(k+1|k)}\,. \tag{21.4.5}$$

Then, the update equation for $\boldsymbol{P}(k+1)$ and the calculation $\boldsymbol{K}(k+1)$ are given as

$$\boldsymbol{P}^{-}(k+1) = \boldsymbol{F}(k)\boldsymbol{P}(k)\boldsymbol{F}^{\mathrm{T}}(k) + \boldsymbol{V}(k)\boldsymbol{M}(k)\boldsymbol{V}^{\mathrm{T}}(k) \tag{21.4.6}$$

$$\boldsymbol{K}(k+1) = \boldsymbol{P}^{-}(k+1)\boldsymbol{G}(k+1)\big(\boldsymbol{G}(k+1)\boldsymbol{P}^{-}(k+1)\boldsymbol{G}^{\mathrm{T}}(k+1) + \boldsymbol{N}(k+1)\big)^{-1} \tag{21.4.7}$$

and

$$\boldsymbol{P}(k+1) = \big(\boldsymbol{I} - \boldsymbol{K}(k+1)\boldsymbol{G}(k+1)\big)\boldsymbol{P}^{-}(k+1). \tag{21.4.8}$$

The state estimates are corrected using the true non-linear relation as

$$\hat{x}(k+1|k+1) = \hat{x}(k+1|k) + \boldsymbol{K}(k+1)\big(y(k+1) - g_{k+1}\big(\hat{\boldsymbol{x}}(k+1|k)\big)\big)\,. \tag{21.4.9}$$

While the derivation of the EKF seems quite simple, it must be stressed at this point that the EKF does *not provide optimal estimates*. While the random variables were remaining Gaussian at all times for the Kalman filter, the distribution of the random variables will change after going through the non-linear transformations in the EKF. Furthermore, one should be aware that the filter can quickly diverge due to the linearization around false operating points, if for example the initial conditions are chosen wrongly. While these points seem to be severe drawbacks of the EKF, it is still used in many applications, the most prominent being navigation systems and GPS devices.

The final extended Kalman filter is then given as

Prediction:

$$\hat{\boldsymbol{x}}(k+1|k) = \boldsymbol{f}_k\big(\hat{\boldsymbol{x}}(k), \boldsymbol{u}(k)\big) \tag{21.4.10}$$

$$\boldsymbol{F}(k) = \left.\frac{\partial \boldsymbol{f}_k(\boldsymbol{x}, \boldsymbol{u})}{\partial \boldsymbol{x}}\right|_{\boldsymbol{x}=\hat{\boldsymbol{x}}(k),\, \boldsymbol{u}=\boldsymbol{u}(k)} \tag{21.4.11}$$

$$\boldsymbol{P}^{-}(k+1) = \boldsymbol{F}(k)\boldsymbol{P}(k)\boldsymbol{F}^{\mathrm{T}}(k) + \boldsymbol{V}(k)\boldsymbol{M}(k)\boldsymbol{V}^{\mathrm{T}}(k) \tag{21.4.12}$$

Correction:

$$\boldsymbol{G}(k+1) = \left.\frac{\partial g_{k+1}(\boldsymbol{x})}{\partial \boldsymbol{x}}\right|_{\boldsymbol{x}=\hat{\boldsymbol{x}}(k+1|k)} \tag{21.4.13}$$

$$\boldsymbol{K}(k+1) = \boldsymbol{P}^{-}(k+1)\boldsymbol{G}(k+1)\big(\boldsymbol{G}(k+1)\boldsymbol{P}^{-}(k+1)\boldsymbol{G}^{\mathrm{T}}(k+1) + \boldsymbol{N}(k+1)\big)^{-1} \tag{21.4.14}$$

$$\hat{x}(k+1|k+1) = \hat{x}(k+1|k) + \boldsymbol{K}(k+1)\big(y(k+1) - g_{k+1}\big(\hat{\boldsymbol{x}}(k+1|k)\big)\big) \tag{21.4.15}$$

$$\boldsymbol{P}(k+1) = \big(\boldsymbol{I} - \boldsymbol{K}(k+1)\boldsymbol{G}(k+1)\big)\boldsymbol{P}^{-}(k+1).\,. \tag{21.4.16}$$

21.5 Extended Kalman Filter for Parameter Estimation

The extended Kalman filter can also be used for parameter estimation. Here, the state vector $\boldsymbol{x}(k)$ is augmented with a parameter vector $\boldsymbol{\theta}$, leading to the state space system

$$\begin{pmatrix} \hat{\boldsymbol{x}}(k+1) \\ \hat{\boldsymbol{\theta}}(k+1) \end{pmatrix} = \begin{pmatrix} \boldsymbol{f}\left(\hat{\boldsymbol{x}}(k), \boldsymbol{\theta}(k), \boldsymbol{u}(k)\right) \\ \boldsymbol{\theta}(k) \end{pmatrix} + \begin{pmatrix} \boldsymbol{F}\boldsymbol{n}(k) \\ \boldsymbol{\xi}(k) \end{pmatrix} \tag{21.5.1}$$

$$\boldsymbol{y}(k) = \boldsymbol{g}\left(\hat{\boldsymbol{x}}(k)\right), \tag{21.5.2}$$

where in comparison with (21.4.1), the parameter vector $\boldsymbol{\theta}(k)$ has been introduced and obeys the dynamics

$$\boldsymbol{\theta}(k+1) = \boldsymbol{\theta}(k) + \boldsymbol{\xi}(k), \tag{21.5.3}$$

see (Chen, 1999). One can see that the parameters are modeled as constant quantities. However, the model includes a stochastic disturbance, i.e. the parameters are modeled as being disturbed by a white noise. If this was not the case, the extended Kalman filter would assume that these values are known exactly and would not adjust these values during the filtering.

21.6 Continuous-Time Models

If one is confronted with continuous-time process models, there are in general two approaches. Typically, the Kalman filter is implemented on a computer and hence implemented in discrete-time. In this case, one can bring the continuous-time model of the dynamics of the observed system to discrete time by determining e.g. the transition matrix, (2.1.27). Then, one can use the above equations for the discrete-time case. A formulation that is completely based in the time domain is called the *Kalman-Bucy filter*, which is e.g. treated in (Grewal and Andrews, 2008). It is computationally difficult to implement, as it requires the solution of the matrix Ricatti differential equation.

21.7 Summary

In this chapter, the Kalman filter has first been introduced as a tool to estimate states of a system, which can e.g. be useful for the application of subspace methods, where the states of the system must be known. The Kalman filter was developed for linear time invariant discrete-time systems in this chapter and it has subsequently been shown that the filter can also be applied to time-varying systems. Then, the extended Kalman filter (EKF) was introduced, which works with non-linear system models. It can be shown that the EKF can not only be used to estimate system states, but also system parameters. Here, it was important to note that the parameters must be modeled as being influenced by a stochastic disturbance since otherwise, the parameters will not be manipulated by the filter equations. Finally, the application to continuous-time systems was shortly discussed. One could use the Kalman-Bucy filter, whose formulation is based entirely in the continuous-time domain, but is mathematically complex. Furthermore, as the filter will nowadays be implemented on digital computers, it is typically more appropriate to just discretize the continuous-time model and then use the discrete-time Kalman filter equations. Some caution should be exercised when using the extended Kalman filter as the filter will only linearize the model around the current (estimated) operating point. The filter can diverge, i.e. go away from the true operating point and lead to completely erroneous results. Real-time implementation issues are discussed e.g. in Chui and Chen (2009) and Grewal and Andrews (2008).

Problems

21.1. Kalman Filter I
What are the differences between a state observer and the Kalman filter?

21.2. Kalman Filter II
Write the Kalman filter down for a first order system and develop a signal flow diagram.

21.3. Extended Kalman Filter I
How can the Extended Kalman Filter be used for parameter estimation. Describe the difference equations that governs the "dynamics" of the parameters.

21.4. Extended Kalman Filter II
What are the implications of linearizing the underlying dynamics locally?

References

Arnold WF III, Laub AJ (1984) Generalized eigenproblem algorithms and software for algebraic Riccati equations. Proc IEEE 72(12):1746–1754

Brookes M (2005) The matrix reference manual. URL `http://www.ee.ic.ac.uk/hp/staff/dmb/matrix/intro.html`

Chen CT (1999) Linear system theory and design, 3rd edn. Oxford University Press, New York

Chui CK, Chen G (2009) Kalman filtering with real-time applications, 4th edn. Springer, Berlin

Grewal MS, Andrews AP (2008) Kalman filtering: Theory and practice using MATLAB, 3rd edn. John Wiley & Sons, Hoboken, NJ

Hänsler E (2001) Statistische Signale: Grundlagen und Anwendungen. Springer, Berlin

Heij C, Ran A, Schagen F (2007) Introduction to mathematical systems theory : linear systems, identification and control. Birkhäuser Verlag, Basel

Kalman RE (1960) A new approach to linear filtering and prediction problems. Trans ASME Journal of Basic Engineering Series D 82:35–45

Papoulis A, Pillai SU (2002) Probability, random variables and stochastic processes, 4th edn. McGraw Hill, Boston

Söderström T (2002) Discrete-time stochastic systems: Estimation and control, 2nd edn. Advanced Textbooks in Control and Signal Processing, Springer, London
Tomizuka M (1998) Advanced control systems II, class notes for ME233. University of California at Berkeley, Dept of Mechanical Engineering, Berkeley, CA
Verhaegen M, Verdult V (2007) Filtering and system identification: A least squares approach. Cambridge University Press, Cambridge

Part VII

Miscellaneous Issues

22

Numerical Aspects

To improve some of the properties of basic parameter estimation methods, the corresponding algorithms can be modified. These modifications serve to enhance the numerical accuracy in digital computers or give access to intermediate results. The numerical properties are important if the word length is confined or if the changes of input signals become small, as in adaptive control or fault detection. Such constraints can lead to ill-conditioned system of equations.

22.1 Condition Numbers

As part of the parameter estimation, the system of equations

$$\boldsymbol{A}\boldsymbol{\theta} = \boldsymbol{b} \tag{22.1.1}$$

must be solved. If now $\boldsymbol{b}$ is disturbed e.g. due to fixed word length of the data or noise, then one obtains

$$\boldsymbol{A}\big(\boldsymbol{\theta} + \Delta\boldsymbol{\theta}\big) = \boldsymbol{b} + \Delta\boldsymbol{b} \ . \tag{22.1.2}$$

and hence for the parameter error

$$\Delta\boldsymbol{\theta} = \boldsymbol{A}^{-1}\Delta\boldsymbol{b} \ . \tag{22.1.3}$$

In order to determine the influence of the $\Delta\boldsymbol{b}$ on the parameter estimation error $\Delta\boldsymbol{\theta}$, one can introduce vector norms for $\|\boldsymbol{b}\|$ and $\|\boldsymbol{\theta}\|$ and an appropriate matrix norm for $\|\boldsymbol{A}^{-1}\|$. Since

$$\Delta\boldsymbol{\theta} = \boldsymbol{A}^{-1}\Delta\boldsymbol{b} \ , \tag{22.1.4}$$

one obtains

$$\|\Delta\boldsymbol{\theta}\| = \|\boldsymbol{A}^{-1}\Delta\boldsymbol{b}\| \leq \|\boldsymbol{A}^{-1}\|\|\Delta\boldsymbol{b}\| \ . \tag{22.1.5}$$

Since furthermore

$$\|\boldsymbol{b}\| = \|\boldsymbol{A}\boldsymbol{\theta}\| \leq \|\boldsymbol{A}\|\,\|\boldsymbol{\theta}\| \ , \tag{22.1.6}$$

one obtains

R. Isermann, M. Münchhof, *Identification of Dynamic Systems*,
DOI 10.1007/978-3-540-78879-9_22, © Springer-Verlag Berlin Heidelberg 2011

$$\frac{1}{\|\boldsymbol{\theta}\|} \leq \frac{\|\boldsymbol{A}\|}{\|\boldsymbol{b}\|} . \tag{22.1.7}$$

From there follows with $\|\boldsymbol{b}\| \neq 0$,

$$\frac{\|\Delta\boldsymbol{\theta}\|}{\|\boldsymbol{\theta}\|} \leq \|\boldsymbol{A}^{-1}\| \frac{\|\Delta\boldsymbol{b}\|}{\|\boldsymbol{x}\|} \leq \|\boldsymbol{A}\|\|\boldsymbol{A}^{-1}\| \frac{\|\Delta\boldsymbol{b}\|}{\|\boldsymbol{b}\|} . \tag{22.1.8}$$

This relation now is a measure for the influence of relative errors in $\boldsymbol{b}$ on the relative errors in $\boldsymbol{\theta}$. This is called the *condition number* of a matrix

$$\operatorname{cond}(\boldsymbol{A}) = \|\boldsymbol{A}\|\|\boldsymbol{A}^{-1}\| . \tag{22.1.9}$$

The condition number depends on the matrix norm. If one uses the 2-norm of a matrix, then one obtains the simple relation that the condition number is given as the ratio of the largest to the smallest singular value of $\boldsymbol{A}$, i.e.

$$\operatorname{cond}(\boldsymbol{A}) = \frac{\sigma_{\max}}{\sigma_{\min}} \geq 1 . \tag{22.1.10}$$

The numerical conditions can now be improved by not calculating $\boldsymbol{P}$ as intermediate value, in which the squares of signals appear, but square roots of $\boldsymbol{P}$. This leads to *square root filtering methods* or *factorization methods* (e.g. Biermann, 1977). By this means, forms can be distinguished which start from the covariance matrix $\boldsymbol{P}$ or the information matrix $\boldsymbol{P}^{-1}$ (Kaminski et al, 1971; Biermann, 1977; Kofahl, 1986).

The advantage of this orthogonalization method can be seen from the error sensitivity of the system that determines the parameters (Golub and van Loan, 1996). If the normal equation (22.3.1) are directly solved by the LS method, the parameter error is bounded by

$$\frac{\|\Delta\hat{\boldsymbol{\theta}}\|}{\|\hat{\boldsymbol{\theta}}\|} \leq \operatorname{cond}(\boldsymbol{\Psi}^{\mathrm{T}}\boldsymbol{\Psi}) \frac{\|\Delta\boldsymbol{y}\|}{\|\boldsymbol{y}\|} = \operatorname{cond}^2(\boldsymbol{\Psi}) \frac{\|\Delta\boldsymbol{y}\|}{\|\boldsymbol{y}\|} , \tag{22.1.11}$$

However, if the orthogonalization approach is used, the upper bound for the parameter errors is given by

$$\frac{\|\Delta\hat{\boldsymbol{\theta}}\|}{\|\hat{\boldsymbol{\theta}}\|} \leq \operatorname{cond}(\boldsymbol{\Psi}) \frac{\|\Delta\boldsymbol{b}\|}{\|\boldsymbol{b}\|} , \tag{22.1.12}$$

i.e. the system (22.3.5) is much less sensitive to measurement errors then the normal equations (22.3.1) themselves.

The following treatment leans on (Isermann et al, 1992). All of the following methods try to solve the normal equations

$$\boldsymbol{\Psi}^{\mathrm{T}}\boldsymbol{\Psi}\hat{\boldsymbol{\theta}} = \boldsymbol{\Psi}^{\mathrm{T}}\boldsymbol{y} , \tag{22.1.13}$$

either in one pass or recursively as new data point become available during online identification.

22.2 Factorization Methods for $\boldsymbol{P}$

A simple approach would be to use the Gaussian elimination to solve (22.1.13). However, as $\boldsymbol{\Psi}^{\mathrm{T}}\boldsymbol{\Psi}$ is positive definite, one can decompose the symmetric matrix $\boldsymbol{P}$ into triangular matrices

$$\boldsymbol{P} = \boldsymbol{S}\boldsymbol{S}^{\mathrm{T}}\,, \tag{22.2.1}$$

where $\boldsymbol{S}$ is called the *square root*. Then, one can work directly on th matrix $\boldsymbol{S}$, which leads to the *discrete square root filtering in covariance form* (DSFC) algorithm. For the RLS method, the resulting algorithm then becomes:

$$\begin{aligned}
\hat{\boldsymbol{\theta}}(k+1) &= \hat{\boldsymbol{\theta}}(k) + \boldsymbol{\gamma}(k)e(k+1)\\
\boldsymbol{\gamma}(k) &= a(k)\boldsymbol{S}(k)\boldsymbol{f}(k)\\
\boldsymbol{f}(k) &= \boldsymbol{S}^{\mathrm{T}}(k)\boldsymbol{\psi}(k+1)\\
\boldsymbol{S}(k+1) &= (\boldsymbol{S}(k) - g(k)\boldsymbol{\gamma}(k)\boldsymbol{f}^{\mathrm{T}}(k))/\sqrt{\lambda(k)}\\
1/(a(k)) &= \boldsymbol{f}^{\mathrm{T}}(k)\boldsymbol{f}(k) + \lambda(k)\\
g(k) &= 1/(1+\sqrt{\lambda(k)a(k)})\,.
\end{aligned} \tag{22.2.2}$$

The starting values are $\boldsymbol{S}(0) = \sqrt{\alpha}\boldsymbol{I}$ and $\hat{\boldsymbol{\theta}}(0) = \boldsymbol{0}$. λ is the forgetting factor, see Sect. 9.6. A disadvantage is the calculation of the square roots for each recursion.

Another method has been proposed by Biermann (1977), the so-called *UD factorization* (DUDC). Here, the covariance matrix is factorized by

$$\boldsymbol{P} = \boldsymbol{U}\boldsymbol{D}\boldsymbol{U}^{\mathrm{T}}\,, \tag{22.2.3}$$

where $\boldsymbol{D}$ is diagonal and $\boldsymbol{U}$ is an upper triangular matrix with ones on the diagonal. Then the recursions for the covariance matrix are

$$\begin{aligned}
&\boldsymbol{U}(k+1)\boldsymbol{D}(k+1)\boldsymbol{U}^{\mathrm{T}}(k+1) =\\
&\frac{1}{\lambda}\big(\boldsymbol{U}(k)\boldsymbol{D}(k)\boldsymbol{U}^{\mathrm{T}}(k) - \boldsymbol{\gamma}(k)\boldsymbol{\psi}^{\mathrm{T}}(k+1)\boldsymbol{U}(k)\boldsymbol{D}(k)\boldsymbol{U}^{\mathrm{T}}(k)\big)\,.
\end{aligned} \tag{22.2.4}$$

After substitution of (9.4.18) and (9.6.12), the right-hand sides becomes

$$\begin{aligned}
\boldsymbol{U}\boldsymbol{D}\boldsymbol{U}^{\mathrm{T}} &= \frac{1}{\lambda}\boldsymbol{U}(k)\Big(\boldsymbol{D}(k) - \frac{1}{\alpha(k)}\boldsymbol{v}(k)\boldsymbol{f}^{\mathrm{T}}(k)\boldsymbol{D}(k)\Big)\boldsymbol{U}^{\mathrm{T}}(k)\\
&= \frac{1}{\lambda}\boldsymbol{U}(k)\Big(\boldsymbol{D}(k) - \frac{1}{\alpha(k)}\boldsymbol{v}(k)\boldsymbol{v}^{\mathrm{T}}(k)\Big)\boldsymbol{U}^{\mathrm{T}}(k)\,,
\end{aligned} \tag{22.2.5}$$

where

$$\begin{aligned}
\boldsymbol{f}(k) &= \boldsymbol{U}^{\mathrm{T}}(k)\boldsymbol{\psi}(k+1)\\
\boldsymbol{v}(k) &= \boldsymbol{D}(k)\boldsymbol{f}(k)\\
\alpha(k) &= \lambda + \boldsymbol{f}^{\mathrm{T}}(k)\boldsymbol{v}(k)\,.
\end{aligned} \tag{22.2.6}$$

The correcting vector then yields

$$\boldsymbol{\gamma}(k) = \frac{1}{\alpha(k)} \boldsymbol{U}(k) \boldsymbol{v}(k) \,. \tag{22.2.7}$$

If the term $(\boldsymbol{D} - \alpha^{-1} \boldsymbol{v}\boldsymbol{v}^{\mathrm{T}})$ in (22.2.5) is again factorized, the recursion for the elements $\boldsymbol{U}$, $\boldsymbol{P}$, and λ becomes

$$\left.\begin{aligned} \alpha_j &= \alpha_{j-1} + v_f f_j \\ d_j(k+1) &= \frac{d_j(k)\alpha(j-1)}{\alpha_j - \lambda} \\ b_j &= v_j \\ v_j &= \frac{f_j}{\alpha_{j-1}} \end{aligned}\right\} j = 2, \ldots, 2m \,, \tag{22.2.8}$$

see (Biermann, 1977), with the initial values

$$\alpha_1 = \lambda + v_1 f_1,\ d_1(k+1) = \frac{d_1(k)}{\alpha_1 \lambda} \tag{22.2.9}$$

$$b_1 = v_1 \,. \tag{22.2.10}$$

For each j, the following expressions hold for the elements of $\boldsymbol{U}$

$$\left.\begin{aligned} u_{ij}(k+1) &= u_{ij}(k) + r_j b_i \\ b_i &= b_i + u_{ij} v_j \end{aligned}\right\} i = 1, \ldots, j \tag{22.2.11}$$

$$\boldsymbol{\gamma}(k) = \frac{1}{\alpha_{2m}} \boldsymbol{b} \,. \tag{22.2.12}$$

The parameters are finally obtained from (9.4.17) as

$$\hat{\boldsymbol{\theta}}(k+1) = \hat{\boldsymbol{\theta}}(k) + \boldsymbol{\gamma}(k) e(k+1) \tag{22.2.13}$$

$$e(k+1) = y(k+1) - \boldsymbol{\psi}^{\mathrm{T}}(k+1)\hat{\boldsymbol{\theta}}(k) \,. \tag{22.2.14}$$

(22.2.12), (22.2.8), and (22.2.11) are calculated instead of (9.4.18) and (9.4.19). As compared to DSFC, here no routines are required for square root calculations. The computational expense is comparable to that of RLS. The numerical properties are similar to those of DSFC, only the matrix elements of $\boldsymbol{U}$ and $\boldsymbol{D}$ may become larger than those of $\boldsymbol{S}$.

To reduce the calculations after each sampling, invariance properties of the matrices (Ljung et al, 1978), may be used to generate fast algorithms. A saving of calculation time only results for order $m > 5$, but at the cost of greater storage requirements and higher sensitivity for starting values.

22.3 Factorization methods[1] for $\boldsymbol{P}^{-1}$

Discrete square root filtering in information form (DSFI) results from the non-recursive LS method of the form

[1] compiled by Michael Vogt

$$\boldsymbol{P}^{-1}(k+1)\hat{\boldsymbol{\theta}}(k+1) = \boldsymbol{\Psi}^{\mathrm{T}}(k+1)\boldsymbol{y}(k+1) = \boldsymbol{f}(k+1) \tag{22.3.1}$$

with

$$\boldsymbol{P}^{-1}(k+1) = \lambda \boldsymbol{P}^{-1}(k) + \boldsymbol{\psi}(k+1)\boldsymbol{\psi}^{\mathrm{T}}(k+1) \tag{22.3.2}$$

$$\boldsymbol{f}(k+1) = \lambda \boldsymbol{f}(k) + \boldsymbol{\psi}(k+1)y(k+1)\,. \tag{22.3.3}$$

The information matrix $\boldsymbol{P}^{-1}$ is now split into upper triangular matrices $\boldsymbol{R}$:

$$\boldsymbol{P}^{-1} = \boldsymbol{R}^{\mathrm{T}}\boldsymbol{R}\,. \tag{22.3.4}$$

Note that $\boldsymbol{R} = \boldsymbol{S}^{-1}$, cf. (22.2.1). Then $\hat{\boldsymbol{\theta}}(k+1)$ is calculated from (22.3.1) by back-substitution from

$$\boldsymbol{R}(k+1)\hat{\boldsymbol{\theta}}(k+1) = \boldsymbol{b}(k+1)\,. \tag{22.3.5}$$

This equation follows from (22.3.1), introducing an orthonormal transformation matrix $\boldsymbol{Q}$ (with $\boldsymbol{Q}^{\mathrm{T}}\boldsymbol{Q} = \boldsymbol{I}$), such that

$$\boldsymbol{\Psi}^{\mathrm{T}}\boldsymbol{Q}^{\mathrm{T}}\boldsymbol{Q}\boldsymbol{\Psi}\hat{\boldsymbol{\theta}} = \boldsymbol{\Psi}^{\mathrm{T}}\boldsymbol{Q}^{\mathrm{T}}\boldsymbol{Q}\boldsymbol{y}\,. \tag{22.3.6}$$

Here,

$$\boldsymbol{Q}\boldsymbol{\Psi} = \begin{pmatrix} \boldsymbol{R} \\ \boldsymbol{0} \end{pmatrix} \tag{22.3.7}$$

possesses an upper triangular form, and the equation

$$\boldsymbol{Q}\boldsymbol{y} = \begin{pmatrix} \boldsymbol{b} \\ \boldsymbol{w} \end{pmatrix} \tag{22.3.8}$$

holds. With (22.3.6), it follows that

$$\boldsymbol{Q}(k+1)\boldsymbol{\Psi}(k+1)\hat{\boldsymbol{\theta}}(k+1) = \boldsymbol{Q}(k+1)\boldsymbol{y}(k+1)\,. \tag{22.3.9}$$

Actually, DSFI uses a different idea to minimize the sum of errors squared

$$V = \sum e^2(k) = \|\boldsymbol{e}\|_2^2 = \|\boldsymbol{\Psi}\hat{\boldsymbol{\theta}} - \boldsymbol{y}\|_2^2\,. \tag{22.3.10}$$

Whereas the LS method solves the *normal equation* $\nabla V = \boldsymbol{0}$, here the *QR factorization*

$$\boldsymbol{Q}\boldsymbol{\Psi} = \begin{pmatrix} \boldsymbol{R} \\ \boldsymbol{0} \end{pmatrix} \tag{22.3.11}$$

is used to simplify (22.3.10). This relies on the fact that the multiplication with an orthonormal matrix $\boldsymbol{Q}$ does not change the norm of a vector, since

$$\begin{aligned} V &= \|\boldsymbol{\Psi}\hat{\boldsymbol{\theta}} - \boldsymbol{y}\|_2^2 = \|\boldsymbol{Q}\boldsymbol{\Psi}\hat{\boldsymbol{\theta}} - \boldsymbol{Q}\boldsymbol{y}\|_2^2 = \left\|\begin{pmatrix} \boldsymbol{R} \\ \boldsymbol{0} \end{pmatrix}\hat{\boldsymbol{\theta}} - \begin{pmatrix} \boldsymbol{b} \\ \boldsymbol{w} \end{pmatrix}\right\|_2^2 \\ &= \left\|\begin{pmatrix} \boldsymbol{R}\hat{\boldsymbol{\theta}} - \boldsymbol{b} \\ \boldsymbol{0} - \boldsymbol{w} \end{pmatrix}\right\|_2^2 = \|\boldsymbol{R}\hat{\boldsymbol{\theta}} - \boldsymbol{b}\|_2^2 + \|\boldsymbol{w}\|_2^2 = \min_{\hat{\boldsymbol{\theta}}}\,. \end{aligned}$$

As already stated in (22.3.5), the parameters $\hat{\boldsymbol{\theta}}$ are determined by solving the system $\boldsymbol{R}\hat{\boldsymbol{\theta}} - \boldsymbol{b} = \boldsymbol{0}$, whereas $\|\boldsymbol{w}\|_2^2$ is the remaining residual, i.e. the sum of errors squared for the optimal parameters $\hat{\boldsymbol{\theta}}$.

The main effort of the method described above is the computation of $\boldsymbol{R}$ and $\boldsymbol{b}$. This is usually done by applying *Householder transformations* to the matrix $(\boldsymbol{\Psi}\ \ \boldsymbol{y})$ (Golub and van Loan, 1996), so that $\boldsymbol{Q}$ does not need to be computed.

However, DSFI computes $\boldsymbol{R}$ and $\boldsymbol{b}$ *recursively*. Assuming that in each step one row is appended to $(\boldsymbol{\Psi}\ \ \boldsymbol{y})$, (22.3.9) is now transferred to a recursive form (Kaminski et al, 1971),

$$\begin{pmatrix} \boldsymbol{R}(k+1) \\ \boldsymbol{0}^{\mathrm{T}} \end{pmatrix} = \boldsymbol{Q}(k+1) \begin{pmatrix} \sqrt{\lambda}\boldsymbol{R}(k) \\ \boldsymbol{\psi}^{\mathrm{T}}(k+1) \end{pmatrix} \tag{22.3.12}$$

$$\begin{pmatrix} \boldsymbol{b}(k+1) \\ w(k+1) \end{pmatrix} = \boldsymbol{Q}(k+1) \begin{pmatrix} \sqrt{\lambda}\boldsymbol{b}(k) \\ y(k+1) \end{pmatrix} . \tag{22.3.13}$$

Then $\boldsymbol{R}(k+1)$ and $\boldsymbol{b}(k+1)$ are used to calculate $\hat{\boldsymbol{\theta}}(k+1)$ with (22.3.5), whereas $w(k+1)$ is the current residual. The method is especially suitable if the parameters are not required for each sample step. Then, only $\boldsymbol{R}$ and $\boldsymbol{b}$ have to be calculated recursively. This is done by applying *Givens rotations* to the right hand sides of (22.3.12) and (22.3.13). The Givens rotation

$$\boldsymbol{G} = \begin{pmatrix} \gamma & \sigma \\ -\sigma & \gamma \end{pmatrix} \tag{22.3.14}$$

is applied to a $2 \times \mu$ matrix $\boldsymbol{M}$ in order to eliminate the element m'_{21} in the transformed matrix $\boldsymbol{M}' = \boldsymbol{G}\boldsymbol{M}$, i.e. to introduce a zero in the matrix

$$\begin{pmatrix} \gamma & \sigma \\ -\sigma & \gamma \end{pmatrix} \begin{pmatrix} m_{11} & m_{12} & \cdots \\ m_{21} & m_{22} & \cdots \end{pmatrix} = \begin{pmatrix} m'_{11} & m'_{12} & \cdots \\ 0 & m'_{22} & \cdots \end{pmatrix} . \tag{22.3.15}$$

The two conditions

$$\det(\boldsymbol{G}) = \gamma^2 + \sigma^2 = 1 \text{ (normalization)} \tag{22.3.16}$$

$$m'_{21} = -\sigma m_{11} + \gamma m_{21} = 0 \text{ (elimination of } m'_{21}) \tag{22.3.17}$$

yield the rotation parameters

$$\gamma = \frac{m_{11}}{\sqrt{m_{11}^2 + m_{21}^2}} \tag{22.3.18}$$

$$\sigma = \frac{m_{21}}{\sqrt{m_{11}^2 + m_{21}^2}} . \tag{22.3.19}$$

This transformation is now sequentially applied to $\boldsymbol{\psi}^{\mathrm{T}}(k+1)$ and the rows of $\sqrt{\lambda}\boldsymbol{R}$ in (22.3.12), where $\boldsymbol{G}$ is now interpreted as an $(n+1) \times (n+1)$ matrix

Table 22.1. Computational expense of different parameter estimation algorithms. Orthogonal methods include back substitution. n is the number of parameters to be estimated

Method	Add/Sub	Mul	Div	Sqrt
NLMS	$3n$	$3n+1$	1	0
RLS	$1.5n^2+3.5n$	$2n^2+4n$	n	0
RMGS	$1.5n^2+1.5n$	$2n^2+3n$	$2n$	0
FDSFI	$1.5n^2+1.5n$	$2n^2+5n$	$2n$	0
DSFI	$1.5n^2+1.5n$	$2.5n^2+6.5n$	$3n$	n

$$\begin{pmatrix} * & * & * \\ 0 & * & * \\ 0 & 0 & * \\ * & * & * \end{pmatrix} \xrightarrow{G_1} \begin{pmatrix} \bullet & \bullet & \bullet \\ 0 & * & * \\ 0 & 0 & * \\ 0 & * & * \end{pmatrix} \xrightarrow{G_2} \begin{pmatrix} \bullet & \bullet & \bullet \\ 0 & \bullet & \bullet \\ 0 & 0 & * \\ 0 & 0 & * \end{pmatrix} \xrightarrow{G_3} \begin{pmatrix} \bullet & \bullet & \bullet \\ 0 & \bullet & \bullet \\ 0 & 0 & \bullet \\ 0 & 0 & 0 \end{pmatrix}$$

The product of the Givens matrices is the transformation matrix $\boldsymbol{Q}(k+1)$:

$$\begin{pmatrix} \boldsymbol{R}(k+1) \\ \boldsymbol{0}^{\mathrm{T}} \end{pmatrix} = \underbrace{\boldsymbol{G}_n(k+1)\ldots\boldsymbol{G}_1(k+1)}_{\boldsymbol{Q}(k+1)} \begin{pmatrix} \sqrt{\lambda}\boldsymbol{R}(k) \\ \boldsymbol{\psi}^{\mathrm{T}}(k+1) \end{pmatrix} \qquad (22.3.20)$$

that produces $\boldsymbol{R}(k+1)$. The same method is used to compute $\boldsymbol{b}(k+1)$ according to (22.3.13). A complete DSFI update step can now be described as follows:

Compute for $i = 1, \ldots, n$:

$$\begin{aligned} r_{ii}(k+1) &= \sqrt{\lambda r_{ii}^2(k) + (\psi_i^{(i)}(k+1))^2} \\ \gamma &= r_{ii}(k)/r_{ii}(k+1) \\ \sigma &= \psi_i^{(i)}(k+1)/r_{ii}(k+1) \\ \left.\begin{aligned} r_{ij}(k+1) &= \sqrt{\lambda}\gamma r_{ij}(k) + \sigma\psi_j^{(i)}(k+1) \\ \psi_j^{(i+1)}(k+1) &= -\sqrt{\lambda}\sigma r_{ij}(k) + \gamma\psi_j^{(i)}(k+1) \end{aligned}\right\} & \quad j = i+1, \ldots, n \\ b_i(k+1) &= \sqrt{\lambda}\gamma b_i(k) + \sigma y^{(i)}(k+1) \\ y^{(i+1)}(k+1) &= -\sqrt{\lambda}\sigma b_i(k) + \gamma y^{(i)}(k+1) \end{aligned} \qquad (22.3.21)$$

Further discussion of square root filtering may be found in (Peterka, 1975; Goodwin and Payne, 1977; Strejc, 1980).

Table 22.1 shall be used to compare the computational expense of different parameter estimation algorithms. The normalized least mean squares algorithm is a stochastic gradient descent method and hence not very reliable in finding the optimum, see Sect. 10.7. The recursive least squares uses more computations per update step, but is much more precise. On the other end of the table, one finds the discrete square root filter in information form (DSFI) algorithm, which is numerically very

Table 22.2. Total number of floating point operations for $n = 4$ and $n = 6$ parameters to be estimated

Method	$n = 4$				$n = 6$			
	Add/Sub	Mul	Div	Sqrt	Add/Sub	Mul	Div	Sqrt
NLMS	12	13	1	0	18	19	1	0
RLS	38	48	4	0	75	96	6	0
DSFI	30	66	12	4	63	129	18	6

robust, but at the same time requires the most computations per update step, including the calculation of n square roots. The DSFI algorithm can be implemented in an even more efficient form, if the following tricks are used: Store only the upper triangular part of the matrix $\boldsymbol{R}$. Store matrix row-wise, as then the elements can be addressed by incrementing the pointer. Do not calculate the parameter vector in every iteration, but only update the matrix $\boldsymbol{R}$. Table 22.2 illustrates the computational effort for $n = 4$ and $n = 6$ parameters to be estimated.

22.4 Summary

No essential differences in the numerical properties can be observed for DSFC and DSFI. Therefore, also DSFI requires the computation of n square roots in each step. There are also factorizations for $\boldsymbol{P}^{-1}$ that do not require square roots, just like the U-D factorization for $\boldsymbol{P}$. These techniques replace the Givens rotations by *fast Givens rotations* (Golub and van Loan, 1996), or employ recursive forms of the *Gram Schmidt orthogonalization*. These fast orthogonalization methods show the same error sensitivity, but their matrix elements may become larger than those of DSFI.

According to Table 22.1 and the fact that computers nowadays provide high computational power, one should in general use the DSFI algorithm. If this is not possible, then the RLS algorithm is a good choice as it has little computational expense and still provides a much higher precision than stochastic gradient descent algorithms.

Problems

22.1. QR Decomposition and Householder Transform
Show how a 5×3 matrix can be QR decomposed by the application of three Householder transforms $\boldsymbol{Q}_1, \boldsymbol{Q}_2, \boldsymbol{Q}_3$. $*$ denotes elements that may change, $\bullet$ denotes elements that will not change.

22.2. DSFI Algorithm 1
Show that for any orthonormal matrix $\boldsymbol{Q}$ and a vector $\boldsymbol{x}$ results $\|\boldsymbol{Q}\boldsymbol{x}\|_2^2 = \|\boldsymbol{x}\|_2^2$.
Show furthermore that for a vector $\boldsymbol{x}$ with

$$x = \begin{pmatrix} a \\ b \end{pmatrix} ,$$

one can obtain

$$\|x\|_2^2 = \|a\|_2^2 + \|b\|_2^2 .$$

Use these results to minimize the cost function

$$V = \|\Psi \hat{\theta} - y\|_2^2$$

by means of the QR decomposition.

22.3. DSFI Algorithm 2
Develop the parameter estimation algorithms to be programmed for a second order dynamic discrete-time process with

$$y(k) + a_1(y(k-1) + a_2 y(k-2) = b_1 u(k-1) + b_2 u(k-2)$$

References

Biermann GJ (1977) Factorization methods for discrete sequential estimation, Mathematics in Science and Engineering, vol 128. Academic Press, New York

Golub GH, van Loan CF (1996) Matrix computations, 3rd edn. Johns Hopkins studies in the mathematical sciences, Johns Hopkins Univ. Press, Baltimore

Goodwin GC, Payne RL (1977) Dynamic system identification: Experiment design and data analysis, Mathematics in Science and Engineering, vol 136. Academic Press, New York, NY

Isermann R, Lachmann KH, Matko D (1992) Adaptive control systems. Prentice Hall international series in systems and control engineering, Prentice Hall, New York, NY

Kaminski P, Bryson A jr, Schmidt S (1971) Discrete square root filtering: A survey of current techniques. IEEE Trans Autom Control 16(6):727–736

Kofahl R (1986) Self-tuning PID controllers based on process parameter estimation. Journal A 27(3):169–174

Ljung L, Morf M, Falconer D (1978) Fast calculation of gain matrices for recursive estimation schemes. Int J Control 27(1):1–19

Peterka V (1975) A square root filter for real time multivariate regression. Kybernetika 11(1):53–67

Strejc V (1980) Least squares parameter estimation. Automatica 16(5):535–550

23

Practical Aspects of Parameter Estimation

Going back to Fig. 1.7, one sees at once that the preceding chapters mainly described the block *Application of Identification Method* and the resulting *Process Model* that was either *non-parametric* or *parametric*. The other blocks will now be discussed in more detail. Also, some special issues, such as low frequent and high frequent disturbances, which are typically not accounted for by the identification method itself, disturbances at the system input, as well as a special treatment of integral acting systems shall be discussed. Also, all methods are summarized and their most important advantages and disadvantages are discussed, see Sect. 23.4. Finally, means to critically evaluate the identification results are presented at the end of this chapter.

23.1 Choice of Input Signal

If the input signal for the identification of a dynamic process can be chosen freely, then one still has to consider the limitations mentioned in Sect. 1.2, i.e. the

- maximum allowable amplitude and speed of the change of the input signal $u(t)$
- maximum allowable amplitude of the output signal $y(t)$
- maximum measurement time $T_{\mathrm{M,max}}$

From the identifiability conditions in Sect. 9.1.4, one knows that the input signal must be persistently exciting of order m, where m is the process order. However, there is still a magnitude of input signals that satisfy the conditions for a consistent estimation. If one wants to obtain a model of maximum fidelity under the given constraints, then the test signal must be designed such that it also optimizes the cost function of the identified model. It seems obvious to derive a suitable quality criterion from the error covariance matrix of the parameter estimates. As a quality criterion, one can indeed define a scalar cost function Φ as

$$V = \mathrm{E}\{\Phi(\boldsymbol{J})\}\,, \tag{23.1.1}$$

see (Goodwin and Payne, 1977; Gevers, 2005), e.g.

R. Isermann, M. Münchhof, *Identification of Dynamic Systems*,
DOI 10.1007/978-3-540-78879-9_23, © Springer-Verlag Berlin Heidelberg 2011

$$V_1 = \mathrm{E}\{\mathrm{tr}\, \boldsymbol{J}^{-1}\}\,, \tag{23.1.2}$$

which is denoted as *A-optimal* or

$$V_2 = \mathrm{E}\{\mathrm{tr}\, \boldsymbol{W}\boldsymbol{J}^{-1}\}\,, \tag{23.1.3}$$

where $\boldsymbol{W}$ is a suitable weighting matrix (called *L-optimal*) . A further metric is

$$V_3 = \det \boldsymbol{J}\,, \tag{23.1.4}$$

which is referred to as *D-optimal*. Under the assumption of a Gaussian distributed error, one can, for the method of least squares, use

$$\boldsymbol{J} = \frac{1}{\sigma_e^2}\mathrm{E}\{\boldsymbol{\Psi}^{\mathrm{T}}\boldsymbol{\Psi}\} = \frac{1}{\sigma_e^2}\mathrm{E}\{\boldsymbol{P}^{-1}\}\,. \tag{23.1.5}$$

Based on the quality criterion, one can then try to design an optimal test signal either in the time or in the frequency domain (Mehra, 1974; Krolikowski and Eykhoff, 1985).

To resolve the mutual dependency of the input signal optimality and the parameter estimates, one can use a minimax approach as outlined in (Welsh et al, 2006). Here, the input signal is optimized by means of any of the above presented cost functions, but the cost function is not only evaluated for a single parameter set $\boldsymbol{\theta}$, but over an entire compact parameter set $\boldsymbol{\Theta}$. The maximum of the cost function over that entire compact parameter set is then minimized, hence leading to a minimax optimization problem.

Optimal test signals can hence only be stated for special cases, such as e.g. efficient parameter estimation methods and long measurement times. Furthermore, the quality criterion should not only be based on the expected error of the parameters, but also on the final application of the model. In practical applications, things are further impaired by thc fact that the model and the noise are not known a priori, so that one can only iteratively design test signals or as an alternative employ nearly optimal test signals. These nearly optimal test signals shall be referred to as *favorable test signals* from now on. For the choice of favorable test signals, the following guidelines are suggested.

Signals from Normal Operation or Artificial Test Signals

As input signals, one can use the signals that occur during the normal operation or one can inject special test signals. The signals from normal operation are however only suitable, if they excite the process to be identified sufficiently in the range of the interesting process dynamics. Furthermore, they must be stationary and uncorrelated with the disturbances acting on the process. This is only true in very rare cases. One should therefore, whenever possible, use artificial test signals, whose properties are exactly known and can be tuned to obtain models of a high fidelity.

Shape of the Test Signal

The shape of the test signal is first and foremost limited by the actuator (e.g. electrically, pneumatically, or hydraulically powered), as the actuator limits the maximum speed of change and hence the time derivative of the input. This also constrains the maximum frequency in the input signal.

Favorable test signals typically excite the interesting eigenvalues of the process continuously and as strongly as possible compared with the disturbance spectrum. For the design/choice of test signals, one has to take into account, compare Sect. 1.5:

- The height of the test signal u_0 should be as large as possible. Here, one has to take limitations of the input and output signal as well as the system states due to operating limitations or the assumption of linearity into account.
- The steeper the edges, the stronger the excitation of the high frequencies (Gibb's phenomenon).
- The smaller the width of the pulses in the input signal, the stronger the excitation of medium to high frequencies. The broader the pulses, the stronger the excitation of low frequencies.

From these considerations follows that the pseudo-random binary signals and generalized random binary signals (GRBS) are especially well suited for correlation analysis and parameter estimation, see Sect. 6.3. If the PRBS signal should excite the high frequencies, then the cycle time λ has to be chosen equal to the sample time T_0. A choice of $\lambda / T_0 = 2, 3, \ldots$ increases the power spectral density at low frequencies and allows a better estimation of the DC gain at the price of diminishing excitation of the high frequencies. By variation of the cycle time λ, one can hence adjust the excited frequency spectrum by means of a single parameter. For the GRBS, one can influence the signal shape by the probability p. One can also imagine to adjust the cycle time online during the experiment, see Sect. 6.3 for a treatment of PRBS and GRBS signals. For processes of low order and limitation of some system states, it can be advisable to use multi frequency signals, see Sect. 5.3, instead of PRBS signals. In (Bombois et al, 2008) a method is presented to design an optimal test signal for multi-sine excitation. Here, the maximum power (i.e. sum of squared amplitudes) is minimized under constraints on the parameter error covariance.

If there is enough measurement time, the excitation with sine-functions for the determination of the frequency responses is one of the best method for linear processes to determine frequency responses, e.g. by the orthogonal correlation, see Sect. 5.5.2. Non-linear processes require multi-valued test signals like APRBS, which are discussed in Sect. 6.3.

23.2 Choice of Sample Rate

For the identification of processes with discrete-time signals, the sample rate must be chosen prior to the measurement. The sample time cannot be reduced later. On the opposite, an increase to twice or three times the sample time can easily be realized by just using every second, third, etc. sample. Before the downsampling, the data should however be low-pass filtered to avoid aliasing in the downsampled data. The choice of the sample time mainly depends on:

- Sample time of the discrete-time model in later application
- Fidelity of the resulting model
- Numerical problems

These issues shall be discussed in the following sections.

23.2.1 Intended Application

If the model is subsequently used to design a digital controller, the sample time must be chosen according to that of the control algorithm. This quantity in turn depends on many aspects such as e.g. desired control quality, chosen control algorithm, and the target hardware. As a reference value, e.g. for PID control algorithms, one can select

$$\frac{T_0}{T_{95}} \approx \frac{1}{5}, \ldots, \frac{1}{15}, \quad (23.2.1)$$

where T_{95} is the 95% settling time of the step response of a proprotional acting process (Isermann and Freyermuth, 1991). Upon higher demands on the control quality, the sample time can become smaller. Similarly, Verhaegen and Verdult (2007) suggested to sample 8 to 9 times during the rise time of the system.

In order to determine the sample rate for an oscillatory system, it is proposed by Verhaegen and Verdult (2007) to count the number of cycles before the steady-state is reached in the step response. If this number is denoted by n and it is assumed that it takes roughly the time of four time constants of the system to settle down, then the time constant can be estimated as

$$T \approx \frac{n T_{\text{cycle}}}{4}. \quad (23.2.2)$$

With the knowledge of the time constant, one can then choose the sample rate so that the system is sampled between 5 and 15 times during the interval of one time constant. A sample rate that is too high is indicated by one or more of the following effects:

- Bad numerical condition of the matrix $\boldsymbol{\Psi}^{\mathrm{T}}\boldsymbol{\Psi}$ due to nearly linearly dependent rows
- Clustering of the poles of a discrete-time system around $z = 1$
- High frequent noise in the data

23.2.2 Fidelity of the Resulting Model

Table 23.1 shows the influence of the sample time T_0 on the parameter estimates a_i and b_i of the transfer function of the Three-Mass Oscillator. As one can see, for decreasing sample times, the absolute value of the parameters b_i decreases and the

Table 23.1. Parameters of the theoretical model of the transfer function of the Three-Mass Oscillator (see App. B) as a function of the sample time T_0. To reduce the sample time, only every k^{th} sample was retained in the data vector

T_0 [s]	0.003	0.012	0.048	0.144
k	1	4	16	48
b_1	-0.013112	-0.0090007	0.055701	5.2643
b_2	-0.0042292	-0.011311	0.49831	7.5739
b_3	0.0086402	0.020682	0.767	2.3529
b_4	0.0032622	-0.0019679	0.44988	0.73567
b_5	-0.0087436	0.026107	0.0081502	0.12386
b_6	0.023896	0.047981	-0.051817	-0.031231
a_1	-0.73415	-1.4584	-1.955	0.11845
a_2	-0.45075	-0.22564	1.718	0.18661
a_3	-0.21071	0.38383	-0.68648	-0.60705
a_4	-0.01038	0.50713	-0.29154	-0.22439
a_5	0.16337	0.16565	0.7275	-0.094103
a_6	0.2451	-0.36914	-0.47486	-0.022469
$\sum b_i$	0.0097141	0.072491	1.7272	16.0194
$1 + \sum a_i$	0.0024734	0.003428	0.037615	0.35704
K	3.9275	21.1468	45.9185	44.8669

sum of the b_i, which is important e.g. to determine the DC gain, depends severely on thc fourth or fifth digit behind the decimal point of the individual b_i. As one can see, small absolute errors in the parameters can have a significant impact on the input/output behavior of the model (gain, impulse response). On the other hand, if the sample time is chosen too large, then the resulting model order reduces.This can also be seen for the last column of Table 23.1. Here, $a_6 \ll |1 + \sum a_i|$ and $b_6 \ll |\sum b_i|$, hence the model order has reduced as the last coefficients are of negligible size.

23.2.3 Numerical Problems

If the sample time is chosen too small, then a badly conditioned system of equations results, as the difference equations for different values of k become nearly linearly depended. Hence, for a decrease of the sample time, one suddenly witnesses a big increase in the parameter variances.

However, the choice of the sample time is rather uncritical as the range between too small and too large sample times is relatively broad.

23.3 Determination of Structure Parameters for Linear Dynamic Models

The determination of the order of a parametric model with the transfer function

$$G_{\mathrm{P}}(z) = \frac{y(z)}{u(z)} = \frac{b_1 z^{-1} + \ldots + b_{\hat{m}} z^{-\hat{m}}}{1 + a_1 z^{-1} + \ldots + a_{\hat{m}} z^{-\hat{m}}} z^{-\hat{d}} \tag{23.3.1}$$

means to determine the structure parameters $\hat{m}$ and $\hat{d}$ of the model of the process with the true order m_0 and d_0. In the ideal case, one should obtain $\hat{m} = m_0$ and $\hat{d} = d_0$.

In most cases, the structure parameters have to be fixed before the estimation of the model parameters. They hence represent part of the a priori assumptions and must be checked as part of the validation of the results, see Sect. 23.8. Therefore, the model order and dead time can be determined by the methods that will subsequently in this chapter be used for model validation. Also, a number of specific methods has been developed which allow to determine the structure parameters and should be seen together with the respective parameter estimation methods.

The methods are termed *order* or *dead time tests* and can be discerned according to the following properties:

- Deterministic or stochastic approach
- Previous parameter estimation necessary or not
- Process and noise model treated separately or together

In the following, some methods are presented, which allow to determine the model order and dead time. Model order tests have been summarized by Söderström (1977), van den Boom (1982), and Raol et al (2004) among others. Often, it is useful to determine the dead time and the model order consecutively. However, they can also be determined in parallel. These criteria can also be used for frequency domain identification as shown in (Pintelon and Schoukens, 2001). Too large models lead to inefficient estimators, too small models to inconsistent estimators (Heij et al, 2007). A first estimate of the model order can be obtained by applying non-parametric identification techniques (e.g. frequency response) and analyzing their results.

23.3.1 Determination of Dead Time

It is for now assumed that the process order m is known. For the determination of the dead time, it is assumed that the true dead time d_0 is bounded by $0 \leq d_0 \leq d_{\max}$ and that the numerator polynomial of the process model

$$y(z) = \frac{B^*(z^{-1})}{A(z^{-1})} u(z) = G_{\mathrm{P}}^*(z) u(z) \tag{23.3.2}$$

has been augmented as

$$B^*(z^{-1}) = b_1^* z^{-1} + \ldots + b_{m+d_{\max}}^* z^{-m-d_{\max}} \,. \tag{23.3.3}$$

For the process model in (23.3.1), one then obtains

$$\left.\begin{aligned} b_i^* &= 0 && \text{for } i = 1, 2, \ldots, \hat{d} \\ b_i^* &= b_{i-\hat{d}} && \text{for } i = 1 + \hat{d}, 2 + \hat{d}, \ldots, m + \hat{d} \\ b_i^* &= 0 && \text{for } i = m + \hat{d} + 1, \ldots, m + d_{\max} \,. \end{aligned}\right\} \tag{23.3.4}$$

For the parameter estimation, one uses the following vectors

$$\boldsymbol{\psi}^{\mathrm{T}}(k+1) = \big(-y(k) \ldots -y(k-m)\big|-u(k-1) \ldots -u(k-m-d_{\max})\big) \tag{23.3.5}$$

$$\hat{\boldsymbol{\theta}} = \big(\hat{a}_1 \ldots \hat{a}_m \big| \hat{b}_1^* \ldots \hat{b}_{m+d_{max}}^*\big) . \tag{23.3.6}$$

For a consistent parameter estimation method, one can expect that

$$\begin{aligned} \mathrm{E}\{\hat{b}_i^*\} = 0 \text{ for } i &= 1, 2, \ldots, d_0 \\ \text{and } i &= m + d_0 + 1, m + d_0 + 2, \ldots, m + d_{\max} . \end{aligned} \tag{23.3.7}$$

Hence, these coefficients of the numerator polynomial must be small compared to the remaining parameters. As a criterion for the determination of $\hat{d}$, one can use

$$|\hat{b}_i^*| \ll \sum_{i=1}^{m-d_{\max}} \hat{b}_i^* \text{ and } |\hat{b}_{i+1}^*| \gg |\hat{b}_i^*| \; i = 1, 2, \ldots, \hat{d} . \tag{23.3.8}$$

In the ideal case, the first condition of this this criterion is satisfied for all $i = \hat{d} \leq d_0$, whereas the second part is only satisfied for $i = \hat{d} = d_0$ (Isermann, 1974). This simple method however assumes that the influence of disturbances is mostly eliminated, either because the disturbance is small or because the measurement time is sufficiently long.

If the disturbance is larger, one can use the following approach (Kurz and Goedecke, 1981):

- *Step 1*: Determine the largest parameter $|\hat{b}_{d'_{\max}}^*|$ of $B^*(z^{-1})$. Then, the dead time must be in the interval
$$0 \leq \hat{d} \leq d'_{\max} . \tag{23.3.9}$$
- *Step 2*: Determine the error of the impulse responses as
$$\Delta g_d(\tau) = \hat{g}^*(\tau) - \hat{g}_d(\tau), \; d = 0, 1, \ldots, d'_{max} , \tag{23.3.10}$$
where $\hat{g}^*(\tau)$ is the impulse response of $G^*(z)$ and $\hat{g}_d(\tau)$ of $G_{\mathrm{P}d}(z)$, where $G_{\mathrm{P}d}(z)$ is given as
$$G_{\mathrm{P}d}(z) = \frac{\hat{B}(z^{-1})}{\hat{A}(z^{-1})} z^{-\hat{d}} . \tag{23.3.11}$$
The parameters of $\hat{A}(z^{-1})$ are identical for both impulse responses. The parameters $\hat{B}(z^{-1})$ of the model $G_{\mathrm{P}d}(z)$ can be determined from $g^*(\tau)$ as follows:
$$\hat{b}_1 = \hat{g}^*(1 + \hat{d}) \tag{23.3.12}$$
$$\hat{b}_i = \hat{g}^*(i + \hat{d}) + \sum_{j=1}^{i-1} a_j \hat{g}^*(i - j + \hat{d}), \; i = 2, \ldots, m . \tag{23.3.13}$$

A recursive formulation for the calculation of the error $\Delta g_d(\tau)$ is presented by Kurz (1979). Then, the cost function

$$V(d) = \sum_{\tau=1}^{M} \Delta g_d^2(\tau),\ d = 0, 1, \ldots, d'_{max} \tag{23.3.14}$$

is evaluated.

- *Step 3*: The minimum value of $F(\hat{d})$ determines the dead time $\hat{d}$
- *Step 4*: The parameters $\hat{b}_i$ can now be estimated

The computational effort of this method is comparably low and hence the method can also be employed after each sample step for the recursive parameter estimation. Kurz and Goedecke (1981) showed examples of adaptive control of processes with variable dead time.

For a non-parametric model in the form of an impulse response or step response, one can also determine the dead time from the delay between the input and the initial reaction of the output to that input.

23.3.2 Determination of Model Order

In order to determine the unknown model order $\hat{m}$, one can use different criteria such as

- cost function
- rank of information matrix
- residuals
- poles and zeros

The common principle is that any of these quantities, upon variation of the model order, will show a distinct behavior upon passing the true model order. These individual criteria are listed in the following sections.

Cost Function

Since all parameter estimation methods minimize a cost function

$$V(m, N) = \boldsymbol{e}^{\mathrm{T}}(m, N)\boldsymbol{e}(m, N)\,, \tag{23.3.15}$$

a rather obvious approach is to analyze the cost function as a function of the model order estimate $\hat{m}$. $\boldsymbol{e}$ can be the vector of the equation error or the residual of the employed parameter estimation method or can be the output error between model and process. Hence, for a given model order estimate $\hat{m}$, one must determine the parameter vector $\hat{\boldsymbol{\theta}}(N)$ and can then determine the error.

For $m = 1, 2, 3, \ldots, m_0$, the cost function value $V(m, N)$ will decrease as the error will get smaller with an increasing model order. If no disturbances are acting on the process, one would theoretically obtain $V(m_0, N) = 0$. If disturbances are

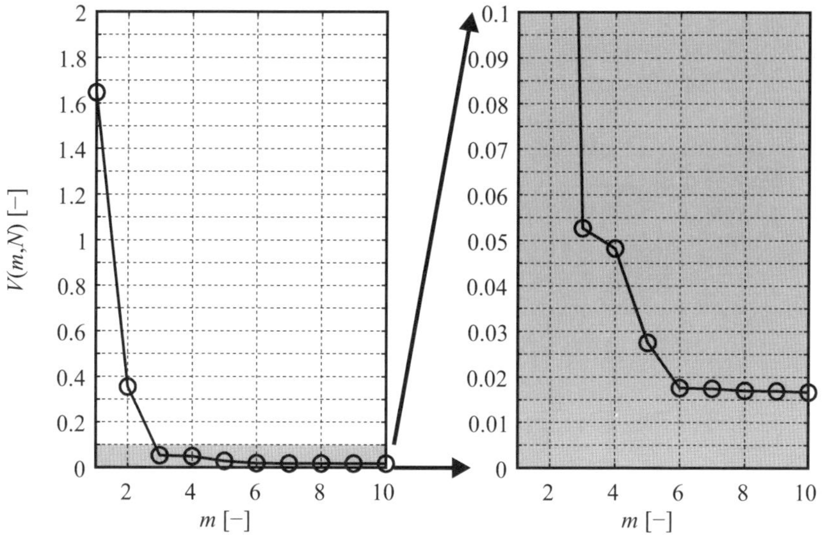

Fig. 23.1. Cost function $V(m, N)$ of the parameter estimation of the Three-Mass Oscillator as a function of model order m

acting on the process, it is expected that $V(m_0, N)$ is a visible turning point in the run of the cost function and that $V(m, N)$ for $m > m_0$ does not change that much. Therefore, a criterion to determine the model order can be the change of $V(m, N)$ upon an increase of the model order as

$$\Delta V(m+1) = V(m) - V(m+1) \tag{23.3.16}$$

and the model order test can be to look for the point

$$\Delta V(\hat{m}+1) \ll \Delta V(\hat{m}) \, , \tag{23.3.17}$$

where no significant improvement of the cost function can be obtained, or basically $V(m)$ does not diminish much more, such that $V(\hat{m}+1) \approx V(\hat{m})$. Then, $\hat{m}$ is the estimate of the model order.

Example 23.1 (Model Order Determination for the Three-Mass Oscillator).

The method of determining the model order by a cost function analysis has been applied to the Three-Mass Oscillator. One can see in Fig. 23.1 that the model order is correctly estimated as $\hat{m} = m_0 = 6$. Noise has been added to the output to show the effect of a very slight reduction for increasing the model order m beyond m_0. □

It has to be added that the calculation of the cost function requires an evaluation of all N data points for each value of m (and also d). Here, methods which employ an intermediate non-parametric model like COR-LS can be an appealing alternative as the number of data points to be processed is much lower than that of the original time sequence. This advantage can be decisive for online estimation of model order and dead time as here all possible candidates of m and d have to be tested simultaneously in parallel during each sample step.

For the automation of the model order estimation, it is advantageous to use statistical hypothesis testing methods to determine whether a significant change in the cost function has taken place if the model order changes from m_1 to m_2. A first possibility is the F-test (Åström, 1968). The test is based on the statistical independence between $V(m_2, N)$ and $V(m_1, N) - V(m_2, N)$, which are χ^2-distributed for normal distributed residuals. To test, whether the cost function changes significantly if the order is increased from m_1 to m_2, i.e. if the number of parameters changes from $2m_1$ to $2m_2$, one uses the test quantity

$$t = \frac{V(m_1) - V(m_2)}{V(m_2)} \frac{N - 2m_2}{2(m_2 - m_1)} . \tag{23.3.18}$$

For large N, the random variable t is asymptotically $F[2(m_2 - m_1), (N - 2m_2)]$ distributed. One can then define a threshold and obtain its value t^* from tables for hypothesis testing using this very distribution (Lehmann and Romano, 2005, e.g.). For $t < t^*$, m_1 is the estimated model order.

In conjunction with the maximum likelihood estimation, Akaike defined some test quantities: The *final prediction error* criterion is given as

$$\mathrm{FPE} = \frac{N + 2m}{N - 2m} \det \frac{1}{N} \sum_{k=1}^{N} \boldsymbol{e}(k, \boldsymbol{\theta}) \boldsymbol{e}^{\mathrm{T}}(k, \boldsymbol{\theta}) , \tag{23.3.19}$$

where the $\boldsymbol{e}(k, \boldsymbol{\theta})$ are the one step ahead prediction errors based on the ML estimates $\boldsymbol{\theta}$. One will then determine the minimum of the FPE, (Akaike, 1970). Another criterion is the *Akaike information criterion (AIC)*

$$\mathrm{AIC} = 2m - 2 \log L(\boldsymbol{\theta}) , \tag{23.3.20}$$

where $L(\boldsymbol{\theta})$ is the likelihood function and $\boldsymbol{\theta}$ are the ML estimates of order m. Once again, one tries to determine the minimum of this criterion. The first term ensures that the cost function increases again if the model is over-parameterized. Another formulation is the Bayesian information criterion, given as

$$\mathrm{BIC} = 2m \log N - 2 \log L(\boldsymbol{\theta}) . \tag{23.3.21}$$

Söderström (1977) has shown that asymptotically, the F-test, the FPE and the AIC are equivalent.

The practical application of the cost function tests requires for each model order m a parameter estimation to determine $\boldsymbol{\theta}(m)$ and an evaluation of the cost function $V(m)$. In order to reduce the computational burden, one can choose between the following approaches:

- Recursive calculation of the covariance matrix $\boldsymbol{P}(m, N)$ for different orders m without matrix inversion (Schumann et al, 1981)
- Successive reduction of the estimates of a model of too high order by parameter estimation employing the DSFI algorithm (smaller computational burden because of triangular matrix) (Kofahl, 1986)

Rank Test of the Information Matrix

For the LS parameter estimation

$$\hat{\theta}(N) = (\boldsymbol{\Psi}^{\mathrm{T}}\boldsymbol{\Psi})^{-1}\boldsymbol{\Psi}^{\mathrm{T}}\boldsymbol{y} , \tag{23.3.22}$$

the behavior of the matrix

$$\boldsymbol{J}' = \boldsymbol{P}^{-1} = \boldsymbol{\Psi}^{\mathrm{T}}\boldsymbol{\Psi} \tag{23.3.23}$$

is investigated. This is part of the *information matrix*, (11.2.56). For the case that no noise is acting on the system, the matrix $\boldsymbol{J}'$ becomes singular for any model order $m > m_0$, i.e.

$$\det \boldsymbol{J}' = \det \boldsymbol{\Psi}^{\mathrm{T}}\boldsymbol{\Psi} = 0 , \tag{23.3.24}$$

since

$$\operatorname{rank} \boldsymbol{J}' = 2m_0 \tag{23.3.25}$$

For disturbances $n(k)$ acting on the system, the determinant will still be non-zero after the transition $\hat{m} = m_0 \rightarrow \hat{m} = m_0 + 1$, but will experience a significant change. Woodside (1971) therefore proposed to evaluate the determinant ratio

$$\mathrm{DR}(m) = \frac{\boldsymbol{J}'_m}{\boldsymbol{J}'_{m+1}} . \tag{23.3.26}$$

For $m = m_0$ and for good signal-to-noise ratios, one will witness a jump. For larger noise levels, it is suggested to use

$$\boldsymbol{J}'' = \boldsymbol{J}' - \sigma^2 \boldsymbol{R} , \tag{23.3.27}$$

where $\sigma^2 \boldsymbol{R}$ is the covariance matrix of the noise, that must then be known.

A method for the determination of the cost function without determining the parameter estimates is given by the following derivation (Hensel, 1987).

$$\hat{\theta} = \left(\boldsymbol{\Psi}^{\mathrm{T}}\boldsymbol{\Psi}\right)^{-1}\boldsymbol{\Psi}^{\mathrm{T}}\boldsymbol{y} = \left(\boldsymbol{\Psi}^{\mathrm{T}}\boldsymbol{\Psi}\right)^{-1}\boldsymbol{q} \tag{23.3.28}$$

$$\boldsymbol{e}^{\mathrm{T}}\boldsymbol{e} = \left(\boldsymbol{y}^{\mathrm{T}} - \hat{\theta}^{\mathrm{T}}\boldsymbol{\Psi}^{\mathrm{T}}\right)\left(\boldsymbol{y} - \boldsymbol{\Psi}\hat{\theta}\right) = \boldsymbol{y}^{\mathrm{T}}\boldsymbol{y} - \boldsymbol{q}^{\mathrm{T}}\left(\boldsymbol{\Psi}^{\mathrm{T}}\boldsymbol{\Psi}\right)^{-1}\boldsymbol{q} \tag{23.3.29}$$

$$= \boldsymbol{y}^{\mathrm{T}}\boldsymbol{y} - \boldsymbol{q}^{\mathrm{T}}\frac{\operatorname{adj} \boldsymbol{\Psi}^{\mathrm{T}}\boldsymbol{\Psi}}{\det \boldsymbol{\Psi}^{\mathrm{T}}\boldsymbol{\Psi}}\boldsymbol{q}$$

$$\boldsymbol{e}^{\mathrm{T}}\boldsymbol{e}\det\left(\boldsymbol{\Psi}^{\mathrm{T}}\boldsymbol{\Psi}\right) = \boldsymbol{y}^{\mathrm{T}}\boldsymbol{y}\det\left(\boldsymbol{\Psi}^{\mathrm{T}}\boldsymbol{\Psi}\right) + \boldsymbol{q}^{\mathrm{T}}\operatorname{adj}\left(\boldsymbol{\Psi}^{\mathrm{T}}\boldsymbol{\Psi}\right)\boldsymbol{q} . \tag{23.3.30}$$

One can now use the fact that

$$\det\begin{pmatrix} 0 & \boldsymbol{x}^{\mathrm{T}} \\ \boldsymbol{w} & \boldsymbol{A} \end{pmatrix} = -\sum_i \sum_j x_i w_j \boldsymbol{A}_{ji} = -\boldsymbol{x}^{\mathrm{T}}(\operatorname{adj} \boldsymbol{A})\boldsymbol{w} \tag{23.3.31}$$

$$\det\begin{pmatrix} 0 & \boldsymbol{q}^{\mathrm{T}} \\ \boldsymbol{q} & \boldsymbol{\Psi}^{\mathrm{T}}\boldsymbol{\Psi} \end{pmatrix} = -\boldsymbol{q}^{\mathrm{T}}\left(\operatorname{adj} \boldsymbol{\Psi}^{\mathrm{T}}\boldsymbol{\Psi}\right)\boldsymbol{q} . \tag{23.3.32}$$

From this follows

$$e^{\mathrm{T}}e \det \boldsymbol{\Psi}^{\mathrm{T}}\boldsymbol{\Psi} = \det \begin{pmatrix} y^{\mathrm{T}}y & q^{\mathrm{T}} \\ q & \boldsymbol{\Psi}^{\mathrm{T}}\boldsymbol{\Psi} \end{pmatrix} = \det \boldsymbol{\Gamma}_m \tag{23.3.33}$$

$$V(m) = e^{\mathrm{T}}e = \frac{\det \boldsymbol{\Gamma}_m}{\det \boldsymbol{J}'_m}\,, \tag{23.3.34}$$

see (Woodside, 1971). The matrix $\boldsymbol{J}'_m$ has the dimension $2m \times 2m$ and the extended information matrix $\boldsymbol{\Gamma}_m$ the dimension $2m+1 \times 2m+1$. Hence, the dimension of $\boldsymbol{\Gamma}_m$ is by 2 smaller than the dimension of the matrix $\boldsymbol{J}'_{m+1}$, which is required for the determinant ratio test.

Mäncher and Hensel (1985) suggested to evaluate the ratio of the cost functions

$$\mathrm{VR}(m) = \frac{V(m-1)}{V(m)} = \frac{\det \boldsymbol{\Gamma}_{m-1}}{\det \boldsymbol{J}'_{m-1}} \frac{\det \boldsymbol{J}'_m}{\det \boldsymbol{\Gamma}_m} \tag{23.3.35}$$

and to test this quantity as a function of m. One can determine the required determinant successively for different m. With this approach, one can use a computationally inexpensive realization of the determination of model order and dead time, which is also suitable for MIMO systems (Mäncher and Hensel, 1985).

The methods just presented test the rank of the information matrix $\boldsymbol{J}'$ or of the extended information matrix $\boldsymbol{\Gamma}$ respectively and establish a link to the theoretical value of the cost function. A big advantage is the fact that one does not have to estimate the parameters, which makes this methods well suited for MIMO processes (Hensel, 1987).

Pole-Zero Test

If a model of higher order m is chosen than is required by the process order m_0, then the identified model will have an additional $(m - m_0)$ poles and zeros that almost cancel each other. This effect can be used to determine the model order. One however has to calculate the roots of the numerator and denominator polynomials.

The following approach is proposed by Pintelon and Schoukens (2001):

- Make an initial guess of the maximum model order, e.g. by looking at the rank of the information matrix. The model order assumed should be conservative, i.e. too high
- Use this model order for an initial parameter estimate
- Cancel poles, zeros, and pole/zero pairs, which do not contribute significantly to the model dynamics. The significance can e.g. be determined from a partial fraction expansion. Check the validity of each reduction and terminate if no further reduction is possible. Then, the model order has been determined and a final parameter estimation should be carried out. Possibly restart this method.

Residual Tests

The parameter estimation methods LS, ELS, GLS, and ML should yield in the case of bias-free estimates and under other idealizing assumptions residuals that are white. Hence, one can test the residuals for whiteness, e.g. by calculating the auto-correlation function. While this method is in general well suited for model validation, it can also be used for model order determination, (van den Boom and van den Enden, 1974; van den Boom, 1982). Starting with a model order estimate of $\hat{m} = 1$, the model order is increased until the residuals are white for the first time, then one expects that $\hat{m} = m_0$.

Concluding Remarks

Practical experience has shown that for processes of higher order, model order identification methods that are based on the cost function or the information matrix yield good results. It might be beneficial to combine several tests. In many cases, however, there might not exist one "best" model order as e.g. several small time constants or dead times could be combined into one time constant, the structure of a distributed parameter system or (weakly) non-linear process cannot be captured exactly by the linear lumped parameter model. Hence, the identified model order can be seen as an approximation. Depending on the type of application, one can make the following suggestions:

Interactive Model Order Determination: The decision is made by the user as part of an offline identification. Here, one can use all methods as the computational burden does not matter. A combination of different methods is suggested, e.g. a cost function test and the pole-zero test.

Automated Model Order Determination: If the model order has to be determined in real time, then the computational effort can play a dominant role in selecting an appropriate algorithm. In combination with recursive parameter estimation methods, it is suggested to carry out a cost function test or a rank test.

23.4 Comparison of Different Parameter Estimation Methods

A comparison of the different methods presented in this book will now be given.

23.4.1 Introductory Remarks

The large number of existing parameter estimation methods shall now be categorized with respect to the model that is to be derived as well as the assumptions on the noise. The main advantages and disadvantages of the respective methods are then summarized.

Before these properties shall be compared in the subsequent sections, some introductory remarks shall now be made with respect to the comparison of identification methods.

For the comparison of the *a priori assumptions*, one just has to compare the different assumptions that had to be made for the mathematical development and for the convergence analysis of the respective method. The same holds true for the *computational effort*. Here, one can determine the number of floating point operations

(FLOPS) and the storage space for the algorithms. For the non-recursive methods, one can compare the total number of floating point operations, whereas for the recursive methods, one can compare the number of FLOPS between two data samples.

Much more difficult is a comparison of the *model fidelity*. One approach is to assume a certain model and noise and then carry out a *simulation* on a computer. One can then e.g. determine the model error as a function of the data sample length or the noise level. A big advantage of this approach is that one can also compare cases where the a priori assumptions or the assumptions that were necessary for the theoretical convergence analysis are violated. This is often encountered in practical applications and hence it is interesting to see, how the methods work in such a case. The major drawback is the fact that all results depend on the simulation model used and are hence in theory only valid for this particular model considered. Although in many cases possible, the results cannot unconditionally be generalized.

For some parameter estimation methods, one can also base a comparison on the *theoretical convergence analysis*. For example, one can compare directly the covariance matrices of the non-recursive methods of LS, WLS, and ML. For a comparison of the recursive methods, one could compare e.g. the trajectories of the parameter estimates. A comparison of the theoretically obtained results for convergence does implicitly only apply to the case of long (possibly even infinitely long) measurement times and under the satisfaction of all a priori assumptions. If one is interested in results for short measurement times and/or in the presence of violations of the a priori assumptions, then one typically has to resort to simulation studies.

A third possibility is to compare the different methods applied to a *real process*. This is especially advisable, if one already has a certain application in mind and can obtain data which are typical for this application. A problem here however is the fact that an *exact* model of the process is in most cases not known. Furthermore, the process behavior and the disturbances can change over time so that the results can often not be generalized.

It can be said that there is no single approach to compare the performance of identification methods in an unambiguous and convincing way. Hence, the results of all three means of comparison, i.e. simulation, theoretical convergence analysis, and real applications must be combined to come to general results.

A further problem is the mathematical quantification of the error between model and process. One can use for example the following errors:

- Parameter error $\Delta\boldsymbol{\theta}_i = \hat{\boldsymbol{\theta}}_i - \boldsymbol{\theta}_{i0}$
- Output error $\Delta y(k) = \hat{y}(k) - y(k)$
- Equation error $e(k) = y(k) - \boldsymbol{\psi}^{\mathrm{T}}(k)\hat{\boldsymbol{\theta}}(k-1)$
- Error of the input/output behavior, for example determined by the impulse response error $\Delta g(\tau) = \hat{g}(\tau) - g(\tau)$

These errors can be evaluated as:

- Absolute values
- Relative values
- Mean values (linear, quadratic)

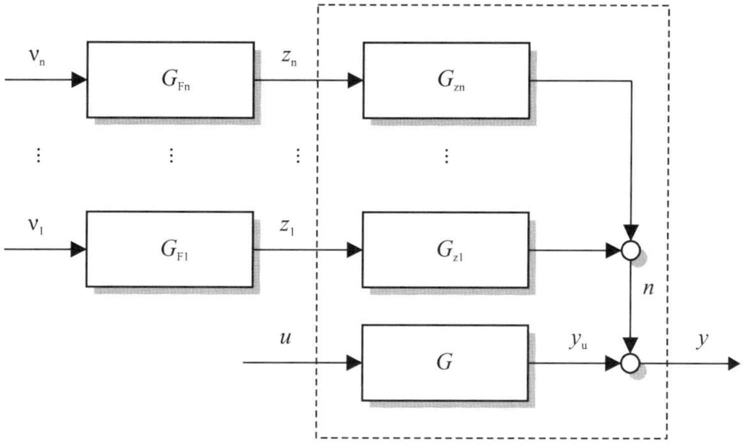

Fig. 23.2. Complex structure of a linear process with disturbances

Due to the possible combinations, there exist a lot of different metrics for the error. Since their information can be quite different, one should always combine some metrics. For the final judgment, however, the application of the model plays the most important role.

23.4.2 Comparison of A Priori Assumptions

For the parameter estimation methods, one had to make certain assumptions to obtain bias-free estimates, especially one had to assume a certain structure of the form filter and the noise or the error. These assumptions shall now be compared and checked, how well these are satisfied in practical parameter estimation applications. Figure 23.2 shows the appropriate model structure. It must in general be assumed that the process is disturbed by several disturbances $z_1, \ldots, z_\nu$. For the disturbances, it is assumed that they act on the measured output signal y by means of the linear disturbance transfer functions $G_{z1}(z), \ldots, G_{z\nu}(z)$, see Fig. 23.2. If the individual z_i are stationary stochastic disturbances, then one can assume that all of them have been generated from white nose by filtering with appropriate form filters. In this case

$$n(z) = G_{z1}(z)G_{\mathrm{F}1}(z)\nu_1(z) + \ldots + G_{z\nu}(z)G_{\mathrm{F}\nu}(z)\nu_\nu(z)\,. \tag{23.4.1}$$

Typically, one uses this model and assumes that the process $G(z)$ is disturbed by one white noise $\nu(z)$ that is filtered by one form filter $G_\nu(z)$, i.e.

$$n(z) = G_\nu \nu(z)\,, \tag{23.4.2}$$

see Fig. 23.3. Hence, one assumes that the individual form filters are linear and that one can replace the individual white noises by one common noise source. At this point, it should be stressed again that the noise signal n is assumed to be stationary.

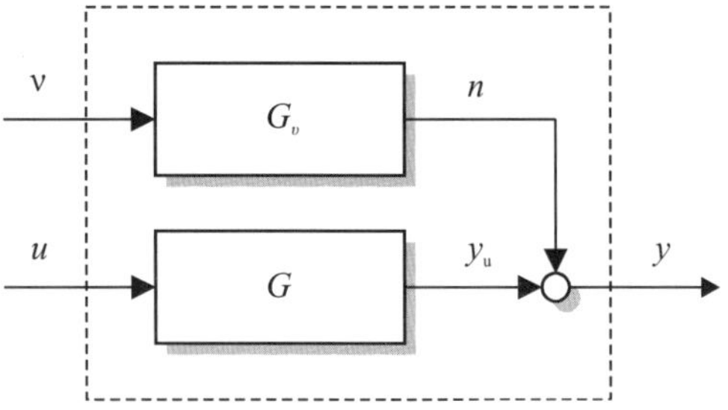

Fig. 23.3. Simplified model structure of the process in Fig. 23.2

Often however, there are also non-stationary noises or noises of unknown character, which must be eliminated by special methods prior to the identification.

Table 23.2 shows the assumptions that have been made about the form filter for the individual parameter estimation methods. In the most generic case, the form filter is given by the structure

$$G_\nu(z) = \frac{D(z^{-1})}{C(z^{-1})} = \frac{d_0 + d_1 z^{-1} + \ldots + d_p z^{-p}}{c_0 + c_1 z^{-1} + \ldots + c_p z^{-p}} \;. \qquad (23.4.3)$$

In order to obtain bias-free estimates with the method LS, the form filter $G_\nu(z)$ must have the form $1/A(z^{-1})$ The form filter hence must have the same denominator polynomial as the process that is identified. Since this is almost never the case, the *method of least squares* will typically provide biased estimates, The same holds true for the *stochastic approximation.*

For the *method of generalized least squares*, the noise filter must have the more general form $1/A(z^{-1})F(z^{-1})$. In most applications, this is not the case, so that this method will also provide biased estimates. However, by an appropriate choice of the model order ν of $F(z^{-1})$, one can reduce the size of the bias compared to the normal method of least squares.

For the *maximum likelihood method* and the *method of extended least squares*, one assumes a form filter as $D(z^{-1})/A(z^{-1})$. This model is not as particular as the models that were assumed for the method of least squares and the method of generalized least squares, but still they cannot fully approximate the generic noise filter $D(z^{-1})/C(z^{-1})$.

The *method of instrumental variables* and the two-stage methods, such as *correlation and least squares* do not require any special assumptions about the noise. They are hence well suited for general purpose applications.

The individual methods are also different with respect to their behavior under the influence of unknown DC values at the input and output. If $\mathrm{E}\{u(k)\} = 0$, then the existence of a DC value Y_{00} has no influence on the parameter estimation while using the method of instrumental variables and the method of correlation and least squares. For the method of least squares, stochastic approximation, and maximum likelihood, the DC value Y_{00} must be identified beforehand or must be introduced into the parameter estimation problem to avoid systematic errors.

Table 23.2. Comparison of noise models for different parameter estimation methods – for the model types, see also Fig. 2.8

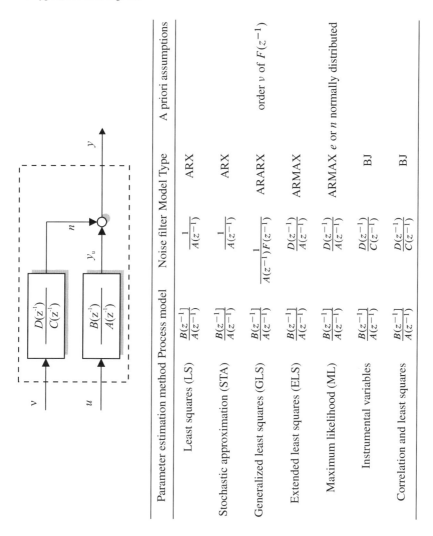

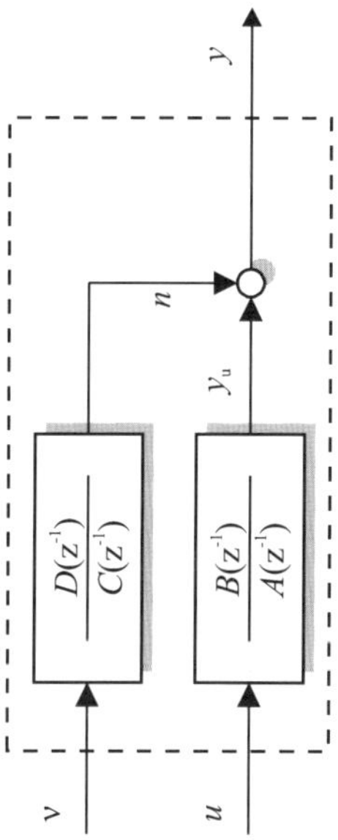

Parameter estimation method	Process model	Noise filter	Model Type	A priori assumptions
Least squares (LS)	$\frac{B(z^{-1})}{A(z^{-1})}$	$\frac{1}{A(z^{-1})}$	ARX	
Stochastic approximation (STA)	$\frac{B(z^{-1})}{A(z^{-1})}$	$\frac{1}{A(z^{-1})}$	ARX	
Generalized least squares (GLS)	$\frac{B(z^{-1})}{A(z^{-1})}$	$\frac{1}{A(z^{-1})F(z^{-1})}$	ARARX	order ν of $F(z^{-1})$
Extended least squares (ELS)	$\frac{B(z^{-1})}{A(z^{-1})}$	$\frac{D(z^{-1})}{A(z^{-1})}$	ARMAX	
Maximum likelihood (ML)	$\frac{B(z^{-1})}{A(z^{-1})}$	$\frac{D(z^{-1})}{A(z^{-1})}$	ARMAX	e or n normally distributed
Instrumental variables	$\frac{B(z^{-1})}{A(z^{-1})}$	$\frac{D(z^{-1})}{C(z^{-1})}$	BJ	
Correlation and least squares	$\frac{B(z^{-1})}{A(z^{-1})}$	$\frac{D(z^{-1})}{C(z^{-1})}$	BJ	

23.4.3 Summary of the Methods Governed in this Book

In this section, some important advantages and disadvantages shall be summarized.

Non-Recursive Parameter Estimation Methods

Method of least squares (LS), Sect. 9.1

– Delivers biased estimates for disturbances

+ However applicable for short measurement times as consistent methods will not provide better results in these cases
- Sensitive to unknown DC value Y_{00}
+ Relatively small computational effort
+ No special a priori assumptions
– Quadratic cost function overemphasizes outliers

Method of least squares for continuous time (LS), Sect. 15.1

– Delivers biased estimates
– Derivatives of (noisy) signals may have to be determined
0 Special filters can be used to calculate derivatives and low-pass filter the signal at the same time
+ Estimated parameters typically directly interpretable

Method of generalized least squares (GLS), Sect. 10.1

– Biased estimates possible because noise filter is very particular
– Relatively large computational effort
+ A noise model is being identified
– A priori assumption about the noise form filter order necessary

Method of extended least squares (ELS), Sect. 10.2

– Biased estimates possible if noise filter does not match noise
– Parameters of noise model converge slower than process parameters
+ A noise model is being identified
+ Little additional effort
– A priori assumption about the noise form filter order necessary
– Fixed denominator polynomial of noise model limits applicability

Method of bias correction (CLS), Sect. 10.3

+ Small computational effort
– Requires that the bias can be determined, but bias can only be determined in special cases
– Limited applicability

Method of total least squares (TLS), Sect. 10.4

+ Can incorporate noise also on the input
– Method of TLS does not account for special structure of data matrix

Method of instrumental variables (IV), Sect. 10.5

+ Good results for a wide range of noises
+ Small/medium computational effort
– Convergence can be problematic
+ Insensitive to unknown DC value Y_{00} provided $\overline{u(k)} = 0$
– A priori assumption about the noise form filter order necessary

Maximum-likelihood method (ML), Sect. 11.2

– Large computational effort
+ A noise model is being identified
– Problematic, if there are local minima of the cost function
+ Detailed theoretical analysis available
± A priori assumptions depend on the optimization method

Correlation and least squares (COR-LS), Sect. 9.3.2

+ Good results for a wide range of noises
+ Small computational effort
+ Intermediate results available which do not depend on the chosen model structure
+ Insensitive to unknown DC value Y_{00} provided $\overline{u(k)} = 0$
+ Small computational effort for model order and dead time determination
+ Easy to validate results
+ A priori assumptions only include the number of correlation function values
– Computationally expensive since correlation functions must first be determined
– Estimation of correlation functions with small error might require long measurement times
0 shares all the advantages and disadvantages associated with the method employed for determination of correlation functions

Frequency response and least squares (FR-LS), Sect. 14.2

+ Good results for a wide range of noises
+ Small computational effort, once frequency response has been determined
+ Intermediate results available which do not depend on the chosen model structure
+ Insensitive to unknown DC value Y_{00} provided $\overline{u(k)} = 0$
+ Small computational effort for model order and dead time determination
+ Easy to validate results
+ A priori assumptions do only include the number of frequency response points to be measured
+ Almost insensitive to noise if orthogonal correlation is used for frequency response measurement
+ Provides a frequency domain model and subsequently identifies a continuous-time model
+ Estimated parameters typically directly interpretable
– Computationally expensive since frequency response must first be determined
– Estimation of frequency response might require long measurement times depending on frequency response identification method used
0 shares all the advantages and disadvantages associated with the method employed for frequency response identification

Recursive Parameter Estimation Methods

Unless noted otherwise, they have the same properties as the non-recursive methods.

Recursive least squares (RLS), Sect. 9.4

+ Robust method with reliable convergence
+ For small identification times or time-varying processes, it should be preferred over consistent methods
+ Implementations with good numerical properties are available and should be preferred whenever possible
+ Has been used in many applications, see Chap. 24

Recursive extended least squares (RELS), Sect. 10.2

+ Good results, if a special noise form filter D/A suits the noise approximately and $1/D(z^{-1})$ has no unstable poles
– Convergence can be problematic
– Slower initial convergence than RLS
– Slower convergence of the parameters of $D(z^{-1})$
+ Rather small computational effort

Recursive instrumental variables (RIV), Sect. 10.5

+ Reliable convergence only if started with other methods, such as RLS
+ Good results for a wide range of noises, if convergence is given

Recursive maximum likelihood (RML), Sect. 11.2

- Basically the same properties as RELS

Stochastic approximation (STA), Sect. 10.6, and normalized least mean square (NLMS), Sect. 10.7

– Control of step-width problematic
– Slow convergence
+ Easy to implement
+ Computationally less expensive than RLS
– RLS can nowadays be implemented in most cases despite larger computational expense

Extended Kalman filter (EKF), Sect. 21.4

+ Allows to estimate states and parameters at the same time
+ Model does not have to be linear in parameters
– Local linearization around operating point hence danger of diverging
– Computationally expensive, since states have to be estimated as well

Other Methods

Characteristic parameter estimation, Sect. 2.5

+ Simple and easy to conduct method that can provide some first ideas about the dominating time constants, process behavior, and such
– Not very precise

Frequency response measurement with non-periodic signals, Chap. 4

+ Fast algorithms for calculation of Fourier transform available
+ Easy to understand method
+ No assumption about model structure necessary
+ Provides a frequency domain model that can easily be converted to a continuous-time model
– Method without averaging over multiple measurements is not consistent

Frequency response measurement with periodic signals (orthogonal correlation), Sect. 5.5.2

+ Extremely strong suppression of noise possible
+ Easy to understand
+ No assumption about model structure necessary
+ Should be used for linear processes whenever experiment time does not play an important role
+ Provides a frequency domain model that can easily be converted to a continuous-time model
– Long measurement times

Deconvolution, Sect. 7.2.1

+ No assumption about model structure necessary
+ Easy to understand
+ Easy evaluation for white noise input
– Impulse response is less informative than frequency response

Iterative optimization, Chap. 19

+ Can use various cost functions, e.g. those that do not overemphasize outliers
+ Model does not have to be linear in parameters
+ Easy inclusion of constraints (e.g. ranges for parameters, stability of the resulting model, etc.)
+ Reliable algorithms readily available
– Large computational expense
– Convergence not guaranteed (local minima)
– Properties of the estimator (such as efficiency) only valid if global minimum is obtained

Subspace methods, Chap. 16

+ Semi-automatic model order determination
+ Formulation suitable for MIMO models from the beginning
+ Reliable methods available for solution
– Large computational effort
– States are typically not interpretable

Neural networks (NN), Chap. 20

+ Universal approximator
+ Model does not have to be linear in parameters
+ Many implementations available (software toolboxes)
– For many nets, limited or even no physical interpretation
– Large computational effort
– Choice of net parameters not always intuitive

The above compilation of the most important advantages and disadvantages has shown that there is no single "best" method. Rather, one has to select among several possibilities the method that is best suited to the experiment conditions, but also to the later application. Hence, the application examples in the following chapter will show, how the combination of different methods will lead to success.

23.5 Parameter Estimation for Processes with Integral Action

Linear integral acting processes have the transfer function

$$G_{\mathrm{P}}(z) = \frac{y(z)}{u(z)} = \frac{B(z^{-1})}{A(z^{-1})} = \frac{B(z^{-1})}{(1-z^{-1})A'(z^{-1})} \tag{23.5.1}$$

with

$$A'(z^{-1}) = 1 + a'_1 z^{-1} + \ldots + a'_{m-1} z^{-(m-1)} \,. \tag{23.5.2}$$

The coefficients a'_i depend on the coefficients of $A(z^{-1})$ as follows

$$\begin{aligned} a_1 &= a'_1 - 1 \\ a_2 &= a'_2 - a'_1 \\ &\vdots \\ a_{m-1} &= a'_{m-1} - a'_{m-2} \\ a_m &= -a'_{m-1} \,. \end{aligned} \tag{23.5.3}$$

These processes hence have a single pole at $z = 1$ and are still stable (critically stable). Because of the unambiguous relation between input and output, in general one can use the same parameter estimation methods as for proportional acting processes. Still, there are some peculiarities, which lead to different methods of estimating the parameters, see Fig. 23.4.

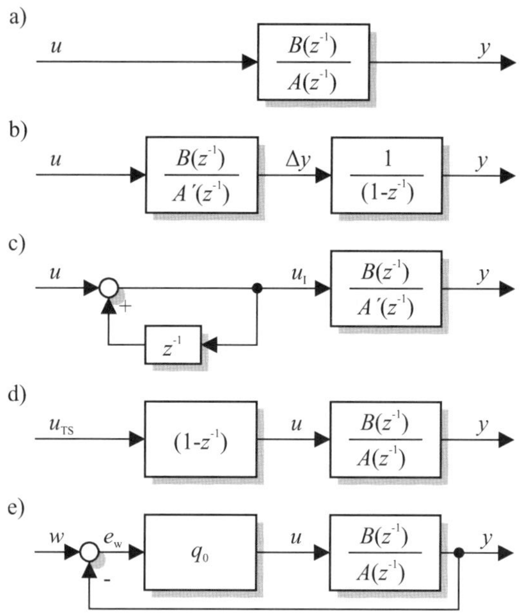

Fig. 23.4. Different setups for the identification of an integral acting process. (**a**) Case 1: Treat like proportional acting system, (**b**) Case 2: Difference of the output signal, (**c**) Case 3: Summation of the input signal, (**d**) Case 4: Difference of the test signal, (**e**) Case 5: Closed loop operation with P controller

Treat like Proportional Acting System

The simplest approach is not to regard the integrating pole at all and hence estimate the parameters of $B(z^{-1})$ and $A(z^{-1})$ based on the model in (23.5.1) and use the measurements of $u(k)$ and $y(k)$ directly, this shall be denoted as case 1.

A Priori Knowledge of the Integral Action

If the presence of a pole at $z = 1$ is known, then one can consider this pole in the signals employed for parameter estimation and hence identify the parameters of $B(z^{-1})$ and $A'(z^{-1})$ for the model in (23.5.1) either based on (case 2)

$$\frac{y(z)(1-z^{-1})}{u(z)} = \frac{\Delta y(z)}{u(z)} = \frac{B(z^{-1})}{A'(z^{-1})} \tag{23.5.4}$$

or (case 3)

$$\frac{y(z)}{u(z)/(1-z^{-1})} = \frac{y(z)}{u_{\mathrm{I}}(z)} = \frac{B(z^{-1})}{A'(z^{-1})} \,. \tag{23.5.5}$$

In case 2, one uses the input signal $u(k)$ and the first difference of the output $\Delta y(k)$. On the contrary, in case 3, one uses the integrated input signal $u_{\mathrm{I}}(k)$ and the output $y(k)$.

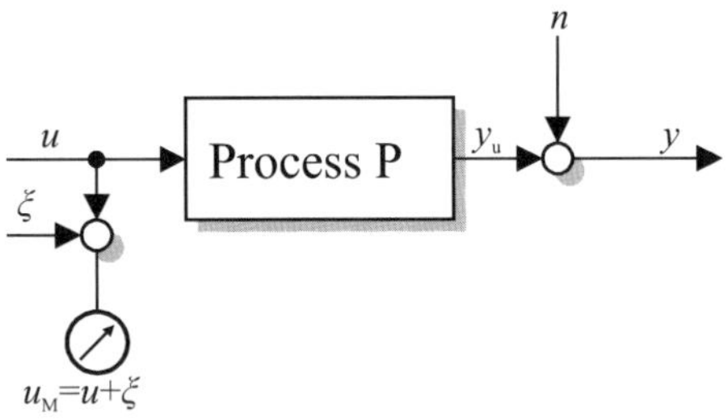

Fig. 23.5. Process with disturbed input and output

Adaptation of the Test Signal

If one uses a standard PRBS signal, one first has to remove the mean of the signal by subtracting the appropriate correction term. In order to excite the high process frequencies sufficiently, one should choose the cycle time of the PRBS as small as possible. For the smallest value, i.e. $\lambda = 1$, the low and high frequencies are excited equally strong. This can lead to large amplitudes of the output of the integral acting process, which might not be admissible. In order to increase the high frequent excitation and at the same time reduce the low frequent excitation, one can use the first difference of the PRBS signal, i.e. the test signal u_{TS}, (case 4) as

$$u(z) = (1 - z^{-1})u_{\mathrm{TS}}(z) = \Delta u_{\mathrm{TS}}(z) . \tag{23.5.6}$$

Closed Loop

In practical applications, it is often difficult to avoid a drift of an integral acting process during the measurement. To overcome this problem, one can operate the process in closed loop control during the experiment and adjust the setpoint according to the test signal (case 5).

Concluding Remarks

The different setups depicted in Fig. 23.4 have been compared on simulated processes of order $m = 2$ and $m = 3$ with and without disturbances by Jordan (1986). As estimation method, the method of recursive extended least squares (RELS) was used. The best convergence was obtained for the adaptation of the test signal (case 4) and by consideration of the integral action in the input or output (cases 2 and 3). Cases 1 and 5 converged most slowly. Methods for identifying integral acting processes are also presented in (Box et al, 2008).

23.6 Disturbances at the System Input

For the parameter estimation methods, it was in most cases assumed that the input signal is known exactly. Now, the case of a disturbed input signal is considered. The input signal can e.g. be falsified by measurement noise as

$$u_{\mathrm{M}}(k) = u(k) + \xi(k) . \tag{23.6.1}$$

The output is also disturbed as

$$y(k) = y_{\mathrm{u}}(k) + n(k) , \tag{23.6.2}$$

see Fig. 23.5. The process to be identified shall be given by

$$y_{\mathrm{u}}(z) = \frac{B(z^{-1})}{A(z^{-1})} u(z) . \tag{23.6.3}$$

The disturbances $\xi(k)$ and $n(k)$ are assumed to be zero mean and not mutually correlated. Furthermore, they are not correlated with the process input $u(k)$. This in particular requires that the process has no external feedback. Calculation of the correlation functions with (23.6.1) and (23.6.2) yields

$$R_{\mathrm{u_M u_M}}(\tau) = \mathrm{E}\{u_{\mathrm{M}}(k)u_{\mathrm{M}}(k+\tau)\} = \mathrm{E}\{u(k)u(k+\tau) + \xi(k)\xi(k+\tau)\} . \tag{23.6.4}$$

If $\xi(k)$ is a white noise, then

$$R_{\mathrm{u_M u_M}}(\tau) = R_{\mathrm{uu}}(\tau) \text{ for } \tau \neq 0 . \tag{23.6.5}$$

Furthermore,

$$R_{\mathrm{u_M y}}(\tau) = \mathrm{E}\{u_{\mathrm{M}}(k-\tau)y(k)\} = \mathrm{E}\{u(k-\tau)y_{\mathrm{u}}(k)\} = R_{\mathrm{uy}_u} . \tag{23.6.6}$$

If the measurable input signal is hence disturbed by a white noise $\xi(k)$, then one can determine the ACF of the input signal for $\tau \neq 0$ and the CCF in the same way as for processes without a disturbance at the input. For parameter estimation, one can hence use the method COR-LS as long as the value $R_{\mathrm{uu}}(0)$ is left out, which however drastically limits the number of equations. Furthermore, the input signal may not be a white noise to ensure that $R_{\mathrm{uu}}(\tau) = 0$ for $\tau \neq 0$. Söderström et al (2009) suggested to estimate the contribution of the disturbing white noise in $u(t)$ on $R_{\mathrm{uu}}(0)$ by estimating the noise variance σ_ξ^2 as part of the system identification. As the noise variance σ_ξ^2 shows up as an additive element for $R_{\mathrm{uu}}(0)$, such a term can be introduced into the problem formulation as an additional parameter.

An overview over different methods for parameter estimation in the presence of noise at the input is given in (Söderström et al, 2002; Söderström, 2007). This problem also subsumes under the name *errors in variables*. It is shown that all methods that are based on the one step prediction do not deliver bias-free estimates, but rather biased estimates as the cost function does not assume a minimum for the true parameters. Under the assumption of a white noise $\xi(k)$, one can employ spectral analysis or *joint input output methods* to obtain unbiased estimates. The last method assumes that the input signal $u(k)$ was generated by a filter from the statistically independent signal $w(k)$. Then, $u(k)$ and $y(k)$ are seen as output signals of a state space model, whose cost function is then minimized by numerical optimization methods. Identifiability issues have been discussed e.g. in (Agüero and Goodwin, 2008).

Another solution of the errors in variables problem is based on the Frisch scheme, see e.g. (Hong et al, 2008). Thil et al (2008) compared two instrumental variable approaches, where the instrumental variables are either chosen as input variables only or as input/output variables, see also Sect. 10.5.

23.7 Elimination of Special Disturbances

In system identification, one is in general only interested in the particular frequency range of the output signal that was excited by the test signal. Henceforth, the identification methods are mostly capable of eliminating disturbances only in this particular frequency band over time. Disturbances with a much higher or much lower frequency have to be eliminated by special measures, as will be discussed in the following. Subsequently, also the elimination of outliers is discussed.

23.7.1 Drifts and High Frequent Noise

High frequent disturbances can best be eliminated by analog low-pass filters prior to sampling. This is also an important issue with respect to anti-aliasing, see Sect. 3.1.2. In this context, one speaks of anti-aliasing filters.

Special attention has to be paid to *low frequent disturbances*, which often manifest themselves as drifts. It had been shown that these low frequent disturbances can have a severe influence, see e.g. Sects. 1.2 and 5.5.2. These low-frequent disturbances can e.g. be described by the following models:

1. *Non-linear drift of order q*

$$d(k) = h_0 + h_1 k + h_2 k^2 + \ldots + h_q k^q \tag{23.7.1}$$

2. *ARIMA process*

$$d(z) = \frac{F(z^{-1})}{E(z^{-1})(1 - z^{-1})^p} \nu(z),\ p = 1, 2\,, \tag{23.7.2}$$

   where $\nu(k)$ is a stationary, uncorrelated signal.

3. *Low-frequent periodic signal*

$$d(k) = \sum_{\nu=0}^{l} \beta_\nu \sin(\omega_\nu T_0 k + \alpha_\nu) \tag{23.7.3}$$

These low-frequent disturbances can be eliminated from the output signal $y(k)$ by the following methods:

Non-Linear Drift of Order q

One estimates the parameters $\hat{h}_0, \hat{h}_1, \ldots, \hat{h}_q$ by the method of least squares and calculates

$$\tilde{y}(k) = y(k) - \hat{d}(k) \,. \tag{23.7.4}$$

This method can however mainly be used for offline identification and long measurement times.

ARIMA Process

The non-stationary part can be eliminated by

$$\tilde{y}(z) = (1 - z^{-1})^p y(z) \,, \tag{23.7.5}$$

i.e. by differentiating the signal $y(k)$ p times (Young et al, 1971). This at the same time however amplifies the high frequent noise so that this method can only be used if the noise $n(k)$ is small compared to the useful signal $y_{\mathrm{u}}(k)$.

Low-Frequent Periodic Signal

For the identification of the low-frequent periodic part, one can use the Fourier analysis. If the frequencies ω_ν of the individual components of the disturbance are unknown, then one can determine them by calculating the amplitude spectrum. From there, one can estimate $\hat{\beta}_\nu$ and $\hat{\alpha}_\nu$ and calculate

$$\tilde{y}(k) = y(k) - \hat{d}(k) \,. \tag{23.7.6}$$

One can also assume a piecewise drift of second order and determine its time-varying parameters.

High-Pass Filtering

All of the methods described so far have the big disadvantage in common that they are only suitable for certain disturbances, whose order q, p, or l must be known a priori or must be determined with large effort. Much easier is the application of a high-pass filter, which blocks the low-frequent signal components, and lets the higher frequent signal components pass. For this, a first order high-pass filter with the transfer function

$$G_{\mathrm{HP}}(s) = \frac{Ts}{1 + Ts} \tag{23.7.7}$$

can be employed. The discrete-time transfer function is given as

$$G_{\mathrm{HP}}(z) = \frac{\tilde{y}(z)}{y(z)} = \frac{1 - z^{-1}}{1 + a_1 z^{-1}} \tag{23.7.8}$$

with

$$a_1 = -e^{-\frac{T_0}{T}}, \tag{23.7.9}$$

see (Isermann and Freyermuth, 1991). To avoid a falsification of the estimation results, the input must be filtered with the same filter.

For the choice of the time constant of the filter, one must ensure that no low-frequent parts of the useful signal are filtered out. Due to this, the choice of T depends both on the spectrum of the low-frequent disturbance as well as the input signal. If a PRBS signal with the cycle time λ and the period time N is used as an input signal, whose lowest frequency is given as $\omega_0 = 2\pi/N\lambda$, then the corner-frequency of the filter should be chosen such that $\omega_{\mathrm{HP}} < \omega_0$, i.e.

$$T > \frac{N\lambda}{2\pi}, \tag{23.7.10}$$

if the dynamics of the process are only of interest for $\omega > \omega_0$.

Pintelon and Schoukens (2001) suggested to simply disregard the low-frequent components of input and output when working with frequency domain identification methods. This is an alternative to high-pass filtering as the amplitude as well as the phase of the higher frequent frequency content remain unchanged.

23.7.2 Outliers

In the following, graphical and analytical methods for the elimination of outliers will be presented. Graphical methods have the big advantage that they can be used very intuitively. No parameters have to be tuned and furthermore, the user can fully control which data points are removed. The major drawback of graphical methods is the fact that they cannot be automated easily and hence are difficult to apply to large data sets, i.e. long measurement times and/or large number of signals.

The elimination of outliers is extremely important for the method of least squares, which uses a quadratic error. Here, outliers are over-emphasized as the error is squared and hence large errors have a large influence on the cost function. The use of alternative cost functions as e.g. discussed in Sect. 19.1 can help to reduce the influence of outliers. Special care has to be taken that only outliers are removed, as there is always the danger that also relevant data are removed from the training data. This can easily happen if the model does not match with the process dynamics.

Furthermore, many estimation methods that estimate dynamic models require an equidistant spacing of the data points. In this case, one cannot just simply remove data points, but must find a suitable replacement. A survey of methods for outlier detection is given by Kötter et al (2007).

x-t Graph

The x-t graph is one of the simplest methods for outlier detection. Here, the measurements are plotted over time. If the measurement is repeated, one can also see drift effects (see Sect. 23.7.1) or trends as well as cross effects such as temperature. One should only remove extraordinarily high or low points as also stepwise excitation etc. can cause sudden jumps of the measured variables.

Measurement vs. Prediction Graph

The measurement vs. prediction graph plots the measured variable on the x axis and the prediction, i.e. the model output, on the y-axis. For a perfect match, all points should lie on the bisectrix between the abscissa and the ordinate. However, the method is very sensitive to unmodeled dynamics, which can lead to a strong deviation from the bisectrix during transients.

Leverage

The leverage is the first analytical method to be presented here. The leverage h_i of a measurement is calculated as

$$h_i = h_{ii} \text{ with } \boldsymbol{H} = \boldsymbol{\Psi}\left(\boldsymbol{\Psi}^{\mathrm{T}}\boldsymbol{\Psi}\right)^{-1}\boldsymbol{\Psi}^{\mathrm{T}} . \tag{23.7.11}$$

The matrix $\boldsymbol{H}$ is termed *hat-matrix* and governs the relation

$$\hat{\boldsymbol{y}} = \boldsymbol{H}\boldsymbol{y} \tag{23.7.12}$$

and allows to calculate the error as

$$\boldsymbol{e} = \boldsymbol{y} - \hat{\boldsymbol{y}} = (\boldsymbol{I} - \boldsymbol{H})\boldsymbol{y} . \tag{23.7.13}$$

The value of h_i is the potential influence of the measurement on the model. Entries with a maximum or minimum value have greater influence on the parameter estimation. Although it is sometimes suggested to remove measurements based on the leverage, this approach seems more than questionable as the leverage only indicates the relevance of a certain data point for the parameter estimation and does not say anything about whether the data point fits well or not with the underlying model, since the output y is not at all considered in the leverage.

Studentized Residuals

The idea of studentized residuals is to normalize the error with respect to the estimated standard deviation of the data set. Furthermore, the error is normalized by the leverage as data points with a high leverage have strong influence on the parameter estimates and will hence "pull" the model output in their respective direction. The studentized residuals are given as

$$e_i' = \frac{e_i}{\hat{\sigma}\sqrt{1 - h_i}} . \tag{23.7.14}$$

The variance is estimated by

$$\hat{\sigma} = \frac{1}{N - m}\sum_{k=1}^{N} e_k^2 , \tag{23.7.15}$$

where N is the number of data points and m the number of model parameters. Thresholds for excluding a data point are typically chosen in the range 2...3 (Kötter et al, 2007; Jann, 2006).

Deleted Studentized Residuals

Deleted studentized residuals differ slightly from the studentized residuals. The only difference is the fact that for the estimation of the variance, the element to be analyzed is not included into the calculation of $\hat{\sigma}$, i.e.

$$\hat{\sigma}_{-i} = \frac{1}{N-m-1} \sum_{\substack{k=1 \\ k \neq i}}^{N} e_k^2 . \tag{23.7.16}$$

This will lead to the calculation of the Deleted Studentized residuals as

$$e_i'' = \frac{e_i}{\hat{\sigma}_{-i}\sqrt{1-h_i}} . \tag{23.7.17}$$

The index $-i$ means that the ith element is not used for the calculation of the respective values. The same thresholds as for the Studentized residuals above can be used.

COOK's D and DFFITS

Other statistics that can be used for outlier detection are COOK's D and DFFITS. The underlying idea is that the occurrence of an outlier becomes likely if high leverage and a high residual occur at the same time. The numbers are calculated as

$$\mathrm{D}_i = \frac{e_i'^2}{p} \frac{h_i}{1-h_i} \tag{23.7.18}$$

and

$$\mathrm{DFFITS}_i = e_i'' \sqrt{\frac{h_i}{1-h_i}} . \tag{23.7.19}$$

Thresholds are suggested as $4/N$ (Jann, 2006) for COOK's D and $2\sqrt{\rho/N}$ for DFFITS, (Myers, 1990).

DFBETAS

The last number that shall be presented here is the DFBETAS. The idea is to calculate the difference in the estimated parameters if one observation is left out. DFBETAS is calculated by

$$\mathrm{DFBETAS}_{il} = \frac{\hat{\theta}_l - \hat{\theta}_{l,-i}}{\hat{\sigma}_{-i}\sqrt{u_{ii}}} , \tag{23.7.20}$$

where u_{ii} is the corresponding element of the matrix $(\boldsymbol{\Psi}\boldsymbol{\Psi})^{-1}$. Here, the parameters are once estimated including the ith measurement, θ_l and once without the ith measurement, $\theta_{l,-i}$. Hence, the measure must be calculated for each data point and for each estimated parameter, leading to $N \times l$ values. As each of these calculations necessitates one full parameter estimation, the calculation of the DFBETAS can become rather time-consuming. A threshold for the rejection of disturbances was given as $2/\sqrt{N}$ by Myers (1990).

23.8 Validation

After the process has been identified and its parameters have been estimated, one has to check, whether the resulting model matches with the process behavior. This check depends on the identification method and the type of evaluation (non-recursive, recursive). One should in all cases check:

- A priori assumptions of the identification method
- Match of the input/output behavior of the identified model with the input/output behavior that has been measured.

The checking of the a priori assumptions can be carried out as follows:

Non-Parametric Identification Methods

- *Linearity*: Comparison of the model identified for different test signal amplitudes, comparison of step-responses for different step heights in both directions
- *Time-Invariance*: Comparison of models identified from different sections of the recorded measurements
- *Disturbance*: Is the noise statistically independent from the test signal and stationary? Is $R_{\mathrm{un}}(\tau) = 0$? Is $\mathrm{E}\{n(k)\} = 0$?
- *Impermissible Disturbances*: Are there any impermissible disturbances? Outliers? Changes in the Mean? Drifts? Check of the recorded signals.
- *Input signal*: Can it be measured without measurement error or noise? Does it excite the process permanently?
- *Steady state values*: Are the steady state values known exactly?

Most methods assume that the system is linear. A simple test for linearity is to scale the input by a factor α and check, whether the output scales accordingly (Pintelon and Schoukens, 2001).

Parametric Identification Methods

In addition to the above stated issues, one should for the individual parameter estimation methods check all other a priori assumptions as well. Depending on the parameter estimation method, these could be

- *Error Signal*: Is it statistically independent? Is $R_{\mathrm{ee}}(\tau) = 0$ for all $|\tau| \neq 0$? Is it statistically independent from the input signal? Is $R_{\mathrm{ue}}(\tau) = 0$? Is $\mathrm{E}\{e(k)\} = 0$?
- *Covariance matrix of the parameter errors*: Are the variances of the error of the parameter estimates decreasing as the measurement time increases? Are they small enough?

A further evaluation of the model can be based on a *comparison of the input/output behavior*, e.g. in an x-t plot. Here, the behavior of the model is compared with the measurements. Also, the measured vs. predicted plot that was introduced in Sect. 23.7.2 can be used for validation. Box et al (2008) also stresses the importance of visual inspection of the results.

1. Comparison of the measured output signal $y(k)$ and the model output $\hat{y}(k)$ for
   - the same input signal $u(k)$ as used for identification
   - other input signals, such as steps or impulses.
2. Comparison of the cross-correlation function $R_{uy}(\tau)$ of the measured signals with the cross-correlation functions determined from the model

For processes with good signal-to-noise ratios, one can directly compare the output signals. In other cases, one has to resort to the correlation functions. As the error $\Delta y(k)$ is given as

$$\Delta y(k) = y(k) - \hat{y}(k) = \left(\frac{B(q^{-1})}{A(q^{-1})} - \frac{\hat{B}(q^{-1})}{\hat{A}(q^{-1})}\right)u(k) + n(k) , \qquad (23.8.1)$$

one can see that for large $n(k)$, this error is dominated by the noise and not by the model error. If a model of the noise has been identified as well, then one can use

$$\Delta y(k) = \left(\frac{B(q^{-1})}{A(q^{-1})} - \frac{\hat{B}(q^{-1})}{\hat{A}(q^{-1})}\right)u(k) + \left(n(k) - \frac{\hat{D}(q^{-1})}{\hat{C}(q^{-1})}\right)\hat{\nu}(k) , \qquad (23.8.2)$$

where one can determine the one step prediction based on the estimate of $\hat{\nu}(k-1)$. A suitable criterion for the comparison of different models is

$$V = \frac{1}{N}\sum_{k=1}^{N} \Delta y^2(k) . \qquad (23.8.3)$$

Other tests can include an analysis of the error. The error must be a Gaussian distributed zero mean white noise sequence. For this test,

$$\hat{\bar{e}} = \frac{1}{N}\sum_{k=0}^{N-1} e(k) \qquad (23.8.4)$$

denotes the mean and

$$\hat{\sigma}_e^2 = \frac{1}{N-1}\sum_{k=0}^{N-1}(e(k) - \hat{\bar{e}})^2 \qquad (23.8.5)$$

the variance of the error sequence. The *skewness* is defined as the standardized third central moment

$$\hat{\mu}_3 = \frac{1}{N}\sum_{k=0}^{N-1}\frac{\left(e(k) - \hat{\bar{e}}\right)^3}{\hat{\sigma}_e^3} . \qquad (23.8.6)$$

The *kurtosis* is then given as the standardized fourth central moment

$$\hat{\mu}_4 = \frac{1}{N}\sum_{k=0}^{N-1}\frac{\left(e(k) - \hat{\bar{e}}\right)^4}{\hat{\sigma}_e^4} . \qquad (23.8.7)$$

For a Gaussian distribution, the skewness is always zero due to the symmetry of the distribution and the kurtosis equal to 3. Hence, one can use these quantities to check for a Gaussian distribution. Statistical independence can be tested be calculating the auto-correlation function, which for statistically independent signal samples should be

$$R_{\mathrm{ee}}(\tau) \approx 0 \text{ for } \tau \neq 0 \,. \tag{23.8.8}$$

For a graphic inspection, one can also integrate the probability density function of the output. This will lead to the cumulative distribution function, which can then be displayed graphically. Upon an appropriate scaling of the y-axis, the resulting line will also be the bisectrix. The plot with the appropriate scaling is called the *normal probability plot*.

Also, one should validate the model on other data than used for system identification. This is termed *cross validation*. Cross validation overcomes the problem of overfitting. The data set can be split up e.g. in $2/3$ that are used for identification, $1/3$ that is used for testing (Ljung, 1999). As was already shown in Chap. 20, the model error can be split up into a bias error and a variance error. Ljung and Lei (1997) pointed out that many models that pass the validation are typically suffering from a dominant variance error, that is, they have already started to adapt to the noise because of too many degrees of freedom.

23.9 Special Devices for Process Identification

For the identification of dynamic systems, one has used process computers since around 1970 and later, starting around 1975, also micro computers. Today, one typically uses computers with appropriate software programs as will be discussed in the next section. Due to this, special devices for parameter estimation are not as frequently used nowadays. With no claim on completeness, some very short remarks shall be made on both special hardware devices as well as the identification with digital computers using software packages.

23.9.1 Hardware Devices

A short reference to the history shall however be made. After the development of the orthogonal correlation, in a time around 1955 special devices entered the market that employed this technique for the identification of the frequency response. Such devices are still available today.

The development of the correlation methods lead to the development of so-called correlators, which were available as mechanic, photoelectric or magnetic correlators. One of these devices used a magnetic tape with varying distance between the recording and reading head. It was completely built in analog technology and was able to output correlation functions and power spectral densities on an x-y plotter.

Of major impact are devices that are based on the digital Fourier transform, which typically employ the Fast Fourier Transform. These are especially well suited for measurements at higher frequencies and also allow to use different representations, such as correlation functions, power spectral densities and such.

For the generation of test signals, there are special signal generators available on the market.

23.9.2 Identification with Digital Computers

Nowadays, processes are typically identified using digital computers. There are several software packages available on the market that provide tools for identification. These include generic numerical toolboxes that allow for example to calculate the Fourier transform or provide tools for a QR decomposition. On the other side, there are also toolboxes that provide complete process identification functionalities, such as the system identification toolbox (Ljung, 1999) or LOLIMOT that provides a toolbox for the LOLIMOT neural net (Nelles, 2001).

23.10 Summary

In this chapter, some aspects of parameter estimation have been discussed. First, the choice of the input signal was outlined. Although it is impossible to determine *the* optimal input signal, one can at least give hints on how the input signal should be chosen. Some methods have been presented that are based on the optimization of a cost function, such as e.g. a function of the information matrix. Unfortunately, most cost functions can only be evaluated once the experiment has been conducted and hence the input signal optimization can only be carried out iteratively with intermediate measurements.

Then, the choice of the sample rate was analyzed. It has been found that the choice of the sample rate is rather uncritical, as a large interval of different sample rates will typically lead to good estimation results. If there are no constraints on the storage space, one should always sample as fast as possible, as one can always reduce the sample time later e.g. by only retaining every second, third, etc. sample in the data vector. Also indicators have been presented that allow to detect if the sample time was chosen wrongly.

After that, methods for the determination of the model order and dead time have been presented. These methods can be based on different test quantities, such as the cost function, a test of the residuals, cancellation of poles and zeros in the resulting model and a rank test of the information matrix.

Integral acting processes can in general be treated just like normal proportional acting processes. However, the estimation results can be improved if a priori knowledge of the integral behavior is exploited. For example, one can specifically adapt the test signal to the integral acting process. In general, the signal must be zero mean to avoid a drift. Furthermore, the derivative of the test signal can be injected into the process to increase the excitation of the high frequent dynamics. Finally, the integral process can be operated in closed loop to take care of drifts caused e.g. by external disturbances acting on the process. The elimination of drifts from the already measured signals has also been discussed in this chapter. Also the case of noise falsifying the input measurement has been discussed shortly.

Also, this method has presented a comparison of all identification methods presented in this book, noting the most important advantages and disadvantages. This allows to select the methods that seem to be appropriate to solve a given identification problem.

Finally, methods to test the identification results and judge their quality have been presented. In most cases, one should compare the process output and the model output for a given input signal. If the influence of noise is too high, one can also compare cross-correlation functions between the input and the output. However, ...

... the best test is the application of the model for the task that it was identified for.

Problems

23.1. Input Signals
What issues have to be taken into consideration for the choice of input signals? What are favorable input signals?

23.2. Sample Time
What happens if the sample time is chosen too small? What happens if the sample time is chosen too large?

23.3. Integral Acting Systems
What methods can be used to account for the a priori knowledge that the process is integrating?

23.4. Special Disturbances
Name examples of special disturbances that should be limited prior to application of the parameter estimation algorithms.
How can these influences be eliminated?

23.5. Model Validation
How can the identification results be validated?

References

Agüero JC, Goodwin GC (2008) Identifiability of errors in variables dynamic systems. Automatica 44(2):371–382

Akaike H (1970) Statistical predictor information. Ann Inst Statist Math 22:203–217

Åström KJ (1968) Lectures on the identification problem: The least squares method. Lund

Bombois X, Barenthin M, van den Hof PMJ (2008) Finite-time experiment design with multisines. In: Proceedings of the 17th IFAC World Congress, Seoul, Korea
van den Boom AJW (1982) System identification - on the variety and coherence in parameter- and order erstimation methods. Ph. D. thesis. TH Eindhoven, Eindhoven
van den Boom AJW, van den Enden A (1974) The determination of the order of process and noise dynamics. Automatica 10(3):245–256
Box GEP, Jenkins GM, Reinsel GC (2008) Time series analysis: Forecasting and control, 4th edn. Wiley Series in Probability and Statistics, John Wiley, Hoboken, NJ
Gevers M (2005) Identification for control: From early achievements to the revival of experimental design. Eur J Cont 2005(11):1–18
Goodwin GC, Payne RL (1977) Dynamic system identification: Experiment design and data analysis, Mathematics in Science and Engineering, vol 136. Academic Press, New York, NY
Heij C, Ran A, Schagen F (2007) Introduction to mathematical systems theory : linear systems, identification and control. Birkhäuser Verlag, Basel
Hensel H (1987) Methoden des rechnergestützten Entwurfs und Echtzeiteinsatzes zeitdiskreter Mehrgrößenregelungen und ihre Realisierung in einem CAD-System. Fortschr.-Ber. VDI Reihe 20 Nr. 4. VDI Verlag, Düsseldorf
Hong M, Söderström T, Soverini U, Diversi R (2008) Comparison of three Frish methods for errors-in-variables identification. In: Proceedings of the 17th IFAC World Congress, Seoul, Korea
Isermann R (1974) Prozessidentifikation: Identifikation und Parameterschätzung dynamischer Prozesse mit diskreten Signalen. Springer, Heidelberg
Isermann R, Freyermuth B (1991) Process fault diagnosis based on process model knowledge. J Dyn Syst Meas Contr 113(4):620–626 & 627–633
Jann B (2006) Diagnostik von Regressionsschätzungen bei kleinen Stichproben. In: Diekmann A (ed) Methoden der Sozialforschung. Sonderheft 44 der Kölner Zeitschrift für Soziologie und Sozielwissenschaften, VS-Verlag für Sozialwissenschaften, Wiesbaden
Jordan M (1986) Strukturen zur Identifikation von Regelstrecken mit integralem Verhalten. Diplomarbeit. Institut für Regelungstechnik, Darmstadt
Kofahl R (1986) Self-tuning PID controllers based on process parameter estimation. Journal A 27(3):169–174
Kötter H, Schneider F, Fang F, Gußner T, Isermann R (2007) Robust regressor and outlier-detection for combustion engine measurements. In: Röpke K (ed) Design of experiments (DoE) in engine development - DoE and other modern development methods, Expert Verlag, Renningen, pp 377–396
Krolikowski A, Eykhoff P (1985) Input signal design for system identification: A comparative analysis. In: Proceedings of the 7th IFAC Symposium Identification, York
Kurz H (1979) Digital parameter-adaptive control of processes with unknown constant or time-varying dead time. In: Proceedings of the 5th IFAC Symposium

on Identification and System Parameter Estimation Darmstadt, Pergamon Press, Darmstadt, Germany

Kurz H, Goedecke W (1981) Digital parameter-adaptive control of processes with unknown deadtime. Automatica 17(1):245–252

Lehmann EL, Romano JP (2005) Testing statistical hypotheses, 3rd edn. Springer texts in statistics, Springer, New York, NY

Ljung L (1999) System identification: Theory for the user, 2nd edn. Prentice Hall Information and System Sciences Series, Prentice Hall PTR, Upper Saddle River, NJ

Ljung L, Lei G (1997) The role of model validation for assessing the size of the unmodeled dynamics. IEEE Trans Autom Control 42(9):1230–1239

Mäncher H, Hensel H (1985) Determination of order and deadtime for multivariable discrete-time parameter estimation methods. In: Proceedings of the 7th IFAC Symposium Identification, York

Mehra R (1974) Optimal input signals for parameter estimation in dynamic systems – Survey and new results. IEEE Trans Autom Control 19(6):753–768

Myers RH (1990) Classical and modern regression with applications (Duxbury Classic), 2nd edn. PWS Kent

Nelles O (2001) Nonlinear system identification: From classical approaches to neural networks and fuzzy models. Springer, Berlin

Pintelon R, Schoukens J (2001) System identification: A frequency domain approach. IEEE Press, Piscataway, NJ

Raol JR, Girija G, Singh J (2004) Modelling and parameter estimation of dynamic systems, IEE control engineering series, vol 65. Institution of Electrical Engineers, London

Schumann R, Lachmann KH, Isermann R (1981) Towards applicability of parameter adaptive control algorithms. In: Proceedings of the 8th IFAC Congress, Kyoto, Japan

Söderström T (1977) On model structure testing in system identification. Int J Control 26(1):1–18

Söderström T (2007) Errors-in-variables methods in system identification. Automatica 43(6):939–958

Söderström T, Soverini U, Kaushik M (2002) Perspectives on errors-in-variables estimation for dynamic systems. Signal Proc 82(8):1139–1154

Söderström T, Mossberg M, Hong M (2009) A covariance matching approach for identifying errors-in-variables systems. Automatica 45(9):2018–2031

Thil S, Gilson M, Garnier H (2008) On instrumental variable-based methods for errors-in-variables model identification. In: Proceedings of the 17th IFAC World Congress, Seoul, Korea

Verhaegen M, Verdult V (2007) Filtering and system identification: A least squares approach. Cambridge University Press, Cambridge

Welsh JS, Goodwin GC, Feuer A (2006) Evaluation and comparison of robust optimal experiment design criteria. In: Proceedings of the 2006 American Control Conference, Minneapolis, MN, USA

Woodside CM (1971) Estimation of the order of linear systems. Automatica 7(6):727–733

Young PC, Shellswell SH, Neethling CG (1971) A recursive approach to time-series analysis. Report CUED/B-Control/TR16. University of Cambridge, Cambridge, UK

Part VIII

Applications

24

Application Examples

In this chapter, examples for the identification of processes from different areas are presented. These examples cover processes as diverse as electric and fluidic actuators, robots and machine tools, internal combustion engines and automotive vehicles up to heat exchangers and illustrate the application of the different methods. However, only a rough overview over the application of the various methods can be given. One will see that optimal results depend on various aspects. These include

- Design of the experiment and here especially the choice of the input signal and the signal conditioning (e.g. low-pass filtering)
- A priori knowledge about the process for the choice of the resulting model type
- Intended application of the derived model

It will become clear in the examples that different combinations of identification methods and filters will be required for different processes, as there will for example be models that have been identified in the time domain and others that have been identified in the frequency domain. Furthermore, there are usually static non-linear models as well as linear dynamic and non-linear dynamic models. Especially with regard to the final application, either continuous-time or discrete-time models are preferred. It will also be shown that the selected model structure and the selected identification method must depend on the physically given properties. Hence, physical (theoretical) modeling of the process is always a good starting point to gather a priori knowledge.

An overview of the examples to be presented in this chapter is found in Table 24.1. For each application example, the special symbols, that are used, are summarized under the corresponding model equations.

24.1 Actuators

This overview shall begin with electric and fluidic actuators, which are found widespread as actuators in automation systems in general and in mechatronic systems in particular. For a detailed treatment of all aspects ranging from construction

R. Isermann, M. Münchhof, *Identification of Dynamic Systems*,
DOI 10.1007/978-3-540-78879-9_24, © Springer-Verlag Berlin Heidelberg 2011

Table 24.1. Overview of examples covered in this chapter

Process (section in parentheses)	Continuous Time	Discrete Time	Time Domain	Frequency Domain	Linear	Non-Linear	Time-Variant	MIMO	Physical Model	Neural Net
Brushless DC Actuators (24.1.1)	✓		✓			✓			✓	
Normal DC motor for electric throttle (24.1.2)	✓		✓			✓			✓	
Hydraulic actuators (24.1.3)	✓		✓			✓			✓	
Machine tool (24.2.1)	✓		✓		✓				✓	
Industrial robot (24.2.2)	✓		✓			✓			✓	
Centrifugal pumps (24.2.3)	✓		✓			✓			✓	
Heat exchangers (24.2.4)	✓		✓		✓	✓	✓		(✓)	✓
Air conditioning (24.2.5)		✓	✓		✓					
Dehumidifier (24.2.6)		✓	✓		✓					
Engine teststand (24.2.7)	✓			✓	✓					
Estimation of vehicle parameters (24.3.1)	✓		✓		✓		✓	✓	✓	
Braking systems (24.3.2)	✓		✓			✓	✓		✓	
Automotive suspension (24.3.3)	✓		✓		✓	(✓)	✓		✓	
Tire pressure (24.3.4)				✓	✓		✓			
Internal combustion engines (24.3.5)		✓	✓			✓		✓		✓

and design features to the specific advantages and disadvantages of individual realizations up to modeling and control, the reader is referred to the book by Isermann (2005) and with regards to applications for fault detection and diagnosis principles to (Isermann, 2010) and given references.

24.1.1 Brushless DC Actuators

As a first example for an electrically actuated system, the cabin outflow valve of a passenger aircraft is presented. The pressure in the fuselage of passenger aircraft flying above 2 000 m is permanently controlled. Fresh bleed air from the engines is constantly released at the front of the fuselage. At the fuselage tail, so-called cabin outflow valves are mounted, which allow the air from the cabin to escape to the surroundings. The valves are driven by brushless DC motors (BLDC) as illustrated in Fig. 24.1 (Moseler et al, 1999). For safety reasons, there is redundancy in the drive and there are two brushless DC motors, which act on a common gear. Another classical DC motor serves as an additional backup.

In the following, a dynamic model of the brushless DC motor and the driven load shall be set up and subsequently, the parameters shall be identified. The technical data of the motor have been summarized in Table 24.2.

As electric motors require a rotating magnetic field to keep up the movement of the rotor, traditional DC motors are equipped with a so-called *commutator*, which is

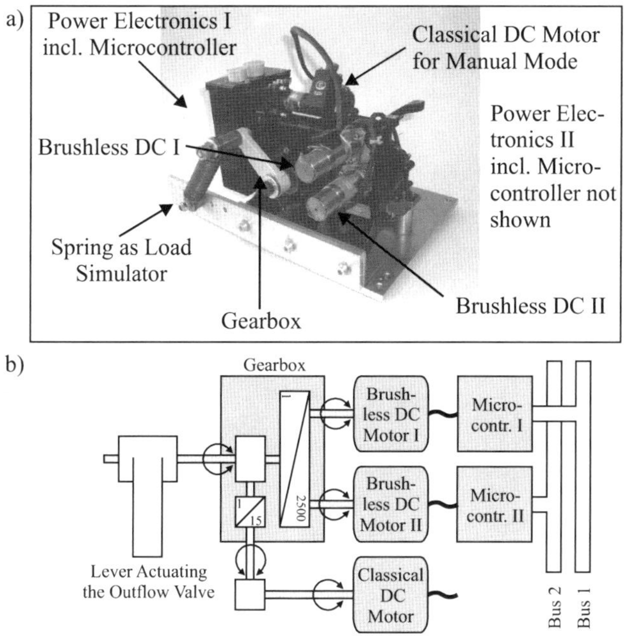

Fig. 24.1. Photo (**a**) and block diagram (**b**) of the cabin outflow valve

Table 24.2. Specifications of the BLDC motor

Parameter	Value
Weight	335 g
Length	87 mm
Diameter	30.5 mm
Supply voltage	28 V
Armature resistance	2.3 Ω
No-load speed	7 400 rpm
No-load current	50 mA
Short-circuit torque	28 mNm
Short-circuit current	1 A (Limited)

a mechanical device that changes the current feed of the rotor windings. A brushless DC motor uses an electronic circuit for this task. Here, the rotor is typically made up of a permanent magnet and the stator windings are supplied with current to generate an appropriately rotating magnetic field. As the rotor consists of permanent magnets only, no electric connection to the rotor is necessary. The position of the rotor is sensed by Hall sensors and is fed back to an electronic circuit that carries out the commutation, see Fig. 24.2. The advantage of the electronic commutation

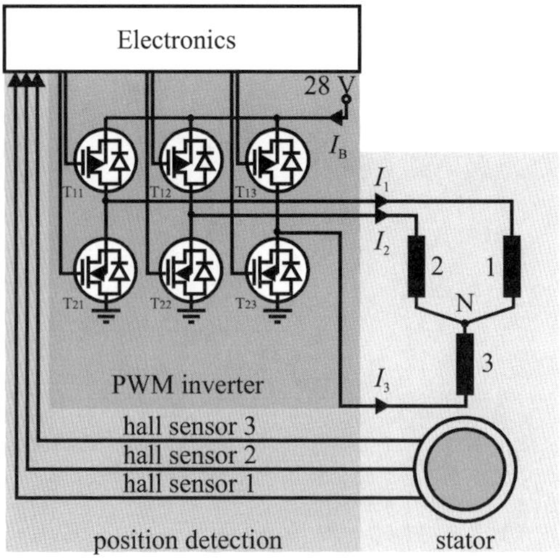

Fig. 24.2. Scheme of a brushless DC motor with electronic commutation realized by an integrated circuit and transistors

is that no brushes exist, which would be subject to wear-and-tear and could cause electromagnetic disturbances. Therefore, the reliability is relatively high.

The stator possesses 3 coils which are Y-connected and are driven by a PWM (pulse width modulation) inverter. The rotor has 4 permanent magnets. The position of the rotor magnets is measured by 3 Hall sensors mounted on the stator. These determine the switching sequence of 6 MOSFET transistors of the PWM inverter. This switching scheme is implemented in a separated programmable logic array. The PWM inverter is supplied with a fixed voltage U_B by the DC power supply and generates square wave voltages through the commutation logic via the 6 transistors to the 3 coils (phases).

Usually, only measurements of the supply voltage U_B, the input current I_B of the 6 phase full bridge circuit, and the angular rotor speed ω_R are available. However, for the detailed modeling that will be carried out in the following, also measurements of the phase current I_A and phase voltage U_A have been utilized. As the brushless DC motor switches the current over from one phase to the next as the rotor moves, the phase voltage and current refer to the *active* pair of phases (in star connection) that correlates with the rotor orientation.

The behavior of the motor can be described by the two differential equations

$$U_A(t) = R_A I_A(t) + L_A \frac{d}{dt} I_A(t) + \Psi \omega_R(t) \tag{24.1.1}$$

$$\Psi I_A(t) = J \frac{d}{dt} \omega_R(t) + M_V \omega_R(t) + M_C \operatorname{sign} \omega_R(t) + M_L^* \,. \tag{24.1.2}$$

The electric and magneto-mechanic subsystem are coupled by the back emf as $U_{emf}(t) = \Psi \omega_R(t)$ and the electric torque $M_{el}(t) = \Psi I(t)$. The mechanic load

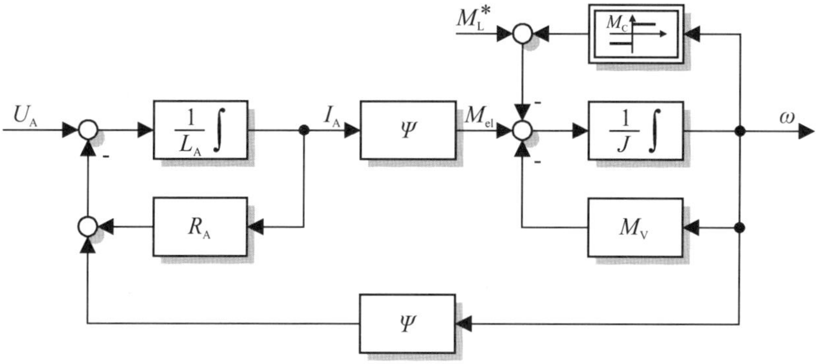

Fig. 24.3. Block diagram of the brushless DC motor model

is modeled as an inertia with both Coulomb and viscous friction. Also, an external load torque M_L is provided for. The drive shaft of the motor is coupled to the flap by means of a gear. The gear ratio ν relates the motor shaft position φ_R to the flap position φ_G

$$\varphi_G = \frac{\varphi_R}{\nu} \tag{24.1.3}$$

with $\nu = 2\,500$. The load torque of the flap is a normal function of the position φ_G

$$M_L = c_s f(\varphi_G) \tag{24.1.4}$$

and the kind of this characteristic is approximately known around the steady-state operation point. (For the experiments, the flap was replaced by a lever with a spring). All quantities have been transferred to the "motor" side, the referred load torque has been denoted as M_L^*. A block diagram is shown in Fig. 24.3. In this context, the following symbols are used:

L_A	armature inductance	J	inertia
R_A	armature resistance	M_C	coefficient of Coulomb friction
Ψ	torque constant	M_V	coefficient of viscous friction
ν	gear ratio		

For the identification, the entire valve was operated in closed-loop position control of the flap angle $\varphi_G(t)$. The setpoint and corresponding measurements are shown in Fig. 24.4. The signals have been sampled with a sampling frequency of 100 Hz. Furthermore, the measurements have been filtered with an FIR realized low-pass filter with $m = 24$ coefficients and a corner frequency of $f_C = 5$ Hz. The rotational velocity has been generated by a differentiating filter from the measurement of the rotational position.

Based on (24.1.1) and (24.1.2), one can now set up a parameter estimation problem. For the parameter estimation, the following measured and calculated signals are available: $U_A(t)$, $I_A(t)$, $\omega_R(t)$, and $\varphi_G(t)$. It was found that the time constant of the RL circuit was so small that the inductance L_A could not be estimated reliably with the given sample rate. Hence, the electric circuit was simplified to

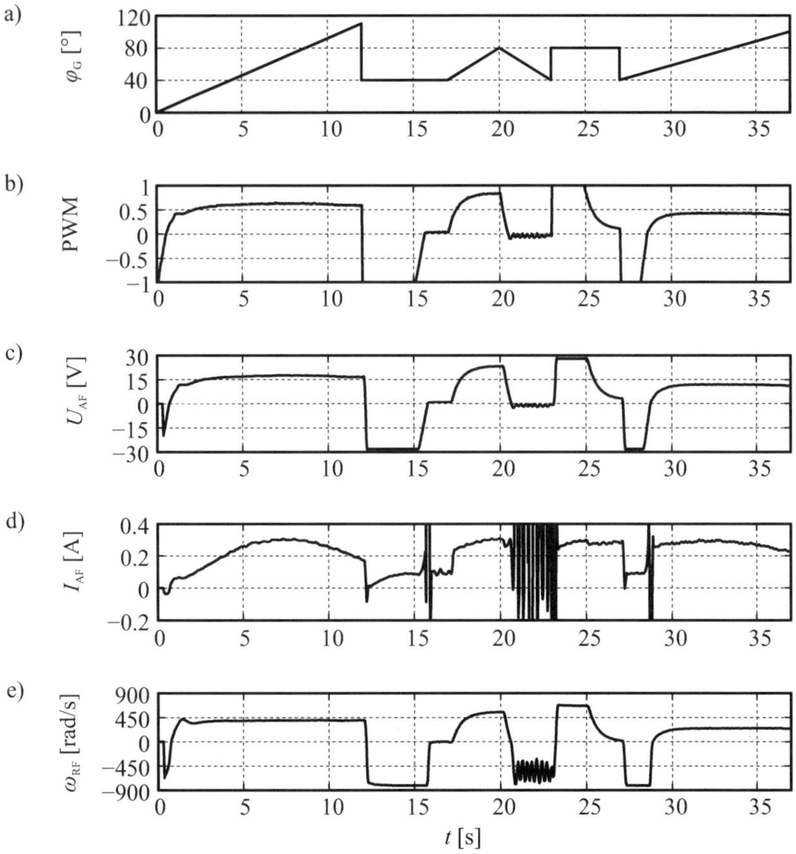

Fig. 24.4. Setpoint and measurements for the brushless DC motor. (**a**) setpoint flap angle, (**b**) PWM duty cycle, (**c**) filtered phase voltage, (**d**) filtered phase current, (**e**) filtered rotational velocity (Vogt, 1998)

$$U_A(t) = R_A I_A(t) + \Psi \omega_R(t) \ . \tag{24.1.5}$$

The signals were sampled and the model for the parameter estimation was then given as

$$y(k) = \boldsymbol{\psi}^T(k)\boldsymbol{\theta} \tag{24.1.6}$$

with

$$y(k) = U_A(k) \tag{24.1.7}$$

$$\boldsymbol{\psi}^T(k) = \big(I_A(k) \ \omega_R(k) \big) \tag{24.1.8}$$

$$\boldsymbol{\theta} = \begin{pmatrix} R_A \\ \Psi \end{pmatrix} , \tag{24.1.9}$$

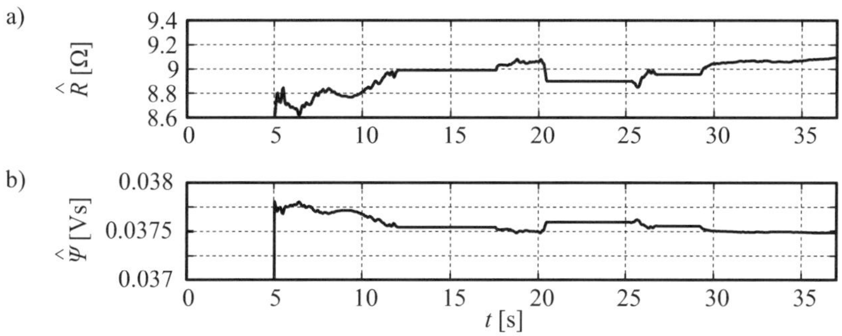

Fig. 24.5. Identification results for the brushless DC Motor. (**a**) Ohmic resistance, (**b**) magnetic flux linkage (Vogt, 1998)

allowing to provide the parameter estimates $\hat{R}_\mathrm{A}$ and $\hat{\Psi}$ by means of the method of least squares.

Also, the magneto-mechanical system had to be simplified, resulting in the model

$$\Psi I_\mathrm{A}(t) = J\frac{\mathrm{d}}{\mathrm{d}t}\omega_\mathrm{R}(t) + M_\mathrm{C}\,\mathrm{sign}\,\omega_\mathrm{R} + c_\mathrm{F}\varphi_\mathrm{G}(t)\;. \tag{24.1.10}$$

A few parameters were assumed to be known, including the rotational inertia and the load characteristics, leading to the model

$$y(k) = \boldsymbol{\psi}^T(k)\boldsymbol{\theta} \tag{24.1.11}$$

with

$$y(k) = \Psi I_\mathrm{A}(k) - c_\mathrm{s}\varphi_\mathrm{G}(k) - J\dot{\omega}_\mathrm{R}(k) \tag{24.1.12}$$

$$\boldsymbol{\psi}^T(k) = \big(\mathrm{sign}\,\omega_\mathrm{R}(k)\big) \tag{24.1.13}$$

$$\boldsymbol{\theta} = \big(M_\mathrm{C}\big)\;, \tag{24.1.14}$$

which allowed to estimate the coefficient of Coulomb friction $\hat{M}_\mathrm{C}$ by means of the method of least squares. Hence, in total, three parameters $\hat{R}_\mathrm{A}$, $\hat{\Psi}$, and $\hat{M}_\mathrm{C}$ are estimated.

Various parameter estimation methods were applied like: RLS (recursive least squares), DSFI (discrete square root filtering), FDSFI (fast DSFI), NLMS (normalized least mean squares) and compared with regard to computational effort in floating point and integer word realization and estimation performance. The floating point implementation is standard for e.g. 16 bit signal processors and in this case RLS, DSFI, or FDFSI can be used. However, integer word implementation is (still) required if reliable and certified low cost microcontrollers like the 16 bit Siemens C167 have to be used. Then, only NLMS is feasible (Moseler, 2001).

The forgetting factor has been chosen as $\lambda = 0.999$. Not all measurements are evaluated for the parameter estimation as the model is not precise enough in certain operating regimes. The estimation algorithm employs only measurements from the

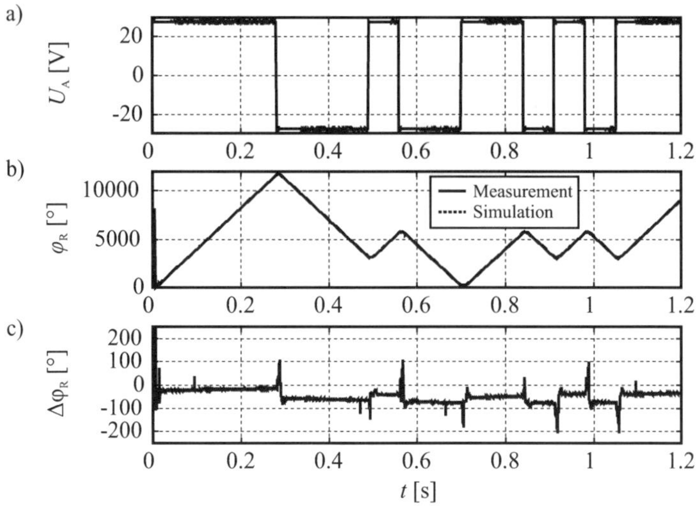

Fig. 24.6. Validation of the model. (**a**) phase voltage, (**b**) rotor angle, note that measured and simulated values lie almost on top of each other, (**c**) error between measured and simulated rotor angle (Vogt, 1998)

interval $0.05 \leq \text{PWM} \leq 0.95$ and $\omega > 10\,\text{rad/s}$. This also ensures that no division by zero can occur in the estimation equation. As FIR filters are used for noise cancellation, the filters can be switched off while no parameters are estimated. Whenever the measurements enter the corridor for parameter estimation, the first m samples are disposed. Upon the time step $m + 1$, the transient effects in the filter have died out and the filter output can safely be used for parameter estimation. The run of the estimated parameters is shown in Fig. 24.5. Furthermore, the first three seconds of the measurements have also been disposed to avoid falsification of the estimation results due to transients. The output of the parameter estimates starts at $t = 5\,\text{s}$ to ensure that the variance of the parameter estimates has already reduced to reasonable values. The high fidelity of the derived model is illustrated in Fig. 24.6. Here, the classical y-t plot has been used for validation of the results.

24.1.2 Electromagnetic Automotive Throttle Valve Actuator

Another example for parameter estimation applied to a DC drive with load is the automotive throttle valve actuator. Since about 1990, electrical driven throttle valves became a standard component for gasoline engines. They control the air mass flow through the intake manifold to the cylinders. The electric throttles are manipulated by the accelerator pedal sensors via an electronic control unit and additional control inputs from idle speed control, traction control, and cruise control. Here, in contrast to the previous example, a traditional DC motor with mechanical commutator is employed.

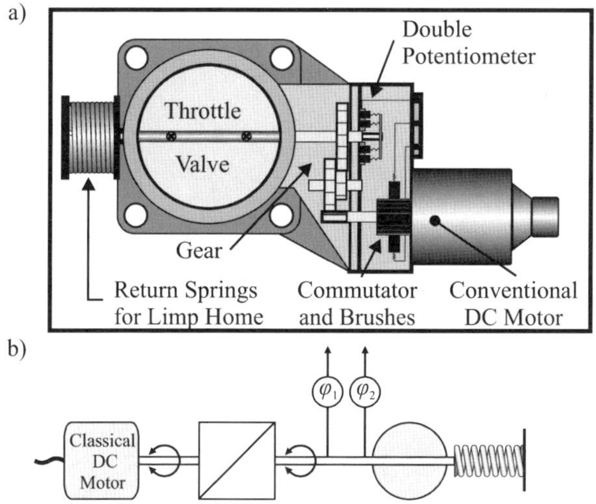

Fig. 24.7. (**a**) Scheme and (**b**) schematic block diagram of an electric throttle of a gasoline engine

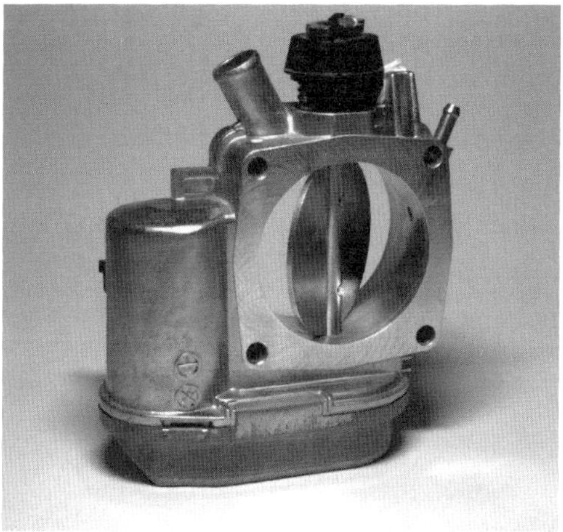

Fig. 24.8. Photo of an electric throttle of a vehicle

Figure 24.7 shows a schematic drawing of the actuator, a photo is shown in Fig. 24.8. A permanently excited DC motor with brush commutation drives the throttle through a two-stage gear in the opening or closing direction. It operates against the main helic spring. A second spring works in the closing region in the opposite direction, in order to open the throttle in the case of a voltage loss into a limp-home position (a mechanical redundancy). The motor is manipulated by a pulse width modulated (PWM) armature voltage $U_{\mathrm{A}} \in (-12\,\mathrm{V} \ldots +12\,\mathrm{V})$. The measured variables are the armature voltage U_{A}, the armature current I_{A}, and the angular throttle posi-

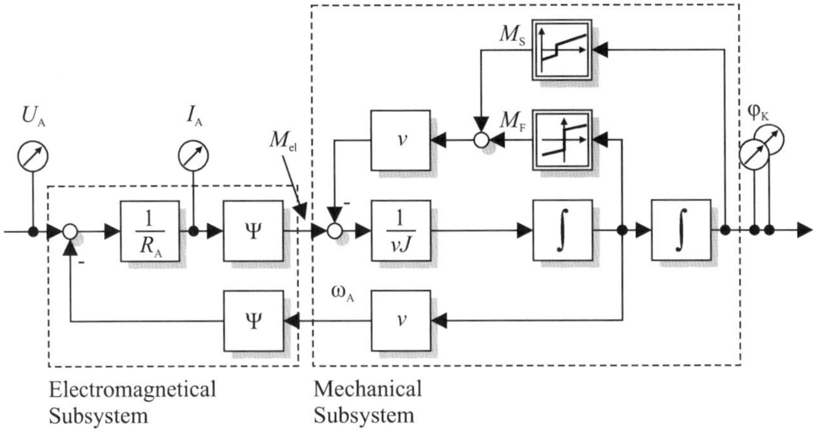

Fig. 24.9. Schematic of the DC drive with load

tion $\varphi_K \in (0° \dots 90°)$. This throttle position is measured by two redundant wiper-potentiometers operating in two different directions. The position controller was a model-based sliding-mode-controller or PID controller with time lag and sampling time $T_0 = 1.5\,\mathrm{ms}$ (Pfeufer, 1997, 1999).

Theoretical modeling of the throttle valve leads to the following basic equations (Isermann, 2005). The DC motor is governed by

$$U_A(t) = R_A I_A(t) + \Psi \omega_A(t) \tag{24.1.15}$$

$$M_{el}(t) = \Psi I_A(t)\,, \tag{24.1.16}$$

while the mechanical part (related to the motor axle) can be modeled as

$$\nu J \dot{\omega}_k(t) = M_{el}(t) - M_{mech}(t) \tag{24.1.17}$$

$$M_{mech}(t) = \frac{1}{\nu}\big(c_{S1}\varphi_k(t) + M_{S0} + M_F\big) \text{ for } \varphi_k > \varphi_{k0} \tag{24.1.18}$$

$$M_F(t) = M_{F0}\,\mathrm{sign}\,\omega_k(t) + M_{F1}\omega_k(t) \tag{24.1.19}$$

In this example, the used symbols are:

R_A	armature resistance	Ψ	magnetic flux linkage
ν	gear ratio ($\nu = 16.42$)	J	moment of inertia of the motor
M_{F0}	Coulomb friction torque	M_{F1}	viscous friction torque
c_{S1}	spring constant	M_{S0}	spring pretension
ω_k	throttle angular speed ($= \dot{\varphi}_k$)	ω_A	motor angular speed, $\omega_A = \nu\omega_k$

The armature inductance can be neglected, because the electrical time constant $M_{el} = L_A/R_A \approx 1\,\mathrm{ms}$ is much smaller than the mechanical dynamics. Depending on the input excitation either the Coulomb friction or the viscous friction turned out to be dominant when measurements were taken at a testbed.

Parameter Estimation for the Dynamic Behavior

The parameter estimation is carried out with recursive least squares estimation in the form of discrete square root filtering (DSFI). The basic model equation is

$$y(t) = \boldsymbol{\psi}^T(t)\boldsymbol{\theta} \tag{24.1.20}$$

and the data vector and the parameter estimation for the electrical part are

$$y(t) = U_A(t) \tag{24.1.21}$$

$$\boldsymbol{\psi}^T(t) = \left(I_A(t)\ \nu\omega_k(t) \right) \tag{24.1.22}$$

$$\boldsymbol{\theta} = \begin{pmatrix} \theta_1 \\ \theta_2 \end{pmatrix} \tag{24.1.23}$$

and for the mechanical part

$$y(t) = \dot{\omega}_k(t) \tag{24.1.24}$$

$$\boldsymbol{\psi}^T(t) = \left(I_A(t)\ \varphi_k(t)\ \omega_k(t)\ 1 \right) \tag{24.1.25}$$

$$\theta = \begin{pmatrix} \theta_4 \\ \theta_5 \\ \theta_6 \\ \theta_7 \end{pmatrix} \tag{24.1.26}$$

Because of a fast input excitation, the Coulomb friction term is neglected and only the viscous friction parameter M_{F1} is estimated under the condition that the speed is sufficiently large, i.e. $|\omega_k| > 1.5\,\mathrm{rad/s}$.

The relation between the physical process coefficients and the parameter estimates are

$$\begin{aligned} &\hat{\theta}_1 = R_A, \quad \hat{\theta}_2 = \Psi, \\ &\hat{\theta}_4 = \frac{\Psi}{\nu J}, \quad \hat{\theta}_5 = -\frac{c_{S1}}{\nu^2 J}, \quad \hat{\theta}_6 = -\frac{M_{F1}}{\nu^2 J}\ \hat{\theta}_7 = -\frac{M_{S0}}{\nu^2 J} \end{aligned} \tag{24.1.27}$$

As the gear ratio ν is known, the rotational inertia follows from

$$J = \frac{\hat{\theta}_2}{\nu \hat{\theta}_4} \tag{24.1.28}$$

All other process coefficients can directly be determined from the parameter estimates $\hat{\theta}_i$.

For the parameter estimation, the actuator operates in closed loop and the setpoint is changed with a PRBS signal between 10° and 70°. The derivatives $\omega_k = \dot{\varphi}_k$ and $\dot{\omega}_k = \ddot{\varphi}_k$ are determined by a state variable filter with the sampling time chosen as $T_{0,SVF} = 2\,\mathrm{ms}$. The sampling time for the parameter estimation is $T_0 = 6\,\mathrm{ms}$. The resulting parameter estimates converge fast and the largest equation error is $\leq 5\%$ or $\leq 3.5°$ for the electrical part and $\leq 7 \ldots 12\%$ for the mechanical part (Pfeufer, 1999).

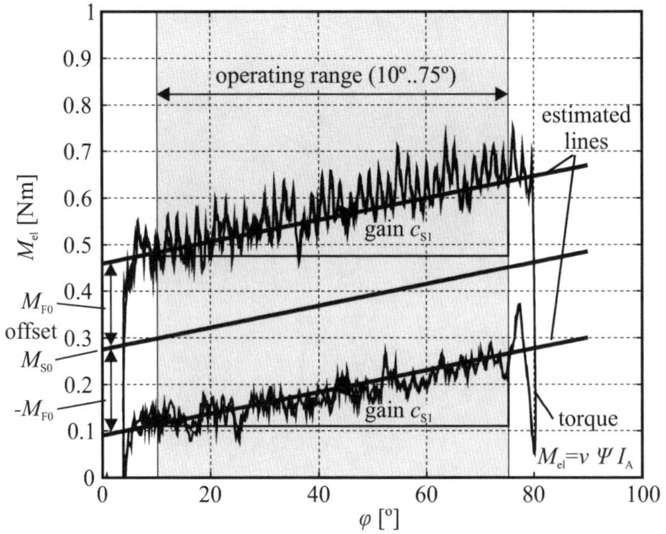

Fig. 24.10. Estimated static behavior of the throttle actuator

Parameter Estimation for the Static Behavior

In order to obtain more precise information about the mechanical part and especially the friction phenomena, only the static behavior is considered for slow continuous input changes.

Setting $\dot{\omega}_k = 0$ and neglecting the viscous friction, (24.1.15) to (24.1.19) yield with $t = kT_0$

$$I_A(k) = \frac{1}{\nu\Psi}\big(c_{S1}\varphi_k(k) + M_{S0} + M_{F0}\,\text{sign}\,\omega_k(k)\big) = \boldsymbol{\psi}^T(k)\boldsymbol{\theta} \qquad (24.1.29)$$

Because of the direction-dependent Coulomb friction for the opening and closing, two separate estimations are made

$$\boldsymbol{\psi}_1^T(k) = \big(\varphi_k^+(k)\ 1\big),\ \boldsymbol{\psi}_2^T(k) = \big(\varphi_k^-(k)\ 1\big)$$

$$\hat{\boldsymbol{\theta}}^+(k) = \begin{pmatrix}\hat{\theta}_1\\ \hat{\theta}_2\end{pmatrix} \qquad \hat{\boldsymbol{\theta}}^-(k) = \begin{pmatrix}\hat{\theta}_3\\ \hat{\theta}_4\end{pmatrix}$$

with

$$\hat{\theta}_1 = \frac{c_{S1}}{\nu\Psi} \qquad \hat{\theta}_2 = \frac{M_{S0}}{+} M_{F0}\nu\Psi$$

$$\hat{\theta}_3 = \frac{c_{S1}}{\nu\Psi} \qquad \hat{\theta}_4 = \frac{M_{S0} - M_{F0}}{\nu\Psi}$$

The magnetic flux linkage Ψ is known from (24.1.27). The physical process parameters then result as

$$c_{S1} = \nu\Psi\frac{\hat{\theta}_1 + \hat{\theta}_3}{2}$$
$$M_{S0} = \nu\Psi\frac{\hat{\theta}_2 + \hat{\theta}_4}{2}$$
$$M_{F0} = \nu\Psi\frac{\hat{\theta}_2 - \hat{\theta}_4}{2}$$

The parameter estimation is performed with recursive DSFI and the sample time $T_0 = 6\,\text{ms}$ for each motion. Figure 24.10 shows the resulting friction characteristics. The spring pretension M_{S0} leads to a positive offset of the linear spring characteristic and the dry friction shifts the friction characteristic by M_{F0}^{+} and M_{F0}^{-} such that a hysteresis characteristic results. A comparison with the electrical torque $M'_{el} = \nu\Psi I_A$ related to the throttle axle indicates a good agreement with the estimated hysteresis characteristic. (The oscillations of the calculated electrical torque are due to the closed loop behavior in connection with adhesive friction or stick-slip effects, which are not modeled. The range around the point of return, where adhesion works, is omitted in the parameter estimation for simplifying reasons).

24.1.3 Hydraulic Actuators

Hydraulic systems are used in manifold applications. They excel especially whenever there is a need for high power density combined with fast actuation capabilities. Furthermore, hydraulic cylinders can easily generate linear motions, whereas e.g. electric motors typically need a conversion gear to generate a linear motion.

Introduction

The typical setup of a hydraulic servo axis is shown in Fig. 24.11: A positive displacement pump acts as a pressure supply. Oil is drawn from a supply tank and is expelled at high pressure into the supply line. From there, the oil flows to the individual hydraulic servo axes, which in the case of a linear motion, typically consist of a proportional acting valve and a hydraulic cylinder. The valve is used to throttle the hydraulic flow and direct it into one of the two chambers of the hydraulic cylinder. For this task, the valve contains a so-called *valve spool*, which can move freely inside the *valve sleeve*. The valve spool can be driven by a variety of forces. By the hand force of a worker (e.g. construction machinery), by a hydraulic force (e.g. second stage of a two-stage valve), by a torque motor (e.g. nozzle-flapper unit of the first stage of a two stage valve), or, as in the case at hand, by the force of two solenoids (direct-driven valve), see Fig. 24.13. As the valve spool moves tiny openings occur and a turbulent flow evolves according to the pressure drop Δp across the tiny opening. The oil flows into the hydraulic cylinder and exerts a pressure on the piston. By means of the piston rod, this force is transferred to drive an external load. As the piston moves, oil from the opposite chamber is pushed back into the supply tank. The two cylinder chambers thus have to be isolated by the piston. Here, one typically uses the setup depicted in Fig. 24.14 for standard hydraulic cylinders. Here,

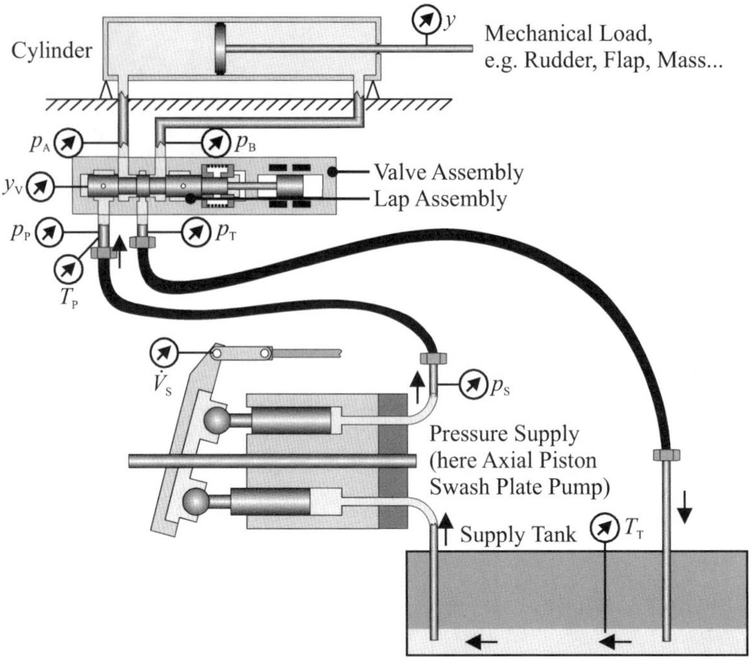

Fig. 24.11. Scheme of a hydraulic servo axis

Fig. 24.12. Photo of the hydraulic servo axis testbed

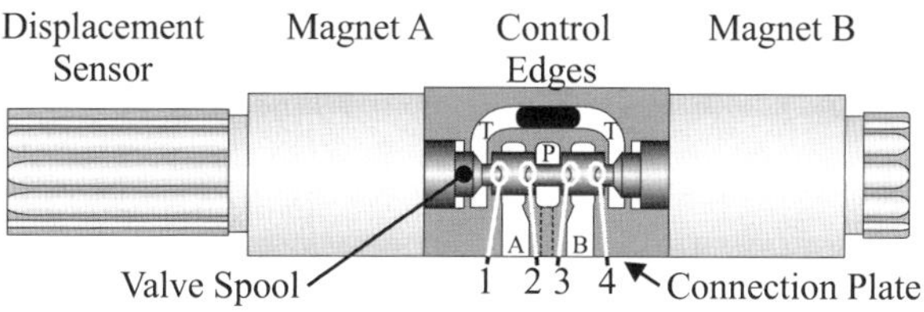

Fig. 24.13. Schematic view of a direct-driven proportional acting valve and control edge numbering

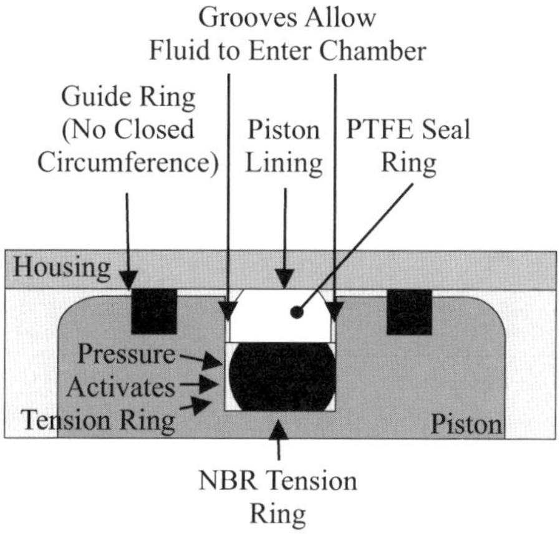

Fig. 24.14. Schematic of the cylinder sealing isolating the two cylinder chambers

one has to obey a certain trade-off: If the cylinder sealing is pushed tightly against the cylinder housing, it would wear out very fast as the piston constantly moves and hence friction will occur between the sealing and the cylinder housing. On the other side, if the sealing is fitted too loosely, a large bypass flow between the two cylinder chambers will result as a combined pressure induced and adhesion induced flow will result. Especially in the area of fault detection and diagnosis, one is interested in determining the coefficient of leakage flow, as it is a good indicator of the sealing health. The measurements and results that are presented in this section stem from the hydraulic servo axis testbed shown in Fig. 24.12. Some technical data are listed in Table 24.3.

Here and in the remainder of this section, the following symbols are used:

b_{V}	coefficient of valve flow	A_{A}	active piston area chamber A
$V_{0\mathrm{A}}$	dead volume chamber A	G_{AB}	laminar leakage coefficient
$E_{0\mathrm{A}}$	bulk modulus chamber A	$\dot{V}$	volume flow rate
T	fluid temperature	$\dot{V}_{\mathrm{A}}$	flow rate into chamber A
y_{V}	valve spool displacement	y	piston displacement
p_{A}	pressure in chamber A	k	spring stiffness of load
m	mass of load	F_0	Coulomb friction force

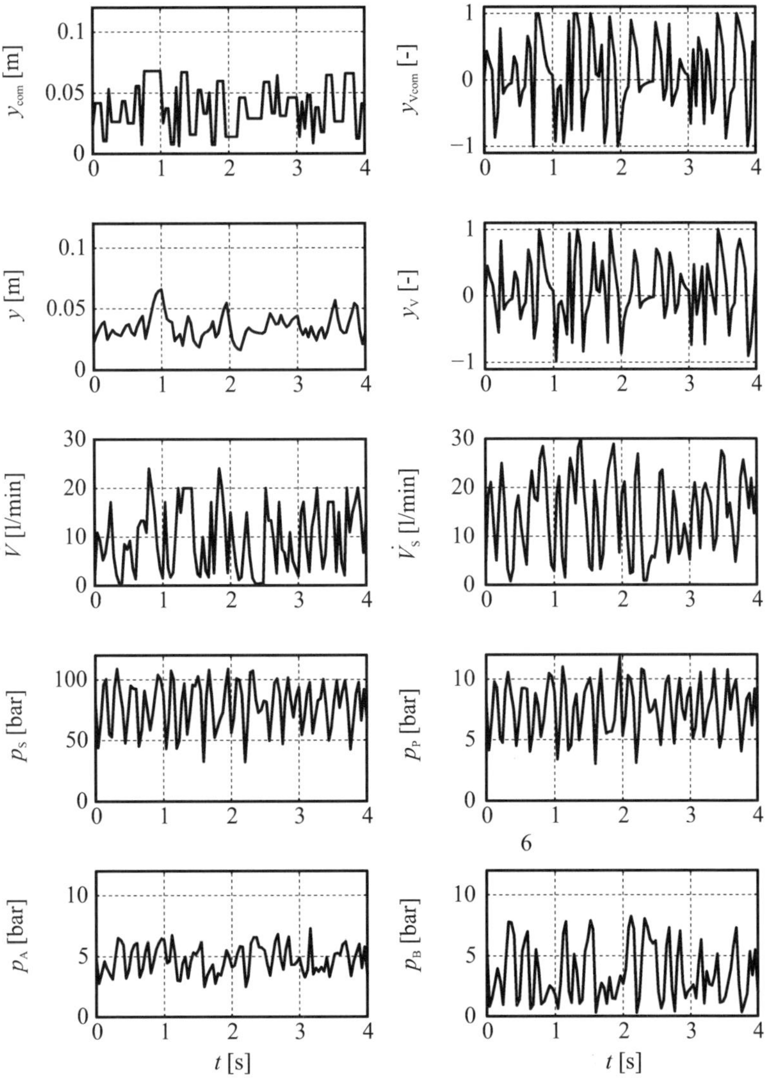

Fig. 24.15. Measurements taken at the hydraulic servo axis. For the sensor location, see Fig. 24.11

Table 24.3. Specifications of the hydraulic servo axis shown in Fig. 24.12 operating with HLP46 as hydraulic oil

Parameter	Value
System pressure	80 bar
Max. pressure supplied by pump	280 bar
Pump displacement	45 cm^3/rev
Cylinder diameter	40 mm
Piston rod diameter	28 mm
Cylinder stroke	300 mm
Max. velocity	1 m/s
Max. force	20 kN
Mass of load	40 kg
Spring stiffness of load	100 000 N/m
Density of fluid at 15°C	0.87 g/cm^3
Mean bulk modulus of fluid	2 GPa

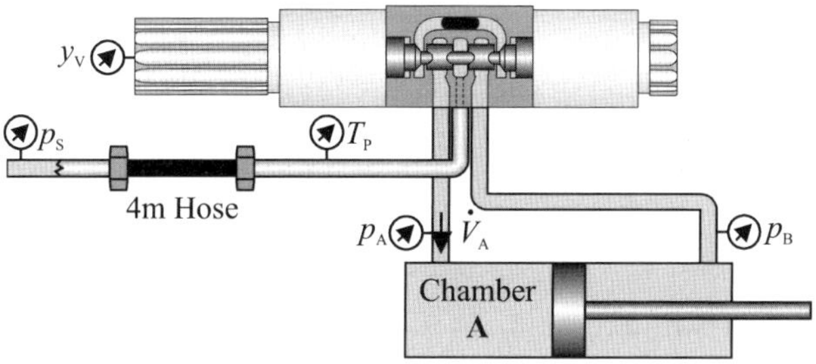

Fig. 24.16. Schematic view of a direct-driven proportional acting valve and hydraulic cylinder

c coefficient of viscous friction

The index denote the following:

A	cylinder chamber A	B	cylinder chamber B
P	pressure line	T	return line or tank
S	pressure supply / pump		

For the sensor location, see Figs. 24.11 and 24.16. A measurement recorded at the hydraulic servo axis is shown in Fig. 24.15.

Identification of the Hydraulics

The flow over the control edge is given as

$$\dot{V}(t) = b_V(y_V(t), T(t))\sqrt{|\Delta p(t)|}\,\mathrm{sign}\,\Delta p(t)\,. \quad (24.1.30)$$

One can see that the coefficient of the flow b_V depends on both the valve spool displacement $y_V(t)$ as well as the fluid temperature. The flow balance for e.g. chamber

A of the hydraulic cylinder is given as

$$\dot{V}_{\mathrm{A}}(t) = A_{\mathrm{A}}\dot{y}(t) - G_{\mathrm{AB}}\big(p_{\mathrm{A}}(t) - p_{\mathrm{B}}(t)\big) - \frac{1}{E_{0\mathrm{A}}}\big(V_{0\mathrm{A}} + A_{\mathrm{A}}y(t)\big)\dot{p}_{\mathrm{A}}(t)\,. \quad (24.1.31)$$

For identification purposes, the following signals will be utilized: $y_{\mathrm{V}}(t)$ valve spool displacement, $p_{\mathrm{A}}(t)$ pressure in chamber A, $p_{\mathrm{B}}(t)$ pressure in chamber B, $p_{\mathrm{S}}(t)$ pressure at the supply, $y(t)$ position of the piston. The fluid temperature $T(t)$ is measured, but can be neglected depending on the requirements of the model fidelity.

The valve opening characteristics $b_{\mathrm{V}}(y_{\mathrm{V}})$ are due to their non-linear behavior modeled as a polynomial as

$$b_{\mathrm{V}i}(y_{\mathrm{V}}) = b_{\mathrm{V}1i}\,y_{\mathrm{V}}(t) + b_{\mathrm{V}2i}\,y_{\mathrm{V}}^2(t) + b_{\mathrm{V}3i}\,y_{\mathrm{V}}^3(t)\,, \quad (24.1.32)$$

where the index i denotes the control edge to be modeled, see Fig. 24.13 for the numbering of the control edges. As there are four possible flow paths, there will also be four independent polynomials.

The individual models will now be combined. This yields

$$\begin{aligned}&\big(V_{0\mathrm{A}} + A_{\mathrm{A}}y(t)\big)\frac{1}{\bar{E}(T)}\dot{p}_{\mathrm{A}}(t) + A_{\mathrm{A}}\dot{y}(t) = \\ &\dot{V}_{\mathrm{PA}}(p_{\mathrm{A}}, p_{\mathrm{P}}, T, y_{\mathrm{V}}) - \dot{V}_{\mathrm{AT}}(p_{\mathrm{A}}, T, y_{\mathrm{V}}) - G_{\mathrm{AB}}(T)\big(p_{\mathrm{A}}(t) - p_{\mathrm{B}}(t)\big)\end{aligned} \quad (24.1.33)$$

with

$$\dot{V}_{\mathrm{PA}}(p_{\mathrm{A}}, p_{\mathrm{S}}, T, y_{\mathrm{V}}) = b_{\mathrm{V}2}(y_{\mathrm{V}}, T)\sqrt{|p_{\mathrm{S}}(t) - p_{\mathrm{A}}(t)|}\,\mathrm{sign}\big(p_{\mathrm{P}}(t) - p_{\mathrm{A}}(t)\big) \quad (24.1.34)$$

$$\dot{V}_{\mathrm{AT}}(p_{\mathrm{A}}, T, y_{\mathrm{V}}) = b_{\mathrm{V}1}(y_{\mathrm{V}}, T)\sqrt{|p_{\mathrm{A}}(t)|}\,\mathrm{sign}\,p_{\mathrm{A}}(t). \quad (24.1.35)$$

These equations can be combined into one model. Since parameter estimation is based on sampled signals, the time t will now be expressed as integer multiples k of a fixed sampling time T_0. The model for chamber A is given as

$$(V_{0\mathrm{A}} + A_{\mathrm{A}}y(k))\frac{1}{\bar{E}(T_{\mathrm{P}})}\dot{p}_{\mathrm{A}}(k) + A_{\mathrm{A}}\dot{y}(k) = \big(\dot{V}_{\mathrm{A}}(p_{\mathrm{A}}, p_{\mathrm{S}}, T_{\mathrm{P}}, y_{\mathrm{V}}) - \dot{V}_{\mathrm{AB}}(p_{\mathrm{A}}, p_{\mathrm{B}}, T_{\mathrm{P}})\big)\,, \quad (24.1.36)$$

where $\bar{E}$ is the mean bulk modulus. The valve flow with a polynomial approximation of the valve flow characteristics is given as

$$\begin{aligned}\dot{V}_{\mathrm{A}}(p_{\mathrm{A}}, p_{\mathrm{S}}, T_{\mathrm{P}}, y_{\mathrm{V}}) =& \left(\sum_{i=k}^{l} b_{1i}(T_{\mathrm{P}})y_{\mathrm{V}}(k)^i\right)\sqrt{|\Delta p_{\mathrm{SA}}(k)|}\,\mathrm{sign}\,\Delta p_{\mathrm{SA}}(k) \\ &-\left(\sum_{i=k}^{l} b_{2i}(T_{\mathrm{P}})y_{\mathrm{V}}(k)^i\right)\sqrt{|\Delta p_{\mathrm{AT}}(k)|}\,\mathrm{sign}\,\Delta p_{\mathrm{AT}}(k)\end{aligned} \quad (24.1.37)$$

where $\Delta p_{\mathrm{SA}}(k) = p_{\mathrm{S}}(k) - p_{\mathrm{A}}(k)$ and $\Delta p_{\mathrm{AT}}(k) = p_{\mathrm{A}}(k)$. The internal leakage is given as

$$\dot{V}_{\mathrm{AB}}(p_{\mathrm{A}}, p_{\mathrm{B}}, T_{\mathrm{P}}) = G_{\mathrm{AB}}(T_{\mathrm{P}})\big(p_{\mathrm{A}}(k) - p_{\mathrm{B}}(k)\big)\,. \tag{24.1.38}$$

Combining the above equations and solving for $\dot{y}(k)$ yields

$$\dot{V}_{\mathrm{A}}(p_{\mathrm{A}}, p_{\mathrm{S}}, T, y_{\mathrm{V}}) - G_{\mathrm{AB}}(T)\big(p_{\mathrm{A}}(k) - p_{\mathrm{B}}(k)\big) - \frac{(V_{0\mathrm{A}} + A_{\mathrm{A}}y(k))}{\bar{E}(T)}\dot{p}_{\mathrm{A}}(k) = A_{\mathrm{A}}\dot{y}(k) \tag{24.1.39}$$

Now, the valve flow can be inserted. The polynomial will be written out to see that this parameter estimation problem is truly linear in parameters,

$$\begin{aligned}&\ldots + b_{1i}(T)y_{\mathrm{V}}(k)^i\sqrt{|\Delta p_{\mathrm{SA}}(k)|}\,\mathrm{sign}\big(\Delta p_{\mathrm{SA}}(k)\big)\big(y_{\mathrm{V}}(k) \geq 0\big) + \ldots\\ &- \ldots - b_{2i}(T)y_{\mathrm{V}}(k)^i\sqrt{|\Delta p_{\mathrm{AT}}(k)|}\,\mathrm{sign}\big(\Delta p_{\mathrm{AT}}(k)\big)\big(y_{\mathrm{V}}(k) < 0\big) - \ldots\\ &- G_{\mathrm{AB}}(T)\big(p_{\mathrm{A}}(k) - p_{\mathrm{B}}(k)\big) - \frac{(V_{0\mathrm{A}} + A_{\mathrm{A}}y(k))}{\bar{E}(T)}\dot{p}_{\mathrm{A}}(k) = A_{\mathrm{A}}\dot{y}(k)\end{aligned} \tag{24.1.40}$$

where the terms $(y_V(k) \geq 0)$ and $(y_V(k) < 0)$ amount to one if the corresponding condition is fulfilled and are zero otherwise.

This entire parameter estimation problem now can be split up into the data matrix

$$\boldsymbol{\Psi}^{\mathrm{T}} = \begin{pmatrix} \vdots & \\ y_{\mathrm{V}}(k)^i\sqrt{|\Delta p_{\mathrm{SA}}(k)|}\,\mathrm{sign}\big(\Delta p_{\mathrm{SA}}(k)\big)\cdot\big(y_{\mathrm{V}}(k) \geq 0\big) & \ldots\\ \vdots & \\ y_{\mathrm{V}}(k)^i\sqrt{|p_{\mathrm{A}}(k)|}\,\mathrm{sign}\,p_{\mathrm{A}}(k)\cdot\big(y_{\mathrm{V}}(k) < 0\big) & \ldots\\ \vdots & \\ p_{\mathrm{A}}(k) - p_{\mathrm{B}}(k) & \ldots\\ A_{\mathrm{A}}\,y(k)\,\dot{p}_{\mathrm{A}}(k) & \ldots \end{pmatrix} \tag{24.1.41}$$

and the output vector

$$\boldsymbol{y}^{\mathrm{T}} = \big(A_{\mathrm{A}}\dot{y}(k), \ldots\big). \tag{24.1.42}$$

The solution of the parameter estimation problem supplies estimates for the parameter vector $\boldsymbol{\theta}$ given as

$$\hat{\boldsymbol{\theta}}^{\mathrm{T}} = \left(\hat{b}_{10}(T)\ldots\hat{b}_{1n}(T)\,\hat{b}_{20}(T)\ldots\hat{b}_{2n}(T)\ldots\hat{G}_{\mathrm{AB}}(T)\,\frac{1}{\hat{\bar{E}}(T)}\right). \tag{24.1.43}$$

This parameter vector contains $2n + 2$ variables. Although such a parameter estimation problem can easily be solved on modern computers, it will only be used for offline identification. For online identification, the parameter estimation problem is split: Two independent parameter estimation problems for $y_{\mathrm{V}}(k) \geq 0$ and $y_{\mathrm{V}}(k) < 0$ are formulated to constrain the number of parameters to be estimated in each iteration. Since the valve spool displacement can only be positive or negative at any instant in time, the parameter estimation problem can be simplified.

The results of the parameter estimation can be seen in Fig. 24.17 and Fig. 24.18. Figure 24.17 shows that the non-linear behavior of the valve flow coefficient as a

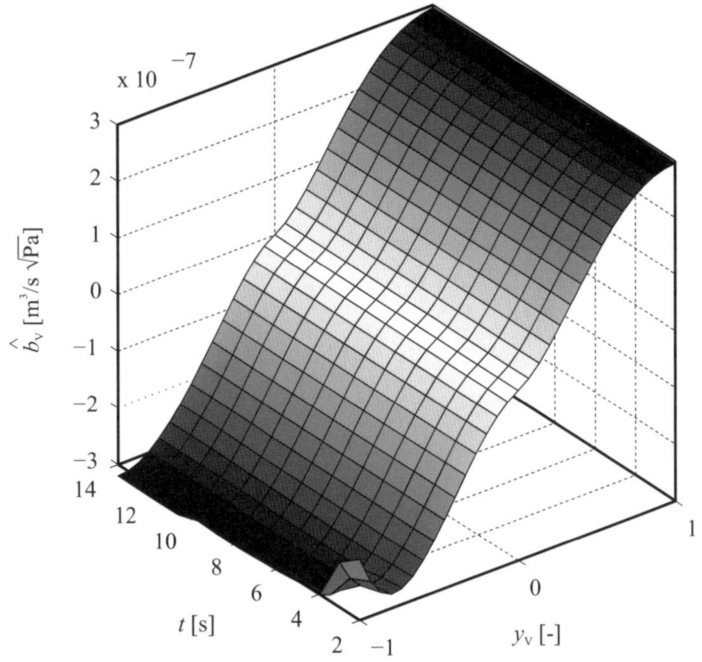

Fig. 24.17. Estimation of non-linear valve characteristics b_V over time. Note that the valve characteristics are non-linear and have been approximated by a polynomial in y_V. The modeled control edges are P $\rightarrow$ A for positive values of y_V and T $\rightarrow$ A for negative values of y_V

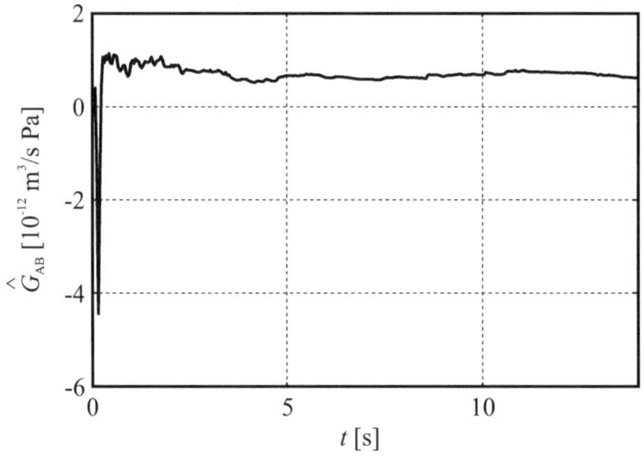

Fig. 24.18. Estimation of laminar leakage coefficient G_AB over time

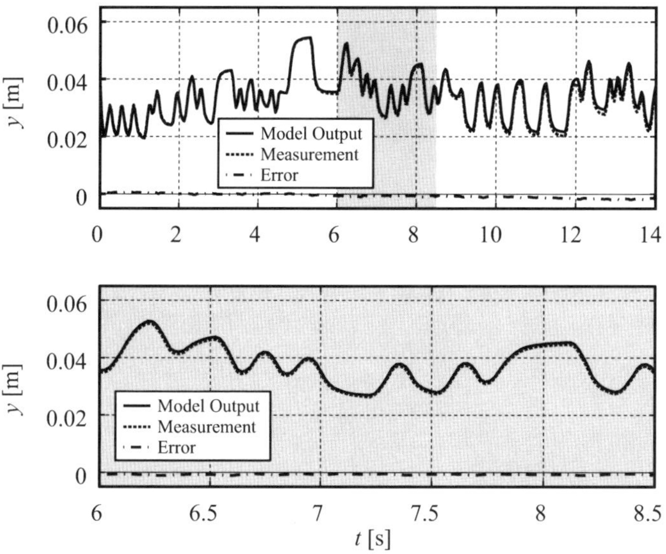

Fig. 24.19. Simulation of the hydraulic subsystem of the hydraulic servo axis

function of the valve opening is identified very well. The identification results of the laminar leakage coefficient in Fig. 24.18 also match very well with the values reported in literature. As was stated in the introduction to this section, knowledge of the laminar leakage coefficient is of special interest in the area of fault detection and diagnosis. Finally, Fig. 24.19 shows a simulation run and comparison with measurements. One can see that the model that has been parameterized by measurements shows a very good model fidelity. This high model fidelity can in large part be attributed to the nonlinear modeling of the valve characteristics.

Hydraulic servo axes typically operate with a large variety of loads. Furthermore, the load can change drastically during operation. To be able to maintain a good control quality despite the large variations in the load and hence the plant parameters, one can use adaptive control algorithms (e.g. Isermann et al, 1992). To be able to update the controller, one needs a good model of the load. Hence, the load driven by the hydraulic servo axis will now be identified based on sensors mounted at the hydraulic servo axis, see Fig. 24.20.

A general load model is shown in Fig. 24.21. This model covers both a mass and a spring as a load and also contains friction effects, which are modeled as a combination of Coulomb and viscous friction. It should be mentioned that even though results have been obtained for a spring-mass load only, the methods will apply to all kinds of loads which can be described by the general load model given in (24.1.44) and Fig. 24.21.

With the piston as the force-generating element of the hydraulic servo axis, the mechanics can be described in general form by

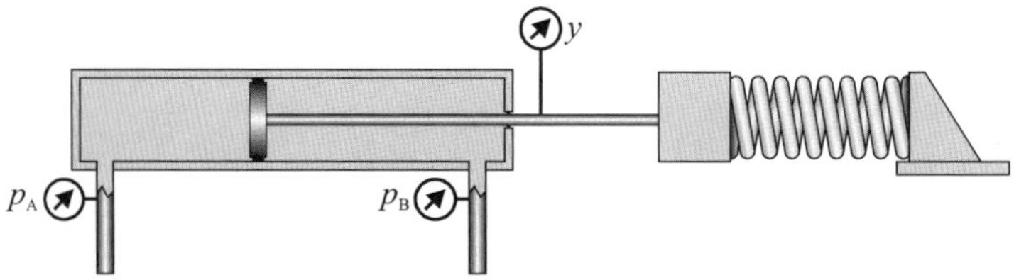

Fig. 24.20. Scheme of the mechanical load consisting of spring and mass

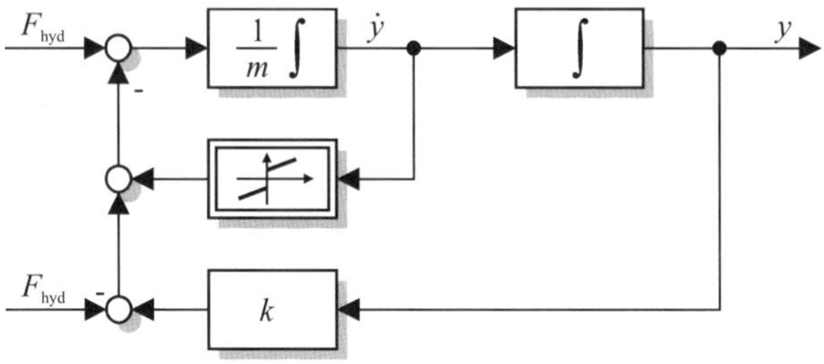

Fig. 24.21. Block diagram of the mechanical load model consisting of spring and mass

$$p_A(t)A_A - p_B(t)A_B = k\big(y(t) - y_0\big) + m\ddot{y}(t) + \begin{cases} -F_0 + c^-\dot{y}(t) \text{ for } \dot{y}(t) < 0 \\ F_0 + c^+\dot{y}(t) \text{ for } \dot{y}(t) \geq 0 \,, \end{cases} \tag{24.1.44}$$

where the friction force $F_F(\dot{y}(t))$ is a combination of Coulomb friction and viscous friction. F_0 is the static friction force and c^- and c^+ are the direction-dependent coefficients of viscous friction.

(24.1.44) is the basis for the parameter estimation problem. The spring pretension is estimated by the spring pre-compression displacement y_0. Based on this equation, the data matrix of the parameter estimation problem becomes

$$\boldsymbol{\Psi} = \begin{pmatrix} y(1) & 1 & \ddot{y}(1) & \operatorname{sign}\dot{y}(1) & \dot{y}(1)\cdot(\dot{y}(1)>0) & \dot{y}(1)\cdot(\dot{y}(1)<0) \\ y(2) & 1 & \ddot{y}(2) & \operatorname{sign}\dot{y}(2) & \dot{y}(2)\cdot(\dot{y}(2)>0) & \dot{y}(2)\cdot(\dot{y}(2)<0) \\ \vdots & \vdots & \vdots & \vdots & \vdots & \vdots \\ y(N) & 1 & \ddot{y}(N) & \operatorname{sign}\dot{y}(N) & \dot{y}(3)\cdot(\dot{y}(N)>0) & \dot{y}(N)\cdot(\dot{y}(N)<0) \end{pmatrix} . \tag{24.1.45}$$

The output vector is

$$\boldsymbol{y} = \begin{pmatrix} p_A(1)A_A - p_B(1)A_B \\ p_A(2)A_A - p_B(2)A_B \\ \vdots \\ p_A(N)A_A - p_B(N)A_B \end{pmatrix} . \tag{24.1.46}$$

The parameter of estimated quantities becomes

$$\boldsymbol{\theta}^T = \big(\hat{k}\ (\hat{k}\cdot\hat{y}_0)\ \hat{m}\ \hat{F}_0\ \hat{c}^-\ \hat{c}^+\big) . \tag{24.1.47}$$

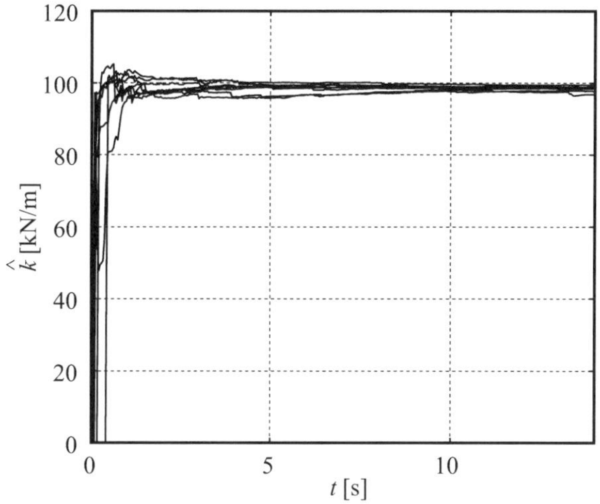

Fig. 24.22. Estimation of spring constant $\hat{k}$ in several test runs

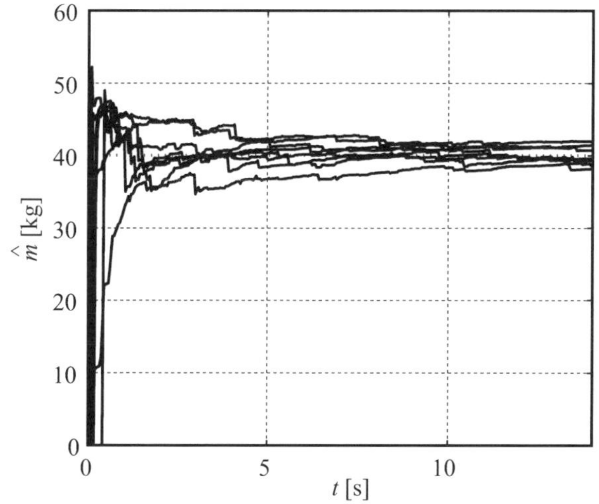

Fig. 24.23. Estimation of load mass $\hat{m}$ in several test runs

With this parameter estimation method, the results depicted in Figs. 24.22 and 24.23 have been obtained. Recalling from the specifications in Table 24.3 and knowing that the mass of load was $m = 40\,\mathrm{kg}$ and the spring stiffness $k = 100\,000\,\mathrm{N/m}$, one can see that the estimates match well with the real values. One can once again run the model in parallel to the measurements to judge the quality of the identification results. This has been done in Fig. 24.24.

The identification method used was DSFI, the sample rate was $f_S = 500\,\mathrm{Hz}$. The first and second derivative of the piston displacement were calculated directly as

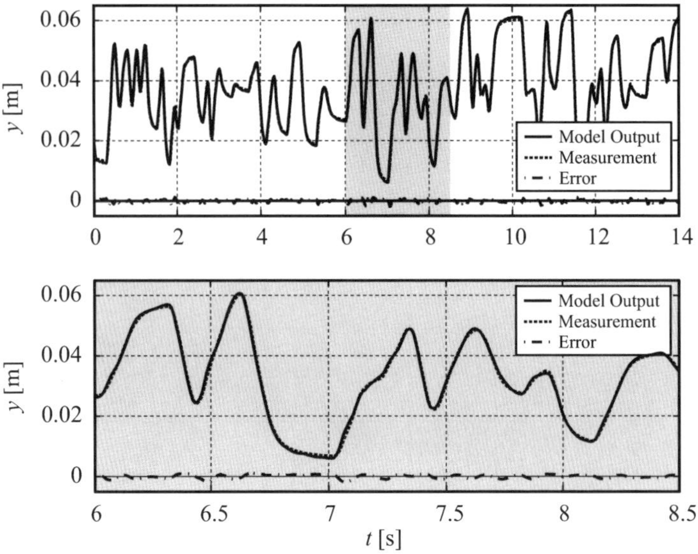

Fig. 24.24. Simulation of the mechanic subsystem of the hydraulic servo axis

the signal was supplied by a digital sensor and had little noise. For this, the central difference quotient was used to ensure that the measurements are not delayed. All methods were implemented as recursive parameter estimation to be able to follow changes of the plant e.g. for fault detection and diagnosis.

The detailed description of these parameter estimation approaches is covered in the dissertation by Muenchhof (2006). There, also the identification using neural nets has been discussed as well as the application of these identification techniques for fault detection and diagnosis.

24.2 Machinery

In the following, the application of identification methods to machinery will be treated. Here, the application of these methods can be advantageous for controller design, adaptive control, up to automatic commissioning of the machines as well as condition monitoring.

24.2.1 Machine Tool

As an example for a main drive, a machining center of type (MAHO MC5) is considered. A speed-controlled DC motor drives a belt, a gear and tool spindle, carrying a cutter or drill. Hence, a multi mass-spring-damper system results with 6 masses. A view of the main drive is shown in Fig. 24.25.

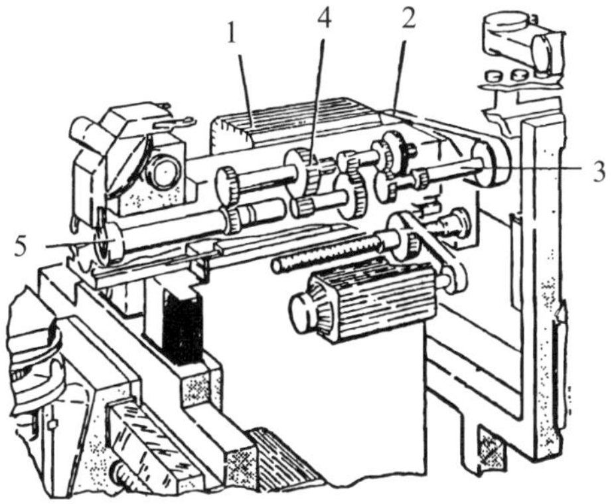

Fig. 24.25. Main drive of machining center MAHO MC5

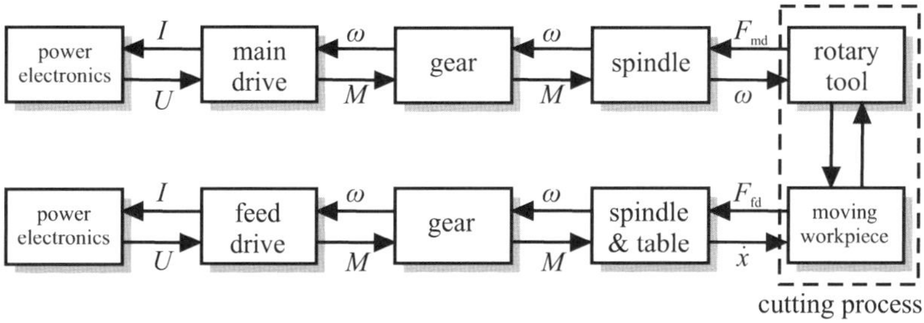

Fig. 24.26. Two-port representation of the signal flow of the main drive and one feed drive of a drilling or milling machine tool

In the following, small deviations of the variables are considered such that linear models can be assumed. The dynamic behavior of the DC motor is like in the previous examples modeled by

$$L_A \frac{\mathrm{d}}{\mathrm{d}t} I_A(t) = -R_A I_A(t) - \Psi \omega_1(t) + U_A(t) \tag{24.2.1}$$

$$J_1 \frac{\mathrm{d}}{\mathrm{d}t} \omega_1(t) = -\Psi I_A(t) - M_1(t) \tag{24.2.2}$$

with

L_A	armature inductance	U_A	armature voltage
R_A	armature resistance	I_A	armature current
Ψ	magnetic flux linkage	ω_1	motor speed ($\omega_1 = \dot{\varphi}$)
J_1	moment of inertia	M_1	load torque

An analysis of the eigenfrequencies of the main drive shows that the motor is able to excite frequencies in open loop $f < 80\,\mathrm{Hz}$ and in closed loop $f < 300\,\mathrm{Hz}$ (Wanke and Isermann, 1992). The eigenfrequency of the belt drive is 123 Hz and those of shaft, gear, and spindle are 706, 412, and 1335 Hz. Hence, the dynamic behavior of the main drive is dominated by the motor and the belt drive and can therefore be modeled by a two-mass system with moment of inertias J_1 (motor plus belt driving pulley) and J_2 (belt driven pulley, shaft, gear, spindle). The mechanical part of the main drive is then described by a linear state space model

$$\dot{\boldsymbol{x}}(t) = \boldsymbol{A}\boldsymbol{x}(t) + \boldsymbol{b}u(t) + \boldsymbol{F}\boldsymbol{z}(t) \tag{24.2.3}$$

with

$$\boldsymbol{x}^{\mathrm{T}}(t) = \big(I_{\mathrm{A}}(t)\ \varphi_1(t)\ \dot{\varphi}_1(t)\ \ldots\ \varphi_5(t)\ \dot{\varphi}_5(t)\big) \tag{24.2.4}$$

$$u(t) = U_{\mathrm{A}}(t) \tag{24.2.5}$$

$$\boldsymbol{z}^{\mathrm{T}}(t) = \big(M_6(t)\ M_{\mathrm{F}}(t)\big) \tag{24.2.6}$$

with M_6 being the load torque and M_{F} the Coulomb friction torque.

The parameters of the main drive can of course be determined from construction data. However, if not all parameters can be determined or for fault detection in normal operation, the estimation of the parameters from measured signals is desired.

To estimate the parameters of the main drive in idle running ($M_6 = 0$) based on measurements of accessible signals $U_{\mathrm{A}}(t)$, $I_{\mathrm{A}}(t)$, $\omega_1(t)$, and spindle speed $\omega_5(t)$, the following equations are used:

$$\begin{aligned} U_{\mathrm{A}}(t) &= \theta_1\omega_1(t) + \theta_2 I_{\mathrm{A}}(t) + \theta_3\dot{I}_{\mathrm{A}}(t) \\ \theta_1 I_{\mathrm{A}}(t) - M_{\mathrm{R}}(t) &= \theta_4\dot{\omega}_1(t) + \theta_5\dot{\omega}_5(t) \\ \omega_5(t) &= \theta_6\dot{\omega}_1(t) + \theta_7\omega_1(t) - \theta_8\dot{\omega}_5(t) - \theta_9\ddot{\omega}_5(t) \end{aligned} \tag{24.2.7}$$

with

$$\begin{aligned} &\theta_1 = \Psi \quad \theta_2 = R_{\mathrm{A}} \quad \theta_3 = L_{\mathrm{A}} \\ &\theta_4 = J_1 \quad \theta_5 = iJ_2 \quad \theta_6 = di/c \\ &\theta_7 = i \quad\ \theta_8 = d/c \quad \theta_9 = J_2 i^2/c \end{aligned} \tag{24.2.8}$$

The armature flux linkage is estimated by the first equation in (24.2.7) beforehand (or known from the data sheet). Then all process coefficients can be determined

$$\begin{aligned} &i = \theta_7\ \text{(gear ratio)} && c = \theta_5\theta_7/\theta_9 \\ &J_1 = \theta_4\ \text{(motor)} && d = \theta_5\theta_7\theta_8/\theta_9 \\ &J_2 = \theta_5/\theta_7\ \text{(spindle)} \end{aligned} \tag{24.2.9}$$

The derivatives of first and second order for continuous-time parameter estimation were determined with state variable filters designed as Butterworth filters of 6th order with corner frequencies of 79.6 Hz and 47.8 Hz. The resolution of the incremental rotation sensors was increased to 4 096 slots for the spindle and 1 024 slots for the motor. Sampling time was $T_0 = 0.5$ ms. The results with the parameter estimation method DSFI (discrete square root filtering information from 40 step functions of the speed over a time interval of 15 s are shown in Figs. 24.27 to 24.30.

The estimated motor coefficients on the motor side $\hat{\Psi}$, $\hat{R}_{\mathrm{A}}$, and $\hat{L}_{\mathrm{A}}$ converge very fast within about 2 s, the mechanical coefficients $\hat{J}_1$, $\hat{J}_2$, $\hat{c}$, and $\hat{d}$ a bit slower within about 5 s. After about 15 s all 8 process coefficients converge to steady state values and agree relatively well with theoretical determined values (Wanke, 1993).

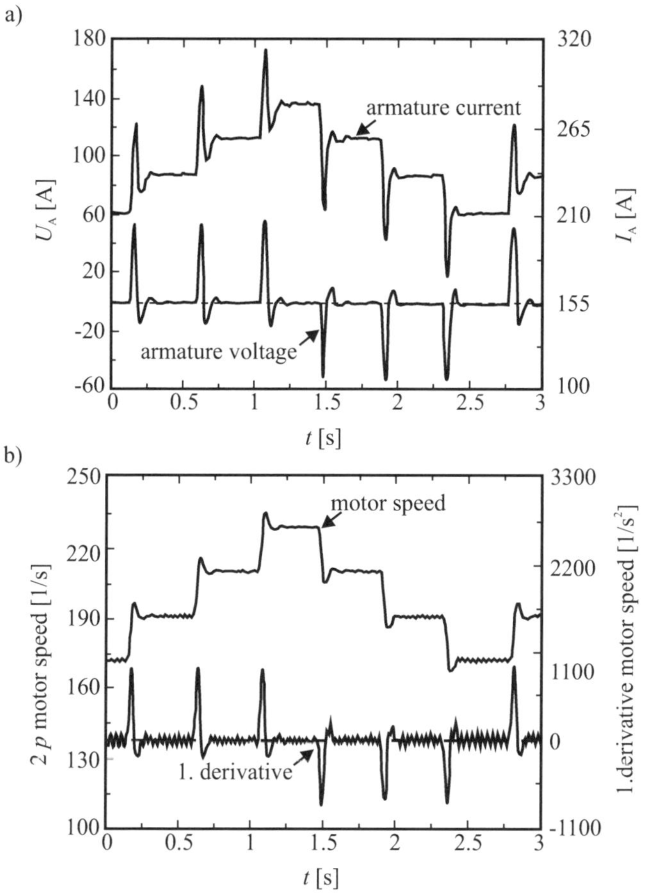

Fig. 24.27. Measured signals for the main drive. (**a**) armature voltage and armature current; (**b**) motor speed ω and first derivative $\dot{\omega}$

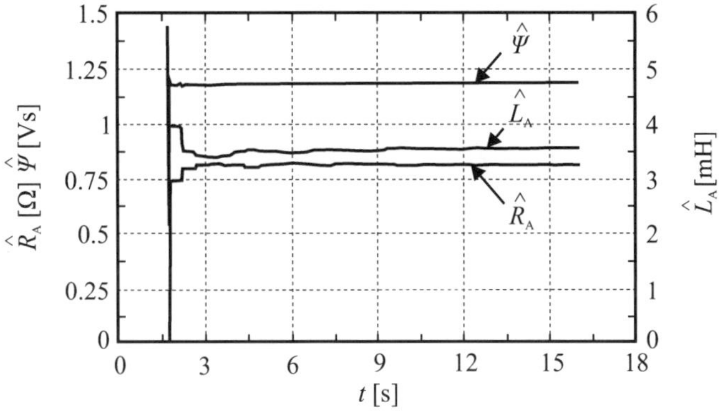

Fig. 24.28. Estimated process coefficients $\hat{R}_A$, $\hat{\Psi}$, and $\hat{L}_A$ of the DC motor

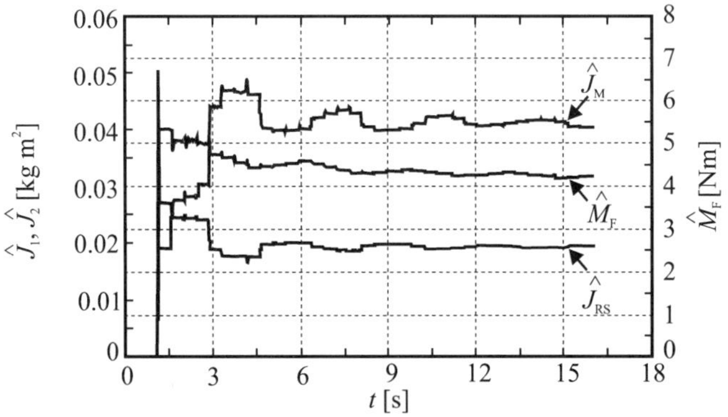

Fig. 24.29. Estimated process coefficients of the main drive. $\hat{J}_1$ and $\hat{J}_2$: moment of inertia of motor and spindle, $\hat{M}_C$: dry friction torque

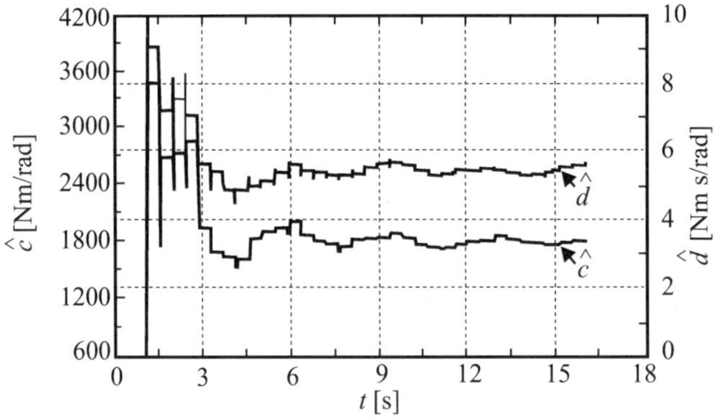

Fig. 24.30. Estimated stiffness $\hat{c}$ and damping $\hat{d}$ of the main drive

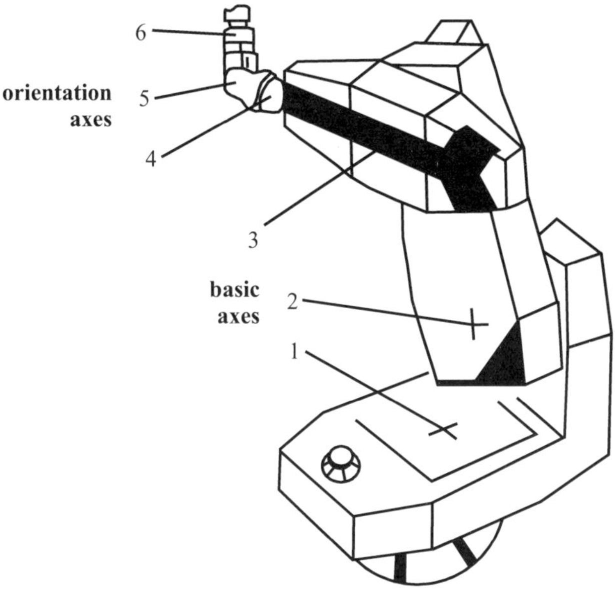

Fig. 24.31. Industrial robot with 6 rotational axes

24.2.2 Industrial Robot

As industrial robots (IR) are usually servo systems with point-to-point movements or trajectory following, they have sufficient dynamic excitation and therefore parameter estimation can be applied very well

Structure of a 6-Axis-Robot

The application of identification methods is in the following described for an industrial robot of type JUNGHEINRICH R106, see Fig. 24.31. The device consists of 6 revolving joints actuated by DC servomotors of high dynamic performance. The following considerations concentrate on the investigation of the mechanical subsystem of the different axes, because a strong demand for parameter estimation techniques exists for applications, such as preventive maintenance and incipient fault diagnosis (Freyermuth, 1991, 1993; Isermann and Freyermuth, 1991).

The mechanical drive chains of the axes consist of different standard machinery elements (gears, bearings, toothed belts, shafts, etc.), transferring torque from the motor to the moved (actuated) arm as shown in Fig. 24.32.

The control of each axis is performed by a cascaded control with an inner speed control of the DC motor and an outer position control of the axis joint. Figure 24.33 depicts the signal flow. The measured variables are the joint position φ, the motor speed ω, and the armature current of the DC motor I_{A}.

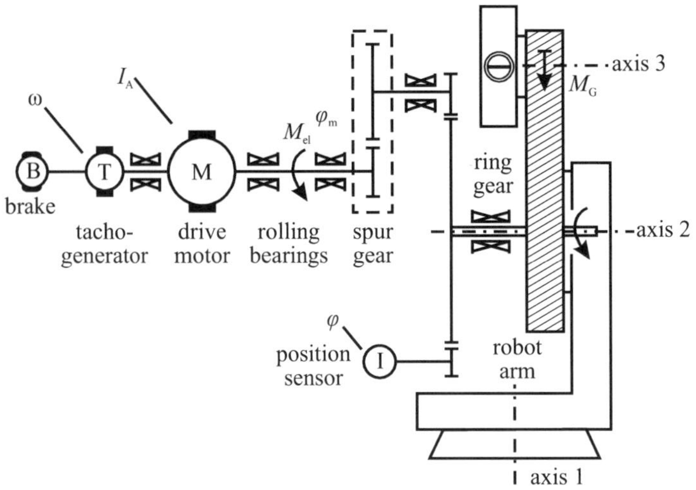

Fig. 24.32. Scheme of the mechanical drive chain (axis) of an industrial robot. Measurable quantities are φ, ω, and I_A

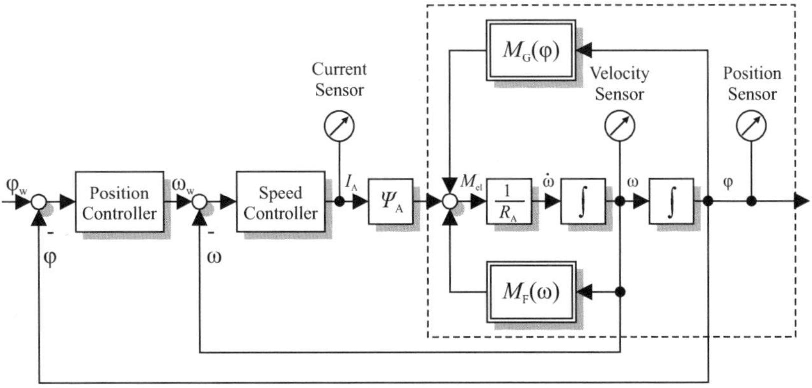

Fig. 24.33. Block diagram of the model of an industrial robot drive unit with conventional cascaded closed control

Assuming the arms as rigid bodies, each joint can be modeled by stating a torque balance related to the joint axis

$$M_{el}(t)/\nu_i = J_L(\varphi_0, m_L)\ddot{\varphi}(t) + M'_{F0}\,\mathrm{sign}\,\dot{\varphi}(t) + M'_{F1}\dot{\varphi}(t) + M'_G(m_L, \varphi_0)\,, \quad (24.2.10)$$

where

$M_{el} = \Psi_A I_A$ electrical torque at motor output axle
Ψ_A armature flux linkage

I_A	armature current
ν	total gear ratio φ/φ_m
J_L	moment of inertia of the arm (position and load dependent)
M'_{F0}	Coulomb friction torque on joint side
M'_{F1}	viscous friction torque on joint side
M'_G	gravitational torque on joint side
m_L	mass of load at end effector
φ	arm position
φ_0	arm base position
$\omega = \dot{\varphi}/\nu$	motor angular speed

The gravitation torque is modeled by

$$M_G(m_L, \varphi_0) = M'_{G0} \cos\varphi \tag{24.2.11}$$

and may be dependent on a kinematic gravitational torque compensation device, e.g. a pneumatic cylinder. The couplings between the axes can be neglected if the movements are not very fast.

Ψ_A is known from the motor data sheet. Discretizing the continuous-time model (24.2.10) with $k = t/T_0$, T_0 sampling time, and relating the parameters to the motor side by multiplying with ν leads to

$$M_{el}(k) = J(\varphi_0, m_L)\dot{\omega}(k) + M_{F0}\,\mathrm{sign}\,\omega(k) + M_{F1}\omega(k) + M_{G0}\cos\varphi(k) \tag{24.2.12}$$

($1/\nu$ is for axis $1, \ldots, 6$: 197, 197, 131, 185, 222, 194)

Then this equation results in vector notation

$$\begin{aligned} M_{el}(k) &= \boldsymbol{\psi}^T(k)\hat{\boldsymbol{\theta}}(k) + e(k) \\ \boldsymbol{\psi}^T(k) &= \big(\dot{\omega}(k)\ \mathrm{sign}\,\omega(k)\ \omega(k)\ \cos\varphi(k)\big) \\ \hat{\boldsymbol{\theta}} &= \begin{pmatrix} \hat{J} \\ \hat{M}_{F0} \\ \hat{M}_{F1} \\ \hat{M}_{G0} \end{pmatrix} \end{aligned} \tag{24.2.13}$$

and is used for recursive parameter estimation in continuous time with

$$\dot{\omega}(k) = \left.\frac{\mathrm{d}\omega(t)}{\mathrm{d}t}\right|_k = \frac{\omega(k) - \omega(k-1)}{T_0},$$

where T_0 is a small sampling time. Note that here, the estimated process parameters are identical to the physical defined process coefficients.

Figure 24.34 illustrates the typical behavior of the measured signals in the case of a point-to-point movement (PTP) of basic axis 1. The end effector did not carry any extra load. Sampling interval $T_0 = 5$ ms is identical to that of the position controller for embedding the parameter estimation software into the robot control system. Analog low-pass filtering is realized at a cut-off frequency $f_C = 40$ Hz. Digital filtering to generate the derivative $\dot{\omega}$ is performed at $f_C = 20$ Hz.

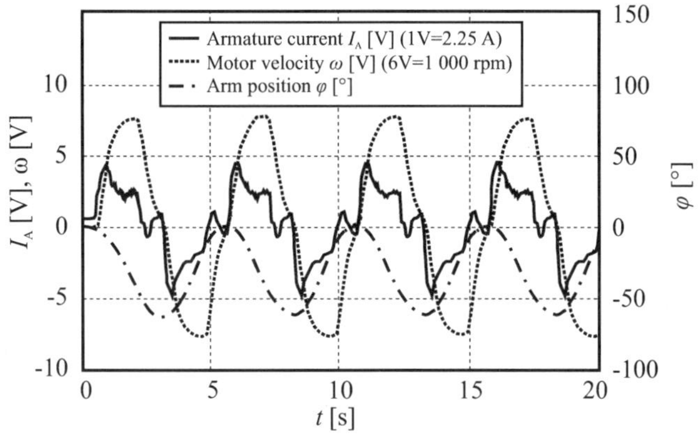

Fig. 24.34. Time history of measurements of basic axis 1 in case of a periodic point-to-point movement

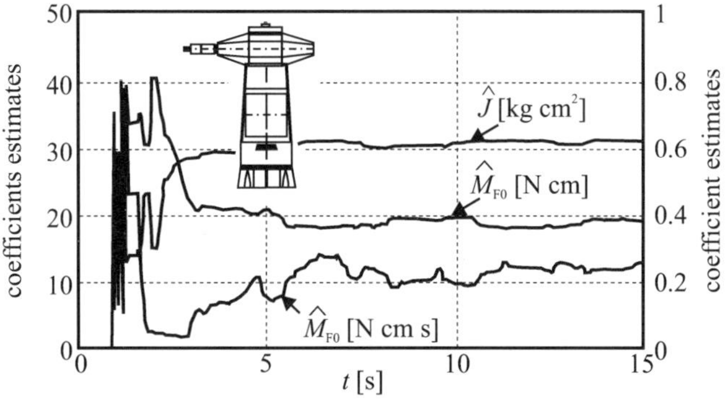

Fig. 24.35. Parameter estimates with the signals of Fig. 24.34

For parameter estimation, the DSFI procedure (discrete square root filtering in information form) is applied, because of its good numerical properties. The forgetting factor λ is set to $\lambda = 0.99$. Figure 24.35 shows the parameter estimates after starting the estimation procedure. They converge within one movement cycle to constant values.

24.2.3 Centrifugal Pumps

Pumps are basic components in most technical processes, like in power and chemical industries, mineral and mining, manufacturing, heating, air conditioning and cooling of engines. They are mostly driven by electrical motors or by combustion engines and consume a high percentage of electrical energy. One distinguishes mainly centrifugal pumps for high deliveries with lower pressures and hydrostatic or positive

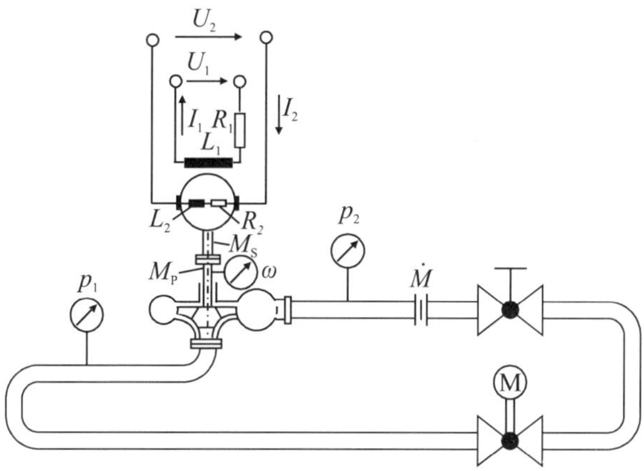

Fig. 24.36. Scheme of the speed-controlled DC motor and centrifugal pump. Closed circuit. Motor: $P_{\max} = 4\,\mathrm{kW}$, $n_{\max} = 3000\,\mathrm{rpm}$, pump: $H = 39\,\mathrm{m}$, $\dot{V}_{\max} = 160\,\mathrm{m}^3/\mathrm{h}$, $n_{\max} = 2600\,\mathrm{rpm}$. An AC motor was used for steady-state operation and a DC motor for dynamic operation.

displacement (reciprocating) pumps for high pressures and small deliveries. They transport pure liquids or mixtures of liquids and solids and herewith increase the pressure to compensate, e.g. for resistance losses or enabling thermodynamic cycles. In the following, this section will concentrate on centrifugal pumps.

In the past, centrifugal pumps were mostly driven with constant speed and the flow rate of liquids was manipulated by valves with corresponding throttling losses. Due to the availability of cheaper speed-controlled induction motors also centrifugal pumps with lower power are now used for directly controlling the flow rate in order to save energy.

The centrifugal pump considered in this example is driven by a speed-controlled DC motor and pumps water through a closed pipe circuit, see Fig. 24.36. Both, the DC motor and the pump are now considered as one unit (Geiger, 1985).

The measured signals are: U_2 armature voltage, I_2 armature current, $\dot{V}$ volume flow rate, ω angular velocity, H pump total head.

The basic equations after some simplifying assumptions are

1. armature circuit

$$L_2 \frac{\mathrm{d}I_2(t)}{\mathrm{d}t} = -R_2 I_2(t) - \Psi\omega(t) + U_2(t)\,. \tag{24.2.14}$$

2. mechanics of motor and pump

$$J_\mathrm{P} \frac{\mathrm{d}\omega}{\mathrm{d}t} = \Psi I_2(t) - M_\mathrm{f0} - \varrho g h_{\mathrm{th}1}\omega(t)\dot{V}(t)\,. \tag{24.2.15}$$

3. hydraulics of the pump (Pfleiderer and Petermann, 2005)

$$H(t) = h_\mathrm{nn}\omega^2(t) - h_\mathrm{nv}\omega(t)\dot{V}(t) - h_\mathrm{vv}\dot{V}^2(t) = h'_\mathrm{nn}\dot{V}^2(t)\,. \tag{24.2.16}$$

In this case, all three terms can again be lumped together, as $\dot{V}$ is proportional to ω.

4. hydraulics of the pipe

$$a_F \frac{d\dot{V}(t)}{dt} = -h_{rr}\dot{V}^2(t) + H(t). \tag{24.2.17}$$

The following symbols are used in this example:

L_2	armature inductance	R_2	armature resistance
U_2	armature voltage	I_2	armature current
Ψ	magnetic flux linkage	ω	rotational velocity
J_P	rotational inertia of pump	M_{f0}	dry friction torque
ϱ	density of fluid	g	gravitational constant
h_{th1}	coeff. theoretical pump head	$\dot{V}$	volume flow rate
H	delivery head	h_{nn}	coefficient of delivery head
h_{nv}	coefficient of delivery head	h_{vv}	coefficient of delivery head
a_F	tube characteristics	h_{rr}	flow friction in tube

The overall model is basically non-linear but linear in the parameters to be estimated. Therefore, least squares parameter estimation can be applied in its direct, explicit form. The models contain 9 process coefficients

$$\boldsymbol{p}^T = \left(L_2\ R_2\ \Psi\ J_P\ M_{f0}\ h_{th1}\ h'_{nn}\ a_F\ h_{rr}\right). \tag{24.2.18}$$

For the parameter estimation, the equations are brought into the form

$$y_j(t) = \boldsymbol{\Psi}_j^T(t)\hat{\boldsymbol{\theta}}_j,\ j = 1, 2, 3, 4, \tag{24.2.19}$$

where

$$\left.\begin{aligned} y_1(t) &= \frac{dI_2(t)}{dt} & y_2(t) &= \frac{d\omega(t)}{dt} \\ y_3(t) &= H(t) & y_4(t) &= \frac{d\dot{V}(t)}{dt} \end{aligned}\right\}. \tag{24.2.20}$$

The model parameters

$$\hat{\boldsymbol{\theta}}^T = \left(\hat{\boldsymbol{\theta}}_1^T\ \hat{\boldsymbol{\theta}}_2^T\ \hat{\boldsymbol{\theta}}_3^T\ \hat{\boldsymbol{\theta}}_4^T\right) \tag{24.2.21}$$

were estimated by the least squares method in the form of discrete square root filtering (DSFI). Based on the model parameter estimates $\hat{\boldsymbol{\theta}}$, all nine process coefficients of $\boldsymbol{p}$ could be calculated uniquely.

The DC motor is controlled by an AC/DC converter with cascade control of the speed and the armature current as auxiliary control variable. The manipulated variable is the armature current U_2. A microcomputer DEC-LSI 11/23 was connected online to the process. For the experiments, the reference value $\omega_S(t)$ of the speed control has been changed stepwise with a magnitude of 750 rpm every 2 min. The operating point was $n = 1\,000$ rpm, $H = 5.4$ m, and $\dot{V} = 6.48\,\text{m}^3/\text{h}$. The signals were sampled with sampling time $T_0 = 5$ ms and 20 ms over a period of 2.5 s and 10 s, so that 500 samples were obtained. These measurements were stored in the core memory before estimation. Hence, one set of parameters and process coefficients was obtained every 120 s. The results of 550 step changes are shown in Table 24.4.

Table 24.4. Estimated parameters of the pump drive

Parameter	Mean	Std. Dev.
L_2 [mH]	57.6	5.6%
R_2 [Ω]	2.20	5.0%
Ψ [Wb]	0.947	1.2%
J [10^{-3} kg m^2]	24.5	3.7%
c_{RO} [Nm]	0.694	13.5%
h_{TH1} [10^{-3} ms^2]	1.27	3.6%
h_{NN} [10^{-3} ms^2]	0.462	0.8%
a_{B} [10^3 s^2/m^2]	0.905	1.46%
h_{RR} [10^6 s^2/m^5]	1.46	3.9%

24.2.4 Heat Exchangers

Heat exchangers are a typical apparatus in the fields of power and chemical engineering, heating, cooling, refrigeration and air conditioning, and are part of all kind of machines and engines. Their task is to transport heat between two or more media, such as e.g. liquids or gases.

Heat Exchanger Types

A large variety of types exists to meet the specific requirements with regard to temperatures, pressures, phase changes, corrosion, efficiency, weight, spacc, and connections. Frequently used types are

- tubular heat exchangers
- plate heat exchangers.

With regard to the flow direction, one distinguishes counter flow, parallel flow and cross flow. The fluids are liquids, gases, or steam, resulting in two media with the combinations:

- liquid - liquid
- gas - liquid
- liquid - steam (condensator, evaporator)
- gas - steam

Steam/Water Heat Exchanger

An industrial size steam-heated heat exchanger, see Fig. 24.37, is considered as an example, which is part of a pilot plant (W. Goedecke, 1987; Isermann and Freyermuth, 1991). This plant consists of an electrically powered steam generator, a steam/condensate circulation (circuit 1), a water circulation (circuit 2), and a cross-flow heat exchanger to transport the heat from water to air.

As inputs and outputs of the considered heat exchanger, the following variables are measured:

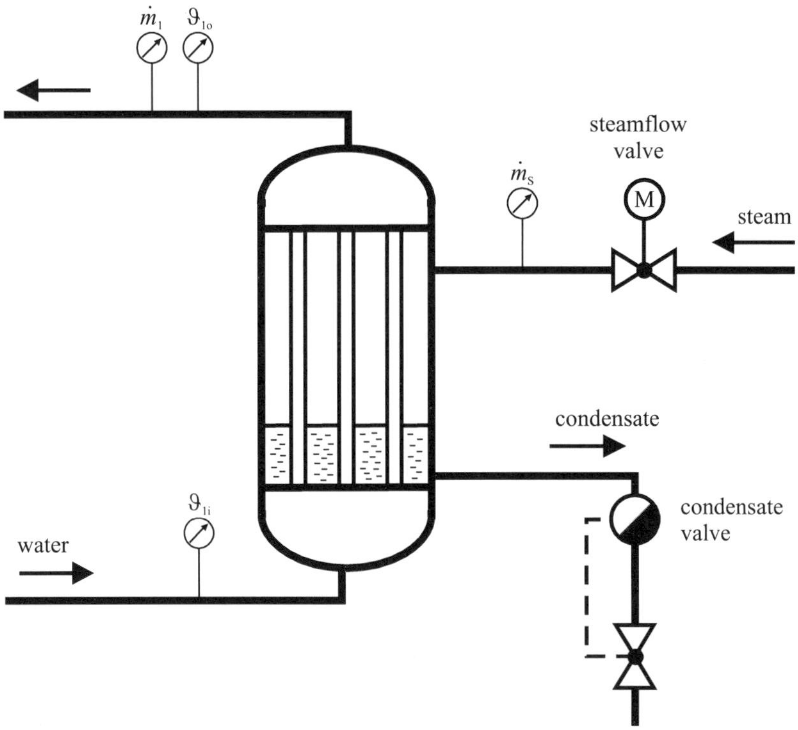

Fig. 24.37. Tubular heat exchanger and measured variables

$\dot{m}_s$	mass flow of the steam
$\dot{m}_1$	mass flow of the liquid fluid (water)
ϑ_{1i}	inlet temperature of the liquid fluid
ϑ_{1o}	outlet temperature of the liquid fluid

The fluid outlet temperature ϑ_{1o} is considered as an output variable, the other three measured variables as input variables.

Linear Models and Parameter Estimation

To model the dynamic behavior, the heat exchanger is subdivided into the tubular section, the water head, a transport delay, and the temperature sensor. The dynamic equations for a heated tube as a distributed parameter process are given e.g. in (Isermann, 2010). In addition, balance equations are stated for the steam space and the shell tube. After approximation of the transcendent transfer function with those for lumped parameters and linearization around the operating point, one obtains for example the approximate transfer function

$$\tilde{G}_{s\vartheta}(s) = \frac{\Delta\vartheta_{1o}(s)}{\Delta\dot{m}_s(s)} = \frac{K_s}{(1+T_{1s}s)(1+T_{2s}s)}e^{-T_{ds}s} \tag{24.2.22}$$

with

$$\left.\begin{aligned} K_{\mathrm{s}} &= \frac{r}{\dot{m}_1 c_1},\; T_{1\mathrm{s}} = \frac{1}{v_1}\left(1+\frac{A_{\mathrm{w}}\varrho_{\mathrm{w}}c_{\mathrm{w}}}{A_1\varrho_1 c_1}\right) \\ T_{2\mathrm{s}} &= \frac{A_{\mathrm{w}}\varrho_{\mathrm{w}}c_{\mathrm{w}}}{\alpha_{\mathrm{w}1}U_1}\frac{1}{\left(1+\frac{A_{\mathrm{w}}\varrho_{\mathrm{w}}c_{\mathrm{w}}}{A_1\varrho_1 c_1}\right)} \end{aligned}\right\} \tag{24.2.23}$$

With the quantities being given as

A	area	c_p	specific heat capacity
$\dot{m}$	mass flow rate	r	vaporization heat
α	heat transfer coefficient	ϑ	temperature
ϱ	density		

and the subscripts

1	primary side of heat exchanger	2	secondary side of heat exchanger
w	wall	s	steam
i	inlet	o	outlet.

In this case, three parameter estimates compare to 10 process coefficients. Therefore, it is not possible to determine all process coefficients uniquely. By assuming some of the process coefficients to be known, however, the following process coefficients and combinations of process coefficients can be determined:

$$\left.\begin{aligned} \alpha_{\mathrm{w}1} &= \frac{A_1\varrho_1 c_1}{T_{2\mathrm{s}}U_1}\left(1-\frac{1}{T_{1\mathrm{s}}v_1}\right) \\ A_{\mathrm{w}}\varrho_{\mathrm{w}}c_{\mathrm{w}} &= T_{1\mathrm{s}}\dot{m}_1 c_1 - A_1\varrho_1 c_1 \\ r &= K_{\mathrm{s}}\dot{m}_1 c_1 \end{aligned}\right\} . \tag{24.2.24}$$

The three parameters $\hat{K}_{\mathrm{s}}$, $\hat{T}_{1\mathrm{s}}$ and $\hat{T}_{2\mathrm{s}}$ are determined by experiments based on transient function measurements of the fluid outlet temperature $\vartheta_{1\mathrm{o}}$ due to changes of the input variables $\vartheta_{1\mathrm{i}}$, $\dot{m}_{\mathrm{s}}$ and $\dot{m}_1$ in the direction of decreasing temperature ϑ_{so}. The operating point was

$$\dot{m}_1 = 3000\,\frac{\mathrm{kg}}{\mathrm{h}},\; \dot{m}_{\mathrm{s}} = 50\frac{\mathrm{kg}}{\mathrm{h}},\; \vartheta_{1\mathrm{i}} = 60°\mathrm{C}, \vartheta_{1\mathrm{o}} \approx 70°\mathrm{C}\;.$$

As sampling time, $T_0 = 500\,\mathrm{ms}$ was selected. The time period of one experiment was 360 s, so that 720 samples were taken. For the parameter estimation, the method of total least squares in a recursive form was applied using a digital state variable filter for the determination of the derivatives.

Figure 24.38a shows one measured transient function and Fig. 24.38b the corresponding time history of the parameter estimates. A good convergence of the parameter estimates was obtained in all cases. A verification of the measured and the calculated transient functions shows a very good agreement.

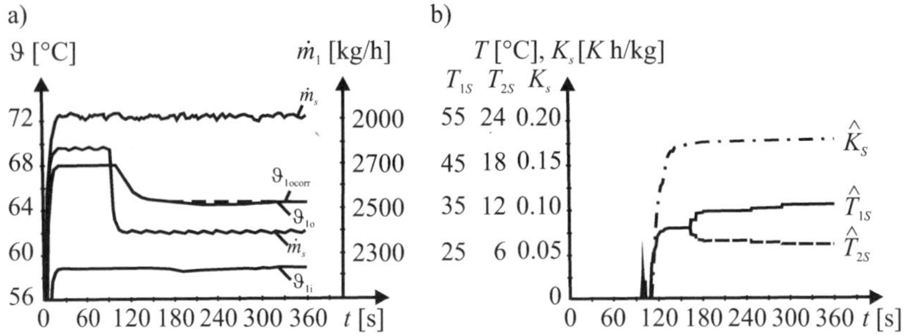

Fig. 24.38. Results for a change in the steam flow $\Delta\dot{m}_S$: (**a**) Measured transient functions for a steam flow change, (**b**) parameter estimates from transient function

Parameter Variable Local Linear Models

As the behavior of heat exchangers depends strongly on the flow-rates, the static and dynamic behavior is non-linear for changing flow-rates. In order to develop models which are applicable over a large operating range, local linear neuronal netmodels of the type LOLIMOT were used to describe first the nominal behavior. This was applied to the steam/water heat exchanger used also for Sect. 24.2.4. By using the LOLIMOT identification method, dynamic models of the water outlet temperature ϑ_{1o} in dependence on the water volume flow $\dot{V}_1$, steam mass flow $\dot{m}_s$, and inlet temperature ϑ_{1i} were determined by simultaneous wide range excitation of the two flows with amplitude modulated PRBS signals (Ballé, 1998).

This resulted in 10 local linear models in dependence on the water flow. Using a sample time of $T_0 = 1$ s, a second order dynamic model was sufficient

$$\begin{aligned}\vartheta_{1o}(k) &= -a_1(z)\vartheta_{1o}(k-1) - a_2(z)\vartheta_{1o}(k-1)\\ &\quad + b_{11}(z)\dot{m}_s(k-1) + b_{12}(z)\dot{m}_s(k-2)\\ &\quad + b_{21}(z)\dot{V}_1(k-1) + b_{31}(z)\vartheta_{1i}(k-1) + c_0(z)\,,\end{aligned} \tag{24.2.25}$$

where the parameters depend on the operating point $z = \dot{V}_1$

$$a_\nu(\dot{V}_1) = \sum_{j=1}^{10} a_\nu \Phi_j(\dot{V}),\ b_{\nu\mu}(\dot{V}_1) = \sum_{j=1}^{10} b_{\nu\mu}(\dot{V}),\ c_0(\dot{V}_1) = \sum_{j=1}^{10} c_0\Phi_j(\dot{V})\,, \tag{24.2.26}$$

where Φ_j is the weighting function within LOLIMOT.

Figure 24.39 shows the resulting stationary outlet temperature in dependence on the two flows. The identified models then allowed to extract three gains and one dominant time constant, partially depicted in Fig. 24.40. The operating point dependence is especially strong for low water flow rates. Static gains and the time constant change with about a factor four.

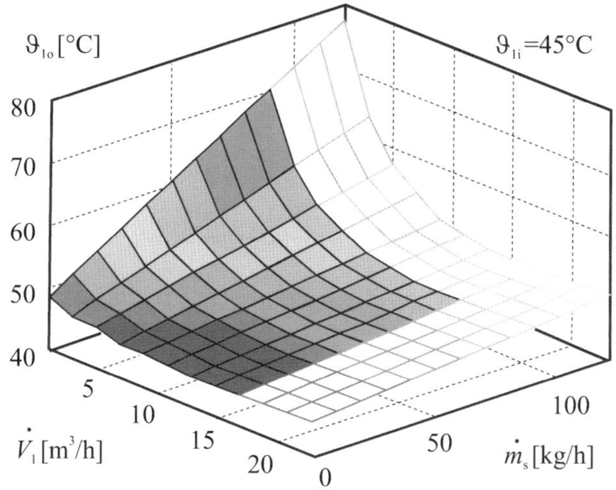

Fig. 24.39. Heat exchanger water outlet temperature static map in dependence on water and steam flow

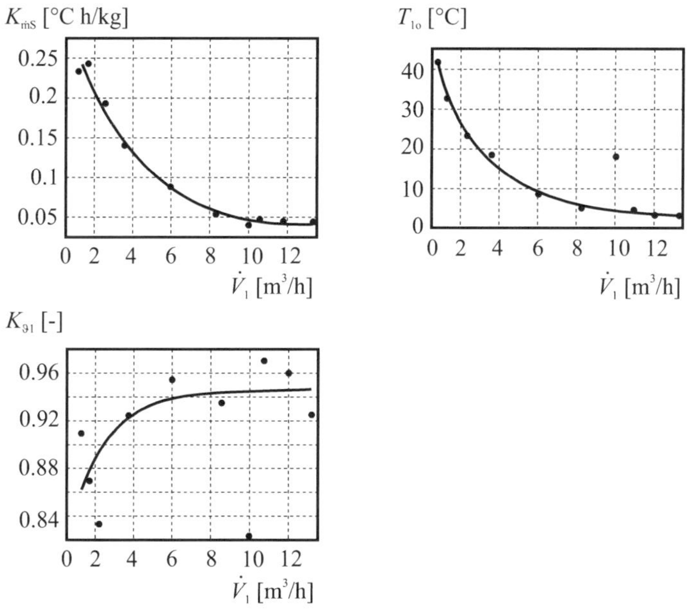

Fig. 24.40. Static gains and time constant for the water outlet temperature in dependence on the water flow rate

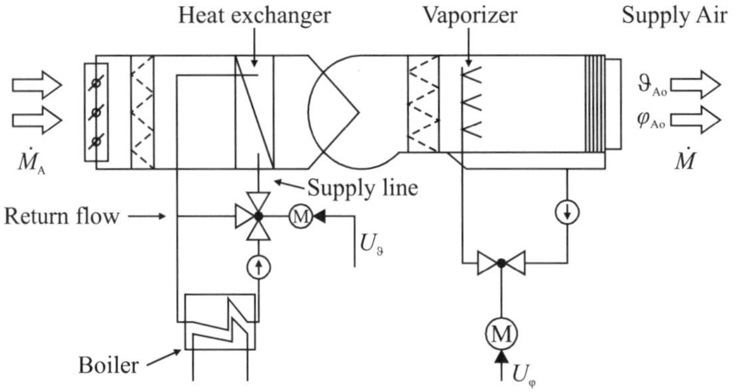

Fig. 24.41. Schematic view of an air condition unit (air mass flow $M_{A,max} = 500\,m^3/s$)

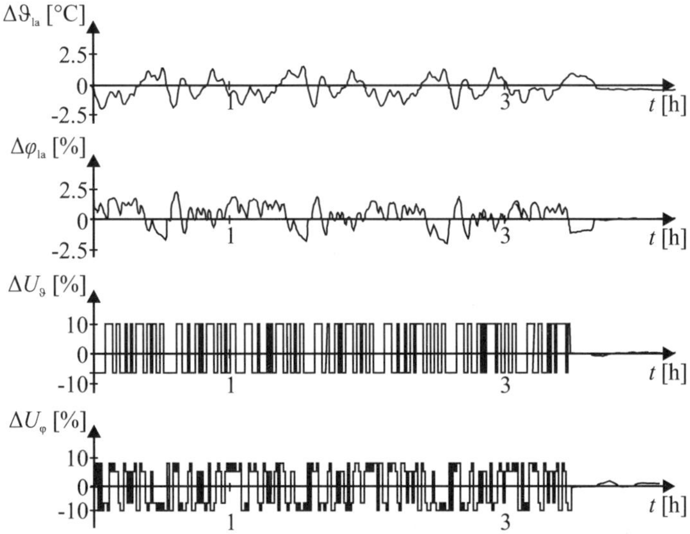

Fig. 24.42. Measurements at an air condition unit. Operating point: $\vartheta_{Ao} = 30°C$, $\varphi_{Aa} = 35°C$, $T_0 = 1\,min$. Test signal amplitudes: $u_1 = 1\,V$ and $u_2 = 0.8\,V$

24.2.5 Air Conditioning

The air condition system that is considered in this section consists of a heater and a humidifier, see Fig. 24.41. By means of the hot water supply, the air temperature after the cross-flow heat exchanger is controlled. Inside the humidifier, the humidity is controlled by means of the water spray flow. Figure 24.42 shows the measurements that have been taken at the process. Since it was a linear MIMO process, a PRBS for the first input was combined with an orthogonal PRMS in order not to apply correlated input signals. The clock time of the base PRBS was $\lambda = 1$, the measurement time was $T_M = 195\,\text{s}$. As a model structure, a simplified P-canonical model was chosen. The model order and dead time estimation by employing a determinant test provided $\hat{m}_1 = 2$, $\hat{d}_{11} = 0$, $\hat{d}_{12} = 0$, and $\hat{m} = 1$, $\hat{d}_{21} = 0$, $\hat{d}_{22} = 0$. By means of COR-LS, the following models (Hensel, 1987) were then identified

$$\hat{G}_{11}(z) = \frac{\Delta\vartheta_{\text{Ao}}}{\Delta U_\vartheta(z)} = \frac{0.0509z^{-1} + 0.0603z^{-2}}{1 - 0.8333z^{-1} + 0.1493z^{-2}}$$
$$\hat{G}_{21}(z) = \frac{\Delta\vartheta_{\text{Ao}}}{\Delta U_\varphi(z)} = \frac{-0.0672z^{-1} - 0.0136z^{-2}}{1 - 0.8333z^{-1} + 0.1493z^{-2}}$$
$$\hat{G}_{22}(z) = \frac{\Delta\varphi_{\text{Ao}}}{\Delta U_\varphi(z)} = \frac{0.2319z^{-1}}{1 - 0.3069z^{-1}}$$
$$\hat{G}_{11}(z) = \frac{\Delta\vartheta_{\text{Ao}}}{\Delta U_\vartheta(z)} = \frac{0.0107z^{-1}}{1 - 0.3069z^{-1}} \,.$$

The estimated static gains result as $\hat{K}_{11} = 0.3520$, $\hat{K}_{12} = -0.2557$, $\hat{K}_{22} = 0.3345$, and $\hat{K}_{12} = 0.0154$. The coupling from $\Delta U_\vartheta(z)$ to $\Delta\varphi_{Ao}(z)$ is hence negligible.

24.2.6 Rotary Dryer

A rotary dryer for sugar beets has been identified as described in (Mann, 1980) and (Isermann, 1987). The input to the process is the amount of fuel supplied, $\dot{m}_F$, and the output is the amount of dry matter, ψ_{DM}. The process shows large disturbances, hence the identification results that had been obtained in the short measurement time are not optimal. Figure 24.43 however shows that the measured output and model output match quite well for longer identification times of roughly 6 hours. The difference in the DC values can be explained by the large disturbances. The process model, which was identified using COR-LS, is given as

$$G(z) = \frac{-1.15z^{-1} + 1.52z^{-2} - 0.54z^{-3} + 0.27z^{-4} + 0.27z^{-5}}{1 - 2.01z^{-1} + 1.27z^{-2} - 0.24z^{-3} + 0.07z^{-4} - 0.07z^{-5}} z^{-2} \,, \tag{24.2.27}$$

see also Fig. 24.45. One can see from the calculated step response in Fig. 24.44 that the resulting model has an all-pass behavior and contains a dead time.

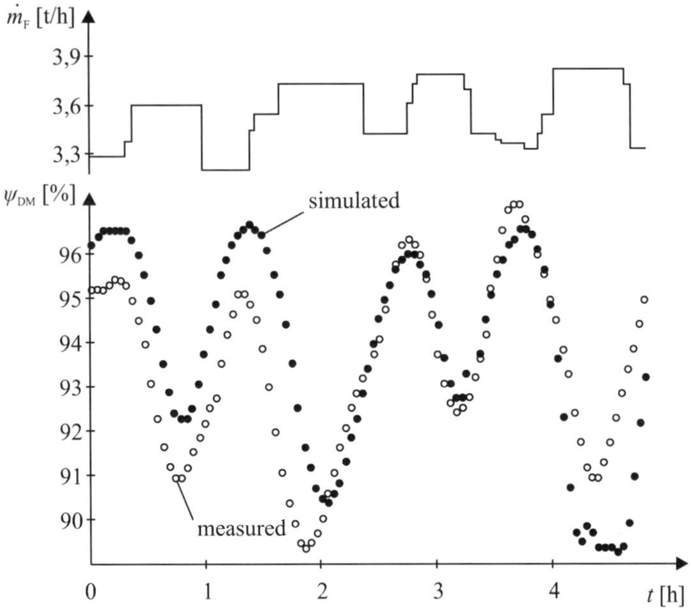

Fig. 24.43. Signals for the identification of a rotary dryer

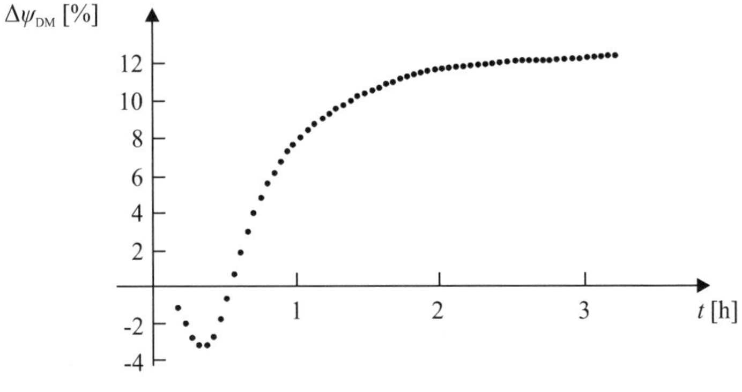

Fig. 24.44. Step response for a change in the manipulated variable fuel mass of the rotary dryer calculated from the identified model

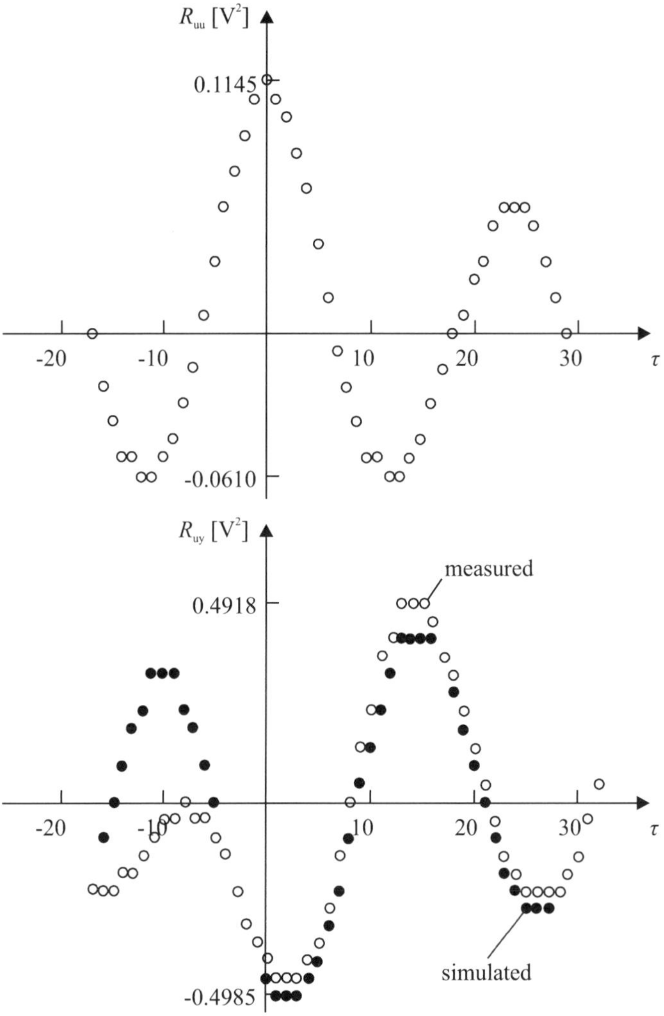

Fig. 24.45. Auto-correlation and cross-correlation function of the rotary dryer

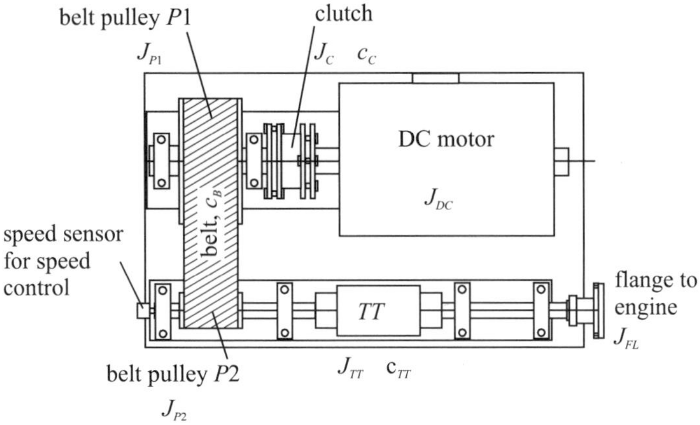

Fig. 24.46. Arrangement of the engine test stand

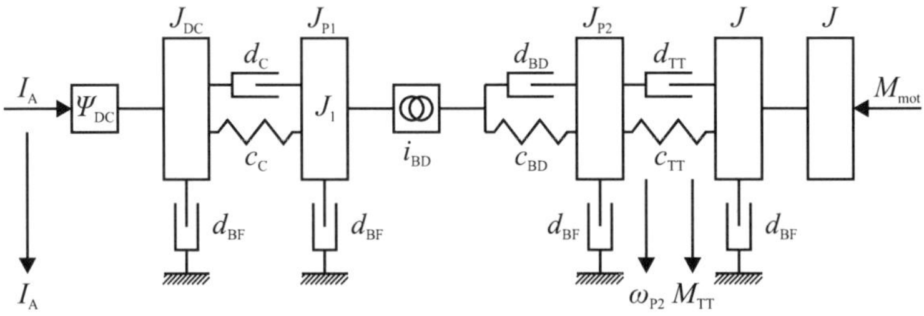

Fig. 24.47. Schematic of the rotational dynamics of the engine test stand

24.2.7 Engine Teststand

Dynamic models of a combustion engine testbench are required for designing the torque or speed control. Figure 24.46 shows the setup. The testbed consists of a DC motor, a spring shackle clutch, a belt transmission, a torque transducer, and a flange to connect the combustion engine to the testbed. The dynamometer applies a dynamic load torque to the combustion engine, which can be shaped to correspond to a certain driving cycle. To design a suitable controller, a precise linear model of the dynamics was required. The input signal of the process is the armature current of the DC motor I_A, the output is the torque at the torque transducer M_{TT} (Voigt, 1991; Pfeiffer, 1997).

Figure 24.47 shows a schematic diagram of the engine testbench with five rotational inertias, which are coupled by spring/damper combinations. The friction in the roller bearings is modeled as viscous friction. The linear behavior can be described by a linear state space model as

$$\dot{\boldsymbol{x}}(t) = \boldsymbol{A}\boldsymbol{x}(t) + \boldsymbol{b}u(t) + \boldsymbol{g}n(t) \tag{24.2.28}$$

$$\boldsymbol{y}(t) = \boldsymbol{C}\boldsymbol{x}(t) \tag{24.2.29}$$

with the input $u(t)$ being the armature current $I_A(t)$, the disturbance $n(t)$ being the torque exerted by the engine $M_{mot}(t)$, and

$$\boldsymbol{x}(t)^T = \big(\omega_{FL}\ \Delta\varphi_{TT}(t)\ \omega_{P2}(t)\ \Delta\varphi_{BD}(t)\ \omega_{P1}(t)\ \Delta\varphi_{C}(t)\ \omega_{DC}(t)\big) \tag{24.2.30}$$

$$\boldsymbol{y}(t) = \big(M_{TT}(t)\ \omega_{P2}(t)\big)\ . \tag{24.2.31}$$

The model parameters are given as

$$\boldsymbol{A} = \begin{pmatrix} -\frac{d_{TT}+d_{BF}}{J_{FL}+J_{mot}} & \frac{c_{TT}}{J_{FL}+J_{mot}} & \frac{d_{TT}}{J_{FL}+J_{mot}} & 0 & 0 & 0 & 0 \\ -1 & 0 & 1 & 0 & 0 & 0 & 0 \\ \frac{d_{TT}}{J_{P2}} & -\frac{c_{TT}}{J_{P2}} & -\frac{d_{BD}+d_{TT}+d_{BF}}{J_{P2}} & \frac{c_{BD}}{J_{P2}} & \frac{i_{BD}d_{BD}}{J_{P2}} & 0 & 0 \\ 0 & 0 & -1 & 0 & i_{BD} & 0 & 0 \\ 0 & 0 & \frac{i_{BD}d_{BD}}{J_{P1}} & -\frac{i_{BD}c_{BD}}{J_{P1}} & \frac{i_{BD}^2 d_{BD}+d_C+d_{BF}}{J_{P1}} & \frac{c_C}{J_{P1}} & \frac{d_C}{J_{P1}} \\ 0 & 0 & 0 & 0 & -1 & 0 & 1 \\ 0 & 0 & 0 & 0\,\frac{d_C}{J_{DC}} & -\frac{c_C}{J_{DC}} & \frac{d_C+d_{BF}}{J_{DC}} & \end{pmatrix} \tag{24.2.32}$$

$$\boldsymbol{b}^T = \left(0\,0\,0\,0\,0\,0\ \frac{\Psi_{DC}}{J_{DC}}\right) \tag{24.2.33}$$

$$\boldsymbol{C} = \begin{pmatrix} 0 & c_{TT} & 0\,0\,0\,0\,0 \\ 0 & 0 & 1\,0\,0\,0\,0 \end{pmatrix}\ . \tag{24.2.34}$$

The transfer function matrix can then be determined by

$$\boldsymbol{G}(s) = \boldsymbol{C}\big(s\boldsymbol{I} - \boldsymbol{A}\big)^{-1}\boldsymbol{b}\ . \tag{24.2.35}$$

Considering $I_A(t)$ as an input and $M_{TT}(t)$ as an output, the model is of seventh order

$$G_{IT}(s) = \frac{b_0 + b_1 s}{a_0 + a_1 s + a_2 s^2 + a_3 s^3 + a_4 s^4 + a_5 s^5 + a_6 s^6 + a_7 s^7}\ . \tag{24.2.36}$$

The parameters of the transfer function depend on the physical parameters of the system. Here, the quantities are given as

Ψ_{DC}	magnetic flux linkage DC	J	rotational moment of inertia
c	spring stiffness	d	damping coefficient
ω	rotational velocity	φ	angle
M	torque	I	current

and the indices denote

BF	bearing friction	P2	belt pulley P2
DC	DC motor	TT	torque transducer
C	clutch	FL	flange
P1	belt pulley P1	mot	test engine
BD	belt drive.		

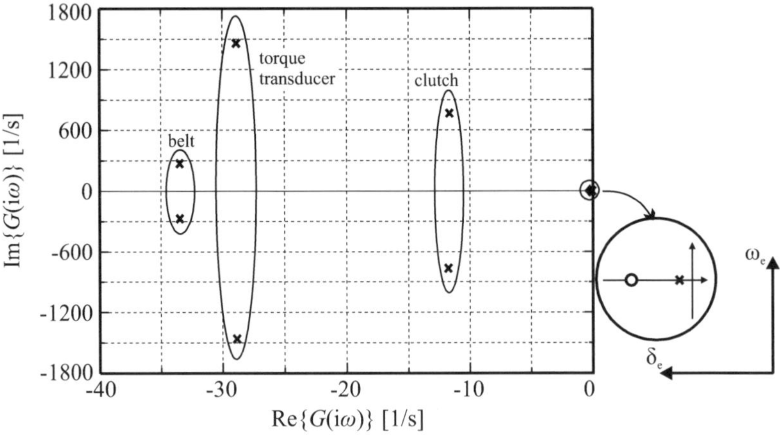

Fig. 24.48. Poles and zeros of the transfer function of the engine test stand

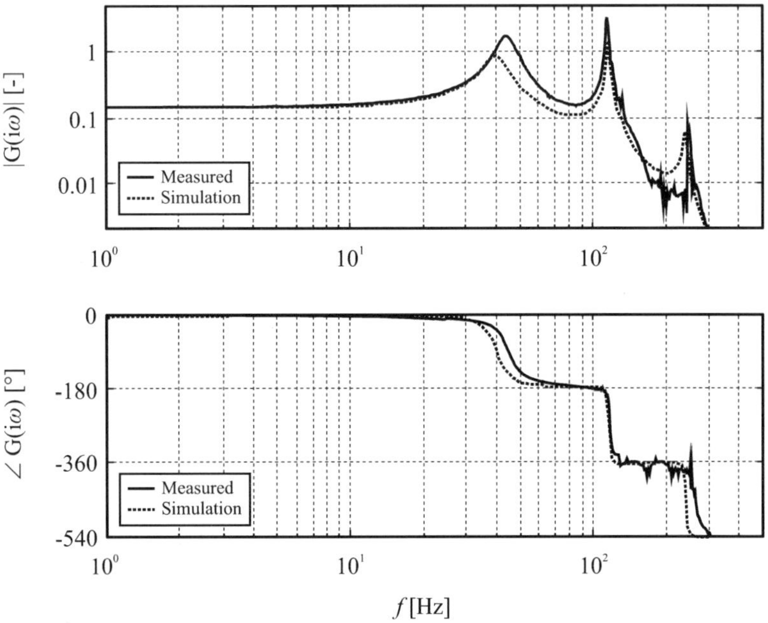

Fig. 24.49. Measured frequency response of the mechanical part of the test stand for current $I_A(t)$ as input and the torque $M_{TT}(t)$ as output

Figure 24.48 shows the poles and zeros of the transfer function. Three pairs of conjugate complex poles are obtained, as well as one real pole and one real zero. Hence, the following natural angular eigenfrequencies can be distinguished $\omega_{\mathrm{e,C}}$ for the clutch, $\omega_{\mathrm{e,TT}}$ for the torque transducer, and $\omega_{\mathrm{e,B}}$ for the belt. Considering the not-coupled elements as undamped second order systems, the characteristic frequencies can be determined as

$$f_{0,\mathrm{C}} = \frac{1}{2\pi}\sqrt{\frac{c_{\mathrm{C}}(J_{\mathrm{DC}}+J_{\mathrm{P1}})}{J_{\mathrm{DC}}J_{\mathrm{P1}}}} = 154.7\,\mathrm{Hz}\ \text{(clutch)} \tag{24.2.37}$$

$$f_{0,\mathrm{BD}} = \frac{1}{2\pi}\sqrt{\frac{c_{\mathrm{BD}}(J_{\mathrm{DC}}+J_{\mathrm{P1}})+i_{\mathrm{BD}}^2(J_{\mathrm{P2}}+J_{\mathrm{FL}})}{(J_{\mathrm{DC}}J_{\mathrm{P1}})(J_{\mathrm{P2}}+J_{\mathrm{FL}})}} = 34.5\,\mathrm{Hz}\ \text{(belt)} \tag{24.2.38}$$

$$f_{0,\mathrm{TT}} = \frac{1}{2\pi}\sqrt{\frac{c_{\mathrm{TT}}(J_{\mathrm{P2}}+J_{\mathrm{FL}})}{J_{\mathrm{P2}}J_{\mathrm{FL}}}} = 154.7\,\mathrm{Hz}\ \text{(clutch)} \tag{24.2.39}$$

Figure 24.49 shows the measured frequency response of the system without an engine attached. The three resonance frequencies of the belt (approximately 45 Hz), clutch (120 Hz), and torque transducer (250 Hz) can clearly be recognized. The results show a good agreement between model and measurement (Isermann et al, 1992; Pfeiffer, 1997). The obtained model allowed the design of a digital torque controller with the connected combustion engine, allowing the compensation of the testbed dynamics with accurate powertrain models up to frequencies of 12 Hz.

24.3 Automotive Vehicles

Automotive vehicles are another interesting area of application, where the use of experimental modeling is of great benefit. While the vehicle dynamics can very well be modeled by a one-track or a two-track model, it is very hard to derive the model parameters analytically. Furthermore, many parameters, describing e.g. the mass distribution and the wheel-road-friction, vary over time as e.g. the load changes or the road surface is dry or wet or icy depending on the weather. Therefore, advanced vehicle dynamics systems have to be adapted to the varying parameters of the vehicle.

Further examples presented in this section include the estimation of wheel suspension system and tire parameters. Both of these are safety critical systems and hence knowledge about the state of these components is of great benefit in terms of supervision of these components. Another safety critical system is the braking system, which will also be modeled and identified in this section.

Finally, combustion engines will be treated. With more and more strict emission requirements, the amount of actuators at the combustion engine increases steadily, making these system true (nonlinear) MIMO systems.

24.3.1 Estimation of Vehicle Parameters

In this section, the estimation of vehicle parameters shall be described. The estimation of vehicle parameters is of interest for advanced vehicle dynamic control

Fig. 24.50. Coordinate system of a road vehicle (Halbe, 2008; Schorn, 2007)

systems, which can employ such an automatically adapted model for model-based control algorithms.

A simple yet sufficiently precise model is the one-track model, (Milliken and Milliken, 1997), which can describe the dynamic behavior of a car up to lateral accelerations of roughly $a_Y \leq 0.4g$. For the one-track model, in contrast to the two-track model, both wheels of each axle are combined into a single wheel at the center of the axle. Furthermore, it is assumed that the center of gravity of the car is on the road surface, i.e. the car does not roll. The force buildup between the tire surface and the road is also assumed to be linearly dependent on the slip angle. Due to these simplifying assumptions, the one-track model is not always of sufficient fidelity, but it can be used in most situations that a normal driver experiences.

Figure 24.50 shows the coordinate system of a passenger car and shall be used to explain the symbols that will be used in the following: x, y, and z describe the three lateral degree of freedom of the car, whereas ϕ denotes the roll, ψ denotes the yaw, and θ denotes the pitch angle.

The characteristics of a tire governing the tire-road surface interaction are shown in Fig. 24.51. Here, one can see that the cornering friction force depends non-linearly on the slip angle, but for small side slip angles can be assumed to be a linear function of the side slip angle.

The one-track model, see Fig. 24.52, can now be derived based on the force and momentum balance of the vehicle,

$$\underbrace{m\frac{v^2}{R}\sin\beta - m\dot{v}\cos\beta}_{\text{D'Alembert inertial forces}} + \underbrace{F_{\text{xr}} + F_{\text{xf}}\cos\delta - F_{\text{yf}}\sin\delta}_{\text{tire forces}} = 0 \tag{24.3.1}$$

$$\underbrace{-m\frac{v^2}{R}\cos\beta - m\dot{v}\sin\beta}_{\text{D'Alembert inertial forces}} + \underbrace{F_{\text{yr}} + F_{\text{xf}}\sin\delta - F_{\text{yf}}\cos\delta}_{\text{tire forces}} = 0 \tag{24.3.2}$$

$$-J_{\text{z}}\ddot{\psi} + \big(F_{\text{yf}}\cos\delta + F_{\text{xf}}\sin\delta\big)l_{\text{f}} - F_{\text{yh}}l_{\text{h}} = 0\,. \tag{24.3.3}$$

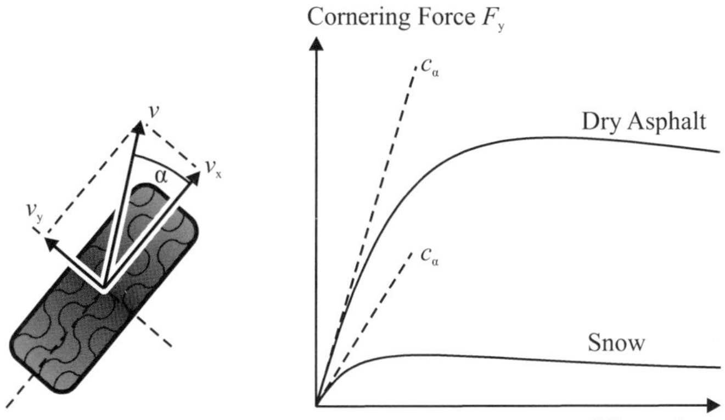

Fig. 24.51. Definition of slip angle and tire characteristics (Halbe, 2008; Schorn, 2007)

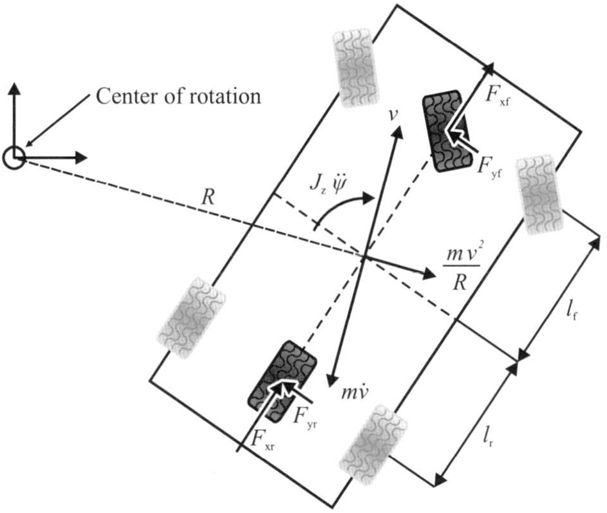

Fig. 24.52. Schematic of the one-track model of a vehicle (Halbe, 2008; Schorn, 2007)

Also, one has the following kinematic relations:

$$\alpha_\mathrm{r} = \arctan\left(\frac{v_\mathrm{yr}}{v_\mathrm{xr}}\right) = \arctan\left(\frac{l_\mathrm{r}\dot{\psi} - v\sin\beta}{v\cos\beta}\right) \approx \left(\frac{l_\mathrm{r}\dot{\psi} - v\sin\beta}{v\cos\beta}\right) \quad (24.3.4)$$

$$\alpha_\mathrm{f} = \delta - \arctan\left(\frac{l_\mathrm{f}\dot{\psi} + v\sin\beta}{v\cos\beta}\right) \approx \delta - \left(\frac{l_\mathrm{f}\dot{\psi} + v\sin\beta}{v\cos\beta}\right), \quad (24.3.5)$$

which provide the slip angle of the front and rear wheel. From there, one can determine the lateral tire forces as

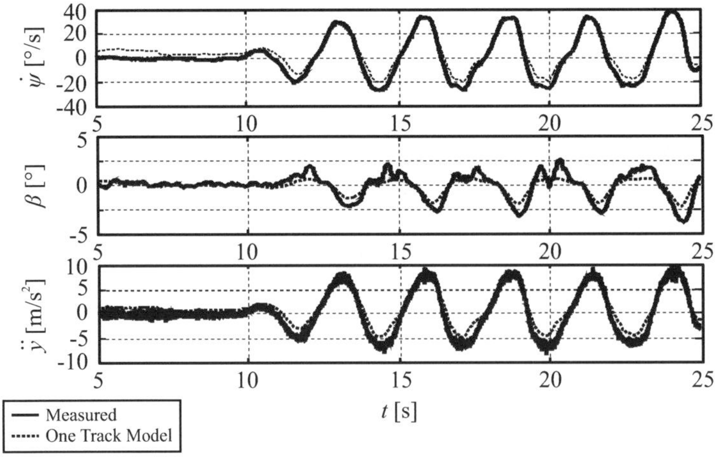

Fig. 24.53. Driving maneuver with a parameterized single track model (Halbe, 2008; Schorn, 2007)

$$f_{\mathrm{yr}} = c_{\alpha\mathrm{r}}\alpha_{\mathrm{r}} \tag{24.3.6}$$

$$f_{\mathrm{yf}} = c_{\alpha\mathrm{f}}\alpha_{\mathrm{f}}\,. \tag{24.3.7}$$

Finally, the one track model can be written in state space form as

$$\begin{pmatrix} \ddot{\psi} \\ \dot{\beta} \end{pmatrix} = \begin{pmatrix} -\dfrac{c_{\alpha\mathrm{f}}l_{\mathrm{f}}^2 + c_{\alpha\mathrm{r}}l_{\mathrm{r}}^2}{J_{\mathrm{z}}v} & -\dfrac{c_{\alpha\mathrm{f}}l_{\mathrm{f}} + c_{\alpha\mathrm{r}}l_{\mathrm{r}}}{J_{\mathrm{z}}} \\ \dfrac{-c_{\alpha\mathrm{f}}l_{\mathrm{f}} - mv^2 + c_{\alpha\mathrm{r}}l_{\mathrm{r}}}{mv^2} & -\dfrac{c_{\alpha\mathrm{f}}l_{\mathrm{f}} + m\dot{v} + c_{\alpha\mathrm{r}}}{mv} \end{pmatrix} \begin{pmatrix} \dot{\psi} \\ \beta \end{pmatrix} + \begin{pmatrix} \dfrac{c_{\alpha\mathrm{f}}l_{\mathrm{f}} + c_{\alpha\mathrm{r}}l_{\mathrm{r}}}{J_{\mathrm{z}}i_{\mathrm{S}}} \\ \dfrac{c_{\alpha\mathrm{f}}}{mvi_{\mathrm{S}}} \end{pmatrix} \delta_{\mathrm{H}}\,. \tag{24.3.8}$$

In these equations, the following symbols are used:

m	mass	v	velocity
R	instantaneous radius	β	slip angle
δ	steering angle	J	moment of inertia
l	length	α	side slip angle
c_α	cornering stiffness		

with the indices

fl	front left	fr	front right
rl	rear left	rr	rear right.

One can now use test drives to estimate the parameters of the one track model. The results for a one track model that has been parameterized by a test drive is shown in Fig. 24.53, where one can see the good match between the model output and the measurement. For more details see (Wesemeier and Isermann, 2007; Halbe, 2008; Schorn, 2007).

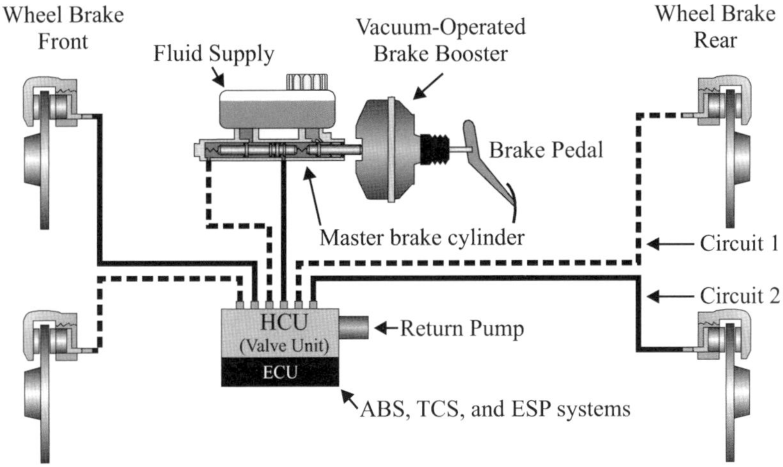

Fig. 24.54. Schematic of a hydraulic vehicle braking system

24.3.2 Braking Systems

This section is concerned with the modeling and identification of components of a hydraulic passenger car braking systems. A schematic view of such a system is shown in Fig. 24.54. The foot force exerted by the driver is amplified by means of a vacuum operated brake booster and is then transmitted to the master brake cylinder, which is typically mounted directly at the vacuum brake booster. The master brake cylinder is divided into two separate chambers by the intermediate piston. From the master brake cylinder, two separate lines run to the hydraulic control unit, which houses all valves for the anti-lock braking system, the traction control system, etc. Four lines connect this block to the four wheel brake cylinders. In the wheel brake cylinders, brake pads can be pushed against the brake disk in order to generate a friction torque which in turn will slow the vehicle down. More detailed information can be found in (Robert Bosch GmbH, 2007; Burckhardt, 1991; Breuer and Bill, 2006).

Hydraulic subsystem

In the following, a detailed model of the hydraulic subsystem of the braking system used at a passenger car braking system testbed (Fig. 24.55) will be derived. This model will be formulated as a non-linear state-space model, with the dynamics of the system being governed by the set of first order non-linear differential equations,

$$\dot{\boldsymbol{x}} = \boldsymbol{a}(\boldsymbol{x}) + \boldsymbol{B}\boldsymbol{u} \tag{24.3.9}$$

and the output given as

$$y = \boldsymbol{c}^{\mathrm{T}}\boldsymbol{x} + \boldsymbol{d}^{\mathrm{T}}\boldsymbol{u} \,. \tag{24.3.10}$$

Fig. 24.55. View of the braking system testbed at IAT, TU Darmstadt (Straky, 2003)

Here and in the remainder of this section, the following symbols are used:

V	volume
$\dot{V}$	volume flow rate
R_{T}	turbulent flow resistance
R_{L}	laminar flow resistance
L	inertia of hydraulic fluid
C	capacity of wheel brake cylinder chamber
p	pressure

and the indices denote

fl	front left
fr	front right
rl	rear left
rr	rear right
I	chamber 1 of master brake cylinder
II	chamber 2 of master brake cylinder
wbc	wheel brake cylinder

Upon operation of the brake pedal, the master brake cylinder will push a certain amount of brake fluid into the individual wheel brake cylinders. This amount of fluid displaced along with the time rate of change of this displacement will be chosen as the states of the system,

$$\boldsymbol{x} = \left(V_{\mathrm{fl}}\ \dot{V}_{\mathrm{fl}}\ V_{\mathrm{fr}}\ \dot{V}_{\mathrm{fr}}\ V_{\mathrm{rl}}\ \dot{V}_{\mathrm{rl}}\ V_{\mathrm{rr}}\ \dot{V}_{\mathrm{rr}} \right)^{\mathrm{T}} . \tag{24.3.11}$$

If subjected to a pressure, the master brake cylinder and the hydraulic connection lines will widen and the hydraulic fluid will be compressed, all of which will lead to a consumption of brake fluid. The flow through the valves inside the hydraulic control unit will give rise to turbulent flow losses. Some of the valves are backflow valves, hence the pressure loss depends on the flow direction and must hence be modeled flow-direction dependent. The long connection lines evoke laminar flow losses. The wheel brake cylinders are modeled as compressible volumes. This once again captures the widening of the cylinder walls and the caliper as well as the compression of the brake pads, brake discs and brake fluid due to a pressure increase.

The dynamics of the system are then governed by

$$a(\boldsymbol{x}) = \begin{pmatrix} \dot{V}_{\mathrm{fl}} \\ -\frac{R_{\mathrm{T,fl}}}{L_{\mathrm{fl}}}\dot{V}_{\mathrm{fl}}^2 - \frac{R_{\mathrm{L,fl}}}{L_{\mathrm{fl}}}\dot{V}_{\mathrm{fl}} - \frac{1}{L_{\mathrm{fl}}}\int \frac{\dot{V}_{\mathrm{fl}}}{C_{\mathrm{fl}}(V_{\mathrm{fl}})}\mathrm{d}t \\ \dot{V}_{\mathrm{fr}} \\ -\frac{R_{\mathrm{T,fr}}}{L_{\mathrm{fr}}}\dot{V}_{\mathrm{fr}}^2 - \frac{R_{\mathrm{L,fr}}}{L_{\mathrm{fr}}}\dot{V}_{\mathrm{fr}} - \frac{1}{L_{\mathrm{fr}}}\int \frac{\dot{V}_{\mathrm{fr}}}{C_{\mathrm{fr}}(V_{\mathrm{fr}})}\mathrm{d}t \\ \dot{V}_{\mathrm{rl}} \\ -\frac{R_{\mathrm{T,rl}}}{L_{\mathrm{rl}}}\dot{V}_{\mathrm{rl}}^2 - \frac{R_{\mathrm{L,rl}}}{L_{\mathrm{rl}}}\dot{V}_{\mathrm{rl}} - \frac{1}{L_{\mathrm{rl}}}\int \frac{\dot{V}_{\mathrm{rl}}}{C_{\mathrm{rl}}(V_{rl})}\mathrm{d}t \\ \dot{V}_{\mathrm{rr}} \\ -\frac{R_{\mathrm{T,rr}}}{L_{\mathrm{rr}}}\dot{V}_{\mathrm{rr}}^2 - \frac{R_{\mathrm{L,rr}}}{L_{\mathrm{rr}}}\dot{V}_{\mathrm{rr}} - \frac{1}{L_{\mathrm{rr}}}\int \frac{\dot{V}_{\mathrm{rr}}}{C_{\mathrm{rr}}(V_{\mathrm{rr}})}\mathrm{d}t \end{pmatrix} . \tag{24.3.12}$$

This input distribution matrix $\boldsymbol{B}$ is defined as

$$\boldsymbol{B} = \begin{pmatrix} 0 & 0 & 0 & \frac{1}{L_{\mathrm{fl}}} & 0 & \frac{1}{L_{\mathrm{rl}}} & 0 & 0 \\ 0 & \frac{1}{L_{\mathrm{fl}}} & 0 & 0 & 0 & 0 & 0 & \frac{1}{L_{\mathrm{rr}}} \end{pmatrix}^{\mathrm{T}} . \tag{24.3.13}$$

with the control input given as the pressure inside the two chambers of the master brake cylinder

$$\boldsymbol{u} = \begin{pmatrix} p_{\mathrm{II}} \\ p_{\mathrm{I}} \end{pmatrix} . \tag{24.3.14}$$

The output of the model is chosen as the total displaced volume, thus it is the sum of the individually displaced volumes, calculated by the output distribution vector

$$\boldsymbol{c}^{\mathrm{T}} = \begin{pmatrix} 1 & 0 & 1 & 0 & 1 & 0 & 1 & 0 \end{pmatrix} \tag{24.3.15}$$

and the direct feed-through vector

$$\boldsymbol{d}^{\mathrm{T}} = \begin{pmatrix} C_{\mathrm{II}} & C_{\mathrm{I}} \end{pmatrix} . \tag{24.3.16}$$

As was already mentioned, the pressure-flow characteristics of the hydraulic control unit are flow direction dependent due to the presence of backflow valves. Thus, the turbulent resistance is given as

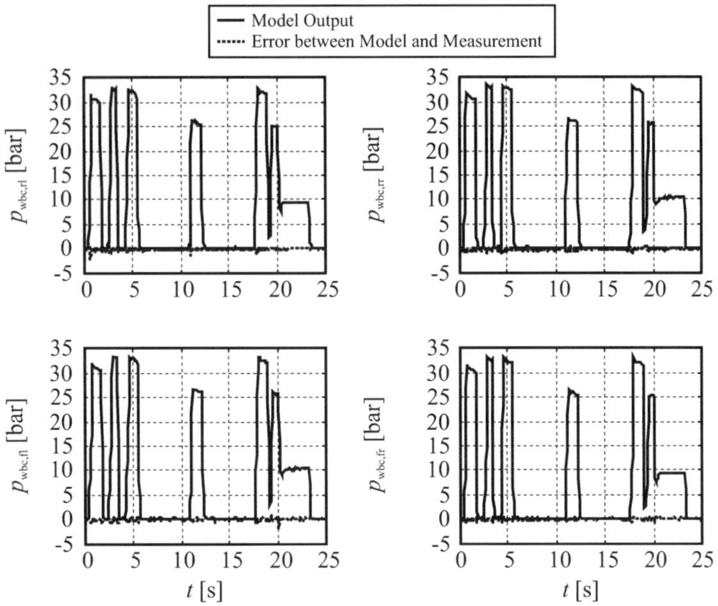

Fig. 24.56. Simulation run of the pressure buildup in the wheel brake cylinders

$$R_{\mathrm{T},i} = \begin{cases} R_{\mathrm{Ta},i} \text{ for } \dot{V}_i \geq 0 \\ R_{\mathrm{Tr},i} \text{ for } \dot{V}_i < 0 \end{cases} \quad \text{for } i \in \{\mathrm{fl}, \mathrm{fr}, \mathrm{rl}, \mathrm{rr}\} \; . \tag{24.3.17}$$

Now, the individual parameters of the model have been determined by iterative optimization based on measurements taken at the testbed in Fig. 24.55. The high fidelity of the derived model can be seen from the simulation in Fig. 24.56. Further details are described in (Straky et al, 2002; Straky, 2003).

Vacuum Brake Booster

The introduction of disc brakes required higher operation forces which in turn mandated some means to amplify the control forces exerted by the driver. Coming up with an easy yet efficient design, the vacuum brake booster is applied. Besides this amplification of the driver input, the vacuum brake booster is also used as an actuation device for the braking system by the brake assistant (Kiesewetter et al, 1997) which initiates a full braking in case of an emergency.

Figure 24.57 shows a cut-away drawing of the vacuum brake booster depicting the different parts of the vacuum brake booster. A photo is shown in Fig. 24.58. The foot-force exerted by the driver is supplied to the vacuum brake booster via the linkage. The vacuum brake booster is divided into two chambers by means of the diaphragm. The vacuum chamber is always kept at a pressure substantially lower than the atmospheric pressure, whereas the pressure in the working chamber is controlled

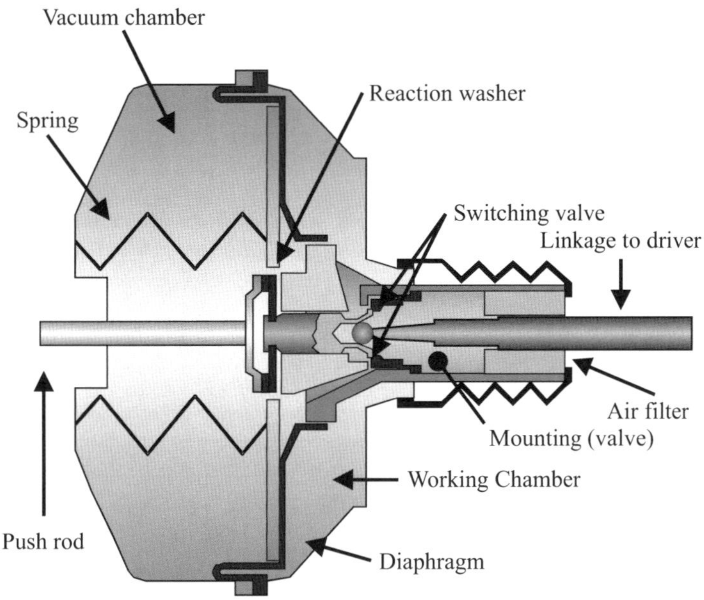

Fig. 24.57. Cut-away drawing of a vacuum brake booster

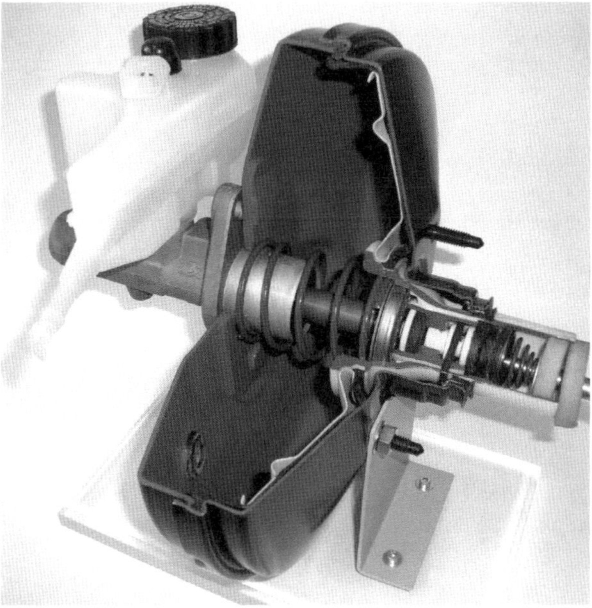

Fig. 24.58. Photo of a vacuum brake booster

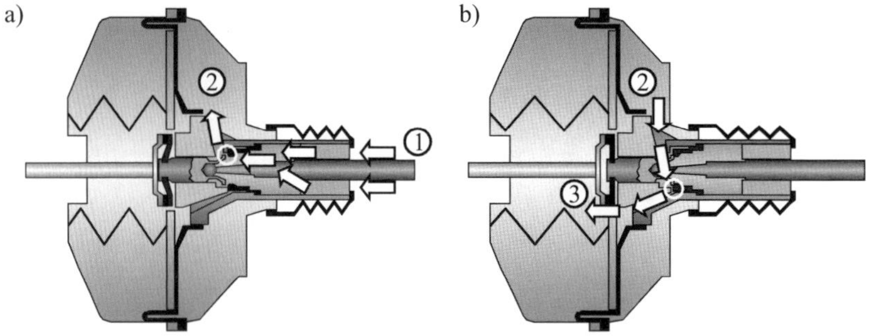

Fig. 24.59. Booster in (**a**) brake stage, (**b**) release stage

by the pneumatic valves located inside the vacuum brake booster. These valves allow to open or shut a flow path from the working chamber to the vacuum chamber (Fig. 24.59a) or from the working chamber to the surroundings (Fig. 24.59b) respectively. Opening and closing of the valves is controlled by the reaction washer, which is basically an elastic rubber disc. The air drawn from the surroundings is traversing an air filter and thereby cleaned. For spark ignition engines, the vacuum pressure is taken from the engine manifold, whereas for Diesel engines, the vacuum pressure is supplied by a vacuum pump. For operation at the testbed (Fig. 24.55), the vacuum brake booster is supplied with vacuum pressure by means of a membrane pump, thus this setup mimics a car equipped with a Diesel engine.

The pressure difference between the two chambers of the vacuum brake booster exerts a force onto the membrane, which is transmitted to the master brake cylinder via a push-rod. For a more detailed description of the vacuum brake booster, the reader is referred to (Robert Bosch GmbH, 2007; Burckhardt, 1991; Breuer and Bill, 2006).

In the following, the quantities are denoted as

p	pressure	V	volume
m	mass	R	gas constant
T	temperature	x	displacement
A	area	A_V	valve opening area
ϱ	density	v	velocity

and indices

vc	vacuum chamber	wc	working chamber
mem	membrane, i.e. diaphragm	amb	ambient
link	linkage		

will be used.

As was indicated in the previous section, the vacuum brake booster consists of two chambers. The chambers are modeled as pneumatic storage devices (Isermann, 2005). The state of such a storage device is described by the equation of state for an ideal gas,

$$p(t)V(t) = m(t)R_{\text{air}}T \;, \tag{24.3.18}$$

where p is the pressure, V the currently enclosed volume, m the mass of the enclosed air, R_{air} the gas constant and T the temperature. The mass m is chosen as the conservative quantity. The volume of the vacuum chamber (vc) is determined by the initial volume of the vacuum chamber, $V_{\text{vc},0}$, and the displacement of the diaphragm, x_{mem},

$$V_{\text{vc}} = V_{\text{vc},0} - x_{\text{mem}}A_{\text{mem}} \;. \tag{24.3.19}$$

Th cross-sectional area of the membrane is written as A_{mem}. Volume and pressure changes are assumed to be isothermal. From the mass of the enclosed air, which was chosen as the conservative quantity, one can calculate the pressure inside the vacuum chamber as

$$p_{\text{vc}}(t) = \frac{m_{\text{vc}}R_{\text{air}}T}{V_{\text{vc}}} \tag{24.3.20}$$

and the actual density as

$$\varrho_{\text{vc}} = \frac{m_{\text{vc}}}{V_{\text{vc}}} \;. \tag{24.3.21}$$

For the working chamber (wc), similar equations can be derived as

$$V_{\text{wc}} = V_{\text{wc},0} + x_{\text{mem}}A_{\text{mem}} \tag{24.3.22}$$

$$p_{\text{wc}} = \frac{m_{\text{wc}}R_{\text{air}}T}{V_{\text{wc}}} \tag{24.3.23}$$

$$\varrho_{\text{wc}} = \frac{m_{\text{wc}}}{V_{\text{wc}}} \;. \tag{24.3.24}$$

Now, the valves are modeled. For this endeavor, Bernoulli's equation (Isermann, 2005) is employed. The fluid is assumed to be barotropic, since the behavior of air can be approximated by a barotropic fluid. In this case, Bernoulli's equation is given as

$$\int_1^2 \frac{\partial v}{\partial t} + \left(P_2 + \frac{v_2^2}{2} + U_2\right) - \left(P_1 + \frac{v_1^2}{2} + U_1\right) = 0 \tag{24.3.25}$$

The influence of the first term, which describes the acceleration induced pressure loss, is neglected. The terms P_i with $i \in \{1, 2\}$ describe the energy expenditure for the compression respectively expansion phases and are calculated as

$$P_I = \int_{p_0}^{p_1} \frac{\mathrm{d}p}{\varrho} = R_{\text{air}}T \int_{p_0}^{p_1} \frac{\mathrm{d}p}{p} = R_{\text{air}}T \ln\left(\frac{p_1}{p_0}\right) . \tag{24.3.26}$$

Bernoulli's equation is now applied to the flow path from point 1 to point 2, as shown in Fig. 24.59, which results in

$$R_{\text{air}}T \ln\left(\frac{p_1}{p_2}\right) + \frac{v_1^2}{2} - \frac{v_2^2}{2} = 0 \;. \tag{24.3.27}$$

Point 1 is in the surroundings, therefore the velocity v_1 is assumed to be negligibly small, and the pressure p_1 is set to the atmospheric pressure p_{amb}. Point 2 is located

a)

b)

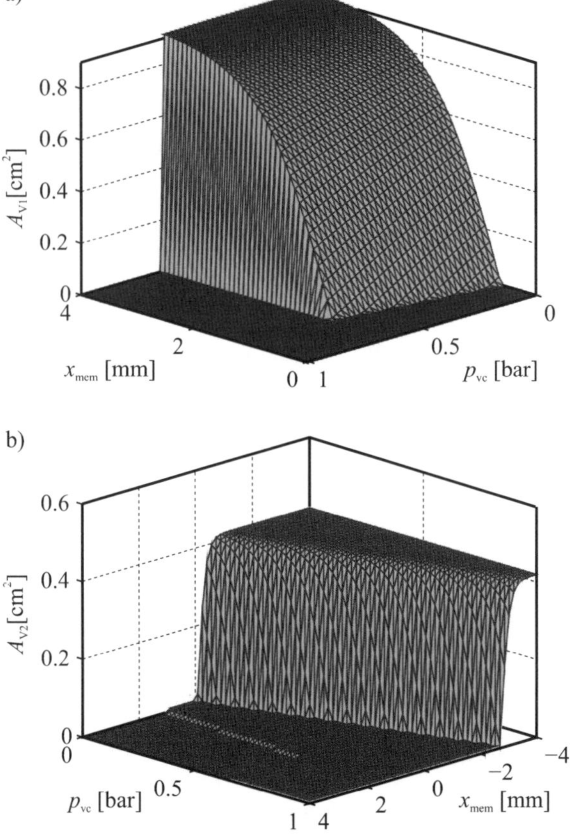

Fig. 24.60. Valve 1 opening function (**a**) and valve 2 opening function (**b**) of the vacuum brake booster

inside the working chamber, thus the pressure p_2 is identical to p_{wc}. One can now solve Eq. 24.3.27 with respect to v_2 as

$$v_2 = \sqrt{2R_{air}T \ln\left(\frac{p_{amb}}{p_{wc}}\right)} . \tag{24.3.28}$$

The mass flow into the working chamber is then determined as

$$\dot{m}_{wc} = A_{V1}\varrho_{wc}v_2 , \tag{24.3.29}$$

where A_{V1} is the active cross-sectional opening area of the aforementioned valve between the surroundings and the working chamber.Similarly, the equation governing the behavior of the valve between the working chamber and the vacuum chamber can now be derived. The flow path is shown in Fig. 24.59 and yields

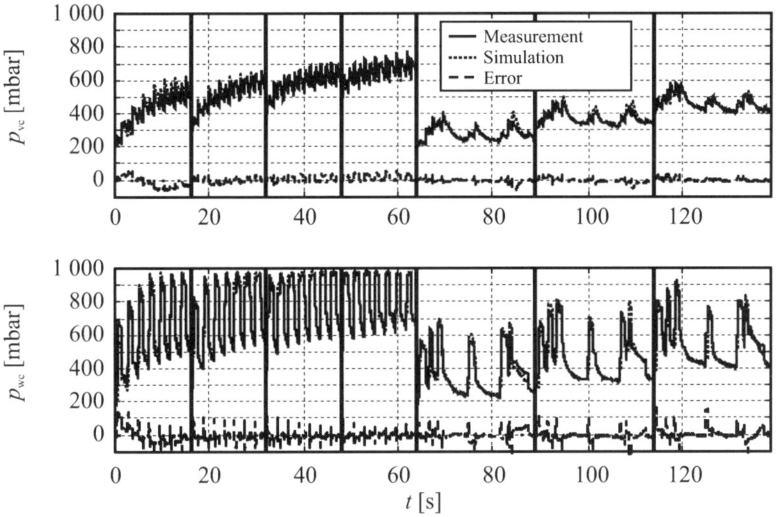

Fig. 24.61. Comparison of model and process output

$$v_3 = \sqrt{2R_{air}T \ln\left(\frac{p_{wc}}{p_{vc}}\right)} \qquad (24.3.30)$$

$$\dot{m}_{vc} = -\dot{m}_{wc} = A_{V2}\varrho_{VC}v_3 \; . \qquad (24.3.31)$$

The active cross-sectional areas of the two valves as a function of the membrane and pedal displacement and the active chamber pressures have been derived from experimental data gathered at the testbed. The resulting valve characteristics are shown in Fig. 24.60. For the identification, an iterative optimization approach was used employing the output error between the simulated and the real chamber pressures and using the Gauss-Newton algorithm with numerically determined derivatives. The model structure was a hybrid physical and black-box model. The pneumatics were modeled physically, whereas the valve opening functions were modeled as black-box functions. Further details are shown in (Muenchhof et al, 2003).

24.3.3 Automotive Suspension

Both, the suspension as well as the tire pressure, covered in the next section, have a large influence on the vehicle dynamics and are thus highly safety critical. In the following, identification techniques will be developed that allow to identify the characteristics of the vehicle suspension and the tire and can serve for supervision of these components.

For application either in a service station, for example for technical inspection, or in a driving state, it is important to use easily measurable variables. If the methods should be used for technical inspection, then the additional sensors must be easily mounted to the car. For on-board identification, the existing variables for suspension

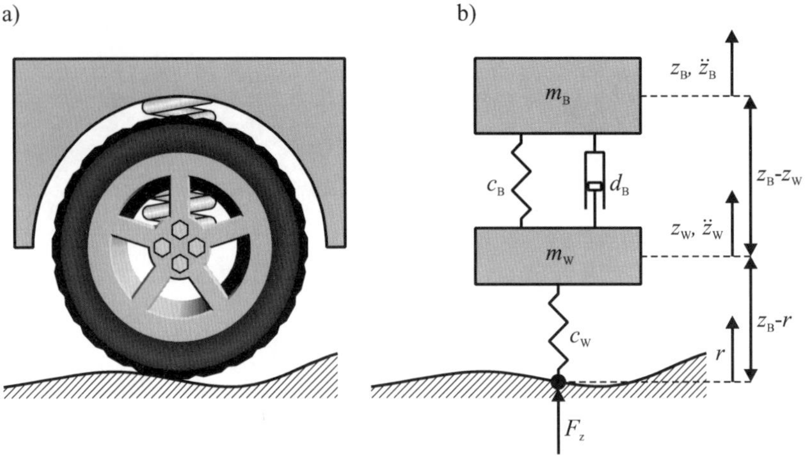

Fig. 24.62. Quarter car model. (**a**) Quarter car (**b**) schematic of the mechanical system

control should be used. Variables which meet these requirements are the vertical accelerations of body and wheel, $\ddot{z}_B$ and $\ddot{z}_W$, and the suspension deflection $z_W - z_B$. Another important point is that the methods should require only little a priori knowledge about the type of car.

A scheme for a simplified model of a car suspension system, a quarter car model, is shown in Fig. 24.62. The following equations follow from force balances

$$m_B \ddot{z}_B(t) = c_B\big(z_W(t) - z_B(t)\big) + d_B\big(\dot{z}_W(t) - \dot{z}_B(t)\big) \qquad (24.3.32)$$

$$m_W \ddot{z}_W(t) = -c_B\big(z_W(t) - z_B(t)\big) - d_B\big(\dot{z}_W(t) - \dot{z}_B(t)\big) + c_W\big(r(t) - z_W(t)\big) \qquad (24.3.33)$$

In this section, the following symbols are used:

a_1, b_1	parameters of transfer functions	F_W	wheel force
c_B	stiffness of body spring	m_B	body mass
c_W	tire stiffness	p_W	wheel pressure
d_B	body damping coefficient	r	road displacement
f_r	resonance frequency	z_B	vert. body displacement
F_C	Coulomb friction force	z_W	vert. wheel displacement
F_D	damper force	$\Delta z_{WB} = z_B - z_W$	
F_S	spring and damper force		suspension deflection

The small damping of the wheel is usually negligible. A survey of passive and semi-active suspensions and their models is given in (Isermann, 2005).

In general, the relationship between force and velocity of a shock absorber is nonlinear. It is usually degressive and depends strongly on the direction of motion of the piston. In addition, the Coulomb friction of the damper should be taken into account. To approximate this behavior, the characteristic damper curve can be divided into m sections as a function of the piston velocity. Considering m sections, the following equation can be obtained.

$$\ddot{z}_B = \frac{d_{B,i}}{m_B(\dot{z}_W - \dot{z}_B) + \frac{c_B}{m_B}(z_W - z_B) + \frac{1}{m_B}F_{C,i}} \, , \quad i = 1, \ldots, m \, , \qquad (24.3.34)$$

compare (24.3.32). $F_{C,i}$ denotes the force generated by Coulomb friction and $d_{B,i}$ the damping coefficient for each section. Using (24.3.34) the damping curve can be estimated with a standard parameter estimation algorithm measuring the body acceleration $\ddot{z}_B$ and suspension deflection $z_W - z_B$. The velocity $\dot{z}_W - \dot{z}_B$ can be obtained by numerical differentiation. In addition, either the body mass m_B or the spring stiffness c_B can be estimated. One of both variables must be known a priori. Using (24.3.32) and (24.3.33), other equations for parameter estimation can be obtained, e.g. (24.3.35) which can be used to estimate the tire stiffness c_W additionally

$$z_W - z_B = -\frac{d_{B,i}}{m_B}(\dot{z}_W - \dot{z}_B) - \frac{c_B}{m_B}(z_W - z_B) + \frac{c_W}{c_B}(r - z_W) - \frac{1}{c_W}F_{C,i} \quad (24.3.35)$$

The disadvantage of this equation is the necessity to measure the distance between road and wheel $(r - z_W)$, see also (Bußhardt, 1995; Weispfenning and Leonhardt, 1996) for modeling and identification of the automotive suspension.

To test the above method in a driving car, a medium class car, an Opel Omega, Fig. 24.63, was equipped with sensors to measure the vertical acceleration of body and wheel as well as the suspension deflections. To realize different damping coefficients, the car is equipped with adjustable shock absorbers at the rear axle, which can be varied in three steps. In Fig. 24.65, the course of the estimated damping coefficients at different damper settings is given for driving over boards of height 2 cm, see Fig. 24.64.

After approximately 2.5 s, the estimated values converge to their final values. The estimated damping coefficients differ approximately 10% from the directly measured ones. In Fig. 24.66, the estimated characteristic curves at the different damper settings are shown. The different settings are separable and the different damping characteristics in compression and rebound is clearly visible, although the effect is not as strong as in the directly measured characteristic curve. More results are given in (Börner et al, 2001).

Next, the damping characteristics of the shock absorber were adjusted during a driving maneuver. Recursive parameter estimation then allow to adapt the damping coefficient accordingly. Figure 24.67 illustrates the suspension deflection $z_W - z_B$, the first derivative of the suspension deflection calculated with a state variable filter $\dot{z}_W - \dot{z}_B$, and the wheel acceleration $\ddot{z}_W$ for the right rear wheel during a highway test drive. After 30, 60, 90, and 120 s, a change of the shock absorber damping was made.

Several estimations have shown that the recursive least squares algorithm (RLS) with exponential forgetting factor received very good results. This recursive parameter estimation is able to adapt to the different damping settings in about 10 s, see Fig. 24.68.

Fig. 24.63. Car for driving experimental for model validation and parameter estimation

Fig. 24.64. Driving experiment with wooden plates to excite the vertical dynamics

Table 24.5. Specifications of the Opel Omega for driving experiments

Parameter	Value
Type	Opel Omega A 2.0i
Year	1993
Drive	Rear wheel drive
Top speed	190 km/h
Engine	4 cylinder OHC spark ignition engine
Engine displacement	$1.998 \times 10^{-3}\,\mathrm{m}^3$
Rated power	85×10^{-3}Nm/s at 5 200 rpm
Max torque	170 Nm at 2 600 rpm
Max drag torque	−49 Nm
Max rpm	5 800 rpm
Transmission	5 speed manual transmission
Brakes	Disc brakes at all four wheels
Steering	Continuous ball race power steering
Steering transmission ratio	13.5
Tire size	195/65 R15 91H
Rim size	6 J × 15
Tire radius of unloaded wheel	0.320 m
Tire rotational inertia	$0.9\,\mathrm{kg\,m}^2$
Unsprung mass front wheel	44 kg
Unsprung mass rear wheel	46 kg
Length	4.738 m
Width	1.760 m
Wheel base	2.730 m
Height of cog	0.58 m

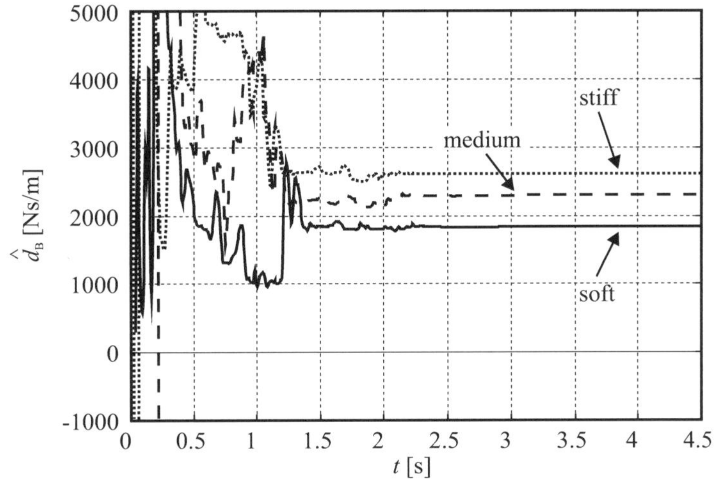

Fig. 24.65. Estimated damping coefficients for different damper settings (speed about 30 km/h)

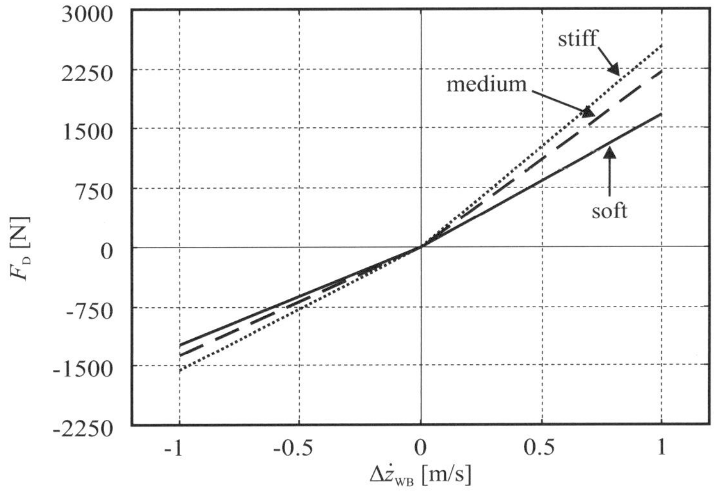

Fig. 24.66. Estimated characteristic damping characteristics for different damper settings

24.3.4 Tire Pressure

The tire pressure is also a very important quantity for vehicle safety. A survey conducted by Michelin in 2006 within the safety campaign "Think Before You Drive"has shown that only 6.5% of the 20 300 inspected cars had the required tire pressure at all four wheels. More than 39.5% of the cars had at least one extremely underinflated tire (< 1.5 bar) (Bridgestone, 2007). It is well known that it is dangerous to drive with underinflated tires. First of all, the risk of an accident increases due to the worse

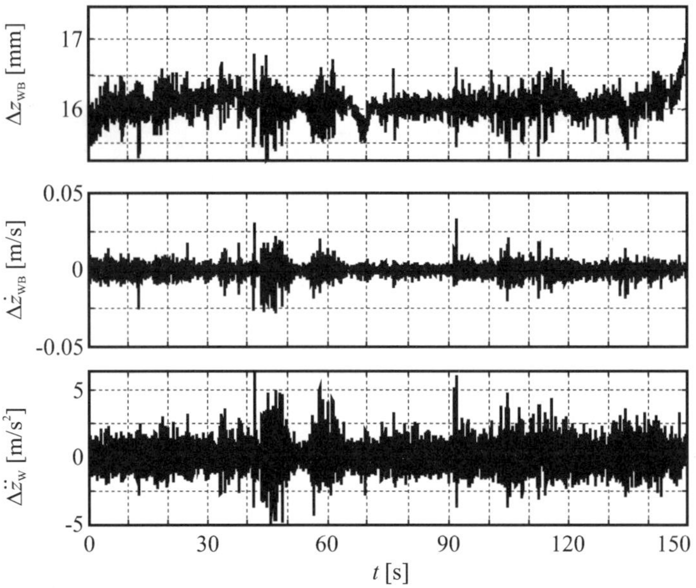

Fig. 24.67. Measured signals on a highway with variation of the damper configuration

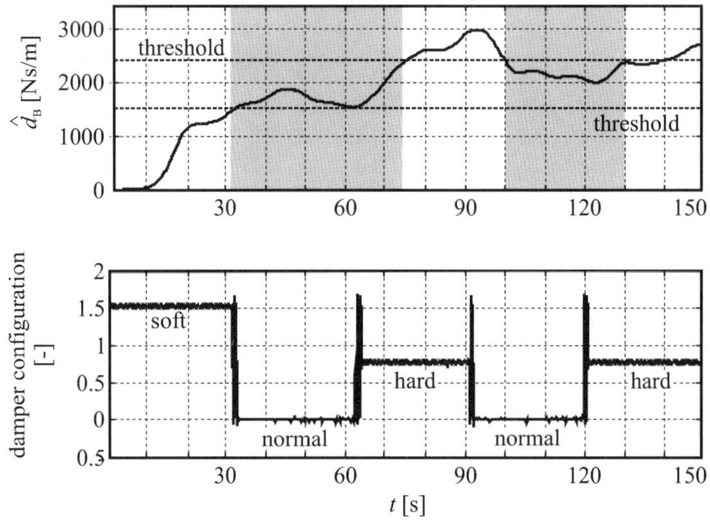

Fig. 24.68. Parameter estimate of the damping coefficient with RLS and forgetting factor. $0 < t < 3$ s: soft damping configuration, $30\,\text{s} < t < 60\,\text{s}$ and $90\,\text{s} < t < 120\,\text{s}$: medium damping configuration (normal situation), $60\,\text{s} < t < 90\,\text{s}$ and $120\,\text{s} < t < 150\,\text{s}$: hard damping coefficient

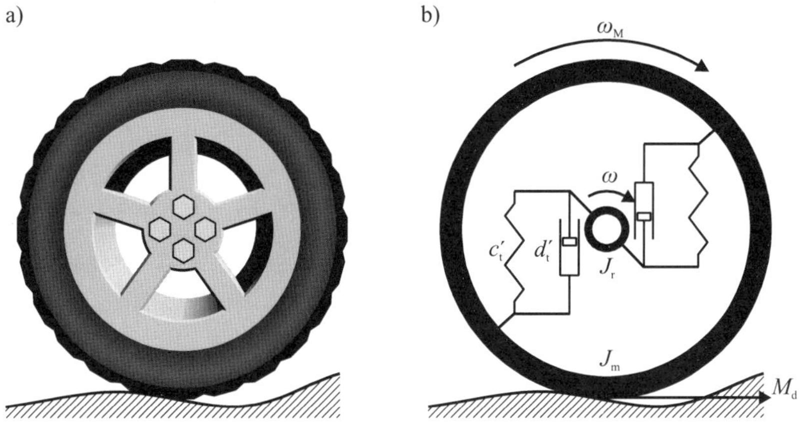

Fig. 24.69. Model of Torsional Wheel Oscillations

vehicle dynamic properties and the increased probability of a tire burst. Because of the increasing deformation, the tire heats up and its structure destabilizes (Normann, 2000). Furthermore, tire deflation increases fuel consumption and tire wear.

One can in general discern two approaches in tire pressure measurement systems. Direct measurement and indirect measurement. Direct tire pressure measurement systems use dedicated pressure sensors (Normann, 2000; Maté and Zittlau, 2006; Wagner, 2004). As the sensor is mounted directly at the tire, it is exposed to extreme environmental conditions, such as a wide temperature range and large accelerations. Also, the power supply of the sensor and the data transmission increase the cost and complexity of the systems. Therefore, one is interested in employing alternative measurement principles.

These are the basis of indirect tire pressure measurement systems. Here, one uses measured signals of the wheel or suspension that are already measured for the use by other vehicle dynamics control systems. One example is the wheel speed that is measured by the wheel speed sensors for use by the ABS system. Besides the wheel speed ω, also the vertical wheel acceleration $\ddot{z}_\mathrm{W}$ can be used to determine the tire pressure, as will be shown in the following.

Torsional Wheel Speed Oscillations

At the mantle of the wheel, a disturbing torque M_d attacks. This torque is caused by variations in the friction coefficient as well as variations of the height of the road surface. This and the elasto-kinematic bearings cause oscillations, which are transferred from the wheel speed at the mantle, ω_m to the wheel speed of the rim ω (Persson et al, 2002; Prokhorov, 2005).

Figure 24.69 shows a schematic diagram of the dynamics of these torsional wheel oscillations. The elasto-kinematics between the wheel and the rim are characterized

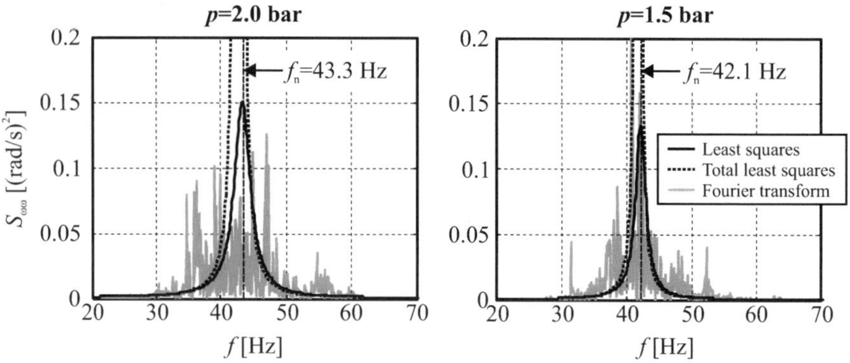

Fig. 24.70. Estimated power spectral density of the wheel speed signal for a straight driving maneuver with $v = 50\,\mathrm{km/h}$ and a pressure of $p = 2.0\,\mathrm{bar}$ and $p = 1.5\,\mathrm{bar}$ respectively of the rear left tire

by a torsional stiffness c'_t and a torsional damping d'_t. The moments of inertia are denoted by J_m and J_r respectively. With these quantities, the transfer function from the disturbance torque M_d to the wheel speed ω is given as

$$G(s) = \frac{\omega(s)}{M_\mathrm{d}(s)} = \frac{d'_\mathrm{t}s + c'_\mathrm{t}}{j_\mathrm{r}J_\mathrm{m}s^3 + (J_\mathrm{r} + J_\mathrm{m})d'_\mathrm{t}s^2 + (J_\mathrm{r} + J_\mathrm{m})c'_\mathrm{t}s} \,. \tag{24.3.36}$$

The torsional stiffness c'_t depends on the tire pressure. Changes in the tire pressure can be detected by analyzing the wheel speed signals. In order to detect changes in the tire pressure, one determines the spectrum of the wheel speed ω. This is done employing parametric spectral analysis methods.

The filter $G(z)$ was modeled as an auto-regressive process, i.e.

$$y(k) = q(k) - a_1 y(k-1) - a_2 y(k-2) - \ldots - a_n y(k-n) \,. \tag{24.3.37}$$

The model parameters can then be determined e.g. by the method of least squares or the method of total least squares.

For the wheel speed analysis, the majority of all spectral components is concentrated in the low frequency range ($< 10\,\mathrm{Hz}$). Only small peaks exist in the frequency range above 10 Hz. As the spectral components that are influenced by the tire pressure are expected between 40 Hz and 50 Hz (Persson et al, 2002; Prokhorov, 2005), a bandpass filter is applied to attenuate spectral components outside of this frequency band. The experimental results of test drives with different tire pressure are shown in Fig. 24.70. One can see that the estimation of the power spectral density of the wheel speed signal by means of the Fourier transform has many peaks and is very disturbed. Hence, the resonance maximum is difficult to detect. Application of the parametric spectral analysis results in a much smoother spectrum. The power spectral density obtained by the spectral analysis with the method of least squares matches well with the general shape of the spectrum as estimated using the Fourier transform.

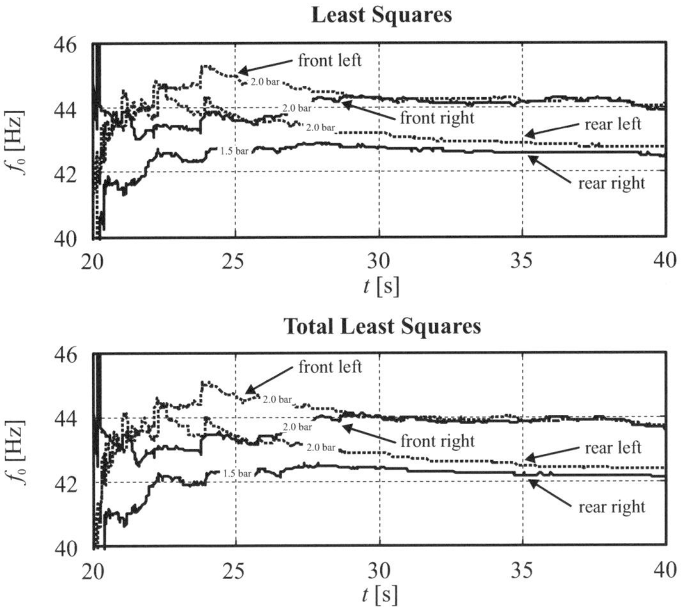

Fig. 24.71. Recursive estimation of the power spectral density of the wheel speed signal for a straight driving maneuver with $v = 50\,\text{km/h}$. Rear right tire underinflated at $p = 1.5\,\text{bar}$, all other tires correctly inflated at $p = 2.0\,\text{bar}$. Power spectral density characterized by the natural frequency f_0

Although the maximum of the power spectral density obtained by using the method of total least squares is far too large, the resonance frequency is still identified well. By the method of least squares, the resonance of the correctly inflated tire is located at 43.3 Hz and for the deflated tire, reduces to 42.1 Hz.

To be able to detect changes in the tire pressure during operation of the vehicle, recursive parameter estimation methods will now be used to estimate the power spectral density of the wheel speed signal online. Instead of identifying the resonance frequency, the natural frequency f_0 will now be identified as it is less affected by wear of the suspension system (Isermann, 2002).

Figures 24.71 and 24.72 show experimental results. In Fig 24.71, the rear right tire was underinflated at $p = 1.5\,\text{bar}$, whereas all other tires correctly inflated at $p = 2.0\,\text{bar}$. One can see that the detection of changes in tire pressure is possible. Note that the rear axle is the driven axle of the car and equipped with a differential gear. Due to the coupling by differential gear, the rear left wheel is also affected although inflated properly.

Vertical Wheel Acceleration

Signals from the suspension system are also influenced by the tire pressure (Börner et al, 2002; Weispfenning and Isermann, 1995; Börner et al, 2000), such as e.g. the

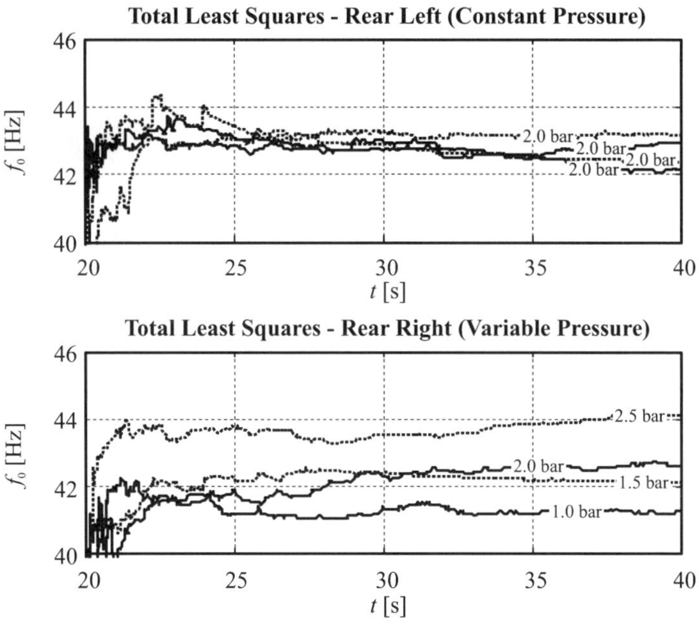

Fig. 24.72. Recursive estimation of the power spectral density of the torsional wheel oscillations for a straight driving maneuver with $v = 50\,\mathrm{km/h}$. Pressure in rear right tire varied from 1.0 bar up to 2.5 bar

spring deflection z_{wb}, the vertical body acceleration $\ddot{z}_{\mathrm{b}}$, or the vertical wheel acceleration $\ddot{z}_{\mathrm{w}}$. As the body motion z_{b}, $\ddot{z}_{\mathrm{bb}}$ is much slower than the wheel motion z_{w}, $\ddot{z}_{\mathrm{w}}$, one cannot expect to see the effect of varying tire pressure in the body motion (Börner et al, 2000). Considering the quarter car model in Fig. 24.62, one can derive a transfer function from the road height r to the wheel acceleration $\ddot{z}_{\mathrm{w}}$ neglecting the movement of the body, i.e. $z_{\mathrm{b}} = 0$, as

$$G(s) = \frac{\ddot{z}_{\mathrm{w}}(s)}{z_{\mathrm{h}}(s)} = \frac{\frac{c_{\mathrm{w}}}{c_{\mathrm{b}}+c_{\mathrm{w}}}s^2}{\frac{m_{\mathrm{w}}}{c_{\mathrm{b}}+c_{\mathrm{w}}}s^2 + \frac{d_{\mathrm{b}}}{c_{\mathrm{b}}+c_{\mathrm{w}}}s + 1}\,. \tag{24.3.38}$$

Since the tire stiffness depends on the tire pressure, one can observe changes in the tire pressure in the spectrum of $\ddot{z}_{\mathrm{w}}$.

Experimental results for this method are shown in Figs. 24.73 through 24.75. For the analysis of the vertical wheel acceleration spectrum, there are few neighboring frequency peaks from other effects and hence no filter is required. Figure 24.73 shows that resonance frequency can be clearly detected and the spectrum is strongly sensitive to changes in the tire pressure. The difference of f_0 between the correctly inflated tires and the underinflated tire is larger when the total least squares spectral analysis is used instead of the least squares spectral analysis. In Fig. 24.75, a strong dependence between the deflation of the left tire and the respective tires natural frequency f_0 is observable. Hence, the vertical wheel acceleration shows stronger

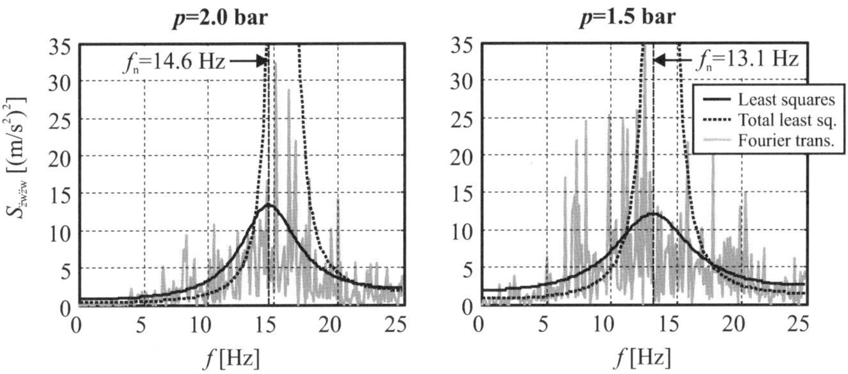

Fig. 24.73. Estimated power spectral density of the vertical wheel acceleration for a straight driving maneuver with $v = 50\,\mathrm{km/h}$ and a pressure of $p = 2.0\,\mathrm{bar}$ and $p = 1.5\,\mathrm{bar}$ respectively of the rear left tire

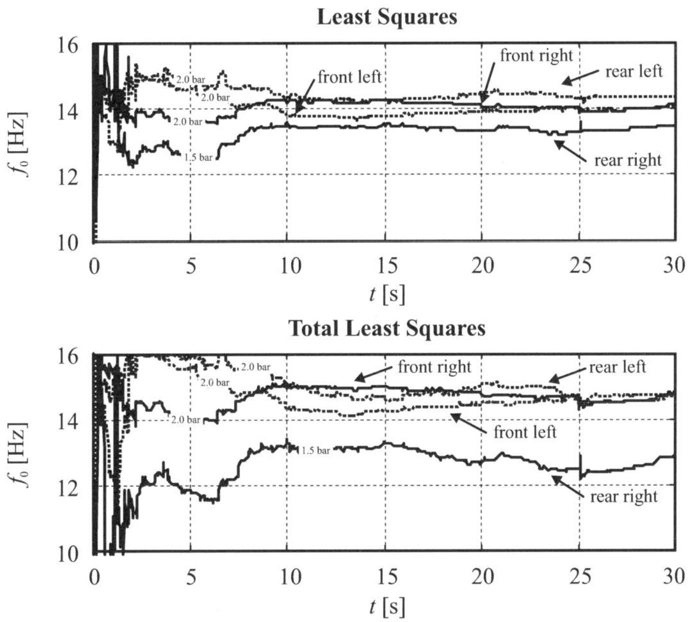

Fig. 24.74. Recursive estimation of the power spectral density of the vertical wheel acceleration for a straight driving maneuver with $v = 50\,\mathrm{km/h}$. Rear right tire underinflated at $p = 1.5\,\mathrm{bar}$, all other tires correctly inflated at $p = 2.0\,\mathrm{bar}$

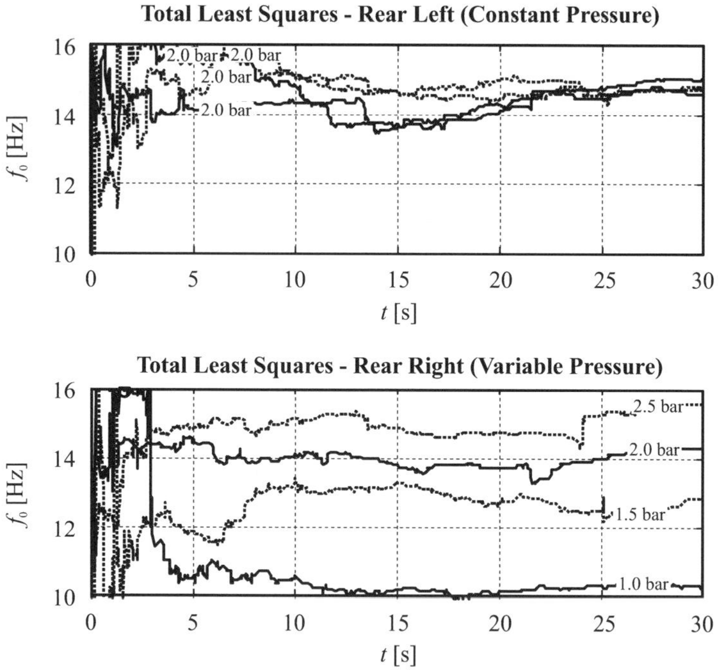

Fig. 24.75. Recursive estimation of the power spectral density of the vertical wheel acceleration for a straight driving maneuver with $v = 50\,\mathrm{km/h}$. Pressure in rear right tire varied from 1.0 bar up to 2.5 bar

dependence on the tire pressure than the torsional wheel oscillations. A survey on tire pressure monitoring with direct and indirect measurements is given in (Fischer, 2003).

24.3.5 Internal Combustion Engines

In the following, the identification of models of internal combustion engines for passenger cars shall be treated. The results can be applied to manifold areas of application for internal combustion engines. In contrast to earlier combustion engines as well as modern gasoline engines, Diesel engines require up to 8 manipulated variables and 8 control variables to meet the ambitious perspectives for low fuel consumption and emissions. Therefore, multi-variable non-linear control is necessary, where the underlying non-linear MIMO models are identified on testbenches. In the following, a Diesel engine is considered as an example of identification of combustion engine models.

Modern Diesel engines are equipped with the following mechatronic actuators:

- High pressure injection system with multiple injections
- Variable camshaft

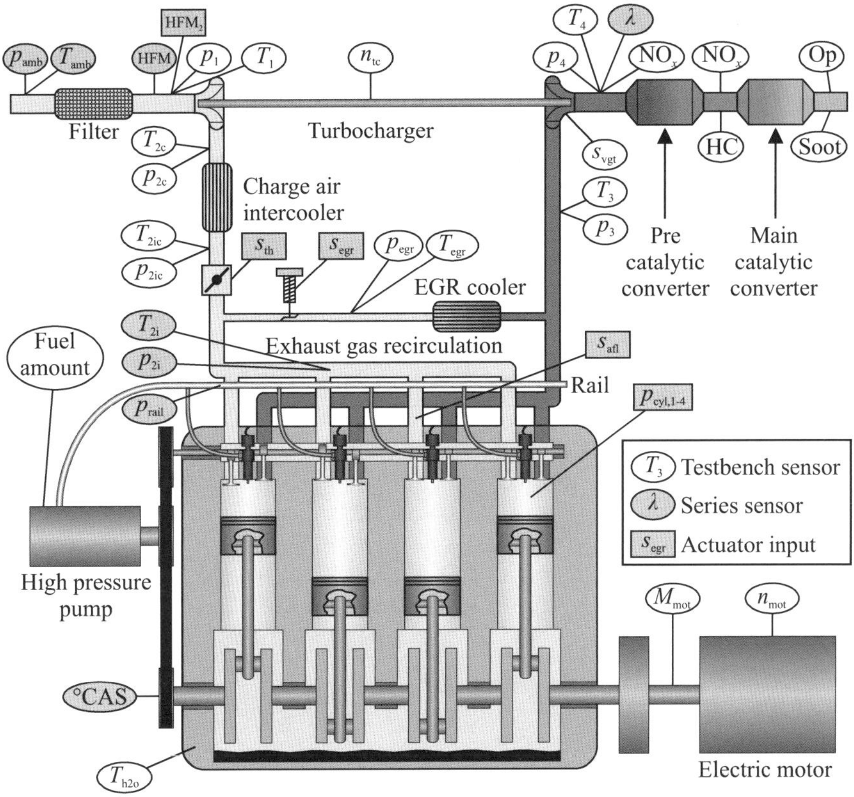

Fig. 24.76. Schematic view of a modern Diesel internal combustion engine with (HFM denotes hot film air mass flow sensor)

- Variable geometry turbo charger
- Exhaust gas recirculation

As was already stated above, these actuators lead to an increase in the number of controlled variables, which all influence the static and dynamic behavior. In a strive for reduced fuel consumption and reduced emissions, one is interested in precise static or also dynamic models of the Diesel engine, which then allow to optimize the engine controllers employed in the engine control unit. The derivation of dynamic engine models becomes more important as recent studies have shown that up to 50% of the emissions of a dynamic driving cycle are caused by accelerations, (Gschweitl and Martini, 2004). The key question in identification of engine models is how the measurement time can be reduced as the huge increase in the number of parameters to be varied has lead to an exponential increase in required measurement time. Here, special attention is paid to the design of the input sequence as to minimize the measurement time while still obtaining models of high fidelity, (Schreiber and Isermann, 2009). The symbols used in this section are:

Fig. 24.77. Control stand

Fig. 24.78. Internal combustion engine on a testbench

m_{air}	air mass in combustion chamber
p_2	boost pressure
φ_{PI}	crank angle of pilot injection
Δt_{PI}	duration of pilot injection
q_{PI}	quantity of fuel of pilot injection
q_{MI}	quantity of fuel of main injection
NO_x	nitrogen-oxides

Table 24.6. Specifications of the vgt turbo-charged Diesel internal combustion engine, type Opel Z19DTH

Parameter	Value
Engine displacement	1.9 l
Engine power	110 kW
Number of cylinders	4
Torque	315 Nm at 2 000 rev
Bore × stroke	82 mm × 90.4 mm
Emission level	Euro 4

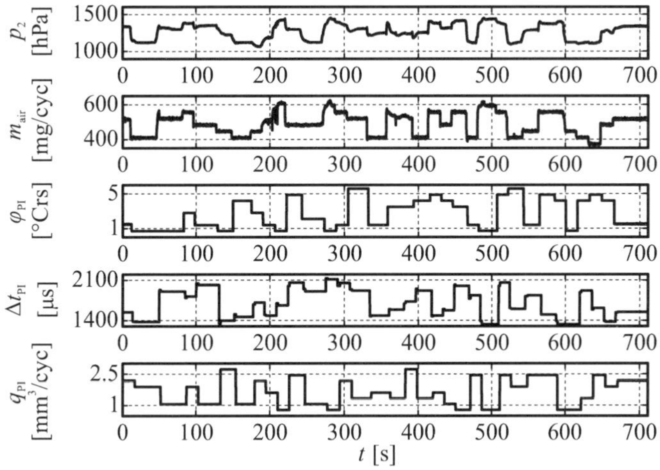

Fig. 24.79. Input signals for internal combustion engine excitation. APRBS test signals for five inputs (simultaneous excitation of the engine at the operating point, $n_{\mathrm{Mot}} = 2000\,\mathrm{rpm}$ and $q_{\mathrm{MI}} = 20\,\mathrm{mm}^3/\mathrm{cyc}$ (Isermann and Schreiber, 2010)

In the following, some experimental results will be presented that have been obtained at an internal combustion engine testbed, see Fig. 24.77. These examples are taken from (Schreiber and Isermann, 2007) and (Isermann and Schreiber, 2010).

The internal combustion engine is mounted on a cart (Fig. 24.78) and can be connected to an asynchronous motor that allows to put a certain torque load on the engine to conduct measurements in different load regimes. The experiments have been conducted on an Opel Z19DTH Diesel engine, whose specifications can be found in Table 24.6.

Figure 24.79 shows a test signal sequence that has been used to conduct measurements at the process. As the behavior of the internal combustion engine is highly non-linear, an APRBS signal has been used for identification, where also the amplitude of the signal is varied in contrast to the two-valued binary PRBS signal. The input signal was furthermore designed to be D-optimal.

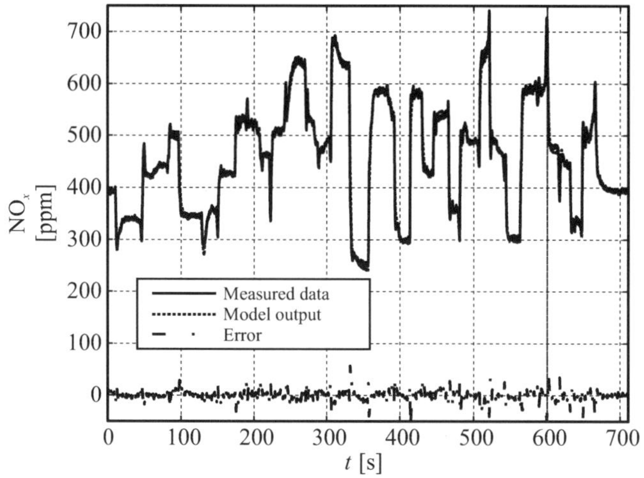

Fig. 24.80. Measured data and model output for NO_x model (Isermann and Schreiber, 2010)

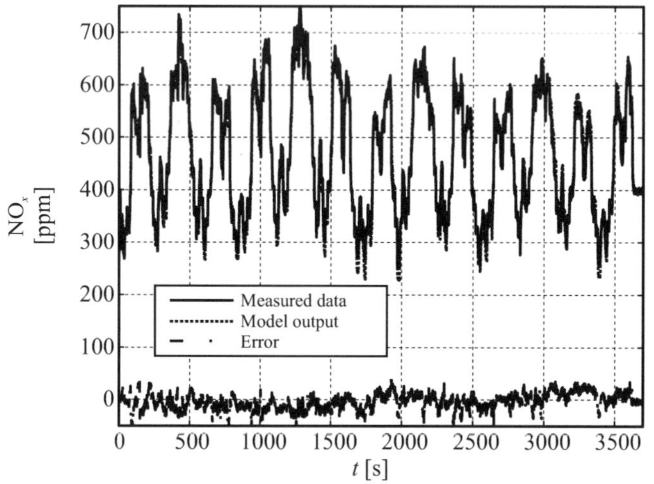

Fig. 24.81. Generalization results for NO_x model for $n_{\text{Mot}} = 2000\,\text{rpm}$ and $q_{\text{MI}} = 20\,\text{mm}^3/\text{cyc}$ (Isermann and Schreiber, 2010)

As an example of a non-linear dynamic MISO model, a model of NO_x emissions has been derived using the LOLIMOT neural network. The output of the neural net and the measurements used for the training are shown in Fig. 24.80. Generalization results are then presented in Fig. 24.81.

The identification of dynamic models has several advantages. First of all, in contrast to stationary measurements, one does not have to wait until the system has settled. Secondly, one can easily deduce a static model from the dynamic model by just

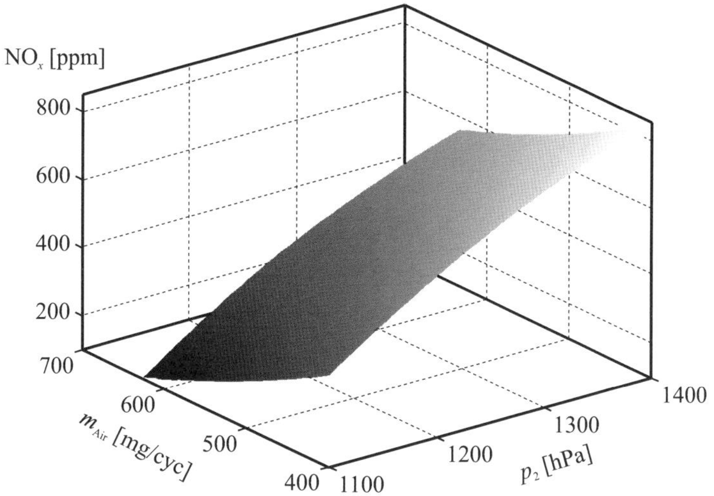

Fig. 24.82. Calculated stationary map for NO_x model for $n_{Mot} = 2000\,\mathrm{rpm}$ and $q_{MI} = 20\,\mathrm{mm}^3/\mathrm{cyc}$ depending on air mass and boost pressure (Isermann and Schreiber, 2010)

calculating the static gain and neglecting the dynamics. The resulting stationary characteristics have been displayed in Fig. 24.82. With the same approach, other models of important dynamics have been identified as well. For the highly non-linear characteristics, the local linear net models are especially well suited for identification. A model of an internal combustion engine had also been the subject of Examples 20.1 and 20.3.

24.4 Summary

This chapter illustrated the use of identification techniques applied to different processes. As one can see from the wide area of applications that has been covered in this chapter, identification methods are a quite universal tools for the extraction of physical parameters and dynamics of processes and haven proven well in many applications. The successful application of identification methods necessitates a certain knowledge of the underlying process dynamics as well as the choice of the appropriate identification method.

For many applications, it is possible to provide physics-based equations that govern the static and dynamic behavior of the system. Then, the method of least squares has been applied for the parameterization based on input/output data. If one cannot provide (simple) equations for the dominant physical effects, one can use (selected) neural nets for identification and modeling. This has been shown for a combustion engine model and the heat exchanger with the LOLIMOT approach.

Other aspects that have been illustrated by the choice of examples are continuous-time and discrete-time models, non-linear models, time domain and frequency domain identification among others, see also Table 24.1. Further application examples can be found in the books (Isermann, 1992, 2005, 2006, 2010).

References

Ballé P (1998) Fuzzy-model-based parity equations for fault isolation. Control Eng Pract 7(2):261–270

Börner M, Straky H, Weispfenning T, Isermann R (2000) Model-based fault detection of vehicle suspension and hydraulik brake systems. In: Proceedings of the 1st IFAC Conference on Mechatronic Systems, Darmstadt

Börner M, Zele M, Isermann R (2001) Comparison of different fault-detection algorithms for active body control components: automotive suspension system. In: Proceedings of the 2001 ACC, Washington, DC

Börner M, Weispfenning T, Isermann R (2002) Überwachung und Diagnose von Radaufhängungen. In: Isermann R (ed) Mechatronische Systeme für den Maschinenbau, Wiley-VCH-Verlag, Weinheim

Breuer B, Bill KH (2006) Bremsenhandbuch: Grundlagen, Komponenten, Systeme, Fahrdynamik, 3rd edn. ATZ/MTZ-Fachbuch, Vieweg, Wiesbaden

Bridgestone (2007) Umweltbelastung durch Reifen mit zu wenig Druck. URL `www.bridgestone.de`

Burckhardt M (1991) Fahrwerktechnik: Bremsdynamik und PKW-Bremsanlagen. Vogel Buchverlag, Würzburg

Bußhardt J (1995) Selbsteinstellende Feder-Dämpfer-Last-Systeme für Kraftfahrzeuge: Fortschr.-Ber. VDI Reihe 12 Nr. 240. VDI Verlag, Düsseldorf

Fischer M (2003) Tire pressure monitoring (BT 243). Verlag moderne industrie, Landsberg

Freyermuth B (1991) Knowledge-based incipient fault diagnosis of industrial robots. In: Prepr. IFAC Symposium on Fault Detection, Supervision and Safety for Technical Processes (SAFEPROCESS), Pergamon Press, Baden-Baden, Germany, vol 2, pp 31–37

Freyermuth B (1993) Wissensbasierte Fehlerdiagnose am Beispiel eines Industrieroboters: Fortschr.-Ber. VDI Reihe 8 Nr. 315. VDI Verlag, Düsseldorf

Geiger G (1985) Technische Fehlerdiagnose mittels Parameterschätzung und Fehlerklassifikation am Beispiel einer elektrisch angetriebene Kreiselpumpe: Fortschr.-Ber. VDI Reihe 8 Nr. 91. VDI Verlag, Düsseldorf

Gschweitl K, Martini E (2004) Transient experimental design for the estimation of dynamic engine behavior. In: Haus der Technik (ed) Motorenentwicklung auf dynamischen Prüfständen, Wiesbaden, Germany

Halbe I (2008) Modellgestützte Sensorinformationsplattform für die Quer- und Längsdynamik von Kraftfahrzeugen: Anwendungen zur Fehlerdiagnose und Fehlertoleranz: Fortschr.-Ber. VDI Reihe 12 Nr. 680. VDI Verlag, Düsseldorf

Hensel H (1987) Methoden des rechnergestützten Entwurfs und Echtzeiteinsatzes zeitdiskreter Mehrgrößenregelungen und ihre Realisierung in einem CAD-System. Fortschr.-Ber. VDI Reihe 20 Nr. 4. VDI Verlag, Düsseldorf

Isermann R (1987) Digitale Regelsysteme Band 1 und 2. Springer, Heidelberg

Isermann R (1992) Identifikation dynamischer Systeme: Besondere Methoden, Anwendungen (Vol 2). Springer, Berlin

Isermann R (2002) Lecture Regelungstechnik I. Shaker Verlag, Aachen

Isermann R (2005) Mechatronic Systems: Fundamentals. Springer, London

Isermann R (2006) Fault-diagnosis systems: An introduction from fault detection to fault tolerance. Springer, Berlin

Isermann R (2010) Fault diagnosis of technical processes. Springer, Berlin

Isermann R, Freyermuth B (1991) Process fault diagnosis based on process model knowledge. J Dyn Syst Meas Contr 113(4):620–626 & 627–633

Isermann R, Schreiber A (2010) Identification of the nonlinear, multivariable behavior of mechatronic combustion engines. In: 5th IFAC Symposium on Mechatronic Systems, Cambridge, MA, USA

Isermann R, Lachmann KH, Matko D (1992) Adaptive control systems. Prentice Hall international series in systems and control engineering, Prentice Hall, New York, NY

Kiesewetter W, Klinkner W, Reichelt W, Steiner M (1997) Der neue Brake-Assistant von Mercedes Benz. atz p 330 ff.

Mann W (1980) Identifikation und digitale Regelung eines Trommeltrockners: Dissertation. TU Darmstadt, Darmstadt

Maté JL, Zittlau D (2006) Elektronik für mehr Sicherheit – Assistenz- und Sicherheitssysteme zur Unfallvermeidung. atz pp 578–585

Milliken WF, Milliken DR (1997) Race Car Vehicle Dynamics. SAE International, Warrendale, PA

Moseler O (2001) Mikrocontrollerbasierte Fehlererkennung für mechatronische Komponenten am Beispiel eines elektromechanischen Stellantriebs. Fortschr.-Ber. VDI Reihe 8 Nr. 908. VDI Verlag, Düsseldorf

Moseler O, Heller T, Isermann R (1999) Model-based fault detection for an actuator driven by a brushless DC motor. In: Proceedings of the 14th IFAC World Congress, Beijing, China

Muenchhof M (2006) Model-based fault detection for a hydraulic servo axis: Fortschr.-Ber. VDI Reihe 8 Nr. 1105. VDI Verlag, Düsseldorf

Muenchhof M, Straky H, Isermann R (2003) Model-based supervision of a vacuum brake booster. In: Proceedings of the 2003 SAFEPROCESS, Washington, DC

Normann N (2000) Reifendruck-Kontrollsystem für alle Fahrzeugklassen. atz (11):950–956

Persson N, Gustafsson F, Drevo M (2002) Indirect tire pressure monitoring system using sensor fusion. In: Proceedings of the Society of Automotive Engineers World Congress, Detroit

Pfeiffer K (1997) Fahrsimulation eines Kraftfahrzeuges mit einem dynamischen Motorenprüfstand: Fortschr.-Ber. VDI Reihe 12 Nr. 336. VDI-Verlag, Düsseldorf

Pfeufer T (1997) Application of model-based fault detection and diagnosis to the quality assurance of an automotive actuator. Control Eng Pract 5(5):703–708

Pfeufer T (1999) Modellgestützte Fehlererkennung und Diagnose am Beispiel eines Fahrzeugaktors. Fortschr.-Ber. VDI Reihe 8 Nr. 749. VDI Verlag, Düsseldorf

Pfleiderer C, Petermann H (2005) Strömungsmaschinen, 7th edn. Springer, Berlin

Prokhorov D (2005) Virtual sensors and their automotive applications. In: Proceedings of the 2005 International Conference on Intelligent Sensors, Sensor Networks and Information Processing, Melbourne

Robert Bosch GmbH (2007) Automotive handbook, 7th edn. Bosch, Plochingen

Schorn M (2007) Quer- und Längsregelung eines Personenkraftwagens für ein Fahrerassistenzsystem zur Unfallvermeidung. Fortschr.-Ber. VDI Reihe 12 Nr. 651. VDI Verlag, Düsseldorf

Schreiber A, Isermann R (2007) Dynamic measurement and modeling of high dimensional combustion processes with dynamic test beds. In: 2. Internationales Symposium für Entwicklungsmethodik, Wiesbaden, Germany

Schreiber A, Isermann R (2009) Methods for stationary and dynamic measurement and modeling of combustion engines. In: Proceedings of the 3rd International Symposium on Development Methodology, Wiesbaden, Germany

Straky H (2003) Modellgestützter Funktionsentwurf für KFZ-Stellglieder: Fortschr.-Ber. VDI Reihe 12 Nr. 546. VDI Verlag, Düsseldorf

Straky H, Muenchhof M, Isermann R (2002) A model based supervision system for the hydraulics of passenger car braking systems. In: Proceedings of the 15th IFAC World Congress, Barcelona, Spain

Vogt M (1998) Weiterentwicklung von Verfahren zur Online-Parameterschätzung und Untersuchung von Methoden zur Erzeugung zeitlicher Ableitungen. Diplomarbeit. Institut für Regelungstechnik, TU Darmstadt, Darmstadt

Voigt KU (1991) Regelung und Steuerung eines dynamischen Motorenprüfstands. In: Proceedings of the 36. Internationales wissenschaftliches Kolloquium, Ilmenau

W Goedecke (1987) Fehlererkennung an einem thermischen Prozess mit Methoden der Parameterschätzung: Fortschritt-Berichte VDI Reihe 8, Nr. 130. VDI Verlag, Düsseldorf

Wagner D (2004) Tire-IQ System – Ein Reifendruckkontrollsystem. atz (660–666)

Wanke P (1993) Modellgestützte Fehlerfrüherkennung am Hauptantrieb von Bearbeitungszentren. Fortschr.-Ber. VDI Reihe 2 Nr. 291. VDI Verlag, Düsseldorf

Wanke P, Isermann R (1992) Modellgestützte Fehlerfrüherkennung am Hauptantrieb eines spanabhebenden Bearbeitungszentrum. at 40(9):349–356

Weispfenning T, Isermann R (1995) Fehlererkennung an semi-aktiven und konventionellen Radaufhängungen. In: Tagung "Aktive Fahrwerkstechnik", Essen

Weispfenning T, Leonhardt S (1996) Model-based identification of a vehicle suspension using parameter estimation and neural networks. In: Proceedings of the 13th IFAC World Congress, San Francisco, CA, USA

Wesemeier D, Isermann R (2007) Identification of vehicle parameters using static driving maneuvers. In: Proceedings of the 5th IFAC Symposium on Advances in Automotive Control, Aptos, CA

Part IX

Appendix

A

Mathematical Aspects

In this appendix, some important fundamental notions of estimation theory shall be repeated. Also, the calculus for vectors and matrices shall very shortly be outlined. A detailed overview of the fundamental notions for estimation theory can e.g. be found in (Papoulis and Pillai, 2002; Doob, 1953; Davenport and Root, 1958; Richter, 1966; Åström, 1970; Fisher, 1922, 1950).

A.1 Convergence for Random Variables

A sequence of random variables x_n is considered with $n = 1, 2, \ldots$. In order to determine whether this sequence converges to a limit random variable x, one can employ different definitions of convergence, which shall shortly by outlines in the following.

Convergence in Distribution

A *very weak form of convergence* is given, if for the cumulative distribution functions $F_n(x)$ of x_n and $F(x)$ of x, the condition

$$\lim_{n\to\infty} F_n(x) = F(x) \tag{A.1.1}$$

is satisfied for all x where $F(x)$ is continuous. This is called convergence in distribution.

Convergence in Probability

The sequence x_n converges in probability to x if

$$\text{For every } \varepsilon > 0 \; \lim_{n\to\infty} P\big(|x_n - x| > \varepsilon\big) = 0 \,. \tag{A.1.2}$$

This is a *weak definition of convergence*. For the convergence in probability, one can also write (Doob, 1953)

R. Isermann, M. Münchhof, *Identification of Dynamic Systems*,
DOI 10.1007/978-3-540-78879-9_25, © Springer-Verlag Berlin Heidelberg 2011

$$\operatorname{plim} x_n = x \text{ or } \operatorname*{plim}_{n\to\infty} x_n = x \,. \tag{A.1.3}$$

Convergence in probability includes convergence in distribution. If x is a constant x_0, then

$$\lim_{n\to\infty} \mathrm{E}\{x_n\} = x_0 \,. \tag{A.1.4}$$

Almost Sure Convergence

A *strong form of convergence* follows from the condition

$$P\left\{\lim_{n\to\infty} x_n = x\right\} = 1 \,, \tag{A.1.5}$$

which also includes convergence in probability and in distribution. This is also called convergence with probability one.

Convergence in the Mean Square

An *even stronger form of convergence* follows from the condition

$$\mathrm{E}\left\{(x_n - x)^2\right\} = 0 \,. \tag{A.1.6}$$

This can be written as

$$\text{l. i. m.}\, x_n = x \,, \tag{A.1.7}$$

(Doob, 1953; Davenport and Root, 1958). Convergence in the mean square includes convergence in probability and in distribution. It does not include almost sure convergence. For the expected value, it includes that

$$\lim_{n\to\infty} \mathrm{E}\{x_n\} = \mathrm{E}\left\{\lim_{n\to\infty} x_n\right\} = \mathrm{E}\{x\} \,. \tag{A.1.8}$$

Slutsky's Theorem

If a sequence of random variables $x_n, n = 1, 2, \ldots$ converges in probability to the constant x_0, i.e. $\operatorname{plim} x_n = x_0$ and $y = g(x_n)$ is a continuous function, then also $\operatorname{plim} y = y_0$ with $y_0 = g(x_0)$. From this follows

$$\operatorname{plim} \boldsymbol{V}\boldsymbol{W} = (\operatorname{plim} \boldsymbol{V})(\operatorname{plim} \boldsymbol{W}) \tag{A.1.9}$$

$$\operatorname{plim} \boldsymbol{V}^{-1} = (\operatorname{plim} \boldsymbol{V})^{-1} \,. \tag{A.1.10}$$

This also includes

$$\lim_{n\to\infty} \mathrm{E}\{\boldsymbol{V}\boldsymbol{W}\} = \lim_{n\to\infty} \mathrm{E}\{\boldsymbol{V}\} \lim_{n\to\infty} \mathrm{E}\{\boldsymbol{W}\} \,, \tag{A.1.11}$$

see also (Goldberger, 1964; Wilks, 1962).

A.2 Properties of Parameter Estimation Methods

In the following, it shall be assumed that a process with the parameters

$$\boldsymbol{\theta}_0^{\mathrm{T}} = \left(\theta_{10}\ \theta_{20}\ \ldots\ \theta_{m0}\right) \tag{A.2.1}$$

is given. These parameters shall not be measurable directly, but can only be estimated based on measurements of an output signal $y(k)$ of the process. The relation between the parameters and the true output variables $y_{\mathrm{M}}(k)$ are known by a model

$$y_{\mathrm{M}}(k) = f(\boldsymbol{\theta}, k) \text{ with } k = 1, 2, \ldots, N\ . \tag{A.2.2}$$

However, the true variables $y_{\mathrm{M}}(k)$ are not known exactly, but only disturbed output variables $y_{\mathrm{P}}(k)$ can be measured. The parameters

$$\hat{\boldsymbol{\theta}}^{\mathrm{T}} = \left(\hat{\theta}_1\ \hat{\theta}_2\ \ldots\ \hat{\theta}_m\right) \tag{A.2.3}$$

shall be estimated such that the outputs of the model $y_{\mathrm{M}}(k)$ agree as good as possible with the recorded measurements y_{P} (Gauss, 1809). The question is now how well the estimates $\hat{\boldsymbol{\theta}}^{\mathrm{T}}$ match with the true values $\boldsymbol{\theta}_0^{\mathrm{T}}$. The following terms were introduced by Fisher (1922, 1950).

Bias

An estimator is termed *unbiased estimator* if

$$\mathrm{E}\{\hat{\boldsymbol{\theta}}\} = \boldsymbol{\theta}_0\ , \tag{A.2.4}$$

or consequently, if it has a systematic error,

$$\mathrm{E}\{\hat{\boldsymbol{\theta}}(N) - \boldsymbol{\theta}_0\} = \mathrm{E}\{\hat{\boldsymbol{\theta}}(N)\} - \boldsymbol{\theta}_0 = \boldsymbol{b} \neq \boldsymbol{0}\ , \tag{A.2.5}$$

then this error is termed *bias*.

Consistent Estimator

An estimator is termed *consistent estimator* if the estimates $\hat{\boldsymbol{\theta}}$ converge in probability to $\boldsymbol{\theta}_0$, hence

$$P\left(\lim_{N\to\infty} \hat{\boldsymbol{\theta}}(N) - \boldsymbol{\theta}_0 = \boldsymbol{0}\right) = 1\ . \tag{A.2.6}$$

If the estimator is consistent, it does only state that for $N \to \infty$ it converges to true values. It does not say anything about the behavior for finite N. Consistent estimators can even be biased for finite N. However, asymptotically, a consistent estimator is bias-free, hence

$$\lim_{N\to\infty} \mathrm{E}\{\hat{\boldsymbol{\theta}}(N)\} = \boldsymbol{\theta}_0\ . \tag{A.2.7}$$

An estimator is termed *consistent in the mean square* if in addition, the variance of the expected value goes to zero

$$\lim_{N\to\infty} \mathrm{E}\left\{\left(\hat{\boldsymbol{\theta}}(N) - \boldsymbol{\theta}_0\right)\left(\hat{\boldsymbol{\theta}}(N) - \boldsymbol{\theta}_0\right)^{\mathrm{T}}\right\} = \boldsymbol{0} \tag{A.2.8}$$

as $N \to \infty$. Then, both the bias as well as the variance tend to zero for $N \to \infty$.

Efficient Estimator

An estimator is termed *efficient* if in a class of estimators, it provides the smallest variance of the estimated parameters, i.e.

$$\lim_{N\to\infty} \operatorname{var} \hat{\boldsymbol{\theta}} = \lim_{N\to\infty} \mathrm{E}\left\{\left(\hat{\boldsymbol{\theta}} - \theta_0\right)\left(\hat{\boldsymbol{\theta}} - \theta_0\right)^{\mathrm{T}}\right\} \to \min \; . \tag{A.2.9}$$

An estimator is called *best linear unbiased estimator (BLUE)* , if it is unbiased and also efficient in the class of all linear unbiased estimators

Sufficient Estimator

An estimator is termed sufficient if it encompasses all information about the observed values, from which the parameters are estimated. A sufficient estimator must have the smallest variance of all estimators and hence is also efficient (Fisher, 1950), see also (Kendall and Stuart, 1961, 1977; Deutsch, 1965).

A.3 Derivatives of Vectors and Matrices

In the derivation of estimators, it is often necessary to determine the first derivative of vector equations to determine the optimum, e.g. the minimum error, the minimum variance or other extremal points. For a vector $\boldsymbol{x}$ and a matrix $\boldsymbol{A}$, one can obtain the following relations

$$\frac{\partial}{\partial \boldsymbol{x}}\left(\boldsymbol{A}\boldsymbol{x}\right) = \boldsymbol{A}^{\mathrm{T}} \tag{A.3.1}$$

$$\frac{\partial}{\partial \boldsymbol{x}}\left(\boldsymbol{x}^{\mathrm{T}}\boldsymbol{A}\right) = \boldsymbol{A} \tag{A.3.2}$$

$$\frac{\partial}{\partial \boldsymbol{x}}\left(\boldsymbol{x}^{\mathrm{T}}\boldsymbol{x}\right) = 2\boldsymbol{x} \tag{A.3.3}$$

$$\frac{\partial}{\partial \boldsymbol{x}}\left(\boldsymbol{x}^{\mathrm{T}}\boldsymbol{A}\boldsymbol{x}\right) = \boldsymbol{A}\boldsymbol{x} + \boldsymbol{A}^{\mathrm{T}}\boldsymbol{x} \tag{A.3.4}$$

$$\frac{\partial}{\partial \boldsymbol{x}}\left(\boldsymbol{x}^{\mathrm{T}}\boldsymbol{A}\boldsymbol{x}\right) = 2\boldsymbol{A}\boldsymbol{x} \text{ if } \boldsymbol{A} \text{ is symmetric, i.e. } \boldsymbol{A}^{\mathrm{T}} = \boldsymbol{A} \; . \tag{A.3.5}$$

Furthermore, the following important rules for the derivative of the trace

$$\frac{\partial}{\partial \boldsymbol{X}} \operatorname{tr}\left(\boldsymbol{A}\boldsymbol{X}\boldsymbol{B}\right) = \boldsymbol{A}^{\mathrm{T}}\boldsymbol{B}^{\mathrm{T}} \tag{A.3.6}$$

$$\frac{\partial}{\partial \boldsymbol{X}} \operatorname{tr}\left(\boldsymbol{A}\boldsymbol{X}^{\mathrm{T}}\boldsymbol{B}\right) = \boldsymbol{B}\boldsymbol{A} \tag{A.3.7}$$

$$\frac{\partial}{\partial \boldsymbol{X}} \operatorname{tr}\left(\boldsymbol{A}\boldsymbol{X}\boldsymbol{B}\boldsymbol{X}^{\mathrm{T}}\boldsymbol{C}\right) = \boldsymbol{A}^{\mathrm{T}}\boldsymbol{C}^{\mathrm{T}}\boldsymbol{X}\boldsymbol{B}^{\mathrm{T}} + \boldsymbol{C}\boldsymbol{A}\boldsymbol{X}\boldsymbol{B} \tag{A.3.8}$$

$$\frac{\partial}{\partial \boldsymbol{X}} \operatorname{tr}\left(\boldsymbol{X}\boldsymbol{A}\boldsymbol{X}^{\mathrm{T}}\right) = \boldsymbol{X}\boldsymbol{A}^{\mathrm{T}} + \boldsymbol{X}\boldsymbol{A} \tag{A.3.9}$$

can be stated (e.g. Brookes, 2005).

A.4 Matrix Inversion Lemma

If $\boldsymbol{A}$, $\boldsymbol{C}$, and $(\boldsymbol{A}^{-1} + \boldsymbol{B}\boldsymbol{C}^{-1}\boldsymbol{D})$ are non-singular quadratic matrices and

$$\boldsymbol{E} = (\boldsymbol{A}^{-1} + \boldsymbol{B}\boldsymbol{C}^{-1}\boldsymbol{D})^{-1} , \tag{A.4.1}$$

then

$$\boldsymbol{E} = \boldsymbol{A} - \boldsymbol{A}\boldsymbol{B}(\boldsymbol{D}\boldsymbol{A}\boldsymbol{B} + \boldsymbol{C})^{-1}\boldsymbol{D}\boldsymbol{A} . \tag{A.4.2}$$

This can be proven as follows:

$$\boldsymbol{E}^{-1} = \boldsymbol{A}^{-1} + \boldsymbol{B}\boldsymbol{C}^{-1}\boldsymbol{D} \tag{A.4.3}$$

upon multiplying $\boldsymbol{E}$ from the left

$$\boldsymbol{I} = \boldsymbol{E}\boldsymbol{A}^{-1} + \boldsymbol{E}\boldsymbol{B}\boldsymbol{C}^{-1}\boldsymbol{D} , \tag{A.4.4}$$

and multiplication of $\boldsymbol{A}$ from the right, one obtains

$$\boldsymbol{A} = \boldsymbol{E} + \boldsymbol{E}\boldsymbol{B}\boldsymbol{C}^{-1}\boldsymbol{D}\boldsymbol{A} . \tag{A.4.5}$$

Furthermore, upon multiplication of $\boldsymbol{B}$ from the right

$$\boldsymbol{A}\boldsymbol{B} = \boldsymbol{E}\boldsymbol{B} + \boldsymbol{E}\boldsymbol{B}\boldsymbol{C}^{-1}\boldsymbol{D}\boldsymbol{A}\boldsymbol{B} \tag{A.4.6}$$

$$= \boldsymbol{E}\boldsymbol{B}\boldsymbol{C}^{-1}(\boldsymbol{C} + \boldsymbol{D}\boldsymbol{A}\boldsymbol{B}) \tag{A.4.7}$$

$$\boldsymbol{A}\boldsymbol{B}(\boldsymbol{C} + \boldsymbol{D}\boldsymbol{A}\boldsymbol{B})^{-1} = \boldsymbol{E}\boldsymbol{B}\boldsymbol{C}^{-1} , \tag{A.4.8}$$

and multiplication of $-\boldsymbol{D}\boldsymbol{A}$ from the right yields

$$-\boldsymbol{A}\boldsymbol{B}(\boldsymbol{D}\boldsymbol{A}\boldsymbol{B} + \boldsymbol{C})^{-1}\boldsymbol{D}\boldsymbol{A} = -\boldsymbol{E}\boldsymbol{B}\boldsymbol{C}^{-1}\boldsymbol{D}\boldsymbol{A} . \tag{A.4.9}$$

Upon introducing (A.4.5),

$$\boldsymbol{A} - \boldsymbol{A}\boldsymbol{B}(\boldsymbol{D}\boldsymbol{A}\boldsymbol{B} + \boldsymbol{C})^{-1}\boldsymbol{D}\boldsymbol{A} = \boldsymbol{E} \tag{A.4.10}$$

results. □

This lemma is also valid for $\boldsymbol{D} = \boldsymbol{B}^{\mathrm{T}}$. The big improvement is that one only needs two matrix inversion in (A.4.2) compared to three in (A.4.1). If $\boldsymbol{D} = \boldsymbol{B}^{\mathrm{T}}$ and $\boldsymbol{B}$ is a column vector and $\boldsymbol{C}$ reduces to a scalar, then one only has to carry out one division instead of two inversions. This lemma can be applied in the progress of the derivation of the recursive method of least squares in (9.4.9) by equating the matrices and vectors as $\boldsymbol{E} = \boldsymbol{P}(k+1)$, $\boldsymbol{A}^{-1} = \boldsymbol{P}^{-1}(k)$, $\boldsymbol{B} = \boldsymbol{\psi}(k+1)$, $\boldsymbol{C}^{-1} = 1$, and $\boldsymbol{D} = \boldsymbol{B}^{\mathrm{T}} = \boldsymbol{\psi}^{\mathrm{T}}(k+1)$, which yields (9.4.15).

References

Åström KJ (1970) Introduction to stochastic control theory. Academic Press, New York

Brookes M (2005) The matrix reference manual. URL http://www.ee.ic.ac.uk/hp/staff/dmb/matrix/intro.html

Davenport W, Root W (1958) An introduction to the theory of random signals and noise. McGraw-Hill, New York

Deutsch R (1965) Estimation theory. Prentice-Hall, Englewood Cliffs, NJ

Doob JL (1953) Stochastic processes. Wiley, New York, NY

Fisher RA (1922) On the mathematical foundation of theoretical statistics. Philos Trans R Soc London, Ser A 222:309–368

Fisher RA (1950) Contributions to mathematical statistics. J. Wiley, New York, NY

Gauss KF (1809) Theory of the motion of the heavenly bodies moving about the sun in conic sections: Reprint 2004. Dover phoenix editions, Dover, Mineola, NY

Goldberger AS (1964) Econometric theory. Wiley Publications in Applied Statistics, John Wiley and Sons Ltd

Kendall MG, Stuart A (1961) The advanced theory of statistics. Volume 2. Griffin, London, UK

Kendall MG, Stuart A (1977) The advanced theory of statistics: Inference and relationship (vol. 2). Charles Griffin, London

Papoulis A, Pillai SU (2002) Probability, random variables and stochastic processes, 4th edn. McGraw Hill, Boston

Richter H (1966) Wahrscheinlichkeitstheorie, 2nd edn. Spinger, Berlin

Wilks SS (1962) Mathematical statistics. Wiley, New York

B

Experimental Systems

In the individual chapters of this book, measurements from a Three-Mass Oscillator are used to illustrate the application of the different identification methods as the system is linearizable in a wide operating range and also easy to understand and model. Furthermore, the resonance frequencies are low and the resonances are very distinct in the Bode plot. The testbed is equipped with couplings so that one or two of the three masses can be disconnected to reduce the system order.

B.1 Three-Mass Oscillator

The Three-Mass Oscillator as shown on the photograph in Fig. B.2 has been realized as a testbed for laboratories on control. It represents a variety of drive trains ranging from the automotive drive train up to tool drives in machine tools. The schematic setup is shown in Fig. B.1. A synchronous motor together with a gear is used to drive the Three-Mass Oscillator. It is fed by a frequency inverter and is operated in torque mode, that means the control signal to the frequency inverter determines the torque generated by the electric motor. Two more masses representing rotational inertias are connected via soft springs. The angular positions $\varphi_1(t)$, $\varphi_2(t)$, and $\varphi_3(t)$ are measured as outputs. By couplings, one can realize three different setups, a one mass rotational inertia, a two and three mass oscillator.

In the following, a physics based, theoretical mathematical model shall be derived. The modeling of multi mass oscillators is also treated by Isermann (2005). For the subsequent modeling of the Three-Mass Oscillator shown in Fig. B.1a, one should first derive a scheme as shown in Fig. B.1b, which shows the dynamic relations between the individual elements and the physical effects that have to be considered. By means of physical relations governing the torque balance at the three rotational masses, see e.g. (Isermann, 2005), one can come up with the following system of differential equations

R. Isermann, M. Münchhof, *Identification of Dynamic Systems*,
DOI 10.1007/978-3-540-78879-9_26, © Springer-Verlag Berlin Heidelberg 2011

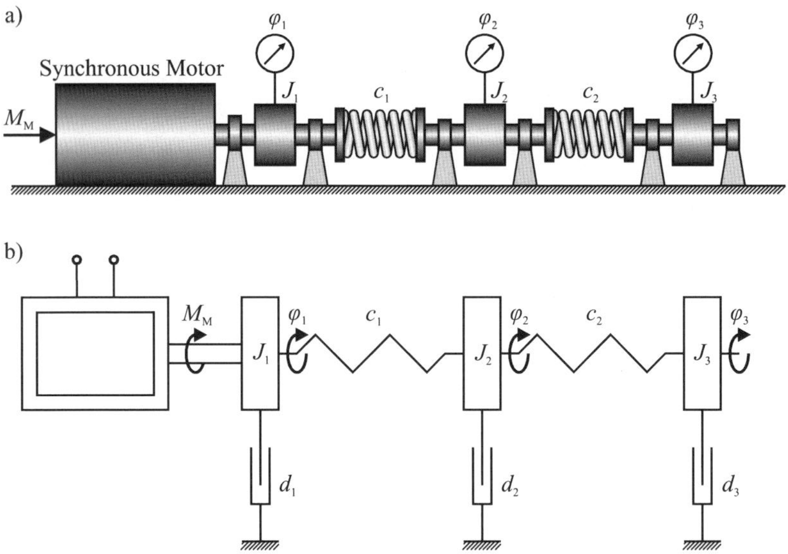

Fig. B.1. Scheme and block diagram of the Three-Mass Oscillator

$$J_1\ddot{\varphi}_1 = -d_1\dot{\varphi}_1 - c_1\varphi_1 + c_1\varphi_2 + M_M \tag{B.1.1}$$

$$J_2\ddot{\varphi}_2 = -d_2\dot{\varphi}_2 + c_1\varphi_1 - (c_1 + c_2)\varphi_2 + c_2\varphi_3 \tag{B.1.2}$$

$$J_3\ddot{\varphi}_3 = -d_3\dot{\varphi}_3 + c_2\varphi_2 - c_2\varphi_3 \,. \tag{B.1.3}$$

These equations of motion can now be written as a system of second order ODEs as

$$\boldsymbol{J}\ddot{\boldsymbol{\varphi}}(t) + \boldsymbol{D}\dot{\boldsymbol{\varphi}}(t) + \boldsymbol{C}\boldsymbol{\varphi} = \boldsymbol{M}M_M(t) \tag{B.1.4}$$

with

$$\boldsymbol{J} = \begin{pmatrix} J_1 & 0 & 0 \\ 0 & J_2 & 0 \\ 0 & 0 & J_3 \end{pmatrix},\ \boldsymbol{D} = \begin{pmatrix} d_1 & 0 & 0 \\ 0 & d_2 & 0 \\ 0 & 0 & d_3 \end{pmatrix},\ \boldsymbol{C} = \begin{pmatrix} c_1 & -c_1 & 0 \\ -c_1 & (c_1 + c_2) & -c_2 \\ 0 & -c_2 & c_2 \end{pmatrix},$$

and

$$\boldsymbol{M} = \begin{pmatrix} 1 \\ 0 \\ 0 \end{pmatrix}.$$

This set of second order ODEs can be rewritten as a system of first oder ODEs with the states chosen as

$$\boldsymbol{x}(t) = \begin{pmatrix} \varphi_1(t) \\ \varphi_2(t) \\ \varphi_3(t) \\ \dot{\varphi}_1(t) \\ \dot{\varphi}_2(t) \\ \dot{\varphi}_3(t) \end{pmatrix}. \tag{B.1.5}$$

Fig. B.2. Photos of the Three-Mass Oscillator. Top photo shows all three rotational inertias and the electric drive. Bottom photo shows a zoom on two rotational inertias with the angular position sensors and the electric drive

The input $u(t)$ is the torque exerted by the electric drive, i.e. $u(t) = M_{\mathrm{M}}(t)$. The output $y(t)$ is the angular position of the third mass, i.e. $y(t) = \varphi_3(t)$. This choice will result directly in the state space representation, (2.1.24), (2.1.25), and Fig. 2.2, with the state matrix $\boldsymbol{A}$ given as

$$\boldsymbol{A} = \begin{pmatrix} \boldsymbol{0} & \boldsymbol{I} \\ -\boldsymbol{J}^{-1}\boldsymbol{C} & -\boldsymbol{J}^{-1}\boldsymbol{D} \end{pmatrix} = \begin{pmatrix} 0 & 0 & 0 & 1 & 0 & 0 \\ 0 & 0 & 0 & 0 & 1 & 0 \\ 0 & 0 & 0 & 0 & 0 & 1 \\ \frac{-c_1}{J_1} & \frac{c_1}{J_1} & 0 & \frac{-d_1}{J_1} & 0 & 0 \\ \frac{c_1}{J_2} & \frac{-(c_1+c_2)}{J_2} & \frac{c_2}{J_2} & 0 & \frac{-d_2}{J_2} & 0 \\ 0 & \frac{c_2}{J_3} & \frac{-c_2}{J_3} & 0 & 0 & \frac{-d_3}{J_3} \end{pmatrix} \tag{B.1.6}$$

the input vector $\boldsymbol{b}$ given as

$$\boldsymbol{b} = \begin{pmatrix} 0 \\ 0 \\ 0 \\ \frac{1}{J_1} \\ 0 \\ 0 \end{pmatrix} \tag{B.1.7}$$

and the output vector $\boldsymbol{c}^{\mathrm{T}}$

$$\boldsymbol{c}^{\mathrm{T}} = \begin{pmatrix} 0 & 0 & 1 & 0 & 0 & 0 \end{pmatrix} . \tag{B.1.8}$$

With the corresponding numerical values, the matrices are given as

$$\boldsymbol{A} = \begin{pmatrix} 0 & 0 & 0 & 1 & 0 & 0 \\ 0 & 0 & 0 & 0 & 1 & 0 \\ 0 & 0 & 0 & 0 & 0 & 1 \\ -73.66 & 73.66 & 0 & -4.995 \times 10^{-3} & 0 & 0 \\ 164.5 & -399.1 & 234.5 & 0 & -1.342 \times 10^{-5} & 0 \\ 0 & 579.9 & -579.9 & 0 & 0 & -5.941 \end{pmatrix} \tag{B.1.9}$$

$$\boldsymbol{b} = \begin{pmatrix} 0 \\ 0 \\ 0 \\ 54.38 \\ 0 \\ 0 \end{pmatrix} \tag{B.1.10}$$

and

$$\boldsymbol{c}^{\mathrm{T}} = \begin{pmatrix} 0 & 0 & 1 & 0 & 0 & 0 \end{pmatrix} . \tag{B.1.11}$$

From the state space representation, one can now determine the transfer function in continuous-time by

$$G(s) = \frac{y(s)}{u(s)} = \boldsymbol{c}^{\mathrm{T}}\left(s\boldsymbol{I} - \boldsymbol{A}\right)^{-1}\boldsymbol{b} . \tag{B.1.12}$$

The transfer functions for the one mass rotational inertia and the two and three mass oscillator will now be derived. The one mass system has the transfer function

$$G(s) = \frac{\varphi_1(s)}{M_{\mathrm{M}}(s)} = \frac{1}{J_1 s^2 + d_1 s} = \frac{\frac{1}{d_1}}{\frac{J_1}{d_1} s^2 + s} = \frac{10.888}{200.21 s^2 + 1 s} \tag{B.1.13}$$

for the angular position and

$$G(s) = \frac{\omega_1(s)}{M_{\mathrm{M}}(s)} = \frac{1}{J_1 s + d_1} = \frac{\frac{1}{d_1}}{\frac{J_1}{d_1} s + 1} = \frac{10.888}{200.21 s + 1} \tag{B.1.14}$$

for the rotational speed.

For the two mass oscillator, a similar approach can be employed. The equations of motion are now given as

$$\begin{aligned} J_1 \ddot{\varphi}_1(t) &= -d_1 \dot{\varphi}_1(t) - c_1 \varphi_1(t) + c_1 \varphi_2(t) + M_{\mathrm{M}}(t) \\ J_2 \ddot{\varphi}_2(t) &= -d_2 \dot{\varphi}_2(t) + c_1 \varphi_1(t) - c_1 \varphi_2(t) \,. \end{aligned}$$

These equations can be brought to the Laplace domain as

$$\begin{aligned} J_1 s^2 \varphi_1(s) &= -d_1 s \varphi_1(s) - c_1 \varphi_1(s) + c_1 \varphi_2(s) + M_{\mathrm{M}}(s) \\ J_2 s^2 \varphi_2(s) &= -d_2 s \varphi_2(s) + c_1 \varphi_1(s) - c_1 \varphi_2(s) \end{aligned}$$

and can then be rewritten in transfer function form as

$$\begin{aligned} G(s) &= \frac{\varphi_2(s)}{M_{\mathrm{M}}(s)} \\ &= \frac{c_1}{J_1 J_2 s^4 + \big(J_2 d_1 + J_1 d_2\big) s^3 + \big(J_2 c_1 + d_1 d_2 + J_1 c_1\big) s^2 + \big(d_2 c_1 + d_1 c_1\big) s} \,. \end{aligned}$$

With the numerical values, one obtains

$$G(s) = \frac{\varphi_2(s)}{M_{\mathrm{M}}(s)} = \frac{1.3545}{1.51 \times 10^{-4} s^4 + 7.58 \times 10^{-7} s^3 + 0.047 s^2 + 1.77 \times 10^{-4} s}$$

for the position of the second mass, φ_2 and

$$G(s) = \frac{\omega_2(s)}{M_{\mathrm{M}}(s)} = \frac{1.3545}{1.51 \times 10^{-4} s^3 + 7.58 \times 10^{-7} s^2 + 0.047 s^1 + 1.77 \times 10^{-4}}$$

for its rotational velocity, ω_2.

For the full three mass system, the transfer function is given as

$$\begin{aligned} G(s) &= \frac{\omega_3(s)}{M_{\mathrm{M}}(s)} \\ &= \frac{5.189 \times 10^6 s - 4.608 \times 10^{-9}}{s^6 + 5.946 s^5 + 1053 s^4 + 2813 s^3 + 155400 s^2 + 103100 s - 3.347 \times 10^{-9}} \,. \end{aligned} \tag{B.1.15}$$

The process coefficients are given in Table B.1. In addition, a dead time, which stems from the signal processing, $T_{\mathrm{D}} = 0.0187\,\mathrm{s}$ could be identified at the testbed.

Table B.1. Process coefficients of the Three-Mass Oscillator

Rotational Inertia J_i $\left[\mathrm{kg\,m^2}\right]$	Damping Constants d_i $\left[\mathrm{Nm\,s}\right]$	Spring Stiffness c_i [Nm]
$J_1 = 18.4 \times 10^{-3}$	$d_1 = 9.18 \times 10^{-5}$	$c_1 = 1.35$
$J_2 = 8.2 \times 10^{-3}$	$d_2 = 1.10 \times 10^{-7}$	$c_1 = 1.93$
$J_3 = 3.3 \times 10^{-3}$	$d_3 = 19.0 \times 10^{-3}$	

References

Isermann R (2005) Mechatronic Systems: Fundamentals. Springer, London

Index

R. Isermann, M. Münchhof, *Identification of Dynamic Systems*,
DOI 10.1007/978-3-540-78879-9, © Springer-Verlag Berlin Heidelberg 2011

Printed by Printforce, the Netherlands